Enzyme Technologies for Pharmaceutical and Biotechnological Applications

Enzyme Technologies for Pharmaceutical and Biotechnological Applications

edited by

Herbert A. Kirst
Eli Lilly and Company
Greenfield, Indiana

Wu-Kuang Yeh
Milton J. Zmijewski, Jr.
Eli Lilly and Company
Indianapolis, Indiana

MARCEL DEKKER, INC. NEW YORK • BASEL

ISBN: 0-8247-0549-1

This book is printed on acid-free paper.

Headquarters
Marcel Dekker, Inc.
270 Madison Avenue, New York, NY 10016
tel: 212-696-9000; fax: 212-685-4540

Eastern Hemisphere Distribution
Marcel Dekker AG
Hutgasse 4, Postfach 812, CH-4001 Basel, Switzerland
tel: 41-61-261-8482; fax: 41-61-261-8896

World Wide Web
http://www.dekker.com

The publisher offers discounts on this book when ordered in bulk quantities. For more information, write to Special Sales/Professional Marketing at the headquarters address above.

Current printing (last digit):
10 9 8 7 6 5 4 3 2 1

PRINTED IN THE UNITED STATES OF AMERICA

Preface

Genes related to human and animal health are being discovered at an ever-increasing rate. As gene products, enzymes are being explored for their function and application in a rapidly emerging field that has been termed *functional genomics*. *Enzyme Technologies for Pharmaceutical and Biotechnological Applications* fills a unique niche for a comprehensive account of certain important enzymes in human and animal health. Readers can also gain important insights into enzyme technologies in both the pharmaceutical and biotechnological industries. The primary aim of this book is to highlight how, what, and where enzymes have become critically important or are rapidly emerging in these two overlapping and interdependent industries.

As a state-of-the-art work on enzyme technologies, the book covers four basic principles and applications in (1) antibiotic biosynthesis, (2) biocatalysis, (3) modern screening/optimization, and (4) emerging new technologies. In Part I, on biosynthesis, the emphasis is placed on both improvements in antibiotic yield and ways to increase antibiotic structural diversity by modifications of the biosynthetic pathways from diverse microorganisms. Here, the emphasis is on using genes to deliver enzymes and to thereby perform metabolic engineering including precursor-directed biosynthesis or mutasynthesis. The use of recombinant techniques to generate protein products that are unnatural to the microbial world is also discussed, using specific examples of challenging problems in this area.

Part II on biocatalysis, covers the direct application of enzymes as chemical tools in manipulating small- to medium-sized synthetic organic compounds. Manipulation of the enzyme tools by genetic engineering is described. Chapter 8 discusses how a novel form of enzymes, cross-linked enzyme crystals (CLCs), are especially useful as chemical catalysts. An example of a large-scale application of these chemical tools brings the area into focus by leaving the laboratory and entering the manufacturing plant.

Part III on screening for and optimization of enzyme inhibitors describes integrated approaches in therapeutic discovery research using enzyme targets relevant to human and animal diseases. For high-throughput screening, the activity assays for the enzyme targets adopt both conventional (colorimetry, spectrometry, and radioactivity) and contemporary methodologies (fluorescence). A selective enzymatic assay maximizes validated hits from large diversified libraries of samples derived from natural products and synthetic compounds, including those arising from combinatorial chemistry. The chapters on screening concentrate on development of effective enzymatic assays, each of which represents specific, kinetic, and molecular interactions between the enzyme and its substrate as well as inhibitors and thus reflects the pharmacological and chemical interplay at the targeted enzyme. The primary screening goal is the production of manageable numbers of hits that ultimately generate high-quality lead compounds. As a practical rule, optimization of those lead compounds by medicinal chemists is the critical follow-up step required for the discovery of viable drug candidates. The process of lead compound optimization for an enzyme inhibitor, often referred to as structure-activity relationship studies or drug design, is dictated by understanding the molecular and kinetic interactions between the enzyme and its inhibitor. These insights are typically gained by analyzing X-ray crystallographic depictions and by elucidating the kinetic behavior of the enzyme-inhibitor complex to improve potency and selectivity and to understand mechanisms of interactions. The chapters on inhibitor screening/optimization emphasize the synergistic importance of high-throughput screening and structure-function based optimization studies for therapeutic discovery programs.

Finally, Part IV on emerging technologies examines some non-traditional methods by which enzymes may play important new roles in the drug discovery processes of the future. The present ability to completely locate and sequence the gene clusters responsible for the multistep biosyntheses of complex natural products has spawned new technologies. Such technologies can precisely and/or deliberately modify certain parts of gene clusters within organisms or, alternatively, can interchange portions of gene clusters between organisms. In each instance, new unnatural natural products may be formed by fermentation of the new genetically modified microorganisms. The exchange of genetic material can be logically extended into a combinatorial paradigm called *combinatorial biosynthesis* or *combinatorial enzymology*, thereby leading to even larger numbers of new natural products. Extensive interdisciplinary collaboration between new target identification and screening laboratories, medicinal chemists, and molecular modeling/computational chemists will become even more essential in the future for rapid discovery of useful new entities to evaluate in the field or clinic.

The last chapters of Part IV describe essential and overlapping enzyme technologies. With the completion of most of the human genome sequence, assigning a precise function to genes (*functional prediction*) and redesigning the function of enzymes (*enzyme engineering)* can play increasingly significant roles in drug discovery. Also, the utility of functional genomics in identifying disease-relevant enzyme targets depends closely on the molecular understanding of these targets under physiological and pathological conditions (*functional proteomics*).

Enzyme Technologies for Pharmaceutical and Biotechnological Applications is informative, practical, timely, and applicable worldwide to the pharmaceutical and biotechnological industries. Real-world examples provided throughout the book are important for discriminating between the use of enzymes solely for academic studies and the practical use of enzymes in industrial applications. The reader will acquire a better understanding of applied sciences in the field. Areas that have been extensively covered in reviews and the general literature (such as use of natural lipases in organic synthesis) have been minimized here. By focusing on real-world applications, the reader will obtain a clearer understanding of what is new and relevant in the field.

The book is intended primarily for industrial and research scientists with interests in adopting and maximizing enzyme technologies for pharmaceutical discovery, development, and manufacturing. The book can also be used by graduate and postdoctoral students in practical enzymology, biochemistry, microbiology, molecular biology, and biochemical engineering, as well as by students in graduate-level courses covering practical enzymology and enzyme biochemistry.

Herbert A. Kirst
Wu-Kuang Yeh
Milton J. Zmijewski, Jr.

Contents

Contributors

Jose L. Adrio Department of Biochemistry, Antibioticos, S.A.U., León, Spain

Akira Arisawa Biochemistry Laboratory, Central Research Laboratories, Mercian Corporation, Fujisawa, Japan

Gerald W. Becker Lilly Research Laboratories, Eli Lilly and Company, Indianapolis, Indiana

Gary M. Birch Lilly Research Laboratories, Eli Lilly and Company, Indianapolis, Indiana

Ridong Chen Department of Biochemistry, University of Saskatchewan, Saskatoon, Saskatchewan, Canada

Andrew R. Cockshott Lilly Research Laboratories, Eli Lilly and Company, Indianapolis, Indiana

James A. Cook Lilly Research Laboratories, Eli Lilly and Company, Indianapolis, Indiana

Karen L. Cox Lilly Research Laboratories, Eli Lilly and Company, Indianapolis, Indiana

Arnold L. Demain Department of Biology, Massachusetts Institute of Technology, Cambridge, Massachusetts

Edward A. Dennis Department of Chemistry and Biochemistry, Revelle College and School of Medicine, University of California, San Diego, California

Claudio D. Denoya Bioprocess Research, Global Research and Development, Pfizer, Inc., Groton, Connecticut

Ralf Dieckmann Biotechnology Center, Technical University Berlin, Berlin, Germany

Joe E. Dotzlaf Lilly Research Laboratories, Eli Lilly and Company, Indianapolis, Indiana

Timothy G. Geary Discovery Research, Pharmacia Animal Health, Kalamazoo, Michigan

Mirko Glinski Max Volmer Institute for Biophysical Chemistry and Biochemistry, Technical University Berlin, Berlin, Germany

Michael D. Grim Westboro, Massachusetts

John E. Hale Lilly Research Laboratories, Eli Lilly and Company, Indianapolis, Indiana

Stephen A. Hitchcock Lilly Research Laboratories, Eli Lilly and Company, Indianapolis, Indiana

Till Hornbogen Max Volmer Institute for Biophysical Chemistry and Biochemistry, Technical University Berlin, Berlin, Germany

Joann Hoskins Lilly Research Laboratories, Eli Lilly and Company, Indianapolis, Indiana

Robert B. Johnson Lilly Research Laboratories, Eli Lilly and Company, Indianapolis, Indiana

Wibke Kallow AnagnosTec GmbH, Luckenwalde, Germany

Margaret M. Kayser Department of Chemistry, University of New Brunswick, Saint John, New Brunswick, Canada

Chaitan Khosla Departments of Chemistry and Chemical Engineering, Stanford University, Stanford, California

Adam J. Kreuzman Lilly Research Laboratories, Eli Lilly and Company, Indianapolis, Indiana

Srinivasan Krishnan Lilly Research Laboratories, Eli Lilly and Company, Indianapolis, Indiana

Hamish A. I. McArthur Bioprocess Research, Global Research and Development, Pfizer, Inc., Groton, Connecticut

Robert McDaniel Kosan Biosciences, Inc., Hayward, California

Timothy I. Meier Lilly Research Laboratories, Eli Lilly and Company, Indianapolis, Indiana

Marian Mosior Lilly Research Laboratories, Eli Lilly and Company, Indianapolis, Indiana

Satoshi Ōmura Research Center for Biological Function, The Kitasato Institute and Graduate School of Pharmaceutical Sciences, The Kitasato Institute and Kitasato University, Tokyo, Japan

Ramesh N. Patel Process Research and Development, Bristol-Myers Squibb Company, New Brunswick, New Jersey

Robert B. Peery Lilly Research Laboratories, Eli Lilly and Company, Indianapolis, Indiana

Jacqueline M. Piret Biology Department, Northeastern University, Boston, Massachusetts

Steven W. Queener Lilly Research Laboratories, Eli Lilly and Company, Indianapolis, Indiana

Sonia Rodríguez Department of Chemistry, University of Florida, Gainesville, Florida

Siddhartha Roychoudhury Discovery-Biology, Procter & Gamble Pharmaceuticals, Mason, Ohio

Molly B. Schmid Microcide Pharmaceuticals, Inc., Mountain View, California

Torsten Schwecke Biotechnology Center, Technical University Berlin, Berlin, Germany

Paul L. Skatrud Lilly Research Laboratories, Eli Lilly and Company, Indianapolis, Indiana

Michele C. Smith Lilly Research Laboratories, Eli Lilly and Company, Indianapolis, Indiana

James Staunton Department of Chemistry, University of Cambridge, Cambridge, United Kingdom

Jon D. Stewart Department of Chemistry, University of Florida, Gainesville, Florida

Kim J. Stutzman-Engwall Bioprocess Research, Global Research and Development, Pfizer, Inc., Groton, Connecticut

Mary Anne Tavanlar National Institute of Molecular Biology and Biotechnology, University of the Philippines Los Baños, Laguna, Philippines

Hiroshi Tomoda Research Center for Biological Function, The Kitasato Institute and Graduate School of Pharmaceutical Sciences, The Kitasato Institute and Kitasato University, Tokyo, Japan

Joaquim Trias Department of Microbiology, Versicor, Inc., Fremont, California

Hiroshi Tsunekawa Pharmaceuticals and Chemicals Division, Mercian Corporation, Tokyo, Japan

Volker Uhlmann Department of Histopathology, St. James's Hospital, Dublin, Ireland

Hans von Döhren Biotechnology Center, Technical University Berlin, Berlin, Germany

Q. May Wang Lilly Research Laboratories, Eli Lilly and Company, Indianapolis, Indiana

Kira J. Weissman Department of Biochemistry, University of Cambridge, Cambridge, United Kingdom

Hsiu-Chiung Yang Lilly Research Laboratories, Eli Lilly and Company, Indianapolis, Indiana

Raymond C. Yao Lilly Research Laboratories, Eli Lilly and Company, Indianapolis, Indiana

Wu-Kuang Yeh Lilly Research Laboratories, Eli Lilly and Company, Indianapolis, Indiana

Zhengyu Yuan Versicor, Inc., Fremont, California

Genshi Zhao Lilly Research Laboratories, Eli Lilly and Company, Indianapolis, Indiana

Rainer Zocher Max Volmer Institute for Biophysical Chemistry and Biochemistry, Technical University Berlin, Berlin, Germany

Enzyme Technologies for Pharmaceutical and Biotechnological Applications

edited by

Herbert A. Kirst
Eli Lilly and Company
Greenfield, Indiana

Wu-Kuang Yeh
Milton J. Zmijewski, Jr.
Eli Lilly and Company
Indianapolis, Indiana

MARCEL DEKKER, INC. NEW YORK • BASEL

1

δ-(L-α-Aminoadipyl)-L-Cysteinyl-D-Valine Synthetase as a Model Tripeptide Synthetase

Hans von Döhren, Torsten Schwecke, and Ralf Dieckmann
Technical University Berlin, Berlin, Germany

Wibke Kallow
AnagnosTec GmbH, Luckenwalde, Germany

Mary Anne Tavanlar
University of the Philippines Los Baños, Laguna, Philippines

Volker Uhlmann
St. James's Hospital, Dublin, Ireland

I. INTRODUCTION

A. Nonribosomal Peptide-Forming Systems and Penicillin

As beta-lactam antibiotics continue to be a major contributor to human health preservation, research on the biosynthesis of penicillin, an almost ancient drug, continues to open up roads to new technologies and perspectives. The provision of precursor peptides to be transformed enzymatically with chemically unachieved efficiency into mono- or bicyclic antibiotics has been termed by Jack Baldwin and colleagues ''the irreversible commitment of metabolic carbon to the secondary metabolism'' [1]. The synthesis of such peptides is indeed performed by a remarkable class of synthetases which, in contrast to the protein-synthesizing machinery, have been termed a nonribosomal system or nonribosomal peptide synthetases (NRPS) [2]. These peptide synthetases have been shown to catalyze the irreversible synthesis of peptides differing both in sequence and structural variability, thus extending the scope of directly gene-encoded polypeptides. The synthetic principle, unravelled mainly by Kiyoshi Kurahashi, Sören Laland, Fritz Lipmann, and later Horst Kleinkauf and colleagues (reviewed in [3]), consists of a process of amino acid selection, activation of carboxyl groups as adenylates, formation of stable thioester intermediates, and successive condensations in an assembly-line process with no free intermediates. The enzymatic machinery thus resembles a kind of self-feeding solid-phase process. Similar sequential condensation processes are known from polyketides and related acetate-derived compounds, and many compounds combining the respective building blocks in mixed biosynthetic systems are known.

A major step in our continuing understanding of these remarkable multistep systems goes back to the cloning of the peptide synthetase producing the penicillin precursor tripeptide δ-(L-α-aminoadipyl)-L-cysteinyl-D-valine (ACV). The first sequences of this remarkable and quite large gene became available in 1990, derived from *Penicillium chrysogenum* [4,5], *Aspergillus nidulans* [6], *Acremonium chrysogenum* [7], and *Nocardia lactamdurans* [8]. These efforts immediately led to the revision of the classical thiotemplate model to the multiple-carrier thiotemplate model [9,10], which was later verified experimentally [11,12]. At the same time, the ACV synthetase sequences were the first to introduce us to the multidomain structure of this class of enzymes, and thus initiated the still-continuing efforts to understand the molecular principles of multistep synthetases and synthases.

B. From Tripeptide to Synthetase

The formation of the tripeptide ACV (the Arnstein tripeptide) upon feeding of labeled valine to mycelia of *P. chrysogenum* had already been shown by Arnstein and co-workers in 1960 [13,14], and demonstrated later in cell-free extracts by Bauer [15]. In more refined work, Abraham and Loder established in extracts of *A. chrysogenum* the $MgATP^{2-}$-dependent formation of ACV from δ-(L-α-aminoadipyl)-L-cysteine (AC) and L-valine [16,17], while no product was observed with L-α-aminoadipate (Aad) and L-cysteinyl-L-valine (LL-CV), or AC and D-valine. The stereochemistry of the tripeptide has been established as LLD [18,19], and the biosynthesis from the respective L-configured amino acids was again demonstrated in a beta-lactam-defective mutant by Shirafuji and colleagues [20]. This mutant Takeda N2 of *A. chrysogenum*, later shown to carry a point mutation in the isopenicillin N synthase gene, was shown by Adlington and co-workers to produce both AC and ACV [21]. Among tripeptide analogs isolated from fermentation broth were Aad-Ala-D-Val (D configuration not proven), Aad-Ser-D-Val, and Aad-Ser-isodehydro-Val [22]. At that time it was generally accepted that the Arnstein tripeptide would be the product of two consecutive reactions catalyzed by the hypothetical AC and ACV synthetases [23]. Thus Lara and colleagues, studying the stimulation of penicillin formation in *P. chrysogenum* by L-glutamate, determined the hypothetical level of AC synthetase by phosphate consumption from $MgATP^{2-}$ in the presence of Aad and Cys, without product analysis [24].

The possible analogy of ACV formation to the well-established thiotemplate processes had earlier led Feodor Lynen (H. Bergmeyer, unpublished data) and then Arny Demain to take a closer look at these synthetases. Employing the industrial strain C-10 of *A. chrysogenum* (ATCC 48272), Banko in Demain's laboratory succeeded in isolating an AC-forming enzyme [25], and later showed the rate of ACV production by this preparation from the three amino acids to be faster than from AC and L-valine [26]. Similar results were obtained by Jensen in Westlake's laboratory with the actinomycete *Streptomyces clavuligerus* [27], and ACV synthesis was also achieved with an immobilized enzyme preparation bound to an anion-exchange resin [28]. The isolation and purification of an ACV synthetase has first been reported with the *A. nidulans* enzyme [29], later to be followed by synthetases from *A. chrysogenum* [30–32] and *S. clavuligerus* [33–36]. These studies established a single multienzyme to be responsible for the synthesis of ACV from Aad, Cys, Val, and $MgATP^{2-}$. An associated protein in the *S. clavuligerus* preparation [34] was later shown to be proclavaminate amidinohydrolase [37]. Thus, possible *in vivo* interactions with other proteins in this biosynthetic process are lost during standard purification operations.

II. ACV SYNTHETASES: ORGANIZATION, PURIFICATION, AND CHARACTERIZATION

A. Overall Organization

Protein sequence data of prokaryotic and eukaryotic ACV synthetases reveal a similar organization of three adenylate domains, three thiolation domains (carrier proteins), two condensation domains, one epimerization domain, and one thioesterase domain. These domains have been well characterized from multiple sequence data of various NRPS systems [38–40] and data are easily accessable, with the terms *AMP binding domain* for adenylate domains (IPR000873, PS00012, PF00501); *pp-binding*, *phosphopantetheine*, or *acp domain* for thiolation domains (IPR000255, PS00455, PF00550); condensation and epimerization domains as *DUF4* (IPR001240) with unknown function; and *thioesterase* domains (IPR001031). A primary approach in the identification of domains is the detection of highly conserved core sequences. Such sequences have been derived from selected NRPS data [38]. The respective ACV synthetase cores show an excellent agreement with little deviation (Table 1). Inspection of the primary data shows alterations especially in the C-terminal subdomains of the adenylate domains (motifs A8–A10), and in the condensation domains. Domain boundaries can be predicted from multiple alignments, which help to identify linker regions [41]. The unique availability of both prokaryotic and eukaryotic sequences may provide additional insights not yet evaluated by data analysis. These sequences include the far N-terminal regions of about 40 amino acid residues present in fungal synthetases, which points also to some insecurity with respect to initiation sites and the actual terminal sequences. Another interesting region is the N-terminal region of about 200 amino acids preceding the first adenylation domain, which shows significant similarities to an internal region of condensation domains. The recent identification of terminal and linker-associated sequences involved in domain interactions has not yet been investigated in NRPS systems [42,43].

As the second elementary step of data analysis, the comparative multiple alignment of the identified domains is conducted. Complete adenylate domain sequences provide an ACV synthetase cluster, with subclusters of the individual adenylate domains in positions 1, 2, and 3 [44]. This clustering indicates a similarity of ACV synthetase domains compared to other NRPS systems, and a likewise similarity of the domains in their respective positions. This clustering has been taken as evidence for the evolution of NRPS by gene duplication within individual systems [45], and would as well support the horizontal gene transfer hypothesis of beta-lactam biosynthetic clusters [46]. A different picture, however, emerges when the sequences of thiolation or condensation domains are compared: the carrier domains of the first modules are found in a different cluster than those of the second and third module [47]. Likewise, a CLUSTAL alignment of the two condensation domains places them in different contexts, which have been speculatively correlated with the specific type of condensation reaction (linking a δ-carboxyl group with an α-amino group, or linking a cysteine carboxyl group with an α-amino group) [47]. The condensation domain-related epimerization domain is found within the respective cluster of epimerization domains. Thus, if applied to thiolation, condensation, or epimerization domains, the system-specific clustering observed for adenylation domains is not evident, possibly due to the more diverse differences in functions. More detailed structural investigations are needed to interpret the functional significance of these observations.

The three-module system found in all ACV synthetases correlates well with the sequential condensation, epimerization, and hydrolytic release of the tripeptide. At first sight, the arrangement of domains and modules fits perfectly the collinearity rule found in the majority of NRPS systems, which implies that the linear sequence of domains corresponds to the sequence of reactions of the pathway [40,45].

B. Purification Work and Protein Studies

Studies have been performed on the purification and characterization of synthetases from *A. nidulans*, *A. chrysogenum*, *S. clavuligerus*, *Flavobacterium* sp. [48], *N. lactamdurans* [49], and *P. chrysogenum* [50]. Except for *Flavobacterium* and *S. clavuligerus*, the complete gene sequences are available, which reveal

Table 1 NRPS Core Sequences and Their Conservation in ACV Synthetases

	Adenylate domains[a]		
	A1 **LTYxEL** **S**	**A2** **LKAGxAYLVPLD** **L I**	**A3** **LAYxxTTSGSTGx--PKG** **T**
Module 1[b]			
P. chrysogenum	LTYGEL	**W**K**S**GAAY VPID	LAYVTYTSGTTGF--PKG
A. chrysogenum	LTYGEL	**W**K**S**GAAY VPID	LAYVTYTSGTTGF--PKG
A. nidulans	LTYGEL	**W**K**S**GAAY VPID	LAYVTYTSGTTGF--PKG
Lysobacter sp.	**I**TYEEL	**W**K**S**GAA**H** VPID	LAYAIYTSGTTGR**PV**PK**A**
N. lactamdurans	LSYREL	**W**KAGAAY VPID	**R**AYVTYTSGTTGV--PKG
Module 2			
P. chrysogenum	LTYREL	**W**K**S**GGAY VPID	LAYIIYTSGTTGR--PKG
A. chrysogenum	LTYREL	**W**K**T**GGAY VPID	LAYIIYTSGTTGK--PKG
A. nidulans	LTYREL	**W**K**S**GGAY VPID	LAYIMYTSGTTGN--PKG
Lysobacter sp.	LSYREL	**W**KAGGAY VPID	LAYAIYTSGTTGR--PKG
N. lactamdurans	LTYREL	**W**K**T**GGAY VPID	LAYIIYTSGTTGK--PKG
Module 3			
P. chrysogenum	LSYA**D**L	**W**KAGAAY VPLD	LAYIIFTSGTSGK--PKG
A. chrysogenum	LSYTEL	**W**K**T**GSAY VPLD	LAYIIFTSGTSGK--PKG
A. nidulans	LSYSEL	**W**KAGAAY VPLD	LAYVIFTSGTSGK--PKG
Lysobacter sp.	LSYREL	**W**KAGAAY VPID	LAYAIYTSGTSGR--PK**A**
N. lactamdurans	**T**TYREL	**W**KAGAAY MPLD	LAYAIFTSGTSGK--PKG

Table 1 Continued

	Adenylate domains[a]		
	A4 **FDxS**	**A5** **NxYGPTE**	**A6** **GELxIxGxGVARGYL**
Module 1			**L**
P. chrysogenum	F**E**P**V**	NEYG**F**TE	GELHIGGLG**ISK**GYL
A. chrysogenum	F**E**P**V**	NEYG**F**TE	GELHIGGLG**ISK**GYL
A. nidulans	F**E**P**V**	NEYG**F**TE	GELHIGGLG**IS**RGYL
Lysobacter sp.	FDH**F**	NGYGPTE	GELYIGGLGVA**K**GYL
N. lactamdurans	F**E**P**H**	NEY**AF**TE	GELHIGGCG**IS**PGYL
Module 2			
P. chrysogenum	FDH**F**	NGYGPTE	GELY**L**GGEGV**V**RGY**H**
A. chrysogenum	FDH**F**	NGYGPTE	GELY**L**GGEGVARGY**H**
A. nidulans	FDH**F**	NGYGPTE	GELY**L**GGDGVARGY**H**
Lysobacter sp.	FDH**F**	NGYGPTE	GELYIGGIGVARGY**H**
N. lactamdurans	FDH**F**	NGYGPTE	GELYIGGIGV**T**RGY**H**
Module 3			
P. chrysogenum	FDFS	NAYG**V**TE	GELHIGGLG**ISK**GYL
A. chrysogenum	FDFS	NAYG**V**TE	GELHIGGLG**ISK**GYL
A. nidulans	FDFS	NAYG**V**TE	GELHIGGLG**IS**RGYL
Lysobacter sp.	FDFS	NAYG**T**TE	GELYIGGLGVA**K**GYL
N. lactamdurans	FDFS	NAY**AT**TE	GELHIGGCG**IS**PGYL

	Adenylate domains[a]		
	A7 **YRTGDL** **K**	**A8** **GRxDxQVKIRGxRIELGEIE**	**A9** **LPxYMIP** **V**
Module 1			
P. chrysogenum	YKTGDL	GRADFQIKLRGIRIEPGEIE	LPRYMIP
A. chrysogenum	YKTGDL	GRADFQIKLRGIRIEPGEIE	LPRYMVP
A. nidulans	YKTGDL	GRADFQIKLRGIRIEPGEIE	LPRYMVP
Lysobacter sp.	YKTGDL	GRNDMQVKIRGQRIELGEVE	LPQAIIP
N. lactamdurans	YRTGDL	GRADFQLKLNGVRIEPGEIE	LIRIMVP
Module 2			
P. chrysogenum	YKTGDL	GRNDFQVKIRGLRIELGEIE	LPGYMVP
A. chrysogenum	YRTGDL	GRNDFQVKIRGLRIELGEIE	LPGYMIP
A. nidulans	YKTGDL	GRNDFQVKIRGQRIELGEIE	LPDYMVP
Lysobacter sp.	YKTGDL	GRNDMQVKIRGLRVELGEIE	LPQAIVP
N. lactamdurans	YRTGDL	GRTDLQVKIRGLRIELGEIE	LPESVVP
Module 3			
P. chrysogenum	YKTGDL	GRGDLQIKMRGYRIEISE	LPTYMVP
A. chrysogenum	YKTGDL	GRADLQVKMRGYRIEPSE	LPAYMVP
A. nidulans	YKTGDL	GRKDQQVKLRGFRIELSE	LPPYMVP
Lysobacter sp.	YKTGDL	GRNDHQVKINGVRIELGE	LTPAMMP
N. lactamdurans	YRTGDL	GRNDAQVKINGLRIEPGE	LMPSMVP

Table 1 Continued

	Adenylate domain[a]	Thiolation domain	Condensation domain
	A10	**T**	**C1**
	NGKVDR	**DxFFxxLGGHSI**	**SxAQxRLWxL**
		D L	**MY**
Domain 1			
P. chrysogenum	NGK**A**D**L**	DNFFR LGGHSI	**ANSLQQGFV**
A. chrysogenum	NGK**A**D**L**	DNFFR LGGHSI	**ANSLQQGFV**
A. nidulans	NGK**A**D**L**	SNFFR LGGHSI	**ANSLQQGFV**
Lysobacter sp.	**S**GK**L**D**V**	DD**I**FG **S**GGDSL	**A QERLL**
N. lactamdurans	NGKVD**W**	DD**I**FR LGG**Q**SI	**ANGLQQGFV**
Domain 2			
P. chrysogenum	**S**GK**L**D**T**	DFFFS LGGDSL	SRAQERL**L**F**I**
A. chrysogenum	**S**GK**L**D**L**	DFFFT LGGDSI	SPAQERL**L**F**I**
A. nidulans	**S**GK**L**D**A**	**S**FFFR LGGDSI	SPAQERL**M**F**I**
Lysobacter sp.	**S**GK**L**D**V**	DF**I**FG **S**GGDSL	SLAQERL**L**F**I**
N. lactamdurans	**S**GK**LMA**	DFFAR **A**GGDSI	SLAQERL**L**F**I**
Domain 3			
P. chrysogenum	NGKLD**V**	DD**DL**FKLGGDSI	
A. chrysogenum	NGKLD**V**	DD**DL**FRLGGDSI	
A. nidulans	NGKLD**L**	DD**DL**FA**R**GGDSI	
Lysobacter sp.	**S**GKLD**V**	**H**D**DL**FQ**I**GGDSL	
N. lactamdurans	**T**GKLD**V**	DD**DL**FR**C**GGDSI	

Condensation domains

	C2	C3	C4
	RHExLRTxF	**MHHxISDGWS**	**YxDFAVW**
		V	**Y**
Domain 1			
P. chrysogenum	**SFP**ALR**L**RF	**C**HHAI**L**DGWS	YT**RTQRY**
A. chrysogenum	**MLP**TLR**L**RF	**C**HHAI**L**DGWS	YT**RTQRY**
A. nidulans	**E**HPALR**L**RF	**C**HHAI**L**DGWS	YR**QSQQY**
Lysobacter sp.	RH**A**ALRTL**L**	**F**HHS**CF**DGWS	YL**LGQQY**
N. lactamdurans	**TYP**ALR**L**RF	**C**HHII**L**DGWS	YV**AAQRY**
Domain 2			
P. chrysogenum	RHEALRTR**E**	**F**HHT**CF**D**A**WS	YK**E**YA**LY**
A. chrysogenum	RHEALRTE**R**	**F**HHT**CF**D**A**WS	YK**E**Y**QYA**
A. nidulans	RHEALRTL**I**	**S**HHL**AF**D**A**WS	YK**E**YA**IE**
Lysobacter sp.	RH**A**ALRTL**L**	**F**HHS**CF**DGWS	YADF**S**VW
N. lactamdurans	RHPALRTR**E**	**V**HHT**CF**D**A**WS	YG**EI**AVW

	C5	C6	C7
	IGxFVNTQCxR	**HQDYPFE**	**RDxSRNPL**
	V LA	**N V**	
Domain 1			
P. chrysogenum	VGPY**I**NTL**P**L**V**	**VMNSRGN**	**T**D**LKHGLF**
A. chrysogenum	VGLY**I**NTL**P**L**V**	**VMNSRGN**	**NELKHGLF**
A. nidulans	VGLF**I**NTL**P**M**I**	**AMNSRGN**	**T**D**LKHGLF**
Lysobacter sp.	VGLY**I**NTL**P**S**V**	**RMNSRSA**	**QGSK**R**RLF**
N. lactamdurans	VGLF**I**NTL**P**L**V**	**TMNSKGI**	**QSGEMRLF**
Domain 2			
P. chrysogenum	IGFFVN**LVV**LR	HQD**M**PF**Q**	**N**DPSR**H**PL
A. chrysogenum	VGFFVN**M**L**P**V**V**	HQD**M**PF**Q**	**N**DASR**H**PL
A. nidulans	VGFFVN**L**L**P**LR	HQD**M**PF**Q**	**H**DPSR**H**PL
Lysobacter sp.	IGFFVN**L**LALR	H**E**D**L**PF**E**	**K**DSSR**H**P**V**
N. lactamdurans	VGFF**A**N**L**LALR	H**GEL**PF**E**	**K**DPSR**H**P**V**

[a] Domains and their respective core sequences are taken from Ref. 38; for clarity, multiple residues in one position have been listed on a second line.
[b] Bold residues are altered with respect to the defined core sequences.

extensive similarities discussed in detail below. Surprisingly, the isolation of the synthetase from strains of *P. chrysogenum* proved most difficult. Steps following salting out by ammonium sulfate fractionation were found to be followed by proteolytic cleavage and inactivation (von Döhren et al., unpublished data). These problems were only recently overcome by Theilgaard and colleagues [50].

In perspective, the lessons learned from other NRPS systems have been confirmed. Due to the large size, ACV synthetases may be salted out at fairly low concentrations of ammonium sulfate and are easily enriched by steps such as gel filtration or purified by gradient centrifugation methods. Strains to be applied should be industrial high producers, such as *A. chrysogenum* C-10, or strains containing high levels of enzyme, such as *S. clavuligerus*. So far, overexpressing constructs which might permit easy access to fungal NRPS are missing. An approach for heterologous expression of the *N. lactamdurans* ACV synthetase in *Streptomyces lividans* proved successful, but only about 0.25% of the total protein constituted the synthetase, and about 30% of the activity was recovered after two column purifications [49].

Comparative assessments of purification protocols are generally difficult, and wide ranges of activities have been reported. ACV synthetases have been assayed by either employing a peptide adsorption resin (Porapak) [29,48] or high performance liquid chromatography (HPLC) detection of the derivatized tripeptide [26,27]. The adsorption assay, employing radiolabeled amino acids, has been

Table 2 Purification of ACV Synthetases: Activities and Recovery

Source	Activity[a] (turnover number), min^{-1}	Protein recovery[b] (%)	Activity recovery (%)	Reference
A. nidulans	0.005 (A)	0.025	2	29
A. chrysogenum	10.6 (A)	0.04 (A)	13	30
	1.5	0.0015 (T)	1.45	53
	8 (H)	0.03 (T)	18	52
P. chrysogenum	0.27 (A)			50
S. clavuligerus	4.2 (A)	0.015 (A)	10.4	30
	0.18 (H)	8.8 (T)	8.8	34
	0.36 (H)	0.001	9.5	33
	4.75 (H)	—	—	36
	0.02 (A)	0.0015	1.6	35
N. lactamdurans	0.26 (A)	0.26 (T)	28.7	49

[a] Activity determined by adsorption assay (A) or HPLC (H).
[b] Protein recovery with respect to total protein (T) or ammonium sulfate fraction (A).

designed for the rapid detection of an unstable enzyme, and being substrate limited, it operates well below saturation concentrations. Activity estimates are of limited use, due to traces of unlabeled substrates diluting the label and coadsorption of amino acids and by-products, if no further purification by, e.g., thin-layer chromatography (TLC) is performed. Estimates from optimized assays relying on the HPLC detection of ACV show comparable turnover numbers in the range of 4–5 min^{-1} for the *S. clavuligerus* enzyme, and 8–11 min^{-1} for the *A. chrysogenum* enzyme (Table 2, [51]). These rates compare well with those established for other NRPS systems, such as gramicidin S, cyclosporin, or enniatin synthetases. This rate, however, is the sum of 10 to 40 individual enzymatic reactions leading to a complex product. The purification protocols illustrate well the balance between activity and purity, considering the decline in activity due to proteolytic degradation and oxidation of these large proteins [54,55]. Thus, effective purification should not exceed two column steps, and conditions promoting enzyme damage should be reduced to a minimum.

Table 3 ACV Synthetases: Protein Properties

Source	Molecular mass from gene data[a]	Molecular mass, SDS-PAGE[b]	Molecular mass native[c]	Reference
A. nidulans	422,448	n.d.	420 (P) 460 (G)	6
P. chrysogenum	421,073 (a) 425,914 (b)	440	470 (P)	4,5,50
A. chrysogenum	414,767	350–400	700–800 (G)	55
		300	600 (G)	30
S. clavuligerus	Not available	283	560 (P)	34
		300	600 (G)	30
		350–400	350–400 (G)	33,55
		500	560 (G)	35
N. lactamdurans	404,079	430	480 (G)	8,49
Lysobacter sp.	411,593	Not available	Not available	56
Flavobacterium sp.	Not available	Not done	300 (G)[d]	48

[a] Data are available under the accession numbers: X54853 (*Aspergillus*); (a) X54296 (*Penicillium*, Turner et al.); (b) P26046 (*Penicillium*, Martin et al.); P258464 (*Acremonium*); P27743 (*Nocardia*); D50308 (*Lysobacter*); assumed holo-enzyme data should contain additional 3 × 340 Da for cofactors.
[b] Polyacrylamide gel electrophoresis under denaturating conditions employing dodecylsulfate.
[c] Estimations done either using native gel filtration (G) or native polyacrylamide gel electrophoresis (P).
[d] Reference value has been corrected with respect to gramicidin S synthetase 2 (512 kDa).

Identity of the isolate with the corresponding genes has been verified for the synthetases from *A. nidulans* [6] and *A. chrysogenum* [30] by sequence analysis of tryptic peptides. Estimation of the molecular mass of purified ACV synthetases agree with the predicted reading frames. Only few marker proteins are available in the respective region of 200 to 1000 kDa, and evaluations of electrophoretic mobility as a function of size in the presence of dodecylsulfate have not been clearly established (Table 3). The more essential question of whether the synthetases associate into dimers has been approached by electrophoresis under native conditions, gel filtration, and sedimentation studies. Again, the uncertainties in these migration studies are high, and shape factors matter even more than for protein–dodecylsulfate complexes. Except for the *A. chrysogenum* enzyme, no significant evidence has been seen for dimerization. No enzyme concentration dependence of ACV synthesis has been found, however, for a functional support of this finding [51].

III. BASIC CATALYTIC PROPERTIES

A. Enzyme Catalytic Studies

1. Basic Reaction Scheme

The evaluation of kinetic properties of NRPS systems is a problem of generally underestimated complexity. The basic path was established in 1971, defining activation, thiolation, and peptidyl transfers as basic reactions. The further refinement from structural data to establish the multiple carrier model, and now to tackle domain interactions, has added some precision to the questions asked. However, we have not yet arrived at a complete kinetic description of even the simple tripeptide synthetase. The ACV synthetase operates with four different substrates at six binding sites, releasing 3 moles of AMP and 3 moles of MgPPi for each ACV formed at optimal conditions [51]. A sequence of 10 reactions has been proposed in analogy to other NRPS-systems:

$$\text{E1} + \text{Aad} + \text{ATP} \rightarrow \text{E1(Aad-AMP)} + \text{PPi} \tag{1}$$
$$\rightarrow \text{E1-S}^{\text{p1}}\text{-Aad} + \text{AMP} \tag{2}$$

$$\text{E2} + \text{Cys} + \text{ATP} \rightarrow \text{E2(Cys-AMP)} + \text{PPi} \tag{3}$$
$$\rightarrow \text{E2-S}^{\text{p2}}\text{-Cys} + \text{AMP} \tag{4}$$

$$\text{E1-S}^{\text{pl}}\text{-Aad} + \text{E2-S}^{\text{p2}}\text{-Cys} \overset{\textbf{C1}}{\rightarrow} \text{E2-S}^{\text{p2}}\text{-Cys-Aad} + \text{El-S}^{\text{pl}}\text{H} \tag{5}$$

$$\text{E3} + \text{Val} + \text{ATP} \rightarrow \text{E3(Val-AMP)} + \text{PPi} \tag{6}$$
$$\rightarrow \text{E3-S}^{\text{p3}}\text{-Val} + \text{AMP} \tag{7}$$

$$\text{E2-S}^{p2}\text{-Cys-Aad} + \text{E3-S}^{p3}\text{-Val} \xrightarrow{\textbf{C2}} \text{E3-S}^{p3}\text{-Val-Cys-Aad} + \text{E2-S}^{p2}\text{H} \qquad (8)$$

$$\text{E3-S}^{p3}\text{-Val-Cys-Aad} \xrightarrow{\textbf{Ep3}} \text{E3-S}^{p3}\text{-D-Val-Cys-Aad} \qquad (9)$$

$$\xrightarrow{\textbf{TE}} \text{Aad-Cys-D-Val} + \text{E3-S}^{p3}\text{H} \qquad (10)$$

The synthetase consists of the three modules E1, E2, and E3 (for a complete description, see Sec. II.A). Each module is composed of an activation site forming the acyl or aminoacyl adenylate, a carrier domain which is posttranslationally modified with 4′-phosphopantetheine (S^P), and a condensation domain (C1, C2) or, alternatively, a structurally similar epimerization domain (Ep). Activation of aminoadipate (Aad) leads to an acylated enzyme intermediate, in which Aad is attached to the terminal cysteamine of the cofactor (E1-S^{p1}-Aad) [reactions (1) and (2)]. Likewise, activation of cysteine (Cys) leads to cysteinylated module 2 [reactions (3) and (4)]. For the condensation reaction to occur between aminoadipate as donor and cysteine as acceptor, both intermediates are thought to react at the condensation site of module 1 (C1). Each condensation site is composed, in analogy to ribosomal peptide formation, of an aminoacyl and a peptidyl site. In this case of initiation, the thioester of Aad enters the P-site, while the thioester of Cys enters the A-site. Condensation occurs and leaves the dipeptidyl intermediate Aad-Cys at the carrier protein of the second module [reaction (5)]. The third amino acid valine is activated on module 3, and Val is attached to the carrier protein 3 [reactions (6) and (7)]. Formation of the tripeptide occurs at the second condensation site C2, with the dipeptidyl intermediate entering the P-site and the valinyl-intermediate the A-site [reaction (8)].

Finally, epimerization of the tripeptide (or dipeptide) intermediate occurs at the epimerization site of module 3 (Ep3) [reaction (9)], and the stereospecific peptide release is controlled by the thioesterase (TE) [reaction (10)].

Early work summarized in Sec. I.B had shown that ACV is made from L-Aad, L-Cys, and L-Val, that δ-(L-α-Aad)-L-Cys (AC) may be converted into ACV, but L-Cys-D-Val is not converted. Likewise, D-Val is not a substrate for ACV synthetase, contrary to the first-characterized NRPS systems of gramicidin S and tyrocidine [2,3]. These observations are in agreement with the scheme, except for the incorporation of AC. This dipeptide was later shown to be activated as an adenylate.

2. Adenylate Formation

Activation of the amino acid carboxyl groups is the only set of clearly reversible reactions catalysed by NRPS. So far these reactions have been studied in the case of ACV synthetases only by the amino acid-dependent ATP–PPi exchange reaction [reaction (11)]:

$$E + RCO_2^- + MgATP^{2-} \rightleftharpoons E\{RCOAMP\} + MgPPi^{2-} \quad (11)$$

In case of equilibrium, addition of labeled PPi yields labeled ATP and this product can be employed to detect NRPS or related enzymes. The respective reaction rates provide information on adenylate formation/pyrophosphorylysis, apparent K_m of substrates and substrate analogs, and with some enzyme kinetic efforts, substrate affinities and the patterns of substrate binding may be deduced. The ease and the sensitivity of the procedure makes it the primary method of investigation of NRPS substrate specificity.

There are several remarkable features of this isotope-exchange assay. If the reaction progress is followed, equilibrium of isotope exchange is attained at a defined distribution of the label between ATP and PPi. This has been shown in the case of ACV synthetase by Baldwin and colleagues [57], but is rarely done because of the restrictions of enzyme stability. Evaluation of this equilibrium, which has been reached in the case of the *A. chrysogenum* enzyme in about 3 hr, has never been attempted, but the concentrations roughly reflect the initial rates of the exchange reactions.

Following structure–activity studies, the adenylate is thought to be stabilized within a cleft formed between the two subdomains of the activation domain [41,58]. The rate is thus related to the formation, presence, and stability of this mixed anhydride with respect to PPi, and at high $MgATP^{2-}$ concentrations, with respect to ATP in the formation of diadenosine tetraphosphate (A_2P_4). Thus a high rate of the amino acid-dependent isotope exchange does not necessarily reflect the efficiency of adenylate formation, and certainly not the efficiency of incorporation of an amino acid into peptidyl intermediates or the final product.

As most NRPS multienzymes are multidomain proteins with multiple activation domains, multiple sites may participate in the reactions assayed, and no clear result concerning a single specific site may result. In ACV synthetases, the nonadditivity of the initial rates has been observed in the *S. clavuligerus* enzyme [35] and the *A. chrysogenum* enzyme [1]. Two or more site activations of one substrate amino acid could be expected to depend on different binding constants, and thus be detectable by kinetic analysis. So far, however, no evidence for mixed types of concentration dependence has been found. It is thus not yet clear if nonadditivity results from misactivation or alteration of kinetic properties in the presence of multiple substrates. In the case of gramicidin S synthetase 2, evidence for misactivations has been reported [59].

A fairly large number of potential amino acids has been assayed for adenylate formation. These amino acids can be grouped into compounds presumably being processed by just one of the three domains, or by several of the domains. Some of the data have been compiled in Table 4, and the main questions to be addressed are: (1) Do we find significant differences in fungal and bacterial ACV

Table 4 Activation of Amino Acids and Related Compounds by ACV Synthetases as Determined by Amino Acid-Dependent ATP-[^{32}P]PPi Exchange

Substrate	ACVS *A. chrysogenum*[a]	ACVS *S. clavuligerus*[b]
Group 1: predicted Aad analogs		
L-2-Aad	++++	++++[c]
S-Carboxy-methyl-Cys	++++	n.a.[d]
Glu	+	−
D,L-2-Amino-pimelate	−	−
D-2-Amino-adipate	−	−
Lys	−	−
Asp	−	n.a.
Phenylacetate	−	n.a.
2-Aminocaproate	−	n.a.
5-Aminovalerate	−	−
L-2-Aminopimelate	(+)	n.a.
Group 2: predicted Cys analogs		
L-Cys	++++	++++
D-Cys	(+)	n.a.
L-Ser	(+)	+
DL-O-Methyl-Ser	(+)	n.a.
L-O-Methyl-Thr	−	n.a.
DL-Homo-Cys	(+)	n.a.
L-S-Methyl-Cys	+++	n.a.
L-N-Acetyl-Cys	(+)	n.a.
L-3-Chloro-Ala	++	n.a.
5-(L-2-Aminoadipyl)-L-cysteine	+	−
Group 3: predicted Val analogs		
L-Val	++++	++++
allo-Ile	++	++
2-Aminobutyrate	++++	++
Norvaline	++	++
Allylglycine	++++	n.a.
L-Leu	++	(+)
L-Ile	++	+
Gly	+++	+
Norleucine	++	+
D-Val	−	(+)
DL-2-Amino-3-methyl-butyrate	++	n.a.
L-Thr	−	−
L-Vinyl-glycine	++	n.a.
L-t-Leu	(+)	n.a.
DL-3-Fluoro-valine	(+)	n.a.
L-3-Chloro-alanine	++++	n.a.
L-Ala	++	+

Table 4 Continued

Substrate	ACVS *A. chrysogenum*[a]	ACVS *S. clavuligerus*[b]
2R,3S-2-Amino-3-fluoro-butyrate	++	n.a.
2R,3R-2-Amino-3-fluoro-butyrate	+	n.a.
L-Phe	+	++++
L-Tyr	+	++
L-Trp	+	+
L-His	−	++
L-Pro	(+)	+
L-Met	(+)	−
3-Ala	(+)	+
4-Abu	−	(+)

[a] Activities are given by: −, not detected or significant (<1%); (+), 1–5%; +, 5–10%; ++, 10–50%; +++, 50–100%; ++++, 100% or >100% relative to natural main substrate.
[b] Data have been taken mainly from Ref. 51.
[c] Note that there are large differences in the rates of Aad activation between the fungal and bacterial enzymes.
[d] n.a., not available.
Source: From Refs. 47, 51; von Döhren et al., unpublished.

synthetases with respect to site specificity? and (2) How does adenylate formation reflect the uptake into the peptide products?

Early investigations missed the detection of 2-aminoadipate activation in the *S. clavuligerus* ACV synthetase [1]. Thus adenylate formation had been questioned, and even a study titrating the release of PPi from gamma-labeled ATP [60] showed less than 1% yield compared to the amount of enzyme applied [60]. Employing Aad concentrations in the range of 5–20 mM and substituting Tris buffer with MOPS buffer, Schwecke finally did demonstrate an Aad-dependent ATP-PPi exchange [60] with a K_m of about 10 mM. Another substantial difference of the Aad domains is activation of L-Glu by the fungal enzyme, while no activation was detectable in the bacterial one. Another significant difference concerns the activation of the dipeptide Aad-Cys, which again does not promote the isotope exchange in *S. clavuligerus* ACV synthetase. As a third major difference, aromatic amino acids are well activated by the bacterial ACV synthetase, but not as well by the *A. chrysogenum* enzyme. A surprising result has been reported with ACV synthetase from *N. lactamdurans* expressed in *S. lividans*: the enzyme converted 6-oxo-piperidine-2-carboxylic acid and cystathionine into ACV [49]. As only nanomoles of labeled ACV were formed from millimolar amounts of precursors, the data need to be substantiated in more detail. No adenylate formation has been found with cystathionine and the *A. chrysogenum* enzyme (von Döhren et al., unpublished).

In a recent analysis of adenylate domains, Conti et al. [61] proposed positionally conserved side chains lining the respective amino acid and acid-binding pockets. This analysis has led both Stachelhaus and Challis and colleagues to the proposal of a nonribosomal code [45,62]. Aligning the polypeptide sequences between the core motifs A3 and A6 (see Sec. II.A), eight or nine pocket-lining residues are predicted. Identical or similar residues permit the prediction of amino acid specificity with remarkable accuracy. The bacterial and fungal ACV synthetase domains each show identical sets of residues, which, however, do not match any set from other known NRPS systems (Table 5). Stachelhaus et al. [62] list the derived ACV synthetase codes as defined templates for Aad, Cys (type 2), and Val (type 2). Most striking is the presence of charged side chains in all of these signatures, which would not be neutralized by substrate charges. Challis et al. [45] include only some of the ACV synthetase data in their analysis, and

Table 5 Predicted Amino Acid-Binding Side Chains for ACV Synthetases and Other NRPS Systems

Substrate	Enzyme/source	Signature	Reference
Aad	ACV synthetase, fungal	EPRHIVEA	von Döhren et al.
	ACV synthetase, bacterial	EPRNLVEA	von Döhren et al.
	Lys2, various eukaryotes	DPRHFVMQ	von Döhren et al.
Cys	ACV synthetase, fungal	DHESDVGI	54
	ACV synthetase, bacterial	DHESDIGI	von Döhren et al.
	Bacterial, type 1	DLYNLSLI	54
	Bacterial, type 2	DLYNLSPI	54
	Bacterial, type 3	DLFNLSLI	54
	Bacterial, type 4	DLYNMSNI	54
Val	ACV synthetase, fungal	DFESTAAY	54
	ACV synthetase, bacterial	DFESLAAY	von Döhren et al.
	Bacterial, type 1	DAFWIGGT	54
	Bacterial, type 2	DAYFWAVT	54
	Bacterial, type 3	DVFWIGGT	54
	Fungal, type 1	DAWMFAAI	54
	Fungal, type 2	DAWMFAAV	54
	Fungal, type 3	DGWFIGII	54

comment on the proposed Cys pocket as being close in structure to the Val pocket, and differing from other NRPS due to a speculative isolated evolution by gene duplication. The predictive structure-based model of amino acid recognition, derived largely by comparative analysis of bacterial NRPS domains, clearly is not applicable for ACV synthetases. A possible explanation could be found in a differing architecture of these substrate-binding sites. Thus an alternative selection of the critical substrate-lining residues might define differing contact sites without the problem of unbalanced charges.

To derive binding constants, the multiplicity of substrates has to be considered. Most work has settled for a constant $MgATP^{2-}$ concentration to approximate a K_m with simple Michaelis-Menten kinetics. It is not clear, however, if such K_m values derived for either ACV synthesis or the amino acid-dependent ATP-PPi exchange are comparable. The available data have been compiled in Table 6.

3. *Adenylate Domains*

To definitely assign specific catalytic activities to defined regions, the dissection of ACV synthetases has been attempted by proteolytic and molecular genetic techniques (Fig. 1). Several quite unexpected results have been obtained. As shown in Fig. 1, limited proteolysis of *A. chrysogenum* ACV synthetase by proteinase K and subtilisin produced two fragments of 119 and 95 kDa, respectively [63]. The 119-kDa fragment (fragment 1) contained the second adenylate domain and specifically catalyzed a cysteine-dependent ATP–PPi exchange reaction. The

Table 6 Kinetic Constants (K_m) of ACV Synthetases

Source of enzyme	Aad (mM)	Cys (mM)	Val (mM)	Reference
ACV biosynthesis				
P. chrysogenum	0.046	0.080	0.083	50
A. chrysogenum	0.055	0.11	0.53	51
	0.17	0.026	0.34	26
S. clavuligerus	0.56	0.07	1.14	27
	0.63	0.12	0.30	36
	0.63	0.43	0.38	54
Amino acid-dependent ATP–PPi exchange				
A. chrysogenum	0.1	0.21	0.09	57
S. clavuligerus	10	1.25[a]	2.4[a]	35

[a] K_D determined at variable $MgATP^{2-}$ concentrations.

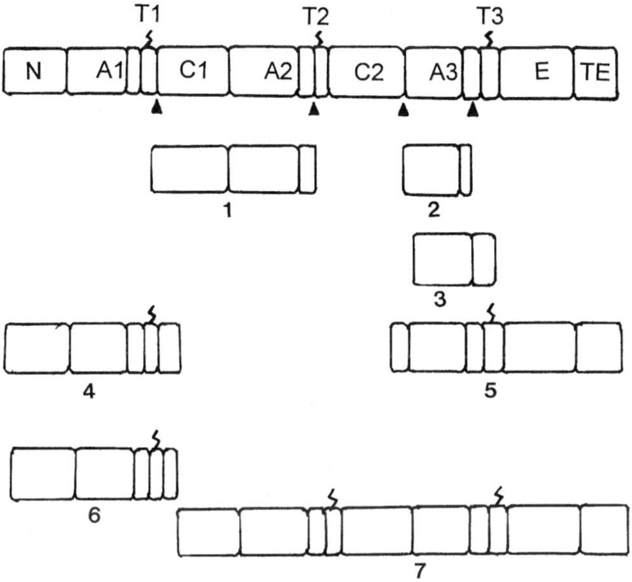

Figure 1 Structural organization of ACV synthetases and proteolytic (1,2) as well as expressed fragments (3–7). Top: domain structure, A1–A3 adenylate domains of the three modules, T1–T3 thiolation domains, C1 and C2 condensation domains, E epimerization domain, TE thioesterase, and N aminoterminal domain. Triangles indicate identified subtilisin and proteinase K cleavage sites of the *A. chrysogenum* enzyme leading to fragment 1 (activating cysteine) and fragment 2 (activating 2-aminoadipate) [63]. Fragment 3 is the complete A3 domain of *P. chrysogenum* expressed in *A. nidulans*, which activates leucine and 2-aminoadipate, but not valine [63]. Fragment 4 is the amino-terminal region of the *S. clavuligerus* enzyme, expressed in insoluble form in *E. coli*, and solubilized. All three substrates, 2-aminoadipate, cysteine, and valine, form adenylates, but exclusively 2-aminoadipate is bound as thioester [60,64]. Fragment 5 is the C-terminal fragment of the *A. chrysogenum* enzyme, expressed in *E. coli* in insoluble form, solubilized, and shown to activate leucine, valine, and 2-aminoadipate (von Döhren et al., unpublished). Fragments 6 and 7 of the *P. chrysogenum* enzyme were expressed in soluble form in *A. nidulans* as β-galactosidase fusion proteins, and activate 2-aminoadipate and valine, and cysteine, valine, and 2-aminoadipate, respectively [65].

95-kDa fragment containing the third adenylate domain yielded a second cleavage product of 47 kDa (fragment 3), which surprisingly activated Aad. Activation of Val was completely lost upon proteinase treatment.

In molecular genetic approaches, a 110-kDa N-terminal fragment of the *S. clavuligerus* ACV synthetase has been expressed in *Escherichia coli* (fragment 4) [60]. The insoluble protein obtained was dissolved in urea and renatured, but activation of all three amino acids was detected, with Val giving the highest reaction rate. Although only a fraction of the expressed protein is posttranslationally modified by pantetheine, the formation of thioesters had been investigated, and only Aad was detected [60,64]. This experiment demonstrated a selective step following adenylate formation for the first time. It further demonstrates that although misactivation may proceed in a fragment of the synthetase, a further selection of intermediates may take place.

As these experiments required the solubilization of denaturated polypeptides, some doubt as to their validity is justified. To overcome the folding and solubility problems, Turner et al. accomplished the homologous expression of ACV synthetase fragments from *P. chrysogenum* in *A. nidulans* [65,66]. A strain carrying an *acvA* deletion was used to express β-galactosidase fusions of the first module and a fragment containing the second and third module, respectively (fragments 6 and 7, β-galactosidase fusion not shown). The respective fusion proteins were obtained only in low yield, and proteolysis precluded their complete purification. However, the first domain was shown to activate Aad, Val, Cys, isoleucine, allo-isoleucine, α-aminobutyrate, S-carboxymethyl-cysteine, and glutamic acid, which is in complete agreement with the activity data of the bacterial fragment. The C-terminal fragment did activate Cys, which also was shown to form a thioester, but in addition produced adenylates with Val, isoleucine, leucine, and α-aminobutyrate. No activation of Aad was detected. These results again supported the assignment of Aad activation to the first adenylate domain.

The third adenylate domain has been further investigated by fragment expression. A C-terminal 136-kDa fragment of the *A. chrysogenum* ACV synthetase was expressed in *E. coli* and the resulting protein pellet solubilized and refolded (von Döhren et al., unpublished, fragment 5). Activation of leucine, Val, α-aminobutyrate, Aad, aminocaproic acid, and norvaline was detected. To achieve expression of a soluble enzyme, the adenylate domain of the *P. chrysogenum* synthetase was excised as a 501-amino acid fragment by polymerase chain reaction (PCR) and expressed in *A. nidulans*, again as a β-galactosidase fusion protein (fragment 3) [63]. The purified protein showed adenylate formation with leucine, 2-amino-ethyl-cysteine, Aad, S-carboxymethyl-Cys, but surprisingly not with Val. The interpretation of the third-domain data implies that the Val-dependent ATP–PPi exchange activity diminished with the reduction of fragment size, and finally disappeared in the fusion protein. Aad and leucine activation have been found in the smaller fragments.

These results show a clear distortion of substrate binding and catalytic activities upon fragmentation of the multienzymes. The substrate pocket architecture seems to depend on the context of adjacent domains as well. Although questions remain, the linearity rule of NRPS holds in ACV synthetases. Open questions remain on the fate of possibly misactivated amino acids in the terminal domains. Could these amino acids be possibly propagated *in trans* to the respective adenylation domains, or are they lost by hydrolytic proofreading?

4. Peptide Synthesis and Adenylate Intermediates

In a remarkable series of experiments, Baldwin and colleagues have studied peptide bond formation in the *A. chrysogenum* ACV synthetase [1,57,67–69]. In the presence of glutamate, Cys, and Val, L-cysteinyl-D-valine is recovered exclusively. As has been stated above, glutamate does form an adenylate, but fails to be incorporated into peptides. The glutamate adenylate is thus not accessable for condensation to proceed at the first condensation domain. Presumably its presence, due to an induced conformational change [41], enhances peptide bond formation at the second condensation domain between the thioesters of Cys and Val, followed by epimerization of the dipeptidyl intermediate and hydrolytic release [reactions (12) and (13), with Cys for A2].

$$\text{E2-S}^{\text{p2}}\text{-A2} + \text{E3-S}^{\text{p3}}\text{-Val} \rightarrow \text{E3-S}^{\text{p3}}\text{-Val-A2} + \text{E2-S}^{\text{p2}}\text{H} \tag{12}$$

$$\text{E3-S}^{\text{p3}}\text{-Val-A2} \xrightarrow{\textbf{Ep3}} \text{E3-S}^{\text{p3}}\text{-D-Val-A2} \xrightarrow{\textbf{TE}} \text{A2-D-Val} + \text{E3-S}^{\text{p3}}\text{H} \tag{13}$$

Likewise, the assumed O-methyl-serinyl-thioester intermediate reacts only slowly with the Aad thioester, but is readily released by aminolysis of free Val [reaction (14), with A2 either Cys or O-methyl-serine] or may react with the Val-thioester intermediate to be further processed [reactions (12) and (13), with O-methyl-serine for A2].

$$\text{E2-S}^{\text{p2}}\text{-A2} + \text{Val} \rightarrow \text{E2-S}^{\text{p2}}\text{H} + \text{A2-Val} \tag{14}$$

Following the dipeptide synthesis from O-methyl-serine and Val by determining the loss ^{18}O from di[^{18}O]valine, evidence for both reactions was obtained, but a direct reaction of Cys-AMP with free Val cannot be excluded [reaction (15)].

$$\text{E2(A2-AMP)} + \text{Val} \rightarrow \text{E2} + \text{A2-Val} + \text{AMP} \tag{15}$$

The dipeptide L,L-O-(methyl-serinyl)valine was formed without significant loss of label, and at the same time, no label was observed in the AMP released. This result thus excludes a thioester intermediate, which by thiolysis of the adenylate would have led to an even distribution of the ^{18}O between AMP and Val. However, the isomeric dipeptide L,D-O-(methyl-serinyl)valine was recovered with all possible labeling patterns of ^{18}O^{18}O, ^{16}O^{18}O, and ^{16}O^{16}O [69]. To explain the retainment of label in the epimerized dipeptide is not easy. Baldwin and colleagues propose an alternative direct acyl transfer mechanism operating with dipeptidyl adenylates of the type Cys-Val, being epimerized, and transferred to a thiol group to undergo peptide bond formation with the Aad thioester followed by hydrolysis [1]. This is a possible interpretation of the data, but the rates of formation of the dipeptide shunt or byproducts in the 1–2% range compared to ACV formation may exclude this path from *in vivo* or optimized *in vitro* conditions. Any direct acyl transfer mechanism would be based on the surface migration of free adenylates between the respective condensation domains, which is both unlikely and not required to interpret the available experimental data [67–69]. Evidence for the collinearity rule in the catalytic sequence has been obtained, omitting Val from the tripeptide biosynthetic assay: the dipeptide Aad-Cys has been shown to accumulate as a thioester intermediate [70].

To demonstrate peptide bond formation from aminoacyl adenylates, Dieckmann [71] has applied the isolated adenylate domain of tyrocidine synthetase 1 to generate various dipeptides. These were obtained from phenylalanyl adenylate with alanine, leucine, leucineamide, and phenylalanine [reaction (16), with AA = amino acid or amine acceptor].

$$\text{E(PheAMP)} + \text{AA} \rightarrow \text{Phe-AA} + \text{E} + \text{AMP} \qquad (16)$$

Further studies are needed on the kinetics of aminolysis of adenylates and thioesters and the possible migration of adenylates to contact reactive intermediates. Such surface diffusion or tunneling has to compete with the effective pantetheine-mediated covalent transport system.

5. *Thiolation*

Aminoacylation or acylation of the "swinging arm" cofactor 4′-phosphopantetheine is considered as the covalent transport principle in NRPS and polyketide synthases (PKS). Experimental procedures to establish the presence of thioester intermediates have largely relied on the demonstration of acid stable and performic acid-cleavable radiolabeled amino acids. This approach has recently been extended by the mass spectrometric detection of cleaved intermediates and cofactor-containing enzyme fragments [12,70]. Any radiolabeling procedure of enzyme-bound intermediates requires free pantetheine thiol groups, but these may be acylated in the respective enzyme preparation. Stabilities of aminoacyl- and peptidylthioesters depend on the type of acyl compound involved. Rates of hydrolytic cleavage have been estimated in the gramicidin S system [72]. At 3°C, half-lives for aminoacyl- or peptidylthioesters were between 1 and 90 hr. Reduced stabilities of 0.4–0.5 hr were observed for thioesters of ornithine or ornithylpeptides due to the cyclization to 3-amino-2-piperidone. The 2-aminoadipyl intermediate apparently does not cyclize effectively, since 6-oxo-piperazine-2-carboxylic acid has not been detected upon incubation of ACV synthetase from *P. chrysogenum* in the presence of Aad and $MgATP^2$ [50,73]. It is thus not surprising that aminoacylation levels are often well below the expected estimates

[69]. Rates of aminoacylation reactions have been determined in the range of more than 100 min^{-1} [74]. The reaction involves the interaction of adjacent adenylate and thiolation domains, or possibly of nonadjacent domains *in trans* [75].

Carrier domains have a conserved structure of four helices with a loop containing the cofactor attachment site [76,108]. 4′-Phosphopantetheine-protein transferases (PPT) from eukaryotes have so far not been identified, but partially purified from *A. nidulans*, *Fusarium scirpi*, and *Tolypocladium niveum* (von Döhren et al., unpublished). Apparently the beta-lactam biosynthetic cluster does not contain the respective gene, but at least some of the producer strains have been shown to contain additional NRPS clusters. The quantitation of the cofactor in isolated synthetases represents an unsolved analytical problem. Microbiological determinations are not reliable, as underestimates are generally obtained [1,30]. A possible approach to determine the state of posttranslational modification, the ratio of apo- to holoenzyme, or the cofactor stoichiometry is to measure the 4′-phosphopantetheine transfer by the pantetheine-protein transferase assay [77–79]. A PPT fraction obtained from *A. nidulans*, which modified apo-tyrocidine synthetase 1 from *Bacillus brevis*, did not transfer significant amounts of pantetheine from labeled CoA to ACV synthetase, indicating the absence of apo-synthetase. So far no fungal PPT-genes have been identified.

The central role of the thiolation domain is evident from their multiple interactions with adjacent and nonadjacent domains. The current multiple-carrier thiotemplate model predicts successive contacts of the first thiolation domain of ACV synthetase with the adjacent aminoadipate adenylate domain and the first condensation domain; the second thiolation domain to interact with the cysteine adenylate domain, and both the adjacent and nonadjacent condensation domains; the third thiolation domain then interacts with the valine adenylate domain, the nonadjacent second condensation domain, the adjacent epimerization domain, and the nonadjacent thioesterase domain (Fig. 2). Protein regions involved in these successive protein–protein interactions involve the highly conserved carrier domain structures of only about 80 amino acids. So far, only two detailed structures of the respective NRPS domains are available [61,108].

6. *Peptide Bond Formation*

Condensation reactions require a domain of about 450 amino acids, which has been functionally identified employing gramicidin S/tyrocidine synthetase systems [80]. The current functional interpretation proposes, in analogy to the ribosomal system, an aminoacyl site and a peptidyl site to enter the activated intermediates [47]. The acylated carrier proteins would thus resemble charged tRNAs, and the condensing site the peptidyl transferase. As condensing domains show structural differences, the question of control of incoming substrates may be relevant. It has been shown by Belshaw and colleagues with a condensing domain

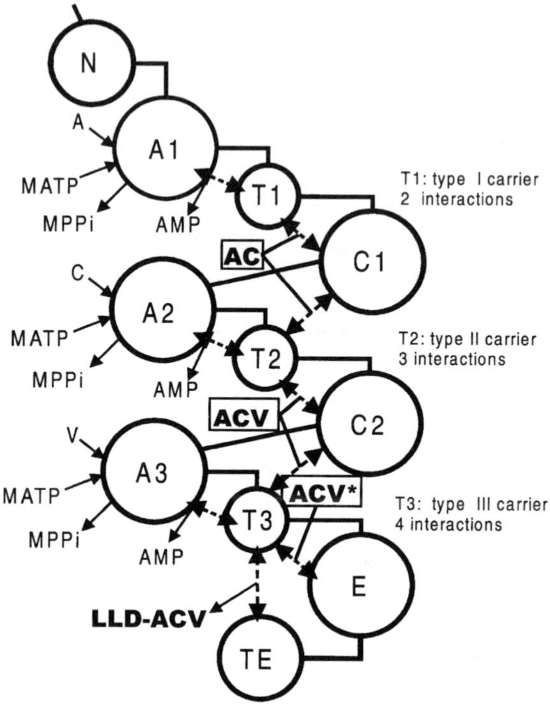

Figure 2 Scheme of domain interactions in ACV synthetases. Domains are presented as circles, linkers are indicated as thin lines, descriptions are as in Fig. 1. Substrate amino acids (A, C, and V) and MATP enter activation domains A1–A3, with MPPi leaving as product. Interactions of the A domains with their adjacent thiolation domains (T1–T3) leads to (amino)acylation with release of AMP. Peptidyl intermediates [AC, LLL-ACV (ACV) and LLD-ACV(ACV*)] are boxed, and are produced upon interactions of the thiolation domains with the respective condensation domains. The third thiolation domain interacts with four domains, and the final interactions lead to epimerization and hydrolytic release.

forming a D-Phe-Pro bond that replacement of proline by alanine dramatically decreases the reaction rate [81]. It is thus likely that the two condensation domains of ACV synthetases are structurally adapted to 2-aminoadipate and cysteine, and cysteine and valine, respectively. Differences between fungal and bacterial condensation domains (see Table 1) may reflect such substrate selection determinants.

A number of substrate analogs have been employed to generate unnatural tripeptides [1,82,83] (Table 7). Some of the work just applying Porapak adsorp-

Table 7 Enzymatic Peptide Synthesis With ACV Synthetase

Product	Amino acids supplied	Identity by	Reference
Aad-δ→Cys→D-Val	Aad, Cys, L-Val	HPLC, MS	36
Aad-δ→Cys→D-Ile	Aad, Cys, L-Ile	HPLC, ESMS, NMR	57
Aad-δ→Cys→D-aIle	Aad, Cys, L-aIle	HPLC	36
Aad-δ→Cys→D-Abu	Aad, Cys, L-Abu	HPLC	36
Aad-δ→Cys→D-Nve	Aad, Cys, L-Nve	HPLC	36
Aad-δ→Cys→D-Leu	Aad, Cys, L-Leu	HPLC	36
Aad-δ→Cys→D-allylGly	Aad, Cys, L-allylGly	HPLC	36
Aad-δ→Cys→Gly	Aad, Cys, Gly	HPLC	36
Aad-δ→Ser→D-Val	Aad, Ser, L-Val	MALDI-MS	von Döhren et al., unpublished
Aad-δ→Hse→D-Val	Aad, Hse, L-Val	MALDI-MS	von Döhren et al., unpublished
Aad-δ→Hcy→D-Val	Aad, D,L-Hcy, L-Val	HPLC, MALDI-MS	36; von Döhren et al., unpublished
Aad-δ→OMeSer→D-Val	Aad, OMeSer, L-Val	MALDI-MS	von Döhren et al., unpublished
CMC-δ→Cys→D-Val	CMC, Cys, L-Val	HPLC, ESMS, NMR	57
Aad-δ→allylGly→D-Val	Aad, allylGly, L-Val	HPLC, ESMS, NMR	57
Aad-δ→vinylGly→D-Val	Aad, vinylGly, L-Val	HPLC, ESMS	57

tion [57] has not been included, since amino acids and nontripeptide products may interfere. Rate estimates for some tripeptides are compiled in Table 8. As with substrates for the adenylate-forming reaction, some differences between prokaryotic and eukaryotic enzymes are obvious. Some rates do not correlate with the reversal of adenylate formation (Table 4).

To approach the kinetic description of the system, concentration dependencies at fixed $MgATP^{2-}$ levels have been determined (Table 6), which resemble simple Michaelis-Menten kinetics. Estimates of the various enzymes and sources are slightly deviating, and some differences can be traced to variable conditions. These K_m values do not correspond to binding constants, as clearly conditions for Michaelis-Menten kinetics are not met. Strikingly different is the binding of 2-aminoadipate in fungal and bacterial ACV synthetases. The difficulty to detect Aad-adenylate by isotope exchange in the *S. clavuligerus* enzyme either points to its stability against pyrophorylysis or a low concentration of enzyme-bound adenylate. The K_m values in the tripeptide assay are only slightly above those in

Table 8 ACV Biosynthesis: Relative Rates[a] of Peptide Formation

Amino acid	*A. chrysogenum*[b]	*S. clavuligerus*[c]
Aad	**100**	**100**
S-carboxymethyl-cysteine	99	52
Cys	**100**	**100**
Serine	4	n.d.
Homocysteine	9	73
Allylglycine	122	n.d.
Homoserine	n.d.[d]	n.d.
O-Methylserine	76	n.d.
Vinylglycine	101	n.d.
Valine	**100**	**100**
Allo-Isoleucine	100	44
2-Aminobutyrate	5	41
Norvaline	n.d.	27
Allylglycine	15	20
Leucine	n.d.	13
Isoleucine	20	11
Glycine	n.d.	10
Vinylglycine	5	n.d.
Norleucine	n.d.	8

[a] Rates are given in percent relative to the original substrates Aad, Cys, or Val.
[b] Adsorption assay (Porapak) [57].
[c] HPLC assay [36].
[d] Not deleted.

fungal systems. This finding is quite a remarkable result, since often in the analysis of NRPS systems, the amino acid-dependent ATP-PPi exchange reaction is tentatively taken as a measure of substrate utilization in peptide synthesis [45,62].

7. Epimerization

The studies on epimerization of the amino acid residue in position 3 have been reviewed in some detail by Bycroft et al. [1], and for NRPS systems by Kleinkauf and von Döhren [84]. From the failure to detect D-valine thioester intermediates in both fungal and bacterial ACV synthetases ([35,60]; Etchegaray, personal communication), epimerization at the peptide stage has been concluded. The epimerization domain located directly next to the valine thiolation domain would then catalyze the reversible epimerization of the pantetheine-attached tripeptidyl intermediate, leading to the enzyme-bound epimers LLD-ACV and LLL-ACV (Fig. 2). Attempts to detect the tripeptide intermediates have been unsuccessful, due to the rapid release of the product LLD-ACV by the C-terminal thioesterase function. In a recent analysis of the epimerization domain of gramicidin S synthetase, Stachelhaus and Walsh [85] have performed a mutational analysis generating mutant enzymes with altered epimer ratios and altered behavior in epimer transfer in the successive condensation reaction. This reaction shows a kinetically controlled stereopreference for the D-aminoacyl intermediate, which is reduced from 98% to 44% in the H753A and Y976A mutations in the motifs E2 (HHxxxDxxSW) and E7 (NY), respectively. The results demonstrate that stereocontrol is not only exerted in the next reaction to follow, but that control may as well be placed in the epimerization domain and presumably the domain-interaction regions. With the idea of blocking the terminating thioesterase function to isolate both tripeptide intermediates, mutagenesis work has been carried out by Kallow et al. ([86]; see Sec. III, A.8).

8. Thioesterase

The carboxy-terminal thioesterase domain of δ- (L-α-aminoadipyl)-L-cysteinyl-D-valine synthetase catalyzes the hydrolytic release of the tripeptide product (LLD-ACV). Following the discussion in Sec. III.A.7 on the epimerization of the peptide, both epimers are thought to be bound as thioesters to the third thiolation domain; upon interaction with the thioesterase, the LLD-peptide would be preferentially hydrolyzed. To abolish hydrolysis, a S3599A change was introduced in the highly conserved GXSXG motif, resulting in a more than 95% decrease of penicillin production [86]. Purification of the modified multienzyme showed surprisingly only a 50% reduction of the peptide formation rate, with the stereoisomer LLL-ACV as the dominating product. This surprising result indicates that the seryl side chain of the classical thioesterase motif is not essential for hydrolysis of the peptidyl intermediate.

Table 9 Core Motifs[a] of Thioesterases Involved in NRPS Systems

	Separate thioesterases	Integrated thioesterases following thiolation domains	Integrated thioesterases following epimerase domains (ACV synthetases)
T1[b]	PxAGG	lFxfxP(a/v)gg	LF(V/L)LPPGEGGAESY
T2	pgr	—	—
T3	pxxxfGHSmGa	GpyxxxG(W/Y)SxGg	QPxGPYxxxGWSFGG
T4	LfiSgxxxAP	—	—
T5	LPxLRAD	—	—
T6	Wr	W	—
T7	GgxHHΦl	gxgxh	—
ET1	—	—	VvFNN
ET2	—	—	LxxiDxΦF
ET3	—	—	LDPI
ET4	—	—	IvLFKA
ET5	—	—	QxxiΦEyy
ET6	—	—	NnLDxiLp

[a] Conserved residues are given in capital letters, not completely conserved positions in small letters, Φ represents aromatic amino acids (F, Y, W).
[b] Motifs designated as T may be found in several types of thioesterases, while motifs defined as ET may be restricted to thioesterases associated with epimerization domains.

Thioesterases of ACV synthetases differ from other thioesterases integrated in nonribosomal peptide synthetases in their direct association with an epimerase domain. A comparison of the primary sequences reveals significant differences to other NRPS thioesterases shown in Table 9. So the GXSXG motif may be involved in the control of tripeptide epimerization by selection of the isomer to be released. Finally, the data support the presence of LLL-ACV as an intermediate in penicillin biosynthesis.

B. *In Vitro* Synthetic Applications

Going beyond product synthesis and just looking into the path of assembly or optimization of enzyme systems to exploit the synthetic potential of NRPS systems was pioneered by Wang and colleagues in the 1970s, mainly on gramicidin S, and later also attempted by Kleinkauf and co-workers with gramicidin S as well as with enniatins, cyclosporins, and ACV-type of tripeptides (reviewed in [87]).

In early studies, Banko et al. used total protein extracts from *A. chrysogenum* C-10 (ATCC 48272) to obtain AC with a linear rate for 6 hr corresponding to about 1 pmol/min*mg [25]. The rate increased to 130 pmol/min*mg for ACV when all three amino acids were added, but decreased to 49% with L-carboxymethylcysteine, to 1.5% with L-glutamate replacing 2-aminoadipate, to 19% with allo-isoleucine, and to 6.5% with L-2-aminobutyrate replacing valine [26]. Extensive purification and stabilization studies of this synthetase have been carried out by Zhang and Demain [55,83], and later by Baldwin and colleagues [30,57]. Final activities of about 10 nmol/min*mg were reached (Table 2). In a study employing an enzyme preparation from the same strain of *A. chrysogenum*, Hadjmalek and Bronstad prepared ACV and the analogs containing L-homocysteine and S-carboxy-methylcysteine at the microgram-to-milligram scale using 20-ml reaction volumes (von Döhren et al., unpublished).

The bacterial ACV synthetase from *S. clavuligerus* had been investigated by Kadima and co-workers [88] with the aim of producing the tripeptide. The enzyme activity was stabilized by glycerol, dithiothreitol, Mg^{2+}, and the substrate amino acids, with a half-life of 6 days at 4°C. Immobilization of the synthetase to DEAE-Trisacryl or aminopropyl-glass resulted in a drastic loss of activity, estimated as 5 pmol ACV/min*mg immobilized enzyme [28]. In an ultrafiltration cell at room temperature and a nitrogen atmosphere, 45–65% activity was retained after 5 uses during a 24-hr period. Extrapolation of the data indicated that 1.3 g of partially purified synthetase would produce 411 mg of ACV in such a period, which corresponds to a specific activity of 600 pmol/min*mg.

IV. PERSPECTIVES

ACV synthetase presents a suitable model system to analyze structure–function relations in NRPS systems [89]. Here we have focused on the current state of understanding and interpretation of structure–function relations of ACV synthetases. These aspects are of importance for the future development of nonribosomal peptide-forming systems, especially in filamentous fungal hosts. Other major aspects of process analysis and metabolic engineering approaches to improve the yield of antibiotic fermentations have not been discussed. These process aspects include the following four approaches.

1. *Alterations of the expression levels of ACV synthetases as possible rate-limiting enzymes.* Several approaches based on the available data conclude that the slow peptide synthetases may be rate-limiting in peptide fermentation processes [90–93]. Thus, expression studies have been conducted altering the bidirectional promoter region between *acvA* and *ipnA*. Inserting the ethanol dehydrogenase promoter, *alcAp*, from *A. nidulans* into the *P. chrysogenum* Wisconsin strain, expression was found to be up to 100 times greater than that from the *acvA* pro-

moter when induced in fermentation conditions. Penicillin yields were found to increase by as much as 30-fold [94]. Such approaches may prove useful, especially in strains that are not extensively selected as industrial penicillin producers.

2. *To evaluate the actual protein levels of peptide synthetases with respect to translation, posttranslational processing, stability, and localization.* These aspects are especially related to the compartmentation of eukaryotic cells. So far, evidence for a vacuolar attachment of ACV synthetase in *P. chrysogenum* has been obtained [95], but it has not been fully substantiated in further studies [96]. The synthetase is now considered to be cytoplasmatic.
3. *To use various approaches of metabolic engineering to increase the levels of other enzymes of the pathway, including pathways to precursors, and proteins involved in transport processes* [97–99].
4. *To introduce pathway alterations to shift the product pattern of, e.g., penicillins to cephalosporins* [100,101].

The principal questions currently studied in NRPS systems concern the understanding, alteration, and design of peptide forming systems. Current engineering approaches which have been performed include the following alterations of domain positions.

1. *The replacement of product amino acid residues by positional alteration of domains* [102,103]. The cysteine- and valine-specific activation-thiolation didomains of *P. chrysogenum* ACV synthetase have been successfully inserted into the terminal position of surfactin synthetase in *Bacillus subtilis* [102]. In the case of ACV synthetase, e.g., specific domains could be replaced to generate new tripeptides or to improve the efficiency of poorly incorporated amino acid analogues.
2. *The alteration of product structures by domain repositioning.* The terminal thioesterase domain of surfactin synthetase has been repositioned to obtain acyl-tetra- and pentapeptide fragments *in vivo* of the cycloheptapeptidolactone [104]. In the case of ACV synthetase, repositioning of the thioesterase domain would be expected to lead to dipeptides of the AC type. If this specific thioesterase would only release peptides of the D-configuration, alternative thioesterases from other NRPS systems might work.
3. *The separation of domains to obtain interacting systems with multiple components.* This approach directs efforts to enzymatic combinatorial approaches and permits selective improvements on single domains [85]. The thioesterase domain of surfactin synthetase has been separately expressed in a thioesterase-deleted surfactin synthetase system and shown to be functional [105]. Thus, specific protein–protein inter-

actions take place, even without covalent linking of the domains. In the case of ACV synthetase, the selective cleavage of domains or modules could be probed to determine how domains interact specifically. Such systems would permit the development of defined interacting modules to exchange certain regions of peptides.

4. *The construction of new synthetases by combining domains from one or various systems to arrive at new peptide synthetases* [106,107]. Work in this direction leads to the development of tools and strategies especially suitable for NRPS systems of filamentous fungi.

ACKNOWLEDGMENTS

Work in the authors' laboratory has been supported by grants from the Deutsche Forschungs-gemeinschaft (K1 148/31-1, *Integrated enzyme systems in peptide synthesis*), and the European Community (BIO4-CT96-0535, *Filamentous fungi as cell factories for the production of proteins and fungal metabolites*).

REFERENCES

1. MF Bycroft, JE Baldwin, C-Y Shiau, CJ Schofield. The mechanism of ACV synthetase. Chem Rev 97:2631–2649, 1997.
2. H Kleinkauf, H von Döhren. A nonribosomal system of peptide biosynthesis. Eur J Biochem 236:335–351, 1996.
3. H Kleinkauf, H von Döhren. Nonribosomal biosynthesis of peptide antibiotics. Eur J Biochem 192:1–15, 1990.
4. DJ Smith, AJ Earl, G Turner. The multifunctional peptide synthetase performing the first step of penicillin biosynthesis in *Penicillium chrysogenum* is a 421,073 dalton protein similar to *Bacillus brevis* peptide antibiotic synthetases. EMBO J 9: 2743–2750, 1990.
5. B Diez, S Gutiérrez, JL Barredo, P van Solingen, LH van der Voort, JF Martin. The cluster of penicillin biosynthetic genes. Identification and characterization of the *pcbAB* gene encoding the alpha-aminoadipyl-cysteinyl-valine synthetase and linkage to the *pcbC* and *penDE* genes. J Biol Chem 265:16358–16365, 1990.
6. A MacCabe, H van Liempt, H Palissa, SE Unkles, MBR Riach, E Pfeifer, H von Döhren, J Kinghorn. Molecular characterization of the *Aspergillus nidulans acvA* gene encoding δ-(L-α-aminoadipyl)-L-cysteinyl-D-valine synthetase, the first enzyme of the penicillin biosynthetic pathway. J Biol Chem 266:12646–12654, 1991.
7. S Gutiérrez, B Díez, E Montenegro, JF Martín. Characterization of the *Cephalosporium acremonium pcbAB* gene encoding alpha-aminoadipyl-cysteinyl-valine synthetase, a large multidomain peptide synthetase: linkage to the *pcbC* gene as a cluster of early cephalosporin biosynthetic genes and evidence of multiple functional domains. J Bacteriol 173:2354–2365, 1991.

8. JJ Coque, JF Martin, JG Calzada, P Liras. The cephamycin biosynthetic genes *pcbAB*, encoding a large multidomain peptide synthetase, and *pcbC* of *Nocardia lactamdurans* are clustered together in an organization different from the same genes in *Acremonium chrysogenum* and *Penicillium chrysogenum*. Mol Microbiol 5:1125–1133, 1991.
9. H van Liempt, H Palissa, E Pfeifer, T Schwecke, R Weckermann, H von Döhren, H Kleinkauf. ACV-synthetase: implications of the amino acid sequence data for the thiotemplate mechanism of peptide biosynthesis. In: H Kleinkauf, H von Döhren, eds. 50 Years of Penicillin Application—History and Trends. Prague: Publica, 1991, pp 136–144.
10. H Kleinkauf, H von Döhren. Thiotemplate mechanism. In: TE Creighton, ed. The Encyclopedia of Molecular Biology. New York: Wiley, 1990, pp 2539–2541.
11. T Stein, J Vater, V Kruft, A Otto, B Wittmann-Liebold, P Franke, M Panico, R McDowell, HR Morris. The multiple carrier model of nonribosomal peptide biosynthesis at modular multienzymatic templates. J Biol Chem 271:15428–15435, 1996.
12. F Leenders, J Vater, T Stein, P Franke. Characterization of the binding site of the tripeptide intermediate D-phenylalanyl L-prolyl-L-valine in gramicidin S biosynthesis. J Biol Chem 273:18011–18014, 1998.
13. HRV Arnstein, N Artman, D Morris, EJ Toms. Sulfur-containing amino acids and peptide in the mycelium of *Penicillium chrysogenum*. Biochem J 76:353–357, 1960.
14. HRV Arnstein, D Morris. The structure of a peptide, containing alpha-aminoadipic acid, cysteine and valine, present in the mycelium of *Penicillium chrysogenum*. Biochem J 76:357–361, 1960.
15. K Bauer. Zur Biosynthese der Penicilline: bildung von 5-(2-aminoadipyl)-cysteinyl-valin in extrakten von *penicillium chrysogenum*. Z Naturforsch [B] 25: 1125–1129, 1970.
16. PB Loder, EP Abraham. Biosynthesis of peptides containing alpha-aminoadipic acid and cysteine in extracts of a *Cephalosporium sp*. Biochem J 123:477–482, 1971.
17. P Fawcett, EP Abraham. Delta-(alpha-aminoadipyl)cysteinylvaline synthetase. Meth Enzymol 43:471–473, 1975.
18. P Adriaens, B Messchaert, W Wuyts, H Vanderhaeghe, H Eyssens. Presence of delta-(L-alpha-aminoadipyl)-L-cysteinyl-D-valine in fermentations of *Penicillium chrysogenum*. Antimicrob Agents Chemother 8:638–642, 1975.
19. JA Chan, F-C Huang, CJ Sih. The absolute configuration of the amino acids in delta-(alpha-aminoadipyl)cysteinylvaline from *Penicillium chrysogenum*. Biochemistry 15:177–180, 1976.
20. H Shirafuji, Y Fujisawa, M Kida, T Kanzaki, M Yoneda. Accumulation of tripeptide derivatives by mutants of *Cephalosporium acremonium*. Agric Biol Chem 43: 155–160, 1979.
21. RM Adlington, JE Baldwin, M Lopez-Nieto, JA Murphy, N Patel. A study of the biosynthesis of the tripeptide delta-(L-alpha-aminoadipyl)-L-cysteinyl-D-valine in a beta-lactam-negative mutant of *Cephalosporium acremonium*. Biochem J 213: 573–576, 1983.
22. N Neuss, RD Miller, CA Affolder, W Nakatsukasa, J Mabe, LL Huckstep, N De

La Higuera, AH Hunt, JL Occolowitz, JH Gilliam. High performance liquid chromatography (HPLC) of natural products. III. Isolation of new tripeptides from the fermentation broth of *P. chrysogenum*. Helv Chim Acta 63:1119–1129, 1980.
23. J Nuesch, J Heim, HJ Treichler. The biosynthesis of sulfur-containing beta-lactam antibiotics. Annu Rev Microbiol 41:51–75, 1987.
24. F Lara, R Del Carmen Mateos, G Vazquez, S Sanchez. Induction of penicillin biosynthesis by L-glutamate in *Penicillium chrysogenum*. Biochem Biophys Res Commun 105:172–178, 1982.
25. G Banko, S Wolfe, AL Demain. Cell-free synthesis of delta-(L-alpha-aminoadipyl)-L-cysteine, the first intermediate of penicillin and cephalosporin biosynthesis. Biochem Biophys Res Commun 137:528–535, 1986.
26. G Banko, AL Demain, S Wolfe. Delta-(L-alpha-aminoadipyl)-L-cysteinyl-D-valine synthetase (ACV synthetase): a multifunctional enzyme with broad substrate specificity for the synthesis of penicillin and cephalosporin precursors. J Am Chem Soc 109:2858–2860, 1987.
27. SE Jensen, DWS Westlake, S Wolfe. Production of the pencillin precursor delta-(L-alpha-aminoadipyl)-L-cysteinyl-D-valine (ACV) by cell-free extracts from *Streptomyces clavuligerus*. FEMS Microbiol Lett 49:213–218, 1988.
28. SE Jensen, S Wolfe, DWS Westlake. Synthesis of the penicillin precursor delta-(L-alpha-aminoadipyl)-L-cysteinyl-D-valine (ACV), by an immobilized enzyme preparation. Appl Microbiol Biotechnol 30:111–114, 1989.
29. H van Liempt, H von Döhren, H Kleinkauf. δ-(L-α-aminoadipyl)-L-cysteinyl-D-valine synthetase from *Aspergillus nidulans*. J Biol Chem 264:3680–3684, 1989.
30. JE Baldwin, JW Bird, RA Fields, NM O'Callaghan, CJ Schofield. Isolation and partial characterization of ACV synthetase from *Cephalosporium acremonium* and *Streptomyces clavuligerus*. J Antibiot 43:1055–1057, 1990.
31. JE Baldwin, JW Bird, RA Fields, NM O'Callaghan, CJ Schofield, A Willis. Isolation and partial characterisation of ACV synthetase from *Cephalosporium acremonium* and *Streptomyces clavuligerus*. Evidence for the presence of phosphopantothenate in ACV synthetase. J Antibiot 44:241–248, 1991.
32. J Zhang, AL Demain. Purification from *Cephalosporium acremonium* of the initial enzyme unique to the biosynthesis of penicillins and cephalosporins. Biochem Biophys Res Commun 169:1145–1152, 1990.
33. J Zhang, AL Demain. Purification of ACV synthetase from *Streptomyces clavuligerus*. Biotechnol Lett 12:649–654, 1990.
34. SE Jensen, A Wong, MJ Rollins, DW Westlake. Purification and partial characterization of delta-(L-alpha-aminoadipyl)-L-cysteinyl-D-valine synthetase from *Streptomyces clavuligerus*. J Bacteriol 172:7269–7271, 1990.
35. T Schwecke, Y Aharonowitz, H Palissa, H von Döhren, H Kleinkauf, H van Liempt. Enzymatic characterisation of the multifunctional enzyme δ-(L-α-aminoadipyl)-L-cysteinyl-D-valine synthetase from *Streptomyces clavuligerus*. Eur J Biochem 205: 687–694, 1992.
36. J Zhang, S Wolfe, AL Demain. Biochemical studies on the activity of δ-(L-α-aminoadipyl)-L-cysteinyl-D-valine synthetase from *Streptomyces clavuligerus*. Biochem J 283:691–698, 1992.
37. KA Aidoo, A Wong, DC Alexander, RA Rittammer, SE Jensen. Cloning, sequenc-

ing and disruption of a gene from *Streptomyces clavuligerus* involved in clavulanic acid biosynthesis. Gene 147:41–46, 1994.
38. MA Marahiel, T Stachelhaus, HD Mootz. Modular peptide synthetases involved in non-ribosomal peptide synthesis. Chem Rev 97:2651–2673, 1997.
39. H von Döhren, U Keller, J Vater, R Zocher. Multifunctional peptide synthetases. Chem Rev 97:2675–2705, 1997.
40. D Konz, MA Marahiel. How do peptide synthetases generate structural diversity? Chem Biol 6:R39–R48, 1999.
41. R Dieckmann, M Pavela-Vrancic, H Kleinkauf, H von Döhren. Probing the domain structure and ligand-induced conformational changes by limited proteolysis of tyrocidine synthetase 1. J Mol Biol 288:129–140, 1999.
42. RS Gokhale, SY Tsuji, DE Cane, C Khosla. Dissecting and exploiting intermodular communication in polyketide synthases. Science 284:482–485, 1999.
43. RS Gokhale, C Khosla. Role of linkers in communication between protein modules. Curr Opin Chem Biol 4:22–27, 2000.
44. K Turgay, M Krause, MA Marahiel. Four homologous domains in the primary structure of GrsB are related to domains in a superfamily of adenylate-forming enzymes. Mol Microbiol 6:529–546, 1992.
45. GL Challis, J Ravel, CA Townsend. Predictive, structure-based model of amino acid recognition by nonribosomal peptide synthetase adenylation domains. Chem Biol 7:211–224, 2000.
46. PL Skatrud. Molecular biology of the beta-lactam producing fungi. In: JW Bennet, L Lasure, eds. More Gene Manipulations in Fungi. New York: Academic Press, 1991, pp 364–395.
47. H von Döhren, R Dieckmann, M Pavela-Vrancic. The nonribosomal code. Chem Biol 6:R273–R279, 1999.
48. H van Liempt. Untersuchungen zur biosynthese der β-lactamvorstufe delta-(L-alpha-aminoadipyl)-L-cysteinyl-D-valine. Ph.D. thesis, Technical University of Berlin, 1988.
49. JJR Coque, JL de la Fuente, P Liras, J Martin. Overexpression of the *Nocardia lactamdurans* alpha-aminoadipyl-cysteinyl-valine synthetase in *Streptomyces lividans*. The purified multienzyme uses cystathionine and 6-oxopiperidine 2-carboxylate as substrates for synthesis of the tripeptide. Eur J Biochem 242:264–270, 1996.
50. HB Theilgaard, KN Kristiansen, CM Henriksen, J Nielsen. Purification and characterization of delta-(L-alpha-aminoadipyl)-L-cysteinyl-D-valine synthetase from *Penicillium chrysogenum*. Biochem J 327:185–191, 1997.
51. W Kallow, H von Döhren, H Kleinkauf. Penicillin biosynthesis: energy requirement for tripeptide precursor formation by delta-(L-alpha-aminoadipyl)-L-cysteinyl-D-valine synthetase from *Acremonium chrysogenum*. Biochemistry 37:5947–5952, 1998.
52. W Kallow. ACV synthetase als modellsystem der nichtribosomalen peptidsynthese. Ph.D. thesis, Technical University of Berlin, 1998.
53. JY Zhang, AL Demain. Purification from *Cephalosporium acremonium* of the initial enzyme unique to the biosynthesis of penicillins and cephalosporins. Biochem Biophys Res Commun 169:1145–1152, 1990.

54. TA Kadima, SE Jensen, MA Pickard. Production of delta-(L-alpha-aminoadipyl)-L-cysteinyl-D-valine by entrapped ACV-synthetase from *Streptomyces clavuligerus*. J Ind Microbiol 14:35–40, 1995.
55. J Zhang, AL Demain. ACV synthetase. Crit Rev Biotechnol 12:245–260, 1992.
56. H Kimura, H Miyashita, Y Sumino. Organization and expression in *Pseudomonas putida* of the gene cluster involved in cephalosporim biosynthesis from *Lysobacter lactamgenus* YK90. Appl Microbiol Biotechnol 45:490–501, 1996.
57. JE Baldwin, C-Y Shiau, MF Byford, CJ Schofield. Substrate specificity of delta-(L-alpha-aminoadipyl)-L-cysteinyl-D-valine formed by ACV synthetase from *Cephalosporium acremonium*: demonstration of the structure of several unnatural tripeptide products. Biochem J 301:367–372, 1994.
58. R Dieckmann, M Pavela-Vrancic, E Pfeifer, H von Döhren, H Kleinkauf. The adenylation domain of tyrocidine synthetase 1: structural and functional role of the interdomain linker region and the (S/T)GT(T/S)GXPKG core sequence. Eur J Biochem 247:1074–1082, 1997.
59. R Kittelberger, M Pavela-Vrancic, H von Döhren. Active site titration of gramicidin S synthetase 2: evidence for misactivation and editing in nonribosomal peptide biosynthesis. FEBS Lett 461:145–148, 1999.
60. T Schwecke. Untersuchungen zur biosynthese der beta-lactamvorstufe delta-(L-alpha-aminoadipyl)-L-cysteinyl-D-valin (ACV) in *Streptomyces clavuligerus*. Ph.D. thesis, Technical University of Berlin, 1992.
61. E Conti, T Stachelhaus, MA Marahiel, P Brick. Structural basis for the activation of phenylalanine in the non-ribosomal biosynthesis of gramicidin S. EMBO J 16: 4174–4183, 1997.
62. T Stachelhaus, HD Mootz, MA Marahiel. The specificity-conferring code of adenylation domains in nonribosomal peptide synthetases. Chem & Biol 6:493–506, 1999.
63. MA Tavanlar. Molecular characterisation of the δ-(L-α-aminoadipyl)-L-cysteinyl-D-valine synthetase (ACVS) from filamentous fungi: determination of the domain arrangement, cloning and expression of the third domain. Ph.D. thesis, University of Los Banos/Technical University of Berlin, 1996.
64. T Schwecke, MB Tobin, S Kovacevic, JR Miller, PL Skatrud, SE Jensen. The A module of ACV synthetase: expression, renaturation and functional characterization. In: RH Baltz, GD Hegeman, PL Skatrud, eds. Industrial Microorganisms: Basic and Applied Molecular Genetics. Washington, DC: American Society for Microbiology, 1993, p. 291.
65. A Etchegaray, R Dieckmann, J Kennedy, G Turner, H von Döhren. ACV synthetase: expression of amino acid activating domains of the *Penicillium chrysogenum* enzyme in *Aspergillus nidulans*. Biochem Biophys Res Commun 237:166–169, 1997.
66. J Kennedy. Genetic manipulation of ACV synthetase. Ph.D. thesis, University of Sheffield, 1995.
67. CY Shiau, JE Baldwin, MF Byford, CJ Schofield. Delta-L-(alpha-aminoadipoyl)-L-cysteinyl-D-valine synthetase: isolation of L-cysteinyl-D-valine, a "shunt" product, and implications for the order of peptide bond formation. FEBS Lett 373: 303–306, 1995.

68. CY Shiau, JE Baldwin, MF Byford, WJ Sobey. CJ Schofield. Delta-L-(alpha-aminoadipoyl)-L-cysteinyl-D-valine synthetase: the order of peptide bond formation and timing of the epimerization reaction. FEBS Lett 358:97–100, 1996.
69. CY Shiau, MF Byford, RT Aplin, JE Baldwin, CJ Schofield. L-Delta-(alpha-aminoadipoyl)-L-cysteinyl-D-valine synthetase: thioesterification of valine is not obligatory for peptide bond formation. Biochemistry 36:8798–8806, 1997.
70. W Kallow, T Neuhof, B Arezi, P Jungblut, H von Doehren. Penicillin biosynthesis: intermediates of biosynthesis of delta-(L-alpha-aminoadipyl)-L-cysteinyl-D-valine formed by ACV synthetase from *Acremonium chrysogenum*. FEBS Lett 414:74–78, 1997.
71. R Dieckmann. Untersuchungen zu struktur/funktionsbeziehungen von peptidsynthetasen am modellsystem tyrocidin synthetase 1. Ph.D. thesis, Technical University of Berlin, 1999.
72. A Gadow, J Vater, W Schlumbohm, Z Palacz, J Salnikow, H Kleinkauf. Gramicidin S synthetase. Stability of reactive thioester intermediates and formation of 3-amino-2-piperidone. Eur J Biochem 132:229–234, 1983.
73. CM Henriksen, J Nielsen, J Villadsen. Cyclization of alpha-aminoadipic acid into the delta-lactam 6-oxo-piperidine-2-carboxylic acid by *Penicillium chrysogenum*. J Antibiot 51:99–105, 1998.
74. PH Weinreb, LE Quadri, CT Walsh, P Zuber. Stoichiometry and specificity of in vitro phosphopantetheinylation and aminoacylation of the valine-activating module of surfactin synthetase. Biochemistry 37:1575–1584, 1998.
75. TA Keating, DA Miller, CT Walsh. Expression, purification, and characterization of HMWP2, a 229 kDa, six domain protein subunit of yersiniabactin synthetase. Biochemistry 39:4729–4739, 2000.
76. R Dieckmann, H von Döhren. Structural model of acyl carrier domains in integrated biosynthetic systems forming peptides, polyketides and fatty acids based on analogy to the *E. coli* acyl carrier protein. In: RH Baltz, GD Hegeman, PL Skatrud, eds. Developments in Industrial Microbiology. Fairfax, VA; Society for Industrial Microbiology, 1997, pp 79–87.
77. CT Walsh, AM Gehring, PH Weinreb, EN Luis, RS Flugel. Post-translational modification of polyketide and nonribosomal peptide synthetases. Curr Opin Chem Biol 1:309–315, 1997.
78. LE Quadri, PH Weinreb, M Lei, MM Nakano, P Zuber, CT Walsh. Characterization of Sfp, a *Bacillus subtilis* phosphopantetheinyl transferase for peptidyl carrier protein domains in peptide synthetases. Biochemistry 37:1585–1595, 1998.
79. RH Lambalot, AM Gehring, RS Flugel, P Zuber, M LaCelle, MA Marahiel, R Reid, C Khosla, CT Walsh. A new enzyme superfamily—the phosphopantetheinyl transferases. Chem Biol 3:923–936, 1996.
80. T Stachelhaus, HD Mootz, V Bergendahl, MA Marahiel. Peptide bond formation in non-ribosomal peptide biosynthesis: catalytic role of the condensation domain. J Biol Chem 273:22773–22781, 1998.
81. P Belshaw, CT Walsh, T Stachelhaus. Aminoacyl-CoAs as probes of condensation domain selectivity in nonribosomal peptide synthesis. Science 284:486–489, 1999.
82. H von Döhren, U Keller, J Vater, R Zocher. Multifunctional peptide synthetases. Chem Rev 97:2675–2705, 1997.

83. J Zhang, S Wolfe, AL Demain. Biochemical studies on the activity of delta-(L-alpha-aminoadipyl)-L-cysteinyl-D-valine synthetase from *Streptomyces clavuligerus*. Biochem J 283:691–698, 1992.
84. H Kleinkauf, H von Döhren. A nonribosomal system of peptide biosynthesis. Eur J Biochem 236:335–351, 1996.
85. T Stachelhaus, CT Walsh. Mutational analysis of the epimerization domain in the initiation module PheATE of gramicidin S synthetase. Biochemistry 39:5775–5787, 2000.
86. W Kallow, J Kennedy, B Arezi, G Turner, H von Döhren. Thioesterase domain of δ-(L-α-aminoadipyl)-L-cysteinyl-D-valine synthetase: alteration of stereospecificity by site-directed mutagenesis. J Mol Biol 297:395–408, 2000.
87. H Kleinkauf, H von Döhren. Applications of peptide synthetases in the synthesis of peptide analogues. Acta Biochim Polonia 44:839–848, 1997.
88. TA Kadima, SE Jensen, MA Pickard. Production kinetics and stability properties of delta(L-alpha-aminoadipyl)-L-cysteinyl-D-valine synthetase from *Streptomyces clavuligerus*. J Ind Microbiol 12:58–65, 1993.
89. Y Aharonowitz, H Bergmeyer, JM Cantoral, G Cohen, AL Demain, U Fink, J Kinghorn, H Kleinkauf, A MacCabe, H Palissa, E Pfeifer, T Schwecke, H van Liempt, H von Döhren, S Wolfe, J Zhang. Delta-(L-alpha-aminoadipyl)-L-cysteinyl-D-valine synthetase, the multienzyme integrating the four primary reactions in β-lactam biosynthesis, as a model peptide synthetase. Biotechnology 11:807–810, 1993.
90. J Nielsen, HS Jørgensen. Metabolic control analysis of the penicillin biosynthetic pathway in a high-yielding strain of *Penicillium chrysogenum*. Biotechnol Prog 11: 299–305, 1995.
91. P de Noronha Pissara, J Nielsen, MJ Bazin. Pathway kinetics and metabolic control analysis of a high-yielding strain of *Penicillium chrysogenum* during fed-batch cultivations. Biotechnol Bioeng 51:168–176, 1996.
92. A Khetan, LH Malmberg, DH Sherman, WS Hu. Metabolic engineering of cephalosporin biosynthesis in *Streptomyces clavuligerus*. Ann NY Acad Sci 782:17–24, 1996.
93. LH Malmberg, WS Hu. Identification of rate-limiting steps in cephalosporin C biosynthesis in *Cephalosporium acremonium*: a theoretical analysis. Appl Microbiol Biotechnol 38:122–128, 1992.
94. J Kennedy, G Turner. Delta-(L-alpha-aminoadipyl)-L-cysteinyl-D-valine synthetase is a rate limiting enzyme for penicillin production in *Aspergillus nidulans*. Mol Gen Genet 253:189–197, 1996.
95. T Lendenfeld, D Ghali, M Wolschek, EM Kubicek-Pranz, CP Kubicek. Subcellular compartmentation of penicillin biosynthesis in *Penicillium chrysogenum*. The amino acid precursors are derived from the vacuole. J Biol Chem 268:665–671, 1993.
96. M van de Kamp, AJ Driessen, WN Konings. Compartmentalization and transport in beta-lactam antibiotic biosynthesis by filamentous fungi. Antonie Van Leeuwenhoek 75:41–78, 1999.
97. M van de Kamp, E Pizzinini, A Vos, TR van der Lende, TA Schuurs, RW Newbert, G Turner, WN Konings, AJ Driessen. Sulfate transport in *Penicillium chryso-*

genum: cloning and characterization of the *sutA* and *sutB* genes. J Bacteriol 181: 7228–7234, 1999.

98. S Gutiérrez, F Fierro, J Casqueiro, JF Martin. Gene organization and plasticity of the beta-lactam genes in different filamentous fungi. Antonie Van Leeuwenhoek 75:81–94, 1999.
99. VK Chary, JL de la Fuente, AL Leitao, P Liras, JF Martin. Overexpression of the *lat* gene in *Nocardia lactamdurans* from strong heterologous promoters results in very high levels of lysine-6-aminotransferase and up to two-fold increase in cephamycin C production. Appl Microbiol Biotechnol 53:282–288, 2000.
100. B Diez, E Mellado, M Rodriguez, R Fouces, JL Barredo. Recombinant microorganisms for industrial production of antibiotics. Biotechnol Bioeng 55:216–226, 1997.
101. JF Martin. New aspects of genes and enzymes for beta-lactam antibiotic biosynthesis. Appl Microbiol Biotechnol 50:1–15, 1998.
102. T Stachelhaus, A Schneider, MA Marahiel. Rational design of peptide antibiotics by targeted replacement of bacterial and fungal domains. Science 269:69–72, 1995.
103. A Schneider, T Stachelhaus, MA Marahiel. Targeted alteration of the substrate specificity of peptide synthetases by rational module swapping. Mol Gen Genet 57:308–318, 1998.
104. F de Ferra, F Rodriguez, O Tortora, C Tosi, G Grandi. Engineering of peptide synthetases. Key role of the thioesterase-like domain for efficient production of recombinant peptides. J Biol Chem 272:25304–25309, 1997.
105. E Guenzi, G Galli, I Grgurina, E Pace, P Ferranti, G Grandi. Coordinate transcription and physical linkage of domains in surfactin synthetase are not essential for proper assembly and activity of the multienzyme complex. J Biol Chem 273: 14403–14410, 1998.
106. F Schauwecker, F Pfennig, N Grammel, U Keller. Construction and in vitro analysis of a new bi-modular polypeptide synthetase for synthesis of N-methylated acyl peptides. Chem Biol 7:287–297, 2000.
107. S Doekel, MA Marahiel. Dipeptide formation on engineered hybrid peptide synthetases. Chem Biol 7:373–384, 2000.
108. T Weber, R Baumgartner, C Renner, MA Marahiel, TA Holak. Solution structure of PCP, a prototype for the peptidyl carrier domains of modular peptide synthetases. Structure 8:407–418, 2000.

2

Metabolic Engineering for Cephalosporin C Yield Improvement and Production of Intermediates

Joe E. Dotzlaf, Steven W. Queener, and Wu-Kuang Yeh
Eli Lilly and Company, Indianapolis, Indiana

I. INTRODUCTION

Improving product yield in an antibiotic fermentation relies on an understanding and exploitation of the basic biology of the producing microorganism. Traditionally, microbiologists and biochemical engineers optimize the productive phase of fermentations by manipulating the nutrition and environment of the producing microorganism [1]. Genetic manipulation of the producing fungus or bacterium through natural selection or deliberate random mutation and screening for enhanced yield or favorable metabolite profile have been companions to fermentation optimization for decades [1,2]. The success of these traditional techniques is typified by the two-orders-of-magnitude increase in the production of penicillin by *Penicillium chrysogenum* since the early 1950s [2].

Biochemical pathways leading from basic metabolic precursors to finished products are deduced through chemical structural analysis of the products and byproducts coupled with biochemical and genetic investigation [3]. Highly sensitive bioanalytical techniques allow biochemists to characterize many of the enzymes responsible for antibiotic production. Similar techniques also help to identify product precursor pools that can indicate rate-limiting steps in the production of an antibiotic. With the introduction of recombinant DNA techniques in the 1970s, the genes coding for the various enzymes responsible for the biochemical steps became targets for antibiotic yield improvement investigations.

The term *metabolic engineering* refers to the deliberate genetic manipulation of one or more steps in a biosynthetic pathway leading to a desired product. Such manipulations may include gene deletions to reduce or eliminate undesirable by-products, an increase in gene copy number to enhance conversion of rate-limiting precursor pools to end products, or the introduction of a non-natural gene resulting in a novel product. In this chapter we will discuss the metabolic engineering of the cephalosporin C (CPC) biosynthetic pathway for yield improvement and production of economically important intermediates for semisynthetic cephalosporin synthesis.

II. ENZYMES OF CEPHALOSPORIN C BIOSYNTHESIS

Cephalosporin C, characterized by a β-lactam-dihydrothiazine fused ring [3], is produced via a multistep biosynthetic pathway by *Cephalosporium acremonium** (Fig. 1). This pathway shares several steps with the pathway that diverges to penicillin produced commercially by *Penicillium chrysogenum* and again diverges and is extended by several steps to yield cephamycin C by *Streptomyces clavuligerus* (Fig. 1; [4]). All of the enzymes in the CPC pathway have been purified and characterized (Table 1), and the genes coding for each enzyme have been cloned using the so-called reverse genetics approach [4,5] and by traditional genetic methodologies. Ingolia and Queener [5] proposed the current nomenclature for the genes in this pathway. Genes common to both penicillin and cephalopsorin biosynthesis are named *pcb*, the genes unique to penicillin are designated *pen*, while those involved only in cephalosporin biosynthesis are named *cef*.

A. ACV Synthetase

In 1971 it was shown that incubating the dipeptide δ-(L-α-aminoadipyl)-L-cysteine (AC) with DL-[^{14}C]-valine and ATP in a particulate fraction from *C. acremonium* protoplast extracts yielded labeled δ-(L-α-aminoadipyl)-L-cysteinyl-D-valine (ACV), implying that AC was an intermediate in ACV biosynthesis [6]. This notion was substantiated when Abraham identified AC from a β-lactam-producing actinomycete [7]. Later, soluble extracts of *C. acremonium* were shown to generate both AC and ACV from labeled amino acid precursors [8]. These data led to the generally accepted conclusion that ACV synthesis was analogous to the two-enzyme process for glutathione biosynthesis [9,10].

* The name of *Cephalosporium acremonium* has been changed to *Acremonium chrysogenum*; however, *Cephalosporium acremonium* is used here for historic continuity, since most of the literature contains this name.

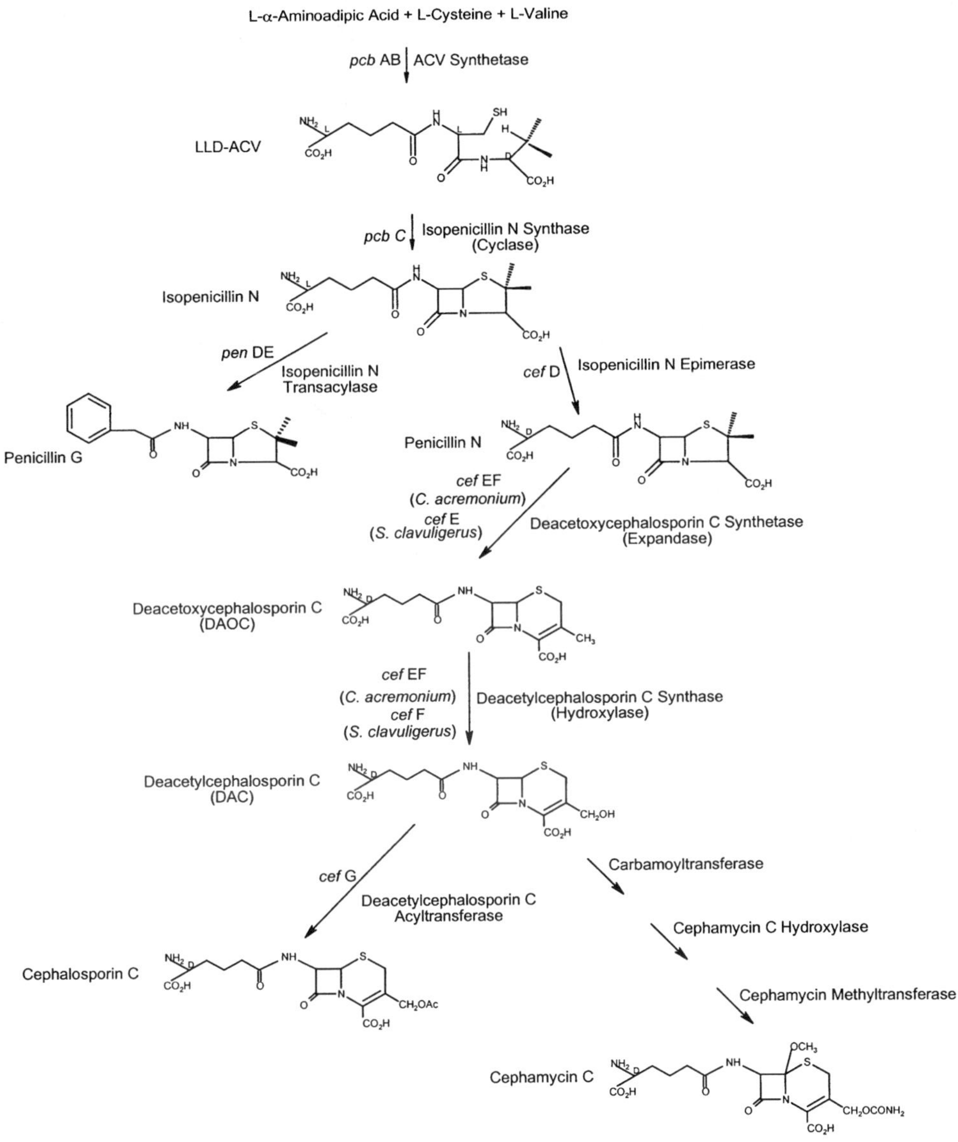

Figure 1 Biosynthetic pathway for cephalosporin C in *C. acremonium*, penicillin G in *P. chrysogenum*, and cephamycin C in *S. clavuligerus*. Gene designations are shown to the left of the arrows and enzyme designations are shown to the right.

Table 1 Physical and Biochemical Properties of Cephalosporin C Biosynthetic Enzymes[a]

Enzyme	Gene	Organism	Functionality	Molecular weight (kDa)	Cofactors	Activity enhancers	References
ACVS	*pcb*AB	*C. acremonium*	Trifunctional	~415	ATP, Mg^{2+}	DTT	13–21
IPNS	*pcb*C	*C. acremonium*	Monofunctional	~37	Electron Donor (e.g., ascorbate), O_2, Fe^{2+}		22–30
IPNE	*cef*D	*S. clavuligerus*	Bifunctional	~48	Pyridoxal 5′-phosphate		31–35
DAOCS/DACS	*cef*EF	*C. acremonium*	Bifunctional	~41	α-Ketoglutarate, O_2, Fe^{2+}	DTT or ascorbate, ATP	36–39,45–47
DAOCS	*cef*E	*S. clavuligerus*	Monofunctional[b]	~35	α-Ketoglutarate, O_2, Fe^{2+}	DTT or ascorbate	40,41,43
DACS	*cef*F	*S. clavuligerus*	Monofunctional[b]	~37	α-Ketoglutarate, O_2, Fe^{2+}	DTT or ascorbate, ATP	40,42,44
DAC-AT	*cef*G	*C. acremonium*	Monofunctional	~50	Acetyl-CoA, (+/−)Mg^{2+}		48–56

[a] Abbreviations: ACVS, δ-(L-α-aminoadipyl)-L-cysteinyl-D-valine Synthetase; IPNS, isopenicillin N synthase; IPNE, isopenicillin N epimerase; DAOCS/DACS, deacetoxycephalosporin C synthetase/deacetylcephalosporin C synthase; DAC-AT, acetyl CoA:deacetylcephalosporin C *O*-acetyltransferase; DTT, dithiothreitol.

[b] DAOCS and DACS of *S. clavuligerus* are, for practical purposes, monofunctional although, under optimized conditions, bifunctionality can be demonstrated *in vitro* [40,41].

Banko et al. [11] speculated that the putative dual-enzyme process was instead catalyzed by a single, multifunctional enzyme. Their data clearly showed that in cell-free extracts of *C. acremonium* the rate of ACV synthesis was significantly greater when the precursor amino acids were provided than it was when AC and valine were used as substrates for the reaction [11]. Jensen and her coworkers [12] demonstrated a similar result using cell-free extracts of *S. clavuligerus*.

The issue was resolved when van Liempt et al. [13] partially purified the enzyme responsible for the multifunctional synthesis of the ACV tripeptide from *Aspergillus nidulans* (Fig. 1). The enzyme was later purified and characterized from *C. acremonium* and *S. clavuligerus* by Baldwin et al. [14,15]. The gene was cloned by reverse genetics from *A. nidulans* [16] and by conventional methods from *P. chrysogenum* [17] and *C. acremonium* [18,19]. The name for the gene coding ACV synthetase, *pcb*AB, is reminiscent of the understanding of the biochemistry prior to 1987 [5]. The biochemistry of ACV formation is thoroughly discussed by von Dohren in Chapter 1. To summarize, the enzyme activates the three component amino acids utilizing ATP, assembles the tripeptide, and epimerizes L-valine to D-valine prior to releasing the tripeptide ([16,20,21]; Table 1).

B. Isopenicillin N Synthase (IPNS)

Isopenicillin N synthase (IPNS, Table 1) or ''ACV cyclase'' (Fig. 1) catalyzes the oxidative cyclization of ACV by the removal of four hydrogens from the tripeptide with the consumption of one molecule of oxygen [22]. The newly formed C—N and C—S bonds of the respective β-lactam and thiazolidine rings are putatively made in a stepwise manner while the intermediates are enzyme bound [22]. IPNS from *C. acremonium* was purified [23a,23b] and its N-terminal sequence reported [24] in the early 1980s. Samson et al. [25] utilized the sequence data to clone the IPNS gene *pcb*C [5]. IPNS has been extensively characterized with respect to its enzymology [23,26,27], substrate specificity, and catalytic mechanism [28–30].

C. Isopenicillin N Epimerase (IPNE)

The generation of cephalosporin from penicillin is dependent on the epimerization of the L-α-aminoadipyl side chain of isopenicillin N to the D-α-aminoadipyl side chain of penicillin N, since penicillin N but not isopenicillin N is the substrate for ring expansion enzymes [31]. This reaction is catalyzed by IPNE ([32]; Table 1) and is coded for by the *cef*D gene (Fig. 1; [5]). While the activity of the IPNE from *C. acremonium* has been studied in cell-free extracts, it has not been purified to date [2]. Jensen et al. [33] described a partial purification and characterization of IPNE from *S. clavuligerus* and noted that it did not share cofactor requirements with other β-lactam biosynthetic enzymes (Table 1). The enzyme was purified to homogeneity by Usui and Yu [34] and was determined to be a racemase, converting isopenicillin N and/or penicillin N to an equimolar ratio of the two substrates/products, shifting to the direction of penicillin N by its further metabolism to DAOC. The gene was cloned by Kovacevic et al. [35].

D. Ring Expansion/Hydroxylation Enzymes (DAOCS/DACS)

The conversion of penicillin N to CPC (Fig. 1) is initiated by the oxidative ring expansion of penicillin N to deacetoxycephalosporin C (DAOC; [36]). Subsequent hydroxylation of the 3′-methyl carbon of DAOC generates deacetylcephalosporin C (DAC; [37]). Kohsaka and Demain [36] described the enzymatic ring expansion activity from cell-free extracts of *C. acremonium.* Two reports of partial purification of the ''expandase'' enzyme by Kupka et al. [38] and Scheidegger et al. [39] and their respective co-workers suggested that ring expansion and hydroxylation activities were catalyzed by a single enzyme. They also described an inherent instability of the enzyme preventing sufficient purification to definitively determine the bifunctional nature of the enzyme [38,39]. In contrast, ring expansion and hydroxylation activities from *S. clavuligerus* were clearly separable by anion-exchange chromatography [40]. *S. clavuligerus* DAOC synthetase (''expandase,'' Table 1) and DAC synthase (''hydroxylase,'' Table 1) were subsequently purified to near homogeneity, biochemically characterized, and cloned and expressed in *E. coli* [41–44]. Dotzlaf and Yeh devised a stabilizing cocktail that was modified as the purification progressed, allowing the purification of DAOC synthetase/DAC synthase from *C. acremonium* to near-homogeneity [45]. The highly purified protein retained its bifunctional nature, an observation also reported by Baldwin et al. [46] in a nearly simultaneous publication. The bifunctionality of the DAOC synthetase/DAC synthase from *C. acremonium* was conclusively demonstrated by cloning and expressing its structural gene in *E. coli* [47]. The gene coding for the two activities in *C. acremonium* is named *cef*EF, while the separate genes coding for ring expansion and hydroxylation activities in *S. clavuligerus* are named *cef*E and *cef*F, respectively [5].

E. DAC Acyltransferase (DAC-AT)

The final step in the biosynthesis of CPC (Fig. 1) is catalyzed by acetyl CoA: deacetylcephalosporin C *O*-acetyltransferase (DAC-AT). Fujisawa et al. [48] showed that cell-free extracts of a CPC-producing strain readily converted DAC and acetyl-l-[^{14}C]-CoA to labeled CPC. They utilized this novel assay system to characterize several mutants that accumulated DAC and correctly proposed that

DAC was an intermediate in the CPC pathway and was converted to CPC by DAC-AT [48,49].

Scheideggar et al. [50] purified DAC-AT approximately 15-fold and estimated the molecular mass of the enzyme to be ~70 kDa. By incorporating 7-aminocephalosporanic acid (7-ACA) in the purification buffers, Matsuyama and his co-workers were able to stabilize DAC-AT, allowing a 1300-fold purification to apparent homogeneity [51]. They estimated the mass to be ~55 kDa based on gel filtration and determined that the enzyme was a heterodimer composed of 14-kDa and 27-kDa subunits based on SDS-PAGE. The difference in mass as measured by the two methods was not discussed by the authors [51]. In a related article, the cloning of the *cef*G gene and expression of DAC-AT activity in *Saccharomyces cerevisiae* using the reverse genetics approach based on the amino acid sequence of the purified protein was reported [52]. In a third paper the disruption of *cef*G gene expression by insertional mutagenesis causing an accumulation of DAC and a lack of CPC production in the mutant *C. acremonium* strain was described [53].

Gutierrez et al. [54] and Mathison et al [55] cloned the *cef*G gene using traditional genetics by first searching for potential open reading frames (ORF) flanking the *pcb*C (IPNS) and *cef*EF (expandase/hydroxylase) genes. A likely ORF was located upstream of the *cef*EF gene by DNA sequence analysis [54,55]. Both groups demonstrated that the cloned gene complemented DAC-AT deficient mutants of *C. acremonium* by restoring CPC production [54,55]. Gutierrez and his co-workers cloned and expressed DAC-AT activity in *P. chrysogenum*, while Mathison and her co-workers reported similar results in *Aspergillus niger*.

Based on DNA sequencing, the mass of the *cef*G gene product was deduced to be 49,269 Da [54], which correlated well with the value reported by Matsuyama et al. [51]. Subsequently, Velasco et al. [56] demonstrated the mass of immunoaffinity purified DAC-AT to be 49–52 kDa based on SDS-PAGE and gel filtration. However, unlike Matsuyama et al. [51], DAC-AT purified from three different *C. acremonium* strains by Velasco et al. [56] was monomeric.

III. METABOLIC ENGINEERING FOR CPC YIELD IMPROVEMENT

Isolation and structural elucidation of the active components from antibiotic-producing microorganisms begins the study of the biosynthetic pathway leading to the desired end product. The abundance of the various intermediate products of an antibiotic biosynthetic pathway can be accurately determined throughout the fermentation cycle by techniques such as high-performance liquid chromatography (HPLC). This and other analytical techniques allow identification of potential rate-limiting steps in the pathway. Three rate-limiting steps in the CPC pathway that are catalyzed by ACV synthetase, DAOC synthetase/DAC synthase, and DAC acetyltransferase are discussed below.

A. ACV Synthetase

Mounting evidence established the formation of ACV to be a rate-limiting step in CPC biosynthesis. Because ACV synthetase catalyzes the transitional step between primary and secondary metabolism in the β-lactam-producing microorganisms, it is likely to be a major regulatory site for β-lactam biosynthesis [20]. Martin and Liras [57] documented that increased ACV synthetase activity paralleled higher levels of cephalosporin or penicillin production in sequential strains. Zhang and Demain [58] showed that ACV synthetase from various cell-free extracts of *C. acremonium* and *S. clavuligerus* possessed from 1% to 10% of the specific activity of isopenicillin N synthase, isopenicillin N epimerase, and deacetoxycephalosporin C synthetase. Finally, analysis of a mathematical model of CPC biosynthesis based on *in vitro* enzyme kinetic data led Malmberg and Hu [59] to conclude that ACV synthesis is the major rate-limiting step in the pathway. At this writing, there are no reports of improved CPC yields in industrial strains by increasing the copy number of *pcb*AB. However, MacCabe et al. [16] described the restoration of penicillin production in *A. nidulans* transformed with the *pcb*AB gene or the *pcb*AB plus *pcb*C genes together. They also claimed that certain transformants expressed a higher level of antimicrobial activity than the corresponding wild-type strain [16] and suggested that extra copies of *pcb*AB might improve yields in the wild-type *A. nidulans*. In related work, Kennedy and Turner [60] overexpressed ACV synthetase when the promoter for *pcb*AB in *A. nidulans* was replaced with the ethanol dehydrogenase promoter and reported up to a 30-fold increase in penicillin yields.

B. Isopenicillin N Synthetase and Epimerase

Isopenicillin N synthase and epimerase have not been implicated as rate-limiting enzymes for CPC biosynthesis. Although ACV was found in the mycelia of *P. chrysogenum* [61] and *C. acremonium* [6], an excessive accumulation of ACV in fermentation broths has not been reported. Furthermore, pools of excreted penicillin were determined to be mostly penicillin N, not isopenicillin N [62,63], favoring the synthesis of the next intermediate, DAOC (Fig. 1). In addition, *in vitro* kinetic data exclude both IPNS and IPNE as potential rate-limiting enzymes [59].

C. DAOC Synthetase/DAC Synthase

Metabolite analysis of ultrafiltered broth from a high-yielding CPC fermentation showed, in addition to CPC, the accumulation of penicillin, DAOC, and DAC

Table 2 Metabolite Analysis of Ultrafiltered Broth from Untransformed *C. acremonium* and Transformed *C. acremonium* Containing One Extra Copy of the *cef*EF Gene Encoding DAOCS/DACS[a]

Strain	Penicillin N (units/liter)	DAOC (units/liter)	DAC (units/liter)	CPC (units/liter)	Total (units/liter)
Untransformed	31	6	6	100	143
Transformed	2	1	6	115	124

[a] The data are from Skatrud et al. [63] and represent values normalized to the concentration of CPC in the control fermentation.

(Table 2; [63]). Further analysis of the accumulated penicillin indicated that >80% was penicillin N, the substrate for the DAOCS [63]. Skatrud et al. [63] described the elegant first use of metabolic engineering of an industrially important antibiotic producing strain when they successfully cloned an additional copy of the *cef*EF gene into *C. acremonium* strain 394-4 (a derivative of *C. acremonium* ATCC 11550). Their cloning efforts resulted in a twofold increase in the DAOCS activity in 150-liter pilot-scale fermentations [63]. Doubling the DAOCS activity translated into a 16-fold reduction in level of penicillin N, a 6-fold decrease of accumulated DAOC, and a 15% increase in CPC yields [63]. There was no effect on the level of DAC [63]. These data predated and confirmed the mathematical model of CPC biosynthesis of Malmberg and Hu [59].

D. DAC Acyltransferase

As stated above, Skatrud et al. clearly showed that, while penicillin N accumulation was significantly reduced and CPC yields were substantially enhanced, the level of DAC was unchanged (Table 2; [63]). The authors suggested that limitation in the carbon flow at the DAC-AT step could be causing the accumulation of DAC; however, they cautioned that DAC present at the end of the fermentation could be the result of degradation of CPC [63]. Cephalosporin C degradation to DAC can result from either or both enzymatic hydrolysis by an extracellular acetylhydrolase [49,64] and chemical hydrolysis due to a rapid rise in pH at the end of the fermentation [65]. Skatrud and his co-workers suggested that cloning and expression of the *cef*G gene (Fig. 1) in a strain similar to 394-4 would be a practical way to address this question, but DAC-AT had not been purified or cloned at that time [63]. By 1993, the means to address this issue were available. The enzyme had been purified, characterized, and cloned in several laboratories by both traditional and reverse genetics [51–55]. Mathison et al. [55] and later Gutierrez et al. [66] successfully cloned and expressed the *cef*G gene in *C. acremonium*. Both groups showed enhanced conversion of DAC to CPC compared to untransformed strains [55,66]. Mathison and her co-workers [55] speculated that, since the DNA fragment used by Skatrud et al. [63] to clone the *cef*EF gene unknowingly contained the *cef*G gene, the reduced penicillin N pool and improved CPC yields reported could have been the result of enhanced DAC-AT rather than enhanced DAOCS activity. They suggested that increased DAC-AT activity might reduce DAC levels to a point below a theoretical repression level for *cef*EF expression [55]. This interesting hypothesis would not account for the lack of reduction in the DAC pool in the Skatrud strain. Velasco et al. [67] showed that the *cef*G gene is poorly transcribed during both early- and late-phase growth in three low-producing *C. acremonium* strains compared to the *cef*EF gene. Gutierrez and his co-authors [66] reported that introduction of extra copies of the *cef*G gene with its native promoter region only marginally improved reduction of the DAC pool with a corresponding increase in the CPC product. However, when the promoter was replaced with four different promoters, including the promoter for the *P. chrysogenum pcb*C gene (IPNS), high steady-state levels of DAC-AT expression and activity for each construct were found [66]. The report by Gutierrez et al. [66] implies that Skatrud et al. [63] correctly interpreted their data. However, to paraphrase Mathison et al. [55], it will be interesting to see if incorporating an altered promoter region will increase *cef*G expression and improve titers in industrial production strains of *C. acremonium.*

IV. METABOLIC ENGINEERING FOR PRODUCTION OF 7-ADCA

The manufacture of several semisynthetic oral cephalosporin antibiotics involves the chemical ring expansion of penicillin V to 7-aminodeacetoxycephalosporanic acid (7-ADCA; Fig. 2; [68]). This is a costly and potentially environmentally damaging process [69]. Cloning of the *C. acremonium cef*EF gene (DAOCS/DACS [47]) and the *S. clavuligerus cef*D (IPNE) and *cef*E gene (DAOCS [43]) opened the possibility for biosynthetic/enzymatic processes for production of 7-ADCA.

A. DAOC Generation

Early *in vitro* data indicated that ring expansion of adipyl-6-APA or penicillin G was either nonexistent or barely detectable [45,70]. Therefore, an alternative route to 7-ADCA through the enzymatic deacylation of DAOC (Fig. 3) requiring the fermentative production of DAOC at economically feasible levels was pursued. Elimination of DACS activity would allow *C. acremonium* to produce DAOC as its end product (Fig. 1). Before it was known that DAOCS and DACS activities in *C. acremonium* were catalyzed by a single bifunctional enzyme,

Penicillin V

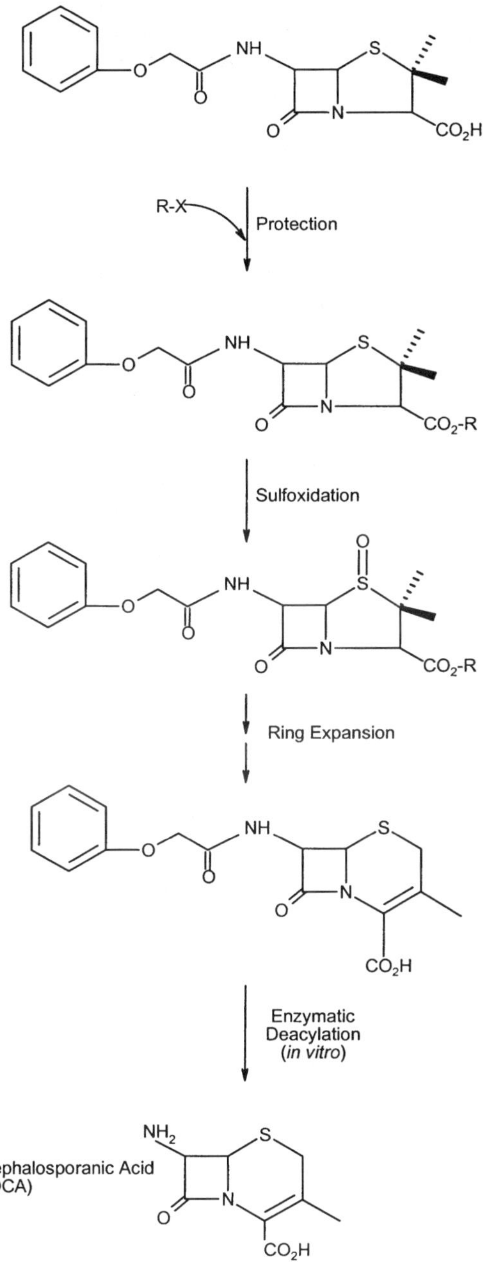

7-Aminodeacetoxycephalosporanic Acid (7-ADCA)

Figure 2 The current chemical process for the production of 7-aminodeacetoxycephalosporanic acid.

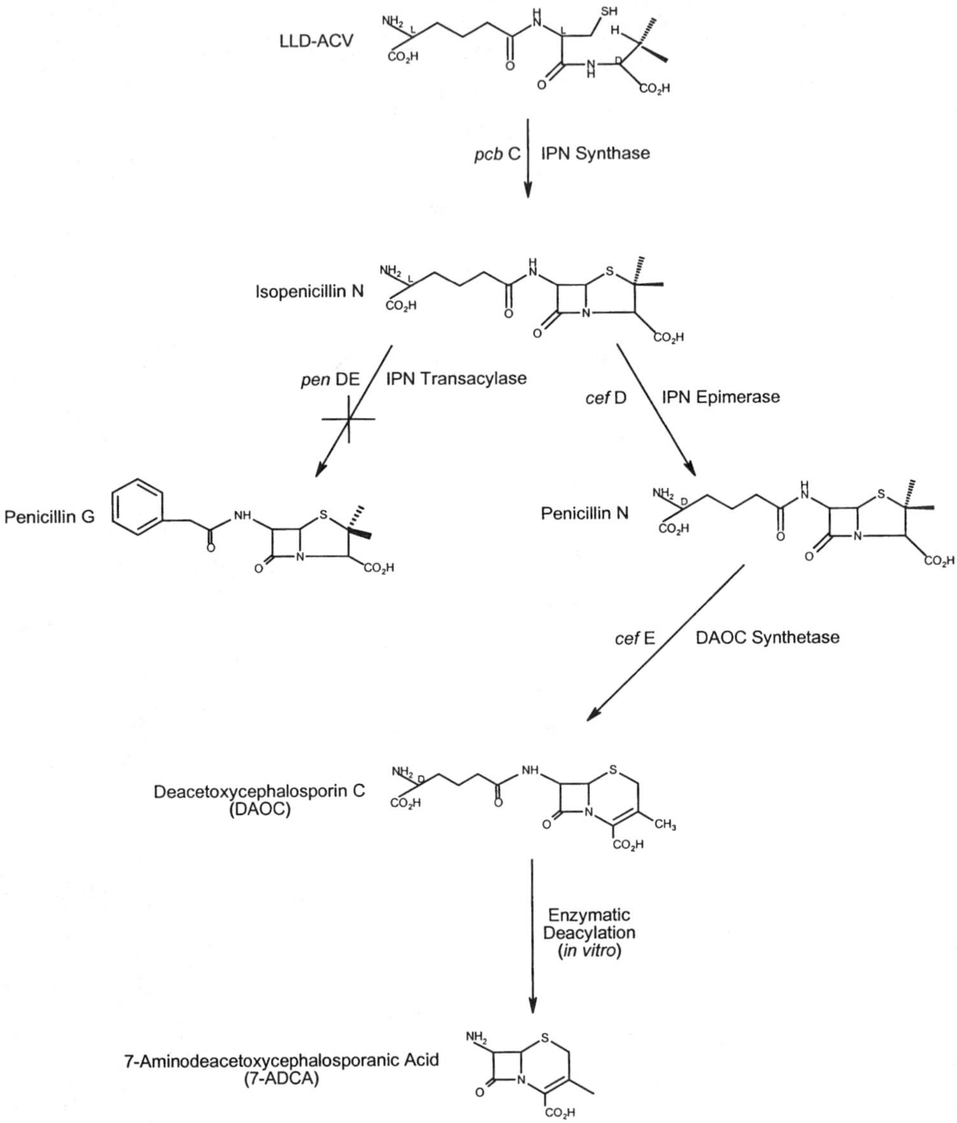

Figure 3 Metabolic engineering for deacetoxycephalosporin production in *P. chrysogenum* expressing the *S. clavuligerus cef*D and *cef*E genes.

Queener et al. [71] conducted mutagenesis and screening for mutants blocked in DACS activity. They effectively eliminated CPC production in some mutants, but DAOC production was only slightly enhanced [71]. Attempts to eliminate DACS activity of the bifunctional DAOCS/DACS by *in vitro* mutagenesis of the cloned *cef*EF gene always resulted in a concomitant loss of DAOCS activity as well [72]. It was later revealed through the use of inhibition kinetics that the two activities probably share a common active site [73]. A simple disruption or deletion of the *cef*F gene in *S. clavuligerus* would result in the desired effect, since DAOCS of that organism produces only minimal amounts of DAC [42]; however, to date this result has not been reported. A second approach involved expressing the *S clavuligerus cef*D and *cef*E genes in *P. chrysogenum.* Cantwell et al. [69] transformed *P. chrysogenum* with a hybrid *cef*E gene containing the promoter sequence from *P. chrysogenum pcb*C gene and reported DAOCS specific activities from 4.3% to 10.3% relative to *S. clavuligerus.* Cantwell et al. [72] later cloned the *S. clavuligerus cef*D gene into the *P. chrysogenum* strain that contained the *S. clavuligerus cef*E gene and demonstrated IPNE activity. This pioneering work was the first demonstration of cephalosporin production by fermentation of *P. chrysogenum.* The DAOC production was at the expense of penicillin V production but much less than the normal end product [72]. The authors concluded that DAOC production would be greatly enhanced if isopenicillin N-to-penicillin V conversion could be prevented [72].

B. DAOG Generation

A second potential biosynthetic/enzymatic route to 7-ADCA involves the enzymatic ring expansion of penicillin G or V *in vivo* followed by *in vitro* deacylation (Fig. 4). Little or no *in vitro* ring expansion activity with these substrates was shown for either the bifunctional DAOCS/DACS or monofunctional DAOCS [46,71]. It was proposed that, through site-directed mutagenesis, DAOCS could be ''enzyme engineered'' to convert penicillin G or V to DAOG or DAOV, which would then be enzymatically deacylated *in vitro* to yield 7-ADCA (Fig. 4 [69,74]). Improved expression of native DAOCS in *P. chrysogenum* was the first goal in the proposed scheme. Queener et al. [74] reconstructed the hybrid gene described by Cantwell et al. [69] by inserting the *cef*E ORF between the *pcb*C promoter and terminator sequences. One *P. chrysogenum* isolate transformed with the new hybrid gene produced DAOCS specific activities 70-fold higher than the best transformant previously reported [69], 4-fold higher than the parent *S. clavuligerus*, and about 75% of the activity of an industrial strain of *C. acremonium* [74]. A similar tactic was used by Bovenberg et al [75] to construct hybrid genes using the promoter and terminator regions of the *P. chrysogenum pen*DE gene and the *cef*E ORF from either *S. clavuligerus* or *Nocardia lactamdurans* (Fig. 1). The *pen*DE promoter and terminator regions ensure expression of

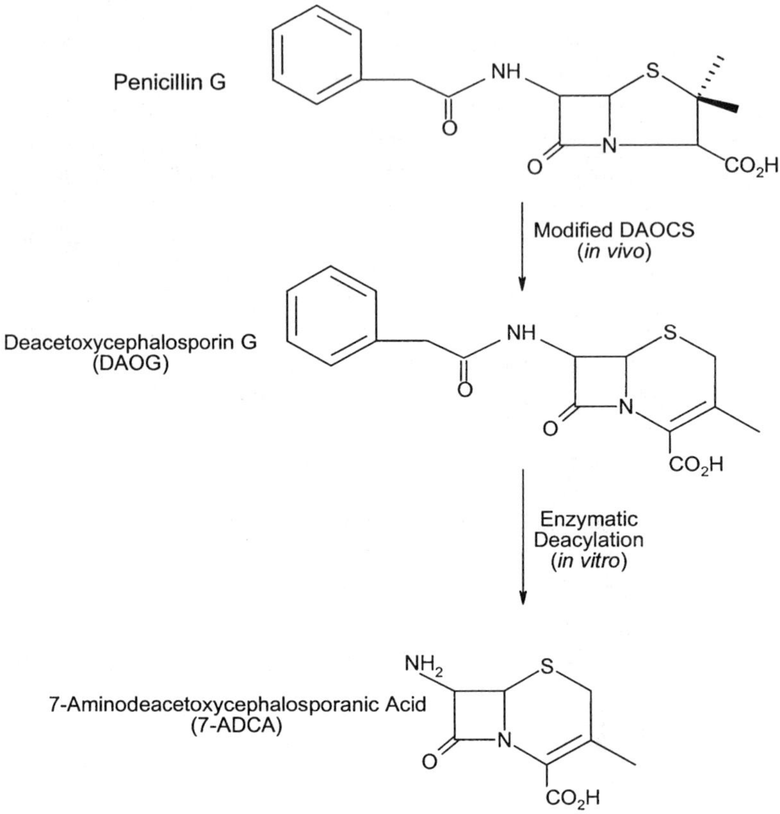

Figure 4 Metabolic engineering for ring expansion of penicillin G in *P. chrysogenum* expressing a modified *S. clavuligerus* DAOCS.

DAOCS after formation of isopenicillin N [75]. Sutherland et al. [76] reasoned that because IPNS and DAOCS were members of the same family of oxidase enzymes, they might share sequence similarity in the binding pockets for their respective substrates. In an elegant set of experiments, they analyzed the binding site of the L-α-aminoadipyl side chain of ACV from the structurally characterized *A. nidulans* IPNS and identified homologous amino acids in *S. clavuligerus* DAOCS responsible for binding penicillin N [76]. Using site-directed mutagenesis, the *S. clavuligerus cef*E gene was modified to accept penicillin G as its primary substrate. *P. chrysogenum*, transformed with the modified gene hybridized as described above, successfully produced DAOG [76].

C. Production of Adipyl-7-ADCA

Yet another process for producing 7-ADCA utilizing expression of CPC pathway genes in *P. chrysogenum* was described by Conder et al. [77] and Crawford et al. [78]. The process involved the generation of adipyl-6-amino penicillanic acid (adipyl-6-APA) by feeding disodium adipate in place of potassium phenoxyacetate to fermentations of *P. chrysogenum, in situ* ring expansion to adipyl-7-ADCA, followed by *in vitro* deacylation to the final product [77,78]. The authors successfully gambled on the expandability of adipyl-6-APA *in vivo*, considering that *in vitro* expansion of the substrate was marginally detectable in one report using purified *C. acremonium* DAOCS/DACS [46] and not detectable in other reports [38,70]. Hybrid genes containing the promoter region from glyceraldehyde-3-phosphate dehydrogenase (GAP), *pcb*C (IPNS) or β-tubulin from *P. chrysogenum* fused to the *cef*E ORF of *S. clavuligerus* were constructed [77,78]. Transformants were screened for the production of adipyl-7-ADCA and, although all isolates expressed mRNA for the hybrid gene and produced adipyl-6-APA, the level of product varied from undetectable to relatively high levels [77,78]. The authors could not correlate product formation to gene copy number, and they concluded that expression was influenced more by site of gene integration than by the promoter sequence fused to the *cef*E ORF adipyl-7-ADCA [77,78]. Finally, traditional strain improvement techniques with the highest-producing transformants significantly improved adipyl-7ADCA production, and the authors predicted a commercially viable process [77,78]. In related work, Crawford et al. [78] described the production of 7-aminocephalosporanic acid (7-ACA), another important intermediate in semisynthetic cephalosporin manufacture, in adipate fed *P. chrysogenum* transformed with a hybrid gene containing the *C. acremonium* *cef*EF gene.

A detailed discussion concerning the conversions of penicillins to cephalosporins is provided by Demain et al. in Chapter 3.

V. CONCLUSIONS

The application of traditional microbiology, mathematical modeling, biochemical engineering, analytical chemistry, mutagenesis, and screening significantly improved the production of CPC by fermentation of *C. acremonium* from the early 1950s through the 1970s. With the advent of recombinant DNA technology combined with new, readily available analytical techniques and sophisticated enzyme biochemistry in the early 1970s, an unparalleled understanding of the biosynthetic process for CPC production was realized. Reducing or eliminating rate limitations in the biosynthesis of CPC through the application of gene cloning resulted in yield increases in production strains where the traditional methods were no longer proving to be successful. Cloning and expression of several CPC biosynthetic genes in *P. chrysogenum* has led to the novel, economical, and environmentally desirable biosynthetic routes to important precursors for the production of semisynthetic cephalosporins. Interestingly, the traditional methods of fermentation optimization and strain improvement are successfully improving production of the semisynthetic cephalosporin intermediates by the newly created cephalosporin-producing *P. chrysogenum* strains. Metabolic engineering for improved biosynthetic production of intermediate and end products for economically important antibiotics has been firmly established as indicated by the pioneering work described in this chapter. Similar advances are discussed in other chapters of this book, and we believe that these gene/enzyme methodologies will be commonly applied to the production of other existing as well as newly discovered antibiotics in the future.

IN MEMORIAL

The authors had the privilege of working with Sir Edward P. Abraham and note with sadness the passing of this historically important scientist. EPA, as we affectionately knew him, helped elucidate the structure of penicillin during World War II. In 1940, with E. Chain, he discovered the β-lactamase enzyme [79]. In the 1950s and early 1960s, with G. G. F. Newton and others, he discovered the cephalosporin antibiotics and described their isolation from the fungus provided to Oxford by Guiseppe Brotzu, culminating in the elucidation of the structure for CPC [80]. EPA provided Lilly scientists with purified isopenicillin N synthetase, thereby enabling, via reverse genetics, the first cloning and characterization of the *pcb*C gene and the expression of recombinant IPNS in *E. coli* [25]. Working with us, he helped demonstrate the transformation of *P. chrysogenum* with hybrid *cef*D and *cef*E genes from *S. clavuligerus* and the production of cephalosporins by the recombinant *P. chrysogenum* [71]. Thus, his scientific career spanned the beta-lactam antibiotic story from first discovery to modern metabolic engineering. Throughout our interactions with him, he exhibited a mentoring, humble manner that inspired us and belied the greatness of this man from Oxford, England.

REFERENCES

1. RP Elander, H Aoki. β-Lactam-producing microorganisms: their biology and fermentation behavior. In: RB Morin, M Gorman, eds. Chemistry and Biology of β-Lactam Antibiotics, Vol. 3. New York: Academic Press, 1982, pp 83–153.
2. B Diez, E Mellado, R Fouces, M Rodriguez, JL Barredo. Recombinant *Acremonium*

chrysogenum strains for the industrial production of cephalosporin. Microbiol Sem 12:359–370, 1996.
3. SW Queener, N Neuss. The biosynthesis of β-lactam antibiotics. In: RB Morin, M Gorman, eds. Chemistry and Biology of β-Lactam Antibiotics, Vol. 3. New York: Academic Press, 1982, pp 1–81.
4. WK Yeh, SW Queener. Potential industrial use of cephalosporin biosynthetic enzymes and genes. An overview. Ann NY Acad Sci 613:128–141, 1990.
5. TD Ingolia, SW Queener. Beta-lactam biosynthetic genes. Med Res Rev 9:245–264, 1989.
6. PB Loder, EP Abraham. Isolation and nature of intracellular peptides from a cephalosporin C-producing *Cephalosporium sp*. Biochem J 123:471–476, 1971.
7. EP Abraham. Beta-lactam antibiotics and related substances. Jpn J Antibiot 30:1–26, 1977.
8. RM Adlington, JE Baldwin, M Lopez-Nieto, JA Murphy, N Patel. A study of the biosynthesis of the tripeptide δ-(L-α-aminoadipyl)-L-cysteinyl-D-valine in a β-lactam-negative mutant of *Cephalosporium acremonium*. Biochem J 213:573–576, 1983.
9. GJM Hersbach, CP van der Beek, PWM van Dijk. The penicillins: properties, biosynthesis, and fermentation. In: EJ Vandamme, ed. Biotechnology of Industrial Antibiotics. New York: Marcel Dekker, 1984, pp 45–140.
10. J Nuesch, J Heim, HJ Treichler. The biosynthesis of sulfur-containing β-lactam antibiotics. Annu Rev Microbiol 41:51–75, 1987.
11. G Banko, AL Demain, S Wolfe. δ-(L-α-aminoadipyl)-L-cysteinyl-D-valine synthetase (ACV synthetase): a multifunctional enzyme with broad substrate specificity for the synthesis of penicillin and cephalosporin precursors. J Am Chem Soc 109: 2858–2860, 1987.
12. SE Jensen, DWS Westlake, S Wolfe. Production of the penicillin precursor δ-(L-α-aminoadipyl)-L-cysteinyl-D-valine (ACV) by cell-free extracts from *Streptomyces clavuligerus*. FEMS Microbiol Lett 49:213–218, 1988.
13. H Van Liempt, H von Dohren, H Kleinkauf. δ-(L-α-aminoadipyl)-L-cysteinyl-D-valine synthetase from *Aspergillus nidulans*. The first enzyme in penicillin biosynthesis is a multifunctional peptide synthetase. J Biol Chem 264:3680–3684, 1989.
14. JE Baldwin, JW Bird, RA Field, NM O'Callaghan, CJ Schofield. Isolation and partial characterization of ACV synthetase from *Cephalosporium acremonium* and *Streptomyces clavuligerus*. J Antibiot 43:1055–1057, 1990.
15. JE Baldwin, JW Bird, RA Field, NM O'Callaghan, CJ Schofield, AC Willis. Isolation and partial characterization of ACV synthetase from *Cephalosporium acremonium* and *Streptomyces clavuligerus*: evidence for the presence of phosphopantothenate in ACV synthetase. J Antibiot 44:241–248, 1991.
16. AP MacCabe, MBR Riach, JR Kinghorn. Identification of the ACV synthetase gene. J Biotechnol 17:91–97, 1991.
17. B Diez, S Gutierrez, JL Barredo, P van Solingen, LHM von der Voort, JF Martin. The cluster of penicillin biosynthetic genes. Identification and characterization of the *pcb*AB gene encoding the α-aminoadipyl-cysteinyl-valine synthetase and linkage to the *pcb*C and *pen*DE genes. J Biol Chem 265:16358–16365, 1990.

18. JA Hoskins, N O'Callaghan, SW Queener, CA Cantwell, JS Wood, VJ Chen, PL Skatrud. Gene disruption of the *pcb*AB gene encoding ACV synthetase in *Cephalosporium acremonium*. Curr Genet 18:523–530, 1990.
19. S Gutierrez, B Diez, E Montegnegro, JF Martin. Characterization of the *Cephalosporium acremonium pcb*AB gene encoding α-aminoadipyl-cysteinyl-valine synthetase, a large multidomain peptide synthetase: linkage to the *pcb*C gene as a cluster of early cephalosporin biosynthetic genes and evidence of multiple functional domains. J Bacteriol 173:2354–2365, 1991.
20. J Zhang, AL Demain. ACV synthetase. Crit Rev Biotechnol 12:245–260, 1992.
21. Y Aharonowitz, J Bergmeyer, JM Cantoral, G Cohen, AL Demain, U Fink, J Kinghorn, H Kleinkauf, A MacCabe, H Palissa, E Pfeifer, T Schwecke, H van Liempt, H von Dohren, S Wolfe, J Zhang. δ-(L-α-aminoadipyl)-L-cysteinyl-D-valine synthetase, the multienzyme integrating the four primary reactions in β-lactam biosynthesis, as a model peptide synthetase. Bio/Technology 11:807–810, 1993.
22. JE Baldwin, EP Abraham. The biosynthesis of penicillins and cephalosporins. Nat Prod Rep 5:129–145, 1988.
23a. CP Pang, B Chakravarti, RM Adlington, HH Ting, RL White, GS Jayatilake, JE Baldwin, EP Abraham. Purification of isopenicillin N synthetase. Biochem J 222: 789–795, 1984.
23b. IJ Hollander, YQ Shen, J Heim, AL Demain, S Wolfe. A pure enzyme catalyzing penicillin biosynthesis. Science 224:610–612, 1984.
24. JE Baldwin, J Gagnon, H Ting. N-terminal amino acid sequence and some properties of isopenicillin-N synthetase from *Cephalosporium acremonium*. FEBS Lett 188: 253–256, 1985.
25. SM Samson, R Belagaje, DT Blankenship, JL Chapman, D Perry, PL Skatrud, RM VanFrank, EP Abraham, JE Baldwin, SW Queener, TD Ingolia. Isolation, sequence determination and expression in *Escherichia coli* of the isopenicillin N synthetase gene from *Cephalosporium acremonium*. Nature 318:191–194, 1985.
26. FR Ramos, MJ Lopez-Nieto, JF Martin. Isopenicillin N synthetase of *Penicillium chrysogenum*, an enzyme that converts δ-(L-α-aminoadipyl)-L-cysteinyl-D-valine to isopenicillin N. Antimicrob Agents Chemother 27:380–387, 1985.
27. SE Jensen, BK Leskiw, LC Vining, Y Aharonowitz, DWS Westlake, S Wolfe. Purification of isopenicillin N synthetase from *Streptomyces clavuligerus*. Can J Microbiol 32:953–958, 1986.
28. JE Baldwin, SJ Killin, AJ Pratt, JD Sutherland, NJ Turner, MJC Crabbe, EP Abraham, AC Willis. Purification and characterization of cloned isopenicillin N synthetase. J Antibiot 40:652–659, 1987.
29. S Wolfe, AL Demain, SE Jensen, DWS Westlake. Enzymatic approach to the synthesis of un-natural beta-lactams. Science 226:1386–1392, 1984.
30. VJ Chen, AM Orville, MR Harpel, CA Frolik, KK Surerus, E Munck, JD Lipscomb. Spectroscopic studies of isopenicillin N synthase: a mononuclear nonheme Fe^{2+} oxidase with metal coordination sites for small molecules and substrate. J Biol Chem 264:21667–21681, 1989.
31. SW Queener. Molecular biology of penicillin and cephalosporin biosynthesis. Antimicrob Agents Chemother 34:943–948, 1990.
32. C Lubbe, S Wolfe, AL Demain. Isopenicillin N epimerase activity in a high cephalo-

sporin-producing strain of *Cephalosporium acremonium*. Appl Microbiol Biotechnol 23:367–368, 1986.
33. SE Jensen, DWS Westlake, S Wolfe. Partial purification and characterization of isopenicillin N epimerase activity from *Streptomyces clavuligerus*. Can J Microbiol 29:1526–1531, 1983.
34. S Usui, CA Yu. Purification and properties of isopenicillin N epimerase from *Streptomyces clavuligerus*. Biochim Biophys Acta 999:78–85, 1989.
35. S Kovacevic, MB Tobin, JR Miller. The β-lactam biosynthesis genes for isopenicillin N epimerase and deacetoxycephalosporin C synthetase are expressed from a single transcript in *Streptomyces clavuligerus*. J Bacteriol 172:3952–3958, 1990.
36. M Kohsaka, AL Demain. Conversion of penicillin N to cephalosporin(s) by cell-free extracts of *Cephalosporium acremonium*. Biochem Biophys Res Commun 70: 465–473, 1976.
37. SJ Brewer, JE Farthing, MK Turner. The oxygenation of the 3-methyl group of 7-β-(5-D-aminoadipamido)-3-methylceph-3-em-4-carboxylic acid (desacetoxycephalosporin C) by extracts of *Acremonium chrysogenum*. Biochem Soc Trans 5:1024–1026, 1977.
38. J Kupka, YQ Shen, S Wolfe, AL Demain. Studies on the ring-cyclization and ring-expansion enzymes of β-lactam biosynthesis in *Cephalosporium acremonium*. Can J Microbiol 29:488–496, 1983.
39. A Scheidegger, MT Kuenzl, J Nuesch. Partial purification and catalytic properties of a bifunctional enzyme in the biosynthetic pathway of β-lactams in *Cephalosporium acremonium*. J Antibiot 37:522–531, 1984.
40. SE Jensen, DWS Westlake, S Wolfe. Deacetoxycephalosporin C synthetase and deacetoxycephalosporin C hydroxylase are two separate enzymes in *Streptomyces clavuligerus*. J Antibiot 38:263–265, 1985.
41. JE Dotzlaf, WK Yeh. Purification and properties of deacetoxycephalosporin C synthase from recombinant *Escherichia coli* and its comparison with the native enzyme purified from *Streptomyces clavuligerus*. J Biol Chem 264:10219–10227, 1989.
42. BJ Baker, JE Dotzlaf, WK Yeh. Deacetoxycephalosporin C hydroxylase of *Streptomyces clavuligerus*: purification, characterization, bifunctionality, and evolutionary implication. J Biol Chem 266:5087–5093, 1991.
43. S Kovacevic, BJ Weigel, MB Tobin, TD Ingolia, JR Miller. Cloning, characterization, and expression in *Escherichia coli* of the *Streptomyces clavuligerus* gene encoding deacetoxycephalosporin C synthetase. J Bacteriol 171:754–760, 1989.
44. S Kovacevic, JR Miller. Cloning and sequencing of the β-lactam hydroxylase gene (*cef*F) from *Streptomyces clavuligerus*: gene duplication may have led to separate hydroxylase and expandase activities in the actinomycetes. J Bacteriol 173:398–400, 1991.
45. JE Dotzlaf, WK Yeh. Copurification and characterization of deacetoxycephalosporin C synthetase/hydroxylase from *Cephalosporium acremonium*. J Bacteriol 169: 1611–1618, 1987.
46. JE Baldwin, RM Adlington, JB Coates, MJC Crabbe, NP Crouch, JW Keeping, GC Knight, CJ Schofield, HH Ting, CA Vallerjo, M Thorniley, EP Abraham. Purification and initial characterization of an enzyme with deacetoxycephalosporin C synthetase and hydroxylase activities. Biochem J 245:831–841, 1987.

47. SM Samson, JE Dotzlaf, ML Slisz, GW Becker, RM Van Frank, LE Veal, WK Yeh, JR Miller, SW Queener, TD Ingolia. Cloning and expression of the fungal expandase/hydroxylase gene involved in cephalosporin biosynthesis. Bio/Technology 5:1207–1214, 1987.
48. Y Fujisawa, T Kanzaki. Role of acetyl CoA: deacetylcephalosporin C acetyltransferase in cephalosporin C biosynthesis by *Cephalosporium acremonium*. Agric Biol Chem 39:2043–2048, 1975.
49. Y Fujisawa, H Shirafuji, M Kida, K Nara, M Yoneda, T Kanzaki. Accumulation of deacetylcephalosporin C by cephalosporin C negative mutants of *Cephalosporium acremonium*. Agric Biol Chem 39:1295–1301, 1975.
50. A Scheidegger, A Gutzwiller, M Kueenzi, A Fiechter, J Nuesch. Investigation of acetyl-CoA: deacetylcephalosporin C *O*-acetyltransferae of *Cephalosporium acremonium*. J Biotechnol 3:109–117, 1985.
51. K Matsuyama, H Matsumoto, A Matsuda, H Sugiura, KI Komatsu, S Ichikawa. Purification of acetyl coenzyme A: deacetylcephalosporin C *O*-acetyltransferase from *Acremonium chrysogenum*. Biosci Biotechnol Biochem 56:1410–1412, 1992.
52. A Matsuda, H Sugiura, K Matsuyama, H Matsumoto, S Ichikawa, KI Komatsu. Molecular cloning of acetyl coenzyme A: deacetylcephalosporin C *O*-acetyltransferase cDNA from *Acremonium chrysogenum*: sequence and expression of catalytic activity in yeast. Biochem Biophys Res Commun 182:995–1001, 1992.
53. A Matsuda, H Sugiura, K Matsuyama, H Matsumoto, S Ichikawa, KI Komatsu. Cloning and disruption of the *cefG* gene encoding acetyl coenzyme A: deacetylcephalosporin C *O*-acetyltransferase from *Acremonium chrysogenum*. Biochem Biophys Res Commun 186:40–46, 1992.
54. S Gutierrez, J Velasco, FJ Fernandez, JF Martin. The *cef*G gene of *Cephalosporium acremonium* is linked to the *cef*EF gene and encodes a deacetylcephalosporin C acetyltransferase closely related to homoserine *O*-acetyltransferase. J Bacteriol 174: 3056–3064, 1992.
55. L Mathison, C Soliday, T Stepan, T Aldrich, J Rambosek. Cloning, characterization, and use in strain improvement of the *Cephalosporium acremonium* gene *cef*G encoding acetyl transferase. Curr Genet 23:33–41, 1993.
56. J Velasco, S Gutierrez, S Campoy, JF Martin. Molecular characterization of the *Acremonium chrysogenum cef*G gene product: the native deacetylcephalosporin C acetyltransferase is not processed into subunits. Biochem J 337:379–385, 1999.
57. JF Martin, P Liras. Enzymes involved in penicillin, cephalosporin and cephamycin biosynthesis. In: A Fiechter, ed. Advances in Biochemical Engineering/Biotechnology, Vol. 39. Berlin: Springer-Verlag, 1989, pp 153–187.
58. J Zhang, AL Demain. Regulation of ACV synthetase in penicillin- and cephalosporin-producing microorganisms. Biochem Adv 9:623, 1992.
59. LH Malmberg, WS Hu. Identification of rate-limiting steps in cephalosporin C biosynthesis in *Cephalosporium acremonium*: a theoretical analysis. Appl Microbiol Biotechnol 38:122–128, 1992.
60. J Kennedy, G Turner. delta-(L-alpha-aminoadipyl)-L-cysteinyl-D-valine synthetase is a rate limiting enzyme for penicillin production in *Aspergillus nidulans*. Mol Gen Genet 253:189–197, 1996.
61. HRV Arnstein, D Morris. The structure of a peptide, containing α-aminoadipic acid,

cysteine and valine, present in the mycelium of *Penicillium chrysogenum*. Biochem J 76:357–361, 1960.
62. EP Abraham, GGF Newton, K Crawford, HS Burton, CW Hale. Cephalosporin *N*: a new type of penicillin. Nature 171:343, 1953.
63. PL Skatrud, AJ Teitz, TD Ingolia, CA Cantwell, DL Fisher, JL Chapman, SW Queener. Use of recombinant DNA to improve production of cephalosporin C by *Cephalosporium acremonium*. Bio/Technology 7:477–485, 1989.
64. A Hinnen, J Nuesch. Enzymatic hydrolysis of cephalosporin C by an extracellular acetylhydrolase of *Cephalosporium acremonium*. Antimicrob Agents Chemother 9: 824–830, 1976.
65. FM Huber, RH Baltz, PG Caltrider. Formation of desacetylcephalosporin C in cephalosporin C fermentation. Appl Microbiol 16:1011–1014, 1968.
66. S Gutierrez, J Velasco, AT Marcos, FJ Fernandez, F Fierro, JL Barredo, B Diez, JF Martin. Expression of the *cef*G gene is limiting for cephalosporin biosynthesis in *Acremonium chrysogenum*. Appl Microbiol Biotechnol 48:606–614, 1997.
67. J Velasco, S Gutierrez, FJ Fernandez, AT Marcos, C Arenos, JF Martin. Exogenous methionine increases levels of mRNAs transcribed from *pcb*AB, *pcb*C, and *cef*EF genes, encoding enzymes of the cephalosporin biosynthetic pathway, in *Acremonium chrysogenum*. J Bacteriol 176:985–991, 1994.
68. CA Bunnell, WD Luke, FM Perry. Industrial manufacture of cephalosporins. In: SF Queener, Weber JA, SW Queener, eds. Beta-lactam Antibiotics for Clinical Use. New York: Marcel Dekker, 1986, pp 255–284.
69. CA Cantwell, RJ Beckmann, JE Dotzlaf, DL Fisher, PL Skatrud, WK Yeh, SW Queener. Cloning and expression of a hybrid *Streptomyces clavuligerus cef*E gene in *Penicillium chrysogenum*. Curr Genet 17:213–221, 1990.
70. WK Yeh, JE Dotzlaf, GW Huffman. Biochemical characterization and evolutionary implication of β-lactam expandase/hydroxylase, expandase and hydroxylase. In: H Kleikauf, H von Dohren, eds. 50 Years of Penicillin Application; History and Trends. Prague: Public, 1994, pp 208–223.
71. SW Queener, JJ Capone, AB Radue, R Nagarajan. Synthesis of deacetoxycephalosporin C by a mutant of *Cephalosporium acremonium*. Antimicrob Agents Chemother 6:334–337, 1974.
72. C Cantwell, R Beckmann, P Whiteman, SW Queener, EP Abraham. Isolation of deacetoxycephalosporin C from fermentation broths of *Penicillium chrysogenum* transformants: construction of a new fungal biosynthetic pathway. Proc R Soc Lond B 248:283–289, 1992.
73. WK Yeh, SK Ghag, SW Queener. Enzymes for epimerization of isopenicillin N, ring expansion of penicillin N, and 3′-hydroxylation of deacetoxycephalosporin C: function, evolution, refolding, and enzyme engineering. Ann NY Acad Sci 672:396–408, 1992.
74. SW Queener, RJ Beckmann, CA Cantwell, RL Hodges, DL Fisher, JE Dotzlaf, WK Yeh, D McGilvray, M Greaney, P Rosteck. Improved expression of a hybrid *Streptomyces clavuligerus cef*E gene in *Penicillium chrysogenum*. Ann NY Acad Sci 721: 178–193, 1994.
75. R Bovenberg, B Koekman, A Hoekema, J van der Laan, J Verweu. Process for the efficient production of 7-ADCA via 2-(carboxyethylthio)acetyl-7-ADCA and

3-(carboxymethylthio)propionyl-7-ADCA. International Patent WO 95/04148, 1995.

76. J Sutherland, R Bovenberg, J van der Laan. Improved process for the production of semi-synthetic cephalosporins via expandase activity on penicillin G. International Patent WO 97/20053, 1997.
77. MJ Conder, PC McAda, JA Rambosek. Recombinant expandase bioprocess for preparing 7-aminodesacetoxycephalosporanic acid (7-ADCA). U.S. Patent 5,318,896, 1994.
78. L Crawford, AM Stepan, PC McAda, JA Ramboesk, MJ Conder, VA Vinci, CD Reeves. Production of cephalosporin intermediates by feeding adipic acid to recombinant *Penicillium chrysogenum* strains expressing ring expansion activity. Bio/Technology 13:58–62, 1995.
79. EP Abraham, E Chain. An enzyme from bacteria able to destroy penicillin. Nature 146:837, 1940.
80. EP Abraham, GGF Newton. The structure of cephalosporin C. Biochem J 79:377–393, 1961.

3

Bioconversion of Penicillins to Cephalosporins

Arnold L. Demain
Massachusetts Institute of Technology, Cambridge, Massachusetts

Jose L. Adrio
Antibioticos, S.A.U., León, Spain

Jacqueline M. Piret
Northeastern University, Boston, Massachusetts

I. INTRODUCTION

Biosynthesis of cephalosporins in *Cephalosporium acremonium* (syn. *Acremonium chrysogenum*) and *Streptomyces clavuligerus* proceeds through a biosynthetic pathway that includes expansion of the five-membered thiazolidine ring of the intermediate penicillin N into the six-membered dihydrothiazine ring of deacetoxycephalosporin C (DAOC; see Fig. 1) [1]. The ring-expansion step is catalyzed by the dioxygenase, DAOC synthase (''expandase'') [2]. Whereas the eukaryotic *C. acremonium* produces penicillin N and cephalosporin C, the prokaryote *S. clavuligerus* produces a variety of β-lactam antibiotics including clavulanic acid, penicillin N, and cephamycin C; cephamycins are 7-α-methoxycephalosporins. Expandase has been purified from both *S. clavuligerus* and *C. acremonium* [3,4]. Its enzyme activity requires α-ketoglutarate, Fe^{2+}, and oxygen. In *C. acremonium*, the activity of expandase resides in a bifunctional enzyme [5,6] that catalyzes not only ring expansion but also the hydroxylation of the methyl group of DAOC to deacetylcephalosporin C (DAC). In *S. clavuligerus* [4,7,8,8a] and in other prokaryotes [9,10], the two activities are associated with

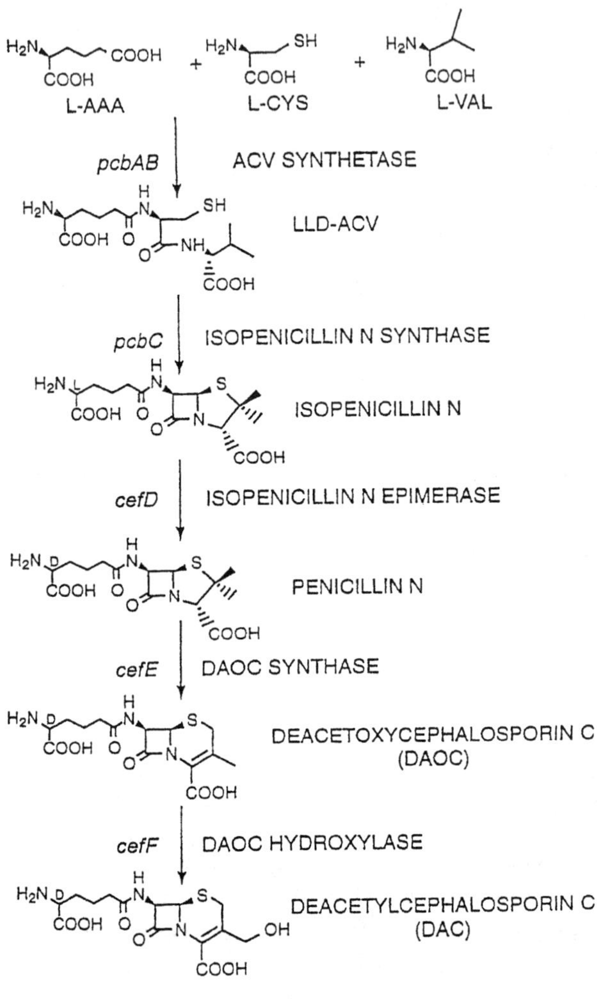

Figure 1 Biosynthetic pathway to DAC. In *Cephalosporium acremonium*, DAC is acetylated by DAC acetyltransferase to cephalosporin C. In *Streptomyces clavuligerus*, it is carbamoylated by *O*-carbamoyltransferase followed by methoxylation by C-7 hydroxylase and methyltransferase.

separate proteins. The properties of this synthase are very similar to those of other α-ketoglutarate-dependent dioxygenases, which incorporate one atom of oxygen into α-ketoglutarate to form succinate and the other atom into the product molecule. However, expandase differs in that no oxygen atom appears in its ring expansion product, DAOC.

Because of the effectiveness of cephalosporins against penicillin-resistant bacterial pathogens, understanding the substrate specificity, action, and regulation of expandase is of great scientific interest and practical importance.

Many investigators [2,3,5,11–13] have reported that the enzyme, in cell-free extracts as well as after purification, has a narrow substrate specificity and no detectable activity on readily available and inexpensive penicillins (such as penicillin G and V) produced by *Penicillium chrysogenum*. Maeda et al. [13] varied temperature, duration of reaction, buffers, and concentration of penicillin G, but without success. They were only able to observe ring expansion when the natural substrate penicillin N or its close relative D-carboxymethylcysteinyl-6-aminopenicillanic acid was used as substrate. Indeed, it is chemical ring expansion of penicillin G plus enzymatic removal of the phenylacetyl side chain that is used in industry to convert penicillin G to 7-aminodeacetoxycephalosporanic acid (7ADCA), an important intermediate for the manufacture of semisynthetic cephalosporins. However, the chemical process requires several steps, is expensive, and polluting (13a). A simple biological route, requiring only two steps, i.e., enzymatic ring expansion and deacylation, might replace the chemical process, thereby reducing costs and environmental problems. Only recently have the conditions that allow some conversion of penicillin G to deacetoxycephalosporin G (DAOG) (phenylacetyl-7ADCA) been identified [14]. Essentially, what was required were increased concentrations of iron and α-ketoglutarate.

II. MOLECULAR STUDIES ON EXPANDASE PROTEINS AND GENES

Ring expansion of penicillin N to DAOC was discovered by Kohsaka and Demain [2] working with cell-free extracts of *C. acremonium*. The product of that reaction was identified as DAOC [15,16]. The enzyme which catalyzed this reaction, expandase, exhibited properties of α-ketoglutarate-linked dioxygenases [17]. Ring expansion activity was found to be markedly increased by ascorbic acid, Fe^{2+} [18,19], α-ketoglutarate [18,20], and oxygen [20,21]. The enzyme showed an absolute requirement for α-ketoglutarate [7,18,19], and this co-substrate could not be replaced by chemically related compounds such as α-ketoadipate, pyruvate, oxaloacetate, glutamate, succinate, α-ketovalerate, or α-ketobutyrate [3,19,22].

The bifunctional expandase/hydroxylase of *C. acremonium* appeared to be a monomer with a molecular weight of 41,000 Da and an isoelectric point of 6.3 [5]. Both catalytic activities require α-ketoglutarate, Fe^{2+}, and oxygen and are stimulated by ascorbate, dithiothreitol (DTT), and ATP. Both activities were evident when Fe^{2+} was replaced by Fe^{3+} in the presence of a reducing agent; however, no activity was observed when Fe^{2+} was replaced by Mg^{2+}, Ca^{2+}, Cu^{2+}, Ni^{+2+} Co^{2+}, Zn^{2+}, Na^{+}, or K^{+}. The K_m value obtained for penicillin N was 29 μM, whereas for α-ketoglutarate it was 22 μM [3,5,12]. Both activities were similarly sensitive to inhibition by metal chelators such as ethylene diaminetetraacetic acid (EDTA) or *o*-phenanthroline, as would be expected with chelators of Fe^{2+}. Inhibition by Zn^{2+} and sulfhydryl reagents such as *p*-hydroxymercuribenzoate (*p*-HMB), 5′-5′-dithiobis-2-nitrobenzoic acid (DTNB), or N-ethylmaleimide (NEM) suggests that at least one sulfhydryl group is essential for ring expansion activity.

When isopenicillin N, penicillin G, penicillin V, ampicillin, carboxy-*n*-butyl penicillin, and 6-amino-penicillanic acid (6APA) were tested as substrate analogs [3,5,12], no ring expansion was observed.

Cloning of the expandase/hydroxylase gene (*cefEF*) from *C. acremonium* in *Escherichia coli* was achieved using probes based on amino acid sequences of fragments of the purified protein [23]. The results obtained revealed an open reading frame (ORF) of 996 nt encoding a protein of 332 amino acids with a molecular weight of 40,000–41,000 Da. In this fungus, the cephalosporin genes are located in two different clusters on separate chromosomes [24–26]. One of them is formed by the genes encoding the first two steps of the pathway, *pcbAB* and *pcbC* [27], found on chromosome VI, whereas the cluster containing the late genes (*cefEF* and *cefG*) was located on chromosome II [28]. The two *cef* genes are oppositely oriented and expressed from divergent promoters located within a 938-bp intergenic region [29].

The *S. clavuligerus* expandase is a monomer of 34,600 Da with two isoelectric points of 6.1 and 5.3 [4]. Its catalytic activity requires α-ketoglutarate, Fe^{2+}, and oxygen, and it is stimulated by DTT and ascorbate but not by ATP [4,12,30]. The synthase was equally active with Fe^{2+} or Fe^{3+} in the presence of ascorbate and DTT; however, these metal ions could not be replaced by Mg^{2+}, Ca^{2+}, Cu^{2+}, Ni^{2+}, Co^{2+}, Na^{+}, or K^{+}. As mentioned above for the fungal synthetase, the *S. clavuligerus* expandase activity is also very sensitive to metals chelators such as EDTA or *o*-phenanthroline, and to sulfhydryl reagents such as *p*-HMB, DTNB, and NEM. As was previously observed by other authors working with cell-free extracts [7,13], purified expandase was unable to expand isopenicillin N, penicillin G, penicillin V, ampicillin, or 6APA [4,12].

The crystal structure of the *S. clavuligerus* expandase has been recently reported [31] and, based on it, a mechanism for the formation of the reactive ferryl species, a catalytic intermediate common to many nonheme oxygenases [32], was proposed. The apo-expandase, a crystallographic trimer, reacts first with the iron, which is ligated by three protein ligands (His 183, His 243, Asp 185) and three solvent molecules, causing dissociation of the apoenzyme into monomers, the catalytically active form. Then, the enzyme–Fe^{2+} complex reacts with α-ketoglutarate, whose binding replaces two solvent molecules around the iron, and with dioxygen. The ferryl form of expandase, which reacts with penicillin N, is created by the splitting of dioxygen and oxidative decarboxylation of α-ketoglutarate to succinate.

Based on a 22-residue amino-terminal sequence of the purified protein, the *cefE* gene coding for the *S. clavuligerus* expandase was cloned and expressed in *E. coli* [33]. The gene contains an ORF of 933 nt encoding a protein of 311 amino acids with a molecular weight of 34,519 Da. Comparison of this gene and the *cefEF* gene of *C. acremonium* showed 67% similarity, whereas at the protein level there was a 57% similarity. The bacterial protein is 21 amino acids shorter than the fungal protein and its amino terminal group is not blocked. In this bacterium, the *cefE* gene is linked to the *cefD* gene (coding for the epimerase converting isopenicillin N to penicillin N) with an 81-bp intergenic region between them [34]. Both genes are located 7 kb upstream of the *lat* gene (encoding lysine ε-aminotransferase) and are expressed from a large transcript [34,35].

The *Nocardia lactamdurans* expandase is a monomer which was initially reported [9] to have a molecular weight of 27,000 Da. Cloning and expression of the *cefE* gene in *Streptomyces lividans* revealed an enzyme of 34,532 Da containing 314 amino acids and an isoelectric point of 4.9 [36]. This synthase behaves very similarly to the expandase of *S. clavuligerus* and requires α-ketoglutarate, Fe^{2+}, and oxygen, although it does not need ATP and ascorbate [9]. A high specificity for the nature of the side chain in the penicillin substrate was also observed and only penicillin N was expanded, not isopenicillin N, penicillin G, or 6APA.

The *N. lactamdurans* gene has a high similarity to the *cefE* gene of *S. clavuligerus* (74.7%) and the *cefEF* gene of *C. acremonium* (69.2%). At the protein level, there is also a high similarity to the expandase of *S. clavuligerus* (70.4%) and with the expandase/hydroxylase of *C. acremonium* (59.5%). In *N. lactamdurans*, the *cefD* and *cefE* genes are clustered 0.7 kb upstream of the *lat* gene, lack any intergenic region, and are translationally coupled [36].

Recently, the gene cluster involved in cephabacin F (a 7-formylamino cephalosporin) biosynthesis in *Lysobacter lactamgenus* was characterized [10]. This *cefE* gene encodes a protein of 319 amino acids with a molecular weight of 35,557 Da that shows 54% and 57% homology with the expandases of *S. clavuligerus* and *C. acremonium*, respectively. The expandase gene is located just downstream of *pcbC* in a simple and compact cluster [37] in which the genes encoding the four enzyme activities converting the tripeptide δ-(L-α-aminoadipyl)-L-cysteinyl-D-valine (ACV) to DAC are tightly linked within a 17-kb region of DNA in the same orientation.

III. BIOCONVERSION OF PENICILLIN G TO DAOG

A. Resting Cells

S. clavuligerus NP1, a mutant that produces only trace levels of cephalosporins [38], was used, since the absence of significant levels of cephalosporins in this strain facilitates detection of cephalosporins produced by ring expansion from added penicillins. Mycelia were developed in two shaken stages of medium MST [39], containing 1% soluble starch, 3% trypticase soy broth without dextrose, and 90 mM 3-(N-morpholino)propanesulfonic acid (MOPS) buffer at pH 7.0. Mycelia were washed twice before being used for the reaction. The standard reaction mixture for expandase action was that described by Maeda et al. [13] except that penicillin G replaced penicillin N. Additions were made in the order established by Shen et al. [22]. At various times during the shaking incubation, samples were taken, centrifuged to remove cells, and supernatant fluids assayed by the paper disk-agar diffusion bioassay, using *E. coli* Ess (a β-lactam supersensitive mutant [15]). The formation of DAOG was determined by inclusion of penicillinase in the assay agar (this narrow-spectrum β-lactamase attacks penicillins but not cephalosporins) and measuring zones of growth inhibition.

Although previous attempts to expand the thiazolidine ring of penicillin G with cell-free extracts of *S. clavuligerus* had failed [2,3,5,7,12,13], success was achieved with resting cells [14]. Table 1 shows the cofactor requirement when ring expansion of penicillin G was carried out with resting cells of *S. clavuligerus* NP1. When $FeSO_4$, α-ketoglutarate, or ascorbic acid was absent, the amount of product was ~30% of that in the control. On the other hand, ATP, $MgSO_4$, KCl, and DTT did not play a significant role in the reaction with resting cells. The omission of DTT actually increased production by 50%.

α-Ketoglutarate was known to be necessary as a cosubstrate for enzymatic ring expansion of penicillin N [3,7,18]. Figure 2 shows that a doubling of activity of resting cells on penicillin G was observed when the α-ketoglutarate concentration was raised from the standard concentration of 0.64 mM [13] to 1.28 mM. Figure 3 reveals that increasing concentrations of Fe^{2+} up to 1.8 mM increased ring expansion of penicillin G. This optimum concentration of Fe^{2+} was 45-fold higher than had been used for conversion of penicillin N by cell-free extracts [14]. The optimum ascorbate concentration for conversion of penicillin G by resting cells was found to be 4–8 mM. Maeda et al. [13] had used 5 mM ascorbate, and Lübbe et al. [40] had used 4 mM ascorbate for cell-free conversion of penicillin N to DAOC. ATP did not improve resting cell activity on penicillin

Table 1 Effect of Cofactors on the Ring Expansion of Penicillin G Using Resting Cells

Cofactor omitted	μg DAOG/ml
None	10.5
DTT (16 mM)	15.5
α-Ketoglutarate (1.28 mM)	3.7
$FeSO_4 \cdot 7H_2O$ (1.8 mM)	3.2
$MgSO_4 \cdot 7H_2O$ (8 mM)	10.1
KCl (8 mM)	11.5
Ascorbic acid (4 mM)	3.0
ATP (0.7 mM)	10.0

All reactions contained 50 mM Tris-HCl pH 7.4 buffer and 2 mg/ml of penicillin G. Dry cell weight: 13 mg/ml. A 1-ml sample from each reaction was taken after 2 hr and centrifuged (14,000 × *g*, 5 min). The supernatants were transferred to new tubes and 200 μl were used in the bioassay.
Source: Ref. 14.

G, whereas cell-free extracts acting on penicillin N had been stimulated by ATP [19]. Increasing cell mass enhanced the concentration of DAOG formed from penicillin G, the optimum concentration being 19 mg/ml. Higher cell concentrations inhibited the reaction, probably because oxygen supply was limiting under such conditions. In the studies of Cho et al. [14], the buffer used for bioconversion was 50 mM Tris-HCl [Tris = tris(hydroxymethyl)aminomethane] at pH 7.4. It was later found that 50 mM MOPS buffer or 50 mM HEPES [N-(2-hydroxyethyl) piperazine-N′-(2-ethanesulfonic acid] buffer at pH 6.5 improved activity [41].

To identify the products of the biotransformation, a system for the separation of penicillin G, DAOG, and deacetylcephalosporin G (DAG) was used [42]. HPLC analysis of the 0-hr sample gave a profile showing several peaks (Fig. 4). The peaks at 3.1 and 3.6 min corresponded to α-ketoglutarate and $FeSO_4$, respectively. The penicillin G peak eluted at 17 min. After 1 hr of incubation, two new peaks (3.65 and 15.3 min) appeared on the chromatogram. The 15.3-min peak was DAOG, but the 3.65-min peak was unidentified. During the subsequent 2 hr of biotransformation, these two peaks increased in size. No peak corresponding to DAG was detected during the reaction. The DAG peak would have eluted at 8 min. Thus, one of the products present in the biotransformation mixture was identified as DAOG. A new unidentified (more polar) peak with a retention time of 3.65 min was also produced. In *S. clavuligerus*, DAC is a precursor of three compounds: O-carbamoyldeacetylcephalosporin C, 7-α-hydroxy-O-carbamoyldeacetylcephalosporin C, and cephamycin C. The new unidentified peak might represent the phenylacetyl version of any one of these products.

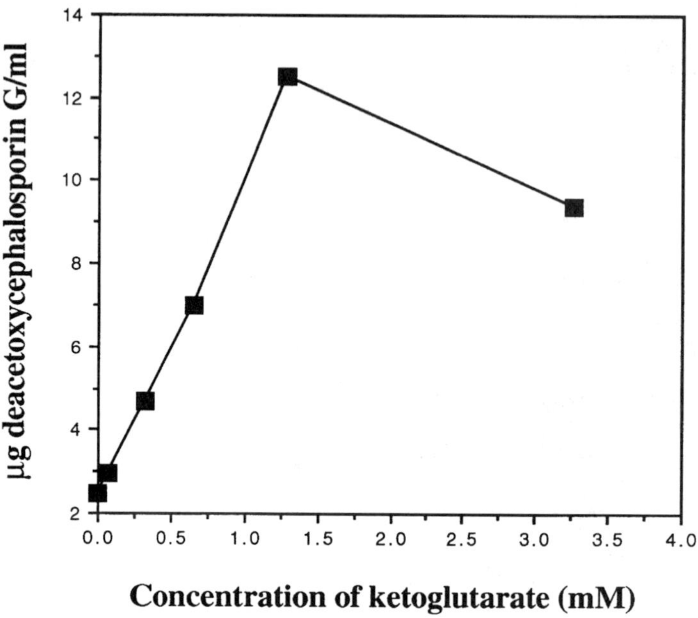

Figure 2 Effect of α-ketoglutarate concentration on ring expansion of penicillin G by resting cells of *Streptomyces clavuligerus*. The concentration of α-ketoglutarate previously used for unsuccessful cell-free ring expansion of penicillin G was 0.64 mM [9]. The reaction mixture contained 50 mM Tris-HCl pH 7.4 buffer, 1.8 mM $FeSO_4$, 4 mM ascorbic acid, and 2 mg/ml penicillin G. Dry cell weight was 12 mg/ml. Samples were taken at 2 hr and centrifuged (14,000 × *g*, 5 min); 200 μl were used in the bioassay. (From Ref. 14.)

B. Cell-Free Extracts

To prepare cell-free extracts, broths were centrifuged and pellets were washed twice in 50 mM Tris-HCl supplemented with 0.1 mM DTT, and disrupted by sonication [14]. After centrifugation, the resulting supernatant fluids were placed on ice and then used for the reaction. Reaction mixtures were incubated in tubes, and the reactions were stopped at various times by the addition of an equal volume of methanol. Using the high concentrations of $FeSO_4$ and α-ketoglutarate established for resting cell conversion of penicillin G to DAOG, activity was observed with cell-free extracts. Furthermore, high protein (6 mg/ml) and high substrate (5 mg/ml) concentrations increased the concentration of product made. Using the conditions employed earlier with penicillin N [14], in which protein concentration in the cell-free extracts was 1–2 mg/ml, only a low level of activity on penicillin G was observed. The penicillin G concentrations previously used by Maeda et al. [13] were only 0.01–0.36 mg/ml. The importance of Fe^{2+},

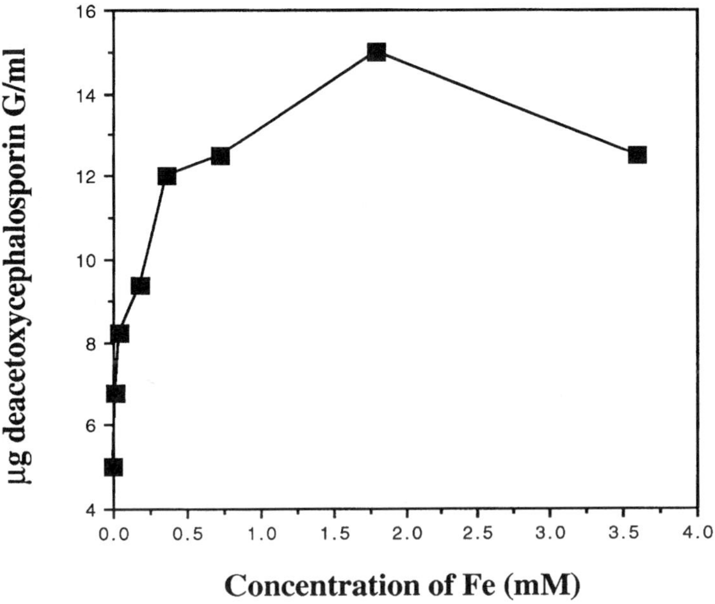

Figure 3 Effect of $FeSO_4$ concentration on ring expansion of penicillin G by resting cells. The concentration of $FeSO_4$ previously used for unsuccessful cell-free ring expansion of penicillin G was 0.04 mM [13]. The reaction mixture contained 50 mM Tris-HCl pH 7.4 buffer, 1.28 mM α-ketoglutarate, 4 mM ascorbic acid, and 2 mg/ml penicillin G. Dry cell weight was 10 mg/ml. Samples were taken at 2 hr and centrifuged (14,000 × *g*, 5 min); 200 µl were used in the bioassay. (From Ref. 14.)

α-ketoglutarate, and ascorbic acid and the lack of importance of ATP, $MgSO_4$, KCl, mentioned above with resting cells, were also observed with cell-free extracts acting on penicillin G. However, DTT did not decrease the activity of extracts on penicillin G, and was included in later experiments to protect the enzyme [40].

The substrate range of the expandase in cell-free extracts using the new conditions included all 15 penicillins tested (Table 2). The largest bioassay zones of inhibition were produced by conversion of penicillins G, X, mX, and 2-thiophenylacetyl-6APA, whereas the smallest zones were obtained with decanoyl-6APA and penicillin V [14].

C. Early Cessation of the Reaction

As mentioned above, the extent of expansion by cell-free extracts increased with increases in concentration of both protein and substrate [14]. However, in all

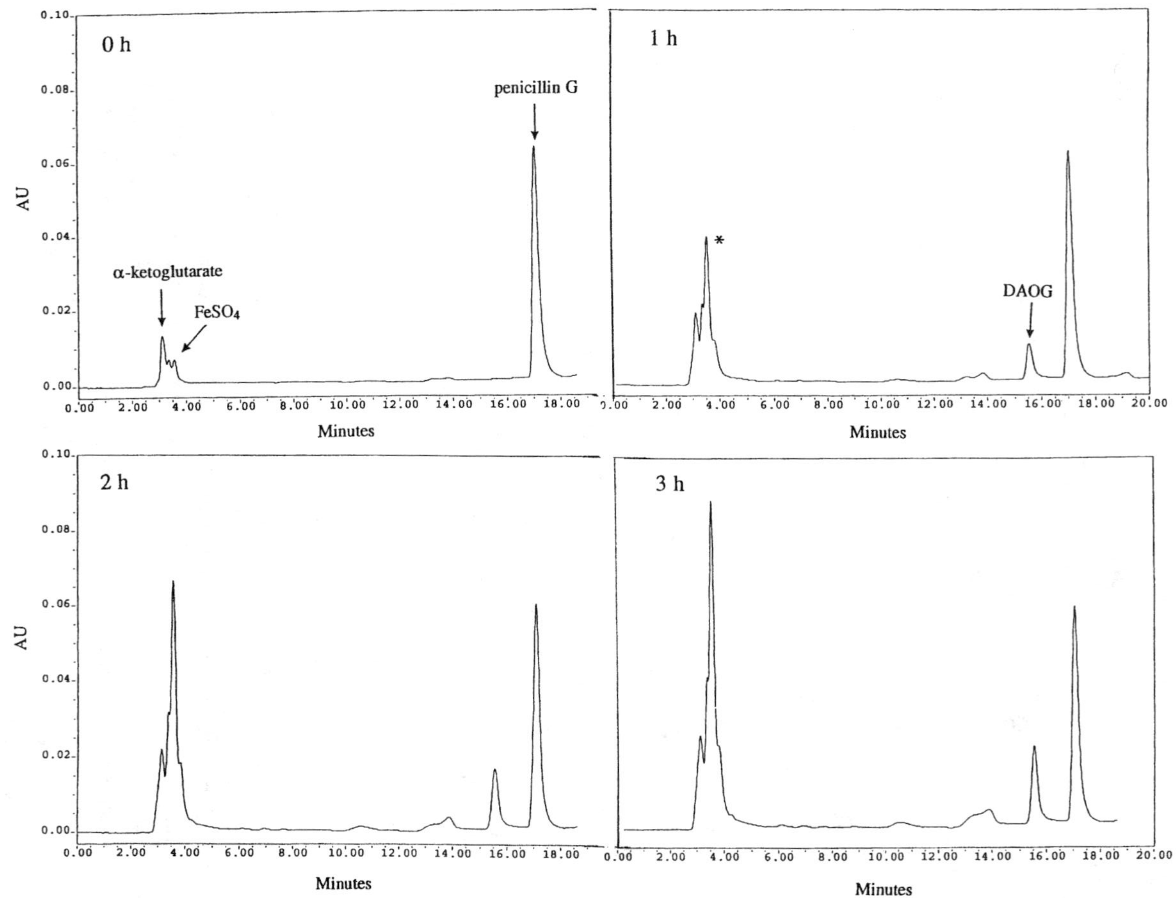

Figure 4 HPLC analysis of biotransformation by resting cells of *S. clavuligerus* NP1. Sensitivity was 0.2 AU (absorbance units). DAOG = deacetoxycephalosporin G, pen G = penicillin G. After 1 hr of incubation, an unidentified peak (*) was present. (From Ref. 42.)

Table 2 Substrate Specificity of Expandase Using Cell-Free Extracts

Substrate	Inhibition zone (mm)
None	0
Adipyl-6APA	20.0
Ampicillin	17.0
Butyryl-6APA	17.5
Decanoyl-6APA	7.0
Heptanoyl-6APA	16.5
Hexanoyl-6APA	19.5
Nonanoyl-6APA	10.5
Octanoyl-6APA	16.0
Penicillin F	21.0
Penicillin G	29.0
Penicillin V	7.5
Penicillin mX	30.5
Penicillin X	30.5
2-Thiophenylacetyl-6APA	32.0
Valeryl-6APA	15.0

Reactions were carried out in 50 mM Tris-HCl pH 7.4 buffer using 4 mg/ml enzyme protein, 1.8 mM $FeSO_4$, 1.28 mM α-ketoglutarate, 8 mM KCl, 8 mM $MgSO_4$, 4 mM ascorbic acid, and 14 mM DTT. All substrates were used at 2 mg/ml, except for penicillin mX and octanoyl-6APA (3 mg/ml). Samples were taken at 2 hr and centrifuged (14,000 × *g*, 5 min). Bioassay was done using 250 μl of reaction mixture using *Escherichia coli* strain Ess, which is supersensitive to β-lactams. Penicillin X is *para*-hydroxypenicillin G and penicillin mX is *meta*-hydroxypenicillin G.
Source: Ref. 14.

cases, the concentration of product increased during the first 1–3 hr and then either remained stable or decreased. The extents of bioconversion were low, amounting to 0.61% at 1 mg/ml penicillin G, 0.32% at 2 mg/ml penicillin G, and 0.30% at 4 mg/ml penicillin G. To determine if the limiting factor after 2 hr was the exhaustion of one or more components of the reaction mixture, different concentrations and combinations of reaction constituents were added to the reaction tubes at 2 hr, but no additive was able to reactivate the system [43]. Even when additional extract protein was provided, no reactivation occurred. Only a very small increment in production was observed when both protein and cofactors were added.

Cell-free extracts were preincubated in buffer in the cold or at 30°C for 2 hr to assess the stability of the expandase over time [43]. After 2 hr, the remaining components of the reaction mixture were added and the tubes were incubated under standard conditions for another 3 hr. In all cases, the initial rates were similar to that of the control (no preincubation) and amounts of product obtained were almost equal to the control value. Thus the lack of activity after 2–3 hr could not be attributed to enzyme instability during shaking at 30°C. When Fe^{3+} (as ferric sulfate) was substituted for Fe^{2+} in the reaction, activity was unaffected. Therefore, inactivation was apparently not due to oxidation of Fe^{2+}.

To determine whether one or more reaction mixture components might inactivate the enzyme, the cell-free extract was preincubated for 2 hr in the presence of different reaction components. After this preincubation, the remaining reaction components, including the penicillin G substrate, were added. The amount of product obtained was markedly different, depending on which component was present during preincubation (Fig. 5). When buffer alone was present, the subsequent reaction yielded up to 95% of the amount of the product formed in the control reaction (Fig. 5b). When individual components were present during incubation, production remained between 64% and 85% of control (Fig. 5a). However, when preincubation was with Fe^{2+} plus ascorbic acid or Fe^{2+} plus α-ketoglutarate, no detectable product was obtained (Figs. 5b, 5c, 5d). Thus, a combination of Fe^{2+} with ascorbate or with α-ketoglutarate inactivated the enzyme during preincubation, and evidently did so during the reaction under standard conditions. The same phenomenon was seen with resting cells [41]. As with extracts (see above), inactivation was not merely due to the presence of oxidized Fe^{2+} because when the normal bioconversion reaction was carried out with Fe^{+3} (ferric sulfate) instead of Fe^{2+}, there was no inhibition.

The above studies done on cell-free extracts and resting cells revealed a very low bioconversion yield (based on added penicillin G), i.e., about 0.6%. To examine the dependence of product formation and bioconversion yield on substrate concentration, different concentrations of penicillin G were tested [41]. As Fig. 6a depicts, the concentration of product formed was higher when higher concentrations of penicillin G were used, but the yield based on substrate added dramatically decreased (Fig. 6b). At 8 mg/ml of penicillin G, bioconversion was about 0.4%; at 2 mg/ml (the standard concentration), the yield was 1.0%; at 0.063 mg/ml, the yield was 9.0%; and at 0.015 mg/ml (data not shown), the yield reached 16.5%.

Inactivation during aerobic incubation with Fe^{2+} plus ascorbic acid and/or α-ketoglutarate might have been due to formation of H_2O_2 [44–46]. As is the case for other α-ketoglutarate-dependent dioxygenases, expandase activity requires a reducing agent in addition to α-ketoglutarate and Fe^{2+}. This requirement can be

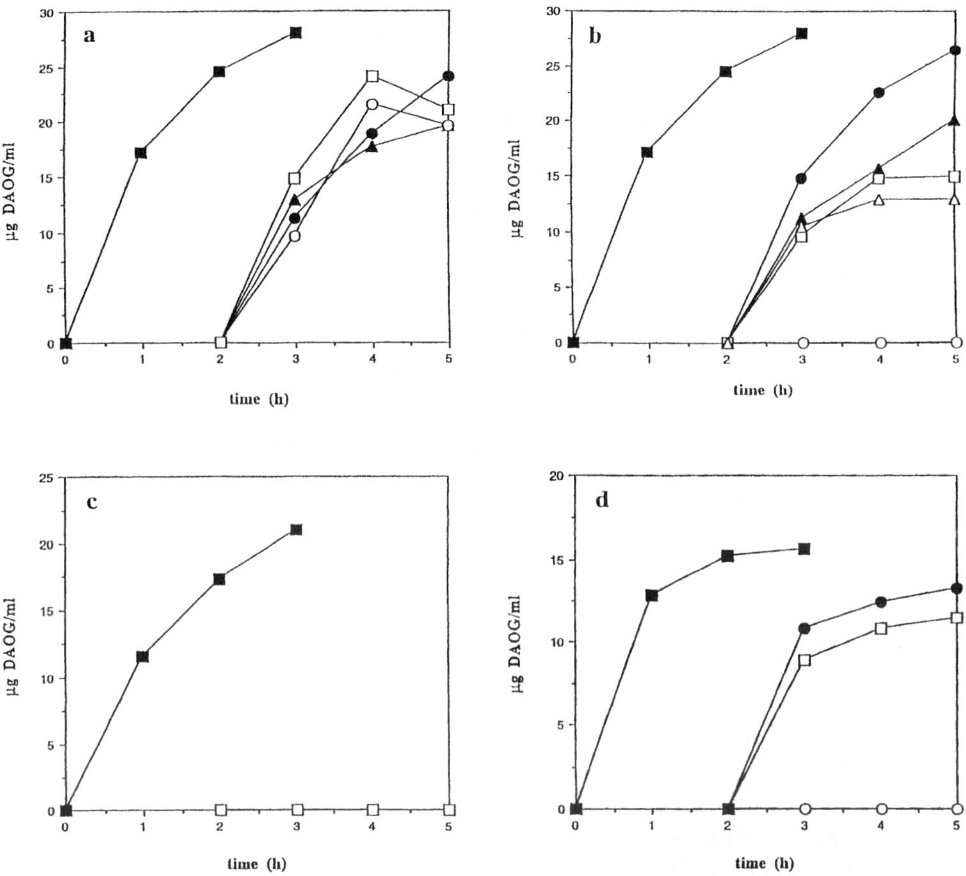

Figure 5 Effect of preincubation of cell-free extract in the presence of different components of the reaction mixture. Cell-free extract in 50 mM Tris-HCl pH 7.4 buffer was preincubated at 30°C, 250 rpm for 2 hr. (a) ■, control (no preincubation); ●, 1.8 mM $FeSO_4$; ▲, 4 mM ascorbate; ○, 1.28 mM α-ketoglutarate; □, 8 mM $MgSO_4$ + 8 mM KCl. (b) ■, control (no preincubation); ●, no additive (only buffer); □, $MgSO_4$ + KCl + $FeSO_4$; ▲, α-ketoglutarate + ascorbate; △, α-ketoglutarate + ascorbate + $MgSO_4$ + KCl; ○, $FeSO_4$ + ascorbate; or $FeSO_4$ + α-ketoglutarate. (c) ■, control (no preincubation); □, $FeSO_4$ + α-ketoglutarate + ascorbate; or $FeSO_4$ + ascorbate + KCl; or $FeSO_4$ + ascorbate + $MgSO_4$; or $FeSO_4$ + α-ketoglutarate + $MgSO_4$; or $FeSO_4$ + α-ketoglutarate + KCl. (d) ■, control (no preincubation); ○, $FeSO_4$ + α-ketoglutarate + ascorbate + $MgSO_4$; or $FeSO_4$ + α-ketoglutarate + ascorbate + KCl; □, ascorbate + $MgSO_4$; ●, ascorbate + KCl. Additive concentrations were the same as in the reaction mixture. (From Ref. 43.)

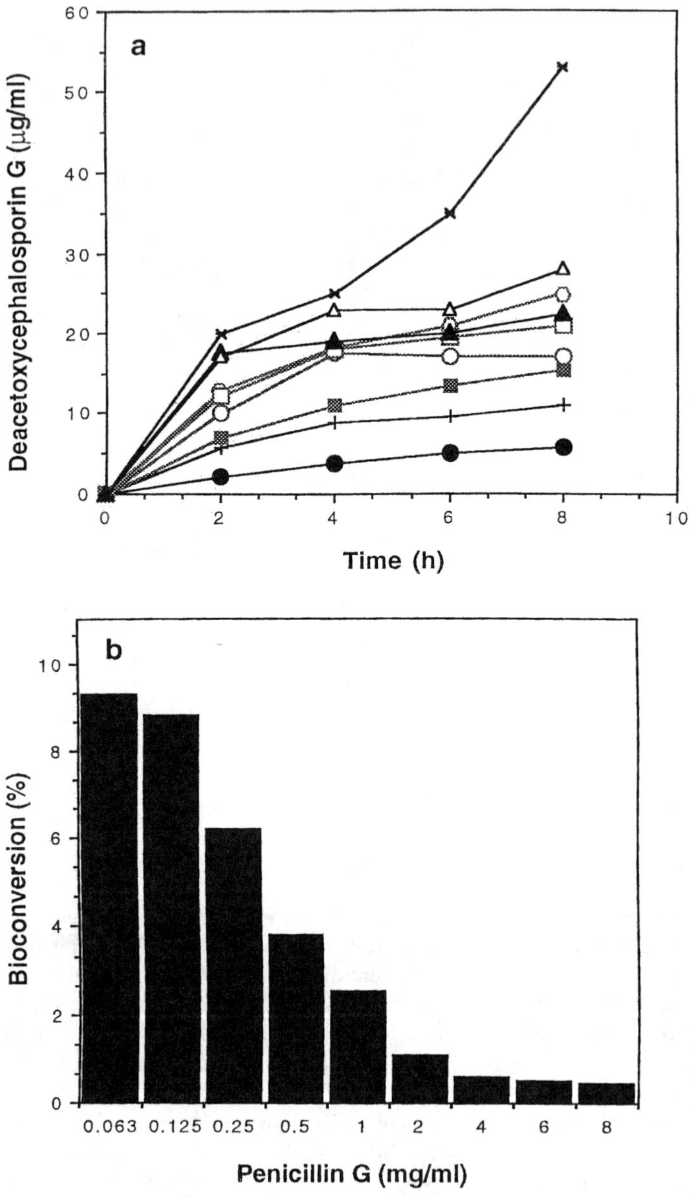

Figure 6 (a) Effect of substrate concentration on product formation. ●, 0.063 mg/ml; ✚, 0.125 mg/ml; ■, 0.25 mg/ml; ○, 0.5 mg/ml; ▲, 1 mg/ml; □, 2 mg/ml; ⬡, 4 mg/ml; △, 6 mg/ml; ×, 8 mg/ml. (b) Yields of bioconversion at different concentrations of added penicillin G. (From Ref. 41.)

fulfilled by several substances, although ascorbate is the most effective and is the reducing agent usually used with this type of enzyme [17]. Its function may be that of keeping iron in the reduced form or, perhaps, to protect sulfhydryl groups on the enzyme [47]. However, ascorbate can have adverse effects, since incubation of some enzymes with ascorbate and oxygen leads to rapid loss of enzymatic activity due to the production of H_2O_2 during autooxidation of this reducing agent [46,48]. Catalase is known to stimulate the activity of such enzymes, and Lübbe et al. [40] reported this effect with expandase in both *C. acremonium* and *S. clavuligerus* using penicillin N as substrate. However, when 0.5–3 mg/ml of catalase was added to the preincubation mixture with cell-free extract, no protection from inactivation was observed [43]. Nevertheless, when 15 mM H_2O_2 was added to a reaction mixture, activity was markedly reduced. If catalase (2 mg/ml) was also added, the inactivation due to peroxide was reversed, indicating that inactivation normally observed with Fe plus ascorbate was not due, at least exclusively, to formation of H_2O_2.

Fucci et al. [44] described the oxygen-dependent inactivation of several key biosynthetic metabolic enzymes that involve mixed-function oxidation systems. These systems catalyze synthesis of H_2O_2 and reduction of Fe^{3+} to Fe^{2+} followed by oxidation of enzyme-bound Fe^{2+} to generate oxygen radicals (such as hydroxyl radicals or singlet oxygen) that attack a histidine (or other oxidizable amino acids) at the metal-binding site of the enzyme. In a recent study, Golan-Goldhirsh et al. [49] reported that preincubation of different proteins in the presence of ascorbic acid and copper (or iron) led to their inactivation due to the autooxidation of ascorbic acid. In this reaction, four strong oxidant species are formed and, after preincubation, changes in the amino acid composition of all proteins studied were observed. In all of them, there was a major loss of histidine (His183 and His243) and methionine residues which ligate ferrous iron (with Asp185) to give a catalytically active form [31]. Because inactivation of the expandase might be due to the formation of such ''activated oxygen'' species, the effects of different radical scavengers such as mannitol, dimethyl sulfoxide, and thiourea were examined [43]. Participation of OH- radicals was contraindicated by the failure of these free-radical scavengers to prevent the inactivation of cell-free extract when preincubated with Fe^{2+} and ascorbate. Histidine inhibition of inactivation is considered presumptive evidence for a role of singlet oxygen in oxidase-catalyzed reactions [44]. However, when this amino acid was added to the preincubation mixture containing Fe^{2+} and ascorbate, it failed to prevent inactivation. Also examined was the possible involvement of superoxide ion ($O_2^{-}\cdot$ formed by a nonenzymatic reaction between ascorbate and iron) [45] by adding different concentrations of superoxide dismutase (SOD) to the preincubation mixture containing extract, buffer, Fe^{2+}, ascorbate, and catalase; again no activity was observed after preincubation. Thus, ''activated oxygen'' species appear not to participate in the inactivation of expandase, since free-radical scavengers and SOD added during preincubation failed to protect enzyme activity.

Although DTT is not necessary in the enzymatic conversion of penicillin G [14], this compound is known to stimulate the expandase activity of frozen crude extracts of *S. clavuligerus* and *C. acremonium* on penicillin N [40]. However, neither this reagent nor β-mercaptoethanol was able to stimulate the subsequent reaction with penicillin G when added during preincubation with $FeSO_4$ and ascorbate [43].

D. Effect of Growth in Alcohols

Production of antibiotic compounds by actinomycetes rarely takes place during periods of rapid growth in rich media. Rather, their biosyntheses occur best under conditions of nutrient imbalance, brought about by limitation of the carbon, nitrogen, or phosphorus source, and at low growth rates [50]. In some cases, the expression of the genetic information for antibiotic biosynthesis has been linked to control mechanisms, such as the Pho system, responsible for detecting and adapting cells to available sources of nutrients present in the environment [51]. The discovery of heat shock proteins (GroEL-like proteins), potentially important in antibiotic export and in the assembly of multienzyme complexes for polyketide antibiotic biosynthesis in a variety of streptomycetes [52], led to the first study of the relationship between non-nutritional stresses, such as heat shock or ethanol treatment, and antibiotic biosynthesis [53]. In that study, jadomycin production was induced by ethanol or heat shock. Thus growth in the presence of alcohols was studied to determine the effect on resting cells of *S. clavuligerus* NP1 with regard to their ability to biotransform penicillin G into cephalosporin-type antibiotics [42]. Cultures grown in control medium MST or MT (MST less starch) formed typical mycelial masses of tangled hyphae. However, in MT supplemented with ethanol, a different morphology was observed, depending on the alcohol concentration in the medium. In 1% ethanol the hyphae were somewhat more dispersed than in control medium, whereas in 2% ethanol the hyphae were extensively fragmented and dispersed.

In the presence of 1% ethanol or 1–2% methanol, growth was slightly less than in MT medium [2.9–3.4 mg/ml dry cell weight (DCW) in alcohol-supplemented cultures versus 3.6 mg/ml in MT]. With 2% ethanol, growth was severely restricted (0.9 mg/ml). Concentrations of alcohol higher that 2% totally inhibited growth. The omission of starch (MT medium) led to a slight increase in specific production of DAOG as compared to cells from MST medium. When MT medium was supplemented with 1% ethanol, there was a marked increase in specific production and a six- to sevenfold increase when 2% ethanol was added [42]; much less of an effect was observed with methanol. Addition of ethanol or methanol at later times during growth (at 2, 6, and 12 hr) did not have any stimulatory effect on the bioconversion. The mechanism of alcohol stimulation of biotransformation of penicillin G to cephalosporins remains unknown, but it could be related to the known ability of ethanol treatment to trigger a heat-shock response [53–56]. Alternatively, growth in alcohol might yield cells with increased membrane permeability, i.e., they might take up penicillin G or release DAOG more easily. The morphological effects observed when cells were grown in alcohol-supplemented media might reflect the latter mechanism.

E. Use of Immobilized Cells

For bioprocessing purposes, increase in the stability of biocatalysts is quite often achieved by immobilization of cells or enzymes [57–60]. This technology is an attractive alternative to the use of expensive free enzymes and cofactors, and can coordinate multistep enzymatic processes into a single operation. Furthermore, fermentative biosynthesis of cephamycin C using immobilized cells of *S. clavuligerus* NRRL 3585 was accomplished by Freeman and Aharonowitz [61]. Jensen et al. [62] reported on the immobilization of β-lactam synthesizing enzymes from the same wild-type culture. None of these early studies used penicillin substrates other than the normal intermediate penicillin N, such as penicillin G.

The oxidative ring expansion of penicillin G by free and entrapped resting cells of *S. clavuligerus* NP1 was compared [63], and immobilized cells were found to perform the expandase reaction albeit more slowly and less extensively than free cells. Both types of cells virtually ceased production at 2 hr. Different biomass amounts (2 g, 4 g, 6 g wet weight) were resuspended in 20 ml of polyethyleneimine-barium alginate solution and tested for ring expansion. Increased resting cell concentrations yielded increased product formation; again, the reaction markedly decreased in rate after 2 hr. To examine multiple cycles of ring expansion, free and immobilized resting cells were allowed to carry out oxidative ring expansion for 2 hr, followed by centrifugation in the cold for 5 min. The cells were washed with 50 mM MOPS (pH 6.5) buffer, and recentrifuged. The expandase reaction was again initiated. Two-hour cycles were carried out for up to four times with assays of product formation done at the end of each cycle. As indicated in Fig. 7, the activity of free cells was reduced by about 60% from the first to the second cycle, and was completely lost after the second cycle. In contrast, immobilized cells experienced only a small reduction of activity at each cycle and still had activity through four cycles. Other polymeric matrices such as agarose (4% w/v) and *k*-carrageenan (3% w/v) did not show activity. This may be due to reaction impedance due to strong diffusional limitations. Such external and internal barriers often reduce catalytic efficiency.

The above results [63] demonstrated the capacity of resting cells entrapped in polyethyleneimine-barium alginate to perform repeated oxidative ring expansion of penicillin G to DAOG. Product formation rate was higher with free than

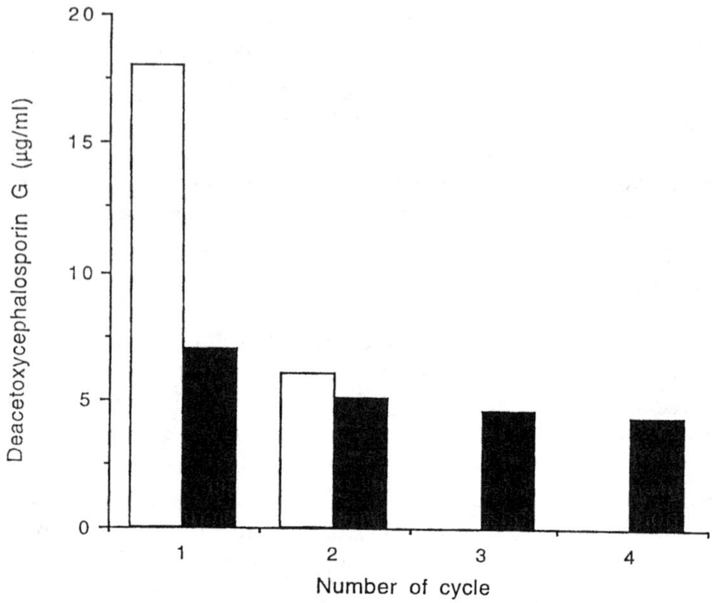

Figure 7 Activity during sequential 2-hr cycles of incubation using (□) free or (■) immobilized cells. Cells were washed between cycles. (From Ref. 63.)

with entrapped resting cells, probably due to strong diffusional limitations which prevent the expandase enzyme from binding the substrate and cofactors and/or preventing oxygen transfer to the interior of the beads. With free cells, diffusional limitation and mass transfer problems are less important than with immobilized systems. Similar behavior was reported by Freeman and Aharonowitz [61], who compared free and immobilized cells of *S. clavuligerus* NRRL 3585 in linear polyacrylamide gels for the fermentative formation of cephamycin C. They observed that free cells produced higher levels of cephamycin C than immobilized cells.

IV. *IN VIVO* CONSTRUCTION OF HYBRID EXPANDASES BY HOMEOLOGOUS RECOMBINATION

Homeologous recombination (recombination between partially homologous sequences) has been successfully used to produce novel antibiotics and functional hybrid proteins in *Streptomyces* [64–66]. If partially homologous sequences are placed in tandem and in the same orientation [67,68], general recombination leads to excision of the duplicated genes as well as of the sequences lying between the sites of recombination. Because the *cefE* genes from *S. clavuligerus* and *N. lactamdurans* have a high degree of similarity at the nucleotide (74%) as well as at the amino acid (70%) sequence levels [36], it was of interest to determine whether recombination could take place at a significant frequency between these genes cloned in the same orientation on a plasmid. If so, hybrid expandases with potentially altered activity/specificity could be generated *in vivo* (Piret, Adrio, and Demain, unpublished data). As described below, chimeric *cefE* genes were obtained by recombination between the *cefE* genes from *S. clavuligerus* and *N. lactamdurans* in both *E. coli* and *Streptomyces* hosts.

A. Recombination of Expandase Genes in *E. coli*

To test whether recombination between the two *cefE* genes could take place on a plasmid, experiments were conducted in the *recA+ E. coli* strain ER1447. Plasmid pJALC was constructed from pCYB160 (derived from pCYB1 [New England Biolabs, Beverly, MA] by F. Feng), which carries the *cefE* gene from *N. lactamdurans*, to which was added a 3-kb *Bam*HI fragment from pULFJ32 (gift of J.-F. Martín, University of León, Spain) containing the *cefE* (and part of *cefD*) gene from *S. clavuligerus* NRRL 3585. Following up to four rounds of growth in liquid culture, plasmid DNA was isolated from *E. coli* ER1447 carrying pJALC, digested with *Bgl*II, dephosphorylated, and used to transform *E. coli* DH5α (*recA*$^-$). Seventeen of the 63 (27%) resulting colonies contained an 8-kb plasmid, the size predicted after excision of the intervening region and one of the duplicated *cefE* genes. Indeed, treatment of these 17 plasmids with *Bgl*II, diagnostic for elimination of the region between the two *cefE* genes in pJALC, gave no digestion. Further restriction analysis revealed two major types of DNA rearrangements in these plasmids. Five of them gave two bands upon digestion with *Pst*I, indicating that both *Pst*I sites in pJALC were still present and that recombination had occurred downstream of the *Pst*I site located in the *N. lactamdurans cefE* gene. All of the remaining 12 plasmids gave a single band when digested with *Pst*I, indicating that one of the original *Pst*I sites had disappeared. Further analysis by restriction digestion showed that the *Pst*I site in the *cefE* gene from *N. lactamdurans* had been eliminated.

To analyze the crossover junctions in *E. coli*, representative recombinant plasmids were sequenced and compared with the native *cefE* genes of *N. lactamdurans* and *S. clavuligerus*. In the five clones analyzed which had retained both *Pst*I sites, recombination had occurred in one of two locations which encode fully conserved sequences of 12 bp in one case and 21 bp in the other. Similarly, in the clones retaining only one *Pst*I site, recombination had taken place in fully conserved sequence stretches of 2, 5, 11, 15, or 17 bp at five different positions within the first 550 bp of the chimeric *cefE* genes. Importantly, all clones analyzed had undergone in-frame recombination events.

B. Recombination of Expandase Genes in *Streptomyces*

Recombination was attempted in *S. lividans* 1326, a well-characterized laboratory host. To screen rapidly for recombination between the two *cefE* genes, the tyrosinase gene *melC* from *Streptomyces glauscescens* was used as a marker and inserted between the two *cefE* genes. Gene *melC* codes for the formation of the black pigment melanin; thus, screening for colonies containing a putative hybrid *cefE* gene was done by examination of colony color, since recombination should lead to loss of pigment production. A 2.3-kb *Kpn*I fragment from pMEA4 [69] containing the tyrosinase gene from *S. glaucescens* was cloned into pSP73 (Promega, Madison, WI) to yield pJA73T-SP6. A 2.3-kb *Eco*RI-*Hind*III fragment from this plasmid was then ligated with pIJ487 [70] to yield pJA487T. pJA487T was digested with *Bcl*I and the largest fragment (about 3 kb) was purified and ligated into the *Bgl*II site of pJALC to give pJALTC. pJALTC was digested with *Eco*RV and *Nsi*I and the 8.6-kb fragment, containing both *cefE* genes as well as the tyrosinase gene, was purified and inserted into the *Pst*I-*Ecl*136II-digested *S. lividans* vector pIJ680 to create pJA680.

Plasmid isolation from colonies grown on supplemented CG medium [69] showed that all colonies producing a halo of black pigment (designated "B") contained a 12.5-kb plasmid, corresponding to pJA680. The 58 melanin-negative colonies (designated "W") analyzed carried plasmids of different sizes. Of these, 37 (64%) strains contained a 6.3-kb plasmid, the size expected to result from recombination between the two *cefE* genes. Plasmids of 5.8 kb (29%), 4 kb (3.5%), and 8 kb (3.5%), presumably due to unexpected, lower-frequency recombination events, were also found.

Plasmids from strains W21 and W64 (carrying 6.3-kb plasmids), W25 and W76 (8-kb plasmids), and B18 (pJA680) were analyzed with several restriction endonucleases. W21 and W64 gave identical restriction patterns and no digestion when treated with *Pst*I or *Xmn*I, meaning that *cefE* from *N. lactamdurans* was partially or fully deleted. Strains W25 and W76 showed no digestion when treated with *Pst*I but were cleaved when incubated with *Xmn*I, *Bam*HI, or *Sca*I, indicating that the *N. lactamdurans cefE* gene was truncated at its 3′ end while *S. clavuligerus cefE* apparently remained intact. Sequencing of the hybrid junctions generated in *S. lividans* is now in progress.

C. Enzymatic Activities of Hybrid Expandases

To study the function of the chimeric *cefE* enzymes generated by recombination in *S. lividans*, the cell-free ring expansion abilities of several recombinants were determined. Three white strains, W21 (6.3-kb plasmid), W25 (8-kb) and W76 (8-kb) as well as the controls B11 and B18 (melanin-positive; pJA680) showed activity, producing inhibition zones in the agar diffusion bioassay. Although *S.*

clavuligerus NP1 showed no activity when grown under the same culture conditions as the *S. lividans* strains, typical bioassay results were obtained when NP1 was grown as previously described [14,43].

To ascertain that the inhibition zones were due to formation of DAOG, the samples were analyzed by HPLC. After 2 hr of incubation, a peak corresponding to DAOG was clearly present on the chromatograms.

V. EXPANSION OF ADIPYLPENICILLIN

Reports on expansion of adipyl-6APA to adipyl-7ADCA in the literature are contradictory. Whereas Baldwin et al. [71] claimed that partially purified extracts from *C. acremonium* were able to expand adipyl-6APA, others [3,12,13] did not observe expansion of this compound using crude extracts, semipurified preparations, or recombinant expandase preparations from *C. acremonium* or *S. clavuligerus*. However, researchers at Panlabs [72] and Merck [73] reported on the expansion of adipyl-6APA to adipyl-7-cephalosporanic acid (adipyl-7ACA) using whole cells. Their work was based on two previous reports which had demonstrated that feeding adipate to *P. chrysogenum* led to production of adipyl-6APA [74] and that DAOC could be produced by *P. chrysogenum* transformants carrying the *S. clavuligerus cefE* (expandase) and *cefD* (epimerase) genes [75]. Fungal plasmids harboring either the *S. clavuligerus cefE* (expandase) gene, the *C. acremonium cefEF* (expandase/hydroxylase) gene, or the fungal *cefEF* and *cefG* (acetyltransferase) genes were used to transform a high-titer penicillin-producing strain of *P. chrysogenum* [72,73]. Transformants were grown in media containing adipic acid (1.7% disodium adipate) as side-chain precursor. When adipate was fed to the untransformed *P. chrysogenum* parent strain, adipyl-6APA was produced at 47% of the penicillin V titer produced by this strain when phenoxyacetate was used as sidechain precursor. One transformant (containing the bacterial expandase) produced up to 14% adipyl-6APA (relative to penicillin V titer) and 17% adipyl-7ADCA. Another recombinant clone containing the eukaryotic expandase/hydroxylase produced 15% adipyl-6APA, 2.6% adipyl-7ADCA, and 6.7% adipyl-7-aminodeacetylcephalosporanic acid. A third clone expressing the fungal expandase/hydroxylase and acetyltranferase activities was able to produce low amounts of adipyl-7ACA. The important cephalosporin intermediates, 7ADCA and 7ACA, were obtained by removing the adipyl side chain using a purified glutaryl acylase derived from a *Pseudomonas* strain [72].

Based on the work described by Crawford et al. [73], researchers at Gist-Brocades [76] transformed *P. chrysogenum* with plasmids containing either the *N. lactamdurans* or *S. clavuligerus cefE* genes under control of different promoters such as *penDE*, *pcbC*, and *gpdA* from *Aspergillus nidulans*. Transformants were grown in media containing adipate (0.5% v/v). Results showed that both expandases were active in *P. chrysogenum* and the highest adipyl-7-ADCA titers were obtained when the *pcbC* promoter was used. When other side-chain precursors were fed into fermentation media, transformants carrying the *cefE* gene from *S. clavuligerus* were able to incorporate these compounds into the corresponding cephalosporanic acids [76]. The titers obtained (relative to the titer obtained with adipate) were 29.6% for carboxymethylpropionic acid, 25.1% for thiodipropionic acid, and 6.2% for phenylacetic acid.

VI. SUMMARY

The *cefE* genes from some cephalosporin-producing bacteria (*S. clavuligerus, N. lactamdurans, L. lactamgenus*) and the *cefEF* gene of *C. acremonium* have been cloned, sequenced, and expressed. All showed similar sizes, homology (ranging from 54 to 74%), and different locations in the cephalosporin gene clusters. The expandases, encoded by those genes, exhibited the properties of α-ketoglutarate-dependent dioxygenases and, in all cases, the presence of α-ketoglutarate, Fe^{2+}, and molecular oxygen was essential for their catalytic activity. All showed a high specificity for the nature of the side chain of the penicillin substrate and only penicillin N, the natural substrate, was expanded under the conditions employed.

Success with resting cells acting on penicillin G allowed Cho et al. [14] to demonstrate the ring expansion of penicillin G. By increasing the concentration of the substrate (penicillin G), cosubstrate (α-ketoglutarate), the chief cofactor, (Fe^{2+}), and cells, the conversion of penicillin G to DAOG was improved. Using such improved conditions, activity with cell-free extracts was also demonstrated on 14 other penicillins.

Incubation of *S. clavuligerus* cell-free extracts in the presence of $FeSO_4$ and either ascorbate or α-ketoglutarate for 2 hr led to elimination of subsequent expansion activity. It is clear that further work will have to be done to understand and overcome this intriguing case of inactivation.

Growth of *S. clavuligerus* NP1 in the presence of ethanol resulted in an increase in DAOG production from penicillin G by resting cells. The mycelium produced in alcohol-supplemented medium was fragmented and dispersed as compared with growth in control (starch) medium.

The reaction of resting cells with penicillin G was rapid only for 1 hr and then proceeded at a very low rate. Preincubation of resting cells for 2 hr in buffer alone or with individual reaction components had only minor effects, whereas preincubation in the presence of combinations of Fe^{2+} and ascorbic acid or Fe^{2+} with α-ketoglutaric acid virtually eliminated activity. Similar results were obtained using cell-free extracts. The reaction yielded less than 1% conversion of added penicillin G. However, conversion yields as high as 16.5% were obtained by reducing the concentration of penicillin G.

Immobilized cells of *S. clavuligerus* NP1, entrapped on a polymeric matrix, were able to perform oxidative ring expansion of penicillin G into DAOG. Cells entrapped in polyethyleneimine barium alginate (1.5%) were able to sustain activity for at least four 2-hr cycles, whereas free resting cells were inactive after the second cycle.

Homeologous recombination (recombination between partially homologous sequences) was used to produce novel, functional expandase enzymes (*in vivo*) which are hybrids of the *S. clavuligerus* and *N. lactamdurans* proteins. DNA sequencing of hybrids obtained in *E. coli* showed that recombination had occurred at several locations displaying conserved sequences as short as 2 bp. Hybrid genes obtained in a *Streptomyces* background produced expandases with altered activity on penicillin G, as confirmed by HPLC analysis.

Additional novel bioprocesses for production of the cephalosporin intermediates 7ACA or 7ADCA have been developed. One process is based on a *P. chrysogenum* strain containing the expandase gene of *S. clavuligerus* or *N. lactamdurans*, or the expandase-hydroxylase gene of *C. acremonium*. The second involves the addition to the fermentation medium of adipic acid leading to the formation of adipylpenicillin, which is used as a substrate by the expandase to yield adipyl-7-cephalosporanic acids, including adipyl-7ADCA and adipyl-7ACA. Removal of the adipyl side chain is then carried out with a glutaryl acylase.

ACKNOWLEDGMENTS

We thank Aiqi Fang for her technical assistance and Fang Fang Traves for her help in the preparation of this manuscript. The MIT/Northeastern University studies reviewed here were funded by Antibioticos S.p.A. (Milan, Italy). We thank Ermanno Bernasconi for encouragement.

REFERENCES

1. SE Jensen, AL Demain. Beta-lactams. In: LC Vining, C Stuttard, eds. Genetics and Biochemistry of Antibiotic Production. Boston: Butterworth-Heinemann, 1994, pp 239–268.
2. M Kohsaka, AL Demain. Conversion of penicillin N to cephalosporin(s) by cell-free extracts of *Cephalosporium acremonium*. Biochem Biophys Res Commun 70: 465–473, 1976.
3. J Kupka, YQ Shen, S Wolfe, AL Demain. Partial purification and properties of the α-ketoglutarate-linked ring expansion enzyme of β-lactam biosynthesis. FEMS Microbiol Lett 16:1–6, 1983.

4. JE Dotzlaf, WK Yeh. Purification and properties of deacetoxycephalosporin C synthase from recombinant *Escherichia coli* and its comparison with the native enzyme purified from *Streptomyces clavuligerus*. J Biol Chem 264:10219–10227, 1989.
5. JE Dotzlaf, WK Yeh. Copurification and characterization of deacetoxycephalosporin C synthetase/hydroxylase from *Cephalosporium acremonium*. J Bacteriol 169:1611–1618, 1987.
6. A Scheidegger, MT Kuenzi, J Nüesch. Partial purification and catalytic properties of a bifunctional enzyme in the biosynthetic pathway of β-lactams in *Cephalosporium acremonium*. J Antibiot 37:522–531, 1984.
7. SE Jensen, DWS Westlake, RJ Bowers, S Wolfe. Cephalosporin formation by cell-free extracts from *Streptomyces clavuligerus*. J Antibiot 10:1351–1360, 1982.
8. SE Jensen, DWS Westlake, S Wolfe. Deacetoxycephalosporin C synthetase and deacetoxycephalosporin hydroxylase are two separate enzymes in *Streptomyces clavuligerus*. J Antibiot 38:263–265, 1985.
8a. BJ Baker, JE Dotzlaf, WK Yeh. Deacetoxycephalosporin C hydroxylase of *Streptomyces clavuligerus*: purification, characterization, bifunctionality and evolutionary implication. J Biol Chem 266:5087–5093, 1991.
9. J Cortés, JF Martín, JM Castro, L Laíz, P Liras. Purification and characterization of a 2-oxoglutarate linked ATP-independent deacetoxycephalosporin C synthase of *Streptomyces lactamdurans*. J Gen Microbiol 133:3165–3174, 1987.
10. H Kimura, M Izawa, Y Sumino. Molecular analysis of the gene cluster involved in cephalosporin biosynthesis from *Lysobacter lactamgenus*. Appl Microbiol Biotechnol 44:589–596, 1996.
11. JE Baldwin, RM Adlington, JB Coates, MJC Crabbe, JW Keeping, GC Knight, T Nomoto, CJ Schofield, H-H Ting. Enzymatic ring expansion of penicillins to cephalosporins: side chain specificity. J Chem Soc Chem Commun 374–375, 1987.
12. WK Yeh, JE Dotzlaf, GW Huffman. Biochemical characterization and evolutionary implication of β-lactam expandase/hydroxylase, expandase and hydroxylase. In: H Kleinkauf, H von Döhren, eds. 50 Years of Penicillin Application: History and Trends. Prague: Public, 1994, pp 208–223.
13. K Maeda, JM Luengo, O Ferrero, S Wolfe, MY Lebedev, A Fang, AL Demain. The substrate specificity of deacetoxycephalosporin C synthase (expandase) is extremely narrow. Enzyme Microb Technol 17:231–234, 1995.
13a. CA Bunnell, WD Luke, RM Perry Jr. Industrial manufacture of cephalosporins. In: SF Queener, JA Webber, SW Queener, eds. Beta-lactam Antibiotics for Clinical Use. New York: Marcel Dekker, 1986, pp 255–284.
14. H Cho, JL Adrio, JM Luengo, S Wolfe, S Ocran, G Hintermann, JM Piret, AL Demain. Elucidation of conditions allowing conversion of penicillin G and other penicillins to deacetoxycephalosporins by resting cells and extracts of *Streptomyces clavuligerus* NP1. Proc Natl Acad Sci (USA) 95:11544–11548, 1998.
15. M Yoshida, T Konomi, M Kohsaka, JE Baldwin, S Herchen, P Singh, NA Hunt, AL Demain. Cell-free ring expansion of penicillin N to deacetoxycephalosporin C by *Cephalosporium acremonium* C-19 and its mutants. Proc Natl Acad Sci (USA) 75:6253–6257, 1978.
16. JE Baldwin, M Jung, P Singh, T Wan, S Haber, S Herchen, J Kitchin, AL Demain, NA Hunt, M Kohsaka, T Konomi, M Yoshida. Recent biosynthetic studies on β-lactam antibiotics. Phil. Trans. R Soc Lond B 298:169–172, 1980.

17. MT Abbott, S Udenfriend. α-Ketoglutarate-coupled dioxygenases. In: O Hayashi, ed. Molecular Mechanisms of Oxygen Activation. New York: Academic Press, 1974, pp 167–214.
18. DJ Hook, LT Chang, RP Elander, RB Morin. Stimulation of the conversion of penicillin N to cephalosporins by ascorbic acid, α-ketoglutarate and ferrous ions in cell-free extracts of strains of *Cephalosporium acremonium*. Biochem Biophys Res Commun 87:258–265, 1979.
19. Y Sawada, NA Hunt, AL Demain. Further studies on microbiological ring-expansion of penicillin N. J Antibiot 32:1303–1310, 1979.
20. HR Felix, HH Peter, HJ Treichler. Microbiological ring expansion of penicillin N. J Antibiot 34:567–575, 1981.
21. MJ Rollins, SE Jensen, S Wolfe, DWS Westlake. Oxygen derepresses deacetoxycephalosporin C synthase and increases the conversion of penicillin N to cephamycin C in *Streptomyces clavuligerus*. Enzyme Microb Technol 12:40–45, 1990.
22. YQ Shen, S Wolfe, AL Demain. Deacetoxycephalosporin C synthetase: importance of order of cofactor/reactant addition. Enzyme Microb Technol 6:402–404, 1984.
23. SM Samson, J Dotzlaf, ML Slisz, GW Becker, RM Van-Frank, LE Veal, WK Yeh, JE Miller, SW Queener, TD Ingolia. Cloning and expression of the fungal expandase/hydroxylase gene involved in cephalosporin biosynthesis. Bio/Technology 5:1207–1214, 1987.
24. Y Aharonowitz, G Cohen, JF Martín. Penicillin and cephalosporin biosynthetic genes: structure, organization, regulation and evolution. Annu Rev Microbiol 46: 461–495, 1992.
25. A Paradkar, SE Jensen, RH Mosher. Comparative genetics and molecular biology of β-lactam biosynthesis. In: WR Strohl, ed. Biotechnology of Antibiotics, 2nd ed. New York: Marcel Dekker, 1997, pp 241–277.
26. JF Martín, S Gutierrez, AL Demain. β-lactams. In: T Anke, ed. Fungal Biotechnology. Weinheim: Chapman & Hall, 1997, pp 91–127.
27. S Gutierrez, B Díez, E Montenegro, JF Martín. Characterization of the *Cephalosporium acremonium pcb*AB gene encoding α-aminoadipylcysteinylvaline synthetase, a large multidomain peptide synthetase: linkage to the *pcb*C gene as a cluster of early cephalosporin-biosynthetic genes and evidence of multiple functional domains. J Bacteriol 173:2354–2365, 1991.
28. PL Skatrud, SW Queener. An electrophoretic molecular karyotype for an industral strain of *Cephalosporium acremonium*. Gene 78:331–338, 1989.
29. S Gutierrez, J Velasco, FJ Fernández, JF Martín. The *cef*G gene of *Cephalosporium acremonium* is linked to the *cef*EF gene and encodes a deacetylcephalosporin acyltransferase closely related to homoserine-O-acetyltransferase. J Bacteriol 174: 3056–3064, 1992.
30. MJ Rollins, DWS Westlake, S Wolfe, SE Jensen. Purification and initial characterization of deacetoxycephalosporin C synthase from *Streptomyces clavuligerus*. Can J Microbiol 34:1196–1202, 1988.
31. K Valegard, ACT van Scheltinga, MD Lloyd, T Hara, S Ramaswamy, A Perrakis, A Thompson, HJ Lee, JE Baldwin, CJ Schofield, J Hadju, I Andersson. Structure of a cephalosporin synthase. Nature 394:805–809, 1998.
32. EL Hegg, L Que. The 2-His-1-carboxylate facial triad-an emerging structural motif in mononuclear non-heme iron (II) enzymes. Eur J Biochem 250:625–629, 1997.

33. S Kovacevic, BJ Weigel, MB Tobin, TD Ingolia, JR Miller. Cloning, characterization and expression of the *Streptomyces clavuligerus* gene encoding deacetoxycephalosporin C synthase. J Bacteriol 171:754–760, 1989.
34. S Kovacevic, MB Tobin, JR Miller. The β-lactam biosynthesis genes for isopenicillin N epimerase and deacetoxycephalosporin C synthase are expressed from a single transcript in *Streptomyces clavuligerus*. J Bacteriol 172:3952–3958, 1990.
35. MB Tobin, S Kovacevic, K Madduri, JA Hoskins, PL Skatrud, LC Vining, C Stuttard, JR Miller. Localization of the lysine ε-aminotransferase (*lat*) and δ-(L-aminoadipyl)-L-(cysteinyl)-D-valine synthetase (*pcb*AB) genes from *Streptomyces clavuligerus* and production of lysine-ε-aminotransferase in *Escherichia coli*. J Bacteriol 173:6223–6229, 1991.
36. JJR Coque, JF Martín, P Liras. Characterization and expression in *Streptomyces lividans* of *cef*D and *cef*E genes from *Nocardia lactamdurans*: the organization of the cephamycin gene cluster differs from that in *Streptomyces clavuligerus*. Mol Gen Genet 236:453–458, 1993.
37. H Kimura, H Miyashita, Y Sumino. Organization and expression in *Pseudomonas putida* of the gene cluster involved in cephalosporin biosynthesis from *Lysobacter lactamgenus*. Appl Microbiol Biotechnol 45:490–501, 1996.
38. B Mahro, AL Demain. *In vivo* conversion of penicillin N into cephalosporin type antibiotics by a non-producing mutant of *Streptomyces clavuligerus*. Appl Microbiol Biotechnol 27:272–275, 1987.
39. SE Jensen, BK Leskiw, LC Vining, Y Aharonowitz, DWS Westlake. Purification of isopenicillin N synthetase from *Streptomyces clavuligerus*. Can J Microbiol 32: 953–958, 1986.
40. C Lübbe, S Wolfe, AL Demain. Dithiothreitol reactivates desacetoxycephalosporin C synthetase after inactivation. Enzyme Microb Technol 7:353–356, 1985.
41. MA Báez-Vásquez, JL Adrio, JM Piret, AL Demain. Further studies on the bioconversion of penicillin G into deacetoxycephalosporin G by resting cells of *Streptomyces clavuligerus* NP-1. Appl Biochem Biotechnol 81:145–152, 1999.
42. M-J Fernández, JL Adrio, JM Piret, S Wolfe, S Ro, AL Demain. Stimulatory effect of growth in the presence of alcohols on biotransformation of penicillin G into cephalosporin-type antibiotics by resting cells of *Streptomyces clavuligerus* NP-1. Appl Microbiol Biotechnol 52:484–488, 1999.
43. JL Adrio, H Cho, JM Piret, AL Demain. Inactivation of deacetoxycephalosporin C synthase in extracts of *Streptomyces clavuligerus* during bioconversion of penicillin G to deacetoxycephalosporin G. Enzyme Microb Technol 25:497–501, 1999.
44. L Fucci, CN Oliver, MJ Coon, ER Stadtman. Inactivation of key enzymes by mixed-function oxidation reactions: possible implication in protein turnover and aging. Proc Natl Acad Sci (USA) 80:1521–1525, 1983.
45. S Nakamura. Initiation of sulfite oxidation by spinach ferredoxin-NADP reductase and ferredoxin system: a model experiment of the superoxide anion radical production by metalloflavoproteins. Biochem Biophys Res Commun 41:177–183, 1970.
46. RL Levine, CN Oliver, RH Fulks, ER Stadtman. Turnover of bacterial glutamine synthetase: oxidative inactivation precedes proteolysis. Proc Natl Acad Sci (USA) 78:2120–2124, 1981.

47. EA Popenoe, RB Aronson, DD Van Slyke. The sulfhydryl nature of collagen proline hydroxylase. Arch Biochem Biophys 133:286–292, 1969.
48. RE Rhoads, S Udenfriend. Purification and properties of collagen proline hydroxylase from newborn rat skin. Arch Biochem Biophys 139:329–339, 1970.
49. A Golan-Goldhirsh, DT Osuga, AO Chen, JR Whitaker. Effect of ascorbic acid and copper on proteins. In: VT D'Souza, J Feder, eds. The Bioorganic Chemistry of Enzymatic Catalysis: An Homage to Myron L. Bender. Boca Raton, FL: CRC Press, 1992, pp 61–76.
50. AL Demain, A Fang. Emerging concepts of secondary metabolism in actinomycetes. Actinomycetologica 9:98–117, 1995.
51. P Liras, JA Asturias, JF Martín. Phosphate control sequences involved in transcriptional regulation of antibiotic biosynthesis. Tibtech 8:184–189, 1990.
52. G Guglielmi, P Mazodier, CJ Thompson, J Davies. A survey of the heat shock response in four *Streptomyces* species reveals two *groEL*-like genes and three GroEL-like proteins in *Streptomyces albus*. J Bacteriol 173:7374–7381, 1991.
53. JL Doull, AK Singh, M Hoare, SW Ayer. Conditions for the production of jadomycin B by *Streptomyces venezuelae* ISP5230: effects of heat shock, ethanol treatment and phage infection. J Ind Microbiol 13:120–125, 1994.
54. PC Lee, BR Bochner, BN Ames. AppppA, heat shock stress and cell oxidation. Proc Natl Acad Sci (USA) 80:7496–7500, 1983.
55. MW Qoronfleh, UN Streips, BJ Wilkinsen. Basic features of the staphylococcal heat shock response. Ant van Leeuwenhoek 58:79–86, 1990.
56. DN Arnosti, VL Singer, MJ Chamberlin. Characterization of heat shock in *Bacillus subtilis*. J Bacteriol 168:1243–1249, 1986.
57. S Fukui, A Tanaka. Immobilized microbial cells. Annu Rev Microbiol 36:145–172, 1982.
58. I Takata, K Kayashima, T Tosa, I Chibata. Improvement of stability of fumarase activity of *Brevibacterium flavum* by immobilization with *k*-carrageenan. J Ferment Technol 60:431–437, 1982.
59. EJ Vandamme. Peptide antibiotic production through immobilized biocatalyst technology. Enzyme Microb Technol 5:403–416, 1983.
60. R Bahulekar, NR Ayangar, S Ponrathnam. Polyethyleneimine in immobilization of biocatalysts. Enzyme Microb Technol 13:858–868, 1991.
61. A Freeman, Y Aharonowitz. Immobilization of microbial cells in crosslinked, polymerized linear polyacrylamide gels; antibiotic production by immobilized *Streptomyces clavuligerus* cells. Biotechnol Bioeng 23:2747–2759, 1981.
62. SE Jensen, DWS Westlake, S Wolfe. Production of penicillins and cephalosporins in an immobilized enzyme reactor. Appl Microbiol Biotechnol 20:155–160, 1984.
63. AL Demain, MA Báez-Vásquez. Immobilized *Streptomyces clavuligerus* NP1 cells for biotransformation of penicillin G into deacetoxycephalosproin G. Appl Biochem Biotechnol 87:135–140, 2000.
64. S Gomi, D Ikeda, H Nakamura, H Naganawa, F Yamashita, K Hotta, S Kondo, Y Okami, H Umezawa. Isolation and structure of a new antibiotic indolizomycin produced by a strain SK2-52 obtained by interspecies fusion treatment. J Antibiot 37:1491–1494, 1984.
65. TJ Hosted, RH Baltz. Mutants of *Streptomyces roseosporus* that express enhanced

recombination within partially homologous genes. Microbiology 142:2803–2813, 1996.
66. RH Baltz. Genetic manipulation of antibiotic-producing *Streptomyces*. Trends Microbiol 6:76–83, 1998.
67. GR Smith. Homologous recombination in prokaryotes. Microbiol Rev 54:1–36, 1988.
68. JJ Perry, JT Staley. Genetic exchange. In: Microbiology: Dynamics and Diversity. New York: Saunders College Publishing, 1997, pp 346–366.
69. G Hintermann. Aspects of the genomic organisation in *Streptomyces glaucescens*. PhD dissertation, ETH, Zürich, Switzerland, 1984.
70. DA Hopwood, MJ Bibb, KF Chater, T Kieser. Plasmid and phage vectors for gene cloning and analysis in *Streptomyces*. Meth Enzymol 153:116–167, 1987.
71. JE Baldwin, RM Adlington, MJ Crabbe, G Knight, T Nomoto, CJ Schofield, HH Ting. The enzymatic ring expansion of penicillins to cephalosporins: side chain specificity. Tetrahedron 43:3009–3014, 1987.
72. M Conder, N Almeida, S Behrens, L Crawford, S Delawder, T Hoerner, P McAda, J Rambosek, C Reeves, T Scimmel, T Stepan, R Stieber, V Vinci. Cephalosporin production in *Penicillium chrysogenum*: the application of metabolic engineering to the development of a new biocatalytic process. Conference on Applied Biocatalysis, Brighton, UK, 1994, pp 20–24.
73. L Crawford, A Stepan, P McAda, J Rambosek, M Conder, V Vinci, C Reeves. Production of cephalosporins intermediates by feeding adipic acid to recombinant *Penicillium chrysogenum* strains expressing ring expansion activity. Bio/Technology 13:58–62, 1995.
74. A Ballio, EB Chain, FD di Accadia, MF Mastropietro-Cancellieri, G Morpurgo, G Serlupi-Crescenzi, G Sermonti. Incorporation of α, ω-dicarboxylic acids as side-chains into the penicillin molecule. Nature 185:97–99, 1960.
75. C Cantwell, R Beckman, P Whiteman, SW Queener, EP Abraham. Isolation of deacetoxycephalosporin C from fermentation broths of *Penicillium chrysogenum* transformants: construction of a new fungal biosynthetic pathway. Proc R Soc Lond 248:283–289, 1992.
76. M Nieboer, A Vollebregt, D Schipper, R Bovenberg. Formation of cephalosporins by *Penicillium chrysogenum*. IV European Conference on Fungal Genetics, León, Spain, 1998, p 201.

4

Direct Fermentative Production of Acyltylosins by Genetically Engineered Strains of *Streptomyces fradiae*

Akira Arisawa
Mercian Corporation, Fujisawa, Japan

Hiroshi Tsunekawa
Mercian Corporation, Tokyo, Japan

I. INTRODUCTION

Streptomyces fradiae produces the 16-membered macrolide antibiotic tylosin (Fig. 1A), which exhibits antimicrobial activities against Gram-positive bacteria and *Mycoplasma* species [1]. Because of its remarkable effect against infections, tylosin has been widely used as a veterinary antibiotic for livestock for the last several decades [2]. However, as with almost all classes of antibiotics in clinical use, the long-term use of tylosin has eventually induced bacteria and *Mycoplasma* species to become resistant to tylosin. The mechanisms of resistance are attributed either to diminished uptake of the antibiotic into cells or to the structural alteration of the 70S ribosome, the target site of macrolide antibiotics, which enables the ribosome to escape from tylosin binding. In our studies to investigate tylosin derivatives with increased antimicrobial activities, one of the acyl derivatives of tylosin, 3-*O*-acetyl-4″-*O*-isovaleryltylosin (AIV) (Fig. 1A), exhibited excellent activity against such resistant microbes [3–8]. Tsuchiya et al. have shown that AIV (1) binds more strongly to the altered 70S ribosome, which has only weak affinity to other macrolides [6,7], and (2) is taken up rapidly even through cell membranes that have acquired low penetrability to other macrolides [7,8]. Higher blood levels of AIV after oral administration to rabbits and mice ensure its strong therapeutic effect [5].

Since tylosin is composed of tylonolide (macrolactone; aglycone ring) and three attached sugars having several chemically reactive hydroxyl groups, it is

A

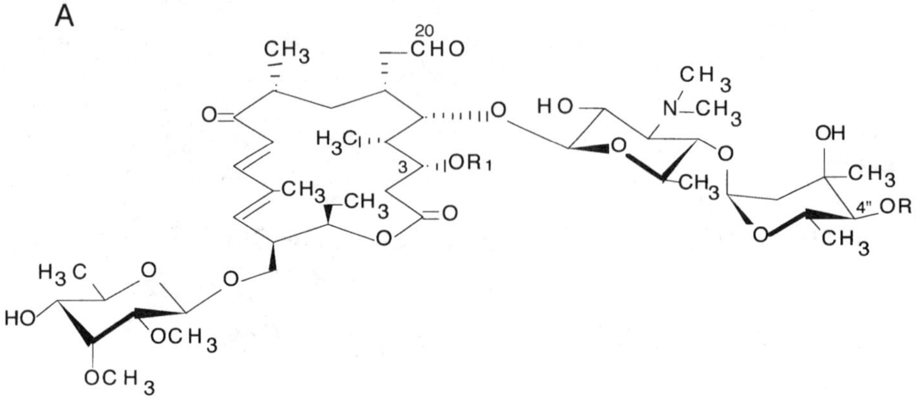

Compound	R_1	R_2
Tylosin	H	H
3-*O*-Acetyltylosin	$COCH_3$	H
4"-*O*-Isovaleryltylosin	H	$COCH_2CH(CH_3)_2$
3-*O*-Acetyl-4"-*O*-isovaleryltylosin (AIV)	$COCH_3$	$COCH_2CH(CH_3)_2$

B

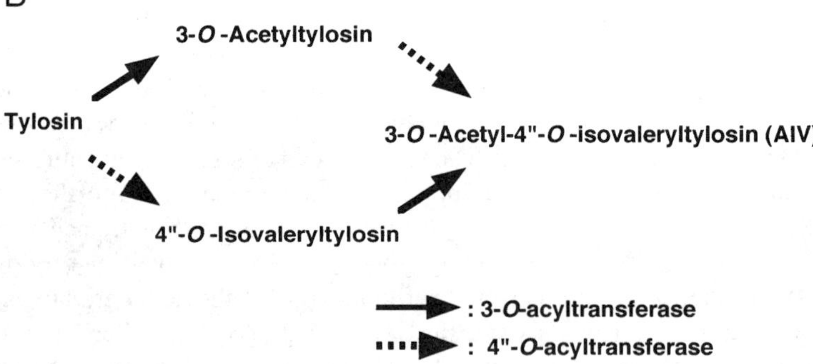

Figure 1 Structural and bioconverting relationships among tylosin, AIV, and its related acyltylosins. (A) Chemical structures of tylosin, AIV, and its related acyltylosins. (B) Biotransformation of tylosin to AIV by two types of macrolide acyltransferases in *S. thermotolerans*: 3-*O*-acyltransferase and 4″-*O*-acyltransferase catalyze acetylation at the 3-hydroxyl group of tylonolide and isovalerylation at the 4″-hydroxyl group of mycarose, respectively.

laborious in terms of industrial production to modify tylosin into AIV by the methods of chemical synthesis [9]. Therefore, an advantageous biotransformation system derived from a streptomycete producing a structurally related macrolide was adopted to produce AIV. In screening for the most suitable converting strain among streptomycetes that have an ability to convert tylosin to AIV, a carbomycin producer, *Streptomyces thermotolerans*, ATCC 11416, was selected [10]. This strain has a 3-*O*-acyltransferase and a 4″-*O*-acyltransferase, both of which are required for the biosynthesis of carbomycin. Tylosin, due to its structural similarity to carbomycin (Fig. 2), can serve as a substrate for these acyltransferases. In practice, AIV was produced using a mutant strain *of S. thermotolerans* (strain TH475, a No. 8254 derivative) [3], which lacks the ability to produce the parent antibiotic carbomycin, but which has both acyltransferase activities, by adding the substrate tylosin exogenously to its growing culture (Fig. 1B). The use of the mutant strain improves recovery of AIV because co-production of carbomycin is eliminated. Utilizing this biotransformation system, industrial manufacturing of AIV was established in 1990 after a number of optimizations of factors influential in large-scale fermentation (Fig. 3). AIV is now on the market with the commercial name of Aivlosin and has achieved some market share of antibiotics used for farm animals (particularly, hogs and fowls) in Southeast Asia, South America, and Eastern Europe as well as Japan.

Although some antibiotics are produced in a single fermentation (e.g., erythromycin, leucomycin, streptomycin, kanamycin), the current manufacturing process for AIV includes two distinct fermentations: first, tylosin fermentation by *S. fradiae*; and second, acyl modification of tylosin by *S. thermotolerans*. This two-step process is relatively laborious and time-consuming, and thus it still needs improvement to reduce manufacturing cost. To solve the problem, we started to work on a strain that produces AIV directly, by expressing cloned acyltransferase genes in *S. fradiae*, which would realize a cost-saving single-

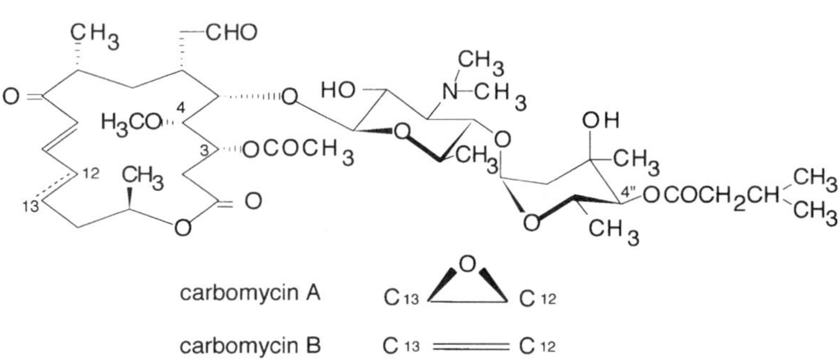

Figure 2 Chemical structures of carbomycins.

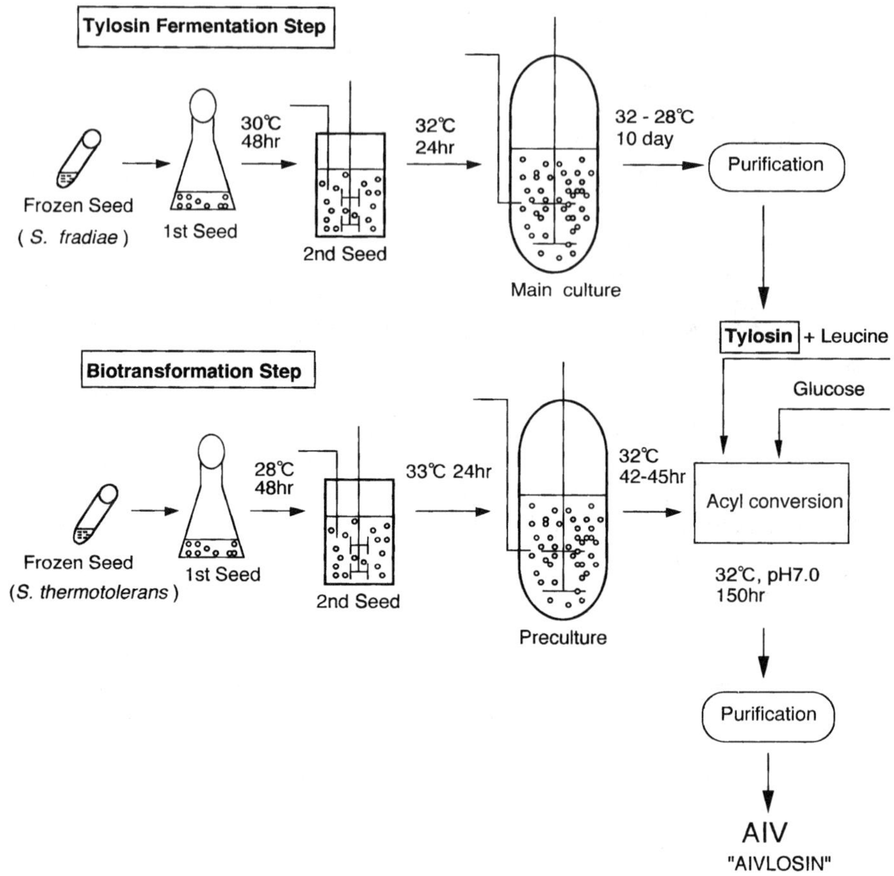

Figure 3 Manufacturing process of AIV. The manufacturing consists of two fermentation steps: tylosin fermentation by *S. fradiae* and tylosin conversion to AIV by *S. thermotolerans*. D-Glucose and L-leucine are fed to the culture for the biotransformation.

step fermentation. Around 1990, when we started the project, the basis for gene manipulation technology in streptomycetes had been established, and the features and roles of genes for some antibiotic biosyntheses were beginning to be clarified [11–13]. These results greatly helped us to move toward the improvement of streptomycetes by DNA recombinant techniques.

In this chapter, we review the cloning of genes involved in the acyl modification of tylosin and the construction of strains that produce acyltylosins, including the important macrolide antibiotic AIV.

II. CLONING GENES INVOLVED IN ACYL MODIFICATIONS OF MACROLIDES FROM *S. THERMOTOLERANS*

A. Cloning of 4″-*O*-Acyltransferase Gene and a Putative Regulatory Gene

Eli Lilly researchers cloned a gene from *S. thermotolerans* for carbomycin 4″-*O*-acyltransferase from the region adjacent to *carB* (carbomycin resistance gene), and designated it *carE* [14–17]. Independently, we cloned the same gene and designated it *acyB1* [18]. Nucleotide sequence analysis of *acyB1* showed that it encodes a protein comprising 388 amino acids with a calculated molecular weight of 43.5 kDa. Interestingly, when we cloned *acyB1*, the cloned fragment included another gene that was found later to activate *acyB1* expression. This putative regulatory gene, called *acyB2*, seemed to start divergently from two possible start codons about 320 bp upstream of *acyB1*. Each protein comprises 374 or 387 amino acids with a calculated molecular weight of 41.7 or 43.0 kDa, respectively. The function of *acyB2* will be discussed in the next section.

B. Cloning of 3-*O*-Acyltransferase Gene

We isolated three contiguous cosmids that cover a 120-kb region containing the carbomycin biosynthetic cluster through extensive cloning by chromosome walking. Since it has been known that genes for antibiotic biosynthesis form a cluster on the genome of its producing actinomycete, this strategy facilitated us to clone a gene for 3-*O*-acyltransferase, which is located in a region approximately 60 kb downstream of *acyB1* [19]. The deduced amino acid sequence of *acyA* (389 amino acids with a calculated molecular weight of 42.8 kDa) [20] showed significant homology to the *acyB1* product (45% identity) and greater homology to the *mdmB* product (midecamycin 3-*O*-acetyltransferase from *Streptomyces mycarofaciens*, 66% identity) [21] (Fig. 4). It is of great interest that comparison of amino acid sequences of those acyltransferases revealed some highly conserved motifs that might provide us with clues to identify domains responsible for regio- and substrate specificity.

A strain of *Streptomyces lividans*, which does not produce macrolide antibiotics, was transformed with plasmids carrying *acyA* and/or *acyB1-acyB2* [18,19, 22]. The resulting transformants were capable of converting exogenously added tylosin to its expected acyl derivatives (i.e., transformant with *acyA*, 3-*O*-acetyltylosin; *acyB1-acyB2*, 4″-*O*-isovaleryltylosin; both sets of genes, AIV).

C. Hybridization Studies with Two Acyltransferase Genes

Southern blot hybridization analysis proved that homologes of *acyA* and *acyB1* were almost perfectly conserved in the genomes of streptomycetes producing the

```
acyA     MESRVERLPSLTGLRWFAALSVFVCHIAQQGIFADPDVASAL-GHLTSLGSLAVSLFFVL    59
mdmB     MPPRVVRLPSLTGLRWFAALAVFACHIAQQQFFADQQVGTAL-LHITTLGSIAVSVFFLL    59
acyB1    MPLP-KHLPALGGMRFISALLVFTSHISTQPFFKNTEINSALQFPLNRLGPLTVSFFFML    59
         *      ** * * *   ** **  **  *  *       **      **   ** ** *

         SGYVLTWSARDGDSVRSFWQRRFAKIYPLHFVTFCIAGVIIFSLSEPVLPGGSTWDGMVP   119
         SGFVLAWSARDKDSVTTFWRRRFAKIYPLHLVTFLIAGVIIFSLAEPTLPGGSVWDGLVP   119
         SGFVLTWAGLPDKSKVNFWRRRTVRAYSLHLPVLLVTLLIVLALNEPNM-GRSVWDGLLT   118
         ** ** *      *   ** **    * **         *   * **   * * ***

         NLLLVHSWLPDAYIVSGFNTPSWSLSCEMAFYLTFPLWYRLLLRIRVSRLW-WYAAALAL   178
         DLLLVQSWLPEPTIIAGFNTPSWSLSCEFAFYLTFPLWYRLVRKIPVRRLW-WCAAGIAA   178
         NLLLIQAWFPDHHEYGSMNPVAWSLSCELFFYAMFPFLFAFFTKVRTDRLWRWAAAVSVA   178
          ***   * *        *   ******  **  **            *** * **

         AVVCMPFVARLLPDS-A-EVVPGMPLRDMWFTYWFPPVRMLEFLLGIVLALIRRQGAWRG   236
         AVICVPFVTSQFPAS-A-ETAPGMPLNELWFACWLPPVRMLEFVLGIVMALILRTGVWRG   236
         AV-SIPLVALLLPASPPLPWDPDMPQLRWWFIYMFPPVRLLEFVLGMLMAQIVIRGRWRG   237
         **   * *    * *      * **    **    **** *** **   * *   * ***

         PGTGTAALLLGGAFALNQVVPPMFTLTATTVVPIALLIAAAADGDLRGRRTGLRAAVLVR   296
         PGVVSSALLLAAAYGVTQVVPPMFTIAACSIVPAALLITALANADVQGLRTGLRSAVLVR   296
         PRPLACVALFSAVFAVTFAVPNHYDPGALTVPVIALLLASVAVGDVRGVRSWLGTRTMVL   297
         *       *          **      *      ***    *  *  * *  *     *

         LGEWSYAFYLIHFLIIRYGHRLLGGDQGYARQWDTLAALGITAAVLGVTIAASAVLHIFV   356
         LGEWSFAFYLVHFMVIRYGHRLMGGELGYARQWSTASAGALALAMLAVAIVAGGLLHTVV   356
         LGELTFAFYLVHYLIIQYGHRFAGGKQGYYRQWDTPAAVGLTLLAFTLALGLSAFLHFFV   357
         ***   **** *   * ****  **  ** *** *  *                 **  *

         ERPCMTLLRGRRPPQGPAPDSGGRPHRAPLERA    389
         ENPCMRLL-GRRRPVATAPDPAT-DEAPKLTRA    387
         EKPVMRTL-GRPRRSPDA-GSTPRSEPAPSGTP    388
         * * *  * **      *
```

Figure 4 Comparison of amino acid sequences of three *O*-acyltransferases for macrolide antibiotics. Amino acid sequences encoded by *acyA* [20], *mdmB* (midecamycin 3-*O*-acetyltransferase) [21], and *acyB1* (*carE*) [15,18] are aligned according to the multi-alignment application program of GENETYX-MAC (Software Development Co., Tokyo, Japan). The residue number is indicated at the right end of each sequence. Identities across all three sequences are denoted by asterisks. (From Ref. 20.)

3,4″-di-*O*-acyl group of 16-membered macrolide antibiotics (Table 1) [18]. The only exception was the *acyB1* probe, which weakly hybridized with the genomic DNA of *Streptomyces ambofaciens* that produces spiramycin, whose 4″-hydroxyl group is not acylated, but whose 3-hydroxyl group is acylated. Hybridization conditions were stringent so that the *acyB1* probe was not allowed to cross-

Table 1 Correlation Between Macrolide Acyltransferase Activities and Presence of *acyA* and *acyB1* Homologs

Strain	Antibiotic	Acyltransferase activities[a]		Hybridization[b]	
		3-*O*-	4″-*O*-	*acyA*	*acyB1*
S. thermotolerans ATCC11416	Carbomycin	+	+	+	+
S. hygroscopicus ATCC21582	Maridomycin	+	+	+	+
S. platensis ATCC29778	Platenomycin	+	+	+	+
S. mycarofaciens ATCC21454	Midecamycin	+	+	+	+
S. kitasatoensis NRRL2487	Leucomycin	+	+	+	+
S. halstedii subsp. *deltae* ATCC33140	Deltamycin	+	+	+	+
S. ambofaciens ATCC1514	Spiramycin	+	−	+	±[c]
S. fradiae ATCC19609	Tylosin	−	−	−	−
Saccharopolyspora erythraea ATCC11635	Erythromycin	−	−	−	−
S. griseus ATCC10137	Streptomycin	−	−	−	−
S. kasugaensis ATCC15715	Kasugamycin	−	−	−	−
S. coelicolor A3(2)	Actinorhodin	−	−	−	Not tested
S. lividans 66	Actinorhodin	−	−	−	−
Streptomyces sp. P9618[d]	Paromomycin	−	−	−	−
S. galilaeus ATCC31133	Aclacinomycin	−	−	−	−

[a] Presence (+) or absence (−) of acyltransferase activity for 16-membered macrolide antibiotics.
[b] Presence (+) or absence (−) of hybridizing band(s) in Southern blot analysis using *acyA* or *acyB1* as a probe.
[c] Weakly hybridizing bands were detected.
[d] A strain from the original culture collection of Mercian Central Research Laboratories.

hybridize with homologs of *acyA*. The *acyB1* probe hybridized with a 3-kb *Sac* I-digested fragment with which the *acyA* probe did not hybridize, suggesting the presence of an *acyB1* homologous gene that does not function in the *S. ambofaciens* genome (A. Arisawa, unpublished data, 1995). The macrolide *O*-acyltransferase genes are conserved among streptomycetes that produce 16-membered acyl macrolides such as midecamycin, leucomycin, and josamycin, but are not homologous to other types of known *O*-acyltransferases for nonmacrolide compounds (e.g., chloramphenicol acetyltransferase [23] and choline acetyltransferase [24]). These results indicate that the genes for macrolide *O*-acyltransferases have a common ancestry and have evolved specifically in acylation of the 3-hydroxyl group or 4″-hydroxyl group in 16-membered macrolides.

A

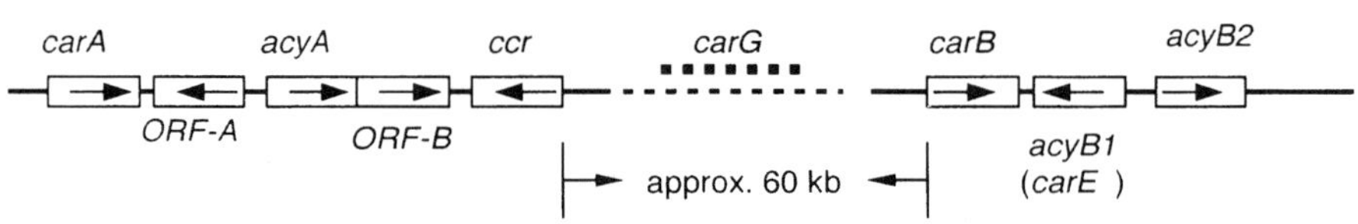

B

Gene	Encoded protein	Function	References
carA	ATP-binding transport protein	carbomycin resistance	(16, 17, 20)
ORF-A	cytochrome P450 monooxygenase	12, 13-epoxidation	(20, 42, 43)
acyA	3-*O*-acyltransferase	acylation at 3-OH	(19, 20)
ORF-B	putative 4-*O*-methyltransferase	methylation at 4-OH	(20, 21)
ccr	crotonyl-CoA reductase	butyryl-CoA synthesis	(A. A., unpublished data. 1999, 44, 45)
carG	polyketide synthase	carbomycin lactone formation	(14, 15, 46)
carB	23S ribosomal RNA methylase	carbomycin resistance	(14, 16, 17)
acyB1 (*carE*)	4"-*O*-acyltransferase	acylation at 4"-OH	(14, 15, 18)
acyB2	tylR-like regulatory factor	positive regulation of *acyA* and *acyB1*	(18, 26)

Figure 5 Genes involved in carbomycin biosynthesis. (A) Structure of the carbomycin biosynthetic cluster of *S. thermotolerans*. Locations and transcriptional directions of nine carbomycin genes identified by the researchers at Eli Lilly and our group are illustrated schematically. (B) List of the carbomycin biosynthetic genes and the functions identified.

D. Other Cloned Genes Involved in Carbomycin Biosynthesis

The regions flanking *acyA* were sequenced. Several genes involved in carbomycin biosynthesis have been identified after sequence homology searches or expression experiments (Ref. 20 and A. Arisawa, unpublished data, 1999). The carbomycin biosynthetic genes identified by us and/or the Eli Lilly research group so far are summarized in Fig. 5.

III. REGULATORY FUNCTION OF *ACYB2*

Expression assays with deletion plasmids revealed that the *acyB2* gene was essential for expression of the 4″-*O*-acyltransferase gene (*acyB1*) [18]. We further characterized the function of *acyB2* in *S. thermotolerans* TH475. When *acyB2* was disrupted by homologous recombination in the strain, the disruptants exhibited quite low activities of both 3-*O*- and 4″-*O*-acyltransferases and failed to convert tylosin to AIV. In contrast, the TH475 transformants in which *acyB2* was overexpressed with a high-copy vector, exhibited elevated activities of both acyltransferases and resulted in a 2.4-fold increased production of AIV [25]. These results suggest that *acyB2* could encode a novel regulatory protein which activates expression of *acyA* and *acyB1* simultaneously.

Bate et al. reported that the deduced product of *tylR* in the biosynthetic gene cluster of tylosin resembled the *acyB2* product (42% identity at amino acid sequence level) [26]. They showed that disruption of *tylR* affected tylosin production, thus concluding that *tylR* encodes a global regulator which controls various aspects of tylosin biosynthesis. Similarly, it is very likely that *acyB2* governs not only two acyltransferases but also other genes governing carbomycin formation. We obtained the *acyB2* disruptants from the TH474 strain of *S. thermotolerans*, which fails to produce carbomycin but expresses acyltransferases. Therefore, the regulatory role of *acyB2* for the carbomycin biosynthesis genes other than *acyA* and *acyB1* has not yet been verified.

IV. PRODUCTION OF ACYLTYLOSINS INCLUDING AIV BY GENETICALLY ENGINEERED *S. FRADIAE*

A. Plasmid Construction and Establishment of Host-Vector System in *S. fradiae*

We screened some strains having an improved productivity of tylosin following a series of conventional mutagenesis on the parent *S. fradiae* ATCC 19609. One of the strains, called MBBF, whose productivity is over 3500 μg/ml, was used as a host for transformation.

Various types of plasmids for producing acyltylosins were designed and constructed. *Streptomyces* high-copy vector pIJ350 [27,28] and pSK1, a plasmid from *Streptomyces kasugaensis* MB273 [29], were used as vectors. The 1.8-kb *acyA* fragment and/or the 3.4-kb *acyB1-acyB2* fragment were inserted into those vectors and the orientation of the insert(s) was verified. On the other hand, a host-vector system of the MBBF strain was established after rounds of transformation and plasmid curing [22]. The resulting strain, in which the original restriction-modification system seemed to be eliminated or markedly diminished, was designated MBBF-c. Tylosin productivity of MBBF-c remained unchanged, whereas its transformation efficiency remarkably improved as compared with the parent MBBF when pIJ350- and pSK-based plasmids were used. After preliminary fermentation tests, the MBBF-c transformants harboring pMAA25 [19], pMAB3 [22], and pAB11ΔEH [22] were selected for subsequent fermentation studies as strains to produce 3-*O*-acetyltylosin, 4″-*O*-isovaleryltylosin and AIV, respectively. Those transformants showed better productivities of acyltylosins than transformants with other types of plasmids. The structures of the three plasmids are shown in Fig. 6. No morphological changes were observed in the selected transformants. Flask fermentation tests were carried out for each transformant. Acyltylosins extracted from the growing culture were analyzed by high-performance liquid chromatography (HPLC) and evaluated for productivities.

B. Production of 3-*O*-Acetyltylosin by *acyA*-Expressing *S. fradiae*

The *S. fradiae* transformant harboring pMAA25 (carrying *acyA*) gave a high titer of 3-*O*-acetyltylosin (3690 μg/ml), while a control strain with the pIJ350 vector produced 3250 μg/ml of tylosin [19] (Fig. 7). During this fermentation, the 3-*O*-acetylation reaction proceeded completely, since no tylosin was left unacylated. Probably a sufficient pool of substrate acetyl-CoA is present in cells to result in the successful conversion. Reduction in total macrolide productivity was not observed after introducing the plasmid into the host strain. Some 3-*O*-acyltranferases from other macrolide producers are likely to utilize propionyl-CoA as well as acetyl-CoA, based on their products, which include 3-*O*-propionyl components. However, no 3-*O*-propionyltylosin was detected from the fermentative products of the *acyA*-expressing strain in the HPLC analysis. Currently, we have no evidence that the *acyA*-encoded 3-*O*-acyltransferase can utilize propionyl-CoA in the reaction with tylosin.

As discussed in the previous section, *acyA* cannot be expressed efficiently without *acyB2*. In the tested strain with pMAA25 carrying *acyA* alone, however, *acyA* was expressed enough to produce a considerable amount of 3-*O*-acetyltylosin under the condition where *acyB2* was not co-expressed. The mRNA constitu-

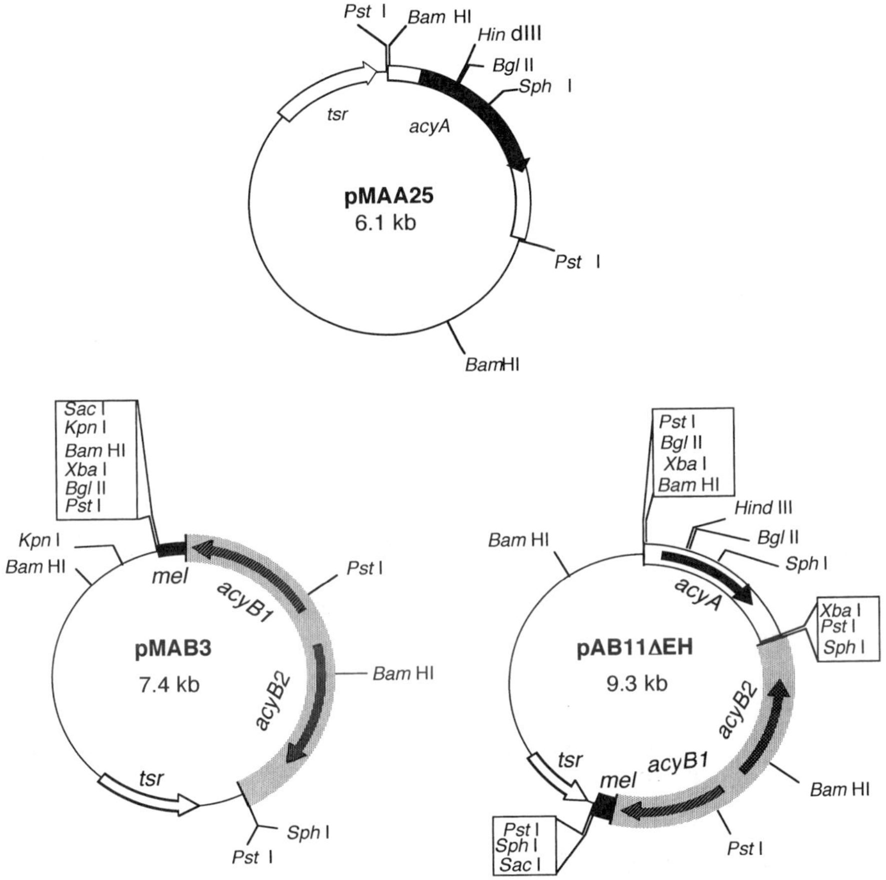

Figure 6 Structures of plasmids for direct production of acyltylosins. pMAA25 carrying the 1.8-kb *acyA* fragment, pMAB3 carrying the 3.4-kb *acyB1-acyB2* fragment, and pAB11ΔEH carrying both fragments were used to produce 3-*O*-acetyltylosin, 4″-*O*-isovaleryltylosin, and AIV, respectively. All plasmids are based on a typical streptomycete high-copy vector pIJ350 derived from pIJ101 [27,28]. *mel*, a *Sac* I-*Bgl* II 0.24-kb portion of tyrosinase gene [47]; *tsr*, thiostrepton-resistance gene [30]. (From Ref. 22.)

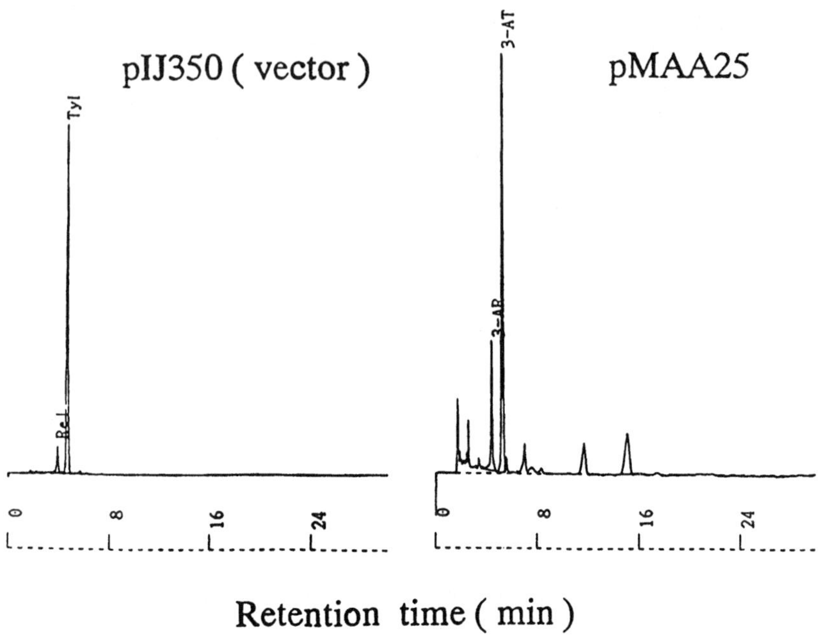

Figure 7 HPLC chromatograms of culture broths of *S. fradiae* harboring pIJ350 and pMAA25. Abbreviations (retention time): Rel, relomycin [20-dihydrotylosin] (3.8 min); Tyl, tylosin (4.6 min); 3-AR, 3-*O*-acetylrelomycin (4.4 min); 3-AT, 3-*O*-acetyltylosin (5.3 min). (From Ref. 19.)

tively transcribed from the *tsr* gene (thiostrepton-resistance gene) [30] lying just upstream of *acyA* is likely to run through *acyA* (Fig. 6; see structure of pMAA25).

C. Production of 4″-*O*-Acyltylosins by *acyB1*-Expressing *S. fradiae*

After 7 days of flask fermentation in a sucrose-based medium, the MBBF-c strain transformed with pMAB3 (carrying *acyB1* and *acyB2*) produced tylosin and several 4″-*O*-acyltylosins in which the 4″-hydroxyl group had undergone various acylations (Fig. 8). In contrast to the *acyA*-encoded 3-*O*-acyltransferase [22], the *acyB1*-encoded 4″-*O*-acyltransferase seemed to utilize several acyl-CoAs as substrates. The productivity of 4″-*O*-acyltylosins, which was determined by HPLC, is 4″-*O*-butyryltylosin (772 μg/ml), 4″-*O*-acetyltylosin (524 μg/ml), 4″-*O*-propionyltylosin (359 μg/ml), and 4″-*O*-isovaleryltylosin (190 μg/ml), in descending order. In this test, unconverted tylosin was also detected, and its production was

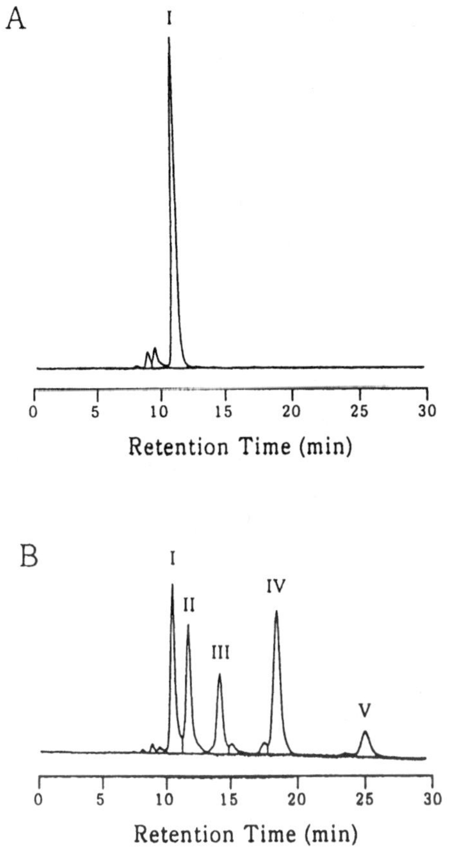

Figure 8 HPLC chromatograms of ethyl acetate extracts from culture broth of *S. fradiae* MBBF-c (A) and *S. fradiae* MBBF-c(pMAB3) (B). Abbreviations: I, tylosin; II, 4″-*O*-acetyltylosin; III, 4″-*O*-propionyltylosin; IV, 4″-*O*-butyryltylosin; V, 4″-*O*-isovaleryltylosin. (From Ref. 22.)

635 μg/ml (25.6% of total macrolide produced). The total macrolide productivity was reduced by 25% as compared with MBBF-c by introducing the plasmid. These results indicated that 74.4% of macrolide produced suffered 4″-*O*-acylation, but that 4″-*O*-isovaleryltylosin was a minor component (7.6% of total macrolide produced). Although 4″-*O*-isovalerylation is the major reaction in *S. thermotolerans* under conditions even without L-leucine, a starting material for isovaleryl-CoA, three other kinds of 4″-*O*-acylation (4″-*O*-acetylation, 4″-*O*-propionylation, and 4″-*O*-butyrylation) are predominant in *S. fradiae*. This finding

is well consistent with differences in the macrolide-producing patterns between *Streptomyces kitasatoensis* and *Streptomyces narbonensis* var. *josamyceticus. S. kitasatoensis* produces many components of leucomycin (a 16-membered macrolide antibiotic related to carbomycin), which differ in their 4″-*O*- and 3-*O*-acyl groups [2]. In contrast, *S. narbonensis* var. *josamyceticus* selectively produces josamycin (identical to the 3-*O*-acetyl-4″-*O*-isovaleryl component of leucomycin, namely, leucomycin A_3) [31]. The production rate of each component of 4″-*O*-acyltylosins is likely to depend on which and how much acyl-CoA is accumulated in the host. The genetic background advantageous for accumulating isovaleryl-CoA in cells will be discussed in the last section.

D. Production of 3-*O*-Acetyl-4″-*O*-acyltylosins Including AIV by *S. fradiae* Expressing *acyA* and *acyB1*

The *S. fradiae* MBBF-c strain transformed with pAB11ΔEH carrying *acyB1-acyB2* and *acyA* produced a variety of 3-*O*-acetyl-4″-*O*-acyltylosins after 7 days of flask cultivation (Fig. 9). As summarized in Table 2, of all the acylated products, AIV was least accumulated at 56 μg/ml (X in Fig. 9B). However, addition of L-leucine in the late period of fermentation increased AIV up to 602 μg/ml (X in Fig. 9C). The introduction of pAB11ΔEH decreased the total macrolide production by 77–78%, which was a larger loss than that caused by the introduction of pMAB3 (i.e., 25% loss) in the same host. Since the host *S. fradiae* exhibits high resistance to various acyltylosins as well as to tylosin, the decreased production was not caused by antimicrobial activities of acyltylosins. Probably some aberrant event caused by pAB11ΔEH, for example, overexpression of the acyltransferases and/or the *acyB2*-encoded regulatory factor, might affect the normal secondary metabolism including macrolide biosynthesis. A similar effect of a plasmid on secondary metabolites was previously reported [32]. Such a negative effect of a plasmid on macrolide production was confirmed by the fact that cured strains (strains that lost pAB11ΔEH) were restored to their original productivity of tylosin.

A time course of AIV production together with other acyltylosins was studied by cultivating the transformant of MBBF-c (pAB11ΔEH) in a 3-liter-jar fermenter for 10 days. Macrolide productivity was determined by HPLC on a daily basis from the second day. As shown in Fig. 10, production of 3-*O*-acetyltylosin

Figure 9 HPLC chromatograms of culture broth of *S. fradiae* MBBF-c (A) and *S. fradiae* MBBF-c(pAB11ΔEH) (B, C). In the fermentation at (C), L-leucine was fed into the culture, and the supernatant was diluted by 20-fold with acetonitrile for analysis. Abbreviations: I, tylosin; VI, 3-*O*-acetyltylosin; VII, 3,4″-di-*O*-acetyltylosin; VIII, 3-*O*-acetyl-4″-*O*-propionyltylosin; IX, 3-*O*-acetyl-4″-*O*-butyryltylosin; X, 3-*O*-acetyl-4″-*O*-isovaleryltylosin (AIV). (From Ref. 22.)

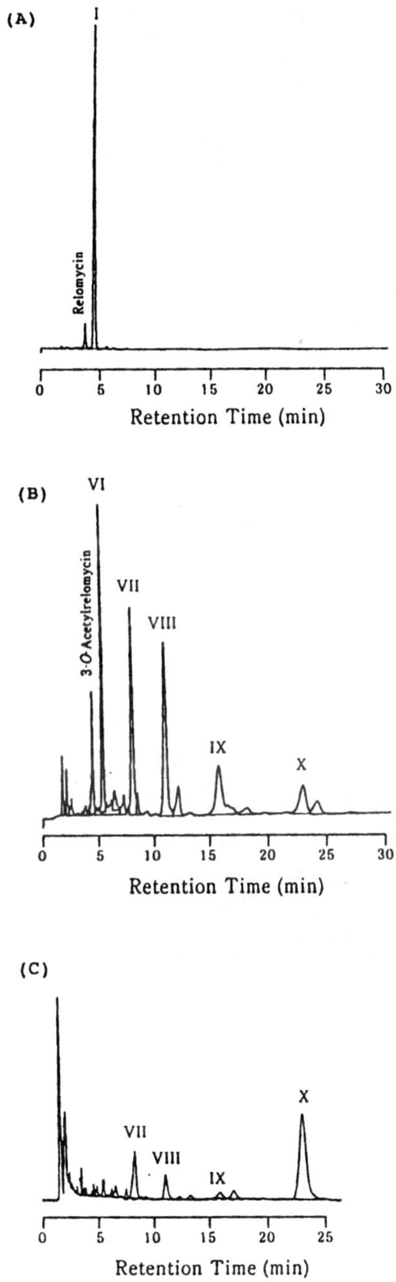
(A)
I
Relomycin
0 5 10 15 20 25 30
Retention Time (min)
(B)
VI
3-O-Acetylrelomycin
VII
VIII
IX
X
0 5 10 15 20 25 30
Retention Time (min)
(C)
X
VII
VIII
IX
0 5 10 15 20 25
Retention Time (min)

Table 2 Productivity of Acyltylosins by *Streptomyces fradiae* (pAB11ΔEH)

Strain	Leucine[a] addition	Macrolide production (μg/ml)[b] TY	RM	3-AT	3-AR	3,4″-AAT	AIV
MBBF-c	−	3320	255	ND	ND	ND	ND
MBBF-c(pAB11ΔEH)	−	ND	7	325	71	329	56
MBBF-c(pAB11ΔEH)	+	ND	ND	29	15	194	602

[a] L-Leucine was added to 0.5 mg/ml on the 5th day of 7 days' culture.
[b] Symbols: TY, tylosin; RM, relomycin (20-dihydrotylosin); 3-AT, 3-*O*-acetyltylosin; 3-AR, 3-*O*-acetylrelomycin; 3,4″-AAT, total amount of 3-*O*-acetyl-4″-*O*-acyltylosins except AIV (VII to IX in Fig. 9); ND, not detected (<1 μg/ml).

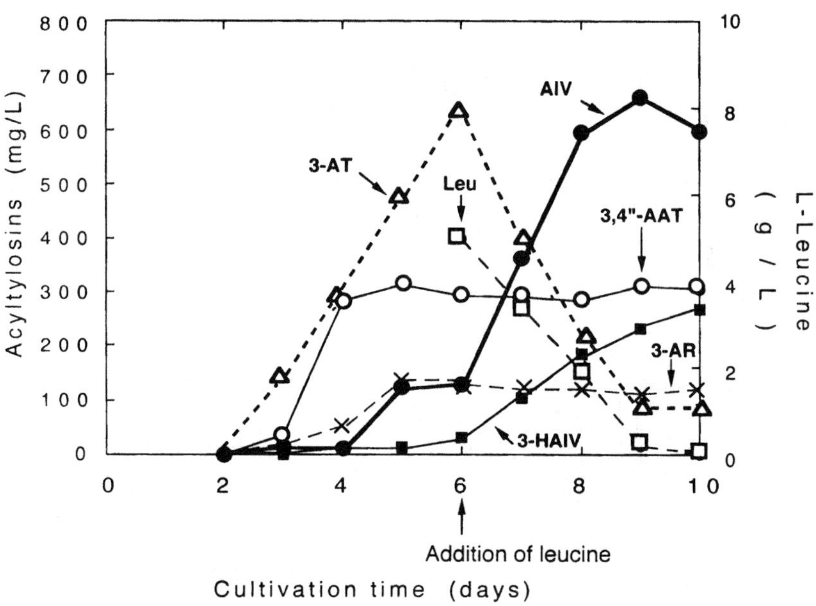

Figure 10 Time course of fermentation with *S. fradiae* harboring pAB11ΔEH using a 3-liter jar fermenter. Titers of AIV (3-*O*-acetyl-4″-*O*-isovaleryltylosin) and other related compounds during 10-day fermentation are presented. Residual L-leucine (Leu) determined is also indicated from the 6th day to the last day of fermentation. Abbreviations for acyltylosins: 3-AT, 3-*O*-acetyltylosin; 3,4″-AAT, total amount of 3-*O*-acetyl-4″-*O*-acyltylosins except AIV; 3-AR, 3-*O*-acetylrelomycin (3-*O*-acetyl-20-dihydrotylosin); 3-HAIV, 3-*O*-acetyl-4″-*O*-(3-hydroxyisovaleryl)tylosin.

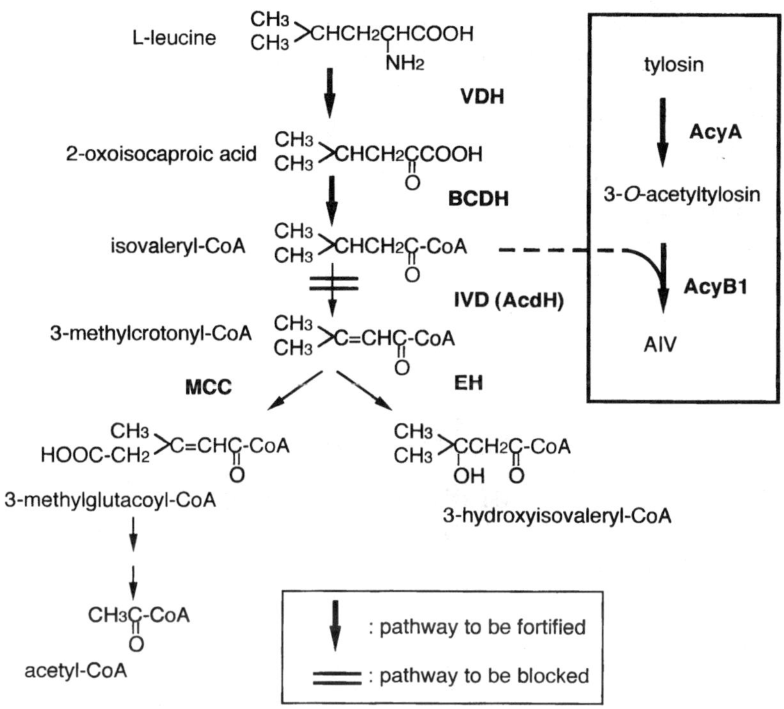

Figure 11 The putative catabolic pathway of L-leucine and its implications for strain improvement. For a promising host strain, the pathway to be blocked is indicated with thick double lines and the pathways to be fortified are indicated with thick arrows. Abbreviations for enzymes participating in the L-leucine catabolism and the acylation of tylosin: VDH, valine (branched-chain amino acid) dehydrogenase; BCDH, branched-chain α-keto acid dehydrogenase; IVD (AcdH), isovaleryl-CoA dehydrogenase (acyl-CoA dehydrogenase); MCC, 3-methylcrotonyl-CoA carboxylase; EH, enoyl-CoA hydratase; AcyA, macrolide 3-*O*-acyltransferase; AcyB1, macrolide 4″-*O*-acyltransferase.

(3-AT) started to rise on day 3 and steadily increased until day 6, when L-leucine was added. Simultaneously, due to acylation at the 4″-hydroxyl group of the intermediate 3-*O*-acetyltylosin, the series of 3-*O*-acetyl-4″-*O*-acyltylosins excluding AIV (3,4″-AAT) began to accumulate and production of AIV followed. The formation of 3,4″-AAT and AIV reached a plateau on day 5, probably because of the shortage of endogenous acyl-CoAs. This assumption is supported by the fact that when exogenous L-leucine was added on day 6, AIV increased exclusively and reached 650 μg/ml on day 9, while the other group of acyltylosins remained unchanged. The AIV productivity was not improved very much in this fermentation experiment using the jar fermenter, even though its control system to keep aeration and pH conditions optimal for macrolide production was expected to ensure better productivity than fermentation in flasks. An unexpected problem appeared upon finding 3-*O*-acetyl-4″-*O*-(3-hydroxyisovaleryl)tylosin (3-HAIV) in the products, which represented a possible further metabolite of AIV that was produced at up to 270 μg/ml. The ratio of 3-HAIV to AIV remained approximately 0.3 throughout the course of the fermentation. Isovaleryl-CoA dehydrogenase and enoyl-CoA hydratase seem to play key roles in the formation of 3-HAIV. It is likely that a series of reactions by both enzymes may convert isovaleryl-CoA to 3-hydroxyisovaleryl-CoA (Fig. 11), which would then be used for 4″-O-acylation of tylosin to form 3-HAIV. Alternatively, the same enzymes may attack AIV directly, to give 3-HAIV.

V. OTHER STUDIES ON CLONING ACYLTRANSFERASE GENES AND HETEROLOGOUS EXPRESSION IN MACROLIDE-PRODUCING STREPTOMYCETES

In 1989, Epp and her co-workers at Eli Lilly first reported success in production of 4″-*O*-isovalerylspiramycin by expressing *carE* (identical to *acyB1*) in spiramycin-producing *S. ambofaciens* [15]. In the study, the *carE* gene was integrated into the chromosome of *S. ambofaciens*. The resulting integrant produced 4″-*O*-isovalerylspiramycin II (3-*O*-acetyl-4″-*O*-isovalerylspiramycin I) and 4″-*O*-isovalerylspiramycin III (3-*O*-propionyl-4″-*O*-isovalerylspiramycin I) as the major components, although their productivities were not clarified. A similar study accomplished by Wang et al. expressed the 4″-*O*-acyltransferase gene from the midecamycin-producer *S. mycarofaciens* in the spiramycin producer *S. ambofaciens* [33]. They obtained strains that produced 4″-*O*-propionylspiramycin II (3-*O*-acetyl-4″-*O*-propionylspiramycin I) and 4″-*O*-propionylspiramycin III (3,4″-di-*O*-propionylspiramycin I). The midecamycin 3-*O*-acyltransferase gene (*mdmB*) was also cloned and expressed in *S. lividans* by Hara and Hutchinson [21], although expression of the gene in heterologous macrolide producers was not mentioned. Researchers at Asahi Chemical Industry have looked for strains that acquired an improved productivity of leucomycin by self-cloning a gene for 4″-*O*-acyltransferase from the leucomycin producer *S. kitasatoensis* [34].

VI. PERSPECTIVES FOR INDUSTRIAL MANUFACTURE OF AIV BY GENETICALLY ENGINEERED *S. FRADIAE*

Reduction in manufacturing cost has been one of the most general and important problems in the fermentation industry. To cope with the problem, strain improvements by both conventional and molecular biology-based strategies have contributed much to increase the productivity of various industrially important bioproducts, including enzymes, chemicals, and pharmaceuticals, especially in the early stage of their development process. Recently, it is becoming acceptable that genetic engineering, incorporating applicable aspects of metabolic engineering [35] and directed enzyme evolution [36], can provide powerful and practical means to improve strains into ones ideal for fermentation with minimum labor and time. In the improvement of AIV-producing strains, as in other cases of antibiotic production, the product must satisfy high qualitative and quantitative demands. In this view, it is more significant to design and alter specific genes in the biosynthetic pathways by genetic engineering than to do random alterations by classical mutagenesis.

The AIV-producing strain that we have constructed and evaluated through the studies mentioned in this chapter is still far from industrial application. As a promising approach to its realization for industrial manufacturing, the use of hosts with highly improved productivity of tylosin is absolutely necessary. Chromosomal integration of *acyA* and *acyB1-acyB2* is thought to be effective to retain the introduced genes stably, and to express enough enzymes for the acyl modifications without a decrease in tylosin productivity during large-scale fermentation. Furthermore, to obtain a host strain which allows accumulation of much endogenous isovaleryl-CoA, we probably need to alter the catabolic pathway of L-leucine. A mutational inactivation of isovaleryl-CoA dehydrogenase gene, together with enhanced expression of valine (branched-chain amino acid) dehydrogenase [37,38] and branched-chain α-keto acid dehydrogenase [39,40], would be most effective (Fig. 11). Lately, genes for acyl-CoA dehydrogenase (*acdH*) were cloned from *Streptomyces avermitilis* and *Streptomyces coelicolor* [41]. The gene product oxidized both short straight- and branched-chain acyl-CoAs including isovaleryl-CoA, and a mutant having a disrupted *acdH* was not able to grow on minimal media containing L-leucine as a sole carbon source. These facts show that this gene is substantially the isovaleryl-CoA dehydrogenase (IVD) gene of our interest. Use of the *acdH* gene (or its homolog in *S. fradiae*) will accelerate the work for the IVD mutant.

It is hoped that these efforts assisted by genetic engineering and genome technology will lead to the direct and selective fermentation of AIV on an industrial scale.

ACKNOWLEDGMENTS

We thank all our colleagues for technical support, including fermentation experiments and product analyses, and constructive discussions particularly with Dr. I. Kojima, Dr. N. Kawamura, T. Narita, K. Takeda, Dr. K. Okamura, Dr. M. Okabe, and Dr. R. Okamoto.

REFERENCES

1. JM McGuire, WS Boniece, CE Higgens, MM Hoehn, WM Stark, J Westhead, RN Wolfe. Tylosin, a new antibiotic I: Microbiological studies. Antibiot Chemother 11: 320–327, 1961.
2. S Ōmura, H Tanaka. Macrolide Antibiotics: Chemistry, Biology, and Practice. New York: Academic Press, 1984, pp 3–338.
3. R Okamoto, T Fukumoto, H Nomura, K Kiyoshima, K Nakamura, A Takamatsu, H Naganawa, T Takeuchi, H Umezawa. Physico-chemical properties of new acyl derivatives of tylosin produced by microbial transformation. J Antibiot 33:1300–1308, 1980.
4. R Okamoto, H Nomura, M Tsuchiya, H Tsunekawa, T Fukumoto, T Inui, T Sawa, T Takeuchi, H Umezawa. The activity of 4″-*O*-acylated tylosin derivatives against macrolide-resistant Gram-positive bacteria. J Antibiot 32:542–544, 1979.
5. R Okamoto, M Tsuchiya, H Nomura, H Iguchi, K Kiyoshima, S Hori, T Inui, T Sawa, T Takeuchi, H Umezawa. Biological properties of new acyl derivatives of tylosin. J Antibiot 33:1309–1315, 1980.
6. M Tsuchiya, T Sawa, T Takeuchi, H Umezawa, R Okamoto. Binding of 3-*O*-acetyl-4″-*O*-isovaleryltylosin to ribosomes from a macrolide-resistant strain of *Staphylococcus aureus*. J Antibiot 35:673–679, 1982.
7. M Tsuchiya, M Hamada, T Takeuchi, H Umezawa, K Yamamoto, H Tanaka, K Kiyoshima, S Mori, R Okamato. Studies of tylosin derivatives effective against macrolide-resistant strains: synthesis and structure-activity relations. J Antibiot 35: 661–672, 1982.
8. M Tsuchiya, K Suzukake, M Hori, T Sawa, T Takeuchi, H Umezawa, R Okamoto, H Nomura, H Tsunekawa, T Inui. Studies on the effects of 3-acetyl-4″-isovaleryltylosin against multi-drug resistant strains of *Staphylococcus aureus*. J Antibiot 34:305–312, 1981.
9. HA Kirst, M Debono, KE Willard, BA Truedell, JE Toth, JR Turner, DR Berry, BB Briggs, DS Fukuda, VM Daupert, AM Felty-Duckworth, JL Ott, FT Counter. Preparation and evaluation of 3,4″-ester derivatives of 16-membered macrolide antibiotics related to tylosin. J Antibiot 39:1724–1735, 1986.
10. R Okamoto, T Fukumoto, K Imafuku, T Kubo, K Kiyoshima, A Takamatsu, T Takeuchi. Screening for 16-membered macrolide-transforming microorganisms. J Ferment Technol 57:519–528, 1979.
11. SE Fishman, K Cox, JL Larson, PA Reynolds, ET Seno, W-K Yeh, R Van Frank, CL Hershberger. Cloning genes for the biosynthesis of a macrolide antibiotic. Proc Natl Acad Sci (USA) 84:8248–8252, 1987.
12. JM Weber, JO Leung, GT Maine, LHB Potenz, TJ Paulus, JP DeWitt. Organization of a cluster of erythromycin genes in *Saccharopolyspora erythaea*. J Bacteriol 172: 2372–2383, 1990.
13. H Kleinkauf, H von Dohren. Antibiotics—cloning of biosynthetic pathways. FEBS Lett 268:405–407, 1990.
14. JK Epp, MLB Huber, JR Turner, B Schoner. Molecular cloning and expression of carbomycin biosynthetic and resistance genes from *S. thermotolerans*. In: Y Okami,

T Beppu, H Ogawara, eds. Biology of Actinomycetes '88: Proceedings of Seventh International Symposium on Biology of Actinomycetes. Tokyo: Japan Scientific Societies Press, 1988, pp 82–85.

15. JK Epp, MLB Huber, JR Turner, T Goodson, and BE Schoner. Production of a hybrid macrolide antibiotic in *Streptomyces ambofaciens* and *Streptomyces lividans* by introduction of a cloned carbomycin biosynthetic gene from *Streptomyces thermotolerans*. Gene 85:293–301, 1989.
16. JK Epp, SG Burgett, BE Schoner. Cloning and nucleotide sequence of a carbomycin resistance gene from *Streptomyces thermotolerans*. Gene 53:73–83, 1987.
17. B Schoner, M Geistlich, P Rosteck Jr, RN Rao, E Seno, P Reynolds, K Cox, S Burgett, C Hershberger. Sequence similarity between macrolide-resistance determinants and ATP-binding transport proteins. Gene 115:93–96, 1992.
18. A Arisawa, N Kawamura, H Tsunekawa, K Okamura, H Tone, R Okamoto. Cloning and nucleotide sequences of two genes involved in the 4″-*O*-acylation of macrolide antibiotics from *Streptomyces thermotolerans*. Biosci Biotechnol Biochem 57:2020–2025, 1993.
19. A Arisawa, N Kawamura, K Takeda, H Tsunekawa, K Okamura, R Okamoto. Cloning of the macrolide antibiotic biosynthesis gene *acyA*, which encodes 3-*O*-acyltransferase, from *Streptomyces thermotolerans* and its use for direct fermentative production of a hybrid macrolide antibiotic. Appl Environ Microbiol 60:2657–2660, 1994.
20. A Arisawa, H Tsunekawa, K Okamura, R Okamoto. Nucleotide sequence analysis of the carbomycin biosynthetic genes including the 3-*O*-acyltransferase gene from *Streptomyces thermotolerans*. Biosci Biotechnol Biochem 59:582–588, 1995.
21. O Hara, CR Hutchinson. A macrolide 3-*O*-acyltransferase gene from the midecamycin-producing species *Streptomyces mycarofaciens*. J Bacteriol 174:5141–5144, 1992.
22. A Arisawa, N Kawamura, T Narita, I Kojima, K Okamura, H Tsunekawa, T Yoshioka, R Okamoto. Direct fermentative production of acyltylosins by genetically-engineered strains of *Streptomyces fradiae*. J Antibiot 49:451–456, 1996.
23. IA Murray, JA Gil, DA Hopwood, WV Shaw. Nucleotide sequence of the chloramphenicol acetyltransferase gene of *Streptomyces acrimycini*. Gene 85:283–291, 1989.
24. N Itoh, JR Slemmon, DH Hawke, R Williamson, E Morita, K Itakura, E Roberts, JE Shively, GD Crawford, PM Salvaterra. Cloning of *Drosophila* choline acetyltransferase cDNA. Proc Natl Acad Sci (USA) 83:4081–4085, 1986.
25. A Arisawa, N Kawamura, T Takeda, T Fujii, T Narita, I Kojima, K Okamura, H Tsunekawa, T Yoshioka, R Okamoto. Cloning of carbomycin biosynthesis genes from *Streptomyces thermotolerans* and its use for strain improvement. Abstracts Book of 10th International Symposium on Biology of Actinomycetes, Beijing, China, 1997, 12S5.
26. N Bate, AR Butler, AR Gandecha, E Cundliffe. Multiple regulatory genes in the tylosin biosynthetic cluster of *Streptomyces fradiae*. Chem Biol 6:617–624, 1999.
27. T Kieser, DA Hopwood, HM Wright, CJ Thompson. pIJ101, a multi-copy broad host-range *Streptomyces* plasmid: functional analysis and development of DNA cloning vecters. Mol Gen Genet 185:223–238, 1982.

28. KJ Kendall, SN Cohen. Complete nucleotide sequence of the *Streptomyces lividans* plasmid pIJ101 and correlation of the sequence with genetic properties. J Bacteriol 170:4634–4651, 1988.
29. H Akagawa, K Kawaguchi, M Ichihara. Plasmids of *Streptomyces kasugaensis* MB273: their pock formation, their dispensable endonuclease cleavage sites for pock formation, and transformation of *S. kasugaensis* MB273 by them. J Antibiot 37: 1016–1025, 1984.
30. DA Hopwood, MJ Bibb, KF Chater, T Kieser, CJ Bruton, HM Kieser, DJ Lydiate, CP Smith, JM Ward, H Schrempf. Genetic manipulation of *Streptomyces*: a laboratory manual. The John Innes Foundation, Norwich, England, 1985.
31. S Ōmura, Y Hironaka, T Hata. Chemistry of leucomycin. IX: Identification of leucomycin A_3 with josamycin. J Antibiot 23:511–513, 1970.
32. DI Thomas, JH Cove, S Baumberg, CA Jones, BAM Rudd. Plasmid effects on secondary metabolite production by a streptomycete synthesizing an anthelmintic macrolide. J Gen Microbiol 137:2331–2337, 1991.
33. Y Wang, L Jin, W Jin, X Zhang, Y Zeng, X Xu, J Yao. Cloning and expression of midecamycin 4″-acylase gene in spiramycin producing strains. Chin J Biotechnol 8:1–13, 1992.
34. K Furuya, T Yamaguchi. DNA containing acyltransferase gene for macrolide antibiotic. Japanese Patent (Kokai Tokkyo Koho) P11-285384, Oct. 19, 1999.
35. J Nielsen. The role of metabolic engineering in the production of secondary metabolites. Curr Opin Microbiol 1:330–336, 1998.
36. FH Arnold, AA Volkov. Directed evolution of biocatalysts. Curr Opin Microbiol 3: 54–59, 1999.
37. A Vancura, I Vancurova, J Volc, SM Fussey, M Flieger, J Neuzil, J Marsalek, V Behal. Valine dehydrogenase from *Streptomyces fradiae*: purification and properties. J Gen Microbiol 138:3213–3219, 1988.
38. L Tang, Y-X Zhang, CR Hutchinson. Amino acid catabolism and antibiotic synthesis: valine is a source of precursors for macrolide biosynthesis in *Streptomyces ambofaciens* and *Streptomyces fradiae*. J Bacteriol 176:6107–6119, 1994.
39. DD Skinner, MR Morgenstern, RW Fedechko, CD Denoya. Cloning and sequencing of a cluster of genes encoding branched chain α-keto acid dehydrogenase from *Streptomyces avermitilis* and production of a functional E1[αβ] component in *Escherichia coli*. J Bacteriol 177:183–190, 1995.
40. CD Denoya, RW Fedechko, EW Hafner, HAI McArthur, MR Morgenstern, DD Skinner, K Stutzman-Engwall, RG Wax, WC Wernau. A second branched-chain α-keto acid dehydrogenase gene cluster (*bkdFGH*) from *Streptomyces avermitilis*: its relationship to avermectin biosynthesis and the construction of a *bkdF* mutant suitable for the production of novel antiparasitic avermectins. J Bacteriol 177:3504–3511, 1995.
41. Y-X Zhang, CD Denoya, DD Skinner, RW Fedechko, HAI McArthur, MR Morgenstern, RA Davies, S Lobo, KA Reynolds, CR Hutchinson. Genes encoding acyl-CoA dehydrogenase (AcdH) homologues from *Streptomyces coelicolor* and *Streptomyces avermitilis* provide insights into the metabolism of small branched-chain fatty acids and macrolide antibiotic production. Microbiology 145:2323–2334, 1999.
42. Y Xue, D Wilson, L Zhao, H-W Liu, DH Sherman. Hydroxylation of macrolactones

YC-17 and narbomycin is mediated by the *pikC*-encoded cytochrome P450 in *Streptomyces venezuelae*. Chem Biol 5:661–667, 1998.
43. M Inouye, Y Takada, N Muto, T Beppu, S Horinouchi. Characterization and expression of a P-450-like mycinamicin biosynthesis gene using a novel *Micromonospora-Escherichia coli* shuttle cosmid vector. Mol Gen Genet 245:456–464, 1994.
44. AR Gandecha, SL Large, E Cundliffe. Analysis of four tylosin biosynthetic genes from the *tylLM* region of the *Streptomyces fradiae* genome. Gene 184:197–203, 1997.
45. SA Fish, E Cundliffe. Stimulation of polyketide metabolism in *Streptomyces fradiae* by tylosin and its glycosylated precursors. Microbiology 143:3871–3876, 1997.
46. JK Epp, BE Schoner. Carbomycin biosynthetic gene, *carG*—used for increasing yields of carbomycin and for producing new antibiotics and antibiotic derivatives. US Patent 4952502, Dec. 15, 1988.
47. V Bernan, D Filpula, W Herber, M Bibb, E Katz. The nucleotide sequence of the tyrosinase gene from *Streptomyces antibioticus* and characterization of the gene product. Gene 37:101–110, 1985.

5

Engineering *Streptomyces avermitilis* for the Production of Novel Avermectins: Mutant Design and Titer Improvement

Claudio D. Denoya, Kim J. Stutzman-Engwall, and Hamish A. I. McArthur
Pfizer, Inc., Groton, Connecticut

I. INTRODUCTION

Streptomycetes are gram-positive, filamentous soil bacteria that have two characteristic features: (1) a complex life cycle that includes mycelial differentiation and sporulation, and (2) the ability to produce a variety of secondary metabolites with therapeutic value [1]. The application of chemical mutagenesis and mutasynthesis have provided important technical tools in the development of industrial cultures for the production of antibiotics and other therapeutic compounds. More recently, the application of recombinant DNA technology in *Streptomyces* spp. has offered the possibility of manipulating the expression of specific genes and the design of new mutants with enhanced ability to produce natural and novel secondary metabolites. Due to the fact that each Streptomycetes soil isolate usually presents distinct characteristics in terms of growth properties, media requirements, genome instability, transformation capability, and recombination proficiency, a significant effort is frequently needed to adapt recombinant DNA techniques to each strain of interest. When successfully developed, genetic engineering strategies can be used to improve the growth characteristics of particular cultures in industrial fermentation, to increase antibiotic production, and to produce novel antibiotic structures. An extremely desirable but challenging objective is the ability to carry out molecular genetic manipulations within high-titer strains already used for the commercial production of secondary metabolites, significantly reducing timelines and effort required to develop novel therapeutic agents, compared to classical process development from environmental isolates. In this chapter, we shall describe how we have applied these approaches to the development of novel avermectins.

Streptomyces avermitilis produces eight closely related polyketide macrocyclic lactones with potent activity against both ecto- and endoparasites. Following the discovery of avermectins in 1979 [2], they have been developed to commercialization and have enjoyed outstanding success in parasite control in the fields of animal health care, agriculture, and human infections. Avermectins were originally isolated as agents which could kill nematodes. It was soon realized that these compounds had a broad spectrum of activity, not only against essentially all roundworms found in the mammalian corpus, including the gastrointestinal tract, lung, heart, and eye, but also against ectoparasites such as mites, biting flies, and ticks. The mode of action of avermectins, and the closely related milbemycins, is believed to be the irreversible opening of chloride channels in muscle membranes [3,4]. These chloride channels were initially believed to be associated with γ-aminobutyric acid (GABA) receptors [5], but more recent data have implicated glutamine-gated chloride channels, a mechanism which would account for the efficacy of the compounds against both roundworms and arthropods [6]. In nematodes, it has been proposed that avermectins block the activities associated with pharyngeal function, feeding, hydrostatic pressure regulation, and/or secretion, thereby causing paralysis of the parasite and facilitating its elimination from the host. In mammals, use of chloride channels as neurotransmitters is restricted to the central nervous system. As a consequence, the blood/brain barrier is believed to prevent the drug from reaching the glutamine-gated chloride channels and accounting for the excellent therapeutic index and safety profile of the avermectins. The combination of such potent activity against a wide variety of parasites and very safe performance in mammals has made avermectins the most successful antiparasitic product ever developed for animal health applications.

II. CHEMICAL STRUCTURE OF AVERMECTINS AND BIOSYNTHESIS

Eight naturally occurring, structurally related avermectins are produced by *Streptomyces avermitilis* [7]. The avermectin polyketide structure is derived from seven acetate and five propionate residues, together with a single 2-methylbutyric acid or isobutyric acid residue which forms the *sec*-butyl or isopropyl group attached to the C25 of the spiroketal moiety [8,9] (Fig. 1). The avermectin aglycone is further modified by glycosylation at C13, with the attachment of two *O*-methylated oleandrose residues and *O*-methylation at C5. Thus, *S. avermitilis*

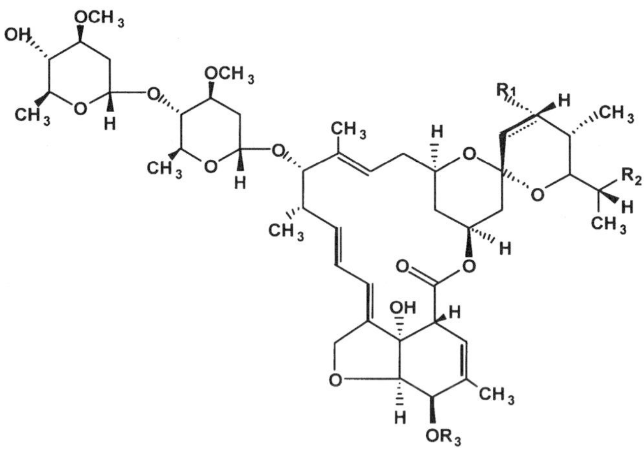

	R_1	R_2	R_3
A_{1a}		C_2H_5	CH_3
A_{1b}		CH_3	CH_3
A_{2a}	OH	C_2H_5	CH_3
A_{2b}	OH	CH_3	CH_3
B_{1a}		C_2H_5	H
B_{1b}		CH_3	H
B_{2a}	OH	C_2H_5	H
B_{2b}	OH	CH_3	H

R_1 is located at C23 of the avermectin aglycone. Where R_1 is absent, the C22,23 double bond is present. R_2 is part of either a *sec*-butyl or an isopropyl group attached to C25. R_3 is part of either a methoxy or a hydroxyl group at C5. The two *O*-methylated oleandrose residues are located at C13.

Figure 1 Structure of the avermectin molecule. As discussed in the text, there are eight naturally occurring, structurally related avermectins, which arise from the presence of (1) either a methoxy or a hydroxyl group at C5 (OR_3), (2) the presence of either double-bond carbons at C22,23 or a hydroxyl group at C23 (R_1), and (3) the presence of either a methyl or an ethyl residue (R_2) located at the isopropyl or the *sec*-butyl group, respectively, that is attached to C25.

produces all the eight different combinations of avermectin resulting from variability at the C5, C22,23, and C25 positions.

The avermectin biosynthetic genes encompass a region of approximately 85 kb of the *S. avermitilis* genome which have been cloned and sequenced by us, by the Merck group [10], and by the Kitasato group [11]. The DNA sequence of the avermectin biosynthetic region was recently published [11]. The avermectin polyketide synthase (PKS) genes have a type I modular organization consisting of four large multifunctional open reading frames (ORFs) that encode a 12-module system (Fig. 2). These PKS genes are required for the biosynthesis of the avermectin aglycone. Additional ORFs are located adjacent to the PKS genes and encode polypeptides involved in postpolyketide modifications. A set of 8 genes to make dTDP-oleandrose from glucose and attach it to the avermectin aglycone is located within an 11-kb *Pst*I fragment as shown in Fig. 2. The deduced gene products have homology to enzymes involved in the biosynthesis of other 6-deoxy sugars and are listed below using the nomenclature from Ikeda et al. [11]:

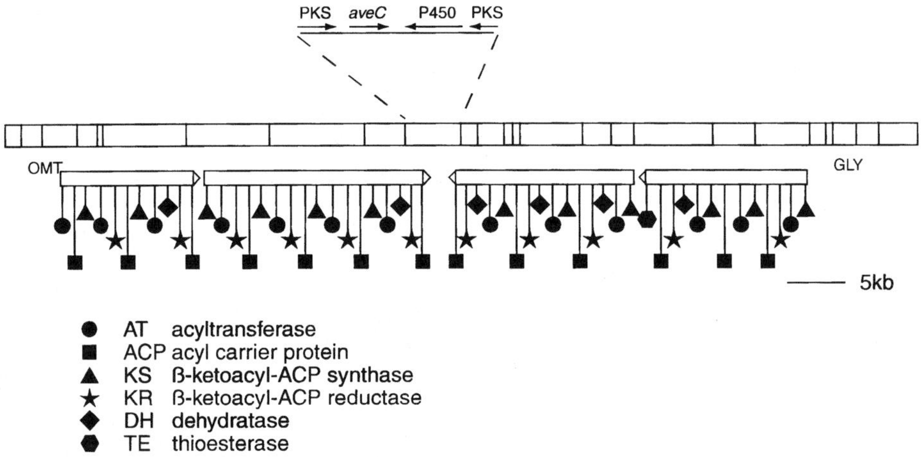

Figure 2 Physical map of the avermectin biosynthetic gene cluster. The location of *Bam*HI restriction sites is indicated by vertical lines in the upper box. There are four long open reading frames (ORFs) encoding very large, multifunctional polypeptides of the avermectin polyketide synthase (PKS) (from left to right: AVES 1, AVES 2, AVES 3, and AVES 4). There are 12 homologous sets of enzyme activities (modules) distributed on the PKS, as follows: modules 1 (AT, ACP, KS, AT, KR, ACP) and 2 (KS, AT, DH, KR, ACP) are located on AVES 1 (far left side). Modules 3 (KS, AT, KR, ACP), 4 (KS, AT, KR, ACP), 5 (KS, AT, KR, ACP), and 6 (KS, AT, DH, KR, ACP) are located on AVES 2 (middle, left). Modules 7 (KS, AT, DH, KR, ACP), 8 (KS, AT, DH, KR, ACP), and 9 (KS, AT, DH, KR, ACP) are located on AVES 3 (middle, right). Modules 10 (KS, AT, KR, ACP), 11 (KS, AT, ACP), and 12 (KS, AT, DH, KR, ACP, TE) are located on AVES 4 (far right side). OMT, C5 *O*-methyltransferase; GLY, genes involved in avermectin glycosylation; *aveC*, a gene associated to mutants that produce only ''A2'' and ''B2'' components of avermectins (C22,23-dehydration) (see Ref. 11).

aveBI	glycosyl transferase (*dnmS*, *eryBV*, *eryCIII*)
aveBII	dTDP-glucose 4,6-dehydratase (*dnmM*, *ery gdh*)
aveBIII	α-D-glucose-1-phosphate thymidylyltransferase (*dnmL*)
aveBIV	dTDP-4-keto-6-deoxy-L-hexose 4-reductase (*dnmV*, *eryBIV*)
aveBV	dTDP-4-keto-6-deoxyhexose 3,5-epimerase (*dnmU*, *eryBVII*)
aveBVI	dTDP-4-keto-6-deoxy-L-hexose 2,3-dehydratase (*dnmT*, *eryBVI*)
aveBVII	dTDP-6-deoxy-L-hexose 3-*O*-methyltransferase (*eryBIII*)
aveBVIII	dTDP-4-keto-6-deoxy-L-hexose 2,3-reductase (*eryBII*)

Although the pathway of 6-deoxysugar biosynthesis in antibiotic-producing bacteria has not been established, recent work by Madduri et al. [12] verifies the function predicted for *aveBIV* and homologs *dnmV* and *eryBIV*. In addition, these authors were able to modify a step in the biosynthesis of dTDP-L-daunosamine and produce the C-4 epimeric form of daunosamine by inactivating the normal *dnmV* gene in *Streptomyces peucetius* and replacing it with the heterologous *aveBIV* or *eryBIV* gene from *S. avermitilis* or *Saccharopolyspora erythraea*, respectively.

Analysis of the avermectin biosynthetic genes has presented the potential to engineer specific changes in the biosynthetic pathway. In the next three sections, the discussion will focus on the genetic basis for the biosynthesis of the eight natural avermectin components and discuss experiments to identify the genes involved in the differences detected at the C5, C22,23, and C25 positions. Later in the chapter, we will discuss how it is possible to manipulate the producing microorganism to direct the biosynthesis of novel avermectins. Finally, there will be a discussion of possible approaches applied to titer improvement.

A. The A and B Forms of Avermectins: The Role of the C5 *O*-Methyltransferase

Avermectins produced by *S. avermitilis* that have a hydroxyl group at C5 are the B components, and avermectins containing a methoxy group at C5 are the A components. The enzyme catalyzing the methylation of the hydroxyl group at C5 is a C5 *O*-methyltransferase (OMT) [13]. The corresponding gene (*aveD*) was identified within a 3.4-kb *Bam*HI fragment by complementing a *S. avermitilis* OMT-deficient mutant previously isolated by screening after random chemical mutagenesis. The gene encoding OMT maps near the left side of the avermectin biosynthetic gene cluster in the *S. avermitilis* genome (Fig. 2). DNA sequence analysis of this chromosomal region by us [14] and by Ikeda et al. [15] revealed a potential open reading frame, and the predicted amino acid sequence of this ORF contains conserved domains found in other methyltransferases (Fig. 3). Ikeda et al. [15] determined that the *aveD* mutant K2034 had a single base-pair

```
CAGCTGGGCCGCCCGGCCCGGATTGCGGAGGTTCCCGACCATGATGGCCGTCCCGGAGCA
                                         M  M  A  V  P  E  Q

GACCCCGCCCTCTTCCCTGGAAGTGGGGACTACTACGACCGTTTGACCGACCTCATGAA
  T  P  P  S  S  L  E  V  G  D  Y  Y  D  R  L  T  D  L  M  N

TCGTGCGCTGGGTGGGAACACCCACCTCGGATACTGGCCGCAGCCGGGGGACGGCAGTTC
  R  A  L  G  G  N  T  H  L  G  Y  W  P  Q  P  G  D  G  S  S

ACCGGGCAAGGCCGCCGACCGGCTCACCGATCTCCTCATCGGCAAGCTGAGAGGCATCAC
  P  G  K  A  A  D  R  L  T  D  L  L  I  G  K  L  R  G  I  T

GGGCCGCCGGGTCCTGGACGTCGGCTGCGGTTCCGGAAAGCCGGCGGTGCGGCTCGCCCT
  G  R  R  V  L  D  V  G  C  G  S  G  K  P  A  V  R  L  A  L

GAGCGCGCCCGTCGATGTCGTCGGCGTGACGGTGAGCGAGGTTCAGGTCGGGCTGGCGAC
  S  A  P  V  D  V  V  G  V  T  V  S  E  V  Q  V  G  L  A  T

CGCTCTCGCGAAGCAGTCGCACGTGGCGGACCGGGTCGTGTTCACCCGTGCCGACGCGAT
  A  L  A  K  Q  S  H  V  A  D  R  V  V  F  T  R  A  D  A  M

GGAACTGCCGTTTCCCGACGGGTCCTTCGACGCGGCGTGGGCCCTGGAGTGTCTCCTGCA
  E  L  P  F  P  D  G  S  F  D  A  A  W  A  L  E  C  L  L  H

CATGCCCAGCCCCGCACAGGTGATCCGGGAGATCGCCCGGGTGCTCCGCCCCGGCGGCCG
  M  P  S  P  A  Q  V  I  R  E  I  A  R  V  L  R  P  G  G  R

GCTGGCCGTCACGGACGTGACGCTGCGCGCCTTCGGGCGGACCGGCATGAAGCGCGGGGA
  L  A  V  T  D  V  T  L  R  A  F  G  R  T  G  M  K  R  G  E

GTGCACGTCCCAGTTGCTCGCGGTCCCGGCCCTGGTGCACATCGACGAGTACGCCGGAAT
  C  T  S  Q  L  L  A  V  P  A  L  V  H  I  D  E  Y  A  G  M

GATCGCCGACGCCGGGCTGGAACTGCATGAGCTGACCGACATCGGCGATCAGGTCGTCGG
  I  A  D  A  G  L  E  L  H  E  L  T  D  I  G  D  Q  V  V  G

CCCCTCTTTCGCCGCGCTGCGTGACCACGTGAACGAGCACCTCGACGAGTACGCGGCGGC
  P  S  F  A  A  L  R  D  H  V  N  E  H  L  D  E  Y  A  A  A

CTTCGGGATCGGCGTCGCGGAGATGCGGAAGGTGGTTGCACAGTGCACGACGCTCCCCTG
  F  G  I  G  V  A  E  M  R  K  V  V  A  Q  C  T  T  L  P  W

GACGCCGGACATCGGCTATGTCGTGCTGACCGCCCGGCGCCCGGGGGAGTGA
  T  P  D  I  G  Y  V  V  L  T  A  R  R  P  G  E  *
```

Figure 3 The nucleotide sequence and the deduced amino acid sequence of the *S. avermitilis* OMT gene. A potential ribosome binding site is in bold type and double-underlined. The stop codon is indicated by an asterisk. Conserved domains found in methyltransferases are underlined.

change which would lead to Thr23→Ile substitution. We examined an *aveD* mutant obtained during mutagenesis of *S. avermitilis* and determined that a single base-pair change resulted in Trp267→termination codon. The AveD protein from this mutant is missing the last 17 amino acids. In addition, disruption of the promoter region of *aveD* eliminated both C5 *O*-methylation activity and C5 keto reduction, presumably because *aveF* (which has a deduced gene product resembling a 3-ketoacyl-ACT/CoA reductase) and *aveD* are part of an operon [16].

B. The 1 and 2 Forms of Avermectins: The Role of the C22,23 Dehydratase

Avermectins produced by *S. avermitilis* that have a hydroxyl group at C23 are the 2 components, and avermectins containing double-bond carbons at C22,23 are the 1 components. Both 1 and 2 forms of avermectin are produced during fermentation. Pioneering studies to characterize the PKS system of the erythromycin biosynthetic gene cluster demonstrated that each biosynthetic step in the assembly of the polyketide chain has a separate active site [17,18]. These steps are organized in sets, one set for each chain extension and (where appropriate) reduction cycle. The order of the structural genes in the biosynthetic gene cluster is related to the order in which the sets of activities act in the polyketide biosynthesis [17,18]. Based on this model, the avermectin dehydratase (DH) domain in Module 2 should be responsible for the reduction at C22,23. To determine if this dehydratase functioned in the production of the 1 and 2 forms of avermectins, we investigated a region of the avermectin PKS which corresponds to the PKS module 2. A cosmid library of *S. avermitilis* was constructed and cosmid clones containing ketosynthase and dehydratase regions were identified by hybridization. A *Bam*HI restriction map was constructed by analysis of overlapping cosmids and hybridization. Sequence analysis identified a region of module 2 that contained >72% homology with the DH oligo probe and a putative conserved DH active site histidine [19].

A comparison of the deduced amino acid sequence of the module 2 DH from *S. avermitilis* with the conserved active site motif of other known DH [17,20] shows that this *S. avermitilis* DH consensus sequence was altered from HxxxGxxxxP to HxxxGxxxxS (Fig. 4) [21]. We designed an amino acid replacement strategy to change the serine (S) residue to a proline (P) residue (since all active DH domains that had been identified contain the conserved sequence HxxxGxxxxP). An *Xho*I restriction site was introduced in the corresponding DNA sequence to follow the gene replacement event. Chromosomal DNA isolated from transformants was analyzed by PCR and *Xho*I digestion to identify gene replacement events. Twenty-three isolates did not produce avermectins, and DNA sequence analysis of 5 avermectin nonproducing isolates confirmed the S-to-P change. Since avermectin production seemed to require the presence of the serine residue in the conserved motif, we investigated gene replacement of a portion of the module 2 DH region with portions of DH region from module 7 or module 12. The DH domain for reduction at C4,C3 was identified, sequenced, and contained the motif HxxxGxxxxP. A gene replacement strategy was designed to exchange 51 amino acids from this region with 51 amino acids from module 2 DH. Two isolates that were confirmed by DNA sequence analysis to have undergone gene replacement produced avermectins at the normal B2:B1 ratio, suggesting that the conserved DH motif HxxxGxxxxP did not provide complete de-

```
           1                                                      50
   Mod10   .HPHHHLDLP TYPFQRQHYW LERSQPGAGD VAAAGLEPAE HPLLAATVQL
    Mod2   PHTHTHLDLP TYPFQHHHYW LESTQPGAGN VSAAGLDPTE HPLLGATLEL
    Mod9   RDRARHLDLP TYAFDHHRYW VDTSAGHPGD LSAAGLGTAG HPLLGSAVAL
     Erm   ....RPAELP TYPFEHQRFW ..PRPHRPAD VSALGVRGAE HPLLLAAVDV

Consensus  -----hldLP TYpF-h--yW -e------gd vsAaGl--ae HPLL-a-v-l


                                 ------ ---------- ---         100
   Mod10   ADTDGCLLTG RLSLRSHPWL GDYEVGGAVL LSGSAFVELA VQVGERVGCT
    Mod2   ATDGGALLAG RLSLRSHPWL ADHAVGGTVL LSGATFLELA LHAGTYVGCD
    Mod9   AESQELLFTG RLSLRTHPWL ADHAIFGTVL LPGTAILELA VRAGDEVDCG
    Erm5   PGHGGAVFTG RLSTDEQPWL AEHVVGGRTL VPGSVLVDLA LAAGEDVGLP

Consensus  a---g-l-tG RLSlr-hPWL adh-vgG-vL l-G----eLA --aG--Vgc-
                                  *   *     *


                                                               150
   Mod10   RIEPLTVH.. .......... .......... .......... ..........
    Mod2   RVDELTLHAP LVVPVDGGVS VQVGVAAADG ......EGRR LVSVYARGGS
    Mod9   TVEELTLRTP LVLPEQGSVI LQLSVGAPQG PQTPEEPERR TFALYAREDD
     Erm   VLEELVLQRP LVLAGAGAL. LRMSVGAPD. .....ESGRR TIDVHAAED.

Consensus  -veeLtl--P LV----G--- ----V-A--- --------RR -----A----


   Mod10   .......... .......... ..........
    Mod2   ACGGGGA..S GGVWTCHASG VLVEAAAGGV
    Mod9   GLSSSSAxxT GTEWTCHATG VLTGTVRPAE
    Erm5   .....VADLA DAQWSQHATG TLAQGVAAGP
Consensus  ------A--- ---W--HA-G -L--------
```

Figure 4 Dehydratase (DH) consensus sequence of avermectin module 10 (inactive), module 9 (active), module 2, and erythromycin DH domains. Amino acids listed in the consensus line in capital letters are shared by all 4 DHs, and those in lower case are shared by 3 of the 4 regions. The underlined module 10 region was inserted into module 2 or the underlined module 9 region was inserted into module 2 for gene replacement experiments. The (*) sequences are the conserved HxxxGxxxxP active domain.

hydratase activity to module 2 DH. The DH domain (nonfunctional) for reduction at C13,C12 was identified, sequenced, and contained the motif YxxxGxxxxS. A gene replacement strategy was designed to exchange 38 amino acids from this region with 38 amino acids from module 2 DH. Three isolates that were confirmed by DNA sequence analysis to have undergone gene replacement produced avermectins at the normal B2:B1 ratio, suggesting that altering the putative C22,23 DH by replacing the active site H (histidine) with a Y (tyrosine) did not affect avermectin production. Previous experiments to alter the active-site histidine in the erythromycin module 4 DH or the fatty acid DH enzyme resulted in the elimination of erythromycin biosynthesis [19] or the disruption of fatty acid biosynthesis [22].

The above results show that the dehydratase reaction that takes place at C22,23 in the fermentation to produce both 1 and 2 avermectins is probably not located in the PKS at the predicted module 2, since replacing a portion of this region with the inactive module 10 DH domain did not affect the amounts of 1 and 2 avermectins produced. Therefore, the identification and mapping of the gene encoding the C22,23 dehydratase remain to be elucidated.

C. The a and b Forms of Avermectins: C25 Substituents

As discussed earlier, the avermectin polyketide backbone is derived from seven acetate and five propionate extender units added to an α branched-chain fatty acid starter, which is either (*S*(+)-α-methylbutyric acid or isobutyric acid. The C25 position of naturally occurring avermectins has two possible substituents: a *sec*-butyl residue derived from the incorporation of *S*(+)-α-methylbutyryl-CoA (''a'' avermectins), or an isopropyl residue derived from the incorporation of isobutyryl-CoA (''b'' avermectins). These α branched-chain fatty acids, which act as starter units in the biosynthesis of the polyketide ring, are derived from the α branched-chain amino acids isoleucine and valine through a branched-chain amino acid transaminase reaction followed by a branched-chain α-keto acid dehydrogenase (BCDH) reaction (Fig. 5) [23].

III. NOVEL AVERMECTINS MODIFIED AT THE C25 SIDE CHAIN: A SUCCESSFUL APPLICATION OF MUTASYNTHESIS

Bu'Lock et al. had shown initially [24] that supplementation of *S. avermitilis* fermentations with a range of fatty acids resulted in their uptake and incorporation to generate novel avermectins modified at the C25 side chain of the molecule. This approach, described as precursor-directed biosynthesis, has been employed to produce many new antibiotics [25], but the co-expression of the parent molecule interferes with the detection and isolation of the novel analogs. To circumvent these difficulties, the elegant technique of ''mutational biosynthesis'' [26] or ''mutasynthesis'' [27,28] was developed. In this approach, a mutant of an organism, deficient in the production of an essential precursor for the secondary metabolite of choice, is isolated, and precursor-directed biosynthesis is then employed to generate only the novel analogs.

In applying mutasynthesis to the production of novel avermectins, the elimination of branched-chain α-keto acid dehydrogenase (BCDH) was targeted. This multienzyme complex is responsible for supplying the 2-methylbutyryl- and isobutyryl-CoA ''starter units'' that initiate natural avermectin biosynthesis [21,29],

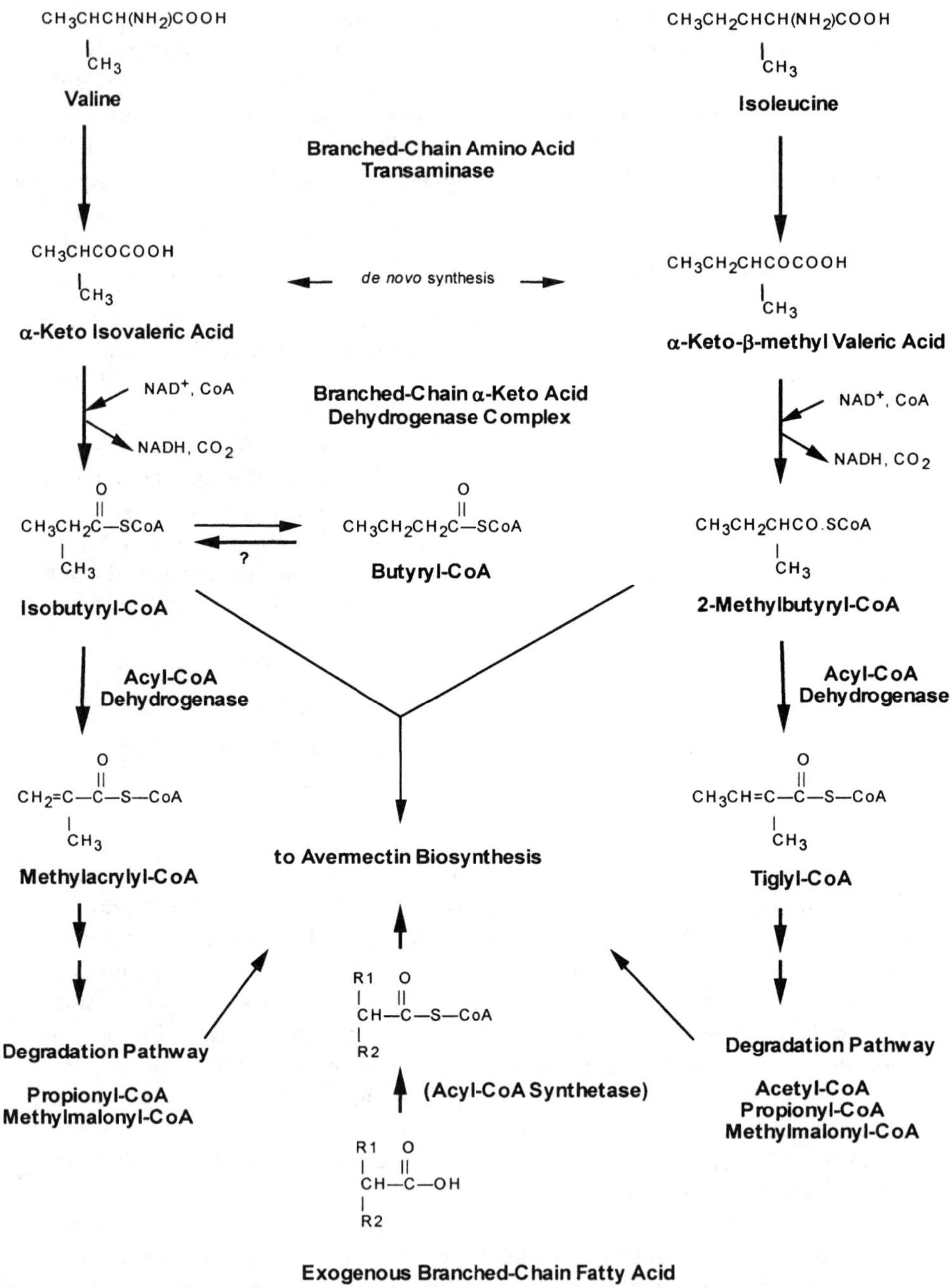

Figure 5 Pathways of valine and isoleucine catabolism and their postulated relationship to avermectin biosynthesis.

and which are derived from branched-chain amino acid metabolism (Fig. 5). However, it was unclear whether this approach would be feasible, for several reasons. Thus, labeling studies [9] had demonstrated that not only were the "starter units" incorporated intact at the C25 position initiating avermectin biosynthesis, but that they were also degraded by subsequent enzymes of branched-chain amino acid catabolism, and incorporated into the 8 "propionate" and 6 "acetate" units of the macrolide. Consequently, if branched-chain amino acid catabolism was a major source of avermectin's "propionate" and "acetate" extender units, disruption of this pathway could compromise novel avermectin expression. In addition, 2-methylbutyryl- and isobutyryl-CoA (along with isovaleryl-CoA produced by the comparable metabolism of leucine) also initiate the biosynthesis of the long, branched-chain fatty acids that are the major cellular fatty acids present in *Streptomyces* and other bacterial genera including *Bacillus* [30]. Previous studies had shown an obligate growth requirement for 2-methylbutyric, isobutyric, and isovaleric acids for *bkd*-deficient mutants of *B. subtilis* [31]. Furthermore, the branched-chain α-keto acid dehydrogenase of *B. subtilis* had been shown to be responsible not only for the oxidative decarboxylation of branched-chain α-keto acids, but also for the oxidative decarboxylation of pyruvic acid [32]. Labeling studies [29] had shown that every carbon atom in avermectin can be derived from glucose; thus, concomitant inactivation of pyruvate dehydrogenase, a key enzyme of glucose metabolism, could also compromise novel avermectin co-expression.

IV. *bkd* MUTANTS ISOLATED AFTER CHEMICAL MUTAGENESIS

Some *bkd* mutants of *S. avermitilis* were isolated using a radioactive screening approach [33,34]. Unlike comparable mutants of *B. subtilis*, pyruvate dehydrogenase was not concomitantly inactivated, nor did the *S. avermitilis* mutants demonstrate an obligate requirement for 2-methylbutyric, isobutyric, and isovaleric acid supplementation. Analysis of these mutants confirmed the absence of long, branched-chain fatty acids, and indicated that the observed increase in the levels of the C16:1 unsaturated fatty acid ensures the required level of "membrane fluidity" normally provided by the C15 and C17 ante-iso-fatty acids resulting from initiation of fatty acid biosynthesis with 2-methylbutyric acid [35,36]. Most important, the *bkd*-deficient mutants of *S. avermitilis* produced no avermectins in the absence of supplementation with fatty acids. However, natural avermectins were produced upon supplementation with isobutyric acid or (*S*)-2-methylbutyric acid, and novel avermectins with other fatty acids [34,37]. Branched-chain α-keto acid dehydrogenase thus plays a pivotal and singular role in branched-chain precursor supply in *S. avermitilis*. In the absence of exogenous supplementation, *bkd*-deficient mutants can produce neither long, branched-chain fatty acids nor avermectins and, in addition, cannot grow on branched-chain amino acids as sole carbon source.

V. *bkd* MUTANTS CONSTRUCTED BY RECOMBINANT DNA TECHNOLOGY

The BCDH complex is a multienzyme complex composed of four functional components: a branched-chain α-keto acid dehydrogenase and decarboxylase (E1[αβ]), a dihydrolipoamide acyltransferase (E2), and a dihydrolipoamide dehydrogenase (E3) [38]. The BCDH complex catalyzes the oxidative decarboxylations of α-ketoisovalerate, α-keto-β-methylvalerate, and α-ketoisocaproate (the deamination products of the branched-chain amino acids valine, isoleucine, and leucine, respectively), releasing CO_2 and generating the corresponding acyl-CoA analogs and NADH (see Fig. 5) [23]. The genes encoding the components of the BCDH complex of *Pseudomonas putida* [39] and the pyruvate dehydrogenase (PDH) and BCDH dual-purpose complex of *B. subtilis* [40] and *Bacillus stearothermophilus* [41] have been cloned and found to be clustered in the following sequence: gene encoding E1α, gene encoding E1β, gene encoding E2, and gene encoding E3. Recently, the genes for the branched-chain fatty acid-specific BCDH from *B. subtilis* were cloned and sequenced [42]. This operon consisted of only the three genes encoding the E1α, E1β, and E2 components. Additionally, the sequences of several eukaryotic genes encoding either E1α or E1β BCDH subunits have been reported [43–49].

To understand further the importance of the BCDH-catalyzed reaction as a source of precursors for natural avermectin production and to manipulate the production of these antibiotics, we decided to clone and analyze the genes encoding the BCDH multienzyme complex from *S. avermitilis*. PCR primers were designed to encompass conserved regions of known *bkd* genes, and multiple sets of primers were used to amplify *S. avermitilis* genomic DNA. Two strong PCR products were detected and sequenced [50,51]. Computer analysis suggested that the deduced amino acid sequences of the amplified PCR products were similar to several published E1-α subunits. Probes were generated from the PCR products and used to screen a cosmid library of *S. avermitilis* genomic DNA. The hybridizing regions were cloned and restriction maps were generated. First, we identified a cluster of genes encoding the E1α, E1β, and E2 components of a BCDH complex of *S. avermitilis* [50]. These genes were designated *bkdA*, *bkdB*, and *bkdC* by analogy with the nomenclature introduced to describe similar genotypes of *P. putida* [39]. When *bkdA* and *bkdB*, encoding the E1α and E1β BCDH subunits, were coexpressed in *Escherichia coli*, a functional E1(αβ) BCDH activity was detected [50]. However, when the genomic copies of these genes were inactivated by gene disruption, no obvious phenotypic changes were observed, suggesting that these genes were silent or that their functions could be accomplished by other genes.

Soon after, we reported the cloning and characterization of a gene cluster encoding another BCDH in *S. avermitilis*, *bkdFGH* [51]. These genes are located approximately 12 kb downstream of the *bkdABC* gene cluster. An *S. avermitilis* *bkd* mutant was then constructed by deletion of a genomic region comprising the 5′ end of *bkdF*. The mutant exhibited a typical *bkd* phenotype, lacking BCDH activity, and it was unable to make natural avermectins in a medium lacking both *S*(+)-α-methylbutyrate and isobutyrate. However, supplementation with *S*(+)-α-methylbutyrate restores production of the corresponding ''a'' avermectins, while supplementation with cyclohexanecarboxylic acid results in the formation of a novel cyclohexyl avermectin without the co-expression of the natural analogs.

These results confirmed that branched-chain amino acid catabolism via the BCDH reaction provides the fatty acid precursors for natural avermectin biosynthesis in *S. avermitilis*. In contrast, *B. subtilis* appears to possess two mechanisms for branched-chain precursor supply. The dual substrate pyruvate/branched-chain α-keto acid dehydrogenase (*aceA*) and an α-keto acid dehydrogenase (*bfmB*), which also has some ability to metabolize pyruvate, appears to be primarily involved in supplying the branched-chain initiators of long, branched-chain fatty acid biosynthesis [32,42]. Two mutations are therefore required to generate the *bkd* phenotype in *B. subtilis* [31,42].

VI. MUTASYNTHESIS

The *bkd* mutants were subjected to an intensive effort to generate novel avermectins by mutasynthesis. More than 800 potential precursors were fed [37] and more than 60 novel avermectins were obtained covering a wide range of functional groups which are exemplified in Figure 6. These include carbon–carbon double bonds, triple bonds, and S or O atoms in the form of ethers. Cyclic fatty acids are also incorporated along with benzoic acid, and thienoic and furoic acids. Alkyl or alkoxy substituents are accepted, but not polar substituents such as hydroxy or amino groups. In addition to these α-branched carboxylic acids, straight-chain acids such as acetic and propionic acid and β-branched acids such as isovaleric acid are also incorporated. The amazing promiscuity of the loading domain of the avermectin PKS has been more recently exploited with its replacement into a wide range of polyketide synthases and triketide lactone model systems [52,53]. The most intensely studied system, to date, involves the substitution of the avermectin loading domain into the erythromycin PKS. Thus, a wide range of novel erythromycins were produced when these chimeric constructs of *S. erythraea* were supplemented exogenously with fatty acids [54].

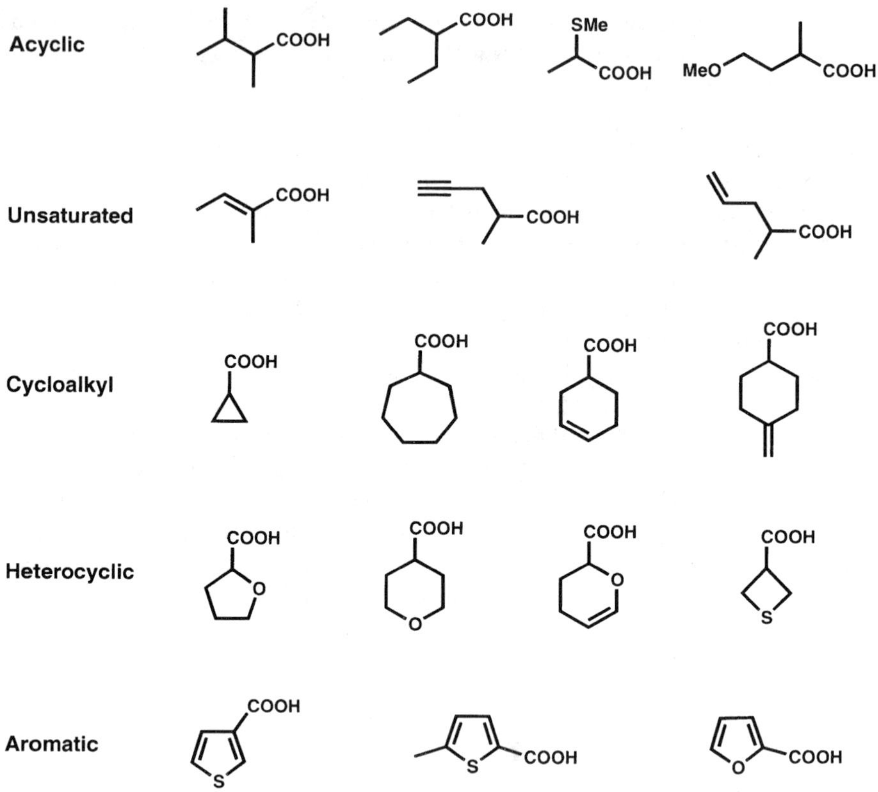

Figure 6 Examples of carboxylic acids incorporated into novel avermectins.

VII. THE NOVEL AVERMECTIN, DORAMECTIN

The potency and spectrum of the novel avermectins was initially profiled *in vitro* using a range of nematode and arthropod parasites. Promising candidates were then tested *in vivo* using endoparasite (nematode-infected rats) and ectoparasite (rabbit ear mite) models. The best candidates were subsequently evaluated for efficacy, persistence, and desirable pharmacokinetics in cattle infected with three species of nematodes (*Dictyocaulus viviparus*, *Cooperia oncophera*, *Ostertagia ostertagi*). The results of these studies suggested that the novel cyclohexyl B1 avermectin, doramectin (Fig. 7), obtained by the incorporation of cyclohexanecarboxylic acid, appeared to offer advantages over Ivermectin, the existing article of commerce [55].

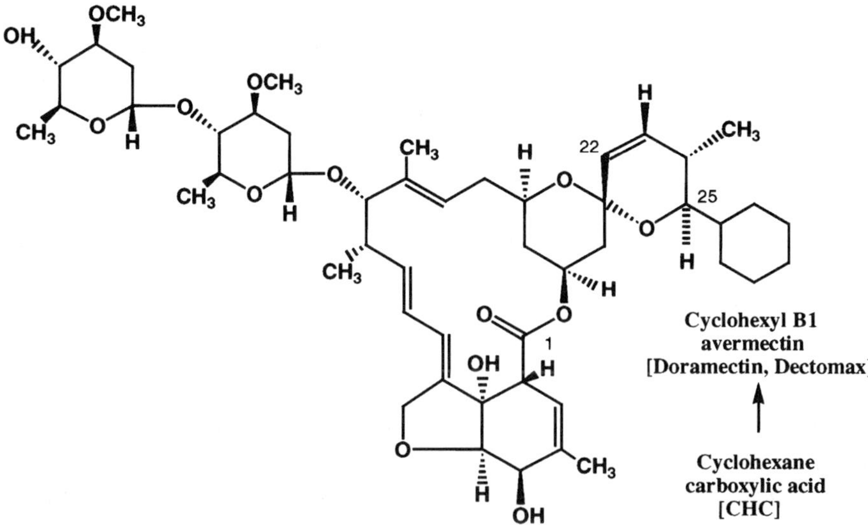

Figure 7 Structure of the doramectin molecule.

Formulation evaluation studies [56] showed that there was considerable scope for manipulating the pharmacokinetic profile of doramectin via the formulation vehicle. An oil-based vehicle of sesame oil:ethyl oleate (90:10) was found to provide high therapeutic and persistent efficacy against a wide range of endo- and ectoparasites [57–59], with excellent injection-site toleration, and so doramectin was selected for development.

VIII. TITER IMPROVEMENT: MANIPULATIONS OF THE PRECURSOR FLOW IN THE PRODUCTION OF DORAMECTIN

As discussed above, when a *S. avermitilis bkd* mutant (lacking BCDH activity) is grown in a fermentation medium supplemented with the unnatural branched-chain fatty acid, cyclohexanecarboxylic acid (CHC), the novel CHC-derived avermectin, doramectin, is produced [34].

Initial studies using radiolabeled CHC showed that the efficiency of incorporation of CHC into cyclohexyl-avermectins was low, accounting for less than 10% of the total precursor added. As CHC is one of the most expensive components of the doramectin fermentation, a more detailed analysis of the CHC flux was undertaken. Two pathways have previously been identified for the metabolism of CHC by microorganisms. The less common route (4-hydroxy pathway) involves the aromatization of CHC to protocatechuic acid via trans-4-OH-CHC, 4-keto-CHC, and *para*-hydroxybenzoic acid [60]. The primary metabolic route (2-hydroxy pathway) is an initial β-oxidation of CHC to 2-OH-CHC, and thence to 2-keto-CHC and pimelic acid [61]. The 4-hydroxy pathway utilizes P450 hydroxylases, while the 2-hydroxy pathway has been proposed to involve coenzyme-A-activated intermediates (*vide infra*). In addition to the catabolism of CHC, *B. subtilis* has been shown to incorporate exogenously supplied CHC to ω-cylohexane fatty acids [62]. Such fatty acid biosynthesis would be anticipated to initiate from cyclohexyl-CoA, the presumptive starter unit for doramectin biosynthesis. A detailed analysis of products derived from 14C-labeled CHC showed the presence of key intermediates from the 2- and 4-hydroxy pathways, ω-cylohexane fatty acids, and cyclohexyl-avermectins, indicating that all presumed routes of CHC flux were operating in *S. avermitilis*.

In order to enhance CHC incorporation into doramectin, we initiated efforts to decrease CHC flux in other pathways. Since a strategy to alter CHC incorporation into lipids would be more difficult to achieve, our initial focus was on decreasing the oxidative catabolism of CHC. Although most of the metabolites of the 4-hydroxy pathway were present at too low a level for effective quantitation, ^{14}C-4-OH-CHC was readily detected by scanning radioautography after thin-layer chromatography. This offered an useful method of screening mutagenized cultures for blocked mutants, and such mutants were easily isolated using this screen (Fig. 8).

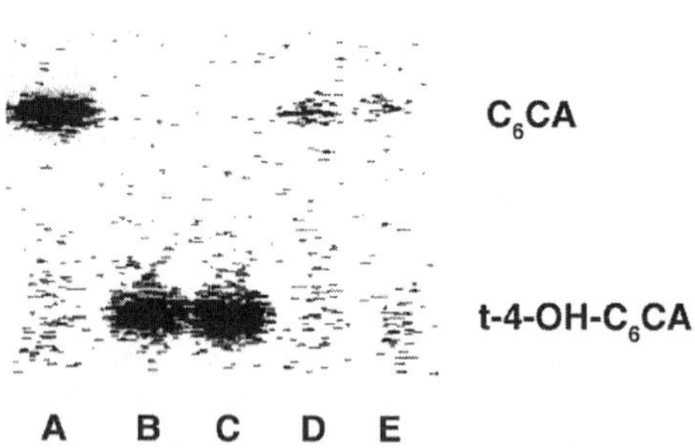

Figure 8 Isolation of mutants deficient in the 4-OH cyclohexanecarboxylic acid metabolic pathway. Thin-layer chromatography radioautogram of ethyl acetate extractable products from doramectin fermentation dosed with [^{14}C]-labeled cyclohexanecarboxylic acid. TLC was done on Kieselgel 60 F254 20 × 20 cm plates (E. Merck) and developed using benzene, ethyl acetate, and formic acid (25:25:2). Lanes: A, uninoculated medium; B and C, culture 109–123; D and E, mutant 324–52.

The 2-hydroxy pathway is analogous to the latter stages of branched-chain amino acid catabolism and the β-oxidation of fatty acids which utilize coenzyme A derivatives (see Fig. 5) [23]. Branched-chain acyl-CoA dehydrogenase acts at the third step of the branched-chain amino acid catabolic pathway, catalyzing the conversion of 2-methylbutyryl-CoA and isobutyryl-CoA into tiglyl-CoA and methacrylyl-CoA, respectively. We recently reported the cloning and nucleotide sequencing of *acdH*, a gene encoding an acyl-CoA dehydrogenase (AcdH), from *S. avermitilis* [63]. AcdH protein expressed in *E. coli* showed a broad range of substrate specificity, oxidizing isobutyryl-CoA, *n*-butyryl-CoA, *n*-valeryl-CoA, isovaleryl-CoA, and cyclohexylcarbonyl-CoA, to their respective 2,3-unsaturated derivatives. In addition, a *S. avermitilis acdH* mutant constructed by insertional inactivation of the *acdH* gene was unable to grow on solid minimal medium containing valine, isoleucine, or leucine as sole carbon sources and had a decreased ability to process isobutyryl-CoA via a methacrylyl-CoA intermediate to methylmalonyl-CoA. These results were consistent with the *acdH* gene encoding an acyl-CoA dehydrogenase with a broad substrate specificity that has a role in the catabolism of branched-chain amino acids in *S. avermitilis*. We speculated that the *S. avermitilis* AcdH is also responsible for the catabolic conversion of

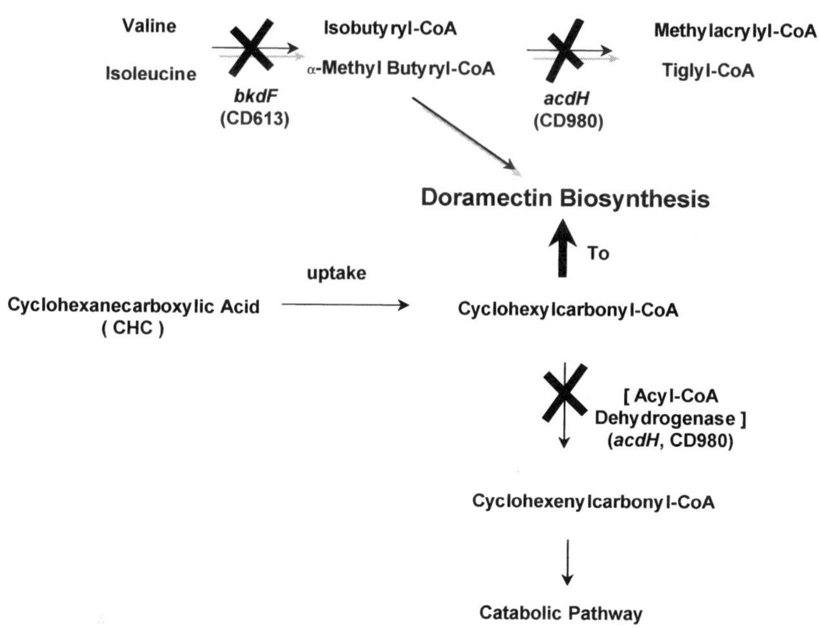

Figure 9 Doramectin titer improvement through disruption of the *acdH* gene in *S. avermitilis*.

the doramectin precursor cyclohexylcarbonyl-CoA into cyclohexenylcarbonyl CoA, in a similar way as the acyl CoA analogs of 2-methylbutyrate and isobutyrate are catabolized in the third step of the isoleucine and valine catabolic pathways, respectively [23]. The ultimate objective of our work was to attain a doramectin titer improvement by optimizing the amount of intracellular cyclohexylcarbonyl-CoA available for doramectin production (Fig. 9). To that end, the *acdH* gene was disrupted by gene replacement in a high-producer *S. avermitilis bkd* mutant. As a result of this manipulation a doramectin titer increase of approximately 20% was observed.

IX. CONCLUSIONS

S. avermitilis and the biosynthesis of avermectins constitute an interesting example where traditional techniques such as chemical mutagenesis and protoplast fusion combined with recombinant DNA technology have been successfully applied in mutant isolation and strain improvement. In addition, this system offered the first opportunity to apply mutasynthesis to the production of better analogs, an application that had never before been exploited commercially. An intense doramectin development effort was therefore initiated with the *bkd*-deficient mutants of *S. avermitilis*. The first step in this process involved the isolation of mutants deficient in 5-*O*-methyltransferase activity to maximize levels of the more bioactive class B avermectins [13]. Thereafter, a combination of strain im-

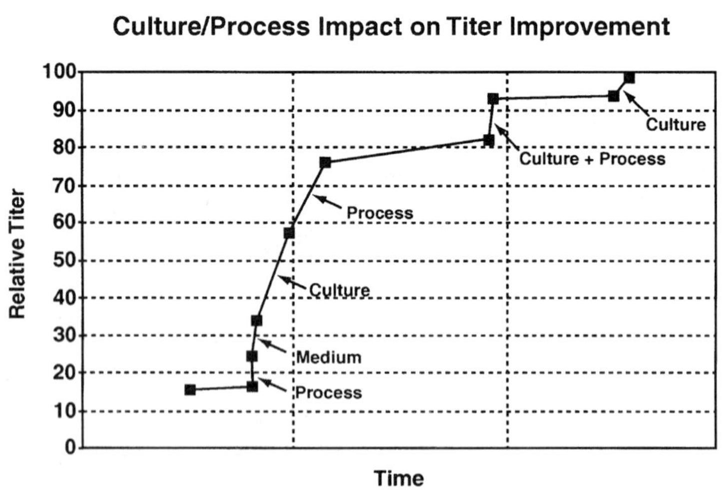

Figure 10 Impact of culture and process on doramectin titer improvement.

provement, employing random mutagenesis, directed enhancements through protoplast fusion approaches, genetic engineering of knockout strains, and process improvement efforts, involving optimization of medium and fermentation (batch feeding, control parameters, etc.), were pursued. In common with development efforts for most secondary metabolites, culture and process improvement contributed almost equally to potency enhancements (Fig. 10). These efforts in concert with parallel development of an optimal recovery process have permitted commercially viable expression levels to be achieved, and doramectin has been launched in all major global markets under the trade name of Dectomax. Thus, doramectin represents the first commercially successful application of mutasynthesis plus a complex combination of strain design through genetic engineering, and fermentation and recovery process technologies.

REFERENCES

1. KF Chater, DA Hopwood. *Streptomyces*. In: AL Sonenshein, JA Hoch, R Losick, eds. *Bacillus subtilis* and Other Gram-Positive Bacteria: Biochemistry, Physiology, and Molecular Genetics. Washington, DC: American Society for Microbiology, 1993, pp 83–99.
2. RW Burg, BM Miller, EE Baker, J Birnbaum, SA Currie, R Hartman, YL Kong, RL Monaghan, G Olson, I Putter, JB Tunac, H Wallick, EO Stapley, R Oiwa, S Omura. Avermectins, new family of potent anthelmintic agents: producing organism and fermentation. Antimicrob Agents Chemother 15:361–367, 1979.
3. DF Cully, DK Vassilatis, KK Liu, PS Paress, LHT Van der Ploeg, JM Schaeffer, JP Arena. Cloning of an avermectin-sensitive glutamate-gated chloride channel from *Caenorhabditis elegans*. Nature (Lond) 371:707–711, 1994.
4. JP Arena, KK Liu, PS Paress, EG Frazier, DF Cully, H Mrozik, JM Schaeffer. The mechanism of action of avermectins in *Caenorhabditis elegans*: correlation between activation of glutamate-sensitive chloride current, membrane binding, and biological activity. J Parasitol 81:286–294, 1995.
5. MJ Turner, JM Schaeffer. Mode of Action of Ivermectin. In: WC Campbell, ed. Ivermectin and Abamectin. New York: Springer-Verlag, 1989, pp 73–88.
6. TG Geary, RD Klein, L Vanover, JW Bowman, DP Thompson. The nervous system of helminths as targets for drugs. J Parasitol 78:215–230, 1992.
7. S Omura, Y Tanaka. Biochemistry, regulation, and genetics of macrolide production. In: S. Omura, ed. Macrolide Antibiotics. Chemistry, Biology, and Practice. New York: Academic Press, 1984, pp 199–229.
8. DE Cane, TC Liang, LK Kaplan, MK Nallin, MD Schulman, OD Hensens, AW Douglas, G Alters-Schonberg. Biosynthetic origin of the carbon skeleton and oxygen atoms of the avermectins. J Am Chem Soc 105:4110–4112, 1983.
9. TS Chen, BH Arison, VP Gullo, ES Inamine. Further studies on the biosynthesis of the avermectins. J Ind Microbiol 4:231–238, 1989.
10. DJ MacNeil, JL Occi, KM Gewain, T MacNeil, PH Gibbons, CL Ruby, SJ Danis.

Complex organization of the *Streptomyces avermitilis* genes encoding the avermectin polyketide synthase. Gene 115:119–125, 1992.
11. H Ikeda, T Nonomiya, M Usami, T Ohta, S Omura. Organization of the biosynthetic gene cluster for the polyketide anthelmintic macrolide avermectin in *Streptomyces avermitilis*. Proc Natl Acad Sci (USA) 96:9509–9514, 1999.
12. K Madduri, J Kennedy, G Rivola, A Inventi-solari, S Filippini, G Zanuso, AL Colombo, KM Gewain, JL Occi, DJ MacNeil, CR Hutchinson. Production of the antitumor drug epirubicin (4′-epidoxorubicin) and its precursor by a genetically engineered strain of *Streptomyces peucetius*. Nature Biotechnol 16:69–74, 1998.
13. MD Schulman, D Valentino, C Ruby. Avermectin B *O*-methyltransferase of *Streptomyces avermitilis*. Fed Proc 44:931, 1985.
14. KJ Stutzman-Engwall, SW Conlon, RW Fedechko, FS Kaczmarek. Engineering *Streptomyces avermitilis* biosynthetic genes. In: CR Hutchinson, J McAlpine, eds. Proceedings from the 1997 Fifth International Conference on the Biotechnology of Microbial Products (Developments in Industrial Microbiology Series, Vol 35). Fairfax, VA: Society for Industrial Microbiology, 1997, pp 7–13.
15. H Ikeda, LR Wang, T Ohta, J Inokoshi, S Omura. Cloning of the gene encoding avermectin B 5-O-methyltransferase in avermectin-producing *Streptomyces avermitilis*. Gene 206:175–180, 1998.
16. H Ikeda, Y Takada, CH Pang, K Matsuzaki, H Tanaka, S Omura. Direct production of 5-oxo derivatives of avermectins by a recombinant strain of *Streptomyces avermitilis*. J Antibiot 48:95–97, 1995.
17. S Donadio, MJ Staver, JB McAlpine, SJ Swanson, L Katz. Modular organization of the genes required for complex polyketide biosynthesis. Science 252:675–679, 1991.
18. J Cortes, SH Haydock, GA Roberts, DJ Bevitt, PF Leadlay. An unusually large multifunctional polypeptide in the erythromycin-producing polyketide synthase of *Saccharopolyspora erythraea*. Nature 348:176–178, 1990.
19. DJ Bevitt, J Staunton, PF Leadlay. Mutagenesis of the dehydratase active site in the erythromycin-producing polyketide synthase. Biochem Soc Trans 21:30S, 1992.
20. DJ Bevitt, J Cortes, SF Haydock, PF Leadlay. 6-Deoxyerythronolide-B synthase 2 from *Saccharopolyspora erythraea*. Cloning of the structural gene, sequence analysis and inferred domain structure of the multifunctional enzyme. Eur J Biochem 204: 39–49, 1992.
21. DJ MacNeil, JL Occi, KM Gewain, T MacNeil. Correlation of the avermectin polyketide synthase genes to the avermectin structure. Implications for designing novel avermectins. Ann NY Acad Sci 123–132, 1994.
22. S Smith. The animal fatty acid synthase: one gene, one polypeptide, seven enzymes. FASEB J 8:1248–1259, 1994.
23. AL Lehninger. Oxidative degradation of amino acids. In: AL Lehninger, ed. Biochemistry. New York: Worth, 1975, pp 559–586.
24. JD Bu'Lock, AC Goudie, KS Holdom, SP Gibson. New antiparasitic avermectin and milbemycin derivatives. Eur Pat Appl 214 731A, 1986.
25. Y Okami, K Hotta. Search and discovery of new antibiotics. In: M Goodfellow, ST Williams, M Mordarski, eds. Actinomycetes in Biotechnology. London: Academic Press, 1988, pp 33–67.

26. K Nagaoka, AL Demain. Mutational biosynthesis of a new antibiotic, streptomutin A, by an idiotroph of *Streptomyces griseus*. J Antibiot 28:627–635, 1975.
27. WT Shier, KL Reinhart Jr, D Gottlieb. Preparation of four new antibiotics from a mutant of *Streptomyces fradiae*. Proc Natl Acad Sci (USA) 63:198–204, 1969.
28. KL Reinhart Jr. Mutasynthesis of new antibiotics. Pure Appl Chem 49:1361–1364, 1977.
29. TS Chen, OD Hensens, MD Schulman. Biosynthesis. In: WC Campbell, ed. Ivermectin and Avermectin. New York: Springer-Verlag, pp 55–72, 1989.
30. T Kaneda. Iso- and anteiso-fatty acids in bacteria: biosynthesis, function and taxonomic significance. Microbiol Rev 55:288–302, 1991.
31. K Willecke, AB Pardee. Fatty acid-requiring mutant of *Bacillus subtilis* defective in branched chain α-keto acid dehydrogenase. J Biol Chem 246:5264–5272, 1971.
32. PN Lowe, JA Hodgson, RN Perham. Dual role of single multienzyme complex in the oxidative decarboxylation of pyruvate and branched-chain 2-oxo acids in *Bacillus subtilis*. Biochem J 215:133–140, 1983.
33. H Tabor, CW Tabor, EW Hafner. Convenient method for detecting $^{14}CO_2$ in multiple samples: application to rapid screening for mutants. J Bacteriol 128:485–486, 1976.
34. EW Hafner, BW Holley, KS Holdom, SE Lee, RG Wax, D Beck, HAI McArthur, WC Wernau. Branched-chain fatty acid requirement for avermectin production by a mutant of *Streptomyces avermitilis* lacking branched-chain 2-oxo acid dehydrogenase activity. J Antibiot 44:349–356, 1991.
35. T Kaneda. Fatty acids in the genus *Bacillus*: an example of branched-chain preference. Bacteriol Rev 41:391–418, 1977.
36. HAI McArthur. The novel avermectin, Doramectin—a successful application of mutasynthesis. In: CR Hutchinson, J McAlpine, eds. Developments in Industrial Microbiology—BMP'97. Fairfax, VA: Soc for Industrial Microbiology, 1998, pp 43–48.
37. CJ Dutton, SP Gibson, AC Goudie, KS Holdom, MS Pacey, JC Ruddock, JD Bu'Lock, MS Richards. Novel avermectins produced by mutational biosynthesis. J Antibiot 44:357–365, 1991.
38. RN Perham. Domains, motifs, and linkers in 2-oxo acid dehydrogenase multienzyme complexes: a paradigm in the design of a functional protein. Biochemistry 30:8501–8512, 1991.
39. PJ Sykes, G Burns, J Menard, K Hatter, JR Sokatch. Molecular cloning of genes encoding branched-chain keto acid dehydrogenase of *Pseudomonas putida*. J Bacteriol 169:1619–1625, 1987.
40. H Hemilä, A Palva, L Paulin, S Arvidson, I Palva. Secretory complex of *Bacillus subtilis*: sequence analysis and identity to pyruvate dehydrogenase. J Bacteriol 172: 5052–5063, 1990.
41. CF Hawkins, A Borges, RN Perham. Cloning and sequence analysis of the genes encoding the α and β subunits of the E1 component of the pyruvate dehydrogenase multienzyme complex of *Bacillus stearothermophilus*. Eur J Biochem 191:337–346, 1990.
42. G Wang, T Kuriki, KL Roy, T Kaneda. The primary structure of branched-chain α-oxo acid dehydrogenase from *Bacillus subtilis* and its similarity to other α-oxo acid dehydrogenases. Eur J Biochem 213:1091–1099, 1993.
43. B Zhang, MJ Kuntz, GW Goodwin, RA Harris, DW Crabb. Molecular cloning of

a cDNA for the E1-α subunit of rat liver branched chain α-ketoacid dehydrogenase. J Biol Chem 262:15220–15224, 1987.
44. B Zhang, DW Crabb, RA Harris. Nucleotide and deduced amino acid sequence of the E1-α subunit of human liver branched-chain α-ketoacid dehydrogenase. Gene 69:159–164, 1988.
45. CWC Hu, KS Lau, TA Griffin, JL Chuang, CW Fisher, RP Cox, DT Chuang. Isolation and sequencing of a cDNA encoding the decarboxylase (E1)-α precursor of bovine branched-chain α-keto acid dehydrogenase complex: expression of E1-α mRNA and subunit in maple-syrup-urine-disease and 3T3-L1 cells. J Biol Chem 263:9007–9014, 1988.
46. CW Fisher, JL Chuang, TA Griffin, KS Lau, RP Cox, DT Chuang. Molecular phenotypes in cultured maple syrup urine disease cells: complete E1-α cDNA sequence and mRNA and subunit contents of the human branched chain α-keto acid dehydrogenase complex. J Biol Chem 264:3448–3453, 1989.
47. I Matsuda, J Asaka, I Akaboshi, F Endo, H Mitsubuchi, Y Nobukuni. Maple syrup urine disease: complete primary structure of the E1-β subunit of human branched chain α-ketoacid dehydrogenase complex deduced from the nucleotide sequence and a gene analysis of patients with this disease. J Clin Invest 86:242–247, 1990.
48. RM Wynn, JL Chuang, JR Davie, CW Fisher, MA Hale, RP Cox, DT Chuang. Cloning and expression in *E. coli* of mature E1β subunit of bovine mitochondrial branched-chain α-keto acid dehydrogenase complex. Mapping of the E1β-binding region on E2. J Biol Chem 267:1881–1887, 1991.
49. MC McKean, KA Winkeler, DJ Danner. Nucleotide sequence of the 5′ end including the initiation codon of cDNA for the E1α subunit of the human branched chain α-ketoacid dehydrogenase complex. Biochim Biophys Acta 1171:109–112, 1992.
50. DD Skinner, MR Morgenstern, RW Fedechko, CD Denoya. Cloning and sequencing of a cluster of genes encoding branched-chain α-keto acid dehydrogenase from *Streptomyces avermitilis* and the production of a functional E1[αβ] component in *Escherichia coli*. J Bacteriol 177:183–190, 1995.
51. CD Denoya, RW Fedechko, EW Hafner, HAI McArthur, MR Morgenstern, DD Skinner, KJ Stutzman-Engwall, RG Wax, WC Wernau. A second branched-chain α-keto acid dehydrogenase gene cluster (*bkdFGH*) from *Streptomyces avermitilis*: its relationship to avermectin biosynthesis and the construction of a *bkdF* mutant suitable for the production of novel antiparasitic avermectins. J Bacteriol 177:3504–3511, 1995.
52. AFA Marsden, B Wilkinson, J Cortes, NJ Dunster, J Staunton, PF Leadlay. Engineering broader specificity into an antibiotic-producing polyketide synthase. Science 279:199–202, 1998.
53. Biotica Technology Ltd., European Patent 910633, 1999.
54. MS Pacey, JP Dirlam, RW Geldart, PF Leadlay, HAI McArthur, EL McCormick, RA Monday, TN O'Connell, J Staunton, TJ Winchester. Novel erythromycins from a recombinant *Saccharopolyspora erythraea* strain NRRL 2338 pIG1. I. Fermentation, isolation and biological activity. J Antibiot 51:1029–1034, 1998.
55. AC Goudie, NA Evans, KAF Gration, BF Bishop, SP Gibson, KS Holdom, B Kaye, SR Wicks, D Lewis, AJ Weatherley, CI Bruce, A Herbert, DJ Seymour. Doramectin—a potent novel endectocide. Vet Parasitol 49:5–15, 1993.

56. SR Wicks, B Kaye, AJ Weatherley, D Lewis, E Davison, SP Gibson, DG Smith. Effect of formulation on the pharmacokinetics and efficacy of doramectin. Vet Parasitol 49:17–26, 1993.
57. C Eddi, I Bianchin, MR Honer, RA Muniz, J Caracostantogolo, YA do Nascimento. Efficacy of doramectin against field nematode infections of cattle in Latin America. Vet Parasitol 49:39–44, 1993.
58. NB Logan, AJ Weatherley, FE Phillips, CP Wilkins, DJ Shaanks. Spectrum of activity of doramectin against cattle mites and lice. Vet Parasitol 49:67–73, 1993.
59. GE Moya-Borja, CMB Oliveira, RA Muniz, LCB Goncalves. Prophylactic and persistent efficacy of doramectin against *Cochliomyia hominivorax* in cattle. Vet Parasitol 49:95–105, 1993.
60. ER Blakley The microbial degradation of cyclohexane carboxylic acid: a pathway involving aromatization to form *p*-hydroxybenzoic acid. Can J Microbiol 20:1297–1306, 1974.
61. EM Rho, WC Evans. The aerobic metabolism of cyclohexane carboxylic acid by *Acinetobacter antitratum*. Biochem J 148:11–15, 1975.
62. R Dreher, K Poralla, WA Konig. Synthesis of w-alicyclic fatty acids from cyclic precursors by *Bacillus subtilis*. J Bacteriol 127:1136–1140, 1976.
63. Y Zhang, CD Denoya, DD Skinner, RW Fedechko, HAI McArthur, MR Morgenstern, RA Davies, S Lobo, KA Reynolds, and CR Hutchinson. Genes encoding acyl-CoA dehydrogenase (AcdH) homologues from *Streptomyces coelicolor* and *Streptomyces avermitilis* provide insights into the metabolism of small branched-chain fatty acids and macrolide antibiotic production. Microbiology (UK) 145:2323–2334, 1999.

6

Biocatalytic Syntheses of Chiral Intermediates for Antihypertensive Drugs

Ramesh N. Patel
Bristol-Myers Squibb Company, New Brunswick, New Jersey

I. INTRODUCTION

Recently, much attention has been focused on the interaction of small molecules with biological macromolecules. The search for selective enzyme inhibitors and receptor agonists or antagonists is one of the keys for target-oriented research in the pharmaceutical industry. Increased understanding of the mechanism of drug interaction on a molecular level has led to wide awareness of the importance of chirality as the key to the efficacy of many drug products. It is now known that in many cases only one enantiomer of a drug substance is required for efficacy and the other enantiomer is either inactive or exhibits considerably reduced activity [1–4]. Pharmaceutical companies are aware that, where appropriate, new drugs for development should be homochiral to avoid the possibility of unnecessary side effects due to an undesirable enantiomer. In many cases where the switch from racemate drug substance to enantiomerically pure compound is feasible, there is the opportunity to extend the use of an industrial process. The physical characteristics of an enantiomer versus racemic compound may confer processing or formulation advantages.

Chiral drug intermediates can be prepared by different routes. One approach is to obtain them from naturally derived chiral synthons, produced mainly by fermentation processes. The chiral pool refers primarily to inexpensive, readily available, optically active natural products. A second approach is to carry out the resolution of racemic compounds. This approach can be achieved by preferential crystallization of enantiomers or diastereomers and by kinetic resolution of racemic compounds by chemical or biocatalytic methods. Finally, chiral synthons can also be prepared by asymmetric synthesis by either chemical or biocatalytic processes using microbial cells or enzymes derived therefrom. The advantages of microbial or enzyme-catalyzed reactions over chemical reactions are that they are stereoselective and can be carried out at ambient temperature and atmospheric pressure. The biocatalytic approach minimizes problems of isomerization, racemization, epimerization, and rearrangement that may occur during chemical processes. Biocatalytic processes are generally carried out in aqueous solution. These types of processes will avoid the use of environmentally harmful chemicals currently implemented in chemical processes and subsequent solvent waste disposal. Furthermore, microbial cells or enzymes derived therefrom can be immobilized and reused for many cycles. Recently, a number of review articles [1–12] have been published on the use of enzymes in organic synthesis. This chapter provides some specific examples of preparation of chiral drug intermediates required for our antihypertensive agents.

II. VASOPEPTIDASE INHIBITOR

A. Enzymatic Synthesis of L-6-Hydroxynorleucine (1)

L-6-Hydroxynorleucine (**1**) (Fig. 1) is a chiral intermediate that is useful for the synthesis of a vasopeptidase inhibitor now in clinical trial and for the synthesis of C-7–substituted azepinones as potential intermediates for other antihypertensive

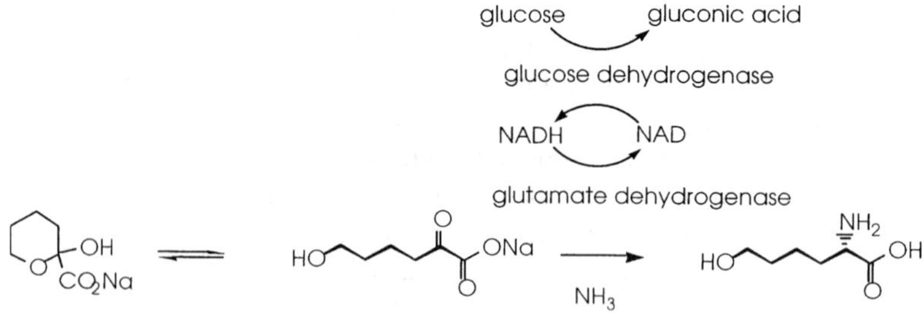

2-hydroxytetrahydropyran 2-carboxylic acid, sodium salt

2-keto-6-hydroxyhexanoic acid, sodium salt

2

L-6-hydroxynorleucine

1

Figure 1 Preparation of chiral synthon for vasopeptidase inhibitor: enzymatic synthesis of L-6-hydroxynorleucine (**1**) using glutamate dehydrogenase.

metalloprotease inhibitors [13,14]. It has also been used for the synthesis of siderophores, indospicines, and peptide hormone analogs [15–17]. Previous synthetically useful methods for obtaining this intermediate have involved synthesis of the racemic compound followed by enzymatic resolution. D-Amino acid oxidase has been used to convert the D-amino acid to the ketoacid, leaving the L-enantiomer which was isolated by ion-exchange chromatography [18]. In a second approach, racemic N-acetyl hydroxynorleucine has been treated with L-amino acid acylase to give the L-enantiomer [13,19]. Both of these resolution methods give a maximum 50% yield and require separation of the desired product. Reductive amination of ketoacids using amino acid dehydrogenases has become a useful method for synthesis of natural and non-natural amino acids [20,21].

We have developed the synthesis and conversion of 2-keto-6-hydroxyhexanoic acid (**2**) to L-6-hydroxynorleucine (**1**) (Fig. 1) by a reductive amination

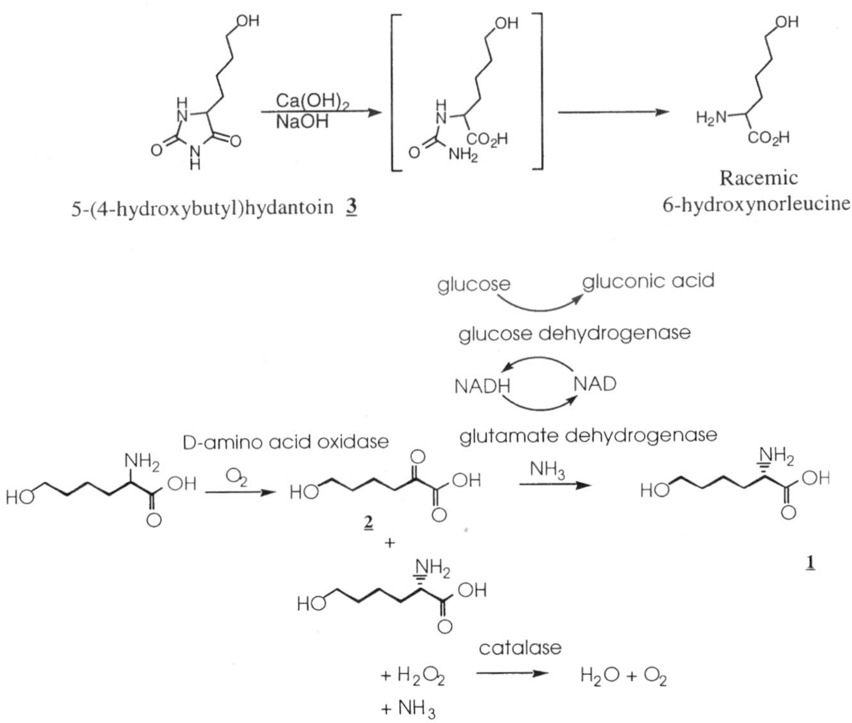

Figure 2 Chemical conversion of 5-(4-hydroxybutyl)hydantoin (**3**) to racemic 6-hydroxynorleucine. Enzymatic conversion of racemic 6-hydroxynorleucine to L-6-hydroxynorleucine (**1**) by D-amino acid oxidase and glutamate dehydrogenase.

process using beef liver glutamate dehydrogenase [22]. 2-Keto-6-hydroxyhexanoic acid (**2**) was converted completely to L-6-hydroxynorleucine (**1**) by beef liver glutamate dehydrogenase. A nicotinamide adenine dinucleotide (NAD^+)-dependent formate dehydrogenase from *Candida boidinii* or glucose dehydrogenase from *Bacillus megaterium* was used for regeneration of reduced nicotinamide adenine dinucleotide (NADH) required for this reaction [22]. The beef liver glutamate dehydrogenase was used for preparative reactions at 100 g/liter substrate concentration. As depicted in Figure 1, 2-keto-6-hydroxyhexanoic acid (**2**), sodium salt, in equilibrium with 2-hydroxytetrahydropyran-2-carboxylic acid, sodium salt, is converted to L-6-hydroxynorleucine. The reaction requires ammonia and NADH. NAD^+ produced during the reaction was recycled to NADH by glucose dehydrogenase from *B. megaterium*. Reaction was completed in about 3 hr with reaction yields of 89–92%, and enantiomeric excess (e.e.) of >98% for L-6-hydroxynorleucine.

Chemical synthesis and isolation of 2-keto-6-hydroxyhexanoic acid required several steps. In a second, more convenient process (Fig. 2), the ketoacid was prepared by treatment of racemic 6-hydroxynorleucine [produced by hydrolysis of 5-(4-hydroxybutyl)hydantoin (**3**)] with D-amino acid oxidase and catalase. After the e.e. of the remaining L-6-hydroxynorleucine had risen to >99%, the reductive amination procedure was used to convert the mixture containing 2-keto-6-hydroxyhexanoic acid and L-6-hydroxynorleucine entirely to L-6-hydroxynorleucine with yields of 91–97% and e.e. of >98%. Sigma porcine kidney D-amino acid oxidase and beef liver catalase or *Trigonopsis variabilis* whole cells (source of oxidase and catalase) were used successfully for this transformation [22].

B. Enzymatic Synthesis of Allysine Ethylene Acetal (4)

(*S*)-2-Amino-5-(1,3-dioxolan-2-yl)-pentanoic acid [allysine ethylene acetal (**4**)] is one of three building blocks used for an alternative synthesis of omapatrilat, a vasopeptidase inhibitor [13,14]. It has previously been prepared in an eight-step synthesis from 3,4-dihydro-2H-pyran [23].

The reductive amination of ketoacid acetal (**5**) to acetal amino acid (**4**) was demonstrated using phenylalanine dehydrogenase from *Thermoactinomyces intermedius* (Fig. 3). The reaction requires ammonia and NADH. NAD^+ produced during the reaction was recycled to NADH by the oxidation of formate to CO_2 using formate dehydrogenase from *C. boidinii*. An initial process was developed using heat-dried cells of *T. intermedius* ATCC 33205 as a source of phenylalanine dehydrogenase, and heat-dried cells of methanol-grown *C. boidinii* as a source of formate dehydrogenase [24].

An improved process was also developed using phenylalanine dehydrogenase from *T. intermedius* expressed in *Escherichia coli BL21 (DE3)* (pPDH155K)

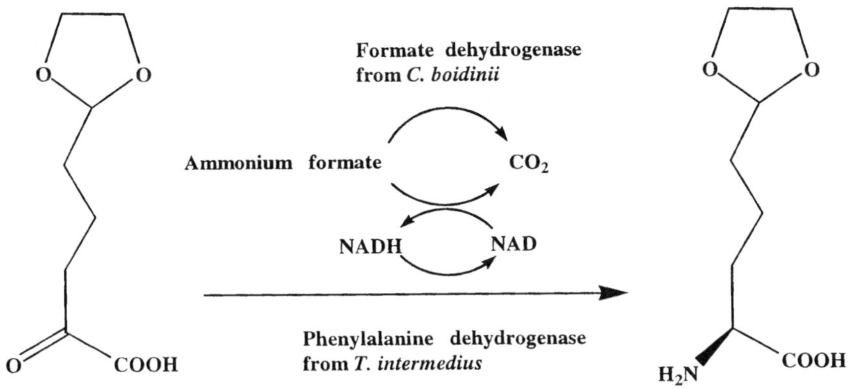

5

4

Figure 3 Preparation of chiral synthon for vasopeptidase inhibitor: enzymatic reductive amination of ketoacid acetal (**5**) to amino acid acetal (**4**).

[SC16144] in combination with *C. boidinii* as a source of formate dehydrogenase. A third-generation process using methanol-grown *Pichia pastoris* as a source of endogenous formate dehydrogenase and *E. coli* SC 16144 expressing *T. intermedius* phenylalanine dehydrogenase was also developed [24].

Glutamate, alanine, leucine, and phenylalanine dehydrogenases (listed in order of increasing effectiveness) converted (**5**) to the desired amino acid (**4**) [24]. Using an extract of *T. intermedius* ATCC 33205 as a source of phenylalanine dehydrogenase and formate dehydrogenase from *C. boidinii* for NADH regeneration, the reaction yield of 80% was obtained, and the process was developed using this enzyme combination. Heat-dried cells of *T. intermedius* and *C. boidinii* SC13822 grown on methanol were used for the reaction.

Phenylalanine dehydrogenase activities in cells recovered from fermentations and fermentor productivities are shown in Table 1. *T. intermedius* gave useful activity on a small scale (15 liters), but lysed soon after the end of the growth period, making recovery of activity difficult or impossible on a large scale (4000 liters). The problem was solved by cloning and expressing the *T. intermedius* phenylalanine dehydrogenase in *E. coli*, inducible by isopropyl thiogalactoside. Fermentation of *T. intermedius* yielded 184 units of phenylalanine dehydrogenase activity per liter of whole broth in 6 hr. At harvest, because the activity was unstable, the fermentor needed to be cooled rapidly. In contrast, the recombinant *E. coli* produced over 19,000 units per liter of whole broth in about 14 hr, and the activity was stable at harvest. *C. boidinii* grown on methanol was a useful source of formate dehydrogenase as described previously [25]. In order

Table 1 Activities and Productivities of Phenylalanine Dehydrogenase and Formate Dehydrogenase for Various Strains Grown in Fermentor

Enzyme	Strain	Specific activity (U/g wet cells)	Volumetric activity (U/liter of broth)	Productivity (U/liter/week)
Phenylalanine dehydrogenase	*T. intermedius*	510	185	900
	E. coli	10,000	24,000	94,000
	P. pastoris	ND	14,500	25,000
Formate dehydrogenase	*C. boidinii*	9	120	350
	P. pastoris	26	1,950	3,200

to recover the cells on a large scale, 0.5% methanol was added to stabilize the cells.

P. pastoris grown on methanol was also a useful source of formate dehydrogenase [26]. Expression of *T. intermedius* phenylalanine dehydrogenase in *P. pastoris*, inducible by methanol, allowed both enzymes to be obtained from a single fermentation. The expression of the two activities during a *P. pastoris* fermentation is shown in Fig. 4. Formate dehydrogenase activity per gram of wet cells was 2.7-fold greater than for *C. boidinii*, and fermentor productivity was increased by 8.7-fold compared to *C. boidinii*. Fermentor productivity for phenylalanine dehydrogenase in *P. pastoris* was about 28% of the recombinant *E. coli* productivity.

Formate dehydrogenase has been reported to have a pH optimum of 7.5–8.5 [25]. The pH optimum for the reductive amination of (**5**) by an extract of *T. intermedius* was found to be about 8.7. Reductive amination reactions were carried out at pH 8.0. A summary of laboratory-scale batches is shown in Table 2. The time course for a representative batch showing conversion of ketoacid (**5**) to amino acid (**4**) is presented in Figure 5 using *E. coli/C. boidinii* heat-dried cells.

The procedure using heat-dried cells of *E. coli* containing cloned phenylalanine dehydrogenase and heat-dried *C. boidinii* was scaled up (Table 3). A total of 197 kg of (**4**) was produced in three 1600-liter batches using a 5% concentration of substrate (**5**) with an average yield of 91 M% and e.e. of >98%.

Third-generation procedure, using dried recombinant *P. pastoris* expressing *T. intermedius* phenylalanine dehydrogenase inducible with methanol, and endogenous formate dehydrogenase induced when *P. pastoris* was grown in medium containing methanol, allowed both enzymes to be produced during a single fermentation. The two enzymes were conveniently produced in about the right ratio that was used for the reaction. The *Pichia* reaction procedure had the follow-

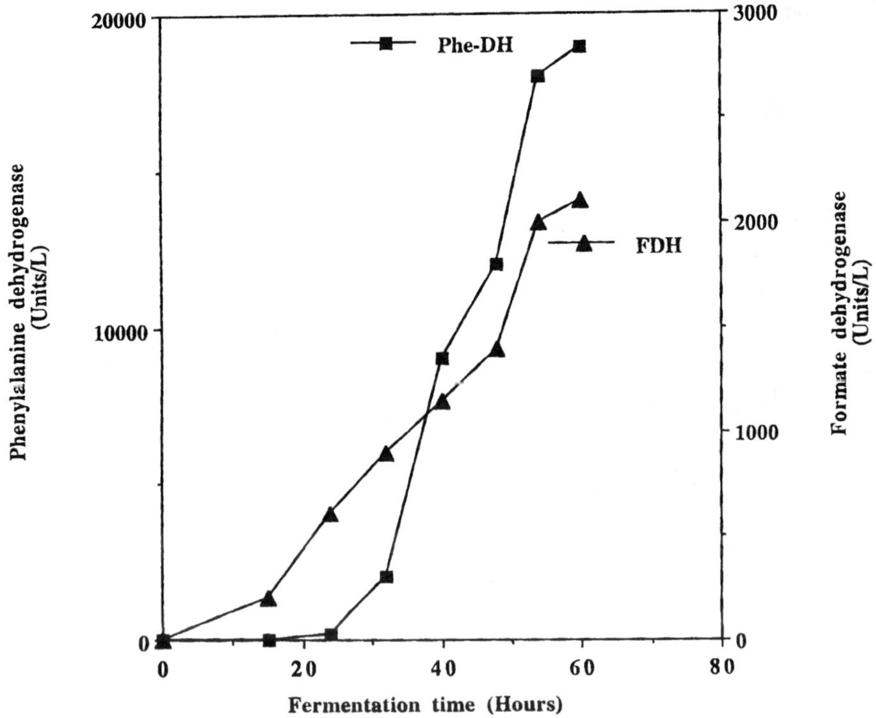

Figure 4 Production of phenylalanine dehydrogenase and formate dehydrogenase in *P. pastoris*.

Table 2 Laboratory-Scale (1-liter) Batches for Reductive Amination Reactions

Phenylalanine dehydrogenase source	Formate dehydrogenase source	Reaction yield of (**4**) (%)	Enantiomeric excess of (**4**) (%)
T. intermedius	*C. boidinii*	85	>99
E. coli	*C. boidinii*	90	>99
P. pastoris	*P. pastoris*	94	>99

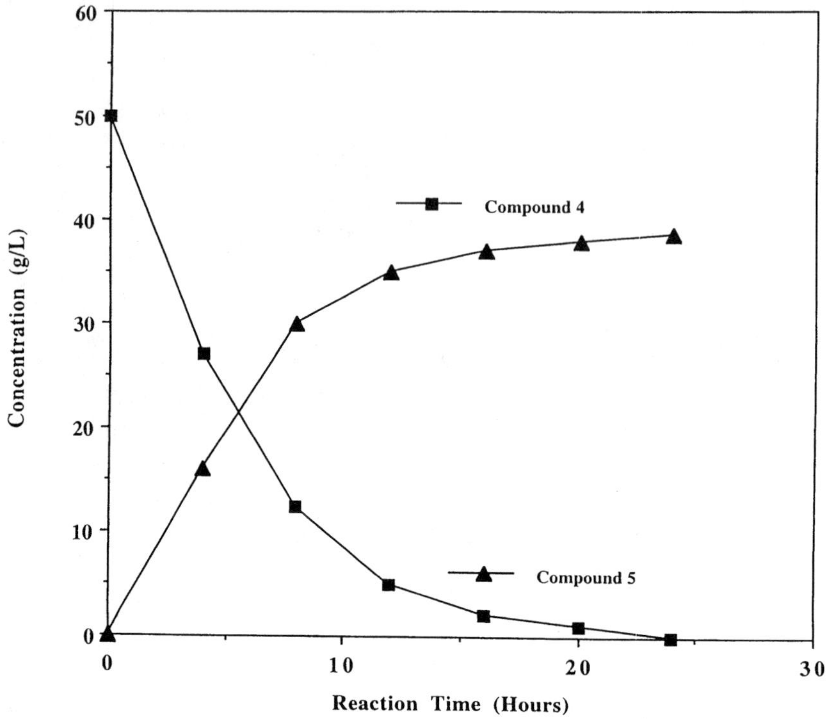

Figure 5 Kinetics of enzymatic conversion of ketoacid acetal (**5**) to amino acid acetal (**4**) by phenylalanine dehydrogenase from recombinant *E. coli* and formate dehydrogenase from *C. boidinii*.

Table 3 Large-Scale Batches for Reductive Amination Reactions

Phenylalanine dehydrogenase source	Formate dehydrogenase source	Keto acid (**5**) input (kg)	Amino acid (**4**) output (kg)	Reaction yield of (**4**) (M%)	Enantiomeric excess amino acid (**4**) (%)
E. coli	*C. boidinii*	80.17	62.4	92	>99
E. coli	*C. boidinii*	79.96	66.75	96	>99
E. coli	*C. boidinii*	89.6	67.61	86	>99
P. pastoris	*P. pastoris*	18.05	15.51	97.5	>99

ing modifications of the *E. coli/C. boidinii* procedure: concentration of substrate was increased to 100 g/liter, one-fourth the amount of NAD^+ was used, and dithiothreitol was omitted. The procedure with *P. pastoris* was also scaled up to produce 15.5 kg of (**4**) with 97 M% yield and e.e. >98% (Table 3) in a 180 liter batch using 10% ketoacid concentration.

For reusability, formate dehydrogenase could be immobilized on Eupergit C and phenylalanine dehydrogenase on Eupergit C250L. The immobilized enzymes were tested for reusability in a jacketed reactor maintained at 40°C, and were used five times for the conversion of (**5**) to (**4**) without much loss of any activity and productivity. At the end of each reaction, the solution was drained from the reactor through a 80/400-mesh stainless steel sieve, which retained the immobilized enzymes, then the reactor was recharged with fresh substrate solution. After the fifth reuse, the reaction rate was decreased; however, the original reaction rate was restored in the seventh-reuse studies by addition of formate dehydrogenase.

III. β-3-RECEPTOR AGONIST

β-Adrenoceptors have been classified as β1 and β2 [27]. Increased heart rate is the primary consequence of β1-receptor stimulation, while bronchodilation and

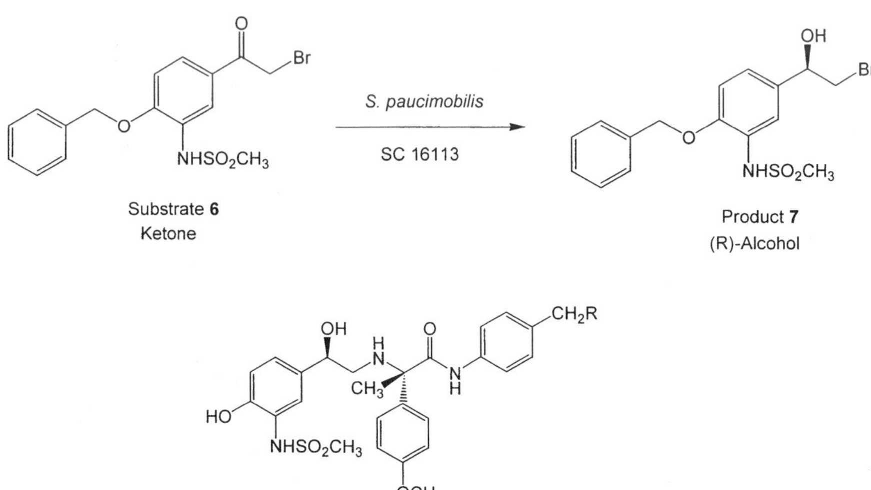

Substrate **6**
Ketone

Product **7**
(R)-Alcohol

BMS-210620

Figure 6 Preparation of chiral synthon for β-3-receptor agonist: microbial reduction of ketone (**6**) to chiral alcohol (**7**).

smooth muscle relaxation are mediated from β-2 receptor stimulation. Rat adipocyte lipolysis was initially thought to be a β-1–mediated process [27]. However, recent results indicate that this type of lipolysis is neither β1 nor β2 receptor-mediated, but is due to "atypical" receptors, later called β3-adrenergic receptors [28]. β3-Adrenergic receptors are found on the cell surface of both white and brown adipocytes and are responsible for lipolysis, thermogenesis, and relaxation of intestinal smooth muscle [29,30]. Consequently, several research groups are engaged in developing selective β-3 agonists for the treatment of gastrointestinal disorders, type II diabetes, and obesity [31–34]. Efficient biocatalytic syntheses of chiral intermediates required for the total chemical syntheses of β-3 receptor agonists have been reported by us [35].

The biocatalytic approaches include (1) the microbial reduction of 4-benzyloxy-3-methanesulfonylamino-2′-bromoacetophenone (**6**) to the corresponding

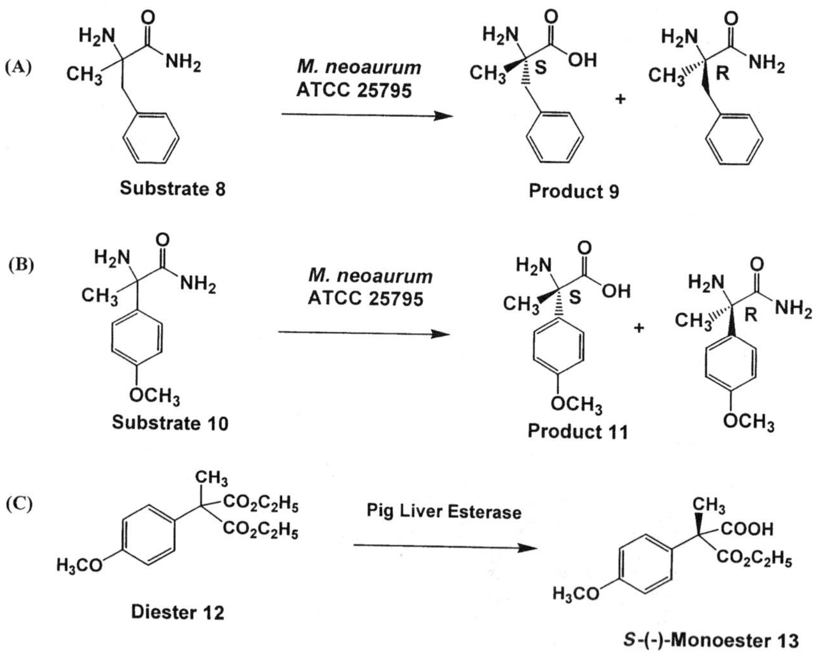

Figure 7 Preparation of chiral synthon for β-3-receptor agonist: (A) enzymatic resolution of racemic amino acid amide (**8**) by amidase from *M. neoaurum* ATCC 25795; (B) enzymatic resolution of racemic amino acid amide (**10**) by amidase from *M. neoaurum* ATCC 25795; (C) enzymatic asymmetric hydrolysis of diester (**12**) to the corresponding (*S*)-monoester (**13**) by pig liver esterase.

(*R*)-alcohol (**7**) by *Sphingomonas paucimobilis* SC 16113 (Fig. 6); (2) the enzymatic resolution of racemic (α-methyl)phenylalanine amide (**8**) and α-(4-methoxyphenyl)alanine amide (**10**) by amidase from *Mycobacterium neoaurum* ATCC 25795 to prepare the corresponding (*S*)-amino acids (**9**) and (**11**), and (3) the asymmetric hydrolysis of methyl-(4-methoxyphenyl)-propanedioic acid, diethyl ester (**12**), to the corresponding (*S*)-monoester (**13**) by pig liver esterase (Fig. 7).

A. Microbial Reduction of 4-Benzyloxy-3-Methanesulfonylamino-2′-Bromoacetophenone

The microbial reduction of 4-benzyloxy-3-methanesulfonylamino-2′-bromoacetophenone (**6**) to the corresponding (*R*)-alcohol (**7**) was demonstrated by *S. paucimobilis* SC 16113 (Fig. 6). Among cultures evaluated, *Hansenula anamola* SC 13833, *H. anamola* SC 16142, *Rhodococcus rhodochrous* ATCC 14347, and *S. paucimobilis* SC 16113, gave desired alcohol (**7**) in >96% e.e. and >15% reaction yield. *S. paucimobilis* SC 16113, in the initial screening, catalyzed the efficient conversion of ketone (**6**) to the desired chiral alcohol (**7**) in 58% reaction yield and >99.5% e.e.

The fermentation of *S. paucimobilis* SC 16113 culture was carried out in a 750-liter fermentor. From each fermentation batch, about 60 kg of wet cell paste was collected. Cells harvested from the fermentor were used to conduct the biotransformation in 1-, 10-, and 210-liter preparative batches under aerobic or anaerobic conditions. The cells were suspended in 80 mM potassium phosphate buffer (pH 6.0) to 20% (w/v, wet cells) concentration. Compound (**6**) (1–2 g/liter) and glucose (25 g/liter) were added to the fermentor and the reduction reaction was carried out at 37°C. In some batches, at the end of the fermentation cycle, the cells were concentrated sevenfold by ceramic crossflow microfiltration using a 0.2-μm filter, diafiltered using 10 mM potassium phosphate buffer (pH 7.0), and used directly in the bioreduction process. In all batches of biotransformation, the reaction yield of >85% and the e.e. of >98% were obtained (Table 4). The isolation of compound (**7**) from the 210-liter preparative batch was carried out to obtain 100 g of product (**7**). The isolated (**7**) gave 83% chemical purity and an e.e. of 99.5%.

In an alternative process, frozen cells of *S. paucimobilis* SC 16113 were used with resin adsorbed (XAD-16 resin) substrate at 5- and 10-g/liter substrate concentrations. In this process, an average reaction yield of 85% and an e.e. of >99% were obtained for product (**7**). At the end of the biotransformation, the reaction mixture was filtered on a 100-mesh (150-m) stainless steel screen, and the resin retained by the screen was washed with 2 liters of water. The product was then desorbed from the resin and crystallized in an overall 75 M% yield with 91% homogeneity and 99.8% e.e.

Table 4 Semipreparative Batches for Microbial Reduction of Ketone (**6**)

Condition	Batch size (liters)	Reaction time (hr)	Substrate input (g/liter)	Reaction yield of (**7**)(%)	Enantiomeric excess of (**7**) (%)
Aerobic	1	4	1	85.3	98.5
Aerobic[a]	2	4	1	87	98.6
Aerobic	1	3	1	87.9	98.5
Anaerobic[a]	1	3	1	89.7	98.7
Anaerobic	1	6	2	87	99.4
Anaerobic	2	3	1	90	98.5
Anaerobic	10	3	1	95	99.4
Anaerobic	200	2	1	84	99.5

Cells were suspended in 80 mM potassium phosphate buffer (pH 6.0) at 20% (w/v, wet cells) concentration.

Cell suspensions were supplemented with substrate (**6**) (in dimethylformamide) and glucose (25 g/liter).

Reactions were carried out at 37°C, 300 rpm, under anaerobic condition.

[a] Microfiltered and diafiltered cells were used.

B. Enzymatic Resolution of Racemic (α-Methyl)phenylalanine Amides

The enzymatic resolution of racemic (α-methyl)phenylalanine amide (**8**) and α-(4-methoxyphenyl)alanine amide (**10**) to the corresponding (*S*)-amino acids (**9**) and (**11**), respectively, by an amidase from *M. neoaurum* ATCC 25795 has been developed (Figs. 7A and 7B). The chiral amino acids are intermediates for the syntheses of β-3-receptor agonists [31–35].

The cells (10%, w/v, wet cells) of *M. neoaurum* ATCC 25795 were evaluated for biotransformation of compound (**8**) to compound (**9**) (Fig. 7A). The reaction was completed in 75 min with a reaction yield of 48 M% (theoretical max. 50%) and an e.e. of 95% for the desired product (**9**). Freeze-dried cells of *M. neoaurum* ATCC 25795 were suspended in 100 mM potassium phosphate buffer (pH 7.0) at 1% concentration and cell suspensions were used for the biotransformation of compound (**8**). The reaction was completed in 60 min with a reaction yield of 49.5 M% (theoretical max. 50%) and an e.e. of 99% for the desired product (**9**). Biotransformation of compound (**8**) was also carried out using a purified amidase. A reaction yield of 49 M% and an e.e. of 99.8% were obtained for desired product (**9**) after 60 min of reaction time.

Freeze-dried cells of *M. neoaurum* ATCC 25795 and partially purified amidase were used for the biotransformation of compound (**10**) (Fig. 7B). A reaction yield of 49 M% and an e.e. of 78% were obtained for the desired product (**11**) using freeze-dried cells. The reaction was completed in 50 hr. Using partially purified amidase, a reaction yield of 49 M% and e.e. of 94% were obtained for desired product (**11**) after a 70-hr reaction time.

C. Asymmetric Hydrolysis of Racemic Methyl-(4-Methoxyphenyl)-Propanedioic Acid, Diethyl Ester (12)

The enzymatic asymmetric hydrolysis of methyl-(4-methoxyphenyl)-propanedioic acid, diethyl ester (**12**) to the corresponding (*S*)-monoester (**13**) by pig liver esterase (PLE) has been demonstrated (Fig. 7C). Chiral (*S*)-monoester is a key intermediate for the syntheses of β-3-receptor agonists.

Various organic solvents were tested for the PLE-catalyzed asymmetric hydrolysis of diester (**12**) in a biphasic system. The results (Table 5) indicate that the reaction yields and e.e. of monoester (**13**) were dependent on the solvent used in the asymmetric hydrolysis. Tetrahydrofuran (THF), methyl isobutyl ketone (MIBK), hexane, and dichloromethane inhibited PLE. Lower reaction yields (28–56 M%) and lower e.e. (59–72%) were obtained using *t*-butyl methyl ether, dimethylformamide (DMF), and dimethylsulfoxide (DMSO) as cosolvent. Higher e.e. (>91%) was obtained using methanol, ethanol, and toluene as cosolvent. Ethanol gave highest reaction yield (96.7%) and e.e. (96%) for monoester (**13**).

Table 5 Effect of Solvent on Asymmetric Hydrolysis of Methyl-(4-Methoxyphenyl)-Propanedioic Acid, Diethyl Ester (**12**)

Solvent	Reaction time (hr)	Diester (**12**) (mg/mliter)	Monoester (**13**) (mg/mliter)	Yield of (**13**) (M%)	Enantiomeric excess of monoester (**13**) (%)
Methanol	22	0	0.65	37	92
Ethanol	22	0	1.7	96.7	96
Acetonitrile	22	0	0.5	28.2	59.3
Dimethylformamide	22	0	0.85	48.3	68.5
Dimethylsulfoxide	22	0.61	1	56.9	72
Acetone	22	0	1.44	81.9	65.1
Methyl ethyl ketone	48	0	1.36	77.3	82.1
Methyl isobutyl ketone	64	2.01	0	0	
t-Butyl methyl ether	22	0.76	0.8	46	64.4
Tetrahydrofuran	48	2	0	0	
Toluene	22	0.18	0.59	33.6	91
Hexane	64	2.05	0	0	

The effect of temperature and pH were evaluated for the PLE-catalyzed hydrolysis of diester (**12**) in a biphasic system using ethanol as cosolvent. It was observed that the e.e. of desired monoester (**13**) was increased with decreasing temperature from 25°C to 10°C. The optimum pH for asymmetric hydrolysis of diester (**12**) in a biphasic system using ethanol as a cosolvent was 7.2 at 10°C.

A semipreparative-scale asymmetric hydrolysis of diester (**12**) was carried out in a biphasic system using 10% ethanol as cosolvent. Substrate (3 g) was used in a 300-ml reaction mixture. The reaction was carried out at 10°C, 125 rpm agitation, and at pH 7.2 for 11 hr. Reaction yield of 96 M% and an e.e. of 96.9% were obtained. From the reaction mixture, 2.6 g of monoester (**13**) were isolated in 86.3 M% overall yield. The e.e. of isolated *S*-(−)-monoester (**13**) was 96.9%.

IV. ANGIOTENSIN CONVERTING ENZYME (ACE) AND NEUTRAL ENDOPEPTIDASE INHIBITORS

Captopril is designated chemically as 1-[(2*S*)-3-mercapto-2-methylpropionyl]-L-proline (**14**) (Fig. 8). It is used as an antihypertensive agent through suppression of the renin–angiotensin–aldosterone system [36,37]. Captopril and other compounds such as enalapril and lisinopril prevent the conversion of angiotensin I to angiotensin II by inhibition of ACE.

The potency of captopril (**14**) as an inhibitor of ACE depends critically on the configuration of the mercaptoalkanoyl moiety; the compound with the S-configuration is about 100 times more active than its corresponding *R*-enantiomer [37]. The required 3-mercapto-(2*S*)-methylpropionic acid moiety has been prepared from microbially derived chiral 3-hydroxy-(2*R*)-methylpropionic acid, which is obtained by the hydroxylation of isobutyric acid [38–40].

The use of extracellular lipases of microbial origin to catalyze the stereoselective hydrolysis of esters of 3-acylthio-2-methylpropionic acid in an aqueous system has been demonstrated to produce optically active 3-acylthio-2-methylpropionic acid [41–43]. The synthesis of the chiral side chain of captopril by the lipase-catalyzed enantioselective hydrolysis of the thioester bond of racemic 3-acetylthio-2-methylpropionic acid (**15**) to yield *S*-(−)-(**15**) has been demonstrated [44]. Among various lipases evaluated, lipase from *Rhizopus oryzae* ATCC 24563 (heat-dried cells), BMS lipase (extracellular lipase derived from the fermentation of *Pseudomonas* sp. SC 13856), and lipase PS-30 from *Pseudomonas cepacia* in an organic solvent system (1,1,2-trichloro-1,2,2-trifluoroethane or toluene) catalyzed the hydrolysis of thioester bond of undesired enantiomer of racemic (**15**) to yield desired *S*-(−) (**15**), *R*-(+)-3-mercapto-2-methylpropionic acid (**16**) and acetic acid (**17**) (Fig. 8A). The reaction yield of

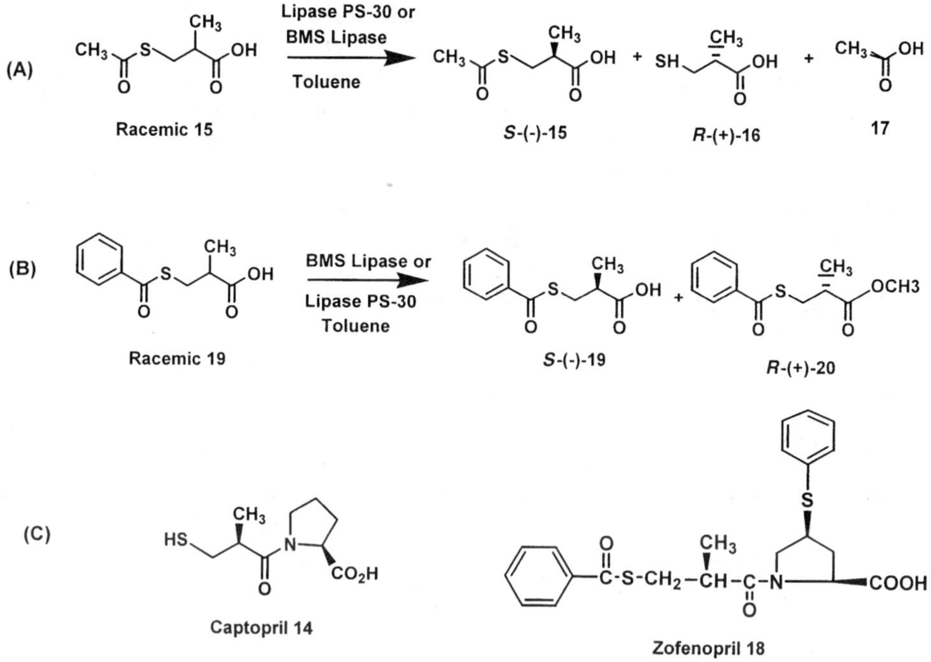

Figure 8 (A) Synthesis of captopril side-chain *S*-(−)-(**15**): stereoselective enzymatic hydrolysis of racemic 3-acylthio-2-methylpropionoic acid (**15**). (B) Synthesis of zofenopril side-chain *S*-(−)-(**19**): stereoselective enzymatic esterification of racemic 3-benzylthio-2-methylpropionic acid (**19**). (C) Structures of captopril (**14**) and zofenopril (**15**).

>24% (theoretical max. 50%) and e.e. of >95% were obtained for *S*-(−)-(**15**) using each lipase in an independent experiment.

In an alternative approach to prepare the chiral side chain of captopril (**14**) and zofenopril (**18**), the lipase-catalyzed stereoselective esterification of racemic 3-benzoylthio-2-methylpropionic acid (**19**) (Fig. 8B) in an organic solvent system was demonstrated to yield *R*-(+)-methyl ester (**20**) and unreacted acid enriched in the desired *S*-(−)-enantiomer (**19**) [45]. Using lipase PS-30 with toluene as solvent and methanol as nucleophile, the desired *S*-(−)-(**19**) was obtained in 37% reaction yield (theoretical max. 50%) and 97% e.e. Substrate was used at 22-g/liter concentration. The amount of water and the concentration of methanol supplied in the reaction mixture was very critical. Water was used at 0.1% concentration in the reaction mixture. More than 1% water led to the aggregation of enzyme in the organic solvent, with a decrease in the rate of reaction which was due to mass transfer limitation. The rate of esterification decreased as the methanol to substrate ratio was increased from 1:1 to 4:1. Higher methanol concentration probably inhibited the esterification reaction by stripping the essential water from the enzyme. Lower e.e. of product S-(−)-(**19**) was obtained at higher methanol concentration (Table 6). Crude lipase PS-30 was immobilized on three different resins, XAD-7, XAD-2 and Accurel polypropylene (Accurel PP) in absorption efficiencies of about 68%, 71%, and 98.5%, respectively. These immobilized lipases were evaluated for the ability to stereoselectively esterify racemic (**19**). Enzyme immobilized on Accurel PP catalyzed efficient esterification, giving 36–45% reaction yield and 97.7% e.e. of *S*(−)-(**19**). The immobilized enzyme under identical conditions gave similar e.e. and yield of product in 23 additional reaction cycles without any loss of activity and productivity. *S*-(−)-(**19**) is a key chiral intermediate for the synthesis of captopril (**14**) [46] or zofenopril (**18**) [47] (Fig. 8C).

The *S*-(−)-α-[(acetylthio)methyl]phenylpropionic acid (**21**) is a key chiral intermediate for the neutral endopeptidase inhibitor (**22**) [48]. We [44] have demonstrated the lipase-catalyzed stereoselective hydrolysis of thioester bond of racemic α-[(acetylthio)methyl]phenylpropionic acid (**21**) in organic solvent to yield *R*-(+)-α-[(mercapto)methyl]phenylpropionic acid (**23**) and *S*-(−)-(**21**). Using lipase PS-30, the *S*-(−)-(**21**) was obtained in 40% reaction yield (theoretical max. 50%) and 98% e.e. (Fig. 9).

The *S*-(−)-2-cyclohexyl-1,3-propanediol monoacetate (**24**) and the *S*-(−)-2-phenyl-1,3-propanediol monoacetate (**25**) are key chiral intermediates for the chemoenzymatic synthesis of Monopril (**26**) (Fig. 10), a new antihypertensive drug which acts as an ACE inhibitor. The asymmetric hydrolysis of 2-cyclohexyl-1,3-propanediol diacetate (**27**) and 2-phenyl-1,3-propanediol diacetate (**28**) to the corresponding *S*-(−)-monoacetate (**24**) and *S*-(−)-monoacetate (**25**) by porcine pancreatic lipase (PPL) and *Chromobacterium viscosum* lipase have been demon-

Table 6 Effect of Methanol Concentration on the Esterification of Racemic 3-Benzoylthio-2-Methylpropionic Acid (**19**)

Methanol concentration (M)	Reaction time (hr)	Conversion to ester (**20**) (%)	Yield of acid *S*-(−)-(**19**) (%)	Enantiomeric excess of acid *S*-(−)-(**19**) (%)
0.5	5.5	55	45	93
1	8	61	39	96
1.5	21	60	40	96.5
2	30	62	38	95.5
3	64	13	87	56
4	64	6	94	56

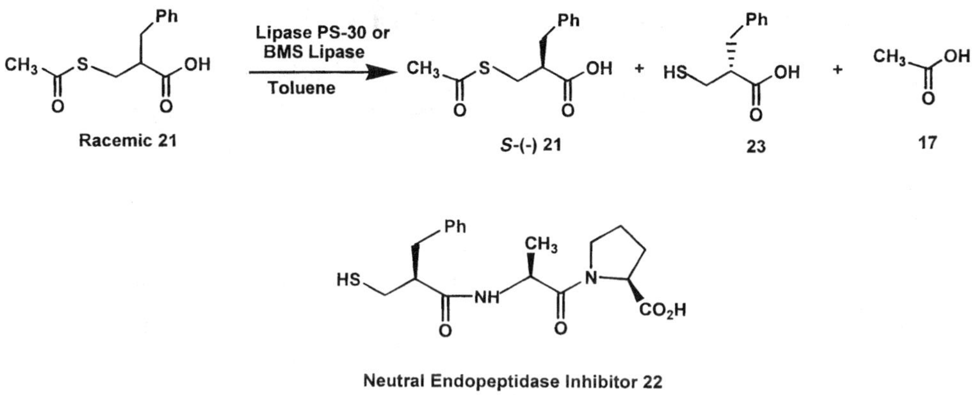

Figure 9 Preparation of chiral synthon for neutral endopeptidase inhibitor: stereoselective enzymatic hydrolysis of racemic α-[(acetylthio)methyl]phenylpropionic acid (**21**).

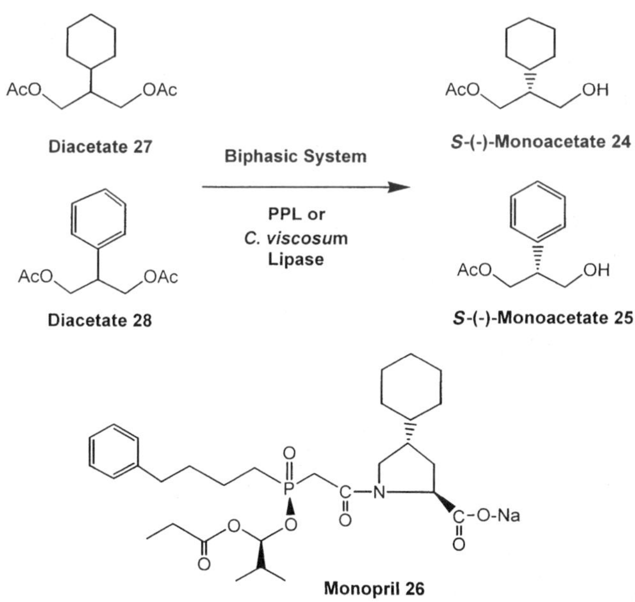

Figure 10 Preparation of chiral synthon for monopril: asymmetric enzymatic hydrolysis of 2-cyclohexyl- and 2-phenyl-1,3-propanediol diacetate (**27**) and (**28**) to the corresponding *S*-(−)-monoacetates.

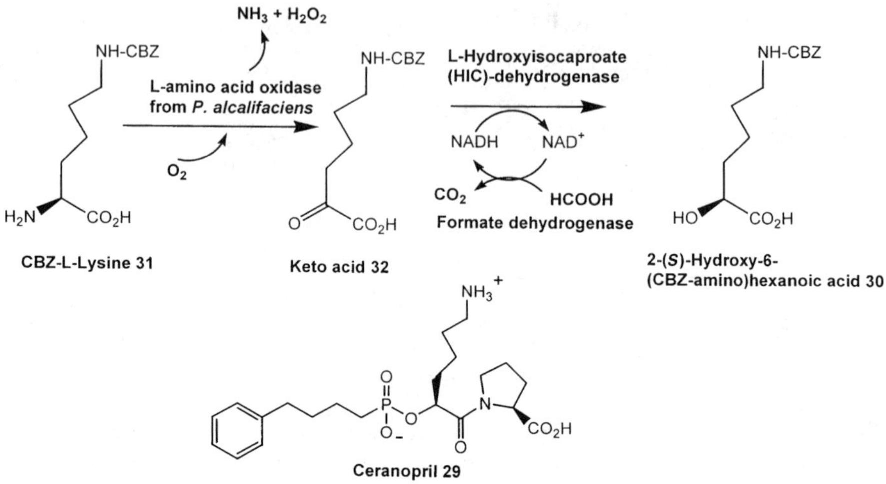

Figure 11 Synthesis of chiral synthon for ceranopril: enzymatic conversion of CBZ-L-lysine (**31**) to (S)-hydroxy-6-(carbobenzyloxyamino)-hexanoic acid (**30**).

strated by Patel et al. [49]. In a biphasic system using 10% toluene, the reaction yield of >65% and e.e. of 99% were obtained for *S*-(−)-(**25**) using each enzyme. *S*-(−)-(**24**) was obtained in 90% reaction yield and 99.8% e.e. using *C. viscosum* lipase under similar conditions.

Ceranopril (**29**) is another ACE inhibitor [50] which requires chiral intermediate 2-(*S*)-hydroxy-6-(carbobenzyloxyamino)-hexanoic acid (**30**) (Fig. 11). A biotransformation process was developed by Hanson et al. [51] to prepare the 2-(*S*)-hydroxy-6-(carbobenzyloxyamino)-hexanoic acid. N-ε-carbobenzoxy(CBZ)-L-lysine (**31**) was first converted to the corresponding keto acid (**32**) by oxidative deamination using cells of *Providencia alcalifaciens* SC 9036 which contained L-amino acid oxidase and catalase. The keto acid (**32**) was subsequently converted to 2-(*S*)-hydroxy-6-(carbobenzyloxyamino)-hexanoic acid (**30**) using L-2-hydroxyisocaproate (HIC) dehydrogenase from *Lactobacillus confusus*. The NADH required for this reaction was regenerated using formate dehydrogenase from *C. boidinii*. The reaction yield of 95% with 98.5% e.e. was obtained in the overall process.

V. THROMBOXANE A2 ANTAGONISTS

Thromboxane A2 (TxA2) is an exceptionally potent pro-aggregatory and vasoconstrictor substance produced by the metabolism of arachidonic acid in blood platelets and other tissues. Together with potent anti-aggregatory and vasodilator compounds, TxA2 plays an important role in the maintenance of vascular homeostasis and contributes to the pathogenesis of a variety of vascular disorders. Approaches toward limiting the effect of TxA2 have focused on either inhibiting its synthesis or blocking its action at its receptor sites by means of an antagonist [52,53]. The lactol (**33**) or lactone (**34**) (Fig. 12A) are key chiral intermediates for the total synthesis of compound (**35**), a new cardiovascular agent useful in the treatment of thrombolic disease [54,55].

Horse liver alcohol dehydrogenase (HLADH) catalyzes the oxidoreduction of a variety of compounds [56,57]. It has been demonstrated that HLADH catalyzes the stereospecific oxidation of only one of the enantiotopic hydroxyl groups of acyclic and monocyclic meso-diols [58,59]. The authors demonstrated the oxidation of meso exo- and endo-7-oxabicyclo [2.2.1]heptane-2,3-dimethanol to the corresponding enantiomerically pure γ-lactones by HLADH. NAD^+ and flavin adenine dinucleotide (FAD) at concentrations of 1 and 20 mmol, respectively,

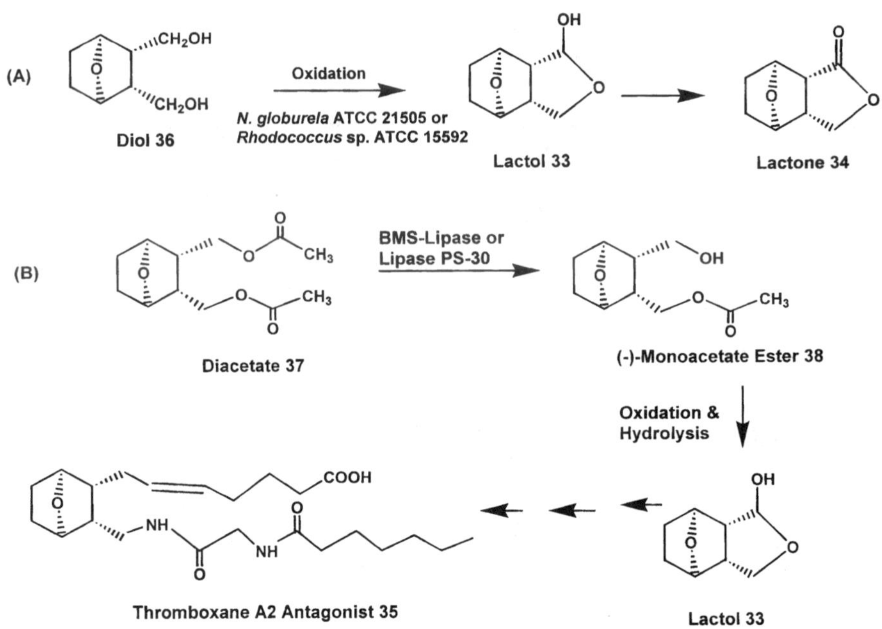

Figure 12 Synthesis of chiral synthon for thromboxane A2 antagonist: (A) stereoselective microbial oxidation of (exo,exo)-7-oxabicyclo[2.2.1]heptane-2,3-dimethanol (**36**) to the corresponding lactol (**33**) and lactone (**34**); (B) asymmetric enzymatic hydrolysis of (exo,exo)-7-oxabicyclo[2.2.1]heptane-2,3-dimethanol, diacetate (**37**) to the corresponding *S*-(−)-monoacetate ester (**38**).

were required for the stereoselective oxidation of 12.7 mmol of substrate. Due to the high cost of enzyme and required cofactors, this process for preparing chiral lactones was economically not feasible for scale-up. Patel et al. [60] described the stereoselective oxidation of (exo,exo)-7-oxabicyclo[2.2.1]heptane-2,3-dimethanol (**36**) to the corresponding chiral lactol (**33**) and lactone (**34**) (Fig. 12A) by cell suspension (10% w/v, wet cells) of *Nocardia globerula* ATCC 21505 or *Rhodococcus* sp. ATCC 15592. The reaction yield of 70 M% and e.e. of 96% were for chiral lactone (**34**) after a 96-hr biotransformation process at 5-g/liter substrate concentration using cell suspensions of *N. globerula* ATCC 21505. An overall reaction yield of 46 M% (lactol and lactone combined) and e.e. of 96.7% and 98.4% were obtained for lactol (**33**) and lactone (**34**), respectively, using cell suspensions of *Rhodococcus* sp. ATCC 15592. Substrate (**36**) was used at 5-g/liter concentration.

The asymmetric hydrolysis of (exo,exo)-7-oxabicyclo[2.2.1]heptane-2,3-dimethanol, diacetate ester (**37**) to the corresponding chiral monoacetate ester (**38**) (Fig. 12B) has been demonstrated with lipases [61]. Lipase PS-30 from *P. cepacia* was most effective in asymmetric hydrolysis to obtain the desired enantiomer of monoacetate ester. The reaction yield of 75 M% and e.e. of >99% were obtained when the reaction was conducted in a biphasic system with 10% toluene at 5 g/liter of the substrate. Lipase PS-30 was immobilized on Accurel PP and the immobilized enzyme was reused (5 cycles) without loss of enzyme activity, productivity, or e.e. of product (**38**). The reaction process was scaled up to 80 liters (400 g of substrate) and monoacetate ester (**38**) was isolated in 80 M% yield with 99.3% e.e. The product was isolated in 99.5% chemical purity. The chiral monoacetate ester (**38**) was oxidized to its corresponding aldehyde and subsequently hydrolyzed to give chiral lactol (**33**) (Fig. 12B). The chiral lactol (**33**) obtained by this enzymatic process was used in chemoenzymatic synthesis of thromboxane A2 antagonist (**35**).

VI. ANTICHOLESTEROL DRUGS

Chiral β-hydroxy esters are versatile synthons in organic synthesis, specifically in the preparation of natural products [62–64]. The asymmetric reduction of carbonyl compounds using Baker's yeast has been demonstrated and reviewed [65,66].

Recently, we [67] have described the reduction of the methyl ester of 4-chloro-3-oxobutanoic acid (**39**) to the methyl ester of *S*-(−)-4-chloro-3-hydroxybutanoic acid (**40**) (Fig. 13) by cell suspensions of *Geotrichum candidum* SC 5469. *S*(−)-(**40**) is a key chiral intermediate in the total chemical synthesis of a cholesterol antagonist (SQ 33600), which acts by inhibiting hydroxymethylglutaryl CoA (HMG CoA) reductase. In the biotransformation process, a reaction

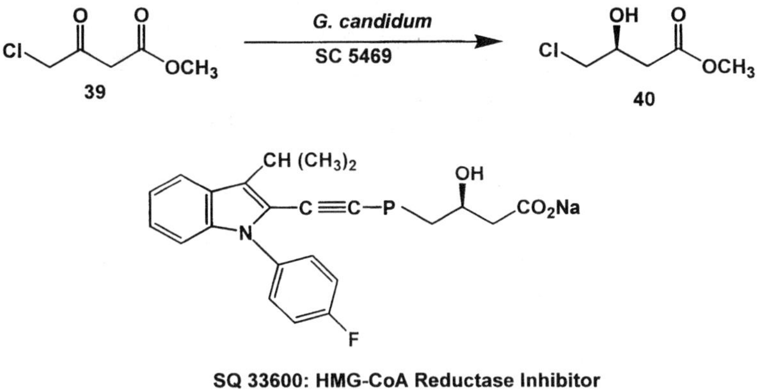

Figure 13 Synthesis of chiral synthon for anticholesterol drug SQ 33600: stereoselective microbial reduction of 4-chloro-3-oxobutanoic acid methyl ester (**39**).

yield of 95% and e.e. of 96% were obtained for *S*-(−)-(**40**) by glucose-, acetate-, or glycerol-grown cells (10% w/v) of *G. candidum* SC 5469. Substrate was used at 10-g/liter concentration. The e.e. of *S*-(−)-(**40**) was increased to 98% by heat treatment of cell suspensions (55°C for 30 min) prior to conducting the bioreduction of (**39**).

Glucose-grown cells of *G. candidum* SC 5469 have also catalyzed the stereoselective reduction of ethyl-, isopropyl-, and tertiary-butyl esters of 4-chloro-3-oxobutanoic acid and methyl and ethyl esters of 4-bromo-3-oxobutanoic acid. A reaction yield of >85% and e.e. of >94% were obtained. NAD^+-dependent oxido-reductase responsible for the stereoselective reduction of β-keto esters of 4-chloro- and 4-bromo-3-oxobutanoic acid was purified 100-fold. The molecular weight of purified enzyme is 950,000. The purified oxido-reductase was immobilized on Eupergit C and used to catalyze the reduction of (**39**) to *S*-(−)-(**40**). The cofactor NAD^+ required for the reduction reaction was regenerated by glucose dehydrogenase.

So far, most microorganisms and enzymes derived therefrom have been used in the reduction of a single keto group of β-keto or α-keto compounds [68–71]. Recently, Patel et al. [72] have demonstrated the stereoselective reduction of 3,5-dioxo-6-(benzyloxy)hexanoic acid, ethyl ester (**41**), to (3*S*,5*R*)-dihydroxy-6-(benzyloxy)hexanoic acid, ethyl ester (**42a**) (Fig. 14). The compound (**42a**) is a key chiral intermediate required for the chemical synthesis of [4-[4α,6β(E)]]-6-[4,4-bis(4-fluorophenyl)-3-(1-methyl-1H-tetrazol-5-yl)-1,3-butadienyl]-tetrahydro-4-hydroxy-2H-pyran-2-one, compound *R*-(+)-(**43**), a new anticholesterol drug that acts by inhibition of HMG CoA reductase [73]. Among various microbial cultures evaluated for the stereoselective reduction of diketone (**41**), cell suspensions of *Aci-*

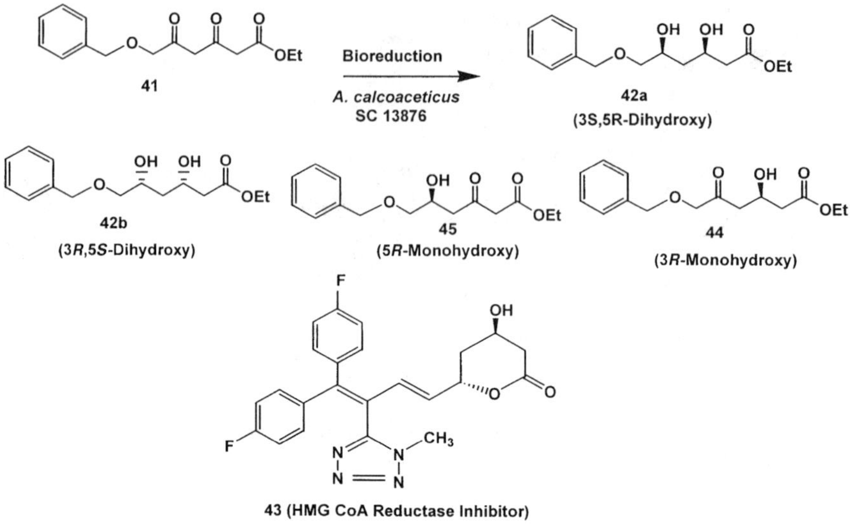

Figure 14 Synthesis of chiral synthon for anticholesterol drug *R*-(+)-(**43**): stereoselective microbial reduction of 3,5-dioxo-6-(benzyloxy)hexanoic acid, ethyl ester (**41**).

netobacter calcoaceticus SC 13876 reduced (**41**) to (**42a**). The reaction yield of 85% and e.e. of 97% were obtained using glycerol-grown cells. The substrate was used at 2 g/liter and cells were used at 20% (w/v, wet cells) concentration.

Cell extracts of *A. calcoaceticus* SC 13876 in the presence of NAD^+, glucose, and glucose dehydrogenase reduced (**41**) to the corresponding monohydroxy compounds (**44**) and (**45**) [3-hydroxy-5-oxo-6-(benzyloxy)hexanoic acid ethyl ester (**44**) and 5-hydroxy-3-oxo-6-(benzyloxy)hexanoic acid ethyl ester (**45**)]. Both (**44**) and (**45**) were further reduced to (3*S*,4*R*)-dihydroxy compound (**42a**) using the cell extracts (Fig. 14). The reaction yield of 92% and the e.e. of 98% were obtained when the reaction was carried out in a 1-liter batch using cell extracts. The substrate was used at 10 g/liter. Product (**42a**) was isolated from the reaction mixture in 72% overall yield. The HPLC area percent purity of the isolated product was 99% and the e.e. was 98.5%. The reductase which converted (**41**) to (**42a**) was purified about 200-fold from cell extracts of *A. calcoaceticus* SC 13876. The purified enzyme gave a single protein band on SDS-PAGE corresponding to 33,000 Da.

Using an enzymatic resolution process, chiral alcohol *R*-(+)-(**43**) was also prepared by the lipase-catalyzed stereoselective acetylation of racemic (**43**) in organic solvent [74]. We evaluated various lipases, among which lipase PS-30 and BMS lipase (produced by fermentation of *Pseudomonas* strain SC 13856) efficiently catalyzed the acetylation of the undesired enantiomer of racemic (**43**) to yield *S*-(−)-acetylated product (**46**) and unreacted desired *R*-(+)-**43** (Fig. 15). A reaction yield of 49 M% (theoretical max. 50 M%) and e.e. of 98.5% were obtained for *R*-(+)-(**43**) when the reaction was conducted in toluene as solvent in the presence of isopropenyl acetate as acyl donor. Substrate was used at 4 g/liter concentration. In methyl ethyl ketone at 50-g/liter substrate concentration, a reaction yield of 46 M% and e.e. of 96% were obtained for *R*-(+)-(**43**).

Lipase PS-30 was immobilized on Accurel PP and the immobilized enzyme was reused five times without any loss of activity or productivity in the resolution process to prepare *R*-(+)-(**43**). The enzymatic process was scaled up to a 640-liter preparative batch using immobilized lipase PS-30 at 4 g/liter racemic substrate (**43**) in toluene as a solvent. From the reaction mixture, *R*-(+)-(**43**) was isolated in 35 M% overall yield with 98.5% e.e. and 99.5% chemical purity. The undesired *S*-(−)-acetate (**46**) produced by this process was enzymatically hydrolyzed by lipase PS-30 in a biphasic system to prepare the corresponding *S*-(−)-alcohol (**43**). Thus both enantiomers of alcohol (**43**) were produced by the enzymatic process.

Pravastatin (**47**) and Mevastatin (**48**) are anticholesterol drugs which act by competitively inhibiting HMG CoA reductase [75]. Pravastatin sodium is produced by two fermentation steps. The first step is the production of compound ML-236B by *Penicillium citrinum* [75–77]. The purified compound was con-

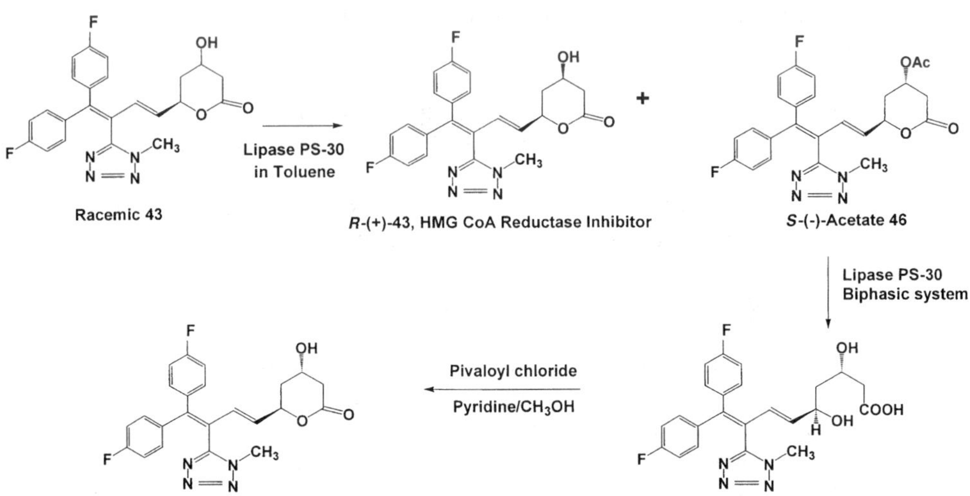

Figure 15 Synthesis of chiral anticholesterol drug *R*-(+)-(**43**): stereoselective enzymatic acetylation of racemic (**43**).

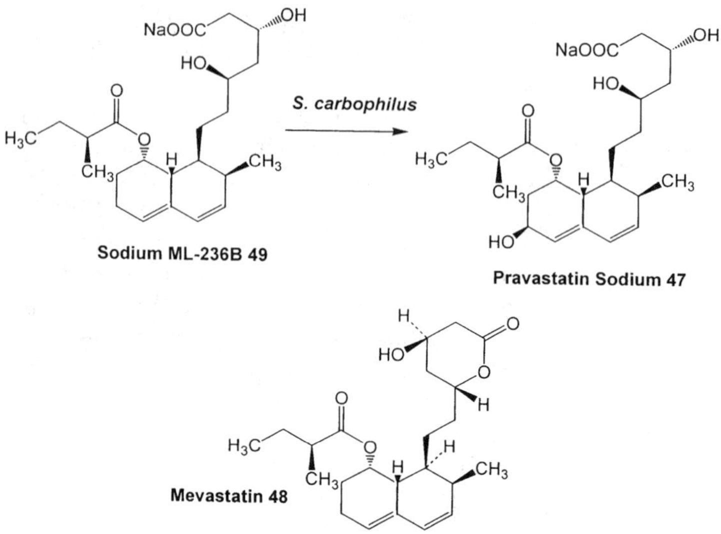

Figure 16 Stereoselective microbial hydroxylation of ML-236B (**49**) to Pravastatin (**47**).

verted to its sodium salt (**49**) with sodium hydroxide and in the second step was hydroxylated to Pravastatin sodium (**47**) (Fig. 16) by *Streptomyces carbophilus* [78]. A cytochrome P-450–containing enzyme system has been demonstrated from *S. carbophilus* which catalyzed the hydroxylation reaction [79].

Squalene synthase is the first pathway-specific enzyme in the biosynthesis of cholesterol and catalyzes the head-to-head condensation of two molecules of farnesyl pyrophosphate (FPP) to form squalene (**50**). It has been implicated in the transformation of FPP into presqualene pyrophosphate (PPP) [80]. FPP analogs are a major class of inhibitors of squalene synthase [81,82]. However, this class of compounds lacks specificity and are potential inhibitors of other FPP consuming transferases such as geranyl-geranyl pyrophosphate synthase. To increase enzyme specificity, analogs of PPP and other mechanism-based enzyme inhibitors have been synthesized [83,84]. BMS-188494 is a potent squalene synthase inhibitor that is effective as an anticholesterol drug [85,86]. (*S*)[1-(acetoxy)-4-(3-phenoxyphenyl)butyl]phosphonic acid, diethyl ester (**51**) is a key chiral intermediate required for the total chemical synthesis of BMS-188494. The stereoselective acetylation of racemic [1-(hydroxy)-4-(3-phenoxyphenyl)butyl]phosphonic acid, diethyl ester (**52**) (Fig. 17), was carried out using *G. candidum* lipase in toluene as solvent and isopropenyl acetate as acyl donor [87]. A reaction

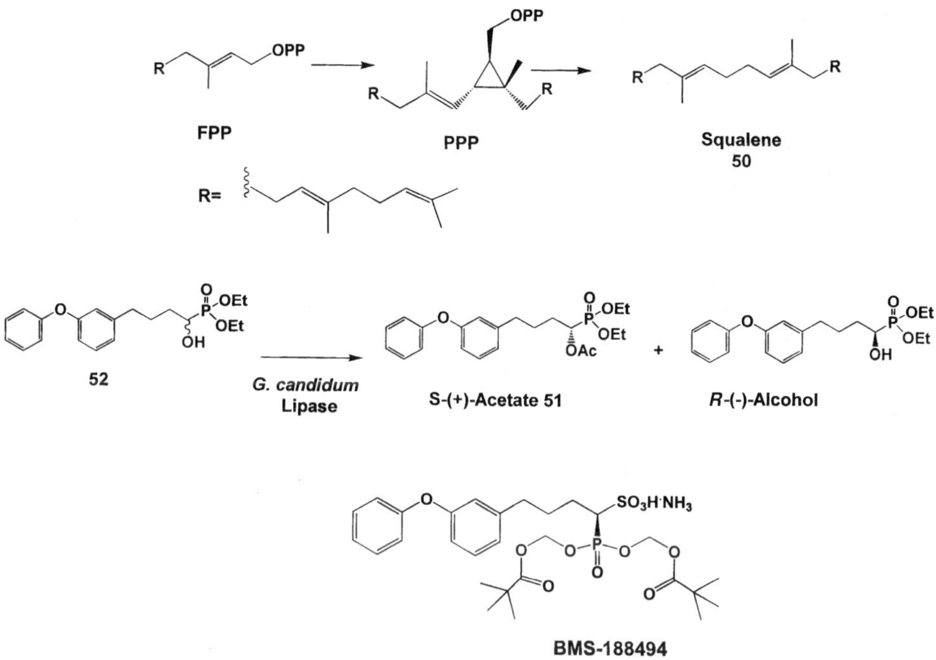

Figure 17 Enzymatic synthesis of chiral synthon for BMS-188494, a squalene synthase inhibitor: stereoselective acetylation of racemic (**52**).

yield of 38% (theoretical max. 50%) and an e.e. of 95% were obtained for chiral (**51**).

VII. CALCIUM CHANNEL BLOCKING AGENTS

Dilthiazem (**53**), a benzothiazepinone calcium channel blocking agent that inhibits influx of extracellular calcium through L-type voltage-operated calcium channels, has been widely used clinically in the treatment of hypertension and angina [88]. Since dilthiazem has a relatively short duration of action [89], recently an 8-chloro derivative has been introduced in the clinic as a more potent analog of dilthiazem [90]. Lack of extended duration of action and little information on structure–activity relationships in this class of compounds led Floyd et al. [91] and Das et al. [92] to prepare isosteric 1-benzazepin-2-ones which resulted in the identification of a 6-trifluoromethyl-1-benzazepin-2-one derivative as a longer-lasting and more potent antihypertensive agent. A key chiral intermediate (**54**) [(3*R*-*cis*)-1,3,4,5-tetrahydro-3-hydroxy-4-(4-methoxyphenyl)-6-(trifluoromethyl)-2H-1-benzazepin-2-one] was required for the total chemical synthesis of the new calcium channel blocking agent [(*cis*)-3-(acetoxy)-1-[2-(dimethylamino)ethyl]-1,3,4,5-tetrahydro-4-(4-methoxyphenyl)-6-(trifluoromethyl)-2H-1-benzazepin-2-one] (**55**). A stereoselective microbial process (Fig. 18A) was developed for the reduction of 4,5-dihydro-4-(4-methoxyphenyl)-6-(trifluoromethyl)-1H-1-benzazepin-2,3-dione (**56**) to chiral (**54**) [93]. Compound (**56**) exists predominantly in the achiral enol form, which is in rapid equilibrium with the two keto-form enantiomers. Reduction of (**56**) could give rise to formation of four possible alcohol stereoisomers. Remarkably, conditions were found under which only the single alcohol isomer (**54**) was obtained by microbial reduction. Among various cultures evaluated, microorganisms from the genera *Nocardia*, *Rhodococcus*, *Corynebacterium*, and *Arthrobacter* reduced compound (**56**) to compound (**54**)

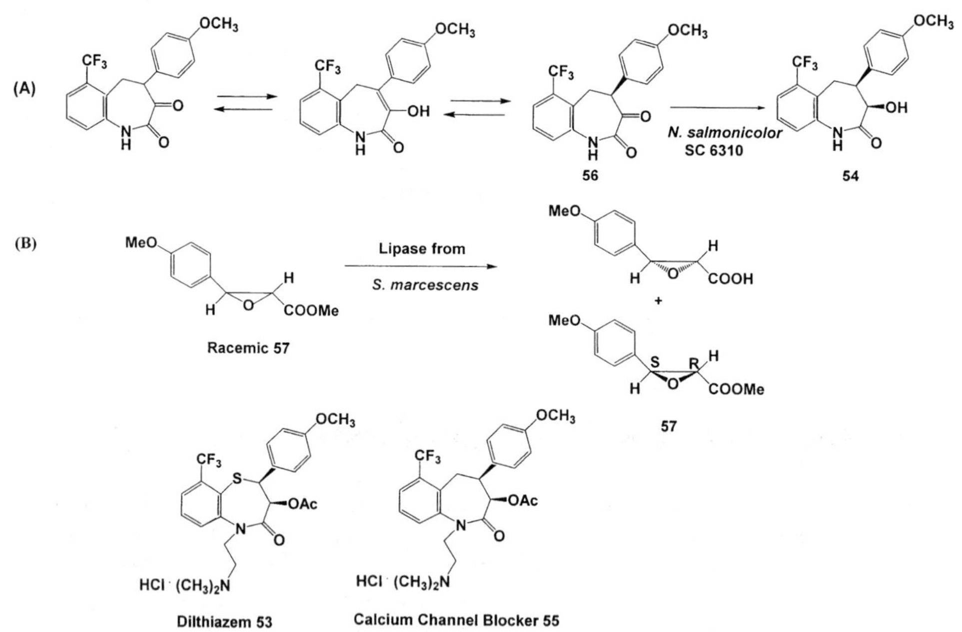

Figure 18 Synthesis of chiral synthon for calcium channel blocker: (A) microbial reduction of 4,5-dihydro-4-(4-methoxyphenyl)-6-(trifluoromethyl)-1H-benzazepin-2,3-dione (**56**); (B) synthesis of chiral synthon for calcium channel blocker, dilthiazem: enantioselective enzymatic hydrolysis of racemic (**57**).

with 60–70% conversion yield at 1-g/liter substrate concentration. The most effective culture, *Nocardia salmonicolor* SC 6310, catalyzed the bioconversion of (**56**) to (**54**) in 96% reaction yield with 99.8% e.e. at 2-g/liter substrate concentration. Product (**54**) was isolated and identified by NMR and MS. A preparative-scale fermentation process for growth of *N. salmonicolor* and a bioreduction process using cell suspensions of the organism were demonstrated.

A chiral intermediate (2*R*,3*S*)-3-(4-methoxyphenyl)glycidic acid methyl ester [(−)-MPGM] (**57**) is required for the synthesis of dilthiazem (**53**). Matsumae et al. [94] screened over 700 microorganisms and identified a lipase from *Serratia marcescens* which catalyzed the enantioselective hydrolysis of racemic MPGM in a biphasic system using toluene as organic phase (Fig. 18B). The reaction yield of 48% and the e.e. of 99.8% were obtained for (−)-MPGM.

VIII. POTASSIUM CHANNEL OPENERS

The study of potassium K-channel biochemistry, physiology, and medicinal chemistry has flourished, and numerous papers and reviews have been published in recent years [95,96]. It has long been known that K-channels play a major role in neuronal excitability and a critical role in the basic electrical and mechanical functions of a wide variety of tissues, including smooth muscle, cardiac muscle, and glands [97]. A new class of highly specific pharmacological compounds has been developed which either open or block K-channels [98,99]. K-channel openers are powerful smooth muscle relaxants with in vivo antihypertensive and bronchodilator activities [97]. Recently, the synthesis and antihypertensive activity of a series of novel K-channel openers [99–102] based on monosubstituted *trans*-4-amino-3,4-dihydro-2,2-dimethyl-2H-l-benzopyran-3-ol (**58**) have been demonstrated. Chiral epoxide (**59**) and diol (**60**) are potential intermediates for the synthesis of K-channel activators that are important as antihypertensive and bronchodilator agents. The stereoselective microbial oxygenation of 2,2-dimethyl-2H-1-benzopyran-6-carbonitrile (**61**) to the corresponding chiral epoxide (**59**) and chiral diol (**60**) (Fig. 19) has been demonstrated [103]. Among microbial cultures evaluated, the best culture, *Mortierella ramanniana* SC 13840, gave reaction yields of 67.5 M% and e.e. of 96% for the (+)-*trans*-diol (**60**). A single-stage process (fermentation/epoxidation) for the biotransformation of (**61**) was developed using *M. ramanniana* SC 13840. In a 25-liter fermentor, the (+)-*trans*-diol (**60**) was obtained in the reaction yield of 60.7 M% and e.e. of 92.5%.

Using a 3-liter cell suspension (10% w/v, wet cells) of *M. ramanniana* SC 13840, the (+)-*trans*-diol (**60**) was obtained in 76 M% yield with an e.e. of 96%. The reaction was carried out in a 5-liter Bioflo fermentor with 2-g/liter substrate and 10-g/liter glucose concentrations. Glucose was supplied to regenerate NADH

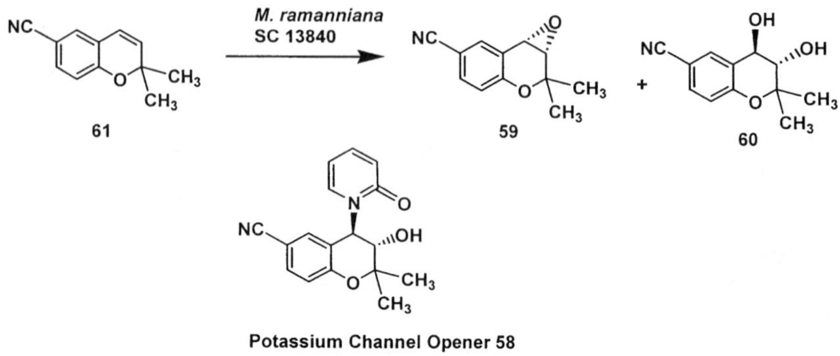

Figure 19 Oxygenation of 2,2-dimethyl-2H-1-benzopyran-6-carbonitrile (**61**) to the corresponding chiral epoxide (**59**) and (+)-*trans*-diol (**60**) by *M. ramanniana* SC 13840.

required for this reaction. From the reaction mixture, (+)-*trans*-diol (**60**) was isolated in 65 M% (4.6 g) overall yield. An enantiomeric excess of 97% and a chemical purity of 98% were obtained for the isolated (+)-*trans*-diol (**60**).

In an enzymatic resolution approach, chiral (+)-*trans*-diol (**60**) was prepared by the stereoselective acetylation of racemic diol with lipases from *Candida cylindraceae* and *P. cepacia*. Both enzymes catalyzed the acetylation of the undesired enantiomer of racemic diol to yield monoacetylated product and unreacted desired (+)-*trans*-diol (**60**). A reaction yield of 40% and an e.e. of >90% were obtained using each lipase [104].

IX. ANTIARRHYTHMIC AGENTS

Larsen and Lish [105] reported the biological activity of a series of phenethanolamine-bearing alkyl sulfonamido groups on the benzene ring. Within this series, some compounds possessed adrenergic and antiadrenergic actions. D-(+)-sotalol (**62**) is a β-blocker [106] that, unlike other β-blockers, has antiarrhythmic properties and has no other peripheral actions [107]. The β-adrenergic blocking drugs such as propranolol (**63**) and sotalol have been separated chemically into the dextrorotatory and levorotatory optical isomers, and it has been demonstrated that the activity of the levo isomer is 50 times that of the corresponding dextro isomer [108]. Chiral alcohol (**64**) is a key intermediate for the chemical synthesis of D-(+)-sotalol (**62**). The stereoselective microbial reduction of N-(4-(2-chloroacetyl)phenyl)methanesulfonamide (**65**) to the corresponding (+)-alcohol (**64**) (Fig. 20) has been demonstrated [109]. Among numbers of microorganisms screened for the transformation of ketone (**65**) to (+)-alcohol (**64**), *Rhodococcus*

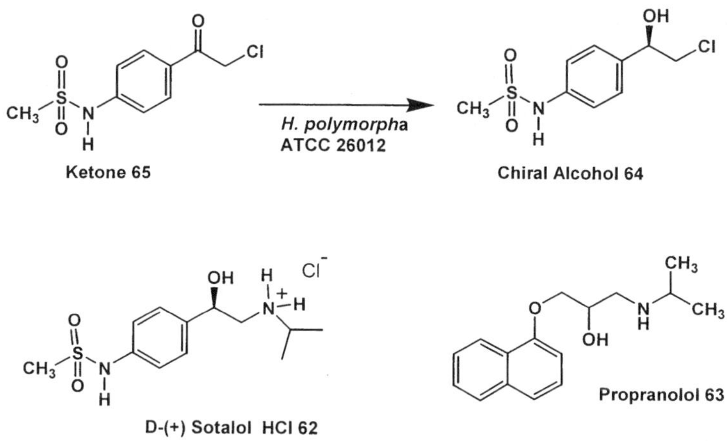

Figure 20 Synthesis of chiral synthon for D-(+)-sotalol, antiarrhythmic agent: stereoselective microbial reduction of N-(4-(2-chloroacetyl)phenyl)methanesulfonamide (**65**).

sp. ATCC 29675, *Rhodococcus rhodochrous* ATCC 21243, *N. salmonicolor* SC 6310, and *Hansenula polymorpha* ATCC 26012 gave the desired (+)-alcohol (**64**) in >90% e.e. *H. polymorpha* ATCC 26012 catalyzed the efficient conversion of ketone (**65**) to (+)-alcohol (**64**) in 95% reaction yield and >99% e.e. Growth of *H. polymorpha* ATCC 26012 culture was carried out in a 380-liter fermentor and cells harvested from the fermentor were used to conduct transformation in a 3-liter preparative batch. Cell suspensions (20% wet cells in 1 liter of 10 mM potassium phosphate buffer, pH 7.0) were supplemented with 12 g of ketone (**65**) and 225 g of glucose and the reduction reaction was carried out at 25°C, 200 rpm, and pH 7. Complete conversion of ketone (**65**) to (+)-alcohol (**64**) was obtained in a 20-hr reaction. Using preparative HPLC, 8.2 g of (+)-alcohol (**64**) were isolated from the reaction mixture in overall 68% yield with >99% e.e.

X. CONCLUSION

The production of optically active chiral intermediates is a subject of increasing importance in the pharmaceutical industry. Increasing regulatory pressure by the Food and Drug Administration to market homochiral drugs has led to the use of alternative approaches, including biocatalysis for the synthesis of chiral compounds. Organic synthesis has been one of the most successful scientific disciplines and has enormous practical utility. One can ask the question, then, why biocatalysis? What does biocatalysis have to offer to synthetic organic chemists?

Biocatalysis gives an added dimension and enormous opportunity to prepare industrially useful chiral compounds. The advantages of biocatalysis over chemical catalysis are that enzyme-catalyzed reactions are stereoselective and regioselective and can be carried out at ambient temperature and atmospheric pressure. In biocatalytic processes, microbial cells and enzymes can be immobilized and an immobilized biocatalyst can be reused for many cycles. In addition, enzymes can be overexpressed to make biocatalytic processes economically efficient. The use of different classes of enzymes in catalysis of different types of chemical reactions is essential to generate a variety of chiral compounds for chemoenzymatic synthesis of pharmaceutical products. This includes the use of hydrolytic enzymes such as lipases, esterases, proteases, dehalogenases, acylases, amidases, nitrilases, lyases, epoxide hydrolases, decarboxylases, and hydantoinases in resolution of racemic compounds and in asymmetric synthesis of optically active compounds. Oxidoreductases and aminotransferases have been used in synthesis of chiral alcohols, aminoalcohols, amino acids, and amines. Aldolases and decarboxylases have been effectively used in asymmetric synthesis by aldol condensation and acyloin condensation reactions. Oxygenases such as monooxygenases have been used in stereoselective and regioselective hydroxylation and epoxidation reactions and dioxygenases in the chemoenzymatic synthesis of chiral diols.

The idea of designing biocatalysts that act specifically in desired reactions of interest can change the face of synthesis. Tailored enzymes made by random and site-directed mutagenesis with modified activity and preparation of thermostable and pH stable enzymes can lead to the production of novel stereoselective biocatalysts. The use of enzymes in organic solvents has led to hundreds of publications on enzyme-catalyzed asymmetric synthesis and resolution processes. Molecular recognition and selective catalysis are key chemical processes in life which are embodied in enzymes. In the course of the last decades, the progress in biochemistry, protein chemistry, molecular cloning, random and site-directed mutagenesis, and fermentation technology has opened up unlimited access to a variety of enzymes and microbial cultures as valuable tools in organic synthesis.

REFERENCES

1. JB Jones. Enzymes in organic synthesis. Tetrahedron 42:3351–3403, 1986.
2. DHG Crout, S Davies, RJ Heath, CO Miles, DR Rathbone, BEP Swoboda. Application of hydrolytic and decarboxylating enzymes in biotransformations. Biocatalysis 9:1–30, 1994.
3. HG Davies, RH Green, DR Kelly, SM Roberts. Recent advances in the generation of chiral intermediates using enzymes. Biotechnology 10:129–152, 1990.
4. J Crosby. Synthesis of optically active compounds: a large scale perspective. Tetrahedron 47:4789–4846, 1991.

5. J Kamphuis, WHJ Boesten, QB Broxterman, HFM Hermes, JAM van Balken, EM Meijer, HE Schoemaker. New developments in chemo-enzymatic production of amino acids. Adv Biochem Eng Biotechnol 42:133–186, 1990.
6. CJ Sih, Q-M Gu, X Holdgrun, K Harris. Optically active compounds via biocatalytic methods. Chirality 4:91–97, 1992.
7. E Santaneillo, P Ferraboschi, P Grisenti, A Manzocchi. The biocatalytic approach to the preparation of enantiomerically pure chiral building blocks. Chem Rev 92: 1071–1140, 1992.
8. AL Margolin. Enzymes in the synthesis of chiral drugs. Enzyme Microb Technol 15:266–280, 1993.
9. DC Cole. Recent stereoselective synthetic approaches to β-amino acids. Tetrahedron 50:9517–9582, 1994.
10. C-H Wong, GM Whitesides. Enzymes in Synthetic Organic Chemistry, Tetrahedron Organic Chemistry Series, Vol. 12. Tarrytown, NY: Pergamon, Elsevier, 1994.
11. K Mori. Biochemical methods in enantioselective synthesis of bioactive natural products. Synlett: 1097–1109, 1995.
12. RN Patel. Stereoselective biotransformations in synthesis of some pharmaceutical intermediates. Adv Appl Microbiol 43:91–140, 1997.
13. JA Robl, C Sun, J Stevenson, DE Ryono, LM Simpkins, MAP Cimarusti, T Dejneka, WA Slusarchyk, S Chao, L Stratton, RN Misra, MS Bednarz, MM Asaad, HS Cheung, BE Aboa-Offei, PL Smith, PD Mathers, M Fox, TR Schaeffer, AA Seymour, NC Trippodo. Dual metalloprotease inhibitors: mercaptoacetyl-based fused heterocyclic dipeptide mimetics as inhibitors of angiotensin-converting enzyme and neutral endopeptidase. J Med Chem 40:1570–1577, 1997.
14. JA Robl, MP Cimarusti. A synthetic route for the generation of C-7 substituted azepinones. Tetrahedron Lett 35:1393–1396, 1994.
15. PJ Maurer, MJ Miller. Microbial iron chelators: total synthesis of aerobactin and its constituent amino acid, N6-acetyl-N6-hydroxylysine. J Am Chem Soc 105:240–245, 1983.
16. CCJ Culvenor, MC Foster, MP Hegarty. Total synthesis of indospicine, 6-amino-hexanoic acid. Austral J Chem 24:371–375, 1971.
17. M Bodanszky, J Martinez, GP Priestly, JD Gardner, V Mutt. Synthesis and properties of a biologically active analog of the C-terminal heptapeptide with ε-hydroxy-norleucine sulfate replacing tyrosine sulfate. J Med Chem 21:1030–1037, 1978.
18. BA Kern, RH Reitz. A general method for the preparation of the D- or L-stereoisomers of 5-substituted hydantoins. Agric Biol Chem 42:1275–1278, 1978.
19. P Dreyfuss. Synthesis and some pharmacological properties of 8-ε-hydroxy norleucine-vasopressin. J Med Chem 17:252–257, 1974.
20. AS Bommarius. In: K Drauz, H Waldmann, eds. Enzyme Catalysis in Organic Synthesis, Vol. II. Weinheim: VCH, 1995, pp 633–641.
21. A Galkin, L Kulakova, T Yoshimura, K Soda, N Esaki. A general method for the preparation of the D- or L-stereoisomers of 5-substituted hydantoins. Appl Environ Microbiol 63:4651–4655, 1997.
22. RL Hanson, MD Schwinden, A Banerjee, DB Brzozowski, B-C Chen, BP Patel, CG McNamee, GA Kodersha, DR Kronenthal, LJ Szarka, RN Patel. Enzymatic synthesis of L-6-hydroxynorleucine. Bioorg Med Chem 7:2247–2252, 1999.

23. A Rumbero, JC Martin, MA Lumbreras, P Liras, C Esmahan. Chemical synthesis of allysine ethylene acetal and conversion in situ into 1-piperidine-6-carboxylic acid: key intermediate of the α-aminoadipic acid for β-lactam antibiotics biosynthesis Bioorg Med Chem 3:1237–1240, 1995.
24. RL Hanson, JM Howell, TL LaPorte, MJ Donovan, DL Cazzulino, V Zannella, MA Montana, VB Nanduri, SR Schwarz, RF Eiring, SC Durand, JM Wasylyk, WL Parker, LJ Szarka, RN Patel. Synthesis of allysine ethylene acetal using phenylalanine dehydrogenase from *Thermoactinomyces intermedius*. Enzyme Microbial Technol 26:348–358, 1999.
25. H Schütte, J Flossdorf, H Sahm, M-R Kula. Purification and properties of formaldehyde dehydrogenase and formate dehydrogenase from *Candida boidinii*. Eur J Biochem 62:151–160, 1976.
26. CT Hou, RN Patel, AI Laskin, N Barnabe. NAD-linked formate dehydrogenase from methanol-grown *Pichia pastoris* NRRL-Y-7556. Arch Biochem Biophys 216: 296–305, 1982.
27. MA Lands, A Arnold, JP McAuliff, FP Luduena, TG Brown. Differentiation of receptor systems activated by sympathomimetic amines. Nature 214:597–598, 1967.
28. JRS Arch. β3-Adrenoceptors and other putative atypical β-adrenoceptors. Pharmacol Rev Commun 9:141–148, 1997.
29. JRS Arch, S Wilson. β3-Adrenoceptors and the regulation of metabolism in adipose tissues. Biochem Soc Trans 24:412–418, 1996.
30. JRS Arch, AT Ainsworth, MA Cawthrone, V Piercy, MV Sennitt, VE Thody, C Wilson, S Wilson. Atypical β-adrenoceptor on brown adipocytes as target for anti-obesity drugs. Nature 309:163–165, 1984.
31. C Wilson, S Wilson, V Piercy, MV Sennitt, JRS Arch. The rat lipolytic β-adrenoceptor: studies using novel β-adrenoceptor agonists. Eur J Pharmacol 100:309–319, 1984.
32. JD Bloom, MD Datta, BD Johnson, A Wissner, MG Bruns, EE Largis, JA Dolan, TH Claus. Disodium (R,R)5-[2-(3-chlorophenyl)-2-hydroxyethyl]amino]propyl]-1,3-benzodioxole-2,2-dicarboxylate. A potent β-adrenergic agonist virtually specific for β3 receptors. J Med Chem 35:3081–3084, 1989.
33. LG Fisher, PM Sher, S Skwish, IM Michel, SM Seiler, KEJ Dickinson. BMS-187257, a potent, selective, and novel heterocyclic β-3 adrenergic receptor agonist. Bioorg Med Chem Lett 6:2253–2258, 1996.
34. PM Sher. Preparation of 5-substituted benzo-1,3-dioxoles as β3 adrenergic receptor agonists. U.S. Patent Appl. 5321036, 1994.
35. RN Patel, A Banerjee, L Chu, D Brzozowski, V Nanduri, LJ Szarka. Microbial synthesis of chiral intermediates for β-3-receptor agonists. J Am Oil Chem Soc 75: 1473–1488, 1998.
36. MA Ondetti, DW Cushman. Inhibition of renin-angiotensin system: a new approach to the theory of hypertension. J Med Chem 24:355–361, 1981.
37. MA Ondetti, B Rubin, DW Cushman. Design of specific inhibitors of angiotensin-converting enzyme: new class of orally active antihypertensive agents. Science 196: 441–444, 1977.
38. CT Goodhue, JR Schaeffer. Preparation of L-(+)-β-hydroxyisobutyric acid

by bacterial oxidation of isobutyric acid. Biotechnol Bioeng 13:203–214, 1971.
39. M Schimazaki, J Hasegawa, K Kan, K Nemura, Y Nose, H Kondo, AA Seymour, TN Swerdel, B Abboa-Offei. Synthesis of captopril from an optically active β-hydroxy acid. J Cardiovasc Pharmacol 17:456–465, 1991.
40. J Hasegawa, M Ogura, H Kanema, N Noda, H Kawaharada, KJ Watanabe. Production of D-β-hydroxyisobutyric acid from isobutyric acid by *Candida rugosa* and its mutant. J Fermen Technol 60:501–508, 1982.
41. QM Gu, CS Chen, CJ Sih. Bifunctional chiral synthons via biochemical methods VIII. Optically active 3-aroyl-thio-2-methylpropionic acid. Tetrahedron Lett 27: 1763–1766, 1986.
42. A Sakimas, K Yuri, N Ryozo, O Hisao. Process for preparing optically active carboxylic acids and antipode esters thereof. European Patent 0172614, 1986.
43. CJ Sih. Process for preparing optically active 3-acylthio-2-methyl propionic acid derivatives. European Patent 87264125, Derwent International Publication W087105328, 1987.
44. RN Patel, JM Howell, CG McNamee, KF Fortney, LJ Szarka. Stereoselective enzymatic hydrolysis of α-[acetylthio)methyl]phenylpropionic acid and 3-acetylthio-2-methylpropionic acid. Biotechnol Appl Biochem 16:34–47, 1992.
45. RN Patel, JM Howell, A Banerjee, KF Fortney, LJ Szarka. Stereoselective enzymatic esterification of 3-benzoylthio-2-methylpropionic acid. Appl Microbiol Biotechnol 36:29–34, 1991.
46. JL Moniot. Preparation of N-[2-(mercaptomethyl)propionyl]-L-prolines. U.S. Patent Appl. CN 88-100862, 1988.
47. MA Ondetti, A Miguel, J Krapcho. Mercaptoacyl derivatives of substituted prolines. U.S. Patent 4316906, 1982.
48. NG Delaney, EN Gordon, JM DeForrest, DW Cushman. Amino acid and peptide derivatives as inhibitors of neutral endopeptidase and their use as antihypertensives and diuretics. European Patent EP361365, 1988.
49. RN Patel, RS Robison, LJ Szarka. Stereoselective enzymic hydrolysis of 2-cyclohexyl- and 2-phenyl-1,3-propanediol diacetate in biphasic systems. Appl Microbiol Biotechnol 34:10–14, 1990.
50. DS Karanewsky, MC Badia, DW Cushman, JM DeForrest, T Dejneka, MJ Loots, MG Perri, EW Petrillo, JR Powell. (Phosphinyloxy)acyl amino acid inhibitors of angiotensin converting enzyme (ACE). 1. Discovery of (S)-1-[6-amino-2-[[hydroxy(4-phenylbutyl)phosphinyl]oxy]-1-oxohexyl]-L-proline, a novel orally active inhibitor of ACE. J Med Chem 31:204–212, 1988.
51. RL Hanson, KS Bembenek, RN Patel, LJ Szarka. Transformation of N-ε-CBZ-L-lysine to CBZ-L-oxylysine using L-amino acid oxidase from *Providencia alcalifaciens* and L-2-hydroxy-isocaproate dehydrogenase from *Lactobacillus confusus*. Appl Microbiol Biotechnol 37:599–603, 1992.
52. M Nakane. Preparation and formulation of 7-oxabicycloheptane substituted sulfonamide prostaglandin analogs useful in treatment of thrombolic disease. U.S. Patent 4,663,336, 1987.
53. AW Ford-Hutchinson. Innovations in drug research: inhibitors of thromboxane and leukotrienes. Clin Exp Allergy 21:272–276, 1991.

54. J Das, MF Haslanger, JZ Gougoutas, MF Malley. Synthesis of optically active 7-oxabicyclo[2.2.1]heptanes and assignment of absolute configuration. Synthesis 12: 1100–1112, 1987.
55. N Hamanaka, T Seko, T Miyazaki, A Kawasaki. Rational design of thromboxane A2 antagonists. Adv Prostaglandin, Thromboxane and Leukotriene Res 21:359–362, 1990.
56. JB Jones, JF Beck. In: JB James, CJ Sih, DE Perlman, eds. Application of Biochemical Systems in Organic Synthesis. New York: Wiley, 1987, pp 248–376.
57. H Yamada, S Shimizu. Microbial and enzymatic processes for the production of valuable biological and chemical compounds. Angew Chem Int Ed Engl 27:622–642, 1988.
58. KP Lok, TJ Jakovac, JB Jones. Enzymes in organic synthesis. 34. Preparation of enantiomerically pure exo- and endo-bridged bicyclic [2.2.1] chiral lactones via stereospecific horse liver alcohol dehydrogenase catalyzed oxidations of meso diols. J Am Chem Soc 107:2521–2526, 1985.
59. AJ Irwin, JB Jones. Stereoselective horse liver alcohol dehydrogenase catalyzed oxidoreductions of some bicyclic [2.2.1] and [3.2.1] ketones and alcohols. J Am Chem Soc 98:8476–8482, 1976.
60. RN Patel, M Liu, A Banerjee, JK Thottathil, J Kloss, LJ Szarka. Stereoselective microbial/enzymatic oxidation of (exo, exo)-7-oxabicyclo[2.2.1]heptane-2,3-dimethanol to the corresponding chiral lactol and lactone. Enzyme Microb Technol 14:778–784, 1992.
61. RN Patel, M Liu, A Banerjee, LJ Szarka. Stereoselective enzymic hydrolysis of (exo, exo)-7-oxabicyclo[2.2.1]heptane-2,3-dimethanol diacetate ester in a biphasic system. Appl Microbiol Biotechnol 37:180–183, 1992.
62. K Mori, H Mori, H. Sugai. Biochemical preparation of both the enantiomers of methyl-3-hydroxypentanoate and their conversion to the enantiomers of 4-hexanolide, the pheromone of *Trogoderma glabrum*. Tetrahedron 41:919–925, 1985.
63. M Hirama, M Uei. A chiral total synthesis of compactin. J Am Chem Soc 104: 4251–4256, 1982.
64. AS Gopalan, CJ Sih. Bifunctional chiral synthons via biochemical methods. 5. Preparation of (S)-ethyl hydrogen-3-hydroxyglutarate, key intermediate to (R)-4-amino-3-hydroxybutyric acid and L-carnitine. Tetrahedron Lett 25:5235–5238, 1984.
65. CJ Sih, BN Zhou, AS Gopalan, WR Shieh, CS Chen, G Girdaukas, F vanMiddlesworth. Enantioselective reduction of β-ketoesters by Baker's yeast. Ann NY Acad Sci 434:186–193, 1984.
66. OP Ward, CS Young. Reductive biotransformations of organic compounds by cells or enzymes of yeast. Enzyme Microb Technol 12:482–493, 1990.
67. RN Patel, CG McNamee, A Banerjee, JM Howell, RS Robison, LJ Szarka. Stereoselective reduction of β-ketoesters by *Geotrichum candidum*. Enzyme Microb Technol 14:731–738, 1992.
68. E Keinan, EK Hafeli, KK Seth, RR Lamed. Thermostable enzymes in organic synthesis. 2. Asymmetric reduction of ketones with alcohol dehydrogenase from *Thermoanaerobium brockii*. J Am Chem Soc 108:162–168, 1986.

69. RN Patel, CT Hou, AI Laskin, P Derelanko. Microbial production of methylketones: properties of purified yeast secondary alcohol dehydrogenase. J Appl Biochem 3:218–232, 1981.
70. CW Bradshaw, H Fu, GJ Shen, C-H Wong. A *Pseudomonas* sp. alcohol dehydrogenase with broad specificity and unusual stereospecificity for organic synthesis. J Org Chem 57:1526–1531, 1992.
71. M Christen, DHG Crout. Biotransformation in organic synthesis: applications of yeast reduction in the synthesis of 3,5-dihydroxy esters of high optical purity. J Chem Soc Chem Commun 45:264–266, 1988.
72. RN Patel, A Banerjee, CG McNamee, D Brzozowski, RL Hanson, LJ Szarka. Enantioselective microbial reduction of 3,5-dioxo-6-(benzyloxy)hexanoic acid, ethyl ester. Enzyme Microb Technol 15:1014–1021, 1993.
73. SY Sit, RA Parker, I Motoe, HW Balsubramanian, CD Cott, PJ Brown, WE Harte, MD Thompson, J Wright. Synthesis, biological profile, and quantitative structure-activity relationship of a series of novel 3-hydroxy-3-methylglutaryl coenzyme A reductase inhibitors. J Med Chem 33:2982–2999, 1990.
74. RN Patel, CG McNamee, LJ Szarka. Enantioselective enzymic acetylation of racemic [4-[4α,6β(E)]]-6-[4,4-bis(4-fluorophenyl)-3-(1-methyl-1H-tetrazol-5-yl)-1,3-butadienyl]tetrahydro-4-hydroxy-2H-pyran-2-one. Appl Microbiol Biotechnol 38: 56–60, 1992.
75. A Endo, M Kuroda, Y Tsujita. ML-236A, ML-236B, and ML-236C, new inhibitors of cholesterogenesis produced by *Penicillum citrinium*. J Antibiot 29:1346–1348, 1976.
76. A Endo, M Kuroda, K Tanzawa. Competitive inhibition of 3-hydroxy-3-methylglutaryl coenzyme A reductase by ML-236A and ML-236B fungal metabolites, having hypocholesterolemic activity. FEBS Lett 72:323–326, 1976.
77. M Hosobuchi, K Kurosawa, H Yoshikawa. Application of computer on monitoring and control of fermentation process: microbial conversion of ML-236B sodium to pravastatin. Biotechnol Bioeng 42:815–820, 1993.
78. N Serizawa, S Serizawa, K Nakagawa, K Furuya, T Okazaki, A Tarahara. Microbial hydroxylation of ML-236B (compactin). Studies on microorganisms capable of 3β-hydroxylation of Ml-236B. J Antibiot 36:887–891, 1983.
79. T Matsuoka, S Miyakoshi, K Tanzawa, K Nakahara, M Hosobuchi, N Serizawa. Purification and characterization of cytochrome P-450sca from *Streptomyces carbophilus*. ML-236B (compactin) induces a cytochrome-P450sca in *Streptomyces carbophilus* that hydroxylates ML-236 B to pravastatin sodium (CS-514), a tissue-selective inhibitor of 3-hydroxy-3-methylglutaryl coenzyme A reductase. Eur J Biochem 184:707–713, 1989.
80. ARPM Valentijn, R de Hann, E de Kant, GA van der Marel, LH Cohen, JH van Boom. Synthesis of a potential enzyme-specific inhibitor of squalene synthase. Rec Trav Chim Pays-Bas 114:332–336, 1995.
81. SA Biller, MJ Sofia, B DeLange, C Foster, EM Gordon, T Harrity, LC Rich, CP Ciosek. The first potent inhibitor of squalene synthase: a profound contribution of an ether oxygen to inhibitor-enzyme interaction. J Am Chem Soc 113:8522–8527, 1991.
82. SA Biller, C Foster, EM Gordon, T Harrity, LC Rich, J Maretta, CP Ciosek. Iso-

prenyl phosphinylformates: new inhibitors of squalene synthase. J Med Chem 34: 1912–1918, 1991.

83. A Steiger, HJ Pyun, RM Coates. Synthesis and characterization of aza analog inhibitors of squalene and geranylgeranyl diphosphate synthases. J Org Chem 57:3444–3450, 1992.
84. C Dufresne, KE Wilson, D Zink, J Smith, JD Bergstrom, M Kurtz, D Rew, M Nallin, R Jenkins. The isolation and structure elucidation of zaragozic acid C, a novel potent squalene synthase inhibitor. Tetrahedron 48:10221–10226, 1992.
85. SA Biller, D Majnin. U.S. Patent 5428028, 1995.
86. RM Lawrence, SA Biller, OM Fryszman. Preparation of α-phosphonosulfinic squalene synthetase inhibitors. U.S. Patent 5447922, 1995.
87. RN Patel, A Banerjee, LJ Szarka. Stereoselective acetylation of [1-(hydroxy)-4-(3-phenoxyphenyl)butyl]phosphonic acid diethyl ester. Tetrahedron Asymmetry 8: 1055–1059, 1997.
88. M Chaffman, RN Brogden. Diltiazem. A review of its pharmacological properties and therapeutic efficacy. Drugs 29:387–390, 1985.
89. C Kawai, T Konishi, E Matsuyama, H Okazaki. Comparative effects of three calcium antagonists, diatiazem, verapamil and nifedipine, on the sinoatrial and atrioventricular nodes. Experimental and clinical studies. Circulation 63:1035–1038, 1981.
90. T Isshiki, B Pegram, E Frohlich. Immediate and prolonged hemodynamic effects of TA-3090 on spontaneously hypertensive (SHR) and normal Wistar-Kyoto (WKY) rats. Cardiovasc Drug Ther 2:539–544, 1988.
91. DM Floyd, RY Moquin, KS Atwal, SZ Ahmed, SH Spergel, JZ Gougoutas, MF Malley. Synthesis of benzazepinone and 3-methylbenzothiazepinone analogs of diltiazem. J Org Chem 55:5572–5575, 1990.
92. J Das, DM Floyd, SM Kimball, RN Patel, JK Thottathil. Chemoenzymatic synthesis of benzazepinone calcium channel blocking agents. Indian J Chem 31B:817–820, 1992.
93. RN Patel, RS Robison, LJ Szarka, J Kloss, JK Thottathil, RH Mueller. Stereospecific microbial reduction of 4,5-dihydro-4-(4-methoxyphenyl)-6-(trifluoromethyl-1H-1)-benzazepin-2-one. Enzyme Microb Technol 13:906–912, 1991.
94. H Matsumae, M Furui, T Sabatani. Lipase-catalyzed asymmetric hydrolysis of 3-phenylglycidic acid ester, the key intermediate in the synthesis of diltiazem hydrochloride. J Ferment Bioeng 75:93–98, 1993.
95. G Edwards, AH Weston. Structure-activity relationships of K+ channel openers. TIPS 11:417–422, 1990.
96. DW Robertson, MI Steinberg. Potassium channel openers: new biological probes. Annu Rep Med Chem 24:91–100, 1989.
97. TC Hamilton, AH Weston. Cromakalim, nicorandil and pinacidil: novel drugs which open potassium channels in smooth muscle. Gen Pharmacol 20:1–9, 1989.
98. VA Ashwood, RE Buckingham, F Cassidy, JM Evans, TC Hamilton, DJ Nash, G Stempo, KJ Willcocks. Synthesis and antihypertensive activity of 4-(cyclic amido)-2H-1-benzopyrans. J Med Chem 29:2194–2201, 1986.
99. EN Jacobsen, W Zhang, AR Muci, JR Ecker, L Deng. Highly enantioselective

epoxidation catalysts derived from 1,2-diaminocyclohexane. J Am Chem Soc 113: 7063–7067, 1991.
100. R Bergmann, V Eiermann, RJ Gericke. 4-Heterocyclyloxy-2H-1-benzopyran potassium channel activators. J Med Chem 33:2759–2761, 1990.
101. K Atwal, GJ Grover, KS Kim. Preparation of benzopyranylcyanoguanidine derivatives. U.S. Patent 5140031, 1991.
102. JM Evans, CS Fake, TC Hamilton, RH Poyser, EA Watts. Synthesis and antihypertensive activity of substituted trans-4-amino-3,4-dihydro-2,2-dimethyl-2H-1-benzopyran-3-ols. J Med Chem 26:1582–1589, 1983.
103. RN Patel, A Banerjee, B Davis, JM Howell, CG McNamee, D Brzozowski, J North, D Kronenthal, LJ Szarka. Stereoselective epoxidation of 2,2-dimethyl-2H-1-benzopyran-6-carbonitrile. Bioorg Med Chem 2:535–542, 1994.
104. RN Patel, A Banerjee, CG McNamee, LJ Szarka. Stereoselective acetylation of 3,4-dihydro-3,4-dihydroxy-2,2-dimethyl-2H-1-benzopyran-6-carbonitrile. Tetrahedron Asymmetry 6:123–130, 1995.
105. AA Larsen, PM Lish. A new bio-isostere: alkylsulfonamidophenethanolamines. Nature 203:1283–1284, 1964.
106. RH Uloth, JR Kirk, WA Gould, AA Larsen. Sulfonanilides. I. Monoalkyl- and arylsulfonamidophenethanolamines. J Med Chem 9:88–96, 1966.
107. PM Lish, JH Weikel, KW Dungan. Pharmacological and toxicological properties of two new β-adrenergic receptor antagonists. J Pharmacol Exp Ther 149:161–173, 1965.
108. P Somani, T Bachand. Blockade of cardiac effects of isoproterenol by the stereoisomers of sotalol. Eur J Pharmacol 7:239–247, 1969.
109. RN Patel, A Banerjee, CG McNamee, LJ Szarka. Stereoselective microbial reduction of N-(4-(1-oxo-2-chloroacetyl ethyl) phenyl methane sulfonamide. Appl Microbiol Biotechnol 40:241–245, 1993.

7

Cloning, Structure, and Activity of Ketone Reductases from Baker's Yeast

Jon D. Stewart and Sonia Rodríguez
University of Florida, Gainesville, Florida

Margaret M. Kayser
University of New Brunswick, Saint John, New Brunswick, Canada

I. INTRODUCTION

While Baker's yeast (*Saccharomyces cerevisiae*) was domesticated over 4000 years ago for bread and beverage production, it was not until the early part of the twentieth century that its potential for asymmetric ketone reductions was appreciated [1]. Since then, baker's yeast, which is inexpensive, simple to handle, nonpathogenic, and reduces a wide range of ketones, has become an important reagent for organic synthesis. Such applications of baker's yeast reductions have been reviewed extensively, and these can provide important precedent for deciding whether baker's yeast is likely to accept a novel substrate as well as guidance in predicting the major product [2–9]. Some of these reviews also comment on the efforts to isolate and characterize enzymes catalyzing asymmetric ketone reductions from baker's yeast; however, there appears to have been no systematic attempt to link the recently completed genome sequence of *S. cerevisiae* [10] to enzymes with synthetic utility, and this is the focus of the present review. The goal of this chapter is to uncover all yeast genes that might participate in the reduction of exogenous ketones, which is the first step toward rational improvements in cofactor and enzyme levels by recombinant DNA techniques, an approach that is still in its infancy [11].

II. SOURCE OF REDUCING EQUIVALENTS

The ability of baker's yeast to reduce exogenously added ketones is closely related to the fermentation process that converts carbohydrates such as glucose or starch to ethanol and carbon dioxide via glycolysis and the subsequent decarboxylation and reduction of pyruvate (Scheme 1). Efficient ethanol and carbon dioxide production directly influences the efficiencies of both bread making and brewing, and *S. cerevisiae* has been subjected to intense selective pressure over the preceding four millennia to optimize carbohydrate assimilation. During fermentation, the metabolic goal is to convert all available carbohydrate to ethanol, a process that does not require oxygen. Glycolysis depletes the cytoplasmic pool of NAD^+, as this cofactor is used to oxidize glyceraldehyde-3-phosphate to 1,3-*bis*-phosphoglycerate (Scheme 1). The NADH-dependent reduction of acetaldehyde yielding ethanol is a major route to regenerating the NAD^+ needed for additional carbohydrate assimilation. On the other hand, any pyruvate decarboxylated to acetyl-coenzyme A (acetyl-CoA) in preparation for oxidative phosphorylation results in an excess of NADH in the cytoplasm. This is believed to be the major source of NADH available to reduce exogenously added ketones by the subset of enzymes that accept this cofactor.

The pentose phosphate pathway is the other major route by which glucose is assimilated by baker's yeast, and this pathway yields excess reducing equivalents in the form of NADPH (Scheme 2). The fate of the ribulose 5-phosphate depends on the cellular requirements for pentose sugars, but it can be converted

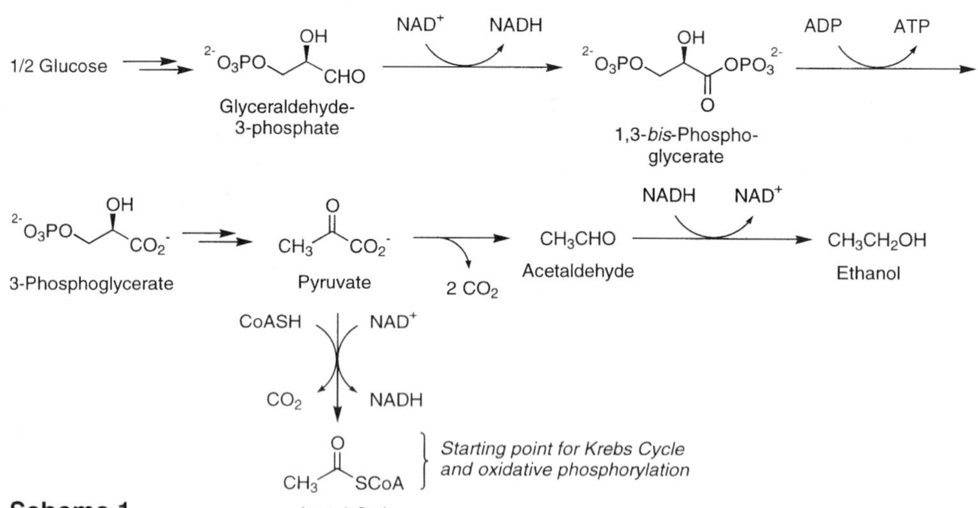

Scheme 1

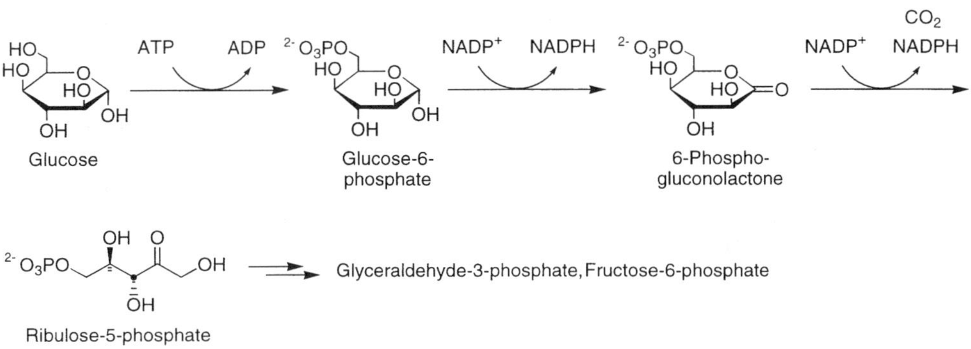

Scheme 2

by a series of equilibria into intermediates in glycolysis (glyceraldehyde 3-phosphate and fructose 6-phosphate). The possibility that the pentose phosphate pathway can operate as a cycle has been debated for some time, although the consensus is that this situation is not normally encountered in sugar-grown baker's yeast [12]. The two oxidative steps at the beginning of the pentose phosphate pathway are believed to be the major source of NADPH required for endogenous biosynthetic pathways when baker's yeast is grown on sugars. When ethanol is used as the carbon source, acetaldehyde dehydrogenase (which catalyzes the $NADP^+$-dependent oxidation of acetaldehyde to acetate) is likely to be the major source of NADPH [12]. This cofactor is required by the majority of yeast reductases that accept exogenous ketones [12].

To facilitate synthetically useful ketone reductions, the quantities of reduced nicotinamides must be kept relatively high. It is experimentally difficult to measure the levels of cellular NADH and NADPH in situ [13], and optimization of this key parameter is therefore generally performed by trial and error using the rate of ketone reduction as the dependent variable. The nature of the carbon source, the supply of oxygen, and other growing conditions affect the levels of reducing equivalents, and variation of these experimental parameters can therefore improve the performance of baker's yeast-mediated ketone reductions [14,15]. For NADH, controlling the partitioning of pyruvate between ethanol and glycerol production (which consumes NADH) and oxidative phosphorylation (which does not regenerate cytoplasmic NAD^+) has the greatest influence on the supply of NADH available for reducing exogenous ketones. For NADPH, the level of carbon flux through the pentose phosphate pathway versus that through glycolysis has the greatest impact on the quantity of cytoplasmic NADPH. The factor(s) controlling these fluxes are not well understood in *S. cerevisiae*. In addition, it appears that the cytoplasmic NADH and NADPH pools are independent in baker's yeast, since this organism lacks both a cytoplasmic pyridine nucleotide transhydrogenase [which interconverts reduced and oxidized NAD(H) and NADP(H)] [16] and NAD(H) kinase [17] activities. This point is important, since many reductases are specific for either NADH or NADPH. Furthermore, Kielland-Brandt and co-workers recently demonstrated that expressing a membrane-bound bacterial transhydrogenase in *S. cerevisiae* resulted in a net transfer of reducing equivalents from NADPH to NADH [18]. While this shift may be useful in certain applications, most previously characterized yeast enzymes that accept synthetically important ketones are specific for NADPH.

III. MULTIPLE BAKER'S YEAST REDUCTASES AND STEREOSELECTIVITY IN WHOLE CELL-MEDIATED REDUCTIONS

Baker's yeast is able to reduce a broad variety of ketones because of the presence of a large complement of reductase enzymes. In nearly all cases, the physiological substrate(s) for these enzymes are not known; instead they have been purified on the basis of activity toward a specific ketone. For this reason, it is important to keep in mind that descriptions such as "important" and "major" contributors to ketone reductions are meaningful only in the context of specific substrates. These practices have also led to ad-hoc naming of yeast reductases that often derive from the order of elution from specific chromatographic columns or reflect the nature of the assay substrate used during the purification, which often leads to confusion when trying to compare results from different laboratories. In many cases, different groups have purified reductases with similar but nonidentical physical properties and it is not clear whether these differences are yeast strain-related or whether the enzymes are distinct. For all of these reasons, the number and properties of ketone reductases produced by baker's yeast is still ambiguous. All of the ketone and keto-ester reductases purified from baker's yeast as of 1998, along with their physical properties, have been tabulated recently by Straathof [12]. An important goal for future studies in this area will be the linking of this biochemical information to the corresponding yeast genes using the complete genome sequence of *S. cerevisiae* [10].

The presence of multiple reductases with overlapping substrate specificities but differing stereoselectivities often leads to mixtures of alcohol products when whole yeast cells are used for ketone reductions. Several ingenious methods have been devised to overcome these problems. These include changing the substrate structure [19,20] or concentration [19,21], altering the carbon source [22], performing reactions in organic solvents or two-phase systems [23–27], immobilization of the yeast cells [28–31], or adding inhibitors [32–37] or sulfur compounds [38,39] to the reductions. All of these strategies are aimed at decreasing the catalytic activities of competing enzymes while leaving that of the target enzyme

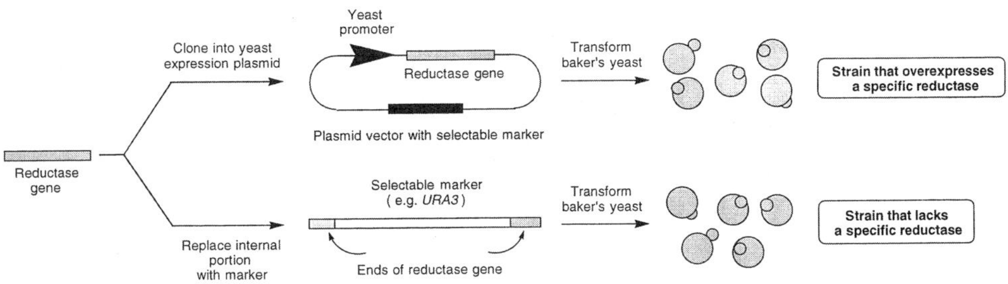

Figure 1 Strategy for tailoring the stereoselectivity of whole-cell baker's yeast-mediated reductions. The level of a yeast reductase can be increased by placing the corresponding gene under control of a strong yeast promoter on an extrachromosomal plasmid. Alternatively, the catalytic activity of a reductase can be eliminated by replacing the corresponding gene on the chromosome with a defective copy prepared in vitro. (Reprinted with permission from Organic Letters 1999, *1*. © 1999 American Chemical Society.)

largely undisturbed. The major difficulty with all of these approaches is that they are empirical and their effects on the reductions of novel substrates are not easily predictable. On the other hand, recombinant DNA technology provides a potentially powerful route to improving the stereoselectivities of yeast reductions rationally. This possibility had been recognized earlier [12,19], although it has only been explored systematically in recent years [11]. In this approach, desirable enzymes are overexpressed while competing reductases are eliminated by knocking out the corresponding genes (Fig. 1). The major advantage of this metabolic engineering strategy over previous methods is that the enzyme whose level is manipulated is known unambiguously. Knocking out individual reductase genes also provides a simple means to determine whether a given enzyme contributes significantly to biotransformations of individual substrates. This approach will become even simpler once construction of a complete library of single-gene knockout strains is completed within the next few years [40].

IV. GENERAL APPROACH TO IDENTIFYING YEAST GENES ENCODING POTENTIAL KETONE REDUCTASES

Using rational metabolic engineering to tailor the stereoselectivity of yeast-mediated reductions is feasible only when all of the relevant genes are known and the relationships between enzyme and acceptable substrates are defined. We are far from this ideal. For a few reductase proteins, amino acid sequence data has

Table 1 *S. cerevisiae* Genes Encoding Potential Ketone Reductases[a]

Protein	Gene	ORF	Mature M_r (Da)	Predicted p*I*	Knockout	Location
Lactate dehydrogenases						
L-Lactate dehydrogenase	*CYB2*	*Yml054c*	56,584	6.36	Viable	Soluble, mitochondria
D-Lactate dehydrogenase	DLD1	Ydl174c	62,156	5.90	Viable	Mitochondrial membrane
D-Lactate dehydrogenase	AIP2	Ydl178w	59,259	5.91	Viable	Soluble, mitochondria
D-Lactate dehydrogenase	—	*Yel071w*	55,089	6.66	Viable	Soluble, cytoplasm
Alcohol dehydrogenases						
Long-chain alcohol dehydrogenase	SFA1	Ydl168w	40,935	6.41	Viable	Unknown
Alcohol dehydrogenase I	ADH1	Yol086c	36,743	6.30	Viable	Soluble, cytoplasm
Alcohol dehydrogenase II	*ADH2*	*Ymr303c*	36,625	6.37	Viable	Soluble, cytoplasm
Alcohol dehydrogenase III	ADH3	Ymr083w	37,164	7.37	Viable	Soluble, mitochondria
Alcohol dehydrogenase V	ADH5	Ybr145w	37,510	6.15	Unknown	Unknown
Alcohol dehydrogenase IV	*ADH4*	*Ygl256w*	50,622	8.71	Viable	Soluble, cytoplasm
Medium-chain alcohol dehydrogenases						
Sorbitol dehydrogenase	SOR1	Yjr159w	38,061	6.52	Unknown	Unknown
Putative oxidoreductase	—	*Ylr070c*	38,461	5.93	Unknown	Unknown
Putative oxidoreductase	—	*Yal061w*	46,084	5.88	Unknown	Unknown
Putative oxidoreductase	—	*Yal060w*	41,531	6.43	Unknown	Unknown
Putative oxidoreductase	—	Ymr318c	39,511	6.38	Unknown	Unknown
Putative oxidoreductase	—	*Ycr105w*	39,348	7.18	Unknown	Unknown
ζ-Crystallin homolog	ZTA1	Ybr046c	37,023	8.89	Viable	Unknown
Short-chain alcohol dehydrogenases						
α-Acetoxyketone reductase	*GRE2*	*Yol151w*	38,060	5.82	Viable	Soluble, cytoplasm
Putative oxidoreductase	—	*Ydr541c*	38,475	6.09	Unknown	Unknown
Putative oxidoreductase	—	Ygl157w	37,944	5.70	Unknown	Unknown
Putative oxidoreductase	—	*Ygl039w*	38,122	5.79	Unknown	Unknown
Putative oxidoreductase	—	Ylr426w	36,716	9.24	Unknown	Unknown
Putative oxidoreductase	—	*Ybr159w*	38,590	9.65	Lethal	Unknown
Putative oxidoreductase	—	*Ybl114w*	34,174	9.64	Unknown	Unknown

Putative oxidoreductase	—	*Ymr226c*	29,054	6.20	Viable	Unknown
Putative oxidoreductase	—	Yir035c	27,376	5.83	Unknown	Unknown
Putative oxidoreductase	—	Yir036c	28,700	5.84	Unknown	Unknown
Putative oxidoreductase	—	Yill24w	32,720	9.34	Unknown	Unknown
Putative oxidoreductase	—	*Ykl107w*	34,520	6.37	Unknown	Unknown
3-Oxoacyl-(acyl-carrier-protein) reductase	*OAR1*	*Ykl055c*	31,191	9.17	Viable	Mitochondria
Aryl alcohol dehydrogenase	*AAD14*	*Ynl331c*	41,856	6.51	Unknown	Soluble, cytoplasm
Putative aryl alcohol dehydrogenase	AAD3	Ycr107w	40,778	6.58	Viable	Soluble, cytoplasm
Putative aryl alcohol dehydrogenase	*AAD4*	*Ydl243c*	36,871	6.40	Unknown	Soluble, cytoplasm
Putative aryl alcohol dehydrogenase	*AAD10*	*Yjr155w*	32,471	9.01	Viable	Soluble, cytoplasm
Putative oxidoreductase	—	Yfl057c	16,493	7.87	Unknown	Unknown
Putative aryl alcohol dehydrogenase	AAD15	Yol165c	15,618	5.78	Unknown	Unknown
Aldose reductase family						
Aldo-keto reductase	*YPR1*	*Ydr368w*	34,622	6.88	Viable	Soluble
Putative oxidoreductase	*GCY1*	*Yor120w*	34,948	7.98	Viable	Soluble
D-Arabinose dehydrogenase	ARA1	Ybr149w	38,411	5.63	Viable	Soluble
Putative oxidoreductase	*GRE3*	*Yhr104w*	37,014	6.70	Unknown	Unknown
Putative oxidoreductase	—	Yjr096w	32,210	6.71	Unknown	Unknown
Putative oxidoreductase	—	Ydl124w	35,454	5.76	Unknown	Unknown
D-Hydroxyacid dehydrogenase family						
Putative oxidoreductase	—	*Ypl113c*	44,880	7.03	Unknown	Unknown
Putative oxidoreductase	—	*Ygl185c*	42,855	8.27	Unknown	Unknown
Putative oxidoreductase	—	*Ynl274c*	38,721	5.98	Unknown	Unknown
Putative oxidoreductase	FDH1	Yor388c	41,606	6.11	Unknown	Unknown
Putative oxidoreductase	—	Yp1275w	26,364	9.41	Unknown	Unknown
Multifunctional						
Fatty acid synthase	*FAS1*	*Ykl182w*	228,678	5.71, 5.27	Lethal	Soluble, cytoplasm
	FAS2	*Ypl231w*	206,895	" "	Lethal	

[a] Yeast proteins are grouped into the appropriate superfamily. Protein names reflect the most common nomenclature in current use. The accepted gene name is given where one has been assigned; in other cases, only the systematic open reading frame code identifies the putative protein. Gene names shown in italics indicate that a strain carrying a deletion in this gene is available as of September 1999.

linked biochemical data with the corresponding gene conclusively. In other cases, reductases have been isolated from yeast cells and their substrate specificities have been defined to a greater or lesser extent, but the absence of amino acid sequence data has precluded matching to the appropriate gene. Given the ability of baker's yeast cells to reduce a seemingly endless variety of ketones, it seems unlikely that all of the yeast ketone reductases have been isolated, and our efforts to identify all genes encoding potential reductases gives a total number of candidates—49—that is much larger than the number of purified reductases [12].

The availability of the genome sequence allowed us to approach the problem of identifying all potential ketone reductases by detecting yeast proteins with sequence similarity to known reductases [41], whether from baker's yeast or other species. These results are presented in Table 1. In some cases, sequence identities of more than 50% were shared with known reductase proteins, making the identification of a putative yeast reductase protein unambiguous. In other cases, the level of sequence identity was much lower (25–35%). In these cases, the yeast protein sequences were grouped according to membership in previously recognized superfamilies—short-chain alcohol dehydrogenases/reductases [42–44], medium-chain alcohol dehydrogenases [44,45], etc.—to allow an informed choice as to their potential as authentic reductase enzymes. Accordingly, these yeast protein sequences were aligned with one another and with a superfamily representative whose catalytic activity and three-dimensional structure had been established (except for short-chain alcohol dehydrogenases, where no related crystal structure is available) [46]. Those yeast proteins that conserved residues known to be critical for catalysis and whose sequence insertions and/or deletions could be reasonably accommodated by the known protein structure were accepted as potential yeast ketone reductases and included in Table 1. We have not attempted to model the yeast reductase proteins using the known X-ray crystal structures, although this may be useful in certain cases. Despite a good deal of effort [47], it is not yet possible to predict reliably the substrate or stereoselectivity of yeast reductases on the basis of computer modeling alone.

One caveat with the genetic approach to identifying yeast reductases is the possibility that their expression may be restricted to specific situations or they may be pseudo-genes that are not expressed under any growth conditions. Gene chip technology offers one approach to addressing these issues by making a genome-wide snapshot of mRNA levels available [48]. While levels of transcription are not always correlated with protein levels, such measurements provide at least some guidance. Clearly, those potential reductases transcribed at reasonable levels are candidates worthy of further investigation. The catalytic efficiency of reductases also plays an important role in determining whether a given candidate is ''important'' for a given substrate, particularly at low substrate concentrations [49].

A. Lactate Dehydrogenases

Baker's yeast contains at least four lactate dehydrogenases, three of which are located in the mitochondrion and one in the cytoplasm (Table 1). Electron flow in the mitochondrial enzymes is linked to cytochromes rather than nicotinamides. Their apparent physiological function is the reversible interconversion of pyruvate and lactate, and it is not clear whether any of these four enzymes accept ketones other than pyruvate, which would limit their importance in organic synthesis.

B. Alcohol Dehydrogenases

Selection for efficient conversion of carbohydrates to ethanol and carbon dioxide has shaped baker's yeast over the past four millennia. One manifestation is the high levels of glycolytic enzymes in the yeast cytoplasm (ca. 30% of total protein). Another is a high level of catalytic activities for the conversion of pyruvate to ethanol by the action of pyruvate decarboxylase and alcohol dehydrogenases. Pyruvate decarboxylation is essentially irreversible under physiological conditions, and it is critical that the cells avoid high concentrations of acetaldehyde given its propensity to react spontaneously with a variety of cellular nucleophiles. It is therefore not surprising that several redundant genes encoding related alcohol dehydrogenases are present in the yeast genome (Table 1).

Five of the yeast alcohol dehydrogenases (encoded by the SFA1, ADH1, ADH2, ADH3, and ADH5 genes) appear to be Zn(II)-dependent enzymes that share a high degree of sequence similarity with mammalian alcohol dehydrogenases. The alcohol dehydrogenase encoded by the ADH4 gene is an iron-dependent enzyme that is a member of a different family. A multiple sequence alignment [46] of the five Zn(II)-dependent enzymes with horse liver alcohol dehydrogenase (HLADH) revealed a number of conserved residues, several of which are known to be critical for structure and function of the horse liver enzyme (Fig. 2A). Horse liver alcohol dehydrogenase contains two Zn(II) ions, a ''structural Zn(II)'' ligated by Cys97, Cys100, Cys103, and Cys111, and a ''catalytic Zn(II)'' ligated by Cys46, His67, and Cys174. The conservation of all seven residues that serve as ligands for the two Zn(II) ions is one important indication that all five yeast genes encode bona-fide alcohol dehydrogenases.

The sequences of these five yeast proteins are also congruent with the experimental three-dimensional structure of the horse liver enzyme, implying that these will have the same overall fold. The length of the Sfal protein (Sfalp) is virtually the same as that of the horse liver enzyme except for a three-amino acid insertion after Asp245 (HLADH numbering), which is located in a surface loop between a β-strand and an α-helix near the entrance to the active site (Fig.

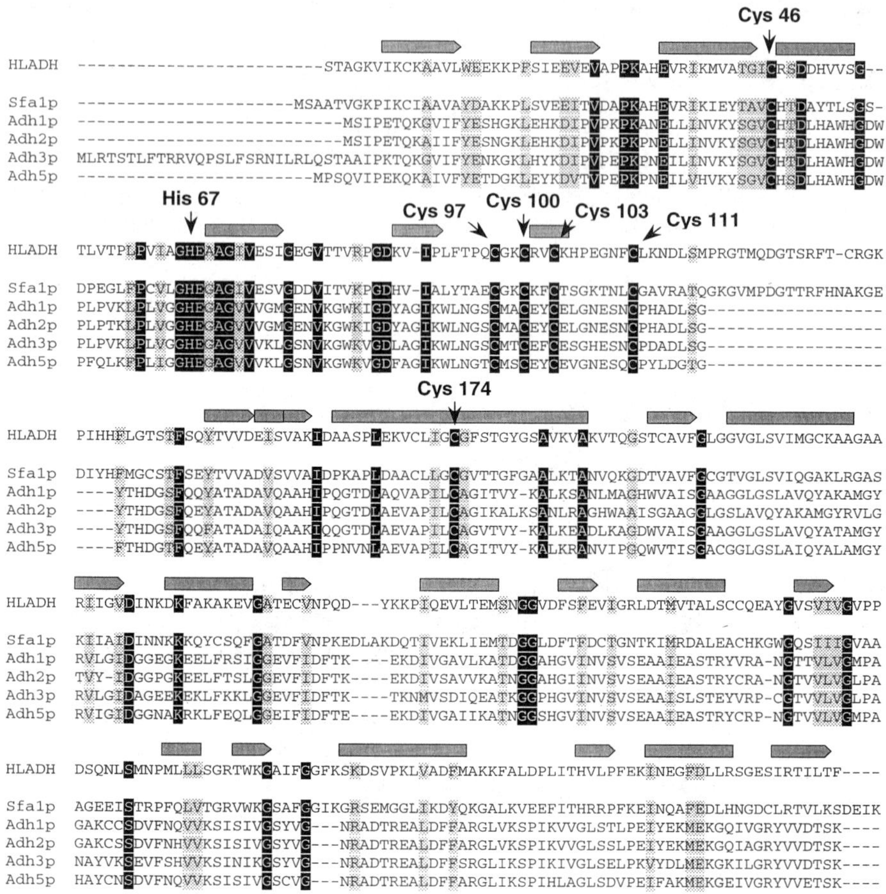

(A)

Figure 2 Alcohol dehydrogenase-like proteins. (A) Aligned sequences of yeast alcohol dehydrogenase-like proteins. The CLUSTAL W program [46] was used to align simultaneously horse liver alcohol dehydrogenase (HLADH) and five related *S. cerevisiae* protein sequences. Residues conserved in all six sequences are shown on a black background, those positions at which a maximum of two closely related amino acids are found are shown on a gray background. Gray arrows above the horse liver alcohol dehydrogenase sequence mark the positions of β-strands, and gray bars represent α-helices from the X-ray crystal structure of this protein (1LDE; [50]). The positions of several key residues important in structure and catalysis are indicated as well. (B) Stereoview of the HLADH structure showing positions of insertions and deletions in similar yeast protein sequences. The active site cleft opens on the right and the location of a three amino acid insertion in Sfa1p (between residues Asp245 and Tyr246) is marked by a dark gray sphere. The locations of two deletions found in yeast proteins Adh1p, Adh2p, Adh3p, and Adh5p relative to HLADH (between Met118 and His139 and between Gly321 and Lys323) are shown in heavy dark lines. Note that only a single monomer of the tetrameric HLADH structure is shown for clarity. This figure was rendered in MolScript [51].

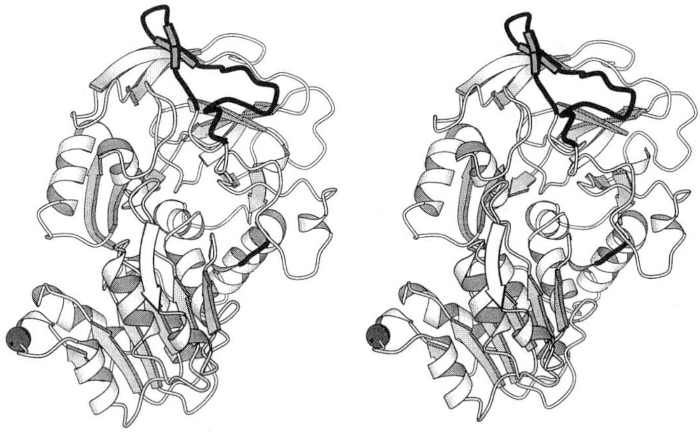

(B)

Figure 2 Continued.

2B)[50]. The proteins encoded by the ADH1, ADH2, ADH3, and ADH5 genes also share a 22-amino acid deletion compared to the HLADH sequence. This region is part of a surface loop in HLADH relatively far from the active site.

The contribution of these alcohol dehydrogenases to reductions of exogenous ketones is unclear. While mammalian alcohol dehydrogenases accept a variety of ketones and aldehydes, yeast alcohol dehydrogenase I has a relatively narrow substrate specificity that is restricted largely to simple ketones [49]. Whether the substrate specificities of enzymes encoded by the ADH2, ADH3, ADH5, and SFA1 genes are similarly restricted awaits experimental investigation, although this seems likely given their high degree of sequence similarity.

C. Medium-Chain Alcohol Dehydrogenases

The *S. cerevisiae* genome also contains a number of other genes related to the above-mentioned alcohol dehydrogenases, and their sequences place all of these proteins in the medium chain alcohol dehydrogenase superfamily (Table 1) [44,45]. Oxidoreductases in this superfamily are generally metal ion-dependent and composed of two domains, a catalytic domain (composed of the N-terminal and C-terminal regions) and a central cofactor-binding domain that contains a Rossman fold. The closest relative of the six putative baker's yeast proteins whose three-dimensional structure is known is human $\chi\chi$ alcohol dehydrogenase [52]. The number of critical residues shared with this enzyme and the locations of sequence insertions and deletions relative to secondary structure elements of the human enzyme suggests that all six of these putative yeast proteins are bona-

```
                                                                     Cys 46
•• ADH    ANEVIKCKAAVAWEAGKPLSIEEIEVAPP---KAHEVRIKIIATAVCHTDAYTLSGADPEG-----------CFPVILG

Ymr318c   -MSYPEKFEGIAIQSHEDWKNPKKTKYDPKPFYDHDIDIKIEACGVCGSDIHCAAGHWGNMK-----------MPLVVG
Ycr105w   -MLYPEKFQGIGISNAKDWKHPKLVSFDPKPFGDHDVDVEIEACGICGSDFHIAVGNWGPVP-----------ENQILG

Yal061w   -------MRALAYFGKGNIRFTN-HLKEPHIVAPDELVIDIEWCGICGTDLHEYTDGPIFFPEDGHTHEISHNPLPQAMG
Yal060w   -------MRALAYFKKGDIHFTN-DIPRPEIQTDDEVIIDVSWCGICGSDLHEYLDGPIFMPKDGECHKLSNAALPLAMG

Sor1p     --MSQNSNPAVVLEKVGDIAIE--QRPIPTIKDPHYVKLAIKATGICGSDIHYYRSGGIGK---------YILKAPMVLG
Ylr070c   -MTDLTTQEAIVLERPGKITLT--NVSIPKISDPNEVIIQIKATGICGSDIHYYTHGRIAN---------YVVESPMVLG

  Glu 68
His 67
•• ADH    HEGAGIVESVGEGVTK-LKAGDTVIPLYIP-------------QCGECKFCLNPKTNLCQKIRVTQGKGLMPDGTSRFT

Ymr318c   HEIVGKVVKLGPKSNSGLKVGQRVGVGAQ------------VFSCLECDRCKNDNEPYCTKFVT-------------T
Ycr105w   HEIIGRVVKVGSKCHTGVKIGDRVGVGAQ------------ALACFECERCKSDNEQYCTNDHVL------------T

Yal061w   HEMAGTVLEVGPGVKN-LKVGDKVVVEPTGTCRDRYRWPLSPNVDKEWCAACKKGYYNICS------------------
Yal060w   HEMSGIVSKVGPKVTK-VKVGDHVVVDAASSCADLHCWPHSKFYNSKPCDACQRGSENLCT------------------

Sor1p     HESSGQVVEVGDAVTR-VKVGDRVAIEPG-------------VPSRYSDETKEGRYNLCP------------------
Ylr070c   HESSGIVALIGENVKT-LKVGDRVALEPG-------------IPDRFSPEMKEGRYNLDP------------------

•• ADH    CKGKTILHYMGTSTFSEYTVVADISVAKIDPLAPLDKVCLLGCGISTGYGAAVNTAKLEPGSVCAVFGLGGVGLAVIMGC

Ymr318c   YSQPYEDGYVSQGGYANYVRVHEHFVVPIPENIPSHLAAPLLCGGLTVYSPLV-RNGCGPGKKVGIVGLGGIGSMGTLIS
Ycr105w   MWTPYKDGYISQGGFASHVRLHEHFAIQIPENIPSPLAAPLLCGGITVFSPLL-RNGCGPGKRVGIVGIGGIGHMGILLA

Yal061w   -YLGLCGAGVQSGGFAERVVMNESHCYKVPDFVP-LDVAALIQPLAVCWHAIR-VCEFKAGSTALIIGAGPIGLGTILAL
Yal060w   -HAGFVGLGVISGGFAEQVVVSQHHIIPVPKEIP-LDVAALVEPLSVTWHAVK-ISGFKKGSSALVLGAGPIGLCTILVL

Sor1p     -HMAFAATPPIDGTLVKYYLSPEDFLVKLPEGVS-YEEGACVEPLSVGVHSNK-LAGVRFGTKVVVFGAGPVGLLTGAVA
Ylr070c   -NLKFAATPPFDGTLTKYYKTMKDFVYKLPDDVS-FEEGALIEPLSVAIHANK-LAKIKFGARCVVFGAGPIGLLAGKVA

•• ADH    KVAGASRIIGVDINKDKFARAKEFGATECINPQD-----FSKPIQEVLIEMTDGGVDYSFECIG-NVKVMRAALEACHKG

Ymr318c   KAMGAETYVISRSSR-KREDAMKMGADHYIATLE----------EGDWGEKYFDTFDLIVVCASSLTDIDFNIMPKAMKV
Ycr105w   KAMGAEVYAFSRGHS-KREDSMKLGADHYIAMLE----------DKGWTEQYSNALDLLVVCSSSLSKVNFDSIVKIMKI

Yal061w   NAAGCKDIVVSEPAKVRRELAEKMGARVYDPTA---HAAKESIDYLRSIADGGDGFDYTFDCSG--LEVTLNAAIQCLTF
Yal060w   KGMGASKIVVSEIAERRIEMAKKLGVEVFNPSK---HGHKS-IEILRGLTKSHDGFDYSYDCSG--IQVTFETSLKALTF

Sor1p     RAFGATDVIFVDVFDNKLQRAKDFGATNTFNSSQFSTDKAQDLADGVQKLLGGNHADVVFECSG--ADVCIDAAVKTTKV
Ylr070c   SVFGAADVVFVDLLENKLETARQFGATHIVNSG--DLPHGVTVDSVIKKAIGKKGADVVFECSG--AEPCVRAGIEVCKA

•• ADH    WGVSVVVGVAASGEEIATRPFQLVTGRTWKGTAFGGWKSVESVPKLVSEYMSKKIKVDEFVTHNLSFDEINKAFELMHSG

Ymr318c   GGRIVSISIP-EQHEMLSLKPYGLKAVSISYSALGSIKELNQLLKLVSEKDIKIW----VETLPVGEAGVHEAFERMEKG
Ycr105w   GGSIVSIAAP-EVNEKLVLKPLGLMGVSISSSAIGSRKEIEQLLKLVSEKNVKIW----VEKLPISEEGVSHAFTRMESG

Yal061w   RGTAVNLAMWGHHKIQFSPMDITLHERKYTGSMCYTHHDFEAVIEALEEGRIDIDRARHMITGRVNIEDGLDGAIMKLIN
Yal060w   KGTATNIAVWGPKPVPFQPMDVTLQEKVMTGSIGYVVEAFEEVVRAIHNGDIAMEDCKQLITGKQRIEDGWEKGFQELMD

Sor1p     GGTMVQVGMG-KNYTNFPIAEVSGKEMKLIGCFRYSFGDYRDAVNLVATGKVNVK---PLITHKFKFEDAAKAYDYNIAH
Ylr070c   GGTIVQVGMG-QEEIQFPISIIPTKELTFQGCFRYCQGDYSDSIELVSSRKLSLK---PFITHRYSFKDAVEAFEETSHH

•• ADH    KSIRTVVKI-----------------------------------------

Ymr318c   DVRYRFTLVGYDKEFSD---------------------------------
Ycr105w   DVKYRFTLVDYDKKFHK---------------------------------

Yal061w   EKESTIKIILTPNNHGELNREADNEKKEISELSSRKDQERLRESINEAKLRHT
Yal060w   HKESNVKILLTPNNHGEMK-------------------------------

Sor1p     GGEVVKTIIFGPE-------------------------------------
Ylr070c   PLNNIKTIIEGPE-------------------------------------
```

(A)

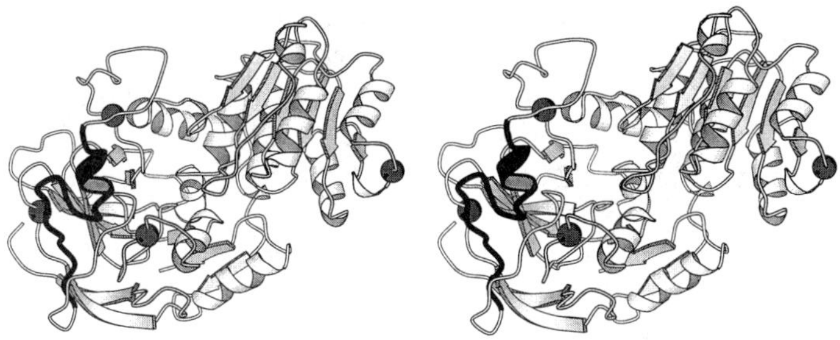

(B)

Figure 3 Medium-chain alcohol dehydrogenase-like proteins. (A) Aligned sequences of yeast medium-chain alcohol dehydrogenase-like proteins. The CLUSTAL W program [46] was used to align simultaneously the sequences of human χχ alcohol dehydrogenase and six related proteins encoded by yeast open reading frames. Coloring of residues and the locations of secondary structure elements from the X-ray crystal structure of χχ alcohol dehydrogenase (1TEH; [51]) are the same as in Fig. 2A. (B) Stereoview of the χχ alcohol dehydrogenase structure showing positions of insertions and deletions in similar yeast protein sequences. The active site cleft opens toward the upper left corner and the locations of four sequence insertions in yeast proteins are marked by dark gray spheres (clockwise from top, after Gly60, Pro31, Pro95, and Asp245). The location of a deletion found in the six yeast proteins relative to χχ alcohol dehydrogenase (between Lys113 and Cys132) is shown in heavy dark lines. Note that only a single monomer of the dimeric χχ alcohol dehydrogenase structure is shown for clarity. This figure was rendered in MolScript [51].

fide dehydrogenases, although their substrates and physiological roles are unknown at this time (Fig. 3A).

The six putative yeast reductases share several key residues that are also conserved in a global sequence alignment of medium-chain alcohol dehydrogenases constructed by Person and co-workers [45] (Table 2). A subset of this superfamily employ an essential metal ion and all six putative yeast enzymes conserve both the Cys46 and His67 residues (χχ alcohol dehydrogenase numbering) that ligate the catalytic Zn(II). The residue that acts as the third ligand to the catalytic Zn(II) (Cys 174, χχ alcohol dehydrogenase numbering) is not conserved between all of the yeast proteins and the human enzyme, however. Both the proteins encoded by the YMR318c and YCR105w genes feature a cysteine at this position; however, this position is occupied by a glutamate in those encoded by the YAL060w, SOR1, and YLR070c genes. Interestingly, the residue at this position in the YAL061wp (glutamine) has limited ability to bind Zn(II), suggesting that the metal ion coordination may be different in this case. Virtually all of superfamily members possess the G–X–G–X–X–G motif characteristic of the Rossman fold, and this is true of the six putative yeast reductases as well. In addition, the complete conservation of several glycine residues in the C-terminal region implies that the overall fold of the χχ alcohol dehydrogenase may be preserved in the putative yeast proteins.

Human χχ alcohol dehydrogenase is also known as formaldehyde dehydrogenase since it oxidizes the spontaneous formaldehyde-glutathione adduct to a thioester that is subsequently hydrolyzed to regenerate reduced glutathione and formate by a separate hydrolase. This sequence may be a detoxification pathway for formaldehyde. Several features of the six putative yeast protein sequences argue that while they are likely to possess alcohol dehydrogenase activity, their substrate specificities may be quite different. Residues known to be important in substrate binding in χχ alcohol dehydrogenase (e.g., Asp57 and Arg115) are not conserved between the human enzyme and the six putative yeast proteins. The yeast open reading frames also contain a variety of insertions and deletions that are likely to affect the substrate-binding pocket (Fig. 3B). Of the six, the proteins encoded by the YMR318c and YCR105w genes are most closely related to human χχ alcohol dehydrogenase. The major difference is a large deletion between residues Lys113 and Cys132 (χχ alcohol dehydrogenase numbering), which form a loop in the human enzyme that makes up part of the substrate-binding pocket. In addition to this deletion, the sequences of putative yeast proteins encoded by the YAL061w and YAL060w genes also differ from the χχ alcohol dehydrogenase by a 14-amino acid insertion after residue Pro95. This insertion—which occurs in a surface loop in χχ alcohol dehydrogenase—would change the substrate-binding site substantially (Fig. 3B).

The final member of the yeast medium-chain alcohol dehydrogenase superfamily, encoded by the ZTA1 gene, differs in several key aspects from the six putative yeast reductases shown in Fig. 3A. Most important, this enzyme lacks the residues that ligate the catalytic Zn(II) ion, in common with other crystallin homologs. Related metal-free proteins are capable of reducing quinones [53]; however, the catalytic activity and substrate specificity of the protein encoded by ZTA1 is currently unknown.

D. Short-Chain Alcohol Dehydrogenases

The short-chain alcohol dehydrogenases constitute a second superfamily whose members are widely distributed across all three phylogenetic kingdoms [42,43]. Sequence analysis of the yeast genome revealed a large number of potential proteins that would be part of this superfamily (Table 1), and we have recently demonstrated that at least one of these is a key player in stereoselective reductions of β-keto esters (*vide infra*) and it is likely that closely related proteins may also participate in similar reductions. Short-chain alcohol dehydrogenases are metal-independent enzymes that can be distinguished by a conserved Y–X–X–X–K motif. Unfortunately, no crystal structure of a dehydrogenase with high sequence similarity to yeast short-chain alcohol dehydrogenases is currently available, so it is not possible to interpret sequence differences in this context. Because the number of potential short-chain alcohol dehydrogenases is quite large, the discussion has been divided into three subsections (potential β-keto ester reductases, distantly related short-chain dehydrogenases and aryl alcohol dehydrogenases).

1. *Potential β-Keto Ester Reductases*

Yeast-mediated reductions of β-keto esters have been the most-utilized biotransformation involving this organism [12]. Several groups have purified yeast enzymes that accept β-keto esters; however, the corresponding gene had not been identified for any β-keto ester reductase member of the short-chain alcohol dehydrogenase superfamily. α-Acetoxyketone reductase reduces a variety of β-keto esters to L-alcohols with very high enantio- and diastereoselectivity [54]. Based on amino acid sequence data [55], it was suggested that this enzyme corresponded to the YJR105w open reading frame [56]. We have shown subsequently that this assignment was incorrect [11]. To identify the correct yeast gene, we purified α-acetoxyketone reductase according to Shieh [57] and determined the amino acid sequences of Lys-C fragments (since the N-terminus was blocked). Data obtained from three separate fragments corresponded precisely to the protein encoded by *S. cerevisiae* GRE2 gene, which has sequence similarity to plant dihydroflavinol and cinnamoyl-CoA reductases (both known members of the short-chain alcohol dehydrogenase superfamily) [58]. The Gre2p sequence was used as a probe in Basic Local Alignment Sequence Tool (BLAST) analysis of the *S. cerevisiae* genome, and three closely related putative proteins were identified (Fig. 4).

The nature of residues conserved between the putative open reading frames and the Gre2p and plant cinnamoyl-CoA reductase argues strongly that the proteins encoded by the YDR541c, YGL157w, and YGL039w genes are all likely to be functional reductases. That all sequences share that consensus Y–X–X–X–K motif places them securely in the short-chain dehydrogenase superfamily. Moreover, their conservation of residues that are also identical between plant dihydroflavinol and cinnamoyl-CoA reductases from different species suggests that their catalytic activities have been identified correctly. On the other hand, whether the substrate specificities of these putative proteins overlap with that of α-acetoxyketone reductase is unknown.

2. *Other Short-Chain Alcohol Dehydrogenases*

Using a variety of known short-chain alcohol dehydrogenases as probes for BLAST searching of the yeast genome revealed an additional eight putative pro-

Table 2 Key Sequence Features of Ketone Reductases and Conservations by *S. cerevisiae* Open Reading Frames

Class	Feature	Conservation by *S. cerevisiae* proteins listed in Table 1	Reference
Medium-chain dehydrogenases/reductases (Figs. 2A, 3A)	Total length of ca. 350 residues; $M_r \sim 40$ kDa	All	44
	Gly-His-Glu sequence ca. 60 residues from N-terminus	All	44
	Gly-X_{1-3}-Gly-X_{1-3}-Gly	Sfa1p, Ymr318c, Ycr105w, Yal061w, Yal060w, Sor1p, Ylor070c	44
	Conserved Gly residues at positions 66, 86, and 201 (HLADH numbering)	Sfa1p, Adh2p, Ymr318c, Ycr105w, Yal061w, Yal060w, Sor1p, Ylor070c	45
Short-chain dehydrogenases/reductases (Figs. 4–6)	Total length of ca. 250 residues, $M_r \sim 29$ kDa	All except Aad10p, Yfl057c and Aad15p	44
	N-terminal Thr-Gly-X_{2-3}-Gly-X-Gly	Ylr426w, Ybr159w, Ydl114w, Ymr226c, Yir035c, Yir036c, Yil124w	44
	Mid-chain Asn-Asn-Ala-Gly	Ylr426w, Ydl114w, Ymr226c, Yil124w; related sequences in Ybr159w, Ylr035c and Ylr036c	44
	Mid-chain Tyr-X-X-X-Lys	All	44
Aldo-keto reductases (Fig. 7A)	Ile-Pro-Lys-Ser-X_5-Arg	Exact match in Gre3p; others have related sequences	63
	Gly38 (human aldose reductase numbering)	All	80
	Asp43 (human aldose reductase numbering)	All	80
	Tyr48 (human aldose reductase numbering)	All	80
	Glu51 (human aldose reductase numbering)	Ypr1p, Gcy1p, Ara1p, Gre3p, Yjr096wp	80
	Gly55 (human aldose reductase numbering)	All	80
	Lys77 (human aldose reductase numbering)	All	80
	Pro112 (human aldose reductase numbering)	All	80
	Gly157 (human aldose reductase numbering)	All	80
	Pro179 (human aldose reductase numbering)	All	80
	Gln185 (human aldose reductase numbering)	All	80
	Ser263 (human aldose reductase numbering)	Ypr1p, Gcy1p, Ara1p, Gre3p, Ydl124w; Thr in Yjr096w	80

D-hydroxyacid dehydrogenases (Fig. 8A)	Gly-X-Gly-X-Ile-Gly	All	64
	Gly78 (D-glycerate dehydrogenase numbering)	Ypl113c, Ygl185c, Ynl274c, Ypl275w; Ala in Yor388c	64
	Asp83 (D-glycerate dehydrogenase numbering)	Ygl185c, Ynl274c, Yor388c	64
	Gly91 (D-glycerate dehydrogenase numbering)	None	64
	Glu107 (D-glycerate dehydrogenase numbering)	Ypl113c, Yor388c; Asp in Ygl185c, Ynl274c	64
	Arg118 (D-glycerate dehydrogenase numbering)	Ygl185c, Ynl274c, Yor388c	64
	Gly151 (D-glycerate dehydrogenase numbering)	Ynl274c	64
	Asp177 (D-glycerate dehydrogenase numbering)	None	64
	Asp205 (D-glycerate dehydrogenase numbering)[a]	Ypl113c, Ynl274c, Yor388c, Ypl275w	64
	Pro212 (D-glycerate dehydrogenase numbering)	All	64
	Lys230 (D-glycerate dehydrogenase numbering)[a]	Ypl113c, Ynl274c, Yor388c, Ypl275w	64
	Asn237 (D-glycerate dehydrogenase numbering)	All	64
	Arg240 (D-glycerate dehydrogenase numbering)	All	64
	Glu269 (D-glycerate dehydrogenase numbering)	Ypl113c, Ygl185c, Ynl274c	64
	Pro286 (D-glycerate dehydrogenase numbering)	Ypl113c, Ygl185c, Ynl274c	64
	His287 (D-glycerate dehydrogenase numbering)	All	64

[a] Not conserved in D-glycerate dehydrogenase.

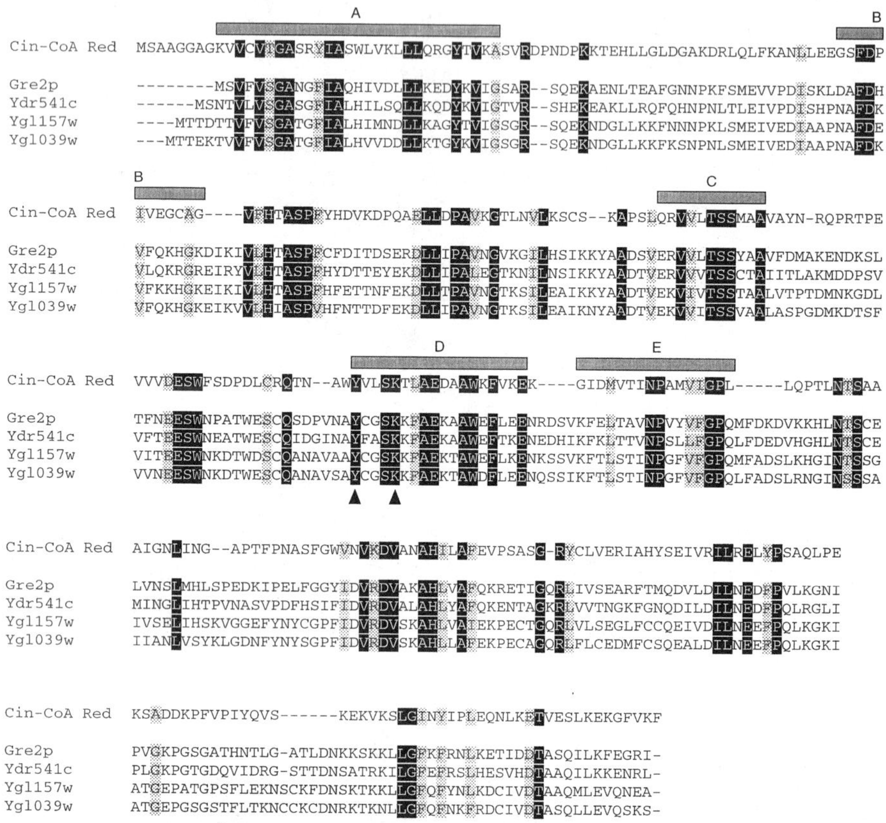

Figure 4 Aligned short-chain alcohol dehydrogenase-like proteins. The CLUSTAL W program [46] was used to align simultaneously the sequences of cider tree cinnamoyl-CoA reductase (Cin-CoA) and four closely related *S. cerevisiae* open reading frames. Residues conserved across all five sequences are shown against a black background, while those positions occupied by two closely related amino acids are shown with a gray background. Sequence blocks labeled with letters correspond to the conserved regions of short-chain alcohol dehydrogenases defined by Krozowski [42]. The location of the Y–X–X–X–K motif is indicated by two black triangles.

teins that are more distantly related (Fig. 5). All eight share the Y–X–X–X–K motif as well as a glycine-rich motif near the N-termini that may represent an NAD(P)H-binding site. In the absence of an X-ray crystal structure of a related dehydrogenase, it is not possible to determine whether the pattern of insertions and deletions would be compatible with a known architecture. The assignment of this group of proteins as potential reductases is tenuous at best.

3. Aryl Alcohol Dehydrogenases

The *S. cerevisiae* genome encodes four very closely related proteins with high sequence similarity to known aryl alcohol dehydrogenases (Fig. 6). These enzymes are characterized by the Y–X–X–X–K motif and are members of the short-chain dehydrogenase family. In addition, two other proteins (encoded by YFL057c and AAD15) correspond to the C-terminal half of the longer sequences. It is particularly interesting that four of the six open reading frames can be derived from alternate start codons in the AAD14 protein sequence. The reason for this apparent genetic redundancy is not known, nor is the physiological role of these proteins. Whether these enzymes participate in reductions of exogenous ketones is also not known.

E. Aldose Reductase Family

Enzymes of the aldo-keto reductase superfamily are also widely distributed within the phylogenetic kingdoms. In baker's yeast, these enzymes may play a role in protection against osmotic stress by producing sugar-derived polyols. On the other hand, at least one of these enzymes (encoded by the YPR1 gene) also plays a major role in the stereoselective reduction of exogenous β-keto esters [55,56,59]. We have shown recently that manipulating the level of the YPR1 protein by recombinant DNA techniques results in useful stereochemical changes in whole-cell-mediated reductions [11], conclusively linking this gene to the aldo-keto reductase isolated by both Sih and Nakamura. Using the Ypr1p and related aldo-keto reductase sequences as probes for BLAST searches of the *S. cerevisiae* genome uncovered five additional putative proteins that were closely related to one another and to human aldose reductase, whose X-ray crystal structure has been determined (Fig. 7A) [60]. It is very likely that these additional five proteins are also ketone reductases, and they may be participants in reductions of β-keto esters [61]. Support for this notion is also provided by the demonstration that the YBR149w open reading frame corresponds to a reductase purified by Nakamura and co-workers on the basis of its ability to reduce a variety of α-carbonyl compounds [56,62]. The structure and function of another member of this family, yeast xylose reductase, has been reviewed recently [63].

A number of key residues are conserved completely between human aldose

```
Ylr426w  MNIDCLCRWVVLPLLRYPLLVALVLRWSLSDSISICLTIYTLLINAFLIANSYIKRSGQVAWKSLREFKNGIVLITGGSK
Ybr159w  --------MTFMQQLQEAGERFRCINGLLWVVFGLGVLKCTTLSLRFLALIFDLFLLPAVNFDKYGAKTGKYCAITGASD
Ydl114w  --------------------------------MVRKNKINRASGTTKHLKDFPSVILSLPSYNPSILSKNATALITGGSS
Ymr226c  ---------------------------------------------------------MSQGRKAAERLAKKTVLITGASA
Yir035c  --------------------------------------------------------------------MGKVILVTGVSR
Yir036c  --------------------------------------------------------------------MGKVILITGASR
Yil124w  -------------------------------------------------------------MSELQSQPKKIAVVTGASG
Ykl071w  ----------------------------------------------------------------MNTSSRITYFIIGGSR

Ylr426w  GLGRAIVSQLLQDYSN----------LT--ILNVDICPSSVRNTRVKDLICDLSDD-EEVAALLNLLKRKYKN-EIRLIV
Ybr159w  GIGKEFARQMAKRGFN----LVLISRTQSKLEALQKELEDQHHVVVKILAIDIAEDKESNYESIKELCAQLP---ITVLV
Ydl114w  GLGFELAKELSRRIN----------KVIVADIQSFPTFAQVEYNNIFYYQCDITSL-DEIKNLKKAIERDHG--NINIII
Ymr226c  GIGKATALEYLEASNGDMKLILAARRLEKLEELKKTIDQEFPNAKVHVAQLDITQA-EKIKPFIENLPQEFK--DIDILV
Yir035c  GIGKSIVDVLFSLDKD-----TVVYGVARSEAPLKKLKEKYG-DRFFYVVGDITED-SVLKQLVNAAVKGHG--KIDSLV
Yir036c  GIGLQLVKTVIEEDDE-----CIVYGVARTEAGLQSLQREYGADKFVYRVLDITDR-SRMEALVEEIRQKHG--KLDGIV
Yil124w  GIGYEVTKELARNG-------YLVYACARRLEPMAQLAIQFGNDSIKPYKLDISKP-EEIVTFSGFLRANLPDGKLDLLY
Ykl071w  GIGFNLVKILSASTGN--TVITSIRGSPSLPKNKQVEDLAKIRKNIHIVQLDLTKD-ESIGNIADEIKKTPFFLGIDIFI

Ylr426w  NNAGVRANFTGFN----GMERDNLDKIFKINTFAPLQFIQELAPSRHSTR---------QCYIVNIASILGILTPAKVAA
Ybr159w  NNVGQSHSIPVPF---LETEEKELRNIITINNTATLLITQIIAPKIVETVKAENKKSGTRGLILTMGSFGGLIPTPLLAT
Ydl114w  NNAGVAHIKKLEH-----MTNKEVEQLIDINLIGAYRIISTFAEDMIDNR---------EGFIINIASVLGELTPARLTS
Ymr226c  NNAGKALGSDRVG----QIATEDIQDVFDTNVTALINITQAVLPIFQAKN---------SGDIVNLGSIAGRDAYPTGSI
Yir035c  ANAGVLEPVQNVN----EIDVNAWKKLYDINFFSIVSLVGIALPELKKT----------NGNVVFVSSDACNMYFSSWGA
Yir036c  ANAGMLEPVKSISQSNSEHDIKQWERLFDVNFFSIVSLVALCLPLLKSSPF--------VGNIVFVSSGASVKPYNGWSA
Yil124w  NNAGQSCTFPALD-----ATDAAVEQCFKVNVFGHINMCRELSEFLIKAK----------GTIVFTGSLAGVVSFPFGSI
Ykl071w  ACSAVSDSYYKVL----ETPKSVWLNHYSTNALGPILALQKVYPLLLLKKTRK------IFFISSVAGSINAFVPLSVSA

Ylr426w  YAASKAALIAFHQSYSFELQNE----GVRNIRTLLVTPGQLNTEMFAGFKPPRQ--FFAPVIDITTLAAKIVRYCELGQR
Ybr159w  YSGSKSFLQGWSNSLAGELSK-------DAIDVELIISYLVTSSMSKIRRSS--LMIPNPQQFVKSTLRSVGRRCGSQER
Ydl114w  YGASKGAMIGFHKCMSRHFRSLSTECNKTGIKTLLVCPGKIKTNMFIDVPTPS--KLLAPDIIPSQLALAIISAMEHNHL
Ymr226c  YCASKFAVGAFTDSLRKELIN-------TKIRVILIAPGLVETEFSLVRYR-----GNEEQ---AKNVYKDTTPLMADDV
Yir035c  YGSSKAALNHFAMTLANEE---------RQVKAIAVAPGIVDTDMQVNIRE-----NVGPSSMSAEQLKMFRGLKENNQL
Yir036c  YGCSKAALNHFAMDIASEEPS-------DKVRAVCIAPGVVDTQMQKDIRE-----TLGPQGMTPKALERFTQLYKTSSL
Yil124w  YSASKAAIHQYARGLHLEMKP-------FNVRVINAITGGVATDIADKRPLPETSIYNFPEGREAFNSRKTMAKDNKPMP
Ykl071w  YGQSKAALNYAVKTLSFELKP-------EGFTVVAFHPGMVSTDMGQYGLD-----HFKEKNIDISGVNIITPEESASAL
         ▲   ▲

Ylr426w  GQLNEPFYCSFAHLLMCVPYS--LQRIVRSFSRIDCCLPDE-----------------------------
Ybr159w  YATMTPYWAHAVYQFVITETFGVYSKIVNSINYSFHKSIRIRALKKAARQVKKE----------------
Ydl114w  QTLNAPYYVNLVPFFKTLSWP--YRHLLKHFSGMDHVTSTSPRAINPKRSA-------------------
Ymr226c  ADLIVYATSRKQNTVIADTLI--FPTNQASPHHIFRG---------------------------------
Yir035c  LDSSVPATVYAKLALHGIPDG--VNGQYLSYNDPALADFMP-----------------------------
Yir036c  LDPKVPAAVLAQLVLKGIPDS--LNGQYLRYNDERLGPVQG-----------------------------
Yil124w  ADAYAKQLVKDILSTSDPVDV--YRGTFANIMRFVMIFVPYWLLEKGLSKKFKLDKVNNALKSKQKNKDD
Ykl071w  IDVFRKILPEDNGKFFNYDGS--EGVF-------------------------------------------
```

Figure 5 Aligned yeast open reading frame sequences with weak similarity to short-chain alcohol dehydrogenases. The CLUSTAL W program [46] was used to align eight putative yeast protein sequences simultaneously against one another. Residues conserved across all eight proteins are shown against a black background, and those positions occupied by two closely related amino acids are shown on a gray background. The location of the Y–X–X–X–K motif is indicated by two black triangles.

reductase and the six yeast protein sequences. A catalytic mechanism in which Tyr48, Lys77, and His110 play key roles was proposed for human aldose reductase by Gabbay and co-workers on the basis of its X-ray crystal structure [60]. In this mechanism, Tyr48 acts as a general acid to protonate the nascent alcohol during carbonyl reduction, and this proton transfer is facilitated by interaction

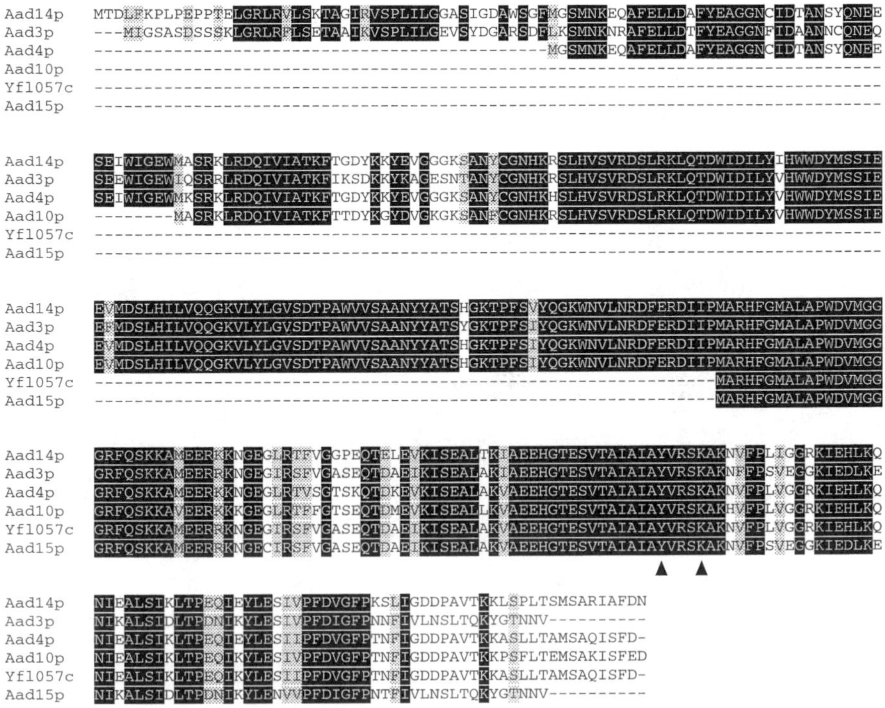

Figure 6 Yeast aryl alcohol dehydrogenases. The sequences of six yeast open reading frames were aligned simultaneously with the CLUSTAL W program [46]. Residues are colored as in Fig. 2A and the Y–X–X–X–K motif is indicated by two black triangles. Because the open reading frames differ dramatically in size, conserved residues are also indicated in regions not present in all six sequences.

with the side chain of Lys77. The side chain of His110 was proposed to influence the stereochemistry of reduction by positioning the substrate correctly. All three of these residues are conserved among all six yeast open reading frames. In addition, other amino acids conserved between the human and yeast sequences match relatively well with residues conserved between mammalian aldose reductases, further suggesting that the yeast sequences are functional reductase enzymes.

The locations of sequence insertions and deletions present in the yeast sequences are congruent with the three-dimensional structure of human aldose reductase. Most occur between secondary structure elements, as would be expected for proteins with the same overall fold but differences in surface-exposed loops

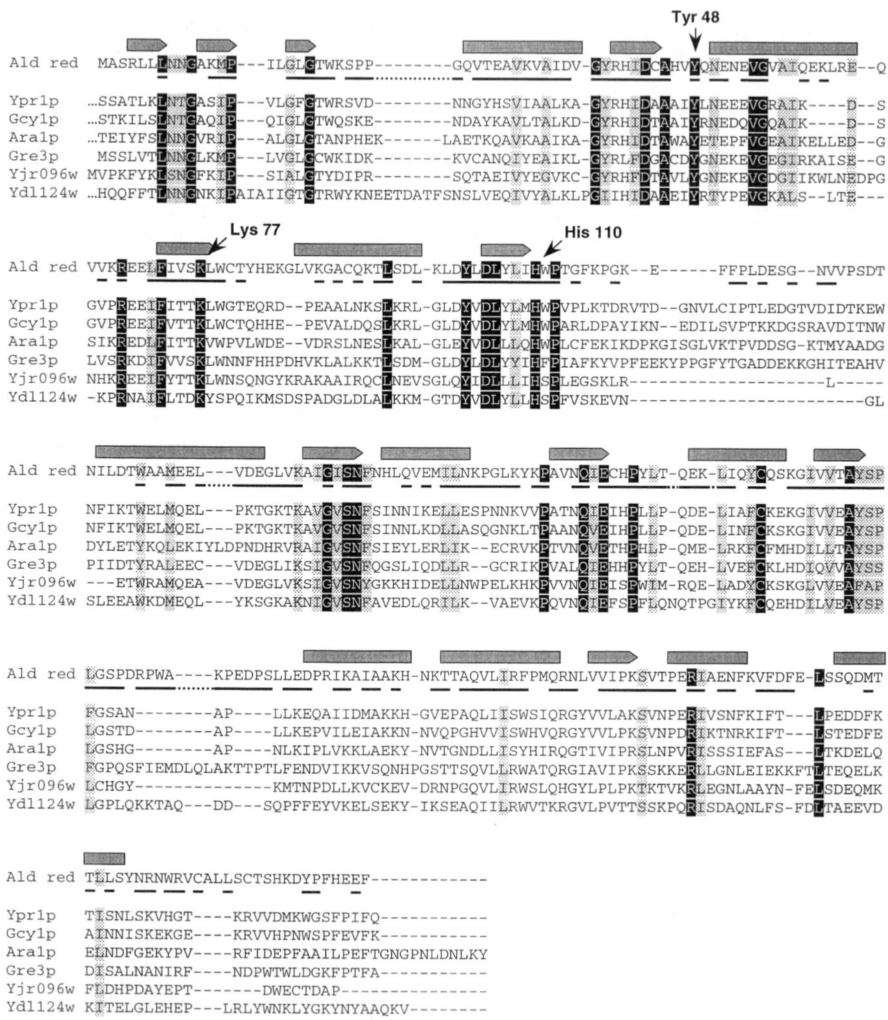
Tyr 48
Ald red MASRLLLNNGAKMP---ILGLGTWKSPP--------GQVTEAVKVAIDV-GYRHIDCAHVYQNENEVGVAIQEKLRE--Q
Ypr1p ...SSATLKLNTGASIP---VLGFGTWRSVD--------NNGYHSVIAALKA-GYRHIDAAAIYLNEEEVGRAIK----D--S
Gcy1p ...STKILSLNTGAQIP---QIGLGTWQSKE--------NDAYKAVLTALKD-GYRHIDTAAIYRNEDQVGQAIK----D--S
Ara1p ...TEIYFSLNNGVRIP---ALGLGTANPHEK------LAETKQAVKAAIKA-GYRHIDTAWAYETEPFVGEAIKELLED--G
Gre3p MSSLVTLNNGLKMP---LVGLGCWKIDK--------KVCANQIYEAIKL-GYRLFDGACDYGNEKEVGEGIRKAISE--G
Yjr096w MVPKFYKLSNGFKIP---SIALGTYDIPR--------SQTAEIVYEGVKC-GYRHFDTAVLYGNEKEVGDGIKWLNEDPG
Ydl124w ...HQQFFTLNNGNKIPAIAIIGTGTRWYKNEETDATFSNSLVEQIVYALKLPGIIHIDAAEIYRTYPEVGKALS--LTE---
Lys 77
His 110
Ald red VVKREELFIVSKLWCTYHEKGLVKGACQKTLSDL-KLDYLDLYLIHWPTGFKPGK--E------FFPLDESG--NVVPSDT
Ypr1p GVPREEIFITTKLWGTEQRD--PEAALNKSLKRL-GLDYVDLYLMHWPVPLKTDRVTD--GNVLCIPTLEDGTVDIDTKEW
Gcy1p GVPREEIFVTTKLWCTQHHE--PEVALDQSLKRL-GLDYVDLYLMHWPARLDPAYIKN--EDILSVPTKKDGSRAVDITNW
Ara1p SIKREDLFITTKVWPVLWDE--VDRSLNESLKAL-GLEYVDLLLQHWPLCFEKIKDPKGISGLVKTPVDDSG-KTMYAADG
Gre3p LVSRKDIFVVSKLWNNFHHPDHVKLALKKTLSDM-GLDYLDLYYIHFPIAFKYVPFEEKYPPGFYTGADDEKKGHITEAHV
Yjr096w NHKREEIFYTTKLWNSQNGYKRAKAAIRQCLNEVSGLQYIDLLLIHSPLEGSKLR------------------L-----
Ydl124w -KPRNAIFLTDKYSPQIKMSDSPADGLDLALKKM-GTDYVDLYLLHSPFVSKEVN-----------------------GL
Ald red NILDTWAAMEEL---VDEGLVKAIGISNFNHLQVEMILNKPGLKYKPAVNQIECHPYLT-QEK-LIQYCQSKGIVVTAYSP
Ypr1p NFIKTWELMQEL---PKTGKTKAVGVSNFSINNIKELLESPNNKVVPATNQIEIHPLLP-QDE-LIAFCKEKGIVVEAYSP
Gcy1p NFIKTWELMQEL---PKTGKTKAVGVSNFSINNLKDLLASQGNKLTPAANQVEIHPLLP-QDE-LINFCKSKGIVVEAYSP
Ara1p DYLETYKQLEKIYLDPNDHRVRAIGVSNFSIEYLERLIK--ECRVKPTVNQVETHPHLP-QME-LRKFCFMHDILLTAYSP
Gre3p PIIDTYRALEEC---VDEGLIKSIGVSNFQGSLIQDLLR--GCRIKPVALQIEHHPYLT-QEH-LVEFCKLHDIQVVAYSS
Yjr096w ---ETWRAMQEA---VDEGLVKSIGVSNYGKKHIDELLNWPELKHKPVVNQIEISPWIM-RQE-LADYCKSKGLVVEAFAP
Ydl124w SLEEAWKDMEQL---YKSGKAKNIGVSNFAVEDLQRILK--VAEVKPQVNQIEFSPFLQNQTPGIYKFCQEHDILVEAYSP
Ald red LGSPDRPWA----KPEDPSLLEDPRIKAIAAKH-NKTTAQVLIRFPMQRNLVVIPKSVTPERIAENFKVFDFE-LSSQDMT
Ypr1p FGSAN--------AP----LLKEQAIIDMAKKH-GVEPAQLIISWSIQRGYVVLAKSVNPERIVSNFKIFT---LPEDDFK
Gcy1p LGSTD--------AP----LLKEPVILEIAKKN-NVQPGHVVISWHVQRGYVVLPKSVNPDRIKTNRKIFT---LSTEDFE
Ara1p LGSHG--------AP----NLKIPLVKKLAEKY-NVTGNDLLISYHIRQGTIVIPRSLNPVRISSSIEFAS---LTKDELQ
Gre3p FGPQSFIEMDLQLAKTTPTLFENDVIKKVSQNHPGSTTSQVLLRWATQRGIAVIPKSSKKERLLGNLEIEKKFTLTEQELK
Yjr096w LCHGY-------------KMTNPDLLKVCKEV-DRNPGQVLIRWSLQHGYLPLPKTKTVKRLEGNLAAYN-FELSDEQMK
Ydl124w LGPLQKKTAQ---DD---SQPFFEYVKELSEKY-IKSEAQIILRWVTKRGVLPVTTSSKPQRISDAQNLFS-FDLTAEEVD
Ald red TLLSYNRNWRVCALLSCTSHKDYPFHEEF-----------
Ypr1p TISNLSKVHGT----KRVVDMKWGSFPIFQ----------
Gcy1p AINNISKEKGE----KRVVHPNWSPFEVFK----------
Ara1p ELNDFGEKYPV----RFIDEPFAAILPEFTGNGPNLDNLKY
Gre3p DISALNANIRF----NDPWTWLDGKFPTFA----------
Yjr096w FLDHPDAYEPT-------DWECTDAP--------------
Ydl124w KITELGLEHEP---LRLYWNKLYGKYNYAAQKV--------

(A)

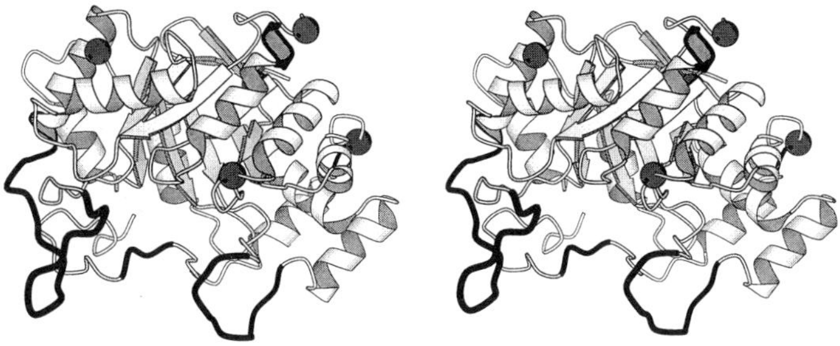

(B)

Figure 7 Yeast protein sequences from the aldo-keto reductase superfamily. (A) Aligned sequences of aldo-keto reductase-like proteins. The sequences of six yeast open reading frames were aligned simultaneously with one another and with human aldose reductase (Ald red) using the CLUSTAL W program [46]. Residues are colored and secondary structure elements from the X-ray crystal structure of human aldose reductase (2ACS; [60]) are indicated as in Fig. 2A. Residues highly conserved between vertebrate aldose reductases are indicated by heavy lines beneath the sequence of human aldose reductase. (B) Stereoview of human aldose reductase showing positions of insertions and deletions in similar yeast protein sequences. The active site cleft opens toward the top and the locations of four sequence insertions in yeast proteins are marked by dark gray spheres (clockwise from top, after Pro24, Leu147, Pro13, and Phe278). The locations of four deletions found in yeast proteins relative to human aldose reductase (clockwise from upper left, Arg217–Ser226, Cys298–Leu301, Lys119–Leu138, and Glu60–Arg63) are shown in heavy dark lines. The first three deletions have large effects on the loops at the entrance to the active site. This figure was rendered in MolScript [51].

(Fig. 7A). The major difference between the human enzyme and those encoded by YPR1, GCY1, and ARA1 is an eight-amino acid deletion between residues Arg217 and Ser226, which forms a loop at the entrance to the active site (Fig. 7B). In addition to this deletion, the sequences of the Yjr096w and Ydl124w proteins also have a large deletion (22 and 16 amino acids, respectively) between residues Lys119 and Leu138 (human enzyme numbering). This region corresponds to a large surface-exposed loop in human aldose reductase, which in principle could be deleted without requiring major changes in the protein architecture.

Taken together, the evidence strongly suggests that all six of the yeast aldose reductase superfamily members are catalytically active. Further explorations of these proteins are likely to yield correlations with previously isolated yeast reductases.

F. D-Hydroxyacid Dehydrogenase Family

Members of the D-hydroxyacid dehydrogenase family catalyze the stereoselective reduction of small α-keto acids such as hydroxypyruvate, pyruvate, etc. [64]. Enantiomerically pure α-hydroxy acids are valuable synthetic intermediates, and a variety of ingenious approaches to these compounds have been described [65–73]. Sequence similarity searching of the *S. cerevisiae* genome revealed five open reading frames that were similar to members of the D-hydroxyacid dehydrogenase family, although one (YPL275w) appeared to lack the N-terminal domain common to this superfamily and may be nonfunctional, at least for this type of reduction. A multiple sequence alignment of the five yeast open reading frames with the sequence of *Hyphomicrobium methylovorum* D-glycerate dehydrogenase revealed that several key residues were conserved and the pattern of sequence insertions and deletions might be reasonably accommodated by the known structure of D-glycerate dehydrogenase (Fig. 8A) [64]. Based on these observations, it appears likely that the five open reading frames encode active reductases; however, it is not clear whether the α-keto ester reductases isolated previously from baker's yeast correspond to one or more of these open reading frames [62].

An alignment of nine authentic D-hydroxyacid dehydrogenases from a variety of sources revealed several conserved residues [64], and many of these were also shared by the five yeast open reading frame sequences. From the crystal structure of D-glycerate dehydrogenase, three residues were suggested to play key roles in catalysis: Arg240, Glu269, and His287 [64]. Both the arginine and histidine residues are conserved in all five yeast open reading frames. A glutamate is present in three of the yeast sequences at the position corresponding to Glu269; however, the remaining two proteins have a glutamine at this position. Such a substitution has also been observed in *Pseudomonas* sp. 101 formate dehydrogenase [74]. In the proposed catalytic mechanism, Glu269 is proposed to hydrogen bond with the conserved histidine (His287), thereby raising its pK_a and facilitating its role in protonating the nascent alcohol. Interestingly, the authors note that the side chain of Asp264 may also participate in catalysis, and this residue is conserved in all five yeast open reading frames (Fig. 8A). The side chain of Arg240 is proposed to interact with the carboxylate of bound substrate and thereby ensure productive substrate binding, and this residue is also present in all five yeast sequences.

The structure of D-glycerate dehydrogenase consists of two domains, an N-terminal catalytic domain and a C-terminal cofactor-binding domain with a typical Rossman fold topology [64]. Insertions in the yeast open reading frame sequences relative to that of D-glycerate dehydrogenase are the major difference in overall primary structure (Fig. 8B). These occur mainly in the catalytic domain and are largely confined to loops between secondary structure elements, suggesting that these would be accommodated by the D-glycerate dehydrogenase architecture.

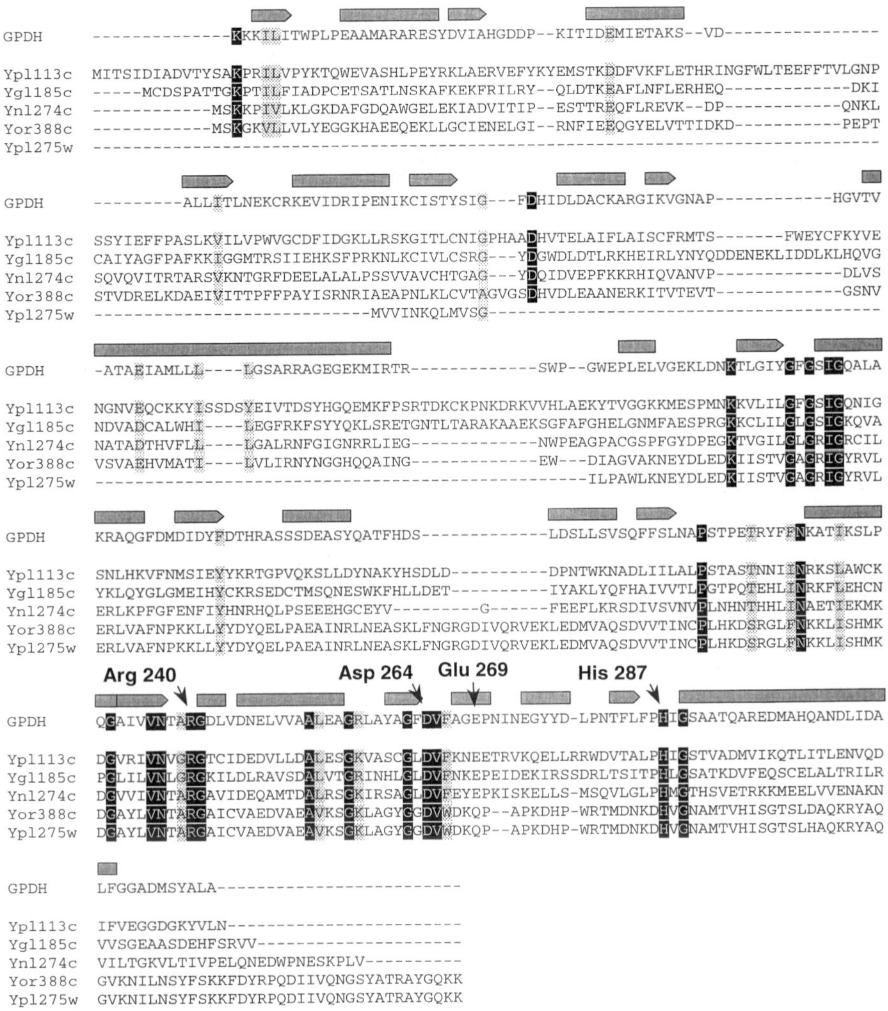

(A)

Figure 8 Yeast D-hydroxyacid dehydrogenase-like proteins. (A) Aligned sequences of D-hydroxyacid dehydrogenase-like proteins. The CLUSTAL W program [46] was used to align simultaneously five related yeast open reading frames with D-glycerate dehydrogenase from *H. methylovorum* (1GDH; [64]). Residues are colored and secondary structure elements are indicated as in Fig. 2A. The locations of critical amino acids in the D-glycerate dehydrogenase are indicated. (B) Stereoview of D-glycerate dehydrogenase showing positions of insertions and deletions in similar yeast protein sequences. The active site lies at the interface between the two domains, and the locations of seven sequence insertions in yeast proteins are marked by dark gray balls (from left, after Asp47, after Pro32, after Gly78, after Pro98, after Arg130, after Leu113, and after Ser196). Note that spheres corresponding to Pro32 and Leu113 are located at the back of the structure and may be difficult to discern. Note that only a single subunit of the dimeric structure is shown for clarity. This figure was rendered in MolScript [51].

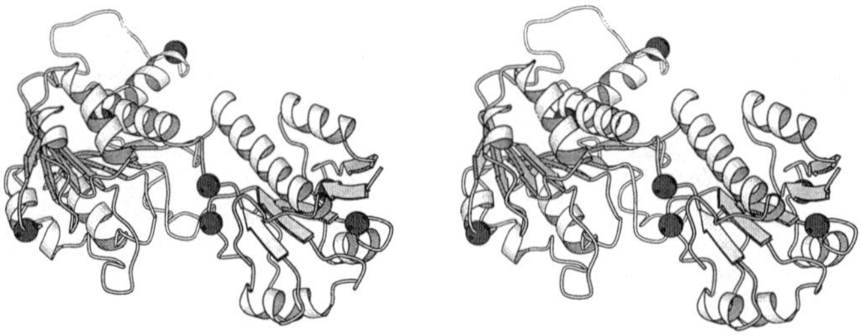

(B)

Figure 8 Continued.

G. Fatty Acid Synthase

Baker's yeast appears to contain only a single pair of genes encoding the subunits of fatty acid synthase (FAS1 and FAS2) [75,76]. This complex reduces a β-keto thioester intermediate stereoselectively to the corresponding D-alcohol during each elongation cycle and also accepts a variety of other β-keto esters as substrates, although it requires an unsubstituted α-position. We have recently shown that the fatty acid synthase complex also reduces an α-keto-β-lactam [77]. While the reduction product did not have the stereochemistry required for the Taxol® side chain, the result suggested that fatty acid synthase has a broader substrate tolerance than might have been anticipated.

Fatty acid biosynthesis occurs in the cytoplasm, whereas degradation is localized to the mitochondrion. No other *S. cerevisiae* open reading frame sequences appear to share high sequence similarity with the FAS2 protein. The only other yeast enzyme that catalyzes a similar reaction is 3-oxoacyl-(acyl-carrier-protein) reductase, the product of the OAR1 gene. This is a mitochondrial enzyme that likely functions in the oxidative direction. Moreover, the OAR1 protein possesses the Y–X–X–X–K motif typical of short-chain alcohol dehydrogenases, a motif that is absent from fatty acid synthase.

V. GENE–PROTEIN RELATIONSHIPS

A major motivation in collecting the information in Table 1 was to facilitate matching biochemical information for yeast reductases with the corresponding gene. Classical genetics has revealed the connection between gene and protein in several cases including yeast alcohol dehydrogenases and fatty acid synthase. In cases where even a fragment of amino acid sequence data from a purified reductase is available, the gene assignment is unequivocal. This method was used to show that the GRE2 gene encodes α-acetoxyketone reductase [58], YPR1 encodes yeast aldo-keto reductase [55,56,59], GRE3 encodes xylose reductase [78], ARA1 encodes D-arabinose dehydrogenase [79], and YBR149w encodes another member of the aldo-keto reductase superfamily [56]. The major disadvantage of this approach is the time and effort required to isolate each protein and determine its amino acid sequence.

Amino acid sequence comparisons with known reductases will continue to play a major role in identifying enzymes in baker's yeast and other species. Several groups have identified key sequence patterns in different reductase superfamilies that allow one to determine whether a protein sequence is consistent with a role in carbonyl reductions [43–45,64,80]. While several of these motifs have been discussed above, a complete collection has been assembled in Table 2, which also indicates the level of conservation of these motifs among the yeast open reading frames discussed previously.

Matching physical properties of purified enzymes to those predicted by computer analysis of yeast open reading frames offers a second approach to matching protein with gene. Molecular weight can be measured easily by gel electrophoresis, and such data are routinely included in all papers in which a yeast reductase has been isolated. Unfortunately, the resolution of this technique is limited to accuracies of ±5%, which is not sufficient to distinguish between related proteins. For example, the six members of the aldo-keto reductase family identified in the *S. cerevisiae* genome are predicted to have an average molecular weight of 35,443 Da, with a standard deviation of 1943 Da (5.5%). A similar situation holds for other families, and an inspection of Table 1 reveals that the vast majority of potential yeast reductases have predicted molecular weights between 30 and 40 kDa. The isoelectric point for purified yeast reductases would be valuable corroborating evidence in assigning the appropriate gene and these values differ significantly, even for proteins in the same superfamily. Unfortunately, such data have not been reported for any of the reductases reported to date.

The behavior of knockout mutants can provide another method to connect a reductase enzyme with the appropriate gene. In most cases, the physiological role for the authentic and putative reductases listed in Table 1 are unknown. Clearly, individual deletions of most of these genes are not lethal, indicating that the proteins are involved in nonessential pathways or there are redundant enzymes. Deletion of open reading frame Ybr159w, however, was lethal. The sequence of this enzyme indicated distant relationship to short-chain alcohol dehydrogenases (Fig. 5). Whether this will be a general observation must await completion of the systematic gene deletion project.

In the short term, the most productive method for identifying yeast reductases will involve testing deletion strains for an inability to reduce ketones that are accepted by the parent strain. This search can be focused on the most likely candidates if a homologous enzyme from another species with similar stereoselectivity has been identified. Ideally, a single knockout mutation would abrogate the reduction of a given ketone. On the other hand, if more than one yeast enzyme is responsible for product formation, gene knockouts may instead alter the stereoselectivity of whole-cell reductions. In fact, such a result provides some of the strongest evidence that incomplete stereoselectivity is the result of multiple enzymes rather than a single reductase with incomplete discrimination. The advantage of this approach is that it is rapid and requires little biochemical or genetic expertise, since all the strains are (or shortly will be) commercially available. One disadvantage of this strategy is that it identifies only those gene products that participate in reducing the specific substrate under investigation. In addition, it will be difficult to identify contributing enzymes if two or more have the same stereoselectivity for a given substrate or if knocking out one gene affects the levels of other enzymes. While knocking out one participant might be expected to reduce the rate of reduction, this analysis can be complicated by changes in overall growth rate as a result of the mutation.

VI. CONCLUSIONS AND FUTURE DIRECTIONS

The increased regulatory demands for optically pure drugs and agrochemicals coupled with pressures to minimize the environmental impacts of chemical processes makes enzyme-mediated processes logical alternatives. The need for chiral ketone reductions often arises in synthetic schemes, and baker's yeast has proven particularly adept at such transformations, even on industrial scales [81,82]. The low cost and experimental simplicity associated with reductions by intact yeast cells provides a strong motivation to employ this organism, particularly because the use of whole yeast cells removes the need to purify enzymes or supply nicotinamide cofactors. Unfortunately, the large number of ketone reductases produced by baker's yeast can lead to side products, and efforts to optimize yeast reductions are therefore focused on minimizing these undesired reactions. The availability of genome sequence information, coupled with knowledge of which genes are likely to encode desirable and competing reductase enzymes, should allow for a rational approach to strain improvement by recombinant DNA techniques.

REFERENCES

1. C Neuberg, A Lewite. Phytochemische Reduktionen. XIV. Hydrierung eines Ketons durch Hefe. (Umwandlung von Methylheptenon in Methylheptenol.) Biochem Z 91: 257–266, 1918.
2. C Neuberg. Biochemical reductions at the expense of sugars. Adv Carb Chem 4: 75–117, 1949.

3. OP Ward, CS Young. Reductive biotransformations of organic compounds by cells or enzymes of yeast. Enzyme Microb Technol 12:482–493, 1990.
4. S Servi. Baker's yeast as a reagent in organic synthesis. Synthesis 1–25, 1990.
5. C Fuganti. Baker's yeast mediated synthesis of natural products. Pure Appl Chem 62:1449–1452, 1990.
6. R Csuk, BI Glaenzer. Baker's yeast mediated transformations in organic chemistry. Chem Rev 91:47–49, 1991.
7. E Santaniello, P Ferraboschi, P Grisenti, A Manzocchi. The biocatalytic approach to the preparation of enantiomerically pure chiral building blocks. Chem Rev 92: 1071–1140, 1992.
8. P D'Arrigo, G Pedrocchi-Fantoni, S Servi. Old and new synthetic capacities of baker's yeast. Adv Appl Microbiol 44:81–123, 1997.
9. R de Souza Pereira. The use of baker's yeast in the generation of asymmetric centers to produce chiral drugs and other compounds. Crit Rev Biotechnol 18:25–83, 1998.
10. HW Mewes, K Albermann, M Bahr, D Frishman, A Gleissner, J Hani, K Heumann, K Kleine, A Maierl, SG Oliver, F Pfeiffer, A Zollner. Overview of the yeast genome. Nature 387:7–8, 1997.
11. S Rodriguez, MM Kayser, JD Stewart. Improving the stereoselectivity of baker's yeast reductions by genetic engineering. Org Lett 8:1153–1155, 1999.
12. WFH Sybesma, AJJ Straathof, JA Jongejan, JT Pronk, JJ Heignen. Reductions of 3-oxo esters by baker's yeast: current status. Biocatal Biotrans 16:95–134, 1998.
13. R de Souza Pereira. Comparison of biochemical effects produced by calcium ions and by monomers of polyacrylamide (acrylamide and bisacrylamide) on strains of *Saccharomyces cerevisiae* used for production of chiral synthons. Mol Cell Biochem 178:33–40, 1998.
14. KC Thomas, SH Hynes, WM Ingledew. Effect of nitrogen limitation on synthesis of enzymes in *Saccharomyces cerevisiae* during fermentation of high concentration of carbohydrates. Biotechnol Lett 18:1165–1168, 1996.
15. AC Dahl, Jø Madsen. Baker's yeast production of D- and L-3-hydroxy esters. Tetrahedron: Asymmetry 9:4395–4417, 1998.
16. R Lagunas, JM Gancedo. Reduced pyridine-nucleotides balance in glucose-growing *Saccharomyces cerevisiae*. Eur J Biochem 37:90–94, 1973.
17. MM Griffiths, C Bernofsky. Activation of the mitochondrial DNP kinase from yeast. FEBS Lett 10:97–100, 1970.
18. M Anderlund, TL Nissen, J Nielsen, J Villadsen, J Rydström, B Hahn-Hägerdal, MC Kielland-Brandt. Expression of the *Escherichia coli pntA* and *pntB* genes, encoding nicotinamide nucleotide transhydrogenase, in *Saccharomyces cerevisiae* and its effect on product formation during anaerobic glucose fermentation. Appl Environ Microbiol 65:2333–2340, 1999.
19. B-N Zhou, AS Gopalin, F VanMiddlesworth, W-R Shieh, CJ Sih. Stereochemical control of yeast reductions. 1. Asymmetric synthesis of L-carnitine. J Am Chem Soc 105:5925–5926, 1983.
20. K Nakamura, T Miyai, K Nozaki, K Ushio, S Oka, A Ohno. Diastereo- and enantioselective reduction of 2-methyl-3-oxobutanoate by baker's yeast. Tetrahedron Lett 3155–3156, 1986.
21. B Wipf, E Kupfer, R Bertazzi, HGW Leuenberger. Production of (+)-(*S*)-ethyl 3-

hydroxybutyrate and (−)- (*R*)-ethyl 3-hydroxybutyrate by microbial reduction of ethyl acetoacetate. Helv Chim Acta 66:485–488, 1983.
22. T Kometani, H Yoshii, Y Takeuchi, R Matsuno. Large-scale preparation of (*S*)-ethyl 3-hydroxybutanoate with a high enantiomeric excess through baker's yeast-mediated bioreduction. J Ferm Bioeng 76:33–37, 1993.
23. K Nakamura, S Kondo, Y Kawai, A Ohno. Reduction by bakers' yeast in benzene. Tetrahedron Lett 7075–7078, 1991.
24. K Nakamura, S Kondo, Y Kawai, A Ohno. Stereochemical control in microbial reduction. XXI. Effect of organic solvents on reduction of α-keto esters mediated by baker's yeast. Bull Chem Soc Jpn 66:2738–2743, 1993.
25. M North. Baker's yeast reduction of β-keto esters in petrol. Tetrahedron Lett 1699–1702, 1996.
26. C Medson, AJ Smallridge, MA Trewhella. The stereoselective preparation of β-hydroxy esters using a yeast reduction in an organic solvent. Tetrahedron: Asymmetry 8:1049–1054, 1997.
27. J-N Cui, R Teraoka, T Ema, T Sakai, M Utaka. Highly regio- and enantioselective reduction of 1-chloro-2,4-alkanediones using baker's yeast: effects of organic solvents as additives. Tetrahedron Lett 3021–3024, 1997.
28. K Nakamura, M Higaki, K Ushio, S Oka, A Ohno. Stereochemical control of microbial reduction. 2. Reduction of β-keto esters by immobilized baker's yeast. Tetrahedron Lett 4213–4216, 1985.
29. K Nakamura, K Inoue, K Ushio, S Oka, A Ohno. Stereochemical control on yeast reduction of α-keto esters. Reduction by immobilized bakers' yeast in hexane. J Org Chem 53:2589–2593, 1988.
30. K Nakamura, Y Kawai, S Oka, A Ohno. A new method for stereochemical control of microbial reduction. Reduction of β-keto esters with baker's yeast immobilized by magnesium alginate. Tetrahedron Lett 2245–2246, 1989.
31. Y Naoshima, J Maeda, Y Munakata, T Nishiyama, M Kamezawa, H Tachibana. Bioreduction with immobilized baker's yeast in hexane using alcohols as an energy source. J Chem Soc Chem Commun 964–965, 1990.
32. K Nakamura, K Inoue, K Oshio, S Oka, A Ohno. Effect of allyl alcohol on reduction of β-keto esters by baker's yeast. Chem Lett 679–682, 1987.
33. Z-W Guo, CJ Sih. Enantioselective inhibition: a strategy for improving the enantioselectivity of biocatalytic systems. J Am Chem Soc 111:6836–6841, 1989.
34. K Nakamura, Y Kawai, S Oka, A Ohno. Stereochemical control in microbial reduction. 8. Stereochemical control in microbial reduction of β-keto esters. Bull Chem Soc Jpn 62:875–879, 1989.
35. K Nakamura, Y Kawai, T Miyai, A Ohno. Stereochemical control in diastereoselective reduction with baker's yeast. Tetrahedron Lett 3631–3632, 1990.
36. K Nakamura, Y Kawai, A Ohno. A novel method to synthesize (*L*)-β-hydroxyl esters by the reduction with bakers' yeast. Tetrahedron Lett 267–270, 1990.
37. Y Kawai, S Kondo, M Tsujimoto, K Nakamura, A Ohno. Stereochemical control in microbial reduction. XXIII. Thermal treatment of baker's yeast for controlling the stereoselectivity of reductions. Bull Chem Soc Jpn 67:2244–2247, 1994.
38. R Hayakawa, M Shimizu, T Fujisawa. The baker's yeast reduction of 1-acetoxy-2-alkanones in the presence of a sulfur compound. Tetrahedron: Asymmetry 8:3201–3204, 1997.

39. R Hayakawa, K Nozawa, M Shimizu, T Fujisawa. Control of enantioselectivity in the baker's yeast reduction of β-keto ester derivatives in the presence of a sulfur compound. Tetrahedron Lett 67–70, 1998.
40. http://sequence-www.stanford.edu/group/yeast_deletion_project/deletions3.html.
41. SF Altschul, TL Madden, AA Schaffer, J Zhang, Z Zhang, W Miller, DJ Lipman. Gapped BLAST and PSI-BLAST: a new generation of protein database search programs. Nucleic Acids Res 25:3389–3402, 1997.
42. Z Krozowski. The short-chain alcohol dehydrogenase superfamily: variations on a common theme. J Steroid Biochem Mole Biol 51:125–130, 1994.
43. H Jörnvall, B Persson, M Krook, S Atrian, R Gonzàlez-Duarte, J Jeffery, D Ghosh. Short-chain dehydrogenases/reductases (SDR). Biochemistry 34: 6003–6013, 1995.
44. H Jörnvall, J-O Höög, B Persson. SDR and MDR: completed genome sequences show these protein families to be large, of old origin, and of complex nature. FEBS Lett 445:261–264, 1999.
45. B Persson, JS Zigler, H Jörnvall. A superfamily of medium-chain dehydrogenases/reductases (MDR). Eur J Biochem 226:15–22, 1994.
46. JD Thompson, DG Higgins, TJ Gibson. CLUSTAL W: improving the sensitivity of progressive multiple sequence alignment through sequence weighting, position-specific gap penalties and weight matrix choice. Nucleic Acids Res 22:4673–4680, 1994.
47. R de Souza Pereira, F Pavão, G Oliva, The use of molecular modelling in the understanding of configurational specificity (*R* or *S*) in asymmetric reactions catalyzed by *Saccharomyces cerevisiae* or isolated dehydrogenases. Mol Cell Biochem 178:27–31, 1998.
48. JL DeRisi, VR Iyer, PO Brown. Exploring the metabolic and genetic control of gene expression on a genomic scale. Science 278:680–686, 1997.
49. CJ Sih, B-N Zhou, AS Gopalan, W-R Shieh, F VanMiddlesworth. Strategies for controlling the stereochemical course of yeast reductions. *Proc. Int. Symp. on Selectivity, a Goal in Synthetic Efficiency, 14th Hoechst Workshop Conf.* 1983, pp. 251–261, 1984.
50. S Ramaswamy, M Scholze, BV Plapp. Binding of formamides to liver alcohol dehydrogenase. Biochemistry 36:3522–3527, 1997.
51. PJ Kraulis. MolScript—a program to produce both detailed and schematic plots of protein structures. J Appl Crystallogr 24:946–950, 1991.
52. Z-N Yang, WF Bosron, TD Hurley. Structure of human χχ alcohol dehydrogenase: a glutathione-dependent formaldehyde dehydrogenase. J Mol Biol 265:330–343, 1997.
53. KJ Edwards, JD Barton, J Rossjohn, JM Thorn, GL Taylor, DL Ollis. Structural and sequence comparisons of quinone oxidoreductase, ζ-crystallin, and glucose and alcohol dehydrogenase. Arch Biochem Biophys 328:173–183, 1996.
54. K Ishihara, N Nakajima, S Tsuboi, M Utaka. Asymmetric reduction of 1-acetoxy-2-alkanones with bakers' yeast: purification and characterization of α-acetoxy ketone reductase. Bull Chem Soc Jpn 67:3314–3319, 1994.
55. N Nakajima, K Ishihara, S Kondo, S Tsuboi, M Utaka, K Nakamura. Differences in protein structure and similarities in catalytic function of two L-stereoselective carbonyl reductases from baker's yeast. Biosci Biotechnol Biochem 58:2080–2081, 1994.

56. K Ishihara, S Kondo, K Nakamura, N Nakajima. Protein sequences of two keto ester reductases: possible identity as hypothetical proteins. Biosci Biotechnol Biochem 60:1538–1539, 1996.
57. W-R Shieh. Part I. Studies on the active site-directed enzyme inhibitors of soybean lipoxygenase. Part II. Purification and studies on the β-keto ester reductases from baker's yeast. Ph.D thesis. University of Wisconsin, Madison, WI, 1987.
58. S Rodriguez Giordano. Baker's yeast β-keto ester reductions: whole cell biocatalysts with improved stereoselectivity by recombinant DNA techniques. Ph.D. thesis, University of Florida, Gainesville, FL, 2000.
59. K Nakamura, S Kondo, Y Kawai, N Nakajima, A Ohno. Amino acid sequence and characterization of aldo-keto reductase from baker's yeast. Biosci Biotechnol Biochem 61:375–377, 1997.
60. DH Harrison, KM Bohren, D Ringe, GA Petsko, KH Gabbay. An anion binding site in human aldose reductase: mechanistic implications for the binding of citrate, cacodylate, and glucose-6-phosphate. Biochemistry 33:2011–2020, 1994.
61. Recent studies have indicated that proteins encoded by GCY1, ARA1, and GRE3 all participate in reductions of exogenous ketones. S Rodrìguez, KT Schroeder, C Charron, MM Kayser, JD Stewart, unpublished results.
62. K Nakamura, S Kondo, Y Kawai, N Nakajima, A Ohno. Purification and characterization of α-keto ester reductases from bakers' yeast. Biosci Biotechnol Biochem 58:2236–2240, 1994.
63. H Lee. The structure and function of yeast xylose (aldose) reductases. Yeast 14: 977–984, 1998.
64. JD Goldberg, T Yoshida, P Brick. Crystal structure of a NAD-dependent D-glycerate dehydrogenase at 2.4 Å resolution. J Mol Biol 236:1123–1140, 1994.
65. DA Evans, MM Morrissey, RL Dorow. Asymmetric oxidation of chiral imide enolates. A general approach to the synthesis of enantiomerically pure α-hydroxy carboxylic acids. J Am Chem Soc 107:4346–4348, 1985.
66. C-H Wong, JR Matos. Enantioselective oxidation of 1,2-diols to L-α-hydroxy acids using co-immobilized alcohol and aldehyde dehydrogenases as catalysts. J Org Chem 50:1992–1994, 1985.
67. T Sugai, H Ohta. Lipase-mediated efficient preparation of both enantiomers of 2-acetoxytetraconsanoic acid, the intermediate for sphingolipid synthesis. Tetrahedron Lett 7063–7064, 1991.
68. K Ushio, S Yamauchi, K Masuda. Preparation of enantiomerically pure (*S*)-2-hydroxy heptanoate via bakers' yeast catalyzed hydrolytic resolution. Biotechnol Lett 13:495–500, 1991.
69. EJ Corey, JO Link, Y Shao. Two effective procedures for the synthesis of trichloromethyl ketones, useful precursors of chiral α-amino and α-hydroxy acids. Tetrahedron Lett 3435–3438, 1992.
70. F Effenberger. Synthesis and reactions of optically active cyanohydrins. Angew Chem Int Ed Engl 33:1555–1564, 1994.
71. W Adam, RT Fell, U Hoch, CR Saha-Möller, P Schreier. Kinetic resolution of chiral α-hydroxy esters by horseradish peroxidase-catalyzed enantioselective reduction to α-hydroxy esters. Tetrahedron: Asymmetry 6:1047–1050, 1995.
72. W Adam, M Lazarus, B Boxx, CR Saha-Möller, H-U Humpf, P Schreier. Enzymatic

resolution of chiral 2-hydroxy carboxylic acids by enantioselective oxidation with molecular oxygen catalyzed by the glycolate oxidase from spinach (*Spinacia oleracea*). J Org Chem 62:7841–7843, 1997.

73. W Adam, W Boland, J Hartmann-Schreier, H-U Humpf, M Lazarus, A Saffert, CR Saha-Möller, P Schreier. α Hydroxylation of carboxylic acids with molecular oxygen catalyzed by the α oxidase of peas (*Pisum sativum*): a novel biocatalytic synthesis of enantiomerically pure (*R*)-2-hydroxy acids. J Am Chem Soc 120:11044–11048, 1998.
74. VO Popov, IA Shumilin, TB Ustinnikova, VS Lamzin, TA Egorov. NAD-dependent formate dehydrogenase from the methylotrophic bacterium *Pseudomonas* sp. 101. I. Amino acid sequence. Bioorg Khim 16:324–335, 1990.
75. L Kühn, H Castorph, E Schweizer. Gene linkage and gene-enzyme relations in the fatty-acid-synthetase system of *Saccharomyces cerevisiae*. Eur J Biochem 24:492–497, 1972.
76. M Schweizer, C Lebert, J Holtke, LM Roberts, E Schweizer. Molecular cloning of the yeast fatty acid synthetase genes, FAS1 and FAS2: illustrating the structure of the FAS1 cluster gene by transcript mapping and transformation studies. Mol Gen Genet 194:457–465, 1984.
77. MM Kayser, MD Mihovilovic, J Kearns, A Feicht, JD Stewart. Baker's yeast-mediated reductions of α-keto-β-lactam. Two routes to the paclitaxel side chain. J Org Chem 64:6603–6608, 1999.
78. A Kuhn, C van Zyl, A van Donder, BA Prior. Purification and partial characterization of an aldo-keto reductase from *Saccharomyces cerevisiae*. Appl Environ Microbiol 61:1580–1585, 1995.
79. ST Kim, WK Huh, BH Lee, SO Kang. *D*-Arabinose dehydrogenase and its gene from *Saccharomyces cerevisiae*. Biochim Biophys Acta 1429:29–39, 1998.
80. JM Jez, TG Flynn, TM Penning. A new nomenclature for the aldo-keto reductase superfamily. Biochem Pharmacol 54:639–647, 1997.
81. JG Keppler. Twenty-five years of flavor research in a food industry. J Am Oil Chem Soc 54:474–477, 1977.
82. V Crocq, C Masson, J Winter, C Richard, G Lemaitre, J Lenay, M Vivat, J Buendia, D Prat. Synthesis of trimegestone: the first industrial application of baker's yeast mediated reduction of a ketone. Org Proc Res Dev 1:2–13, 1997.

8
Cross-Linked Enzyme Crystals: Biocatalysts for the Organic Chemist

Michael D. Grim
Westboro, Massachusetts

I. INTRODUCTION

Cross-linked enzyme crystals, one of the most exciting new developments in biocatalysis to come along since people first began to use enzymes to do chemistry, are biocatalysts that the average organic chemist can use as easily as many other standard chemical reagents. Indeed, this new form of enzyme has the potential to transform biocatalysis from an esoteric specialization into the widely practiced synthetic technique it deserves to be. First, the development of cross-linked enzyme crystals (CLCs) will be discussed from a historical perspective. Next, a summary of the properties of enzymes in this form compared to soluble and immobilized enzymes will be presented. Finally, a review of recent applications of CLCs to show how they are bringing biocatalysis into the mainstream of organic synthesis, particularly with respect to the pharmaceutical industry, will be given.

Throughout the bulk of this chapter, CLC will be used as an abbreviation for cross-linked enzyme crystal. Occasionally, the abbreviation CLEC® will also be used to indicate cross-linked enzyme crystal. This acronym is a registered trademark of Altus Biologics, Inc. (Cambridge, MA) and will be used in discussing work done with various cross-linked enzyme crystals which are commercially available from Altus (Table 1). Finally, the notation CPC will be used to denote cross-linked protein crystal.

Table 1 Commercially Available Cross-Linked Enzyme Crystals[a]

Product	Synthetic utility
ChiroCLEC™-BL (Subtilisin protease)	Resolution of amino acids, amines, and amino acid analogs
ChiroCLEC™-CR (*Candida rugosa* lipase)	Resolution of alcohols, acids and esters
ChiroCLEC™-EC (Penicillin acylase)	Resolution of amino acids and amines; protection/deprotection
ChiroCLEC™-PC (*Burkholderia cepacia* lipase)	Resolution of acids, esters, and alcohols
PeptiCLEC™-BL (Subtilisin protease)	Coupling of α-amino acid esters to amines, amino acids, or peptides
PeptiCLEC™-TR (Thermolysin protease)	Coupling of α-amino acids to amines, amino acids, or peptides
SynthaCLEC™-PA (Penicillin acylase)	Antibiotic synthesis

[a] Available from Altus Biologics, Inc., Cambridge, MA.

II. HISTORY

The first reported preparation of cross-linked enzyme crystals was by Quiocho and Richards in 1964 [1]. They prepared crystals of carboxypeptidase-A and cross-linked them with glutaraldehyde. The material they prepared retained only about 5% of the activity of the soluble enzyme and showed a measurable increase in mechanical stability. The authors quite correctly predicted that ''cross-linked enzyme crystals, particularly ones of small size where the diffusion problem is not serious, may be useful as reagents which can be removed by sedimentation and filtration.'' Two years later the same authors reported a more detailed study of the enzymic behavior of CLCs of carboxypeptidase-A [2]. In this study they reported that only the lysine residues in the protein were modified by the glutaraldehyde cross-linking. The CLCs were packed in a column for a flow-through assay and maintained activity after many uses over a period of 3 months.

Little was done in the area of cross-linked enzyme crystals over the next 10 years. In 1977, the kinetic properties of CLCs of the protease subtilisin were reported by Tuchsen and Ottesen [3]. They reported that cross-linked enzyme crystals of subtilisin were highly effective catalysts with increased thermal stability and increased stability toward acid compared to the soluble enzyme. They further reported that the CLCs of subtilisin showed essentially no autodigestion at 30°C. Like Quiocho and Richards before them, Tuchsen and Ottesen found that only the lysine residues of the enzyme were affected by the glutaraldehyde treatment.

Another gap of almost 10 years occurred before work in the area of CLCs picked up again. In 1985, a group at the Louis Pasteur University in Strasbourg, France, prepared cross-linked crystals of horse liver alcohol dehydrogenase [4]. The activity of the enzyme in CLC form was maintained and the coenzyme was found to be firmly bound to the crystals. The cross-linked crystals could be used as redox catalysts with no addition of coenzyme. The authors also reported the increased stability of CLCs toward organic solvents.

Up to this time CLCs had been prepared largely for academic interest and showed significant stability but generally only very low activity when compared to the soluble enzymes from which they were derived. Cross-linked enzyme crystals were transformed from a mildly interesting, esoteric area of academic research into a viable commercial technology when Navia and St. Clair reported for the first time the development of cross-linked enzyme crystals (CLECs®) which combined high activity with the high stability already associated with CLCs [5]. The use of cross-linked enzyme crystals in biocatalysis finally started to become more common in 1992–1993 when Altus Biologics was formed to commercialize the technology developed by Navia and co-workers [6]. Altus has obtained very broad patent coverage for its technology [7]. Today many different enzymes are available in the cross-linked crystal format exclusively from Altus Biologics under the ChiroCLEC™, PeptiCLEC™, and SynthaCLEC™ brand names (Table 1).

III. PROPERTIES

Having briefly outlined the historical development of cross-linked enzyme crystals, a discussion of their properties as compared to soluble and immobilized enzymes is in order.

A. Structure

First and foremost, cross-linked enzyme crystals are crystals. Within the crystal lattice the concentration of protein approaches the theoretical limit. This is important to the process development chemist, who would much rather use a small quantity of a very active catalyst in a reactor than fill it with an immobilized enzyme. Typically an immobilized enzyme contains only 1–10% by weight enzyme, with the remaining carrier material simply occupying valuable reactor space. The crystallinity is absolutely required to achieve the stability exhibited by CLCs [8]. Cross-linked soluble thermolysin and cross-linked precipitate of thermolysin are no more stable than the soluble enzyme. Crystals of proteins (and enzymes) are not solid structures. They typically contain 30–65%, by volume, open channels, which are usually filled with solvent [9]. In the case of cross-linked crystals of thermolysin, the diameter of those channels is about 25 Å [6]. Cross-linked enzyme crystals and cross-linked protein crystals (CPCs) are microporous materials with pore surface areas between 500 and 2000 m^2/mg [10]. This is the same range as for inorganic zeolites. CLCs, however, have pore diameters in the range of 3–8 nm, as compared to 0.2–1.0 nm for normal inorganic zeolites. Indeed, CLCs can be thought of as bioorganic zeolites. The macroporous nature of CLCs enables small molecules to penetrate easily into the crystals. Diffusion is not rate limiting, as long as microcrystals (10–100 μm) of enzymes are used to catalyze chemical reactions [11].

While it has been well established that glutaraldehyde reacts with the side-chain amino groups of lysine residues, the exact nature of the bonds formed is unclear. The stability exhibited to acid appears to eliminate simple imine formation as the mechanism [12]. Aqueous glutaraldehyde forms a mixture of oligomers of various lengths and structures [13]. It is thought that an enzyme crystal which is treated with such a mixture is able to self-select the best cross-linking agents from the mixture for its unique set of side-chain functionalities and their arrangement in space [14].

B. Purity

Commercially available, soluble enzyme preparations are generally very crude mixtures containing from less than 1% to about 10% active enzyme. The remaining materials present in these mixtures commonly include buffer salts, cell debris from the fermentation process, and even other enzymes, which may or may not have beneficial activities. Pure soluble enzymes tend to be prohibitively expensive due to the often complex chromatography used to achieve the purification, and frequently prove to be less stable in the purified state. Crystallization is an inexpensive, standard purification technique in organic chemistry. Cross-linked enzyme crystals are therefore, by definition, highly purified forms of enzymes. In contrast to crude enzyme preparations, the high purity of CLCs has several benefits when they are used to conduct chemistry. The most obvious benefit is that the chemist is putting fewer impurities into the reaction to begin with, so he or she can expect a cleaner reaction product and easier workup when the reaction is complete.

Another major benefit of the purity of CLCs relates to the fact that many commercially available enzyme products actually contain more than one enzyme, and these contaminating enzymes can often lead to undesired side reactions. In the case of a resolution this can mean the difference between success and failure. For example, in the resolution of racemic ketoprofen by ester hydrolysis (Fig. 1), the enantioselectivity using crude *Candida rugosa* lipase is poor ($E = 5$).

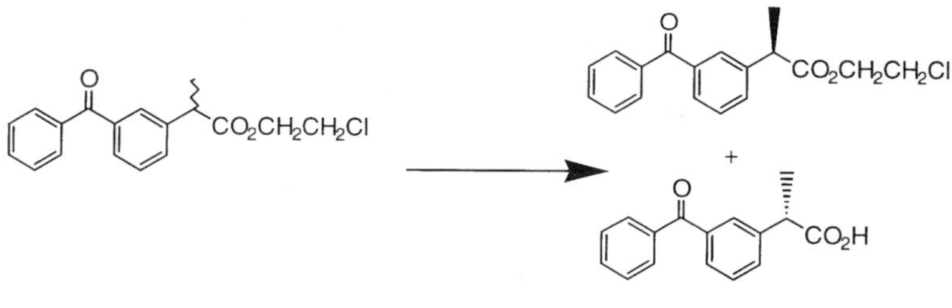

Figure 1 Resolution of racemic ketoprofen by enzyme catalyzed ester hydrolysis.

The E value represents the ratio of the specificity constants of the enzyme for the two enantiomers of the substrate and allows one to compare directly the selectivity of different enzymes in a reaction [15]. In contrast, using the purified CLC form of the enzyme gives an $E = 66$, which makes it a viable method for obtaining pure S-ketoprofen [16].

C. Stability

Probably the most striking and valuable characteristic of cross-linked enzyme crystals is the remarkable stability they exhibit in comparison to soluble and immobilized enzymes. They can withstand exposure to organic solvents, high temperatures, mechanical stress such as shear, extremes of pH, and even exposure to proteases.

1. Organic Solvent Stability

Many of the compounds the organic chemist works with in the pharmaceutical industry are not very soluble in water. Enzymes are generally rapidly denatured in the presence of water-miscible organic solvents. Consequently, enzyme-catalyzed reactions which are typically water based have historically been viewed as a technique which was irrelevant to the organic chemist. The remarkable stability of cross-linked enzyme crystals has changed everything. Figure 2 shows the stability in water/organic solvent mixtures of ChiroCLEC™-CR compared to the soluble *Candida rugosa* lipase from which it is derived.

It is difficult to overstate the significance of the organic solvent stability of CLCs. This stability allows the organic chemist to use enzymes in a much broader set of reaction conditions than was ever possible before [17]. Since it is now possible to use enzymes with the standard type of conditions under which they normally work, chemists are much more likely to try a biocatalytic step

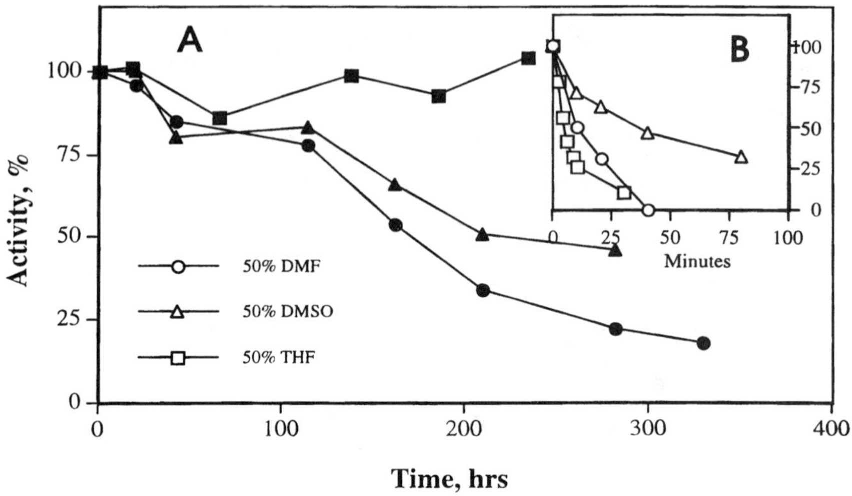

Figure 2 Stability of (A) CRL-CLEC and (B) commercial CRL (inset) in the presence of water-miscible organic solvents. The catalyst was stored in the presence of 50% aqueous solvent at 25°C. CRL is *Candida rugosa* lipase. (Reproduced from Ref. 16. © 1995 American Chemical Society.)

when developing a new synthetic scheme. Thus CLCs are the biocatalysts of choice for the organic chemist, and their commercial availability is spearheading an increase in the use of enzymes in the pharmaceutical industry.

2. *Thermal Stability*

Soluble enzymes are typically easily inactivated by heat. It does not require much energy to disrupt the hydrogen bonds and other weak forces which hold an enzyme in its active conformation. Like the organic solvent stability, the thermal stability of CLCs can be two to three orders of magnitude greater than that of the soluble protein. When an enzyme forms a crystal, a very large number of stabilizing contacts are formed between individual enzyme molecules. Energy must be put into the system in order to disrupt these new contacts. On top of that, additional energy is required to break the covalent cross-links before the CLC begins to dissolve and then denature. CLCs therefore are much more stable than the same enzyme in solution. Figure 3 illustrates the thermal stability of PeptiCLEC™-TR, the CLC form of the protease thermolysin, as compared to soluble enzyme.

The increased thermal stability offers several major benefits to the organic chemist using CLCs as catalysts. First, no special storage conditions are required

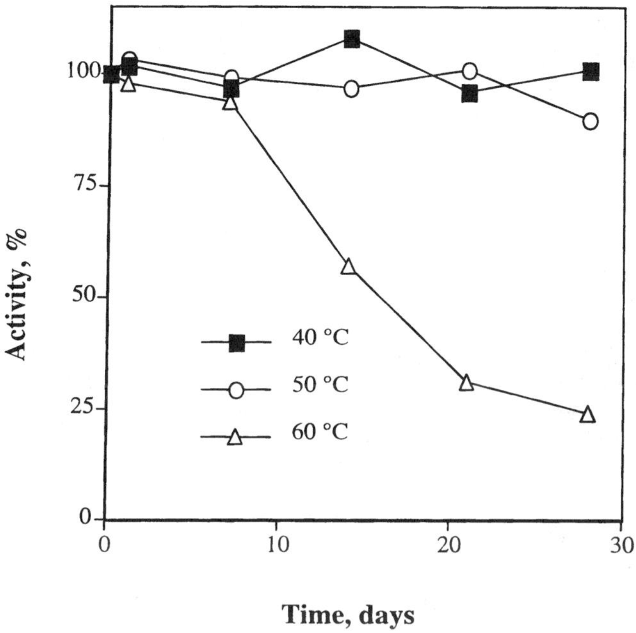

Figure 3 Stability of thermolysin CLEC at elevated temperatures. In comparison, the half-life of soluble thermolysin is only 10 hr at 50°C.

for CLCs. Unlike soluble and immobilized forms, which must be shipped and stored cold, CLCs are stable for years at room temperature. Second, since enzyme catalyzed reactions are generally slower in organic solvents and with non-natural substrates, the ability to run such reactions at higher temperatures to increase the rate is a significant benefit to the chemist.

3. Mechanical Stability

Testing has shown that CLEC® catalysts are stable for long periods of time under standard fine chemical production conditions such as agitation, pumping, and filtration [18]. As a process is scaled up from the bench, through pilot scale and production, there is an increase in the mechanical shear and pressure to which the catalyst is exposed to. CLCs of thermolysin and *Candida rugosa* lipase under high or moderate shear mixing showed little or no loss in average particle size. Extended agitation of an aqueous CLEC® suspension at high mixing speeds using a flat turbine (high shear) and an A310 propeller blade (moderate shear) showed virtually no particle breakage. Under the more drastic treatment of continuous pumping using a high-shear centrifugal pump for 24 hr, particle fracturing for ChiroCLEC™-CR was seen only at the highest speed. CLEC® particles are mechanically stable under the highest shear conditions seen in fluid transfer and agitation.

4. *Stability to Proteolysis*

Enzymes in the cross-linked crystal form are essentially impervious to degradation by exogenous proteases and from autolysis, in the case of CLCs of proteases themselves [5]. This stability makes the enzyme-catalyzed preparation of peptides and peptide mimics truly practical [6]. Examples will be discussed in more detail in Sec. IV. Further, one could conceive of using multiple enzymes in ''one-pot'' reaction systems mimicking natural biosynthetic cascades. Indeed, the application of this concept has been reported for a mixture of lipoamide dehydrogenase and lactate dehydrogenase [19].

IV. GENERAL APPLICATIONS

The primary focus of this section will be the use of cross-linked enzyme crystals as catalysts of chemical reactions and particularly those reactions of relevance to the pharmaceutical industry. Before beginning this discussion, however, a brief look at the use of CLCs in chromatography is given, since this application also has possible uses in the pharmaceutical industry.

A. Chromatography

As discussed in Sec. III.A, cross-linked enzyme crystals and cross-linked protein crystals can be thought of as bioorganic zeolites. In addition, CLCs and CPCs are chiral in nature and in the case of cross-linked enzyme crystals also contain special binding sites. It has recently been shown that CLCs and CPCs can be packed in high-performance liquid chromatography (HPLC) columns and used as unique stationary phases. These ''bioorganic zeolite'' columns are able to perform separations via size exclusion, adsorption, and chiral discrimination [20]. The materials showed excellent stability over 500 runs cycling between water and 50% acetonitrile. This application of CLCs may become much more important in the future if large-scale chromatography becomes more widely accepted and utilized in the manufacture of pharmaceutical active ingredients and intermediates.

B. Biocatalysis

The major application of cross-linked enzyme crystals in the pharmaceutical and fine chemicals industry is in biocatalysis or the use of CLCs to catalyze various

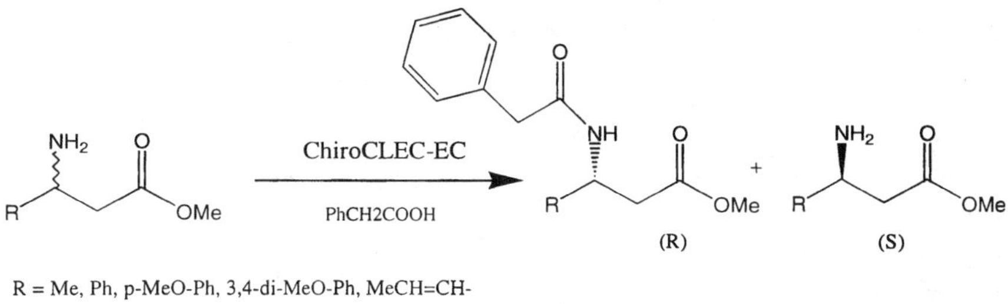

Figure 4 Resolution of β-amino esters via selective acylation.

chemical reactions. In this section we will review recent applications, grouped by reaction type.

1. Resolutions

Resolutions of racemic mixtures are by far the most frequent applications of biocatalysts in the pharmaceutical industry. Repic et al., of the process research and development group at Novartis, recently published work to develop a method for the resolution of racemic β-amino acid esters, an important class of intermediates for the preparation of peptidomimetics [21]. The Novartis group used Chiro-CLEC™-EC [22] in 2% aqueous toluene to selectively acylate several different β-amino esters (Fig. 4). The authors were able to isolate the desired (*S*) isomer of the amino esters in >95% ee in a simple one-step reaction and described it as a method which could be amenable for large-scale preparation.

The bicyclic aminoalcohol, 3-quinuclidinol, is an important synthon for the preparation of cholinergic receptor ligands [23], anesthetics [24], and drugs for the treatment of Alzheimer's disease and asthma [5]. P. Bossard at Lonza AG developed and patented an enantioselective acylation of racemic 3-quinuclidinol using ChiroCLEC™-BL, the CLC form of subtilisin (Fig. 5) [25]. The reaction was run in 2-methyl-3-butanol with vinyl butyrate used as the acylating agent.

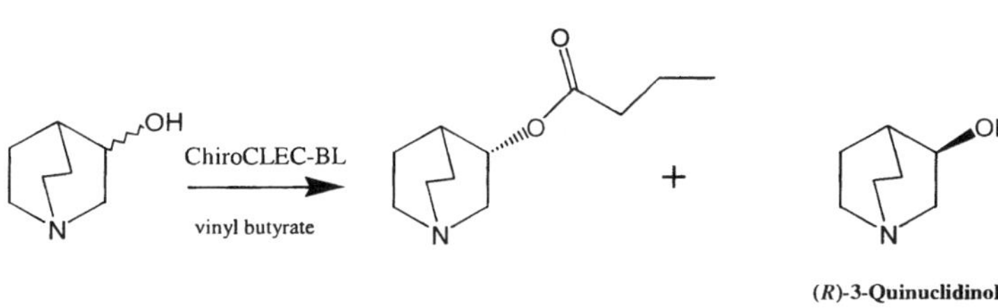

Figure 5 Resolution of (*R*,*S*)-3-quinuclidinol via selective acylation.

The pure (*R*) enantiomer was obtained in 68% yield and 96.2% ee after crystallization.

The natural products epothilone A and B are structurally different from taxol but have similar anticancer activity. Significantly, they have been reported to be much more active against cell lines exhibiting multiple-drug resistance [26]. Taylor and co-workers at the University of Notre Dame have recently published an elegant, formal total synthesis of epothilone A [27]. In this work, the authors used the CLC form of *Burkholderia cepacia* (formerly *Pseudomonas cepacia*) lipase (ChiroCLEC™-PC) to resolve a key alcohol intermediate by selective acylation with vinyl acetate in *t*-butyl methyl ether (Fig. 6). The enantioselectivity was >20:1 at 47% conversion and efficiently provided gram quantities of the desired (*R*) alcohol. Since the unreacted (*S*) alcohol can easily be epimerized by a simple oxidation–reduction sequence and the catalyst reused without significant loss in activity, the method is ideally suited for scale-up.

Researchers at the biotech company EntreMed, Inc., have recently prepared and tested 2-phthalimidino-glutaric acid analogs of thalidomide and found them to be potent inhibitors of tumor metastasis [28]. The key to the success of their synthesis was a resolution via enantioselective ester hydrolysis catalyzed by ChiroCLEC™-BL, the CLC form of the protease subtilisin. The authors were able to isolate both enantiomers of the desired product with good optical purity (95% ee) (see Fig. 7).

2. *Peptide Synthesis*

Peptide synthesis is an extremely important area of chemistry for the pharmaceutical industry, and like any specialized area of chemistry, has its own set of unique problems associated with it. Racemization and purification of final products are two of the most difficult problems in this area. The use of enzymes has been explored as a possible answer to these problems since 1938 [29]. However, proteases needed to catalyze peptide synthesis are subject to rapid autolysis under the conditions needed to affect peptide coupling, so this has generally not been a practical approach until cross-linked enzyme crystals of proteases became available. The synthetic utility of protease-CLCs was demonstrated by the thermolysin CLC (PeptiCLEC™-TR)-catalyzed preparation of the aspartame precursor Z-

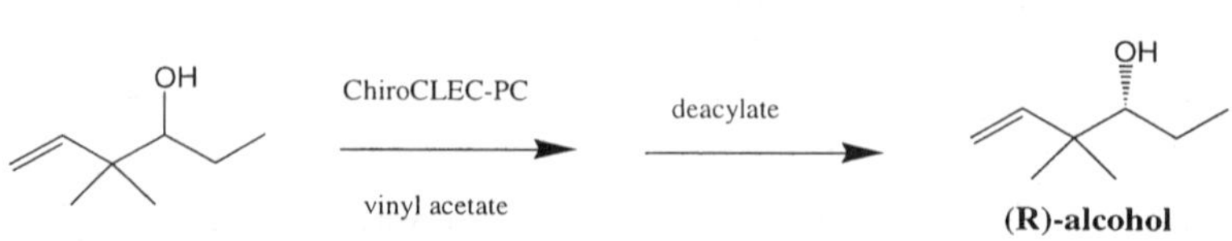

Figure 6 Resolution of key alcohol intermediate for Epothilone A and B.

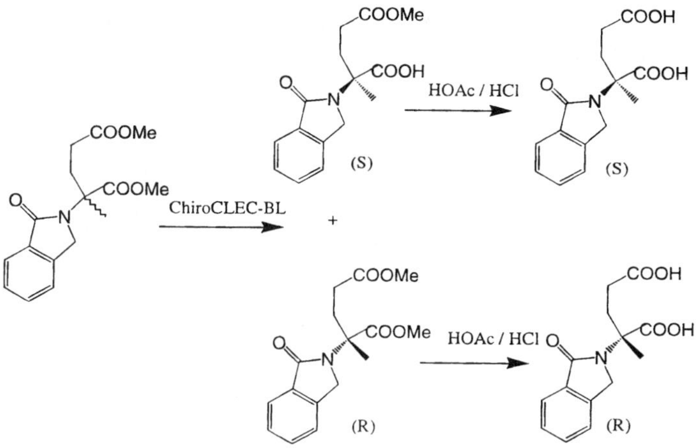

Figure 7 Synthesis of 2-phthalimidino-glutaric acid analogs.

Asp-Phe-OMe shown in Figure 8. The product was obtained in good yield (95%) and the reaction was repeated over 19 cycles in pure ethyl acetate at 55°C with no loss in enzyme activity [5].

Biologically active peptides are generally metabolized very quickly by the body and often present bioavailability problems. Pharmaceutical companies circumvent these problems by designing peptide mimics which might contain non-natural amino acids and/or functional isosteres. The synthetic potential for using cross-linked enzyme crystals to prepare peptidomimetics was demonstrated in recent work by Margolin et al. [30]. A variety of chiral alkylamide derivatives of amino acids and peptides were prepared by CLC-catalyzed (PeptiCLEC™-BL) coupling reactions. The power in this approach is that racemic carboxy and/or amine components can be stereoselectively coupled to give the desired (*S*,*S*) products and thereby eliminate several steps from the process. In the coming years,

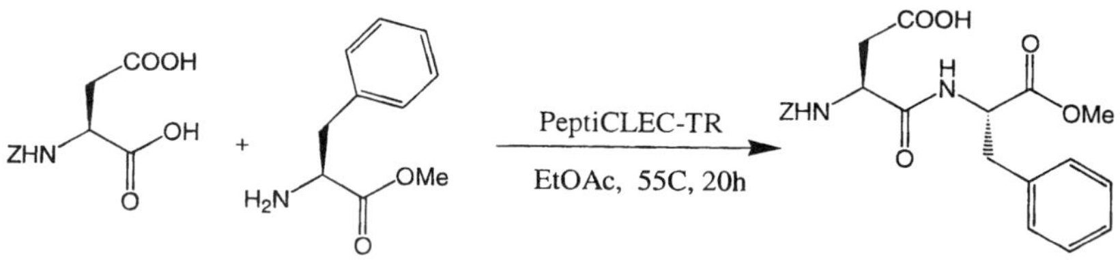

Figure 8 Synthesis of aspartame precursor.

we will no doubt see many new peptidomimetic drugs come to market which are produced via CLC-catalyzed coupling reactions.

3. Regioselective Reactions

Regioselective reactions in systems which contain multiple functional groups are an area ideally suited for biocatalysis. Linhardt and co-workers at the University of Iowa's Division of Medicinal and Natural Products Chemistry recently published the synthesis of a series of 1′-O-acyl sucrose derivatives [31]. Using ChiroCLEC™-BL (the CLC of subtilisin) and vinyl esters of the acylating agent in pyridine as solvent, the authors prepared 1′-O-lauryl sucrose, 1′-O-myristyl sucrose, and 1′-O-stearyl sucrose in 80–90% yield (Fig. 9). Their method represents a ''green'' alternative to the tin chemistry previously used [32].

4. Carbon–Carbon Bond-Forming Reactions

Carbon–carbon bond-forming reactions are some of the most important transformations in organic chemistry. Sobolov et al. [33] reported that CLCs of fructose 1,6-diphosphate aldolase from rabbit muscle are much more stable than the soluble enzyme. The synthetic potential of these CLCs was demonstrated by the preparation of a series of compounds shown in Fig. 10.

5. Reductions

Reduction of achiral precursors is often used to produce chiral products. The advantage of this approach is that the theoretical yield of product is 100% compared to the 50% theoretical maximum for the resolution of racemates. Cross-linked crystals of lactate dehydrogenase have been used to prepare L-lactic acid from pyruvic acid in an electrolytic cell. The LDH CLCs maintained constant

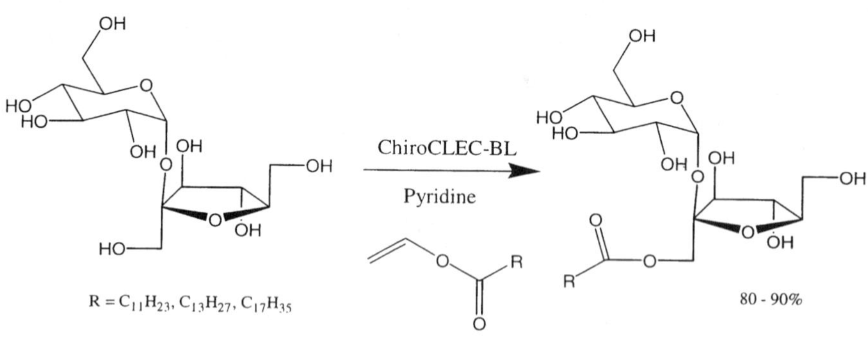

Figure 9 Regioselective acylation using ChiroCLEC™-BL.

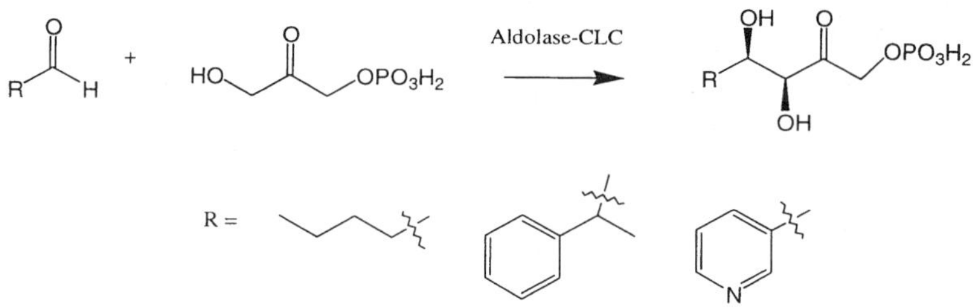

Figure 10 Aldolase-CLC-catalyzed C–C bond formation.

activity over 25 days and were much less sensitive to pH than the soluble enzyme [34]. Horse liver alcohol dehydrogenase has been crystallized and cross-linked with the cofactor bound to the enzyme and the resulting CLCs used to reduce a number of ketones to chiral alcohols (Fig. 11) [35]. The CLCs exhibited activity similar to the soluble enzyme and were more stable toward heat and the presence of alcohols. The cofactor remained tightly bound to the CLC and could be regenerated in situ with a high turnover number.

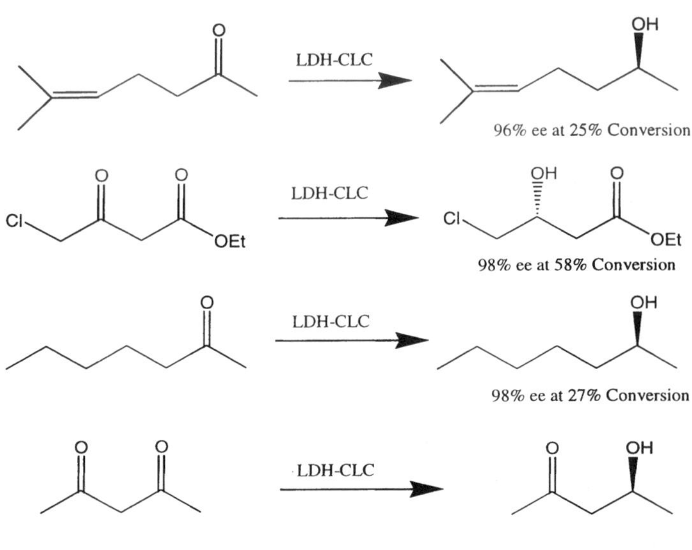

Figure 11 Reduction of ketones with LDH-CLC. Conversions shown were reached after 20 hr.

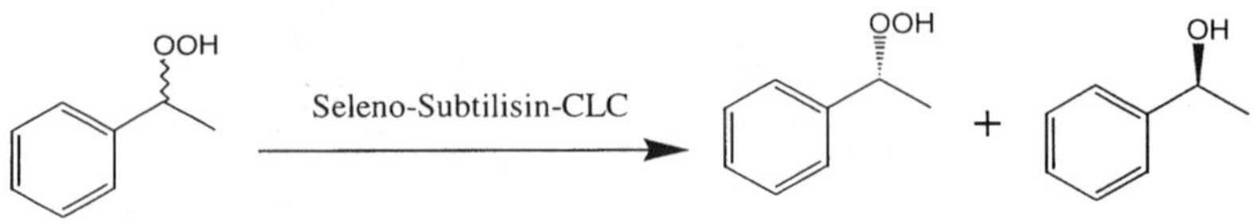

Figure 12 Resolution using semisynthetic seleno-subtilisin CLC.

Haring and Schreier have modified the active site of subtilisin cross-linked enzyme crystals by introducing selenium into it and thereby converting the enzyme into a peroxidase [36]. The rigid CLC matrix allowed them to chemically modify subtilisin without loss of the tertiary structure. The kinetic resolution of racemic 2-hydroxy-1-phenylethyl hydroperoxide was demonstrated using the semisynthetic CLC (Fig. 12). The reaction time was 25–30 min with an ee of 97%. The authors demonstrated the stability of these semisynthetic CLCs by cycling their enzyme 10 times.

V. INDUSTRIAL APPLICATIONS

The true value of cross-linked enzyme crystals is that this technology minimizes many if not all of the problems which have limited the industrial use of enzymes to date. Issues of stability, purity, and cost are all addressed favorably by cross-linked enzyme crystal technology.

The first large-scale commercial application of cross-linked enzyme crystals was the use of glucose isomerase CLCs to produce high-fructose corn syrup. While this is not a pharmaceutical or a biotechnological application, it is included here because it serves to demonstrate the economic viability of the technology in a very cost-sensitive business. In this application the CLCs were attached to the surface of a polystyrene-cellulose-titanium oxide composite carrier in a ratio of 9:1 carrier:enzyme. The catalyst had a half-life of 150 days at 57°C, and 12–18 tons of dry sugar product could be produced per kilogram of enzyme [37].

A group at Industrial Research Limited in New Zealand recently reported the results of a study to determine if a cross-linked enzyme crystal-catalyzed resolution can compete with alternative chiral technologies in the pharmaceutical and chemical process industries. The group used the enantioselective hydrolysis of α-phenylethyl acetate catalyzed by ChiroCLEC™-PC as a model system (Fig. 13). Based on their results with 270 kg of racemate, they evaluated the economics of running the process at the 600-kg batch scale and concluded that ''this process is economically feasible'' [38].

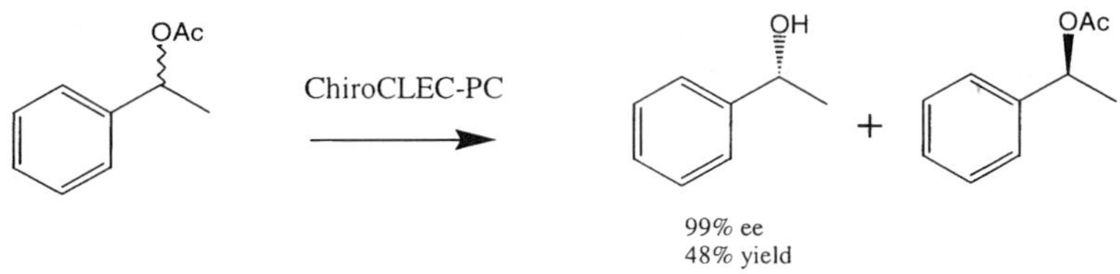

Figure 13 Resolution of racemic α-phenylethyl acetate via ChiroCLEC™-PC.

As discussed earlier in the chapter, the use of CLCs did not begin to take off until Altus Biologics was formed and commercialized the technology in 1993. Since it takes several years for a new drug to be developed and commercialized and many do not make it all the way to the market, there are few examples in the public domain of recent drug introductions which are produced using a CLC step. The syntheses of Chiroscience's single-isomer MMPIs (matrix metalloprotease inhibitors) D1927 and D2163 were elegantly achieved via the use of PeptiCLEC™-TR. In these cases the key molecular structure and a final element of asymmetry were introduced via a PeptiCLEC™-TR-mediated amide bond formation [39].

In the very important therapeutic class of antibiotics, both 6-APA (6-aminopenicillinic acid) and 7-ADCA (7-aminodeacetoxycephalosporanic acid) are being produced today on multi-ton scale using the cross-linked crystal form of penicillin-G amidase (SynthaCLEC™-PA) (P. Lanciano, President, Altus Biologics,

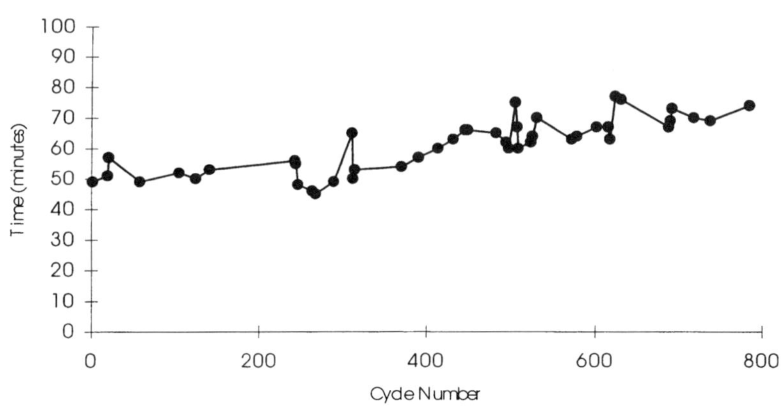

Figure 14 Repetitive batch hydrolysis of penicillin-G to 6-aminopenicillanic acid. Plot shows batch number versus time required for complete hydrolysis of starting material.

personal communication, 1999). The stability of enzymes in the cross-linked crystal form is again clearly demonstrated by the multicycle piloting reaction run for the hydrolysis of penicillin-G shown in Figure 14.

VI. SUMMARY

Like all "new technologies," cross-linked enzyme crystals were not developed overnight. Almost 30 years separate the first report of cross-linked enzyme crystals from the introduction of truly useful, commercially available CLEC® products.

The tremendous advantages of purity, stability, and ultimately economy of using enzymes in CLC form have only very recently been discovered by the organic chemist and in fact is still an ongoing process. As the examples in this chapter illustrate, a wide variety of synthetic targets, especially in the pharmaceutical area, are being successfully prepared using cross-linked enzyme crystals. Further, processes for large-scale manufacturing employing CLC steps have been developed and successfully employed.

It is a certainty that many of the exciting new therapeutic agents which will reach the market in the coming years will be manufactured by processes which include the use of cross-linked enzyme crystals.

REFERENCES

1. FA Quiocho, FM Richards. Intermolecular cross-linking of a protein in the crystalline state: carboxypeptidase-A. Proc Natl Acad Sci (USA) 52:833–839, 1964.
2. FA Quiocho, FM Richards. The enzymic behavior of carboxypeptidase-A in the solid state. Biochemistry 5:4062–4075, 1966.
3. E Tuchsen, M Ottesen. Kinetic properties of subtilisin type Carlsberg in the crystalline state. Carlsberg Res Commun 42:407–420, 1977.
4. KM Lee, M Blaghen, J-P Samama, J-F Biellmann. Crosslinked crystalline horse liver alcohol dehydrogenase as a redox catalyst: activity and stability toward organic solvent. Bioorg Chem 14:202–210, 1986.
5. NL St. Clair, M Navia. Cross-linked enzyme crystals as robust biocatalysts. J Am Chem Soc 114:7314–7316, 1992.
6. D Haring, P Schreier. Cross-linked enzyme crystals. Curr Opin Chem Biol 3:35–38, 1999.
7. MA Navia, NL St. Clair. U.S. Patent 5,618,710, 1997.
8. RA Persichetti, N St. Clair, JP Griffith, MA Navia, AL Margolin. Cross-linked enzyme crystals (CLECs) of thermolysin in the synthesis of peptides. J Am Chem Soc 117:2732–2737, 1995.

9. BW Matthews. Solvent content of protein crystals. J Mol Biol 33:491–497, 1968.
10. LZ Vilenchik, JP Griffith, NL St. Clair, MA Navia, AL Margolin. Protein crystals as novel microporous materials. J Am Chem Soc 120:4290–4294, 1998.
11. JL Schmitke, CR Wescott, AM Klibanov. The mechanistic dissection of the plunge in enzymatic activity upon transition from water to anhydrous solvents. J Am Chem Soc 118:3360–3365, 1996.
12. PM Hardy, AC Nicholls, HN Rydon. Nature of cross-linking of proteins by glutaraldehyde. 1. Interaction of glutaraldehyde with amino groups of 6-aminohexanoic acid and of α-N-acetyl-L-lysine. J Chem Soc, Perkin Trans I 958–962, 1976.
13. T Tashima, M Imai, YY Kuroda, S Yagi, T Nakagawa. Structure of a new oligomer of glutaraldehyde produced by aldol condensation reaction. J Org Chem 56:694–697, 1991.
14. K Martinek, VP Torchilin. Stabilization of enzymes by intramolecular cross-linking using bi-functional reagents. Meth Enzymol 137:615–624, 1983.
15. CS Chen, CJ Sih, Y Fujimoto, G Girdaukus. Quantitative analyses of biochemical kinetic resolutions of enantiomers. J Am Chem Soc 104:7294–7299, 1982.
16. JJ Lalonde, C Govardhan, N Khalaf, AG Martinez, K Visuri, AL Margolin. Cross-linked crystals of *Candida rugosa* lipase: highly efficient catalysts for the resolution of chiral esters. J Am Chem Soc 117:6845–6852, 1995.
17. Y-F Wang, K Yakovlevsky, B Zhang, AL Margolin. Cross-linked crystals of subtilisin: versatile catalyst for organic synthesis. J Org Chem 62:3488–3495, 1997.
18. JJ Lalonde, C Govardhan, RA Persichetti, Y-F Wang, AL Margolin. Process-scale application of CLECs: the processing and economic benefits of using cross-linked enzyme crystals as biocatalysts. Proc. Industrial Biocatalysis: InBio '96, Manchester, UK, 1996.
19. AJ Fry, SB Sobolov, MD Leonida, KI Voivodov, J Fenton. New biopolymers for electroenzymatic synthesis of α-hydroxy acids. Electrochem Soc Proc 97:115–122, 1997.
20. AL Margolin, LZ Vilenchik. Crosslinked protein crystals as universal separation media. 24 September 1996, WO9813119.
21. D Roche, K Prassad, O Repic. Enantioselective acylation of β-amino esters using penicillin-G acylase in organic solvents. Tetrahedron Lett 40:3665–3668, 1999.
22. ChiroCLEC™-EC is cross-linked crystals of penicillin-G acylase and is available from Altus Biologics, Inc., Cambridge, MA.
23. VI Cohen, RE Gibson, LH Fan, R De la Cruz, MS Gitler, E Hariman, RC Reba. Synthesis and muscarinic cholinergic receptor affinities of 3-quinuclidinyl-1-α-(alkoxyalkyl)-α-aryl-α-hydroxyacetates. J Med Chem 34:2984–2993, 1991.
24. F Gregan, J Durinda, E Racanska, J Zamocka. Synthesis and local anaesthetic activities of 3-(2-alkoxyphenylcarbamoyloxy)chinuclidinium chlorides. Pharmazie 48: 465–466, 1993.
25. P Bossard. Patent, DE 19715465, 1997.
26. G Hoefle, N Bedorf, K Gerth, H Reichenbach. Patent DE 4138042, 1993.
27. RE Taylor, GM Galvin, KA Hilfiker, Y Chen. A formal total synthesis of epothilone A: enantioselective preparation of the C1-C6 and C7-C12 fragments. J Org Chem 63:9580–9583, 1998.

28. JH Shah, GM Swartz, AE Papathanassiu, AM Treston, WE Fogler, JW Madsen, SJ Green. Synthesis and enantiomeric separation of 2-phthalimido-glutaric acid analogues: potent inhibitors of tumor metastasis. J Med Chem 42:3012–3017, 1999.
29. CH Wong, GM Whitesides. Enzymes in Synthetic Organic Chemistry. Tarrytown, NY: Pergamon, 1994, p 46.
30. Y-F Wang, K Yakovlevsky, AL Margolin. An efficient synthesis of chiral amino acid and peptide alkylamides via CLEC-subtilisin catalyzed coupling and in situ resolution. Tetrahedron Lett 37:5317–5320, 1996.
31. T Polat, HG Bazin, RJ Linhardt. Enzyme catalyzed regioselective synthesis of sucrose fatty acid ester surfactants. J Carbohydr Chem 16:1319–1325, 1997.
32. U.S. Patent 4,350,746, 1990. U.S. Patent 5,023,329, 1991. U.S. Patent 5,089,608, 1992. U.S. Patent 5,470,969, 1995.
33. SB Sobolov, A Bartoszko-Malik, TR Oeschger, MM Montelbano. Cross-linked enzyme crystals of fructose diphosphate aldolase: development as a biocatalyst for synthesis. Tetrahedron Lett 35:7751–7754, 1994.
34. SB Sobolov, MD Leonida, A Bartoszko-Malik, KI Voivodov, F McKinney, AJ Fry. Cross-linked LDH crystals for lactate synthesis coupled to electroenzymatic regeneration of NADH. J Org Chem 61:2125–2128, 1996.
35. N St. Clair, Y-F Wang, AL Margolin. Cofactor bound cross-linked enzyme crystals (CLEC) of alcohol dehydrogenase. Angew Chem Int Ed 39:380–383, 2000.
36. D Haring, P Schreier. Novel biocatalysts by chemical modification of known enzymes: cross-linked microcrystals of semi-synthetic peroxidase seleno-subtilisin. Angew Chem Int Ed 37:2471–2473, 1998.
37. (a) KJ Visuri. U.S. Patent 4,699,882, 1987. (b) K Visuri. U.S. Patent 5,120,650, 1992. (c) K Visuri. U.S. Patent 5,437,993, 1995. (d) Personal communication with K Visuri.
38. AM Collins, C Maslin, RJ Davies. Scale-up of a chiral resolution using cross-linked enzyme crystals. Org Process Res Dev 2:400–406, 1998.
39. AD Baxter, JB Bird, R Bannister, R Bhogal, DT Manallack, RW Watson, DA Owen, J Montana, J Henshilwood, RC Jackson. Matrix metalloproteinase inhibitors in cancer therapy. Totawa, NJ: Humana Press, 2000, pp 193–222.

9
Enzymatic Deacylation of Echinocandins and Related Antifungal Agents

Andrew R. Cockshott, Adam J. Kreuzman, and Wu-Kuang Yeh
Eli Lilly and Company, Indianapolis, Indiana

I. INTRODUCTION

Fungal infections cover a wide range from superficial conditions (e.g., athlete's foot) to life-threatening diseases (e.g., HIV-mediated infection). The types and frequencies of life-threatening fungal infections have increased dramatically within the last two decades [1,2]. The major opportunistic fungal pathogen in the immuno-compromised host is *Candida albicans* [2,3]. However, *Aspergillus* sp. have caused a high mortality in bone marrow transplant recipients, and *Pneumocystis carinii* is a leading cause of death in HIV-infected patients in North America and Europe [2,4].

Treatment therapies for invasive fungal infections are essentially limited to the polyenes and the azoles [2]. Both classes of compounds have mechanisms of action that target ergosterol, the major sterol in fungal membranes. Amphotericin B, a polyene that has potent broad-spectrum fungicidal activity [2,5], is still the "gold standard" for the most severe fungal infections, but its utility is limited by nephrotoxicity [1]. Newer azoles such as fluconazole and itraconazole are fungistatic agents with limited spectrum compared to amphotericin B. These newer azoles have fewer side effects, but fungal resistance is emerging [3,6,7].

There is an urgent unmet medical need for antifungal agents that are safe, broad-spectrum, and fungicidal. Recent reviews suggest that the most effective antifungal agents may come from those targeting cell wall biosynthesis [2,3,8]. Among the known cell wall-active agents, the echinocandin and pneumocandin analogs (both are cyclic lipopetides) that inhibit 1,3-β-D-glucan synthase [1,9] appear most promising as new broad-spectrum and fungicidal agents [1,2,9]. Ad-

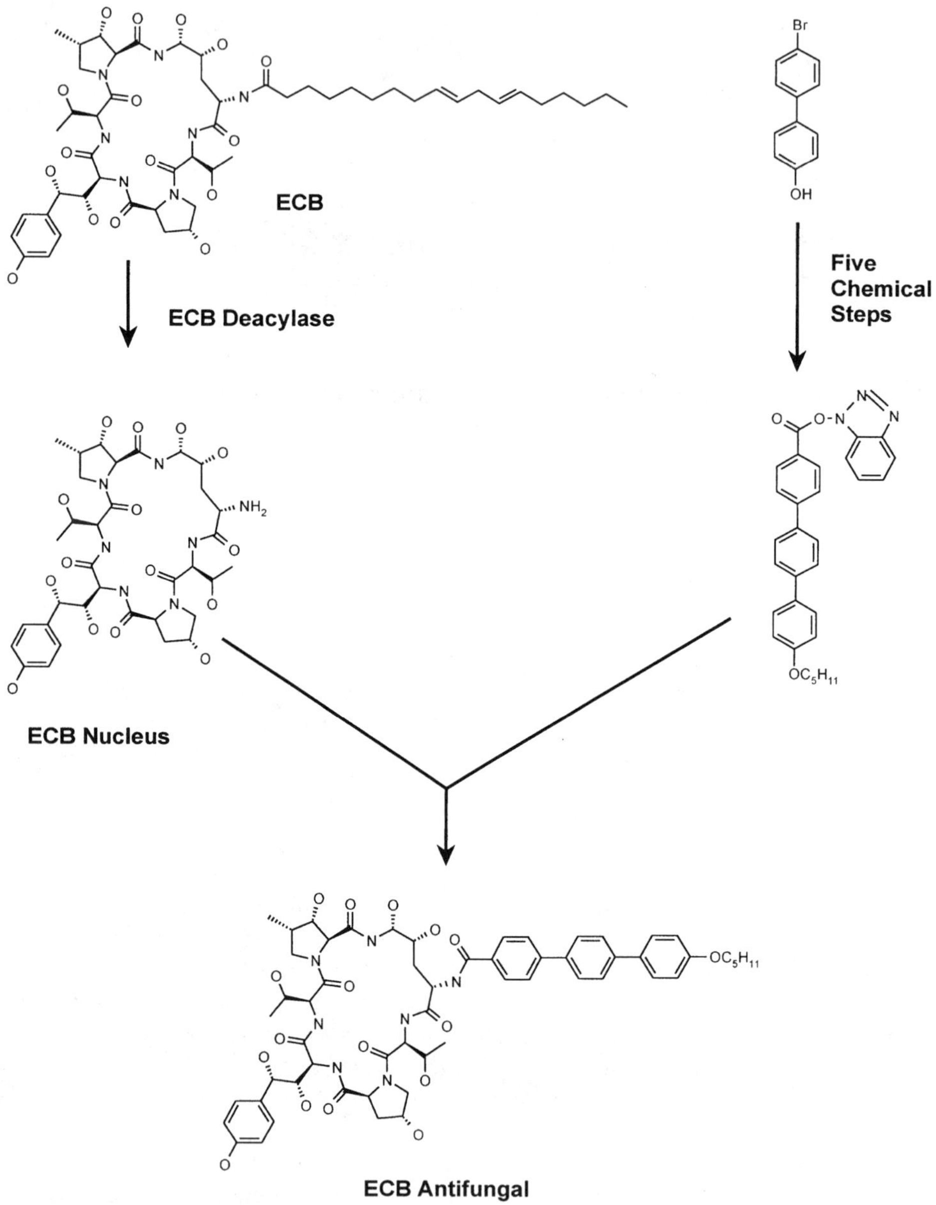

Figure 1 Chemo-enzymatic synthesis of an ECB antifungal agent.

Table 1 Substrate Specificity of ECB Deacylase

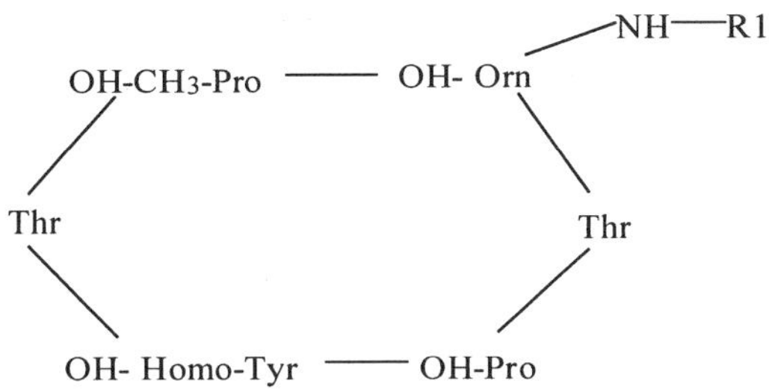

OH-Orn: Hydroxyornithine
OH-Pro: Hydroxyproline
$OH-CH_3-Pro$: Hydroxymethylproline
OH-Homo-Tyr: Hydroxyhomotyrosine

Substrate	R1	Activity (%)[a]
ECB		100
ECB Analog 1	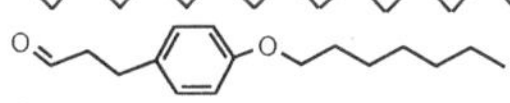	30
ECB Analog 2	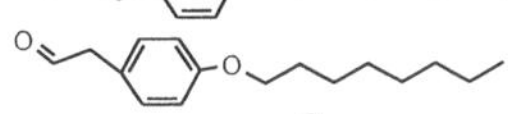	30
ECB Analog 3	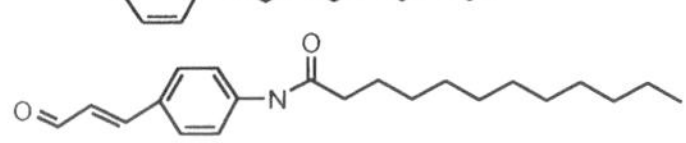	15
ECB Analog 4	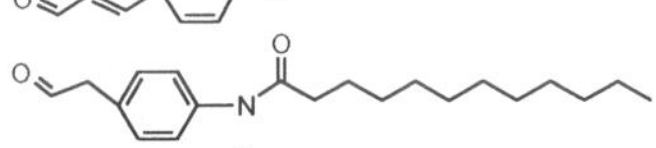	5
ECB Analog 5	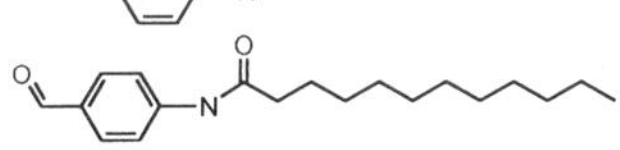	5
Cilofungin	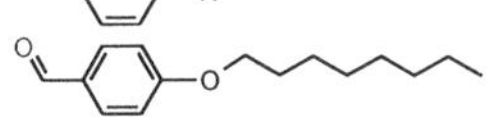	0
ECB Antifungal	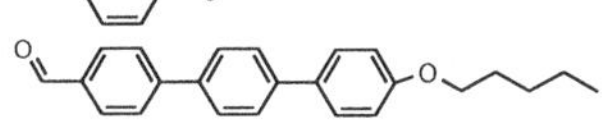	0
Aculeacin	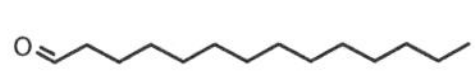	208

[a] Monitored by formation of ECB nucleus with HPLC as described previously [25].

ditional promising cell wall-active agents are cyclic peptides such as pseudomycin and syringomycin [10,11].

The promising echinocandin and pneumocandin analogs, which include FK463 [12], are semisynthetic. Pneumocandin A0 was produced by bioconversion of pneumocandin to its cyclic peptide [13] and chemical modification of the cyclic peptide. In contrast, cilofungin, the first echnocandin analog to reach clini-

Table 2 Substrate Specificity of ECB Deacylase

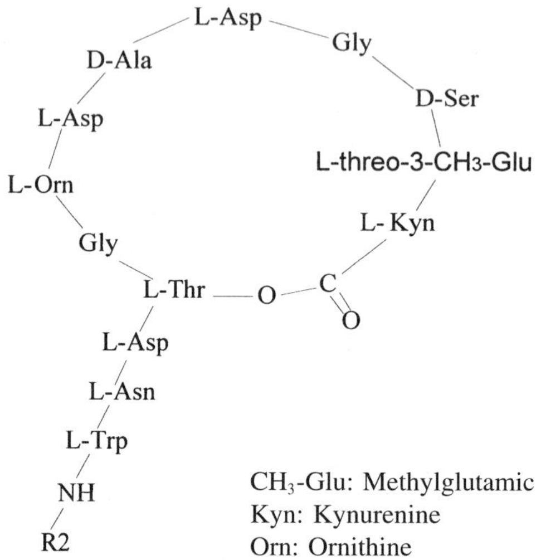

CH_3-Glu: Methylglutamic acid
Kyn: Kynurenine
Orn: Ornithine

Substrate	R2	Activity (%)[a]
Daptomycin		437
A21978C1	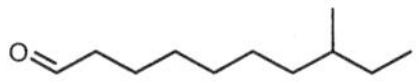	278
A21978C2	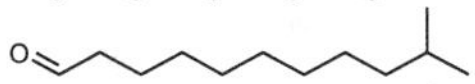	282
A21978C3	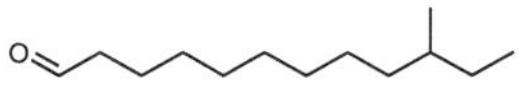	208

[a] Monitored by formation of daptomycin nucleus with HPLC and expressed relative to ECB as the substrate as indicated in the footnote to Table 1.

cal trials, was produced by bioconversion of echinocandin B (ECB) to its cyclic peptide and chemical addition of a new side chain to the cyclic peptide [14–17]. LY303366 and related analogs, generated by a two-step process similar to cilofungin (Fig. 1 and Table 1; [18–20]), are potentially useful fungicidal agents for the treatment of life-threatening fungal infections caused by important opportunistic pathogens including *Candida* sp., *Aspergillus* sp., and *Pneumocystis carinii*. For comparison, daptomycin (Table 2), an acidic cyclic lipopeptide, is effective against Gram-positive resistant bacteria [21]. Interestingly, daptomycin analogs were produced directly by feeding different side chains during fermentation [22].

ECB can be cleaved by a deacylating enzyme (''ECB deacylase'') from a *Pseudomonas* sp. [23] or from *Actinoplanes utahensis* [24], generating ECB nucleus (Fig. 1), an important antifungal intermediate. Whole cells containing the biocatalyst were used previously with ECB as the substrate [24]. In this chapter, we will describe a highly effective and complete large-scale enzyme process for generation of the antifungal intermediate. The chapter includes purification and characterization of native ECB deacylase, heterologous gene cloning and overexpression of the enzyme, as well as optimization in both ECB and the deacylase fermentations, and new technology for the deacylase-catalyzed conversion of ECB to the antifungal intermediate.

II. ECB DEACYLASE: ENZYMOLOGY, CLONING, AND SIGNIFICANCE

In this section we will describe the purification, properties, and substrate specificity of native ECB deacylase, its heterologous cloning and overexpression, as well as potential applications of the deacylase in producing antifungal intermediates. Deacylase activity was determined by using ECB as the substrate and by monitoring formation of the cyclic hexapeptide (ECB nucleus) at 225 nm using high-performance liquid chromatography (HPLC) [25]. The reaction mixture for the deacylase assay contained dimethyl sulfoxide (DMSO) to solubilize ECB and KCl for maximal activity. An alternative method was used where activity was determined by using daptomycin as the substrate in the absence of DMSO and by monitoring formation of daptomycin nucleus at 225 nm using HPLC [25].

A. Native ECB Deacylase: Solubilization and Purification

ECB deacylase is produced naturally by *Actinoplanes utahensis*. However, very little (~0.2%) deacylase activity was detected in the culture filtrate of *Actinoplanes utahensis*. Less than 5% of the cell-associated deacylase activity was released by incubation of the cells at 0.01 M KH_2PO_4, pH 6, for 1 day. A simple salt treatment with 0.8 M KCl resulted in 60–80% recovery of soluble deacylase at a high specific activity [25]. This salt-induced solubilization suggests that the deacylase is loosely bound to the membrane of *A. utahensis* and can be released by disruption of ionic interactions [26]. The solubilized enzyme was stable and purified to apparent homogeneity by a four-step conventional procedure [25].

B. ECB Deacylase Properties

1. Catalytic Properties

ECB deacylase is a heterodimer consisting of 63-kDa and 18–20-kDa subunits [25]. The purified ECB deacylase requires no external cofactor, metal ion, or

Table 3 Comparison of ECB Deacylase Versus Aculeacin A Acylase

Parameter	ECB deacylase[a]	Aculeacin A acylase[b]
Producing organism	*A. utahensis*	*A. utahensis*
Enzyme location	Cell-associated	Extracellular
Substrate		
Cyclic peptide	Six-membered	Six-membered
Acyl side-chain	Linoleoyl	Palmitoyl
Effect of salt (KCl or NaCl)		
Enzyme solubility	Required	Not required
Enzyme stimulation	3-Fold	0.6-Fold
Oligomeric structure	Dimer	Dimer
Large subunit	63 kDa	55 kDa
Small subunit	18–20 kDa	19 kDa
Amino-terminal sequence		
Large subunit	SNAYG	SNAYG
Small subunit	HDGGY	GGY

[a] From Ref. 25.
[b] From Refs. 30 and 31.

reducing agent for maximal activity. Also, the activity was not affected by addition of a metal chelator or sulfhydryl reagent. The substrate, ECB, was practically insoluble in aqueous solutions. Among several organic solvents tested, DMSO at a low concentration (i.e., 15%) was compatible with the deacylase-catalyzed reaction and a subsequent enzyme activity analysis with HPLC. The deacylase was optimally active at pH 6 in 0.05 KH_2PO_4 and 60°C. A stimulation of the deacylase up to threefold was observed by addition of a salt, and the salt stimulation was general with respect to several mono- and divalent metal salts examined (Table 3; [25]). The deacylase was at least twice as active when the reaction was initiated with the enzyme rather than when it was initiated with the substrate. This suggests that the substrate might protect the deacylase from its subunit dissociation at a low enzyme concentration. The K_m of the enzyme for ECB was 50 μM and the V_{max} for the deacylase-catalyzed reaction was ~15 μmol of the ECB nucleus formed/min/mg protein [25].

2. *Substrate Specificity*

Enzymatic deacylation of acyl side-chain and peptide nucleus analogs of ECB, containing a cyclic hexapeptide, was examined with the purified native deacylase under the reaction conditions optimized with ECB, as described previously [25]. Unless specified, the deacylation reaction was conducted in the presence of

DMSO. The deacylation rate for ECB analogs was analyzed by both substrate disappearance and product formation with HPLC. As shown in Table 1 [20,25], the deacylase catalyzed effectively the cleavage of aculeacin, a biological product that differs from ECB by having a palmitoyl (instead of linoleoyl) side chain. In comparison to ECB, the enzyme catalyzed a moderate (30%) or a low-level cleavage (5–15%) of several chemical side-chain analogs of ECB (Table 1). However, neither cilofungin nor the ECB antifungal, known as a potentially important chemical ECB analog, was a substrate for the enzyme. The deacylase also catalyzed effectively the cleavage of daptomycin, a cyclic decapeptide analog of ECB, and its three biological side-chain analogs (Table 2; [20,25]). In addition, the deacylase-catalyzed cleavage was also observed with teicoplanin and pseudomycin A [25,27]; both are natural products. In contrast, no enzymatic deacylation was observed for the following β-lactams: penicillin N, penicillin G, penicillin V, ampicillin, deacetoxycephalosporin C, deacetylcephalosporin C, and cephalosporin C [20,25]. Thus, ECB deacylase exhibited a broad side-chain specificity toward the cyclic hexapeptides (ECB/aculeacin A) and more complex cyclic peptides (daptomycin/pseudomycin), but not for β-lactams (penicillin/cephalosporin).

C. Heterologous Cloning, Overexpression, Subunit Stability, and Purification of Recombinant ECB Deacylase

The ECB deacylase genes were isolated from *A. utahensis* (NRRL 12052) using standard methods [28]. Radiolabeled degenerate oligonucleotides were synthesized based on the N-terminal amino acid sequences of the purified native ECB deacylase [25] and used to probe restriction digests of *A. utahensis* genomic DNA. A PstI fragment of approximately 4 kb was observed following Southern hybridization. The hybridizing fragment was isolated from low-melting-point gel and cloned into the pUC19 cloning vector. The resulting correct plasmid, pSHP100 [25], was sequenced and shown to contain the entire ECB deacylase coding region with an additional 1 kb of DNA upstream and approximately 0.5 kb of DNA downstream of the ECB deacylase coding region.

Increased expression of the ECB deacylase genes was achieved using an autonomously replicating *Streptomyces* plasmid, pSHP150 [25]. This plasmid was constructed by insertion of the 4 kb PstI-BamHI fragment of pSHP100 containing the ECB deacylase genes into a 5.8 kb PstI-BglII fragment of the autonomously replicating *Streptomyces* cloning vector pIJ702 [29]. The resulting expression vector was introduced into *Streptomyces lividans* TK23 using polyethylene glycol (PEG)-mediated transformation of protoplasts. Thiostrepton resistant transformants were subsequently tested for activity. Expression of the

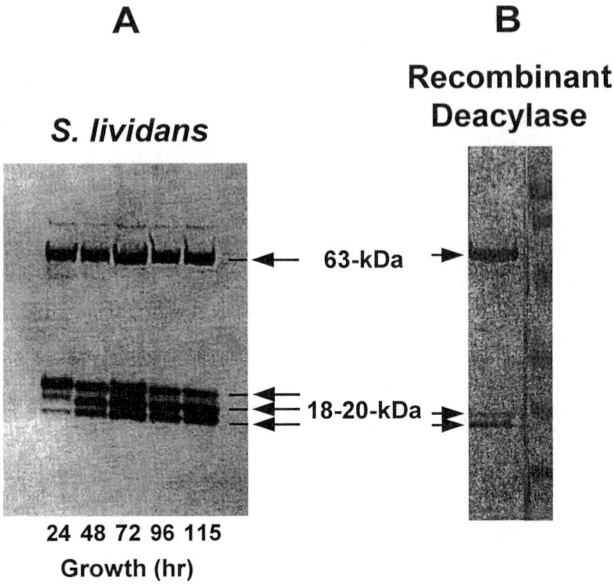

Figure 2 (A) Western blot of the culture filtrates of *S. lividans* using the antisera prepared against purified native ECB deacylase subunits (two antisera were combined). (B) SDS-PAGE of purified recombinant ECB deacylase (20 μg).

ECB deacylase genes by pSHP150 is regulated by the native *A. utahensis* 5′ and 3′ regulatory sequences.

The deacylase activity from the culture filtrate of *S. lividans* was analyzed by using ECB as the substrate. The soluble activity from recombinant *S. lividans* was 180-fold higher than that of *A. utahensis* [25]. The soluble activity from recombinant *S. lividans* was 20-fold higher than that reported previously [30]. The two subunits from *S. lividans* appeared stable during the 115-hr growth period (Fig. 2A). In contrast, the deacylase genes were also cloned and overexpressed in *Streptomyces griseus*, but the large (63-kDa) subunit appeared unstable during growth [25]. A stepwise processing of the small subunits (i.e., from 20 to 18 kDa) from *S. lividans* was observed. Due to its high activity and relative protein stability, *S. lividans* has become our preferred source of enzyme for purification and enzymatic deacylation studies.

The two genes encoding ECB deacylase subunits from *A. utahensis* were cloned and overexpressed by *S. lividans* as soluble extracellular proteins [25]. The recombinant *S. lividans* was grown in a 10-liter fermenter under conditions described previously [30]. After 2 days of growth, the culture was filtered through #1 Whatman filter paper. The recombinant ECB deacylase from the culture filtrate was purified to apparent homogeneity by a single-step cation-exchange (S-Sepharose) chromatography (Fig. 2B; [25]).

D. ECB Deacylase: Potential Pharmaceutical Applications

The broad substrate specificity of ECB deacylase (Tables 1 and 2) suggests that the native enzyme and any modified form may play a major role toward generating important antifungal intermediates. The following two comparative enzyme cases can serve to alert the readers of possible biotechnological utility of ECB deacylase.

1. Comparison of ECB Deacylase and Aculeacin A Acylase

Both ECB deacylase and aculeacin A acylase were purified from *A. utahensis* [25,31], but whether they are from a single bacterial strain remains unclear. They share similar catalytic properties with regard to optimal pH and temperature, salt stimulation, as well as lack of requirement or stimulation by cofactor, metal ion, or reducing agent (Table 3; [31]). They also appear to catalyze deacylation with similar substrate specificity [20,25,31]. However, ECB deacylase can be differentiated from aculeacin A acylase by the enzyme location, salt-induced solubilization, and stimulation extent as well as the large-subunit size (Table 3). The two different subunits of ECB deacylase, based on their amino-terminal sequences [25], are identical to those reported for aculeacin A acylase except that the first two residues of the small subunit of ECB deacylase are absent from aculeacin A acylase [30]. The overall sequence homology of the two enzymes is possibly quite high but has not yet been reported. Thus, both ECB deacylase and aculeacin A acylase will likely have similar biocatalytic activity for pharmaceutical applications.

2. ECB Deacylase and Penicillin Acylase: Evolutionary Implications

ECB deacylase is an 81–83-kDa heterodimer consisting of 63- and 18–20-kDa subunits. Penicillin G acylase from *Escherichia coli* is an 87-kDa heterodimer with 65- and 22-kDa subunits [32]. For comparison, cephalosporin acylase from a *Pseudomonas* strain is an 83-kDa heterodimer consisting of 57- and 26-kDa subunits [33]. The essential absence of any external catalytic requirement, cofactor stimulation, or product inhibition of ECB deacylase is also an intrinsic property of penicillin acylase [34]. Based on the amino-terminal sequences of the two subunits of ECB deacylase, a 48% sequence similarity has been observed between the small subunit of ECB deacylase and a penicillin acylase [25]. This statistically significant albeit moderate sequence similarity from two short segments of the enzymes suggests an evolutionary relationship between ECB deacylase and penicillin G acylase. Such sequence similarity indicates convergent evolution of the enzymatic functions toward similar cyclic peptide substrates. The similar catalytic functions of two enzymes can be defined by just a few common amino acid residues (e.g., for a moderate sequence similarity); this observation is described eloquently in a latter chapter by Ridong Chen on "Assigning Precise Function to Genes." The broad substrate specificity of penicillin and cephalosporin acylases using β-lactams as substrates [33,35,36] has allowed them to play a critical role in generation of key intermediates for manufacturing of therapeutic antibacterial β-lactam compounds. Although ECB deacylase has some biochemical and structural similarities to penicillin/cephalosporin acylases, as described above, ECB deacylase has displayed broad substrate specificity for the cyclic hexapeptides and more complex cyclic peptides such as daptomycin. In analogy to penicillin acylase in β-lactams as substrates and practical significance, ECB deacylase may affect, by its enzymatic deacylation and acylation, the development of therapeutic antifungal cyclic peptides. The broad substrate specificity of the native enzyme may be further extended, e.g., in acquiring a novel deacylation or acylation, by enzyme engineering. Toward this direction, future elucidation of the active site of ECB deacylase appears desirable for potential industrial applications.

III. ECB AND ECB DEACYLASE: FERMENTATION IMPROVEMENT

As described above, ECB deacylase catalyzes the cleavage of ECB to produce the important antifungal intermediate ECB nucleus. Both ECB and ECB deacylase are made by fermentation. This section outlines how these fermentations were optimized to create a commercially viable process. Section IV looks at the optimization of the bioconversion process itself.

Cloning the gene for ECB deacylase from *A. utahensis* into *S. lividans*, as described in the previous section, increased total enzyme production of the soluble enzyme sixfold (Fig. 3). Improvements in the fermentation conditions, adding glucose and ammonia feeding, lead to a further sevenfold increase in enzyme production (Fig. 3).

For ECB, produced by *Aspergillus nidulans*, genetic engineering was not a viable approach since the genes that influence productivity were not known. Therefore a classical screening approach was used [37]. Successive rounds of mutagenesis and screening led to over a 50-fold increase in titer and to strains that were unable to produce sterigmatocystin (a potent carcinogen that can be made by this organism) [38]. Further yield improvements were made by optimizing the fermentation medium [39,40] and by using glucose and ammonia feeding.

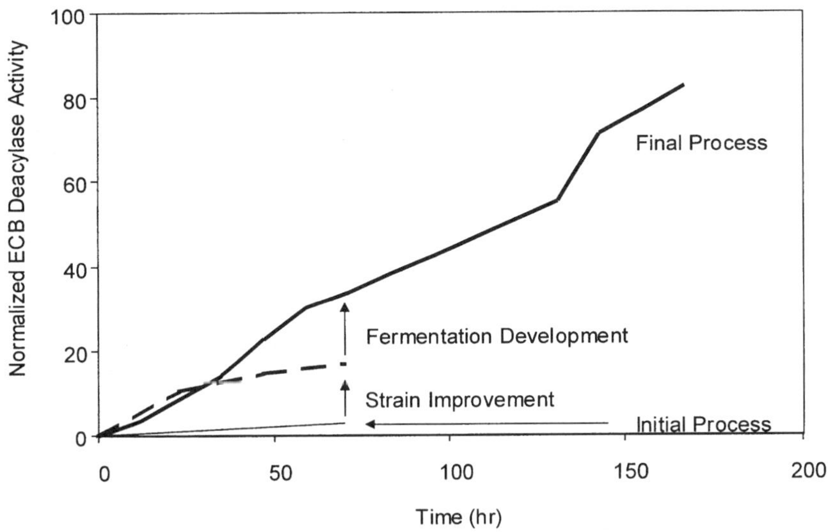

Figure 3 Improvement of ECB deacylase fermentation process due to strain improvement and fermentation development.

IV. ECB AND ECB DEACYLASE: BIOCONVERSION OPTIMIZATION

The development of the fermentation processes for making the substrate (ECB) and the enzyme (ECB deacylase) has been described briefly in the proceeding section. The development of the bioconversion process itself will be considered in the following section.

A. Choice of Substrate and Enzyme Conditions

The first question to consider is what form the substrate and enzyme should be. Options include whole broth, purified, or immobilized. By answering key questions at an early stage about the nature of the substrate, enzyme, and product, it is possible to eliminate reactor options quickly without expensive experimentation [41].

1. Whole-Broth Bioconversion

The easiest process for a bioconversion involving a substrate and an enzyme made by fermentation is simply to mix the fermentation broths together under suitable conditions and then purify the product. However, there are a number of disadvantages to such an approach. The first, and most obvious, consideration is whether the enzyme and the substrate are both physically available for reaction, i.e., if both were intracellular then no simple mixed-broth bioconversion process would be possible. Another important consideration is whether there are impurities in either broth that would interfere with the bioconversion process. For example, proteases produced during the fermentation of *A. nidulans* to produce ECB might subsequently degrade the ECB deacylase and lower the yield of product or rate of the bioconversion process. The same would be true of inhibitors of the ECB deacylase that might be produced by *A. nidulans*. Another practical problem with the mixed-broth process would be that both the enzyme and substrate fermentation broths would have to be scheduled so that they would both be available at the same time. The alternative would be to hold one of the broths chilled waiting for the other, but such an approach would be suboptimal for equipment utilization.

2. Purified/Immobilized Enzyme or Substrate

One solution to the above problems would be to purify either the substrate or the enzyme. For example, in the case of ECB, the substrate of the enzyme could be extracted from the cells and then purified by chromatography to remove any impurities that would interfere with the bioconversion processes. The ECB could be left as a solution in a solvent or dried to a powder form. ECB deacylase could also be purified to either a concentrated enzyme solution or solid form. Another option is to immobilize the enzyme or substrate on a suitable support. In the case of ECB deacylase, the advantage would be that the enzyme could be reused and the savings gained would obviously have to be compared with the extra costs of immobilization. The next section looks in more detail at the novel approach of immobilizing the ECB substrate.

3. ECB Immobilization

Insoluble ECB can be selectively extracted from *A. nidulans* by an organic solvent such as acetone. The extracted ECB is about 50% pure. Partially purified ECB is dissolved in an organic solvent such as methanol or acetone and mixed with a type of nonfunctional polystyrene HP20 resin such as HP20SS prewetted with methanol. Slow addition of water to the mixture to about 25% methanol and adjustment of the pH to about 4.0 with continuous stirring can cause complete binding of ECB to the resin. Any residual methanol can be removed by washing the ECB-bound resin with water. The binding capacity of HP20 for ECB is 100–200 mg/ml.

This immobilized ECB can then be converted to soluble ECB nucleus (as flow-through) by passing ECB deacylase slowly through the ECB-bound HP20

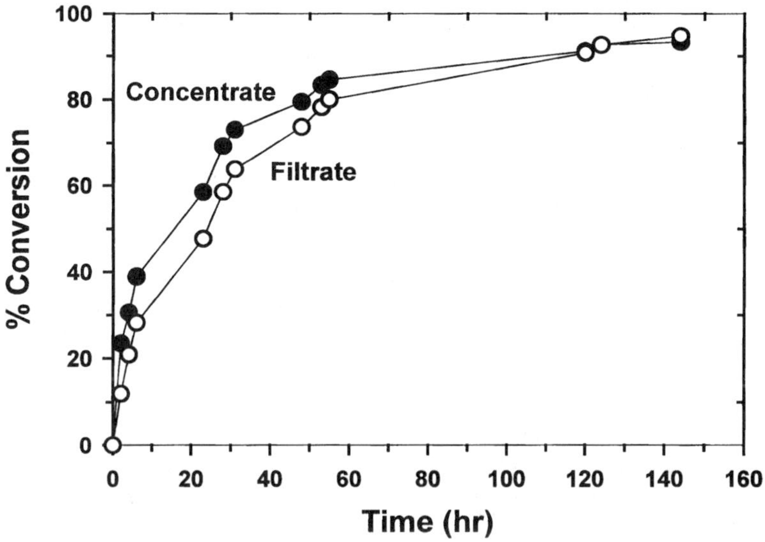

Figure 4 Enzymatic deacylation of immobilized ECB: using filtrate versus concentrate.

resin at pH 6.0 and room temperature. On a small scale (1.2 g ECB), by a single enzyme addition, it was shown that 60% conversion was observed [25]. With multiple enzyme additions, near-complete conversion was achieved (Fig. 4; [25]).

Further improvements could be made to this process. For example, by lowering the pH of the resulting eluent from the enzymatic reaction to pH 4.0, the stability of ECB nucleus could be increased. The low pH would also enhance the nucleus binding to a second column of HP20 resin, thus potentially improving purity and recovery of ECB nucleus from the resin by elution with an organic solvent such as methanol. Also, the enzyme in the column effluent could be separated by ultrafiltration and reused. Reuse of HP20 resin could be easily achieved by washing the used resin with an organic solvent such as methanol.

With some modification, the immobilized process for ECB may also be applicable to other insoluble hydrophobic substrates. Modification of the immobilized ECB process may include specific binding resins for enzymatic conversion and for product recovery.

B. Optimization of Bioconversion Conditions

The substrate, enzyme and product should each be investigated to determine their stability at a range of pH, temperature, and solvent conditions. The enzyme activity should also be measured under the same range of conditions. Such an approach will help to identify suitable operating conditions for the bioconversion and choice of substrate and enzyme form. Final optimization of the bioconversion should be done using statistical design.

For example, in the case of ECB deacylase the initial investigations showed that the enzyme was most active at high pH and temperature (e.g., pH 8, 70°C), but that the product (ECB nucleus) degraded rapidly under these conditions. One solution was to perform the bioconversion at lower pH and temperature, where the product would be stable, and accept suboptimal enzyme performance (i.e., slower reaction time or more enzyme usage).

These investigations also showed that the conversion of ECB to ECB nucleus would proceed more rapidly if ECB were first solubilized in a suitable solvent such as methanol or acetone. However, if the concentration of solvent was too high, the enzyme activity was reduced. Ideally, the enzyme itself could be tailored to suit the industrially preferred conditions (e.g., to make it more resistant to solvent or active at a different pH). One method for achieving this is to use directed evolution [42], whereby genes encoding the enzyme are mutated, screened and then recombined *in vitro*. Although the contributions of individual amino acid mutations are small, the accumulation of multiple mutations by directed evolution allows significant improvement in the biocatalyst for reactions on substrates or under conditions not already optimized in nature. This approach was used by Arnold and Moore [43] to make a 150-fold improvement in the activity of a *p*-nitrobenzyl esterase in the presence of 15% DMSO.

C. Economic Model for the Bioconversion Process

We have discussed how each step in the bioconversion process (e.g., the ECB and ECB deacylase fermentations and the bioconversion) can be improved. The overall improvement in the process is shown in Figure 5. However, with limited resources and time, it is important to focus development efforts on steps that will have the greatest effect on the overall productivity of the process. An excellent way to achieve this is to develop an economic model of the overall process and to use this as shown schematically in Fig. 6. For example, it may be wasteful to focus efforts on improving the yield of the enzyme fermentation if the major cost was due to the ECB fermentation.

V. CONCLUSIONS

The membrane-associated ECB deacylase of *A. utahensis* can be salt-solubilized, heat-treated, and purified to apparent homogeneity by a three-step chromatographic procedure. The deacylase is a heterodimer consisting of 63- and 18–20-kDa subunits. Except for salt stimulation, the deacylase has no external cofactor or metal ion requirement and has broad substrate specificity for the cyclic hexa-

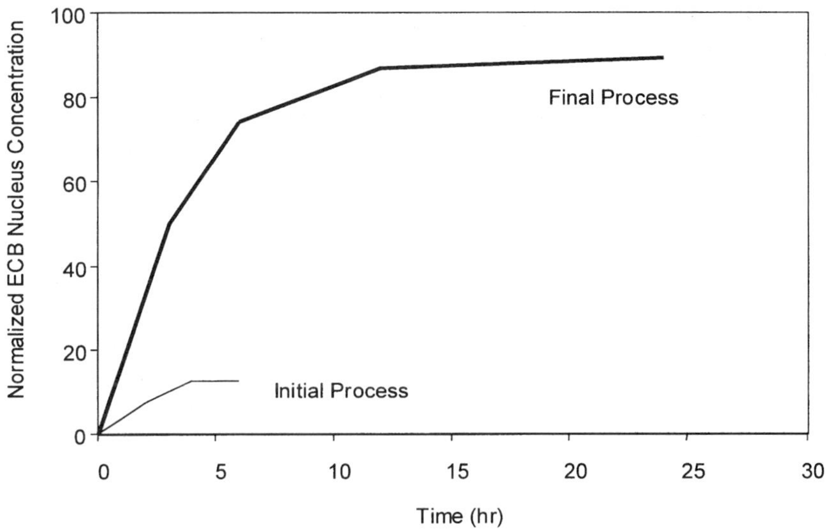

Figure 5 Improvement of ECB bioconversion process due to improvements in the yields of the ECB and ECB deacylase fermentation processes and optimization of the bioconversion process itself.

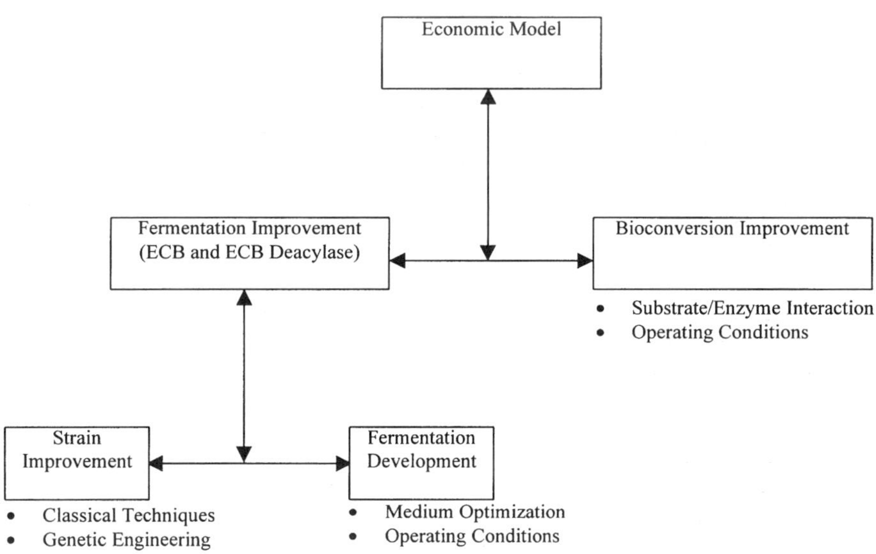

Figure 6 Overview of ECB bioconversion process optimization.

peptide and more complex cyclic peptides. The deacylase genes have been cloned, overexpressed, and stabilized in *S. lividans*. ECB deacylase is similar to, in structure/function, but can be differentiated from, aculeacin A acylase. Also, ECB deacylase shares a detectable sequence similarity with penicillin G acylase. Thus, the recombinant ECB deacylase and its yet to-be-improved versions, e.g., by enzyme engineering, can be potentially useful for pharmaceutical and biotechnological applications.

The development of a commercially viable enzyme process for the production of ECB nucleus was achieved by improving the ECB and ECB deacylase fermentation processes and the bioconversion processes. Increased yields in the fermentation processes were achieved through linked programs for strain improvement and fermentation development. The bioconversion process was improved by the choice of substrate and enzyme conditions and subsequent optimization of operating conditions. An economic model was used to decide where development resources should be focused.

ACKNOWLEDGMENTS

We are grateful to Bill Current, Joe Dotzlaf, and Bob Strobel for critical reviewing of the manuscript.

REFERENCES

1. NH Georgopapadakou, TJ Walsh. Antifungal agents: chemotherapeutic targets and immunologic strategies. Antimicrob Agents Chemother 40:279–291, 1996.
2. WW Turner, WL Current. Echinocandin antifungal agents. In: WR Strohl, ed. Biotechnology of Antibiotics. New York: Marcel Dekker, 1997, pp 315–334.
3. MA Pfaller. Nosocomial candidiasis: emerging species, reservoirs, and modes of transmission. Clin Infect Dis 22(suppl 2): S89–S94, 1996.
4. RA Fromtling. Current antifungal therapy of human mycoses. Drug News Perspectives 10:170–178, 1997.
5. JL Lesher. New antifungal agents. Dermatol Clin 10:799–805, 1992.
6. Stenberg S. The emerging fungal threat. Science 266:1632–1634, 1994.
7. JH Rex, MG Rinaldi, MA Pfaller. Resistance of *Candida* species to fluconazole. Antimicrob Agents Chemother 39:1–8, 1995.
8. MB Kurtz, CM Douglas. Lipopeptide inhibitors of fungal glucan synthase. J Med Vet Mycol 35:79–86, 1997.
9. M Debono, RS Gordee. Antibiotics that inhibit fungal cell wall development. Annu Rev Microbiol 48:471–497, 1994.
10. A Ballio, D Bossa, D DiGiorgio, P Ferranti, M Paci, P Pucci, A Scaloni, A Segre, G Strobel. Novel bioactive lipodepsipeptides from *Pseudomonas syringae*: the pseudomycins. FEBS Lett 355:96–100, 1994.

11. KN Sorensen, AA Wanstrom, SD Allen, JY Takemoto. Efficacy of syringomycin E in a murine model of vaginal candidasis. J Antibiot 51:743–749, 1998.
12. M Tomishima, H Ohki, A Yamada, H Takasugi, K Maki, S Tawara, H Tanaka. FK463, a novel water-soluble echinocandin lipopeptide: synthesis and antifungal activity. J Antibiot 52:674–676, 1999.
13. DM Schmatz, G Abruzzo, MA Powles, DC McFadden, JM Balkovec, RM Black, K Nollstadt, K Bartizal. Pneumocandins from *Zalerion arboricola*. IV. Biological evaluation of natural and semi-synthetic pneumocandins for activity against *Pneumocystis carinii* and *Candida* species. J Antibiot 45:1886–1891, 1992.
14. M Debono, BJ Abbott, DS Fukuda, M Barnhart, KE Willard, RM Molloy, KM Michel, JR Turner, TF Butler, AH Hunt. Synthesis of new analogs of echinocandin B by enzymatic deacylation and chemical reacylation of the echinocandin B peptide: synthesis of the antifungal agent cilofungin (LY121019). J Antibiot 42:389–397, 1988.
15. M Debono, BJ Abbott, JR Turner, LC Howard, RS Gordee, AS Hunt, M Barnhart, RM Molloy, KE Willard, DS Fukuda, TF Butler, DJ Zeckner. Synthesis and evaluation of LY121019, a member of a series of semisynthetic analogues of the antifungal lipopeptide echinocandin B. Ann NY Acad Sci 544:152–167, 1988.
16. LD Boeck, DS Fukuda, BJ Abbott, M Debono. Deacylation of A21978C, an acidic lipopeptide antiobiotic complex, by *Actinoplanes utahensis*. J Antibiot 41:1085–1092, 1988.
17. RS Gordee, DJ Zeckner, LC Howard, WE Alborn, M Debono. Anti-*Candida* activity and toxicology of LY121019, a novel semisynthetic polypeptide antifungal antibiotic. Ann NY Acad Sci 544:294–309, 1988.
18. M Debono, WW Turner, L LaGrandeur, FJ Burkhardt, JS Nissen, KH Nichols, MJ Rodriguez, MJ Zweifel, DJ Zeckner, RS Gordee, J Tang, TR Parr Jr. Semisynthetic chemical modification of the antifungal lipopeptide echinocandin B (ECB): structure-activity studies of the lipophilic and geometric parameters of polyarylated acyl analogs of ECB. J Med Chem 38:3271–3281, 1995.
19. JA Jamison, DJ Zeckner, MJ Rodriguez. The synthesis and antifungal activity of *N*-alkylated analogs of echinocandin B. J. Antibiot 50:562–566, 1997.
20. WK Yeh. Evolving enzyme technology for pharmaceutical applications: case studies. J Ind Microbiol Biotechnol 19:334–343, 1997.
21. FP Tally. Researchers reveal ways to defeat ''superbugs.'' Drug Discovery Today 4: 395–398, 1999.
22. FM Huber, DM Berry, RL Pieper, AJ Tietz. The synthesis of A21978C analogs by *Streptomyces roseosporus* cultivated under carbon limitation and fed fatty acids. Biotechnol Lett 12:789–792, 1990.
23. W Pache, C Keller, M Kuhn. Cleavage of echinocandin B with polymyxin acylase liberating the fatty acid and reacylation of the peptide moiety. Experientia 34:1670–1671, 1978.
24. LD Boeck, DS Fukuda, BJ Abbott, M Debono. Deacylation of echinocandin B by *Actinoplanes utahensis*. J Antibiot 42:382–388, 1989.
25. AJ Kreuzman, RL Hodges, JR Swartling, TE Pohl, SK Ghag, PJ Baker, D McGilvray, WK Yeh. Membrane-associated echinocandin B deacylase of *Actinoplanes utahensis*: purification, characterization, heterologous cloning and enzymatic deacylation reaction. J Ind Microbiol Biotechnol 24:173–180, 2000.

26. AB Cubitt, MC Gershengorn. Characterization of a salt-extractable phosphatidylinositol synthase from rat pituitary-tumour membranes. Biochem J 257:639–644, 1989.
27. NJ Snyder, RD Cooper, BS Briggs, M Zmijewski, DL Mullen, RE Kaiser, TI Nicas. Enzymatic deacylation of teicoplanin followed by reductive alkylation: synthesis and antibacterial activity of new glycopeptides. J Antibiot 51:945–951, 1998.
28. J Sambrook, EF Fritsch, T Maniatis. Molecular Cloning: A Laboratory Manual. 2nd ed. Cold Spring Harbor, NY: Cold Spring Harbor Laboratory Press, 1989, pp 50–250.
29. DA Hopwood, MJ Bibb, KF Chater, T Kieser, CJ Bruton, HM Kieser, DJ Lydiate, CP Smith, JM Ward, H Schrempf. Genetic Regulation of *Streptomyces*—A Laboratory Manual. Norwich, England: John Innes Foundation, 1985.
30. J Inokoshi, H Takeshima, H Ikeda, S Omura. Cloning and sequencing of the aculeacin A acylase-encoding gene from *Actinoplanes utahensis* and expression in *Streptomyces lividans*. Gene 119:29–35, 1992.
31. H Takeshima, J Inokoshi, Y Takada, H Tanaka, S Omura. A deacylation enzyme for aculeacin A, a neutral lipopeptide antibiotic, from *Actinoplanes utahensis*: purification and characterization. J Biochem 105:606–610, 1989.
32. W Bruns, J Hoppe, H Tsai, HJ Bruning, F Maywald, J Collins, H Mayer. Structure of the penicillin acylase gene from *Escherichia coli*: a periplasmic enzyme that undergoes multiple proteolytic processing. J Mol Appl Genet 3:36–44, 1985.
33. A Matsuda, K Matsuyama, K Yamamoto, S Ichikawa, KI Komatsu. Cloning and characterization of the genes for two distinct cephalosporin acylases from a *Pseudomonas* strain. J Bacteriol 169:5815–5820, 1987.
34. C Kutzbach, E Rauenbusch. Preparation and general properties of crystalline penicillin acylase from *Escherichia coli* ATCC 11105. Hoppe-Seyler's Z Physiol Chem 354:45–53, 1974.
35. AL Margolin, VK Svedas, IV Berezin. Substrate specificity of penicillin amidase from *E. coli*. Biochim Biophys Acta 616:283–289, 1980.
36. VK Svedas, AL Margolin, IV Berezin. Enzymatic synthesis of β-lactam antibiotics: a thermodynamic background. Enzyme Microb Technol 2:138–144, 1980.
37. RL Hodges, DW Hodges, K Goggans, X Xuei, P Skatrud, D McGilvray. Genetic modification of an echinocandin B producing strain of *Aspergillus nidulans* to produce mutants blocked in sterigmatocystin biosynthesis. J Ind Microbiol 13:372–381, 1994.
38. VA Vinci. Nonrecombinant strain improvement. In: AL Demain, NA Solomon, eds. Manual of Industrial Microbiology and Biotechnology. 2nd ed. Washington DC: American Society for Microbiology, 1996.
39. AR Cockshott, GR Sullivan. Improving the fermentation medium for echinocandin B production Part I: sequential statistical experiment design. Process Biochem 36: 647–660, 2001.
40. AR Cockshott, BE Hartman. Improving the fermentation medium for echinocandin B production part II: particle swarm optimization 36:661–669, 2001.
41. MD Lilly. Advances in biotransformation processes. Chem Eng Sci 49:151–159, 1994.
42. JC Moore, HM Jin, O Kuchner, FH Arnold. Strategies for the *in vitro* evolution of protein function: enzyme evolution by random recombination of improved sequences. J Mol Biol 272:336–347, 1997.
43. FH Arnold, JC Moore. Optimizing industrial enzymes by directed evolution. Adv Biochem Eng Biotechnol 58:1–14, 1997.

10
Roles of Enzymes in Antibacterial Drug Discovery

Siddhartha Roychoudhury
Procter & Gamble Pharmaceuticals, Mason, Ohio

I. INTRODUCTION

More than half a century ago, a revolution in the field of modern medicine was ushered in by the discovery of penicillin [1], the first in a series of antibiotics that fought bacterial infections with unprecedented success. Over the following decades, numerous drugs with progressively higher potency and broader spectrum of antibacterial activity were introduced, leading to the belief that infectious diseases caused by bacteria no longer posed a major threat to human health, especially in the developed world. However, there have been several recent developments that call for a more cautious view. First, it is increasingly clear that bacteria are remarkably adept at becoming resistant, not only to currently utilized antibacterial drugs, but also to many of their structural analogs [2]. Second, certain bacterial species with intrinsic resistance to multiple drugs are causing infections with an increasing frequency [3]. Third, changes in demographics are leading to an increase in the number of people with impaired immunity to bacterial infections. Together, these factors are encouraging renewed efforts toward the discovery and development of new antibacterial drugs.

In light of the fact that effective inhibition of bacterial enzymes is a well-known mechanism by which several currently available drugs elicit their antibacterial activity, enzymes continue to draw attention as potential targets for new antibacterial compounds. Currently, several approaches are being pursued to study and utilize enzymes for the purpose of antibacterial drug discovery. First, genomic approaches are being used to discover previously unknown enzymes that could serve as potential targets. Second, known enzymes, yet to be utilized as targets, are being studied with the goal of identifying inhibitors that could be developed into potential antibacterial drugs. Third, enzyme targets for known classes of drugs are being studied with renewed interest to gain further insight into their molecular mechanisms of action and structure–activity relationships (SAR) in order to develop newer inhibitors with higher potency, altered spectrum of activity, superior pharmacokinetic properties, or lower toxicity profiles. Finally, enzymes involved in the development of resistance to currently available antibacterial agents are being studied to develop inhibitors that could help existing drugs circumvent bacterial resistance.

To focus on the approaches discussed above, the following two sections of this chapter will discuss specific examples without attempting to include all the enzymes that are, or could be, potential targets of antibacterial drug discovery. The first section will focus on examples to discuss the biochemical background that forms the rationale for choosing certain enzymes as targets. The next section will then focus on examples of the methodologies used and recent progress made toward discovering inhibitors and antibacterial leads. Although enzymes also play important roles in down-stream drug development issues such as toxicity, metabolism, and drug–drug interactions, this chapter will be limited to the study of enzymes in antibacterial drug discovery efforts.

II. BACTERIAL ENZYME TARGETS: BIOCHEMICAL BACKGROUND

A. Unknown Enzymes as New Targets

With the advent of genomic and proteomic technologies, it is now possible to simultaneously identify numerous genes and their protein and enzyme products that could be potential targets for new antibacterial compounds. With the DNA sequence of a number of bacterial genomes being known, it is possible to identify proteins and enzymes that could serve as targets for new antibacterial agents with different goals. Enzymes that are ubiquitous in bacteria could be targeted to identify potential inhibitors with broad-spectrum applications, while enzymes that are specific to particular pathogens could be targeted for inhibitors with narrow-spectrum applications. In many cases, enzyme functions can be predicted from homology analyses. In fact, DNA sequence sampling has been effectively used to identify new enzymes as potential targets in *Streptococcus pneumoniae* [4]. While several genomics-based approaches have been described and reviewed elsewhere [5], it is particularly interesting to note that certain functional genomic approaches combine molecular genetics with whole-cell inhibition of enzymes that lead to cell growth inhibition [6]. These approaches could not only identify potential targets, but also their potential inhibitors. The advantages of this approach include identifying those enzymes whose inhibition results in bacterial cell growth inhibition. Bacterial enzymes and protein factors that are specifically expressed *in vivo* during infection are also drawing special attention. The *in vivo* expression technology (IVET) [7,8] and similar approaches have the potential advantage of identifying enzymes whose inhibitors could lead to drugs that interfere with the molecular events involved in bacterial pathogenesis and virulence, thereby leading to potentially novel antibacterial agents with unique mechanism(s) of action.

B. Known Enzymes as New Targets

Several enzymes in this category are of significant interest. Many of these enzymes have been cloned, overexpressed, purified, and biochemically characterized. This facilitates the development of high-throughput screens as well as structure–function-based drug design efforts. In most cases, the enzymes catalyze biochemical reactions involved in pathways that are known to be essential for bacterial survival. For example, lipid A is an essential component of the outer membrane of Gram-negative bacteria and is essential for bacterial growth. Recently, inhibition of a deacetylase, involved in the biosynthesis of lipid A, has been exploited to discover potent inhibitors with *in vitro* and *in vivo* antibacterial activity [9,10]. Removal of N-formyl groups from nascent bacterial polypeptides is an essential step in bacterial protein synthesis. Bacterial peptide deformylases are responsible for this activity and are being studied to identify inhibitors [11]. Bacterial signal peptidases are members of the serine protease class of enzymes and are responsible for the proteolytic removal of N-terminal signal peptides from pre-proteins that are secreted. Recently, the *spsB* gene, coding for the type I signal peptidase in *Staphylococcus aureus*, has been shown to be essential for bacterial growth [12]. Signal peptidases are ubiquitous in Gram-positive and Gram-negative bacteria and appear to be significantly different from their eukaryotic counterparts, based on DNA sequences. Thus, bacterial signal peptidases are considered attractive targets for new antibacterial agents [13].

While there are several known inhibitors of bacterial cell wall biosynthesis with antibacterial activity, it is important to point out that numerous enzymes are involved in this pathway. For example, the *murA*, *murB*, *murG*, and *mraY* gene products are enzymes involved in peptidoglycan synthesis in *Escherichia coli* and are being studied as potential targets for new antibacterial agents [14–16]. Other enzymes involved in bacterial cell wall synthesis, such as the D-alanine-D-alanine ligase and glucosamine 6-phosphate synthase, are also being studied [17,18].

Known classes of enzymes that are involved in regulating sensory signal transduction and virulence in bacteria have drawn attention as targets for novel antibacterial agents that might be especially effective in combating bacterial virulence [19–22]. Bacterial two-component kinases are the best-known class of en-

Table 1 Additional Examples of Enzymes as Antibacterial Drug Targets

Enzyme	Function	Reference
t-RNA Synthetase	Translation	75
ppGpp Degradase	Metabolism of bacterial stress signal ppGpp	76
Dihydroneopterin aldolase	Folate biosynthesis	77
RNAse P	RNA processing	78
SecA	Protein secretion	79
MurA-F	Peptidoglycan synthesis	80
Mur ligases	Peptidoglycan synthesis	81
Coenzyme A reductase	Oxidative stress and redox balance	82

zymes in this category. These histidine protein kinases undergo ATP-dependent autophosphorylation at a conserved histidine residue. The phosphate is subsequently transferred to an aspartate residue of the ''response regulator'' (the second component), which is usually a DNA-binding protein that regulates transcription of specific genes. The phosphorylation status of the ''response regulator'' modulates its DNA-binding activity, thereby regulating bacterial gene expression [23]. Numerous pairs of these two-component systems are widespread in bacteria and are essential for cell differentiation and virulence as well as response to environmental changes [24,25]. Recently, a new type of two-component system (encoded by the *yycFG* genes) that regulates bacterial cell division in *S. aureus* has been found to be essential for bacterial growth [26]. The two-component kinase VanS is known to regulate vancomycin resistance in enterococci [27]. VncS, another two-component kinase, has been linked to vancomycin tolerance in *S. pneumoniae* [28]. In addition, inhibitors of a fungal two-component system are being studied as potential antifungal agents [29]. Inhibitors of bacterial two-component kinases could therefore represent a novel class of agents with multiple modes of antimicrobial activity. Examples of additional targets under investigation are provided in Table 1.

C. Known Enzymes as Known Targets

Enzymes that are established targets of known antibacterial agents are also being studied with the goal of identifying superior inhibitors. The advantage of enzyme targets in this category is their demonstrated essentiality for bacterial growth and survival. Several of these enzymes are being studied with renewed interest for two reasons. On the one hand, a better understanding of the structure–function and biochemical properties of these enzymes could lead to the discovery of new inhibitors that may not be susceptible to existing and emerging mechanisms of resistance. On the other hand, by studying enzyme targets, a better understanding of the mechanisms of action of known inhibitors at the molecular level could be developed and used to design new members of known classes of inhibitors.

Bacterial dihydropteroate synthase, an enzyme involved in folate biosynthesis, is the target for sulfonamides, which were among the earliest antibacterial agents. Sulfonamides are structural analogs of *para*-aminobenzoic acid, one of the substrates of dihydropteroate synthase. The biochemical mechanism involved in the inhibition of this enzyme is being studied with renewed interest [30]. Dihydrofolate reductase is the second enzyme involved in folate biosynthesis to be a target for known antibacterial agents such as trimethoprim [31]. Newer analogs of trimethoprim are being developed [32]. A combination of sulfonamides and trimethoprims is commonly used to treat a variety of infections [33]. This example shows that in certain cases it is advantageous to target two enzymes, such as the synthase and the reductase in this case, in a biosynthetic pathway to accomplish a successful clinical outcome.

Bacterial DNA gyrase is a type II topoisomerase involved in regulating the level of DNA supercoiling. The activity of this enzyme, a target for the quinolone class of antibacterial agents, is essential for bacterial DNA replication and, consequently, for growth and survival [34]. Understanding the molecular aspects of how quinolones inhibit DNA gyrase has played an important role in the design of new-generation quinolones with improved potency and spectrum of antibacterial activity [35]. In addition, during the process of understanding the quinolone mechanism of action, topoisomerase IV, an enzyme involved in bacterial chromosome segregation, was discovered as a second target for the quinolones [36]. Generation of multistep point mutations in the genes of DNA gyrase and topoisomerase IV has been shown to incrementally increase the level of quinolone resistance in several pathogens [35]. Thus, quinolones with comparable potencies toward both of the targets are now believed to be more desirable because of the lower probability of stepwise resistance development [37]. Additional classes of bacterial DNA gyrase inhibitors, such as coumarins and cyclothialidines, have been identified [38,39], although they have yet to be developed as successful antibacterial agents. Unlike quinolones, these inhibitors interact with the *gyrB* subunit of DNA gyrase and therefore have a different mechanism of action relative to quinolones.

Bacterial RNA polymerase, the target for the rifamycin class of antibacterial agents, is the enzyme responsible for transcription of genomic DNA in bacteria [40]. Like DNA gyrase, RNA polymerase is a multifunctional, multisubunit enzyme with multiple active conformations. This increases the number of possible mechanisms of inhibition of RNA polymerase. For example, in addition to the β subunit, which is the apparent target for rifamycin, bacterial transcription initiation is a unique process in which the σ subunit plays a unique role in the recognition of bacterial promoter sequences [41]. Alternative σ subunits, such as σ^S and σ^E, have been implicated in the transcription of virulence genes in pathogens such as *Salmonella typhimurium* and *Pseudomonas aeruginosa* [42,43]. These alternative σ subunits associate with the core RNA polymerase to activate transcription of specific genes that are essential for pathogenic bacteria to cause infection in humans and animals. As part of the bacterial RNA polymerase holoenzyme, the σ subunits therefore represent an as yet unexploited, but attractive set of targets for antibacterial drug discovery.

Enzymes involved in bacterial cell wall biosynthesis are well-known targets of several classes of antibacterial agents including β-lactams, which are among the most commonly used drugs to treat bacterial infections. The targets of β-lactams are the so-called penicillin-binding proteins, which are transpeptidases involved in the cross-linking of peptidogylcans. These enzymes continue to be studied extensively and are discussed in Chapter 11 [44]. Enzymes involved in bacterial fatty acid synthesis are also being studied as targets of known antibacterial agents. The enoyl-acyl carrier protein reductase, the *inhA* gene product, is the target utilized by isoniazid, an antibacterial agent used against *Mycobacterium tuberculosis* [45]. Recently, the enoyl-acyl carrier protein reductase, the *fabI* gene product in *E. coli*, has been identified as the molecular target for the 2-hydroxydiphenyl ether class of agents which include triclosan, a broad-spectrum antibacterial agent that is widely used in a variety of consumer products [46]. Therefore, further studies of bacterial fatty acid biosynthetic enzymes could lead to new antibacterial agents.

D. Resistance-Determining Enzymes as Targets

As mentioned in the Introduction, bacterial resistance to currently used antibacterial agents poses a major threat to their continued effectiveness [2]. Interestingly, a number of enzymes are involved in the development of antibacterial resistance via several mechanisms. For example, there are enzymes that catalyze the inactivation of antibacterial agents [47]. β-Lactamases are the best-known examples of this class of enzymes and have been studied extensively as targets for new inhibitors that are used in combination with β-lactams to prevent their inactivation [48]. Several β-lactamase inhibitors have been developed and are used in combination with β-lactams (e.g., amoxicillin/clavulanic acid, piperacillin/tazobactam, and others) [49,50].

Bacterial resistance to the macrolide class of antibacterial agents is mediated by the *erm* class of gene products, which function as ribosomal RNA methyltransferases [51]. Numerous members of this class of enzymes have been identified [52]. These enzymes methylate the 23S rRNA component of the 50S subunit of bacterial ribosomes, thereby reducing the affinity of macrolides for ribosomes. Attempts at identifying inhibitors of methyltransferases have been made, and potent inhibitors have been identified using high-throughput screening [53]. Some of these inhibitors inhibited bacterial cell growth *in vitro*, although their true mechanism of action against bacterial whole cells remain unclear. More recently, a series of triazine-containing compounds that inhibit Erm methyltransferases have been identified using an NMR-based screen followed by a parallel synthesis approach to optimize inhibition potency [54]. Further investigation of these various classes of Erm methyltransferase inhibitors could lead to therapeutic agents that reverse macrolide resistance in bacteria.

Bacterial resistance to vancomycin is another example of enzymes playing a key role at the molecular level [55]. Vancomycin is a well-known antibiotic that inhibits bacterial peptidoglycan synthesis and is used to treat infections by multiple-drug-resistant Gram-positive pathogens such as methicillin-resistant *S. aureus* (MRSA). Resistance to vancomycin has now emerged as a significant problem in hospital-acquired infections caused by *Enterococcus faecium* and *Enterococcus faecalis*, pathogens that are resistant to multiple antibacterial agents [56]. Studies on the molecular basis for this resistance revealed the involvement of several enzymes. At least three enzymes, encoded by the *vanH*, *vanA*, and *vanX* genes, are involved in converting the D-alanyl-D-alanine dipeptide, the target for vancomycin, to D-alanyl-D-lactate, which is still capable of transpeptidation but insensitive to vancomycin. First, the *vanX* gene product, a dipeptidase cleaves the terminal D-alanine of the D-alanyl-D-alanine dipeptide. Next, the *vanH* gene product, a dehydrogenase reduces D-pyruvate to D-lactate, which is then used to replace the D-alanine of the dipeptide by the *vanA* gene product, a ligase [57]. Recently, the VanX peptidase and the VanA ligase have been the targets for discovery of new inhibitors that could potentially inhibit vancomycin resistance in enterococci and prolong the use of this antibiotic [58,59].

III. METHODOLOGIES: ANTIBACTERIAL DRUG DISCOVERY

While identification of a target is very important in antibacterial drug discovery, it is only the first in a series of steps leading to the identification of a lead compound. As far as enzyme targets are concerned, certain follow-up steps are common to all of these targets, regardless of their role in bacterial growth and physiology. These steps comprise the use of certain methodologies and tools, examples of which are summarized in Fig. 1 and discussed below.

A. Overexpression and Purification of the Enzyme Target: The *murG* Gene Product

As mentioned in the previous section, enzymes involved in bacterial cell wall biosynthesis are being studied with renewed interest. Among these enzymes is

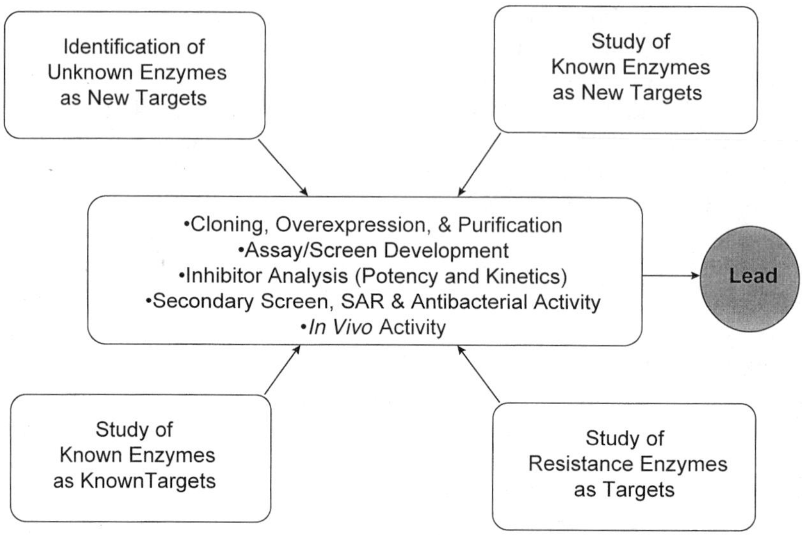

Figure 1 A summary of enzymology-based approaches in antibacterial drug discovery. Enzymes identified with four different approaches are studied using similar methodologies (box at the center) and follow-up steps resulting in the identification of a lead.

the *murG* gene product that catalyzes the formation of lipid II, a precursor of peptidoglycan, which is a primary component of the bacterial cell wall [60]. As the first step toward studying this enzyme as a target for new antibacterial agents, the *murG* gene from *E. coli* has been cloned and expressed from a plasmid. To facilitate purification of the enzyme product, the *murG* gene was tagged with multiple histidine residues. This is a popular technique that usually retains the activity of the enzyme while allowing its efficient purification using metal affinity (nickel) chromatography. Plasmids expressing the cloned gene were used to complement the *murG* mutation in *E. coli*, indicating that the cloned enzyme retained its biological activity. Affinity chromatography (nickel) was then used to purify the enzyme, which demonstrated activity in a peptidoglycan polymerization assay [60].

B. Development of a Functional Assay: The *vanHAX* Gene Products

Once the role of a gene product as a potential enzyme target is identified via genetic techniques, such as cloning and complementation, the next step involves its biochemical characterization including the development of a quantitative assay. Once the vancomycin-resistance genes in enterococci were characterized genetically, suitable enzymatic assays were developed [61–63]. Three distinct enzymatic activities were linked to the products of *vanA*, *vanX*, and *vanH*. Enzymatic characterization of these gene products not only helped elucidate the biochemical mechanism of vancomycin resistance in enterococci, but also provided the biochemical basis for screen development, which will be discussed later in this chapter.

C. Elucidation of the Enzyme Structure via X-Ray Crystallography

1. *The β-Lactamases*

The β-lactamases, a group of resistance-determining enzymes, has been studied extensively via structural analysis [64,65]. Use of β-lactamase inactivators, such as clavulanate, sulbactam, and tazobactam (as discussed in the previous section), has led to the selection of inhibitor-resistant β-lactmases [66]. Recently, X-ray crystallographic studies have been used to elucidate the mechanism of inhibitor insensitivity of an inhibitor-resistant β-lactamase from *E. coli* at the atomic level. Using a high-resolution X-ray structure (2.3 Å), it has been shown that the Asn276-Asp point mutation results in facilitating the turnover of clavulanate, thereby treating it as another β-lactam substrate, rather than a mechanistic inactivator [67]. This finding could facilitate the design of newer β-lactamase inactivators capable of evading the effect of the Asn276-Asp mutation. In a separate study, potent (K_i = 27 nM), non-β-lactam inhibitors were designed using the X-ray crystal structure of AmpC, an intrinsically resistant β-lactamase that was complexed with *m*-aminophenylboronic acid. These inhibitors were also able to potentiate the antibacterial activity of β-lactams [68].

2. *Bacterial Signal Peptidase*

Bacterial signal peptidase is an example of a known enzyme that could serve as a target for new antibacterial agents [13]. Recently, a catalytically active, soluble fragment of signal peptidase from *E. coli* has been crystallized as a complex with a β-lactam inhibitor [69]. This represents a major step in the efforts toward the rational design of inhibitors that can be readily tested for their enzyme inhibition and bacterial growth-inhibition activities.

D. Development of High-Throughput Screening

Even though a biochemical assay for an enzyme target might be available, development of a high-throughput screen can still pose significant challenges. First, the assay has to be miniaturized to a microtiter format (e.g., 96- or 384-well format). Second, steps involving liquid handling, such as the addition of substrate, cofactor, reaction termination buffer, etc., and detection of the reaction product have to be simplified to facilitate automation. Third, the assay has to have adequate detection sensitivity to be quantitatively reliable with acceptable levels of well-to-well, plate-to-plate, and day-to-day variations.

1. *Bacterial RNA Polymerase*

A high-throughput assay for bacterial RNA polymerase has been successfully developed and validated using a 96-well, automated format [70]. The reaction mixture contained a DNA template, nucleotide substrates (NTPs), supplemented with α-^{33}P-labeled CTP in Tris-acetate buffer (pH 6.8). The polymerase reaction was carried out at 34°C for 40 min (providing linear kinetics). The effect of dimethylsulfoxide (DMSO), the usual solvent for test compounds used in a screen, was taken into consideration. The radiolabeled RNA transcripts were allowed to bind diethyl aminoethyl (DEAE) beads, which were then separated via filtration, and radioactivity associated with the wells was quantitated to measure the RNA polymerase activity. The standard deviation of the measured activity was typically $<15\%$ of the average. Use of this assay to screen for RNA polymerase inhibitors from chemical libraries and natural products led to the identification of DNA intercalators (known to inhibit RNA polymerase activity), rifampicin (a known inhibitors of RNA polymerase), and several derivatives of rifampicin from Actinomycetes extracts. Therefore this assay can be reliably utilized to detect novel inhibitors of bacterial RNA polymerase.

2. *Bacterial Two-Component Kinase*

By virtue of their prevalence and importance, eukaryotic tyrosine, threonine, and serine protein kinases continue to be targets of intense research. As a result, reliable and quantitative assays to monitor their activity have been developed [71]. However, in the case of bacterial two-component kinases, which are autophosphorylating enzymes with no known low-molecular-weight substrates or reaction products, traditional assays are limited in terms of their throughput. Recently, a high-throughput autophosphorylation assay was developed using CheA, the two-component kinase that regulates chemotaxis in *E. coli* [72]. Cell extracts from an *E. coli* strain, containing a recombinant plasmid to overexpress the *cheA* gene, were used as the source of the enzyme. Partially purified ($\sim$50%) enzyme preparations, with no contaminating autophosphorylation activity, were used for the assay. The autophosphorylation assay was performed in a 96-well format using DEAE or nitrocellulose filter plates. [γ-^{33}P]ATP was used as a phosphate-donating substrate and ^{33}P-labeled, phosphorylated CheA was detected as the reaction product. Following a 30-min reaction at room temperature, free [γ-^{33}P]ATP was removed by washing the filter wells with high-salt buffer under vacuum.

Radioactivity associated with the washed and dried wells was measured by scintillation counting. Using this screen, several inhibitors of CheA have been identified from combinatorial libraries [73]. Other members of the two-component kinase family, such as NtrB (NR_{II}) and KinA have also been used to screen for inhibitors [20–22].

3. *Coupled VanA/VanX Screen*

To screen for inhibitors of both VanA and VanX enzymes, a coupled high-throughput assay was developed by linking the two reactions catalyzed by these two enzymes [59]. The coupling of the two reactions was facilitated by the fact that the product of the VanX dipeptidase reaction is D-alanine, a substrate utilized by the VanA ligase, along with D-lactate. Since the VanA reaction hydrolyzes ATP, the resulting inorganic phosphate was used for measuring the coupled enzyme activity. The assay was validated using D-cycloserine, a known inhibitor of VanA, and was used in a high-throughput format to screen approximately 250,000 synthetic compounds. While none of the inhibitors detected by this screen was able to reverse vancomycin resistance in *E. faecium (vanA)*, one compound, VAN32, showed synergistic activity with vancomycin against *E. faecalis (vanB)*.

E. Analysis of Inhibitors

Regardless of the properties of a particular enzyme target, numerous preliminary inhibitors are usually identified as a result of a high-throughput screen. However, the number of compounds identified at this step as inhibitors usually depends on the minimum potency deemed necessary for further investigation. A variety of approaches are subsequently used to verify the potency and specificity of the inhibitors. A few of these approaches are discussed below, using the examples of efforts following high-throughput screening for bacterial two-component kinases.

Screening of KinA as a prototypical two-component kinase led to the identification of several classes of inhibitors, such as cyclohexanes, benzimidazoles, bis-phenols, salicylanilides, trityls, and benzoxazines [20,74]. Several members of these inhibitor classes showed significant antibacterial activity against Gram-positive pathogens, including methicillin-resistant *S. aureus* (MRSA) and vancomycin-resistant *E. faecium* (VRE). In addition, a correlation between the KinA inhibition potency (IC_{50}) and antibacterial activity against *S. aureus* was observed [21]. RWJ-49815, a lead compound identified in this study, also showed bactericidal activity against MRSA. In serial passage experiments, MRSA isolates showed a lower tendency to develop resistance to RWJ-49815 than to ciprofloxacin, a known fluoroquinolone. These findings were followed with a series of whole-cell analyses designed to establish a link between the two-component kinase inhibition activity of these compounds and their antibacterial activity [74]. Using assays to determine membrane damage, bacterial viability, hemolysis, and macromolecular synthesis, this study found that the leads with potent enzyme inhibition activity either damaged *S. aureus* membrane or caused hemolysis of equine erythrocytes. Most of these compounds also caused a generalized inhibition of macromolecular (DNA, RNA, and protein) synthesis in bacteria. The authors therefore concluded that the antibacterial activity of the leads identified by the KinA screen was most likely attributable to a nonspecific effect, rather than to the inhibition of two-component kinases.

F. Whole-Cell Screening of Enzyme Inhibitors: Lipid A Biosynthesis

It is important to note that potent enzyme inhibitors can be identified by screening for whole-cell inhibitors of biosynthetic pathways. A successful example of this strategy is the discovery of L-573,655, a member of the carboxyamido-oxazolidines [9,10]. This compound was identified using a whole-cell screen to identify inhibitors of lipid A biosynthesis, an essential pathway for Gram-negative bacteria. Unlike the two-component kinase inhibitors discussed above, the specificity of L-573,655 was demonstrated by its inability to inhibit bacterial DNA, RNA, protein, and phospholipid synthesis. Secondary screening of the nine enzymes involved in *E. coli* lipid A biosynthesis identified the second enzyme in the pathway, UDP-3-O-[*R*-3-hydroxymyristoyl]-GlcNAc deacetylase, as the molecular target for this inhibitor. Subsequent SAR studies of over 200 analogs of L-573,655 led to the identification of L-161,240, which had a K_i of 50 nM for the enzyme target and significant antibacterial activity against several Gram-negative pathogens, such as *E. coli*, *Enterobacter cloacae*, and *Klebsiella pneumoniae*. In addition, this compound demonstrated *in vivo* efficacy by protecting mice from lethal bacterial infection.

IV. CONCLUDING REMARKS

Using specific examples, this chapter attempted to provide a broad overview of the roles of enzymes as targets of antibacterial drug discovery. On the one hand, more detailed knowledge of bacterial pathogenesis and physiology, together with genomics, will lead to the identification of additional enzymes as potential targets for new drugs. On the other hand, advances in high-throughput screening, combinatorial chemistry, parallel synthesis, and automated data processing will undoubtedly enrich the field of antibacterial drug discovery. Developments in molecular diagnostics might also open up new possibilities for targeted, narrow-spectrum therapies based on specific inhibition of enzymes. While the area of antibacterial drugs is a mature one, with numerous potent and effective enzyme inhibitors, the targets were usually identified following the discovery of the inhibitors. Use of enzyme targets for *de novo* discovery, design, and development of antibacterial agents is a field that is still in its early days.

ACKNOWLEDGMENTS

The author would like to thank C. N. Parker, K. S. Howard-Nordan, and C. C. McOsker for critical reading of the manuscript.

REFERENCES

1. AJ Wright. The penicillins. Mayo Clinic Proc 74:290–307, 1999.
2. RN Jones, MA Pfaller. Bacterial resistance: a worldwide problem. Diagn Microbiol Infect Dis 31:379–388, 1998.
3. JP Quinn. Clinical problems posed by multiresistant nonfermenting gram-negative pathogens. Clin Infect Dis 27 (Suppl) 1:S117–S124, 1998.
4. RH Baltz, FH Norris, P Matsushima, BS DeHoff, P Rockey, G Porter, S Burgett, R Peery, J Hoskins, L Braverman, I Jenkins, P Solenberg, M Young, MA McHenney, PL Skatrud, PR Rosteck Jr. DNA sequence sampling of the *Streptococcus pneumoniae* genome to identify novel targets for antibiotic development. Microb Drug Res 4:1–9, 1998.
5. DT Moir, KJ Shaw, RS Hare, GF Vovis. Genomic and antimicrobial drug discovery. Antimicrob Agents Chemother 43:439–446, 1999.
6. DC Young, NG Rafanan, CI Rosenow, D Chen, Z Yuan, R White, J Trias. A novel means of creating selective hypersusceptible screening strains by down regulation of essential genes. Abstracts of the 99th General Meeting of the American Society for Microbiology, Washington DC, 1999, p 23.
7. MJ Mahan, JW Tobias, JM Slouch, PC Hanna, RJ Collier, JJ Mekalanos. Antibiotic-based selection for bacterial genes that are specifically induced during infection of a host. Proc Natl Acad Sci (USA) 92:669–673, 1995.
8. DM Heithoff, CP Conner, PC Hanna, SM Julio, U Hentschel, MJ Mahan. Bacterial infection as assessed by *in vivo* gene expression. Proc Natl Acad Sci (USA) 94: 934–939, 1997.
9. HR Onishi, BA Pelak, LS Gerckens, LL Silver, FM Kahan, MH Chen, AA Patchett, SM Galloway, SA Hyland, MS Anderson, CRH Raetz. Antibacterial agents that inhibit lipid A biosynthesis. Science 274:980–982, 1996.
10. MH Chen, MG Steiner, SE de Laszlo, AA Patchett, MS Anderson, SA Hyland, HR Onishi, LL Silver, CRH Raetz. Carbohydroxamido-oxazolidines: antibacterial agents that target lipid A biosynthesis. Bioorg Med Chem Lett 9:313–318, 1999.
11. DJ Durand, GB Gordon, JF O'Connell, SK Grant. Peptide aldehyde inhibitors of bacterial peptide deformylases. Arch Biochem Biophys 367:297–302, 1999.

12. KM Cregg, I Wilding, MT Black. Molecular cloning and expression of the *spsB* gene encoding an essential type I signal peptidase from *Staphylococcus aureus*. J Bacteriol 178:5712–5718, 1996.
13. MT Black, G Burton. Inhibitors of bacterial signal peptidases. Curr Pharm Des 2: 133–154, 1998.
14. T Skarzynski, A Mistry, A Wonacott, SE Hutchinson, VA Kelly, K Duncan. Structure of UDP-N-acetylglucosamine enolpyruvyl transferase, an enzyme essential for the synthesis of bacterial peptidoglycan, complexed with substrate UDP-N-acetylglucosamine and the drug fosfomycin. Structure 4:1465–1474, 1996.
15. TE Benson, CT Walsh, JM Hogle. The structure of the substrate-free form of MurA, an essential enzyme for the synthesis of bacterial cell wall. Structure 4:47–54, 1996.
16. AA Branstrom, S Midha, CS Mintz, CB Longley, K Han, ER Baizman, HR Axelrod. Purification and utilization of the gene products encoded by *mraY* and *murG* in the development of assays for high throughput screening. Abstracts of the 37th Interscience Conference on Antimicrobial Agents and Chemotherapy, American Society for Microbiology, Washington, D.C., 1997, p 186.
17. PK Chakravarty, WJ Greenleem, WH Parsons, AA Patchett, P Combs, A Roth, RD Busch, TN Mellin. (3-Amino-2-oxoalkyl)phosphponic acids and their analogs as novel inhibitors of D-alanine:D-alanine ligase. J Med Chem 32:1886–1890, 1989.
18. SL Bearne. Active site-directed inactivation of *Escherichia coli* glucosamine-6-phosphate synthase. Determination of the fructose 6-phosphate binding constant using a carbohydrate-based inactivator. J Biol Chem 271:3052–3057, 1996.
19. S Roychoudhury, NA Zielinski, AJ Ninfa, NE Allen, LN Jungheim, TI Nicas, AM Chakrabarty. Inhibitors of two-component signal transduction systems: inhibition of alginate gene activation in *Pseudomonas aeruginosa*. Proc Natl Acad Sci (USA) 90: 965–969, 1993.
20. JM Domagala, D Alessi, M Cummings, S Gracheck, L Huang, M Huband, G Johnson, E Olson, M Shapiro, R Singh, Y Song, R Van Bogelen, D Vo, S World. Bacterial two-component signaling as a therapeutic target in drug design: Inhibition of NRII by the diphenolic methanes (bisphenols). Adv Exp Med Biol 456:269–286, 1998.
21. JF Barrett, RM Goldschmidt, LE Lawrence, B Foleno, R Chen, JP Demers, S Johnson, R Kanojia, J Fernandez, J Bernstein, L Licata, A Donetz, S Huang, DJ Hlasta, MJ Macielag, K Ohemeng, R Frechette, MB Frosco, DH Klaubert, JM Whiteley, L Wang, JA Hoch. Antibacterial agents that inhibit two-component signal transduction systems. Proc Natl Acad Sci (USA) 95:5317–5322, 1998.
22. JF Barrett, JA Hoch. Two-component signal transduction as a target for microbial anti-infective therapy. Antimicrob Agents Chemother 42:1529–1536, 1998.
23. JS Parkinson. Genetic approaches for signaling pathways and proteins. In: JA Hoch, TJ Silhavy, eds. Two-component signal transduction. Washington, DC: ASM Press, 1995, pp 9–23.
24. T Lane, A Benson, GB Hecht, GJ Burton, A Newton. Switches and signal transduction networks in the *Coulobacter crescentus* cell cycle. In: JA Hoch, TJ Silhavy, eds. Two-component signal transduction. Washington, DC: ASM Press, 1995, pp 403–430.

25. M Dziejman, JJ Mekalanos. Two-component signal transduction and its role in the expression of bacterial virulence factors. In: JA Hoch, TJ Silhavy, eds. Two-component signal transduction. Washington, DC: ASM Press, 1995, pp 305–317.
26. PK Martin, T Li, D Sun, DP Biek, MB Schmid. Role in cell permeability of an essential two-component system in *Staphylococcus aureus*. J Bacteriol 181:3666–3673, 1999.
27. M Arthur, C Molinas, P Courvalin. The VanS-VanR two-component regulatory system controls synthesis of depsipeptide peptidoglycan precursors in *Enterococcus faecium*. J Bacteriol 174:2582–2591, 1992.
28. R Novak, B Henriques, E Charpentier, S Normark, E Tuomanen. Emergence of vancomycin tolerance in *Streptococcus pneumoniae*. Nature 399:590–593, 1999.
29. RJ Deschenes, H Lin, AD Ault, JS Fassler. Antifungal properties and target evaluation of three putative bacterial histidine kinase inhibitors. Antimicrob Agents Chemother 43:1700–1703, 1999.
30. HG Vinnicombe, JP Derrick. Dihydropteroate synthase from *Streptococcus pneumoniae*: characterization of substrate binding order and sulfonamide inhibition. Biochem Biophys Res Commun 258:752–757, 1999.
31. RL Then. History and future of antimicrobial diaminopyrimidines. J Chemother 6: 361–368, 1993.
32. HH Locher, H Schlunegger, PG Harman, P Angern, RL Then. Antibacterial activities of epiroprim, a new dihydrofolate reductase inhibitor, alone and in combination with dapsone. Antimicrob Agents Chemother 40:1376–1381, 1996.
33. AJ Salter. Trimethoprim-sulfamethoxazole in treatment of severe infections. Rev Infect Dis 4:338–350, 1982.
34. TD Gootz, KE Brighty. Chemistry and mechanism of action of the quinolone antibacterials. In: VT Andriole, ed. The Quinolones. 2nd ed. San Diego, CA: Academic Press, 1998, pp 29–80.
35. K Drlica, X Zhao. DNA gyrase, topoisomerase IV, and the 4-quinolones. Microbiol Mol Biol Rev 61:377–392, 1997.
36. JI Kato, Y Nishimura, R Imamura, H Niki, S Hiraga, H Suzuki. New topoisomerase essential for chromosome segregation in *E. coli*. Cell 63:393–404, 1990.
37. XS Pan, LM Fisher. DNA gyrase and topoisomerase IV are dual targets of clinafloxacin action in *Streptococcus pneumoniae*. Antimicrob Agents Chemother 42:2810–2816, 1998.
38. P Laurin, D Ferroud, M Klich, C Dupuis-Hamelin, P Mauvais, P Lassaigne, A Bonnefoy, B Musicki. Synthesis and *in vitro* evaluation of novel highly potent coumarin inhibitors of gyrase B. Bioorg Med Chem Lett 9:2079–2084, 1999.
39. N Nakada, H Shimada, T Hirata, Y Aoki, T Kariyama, J Watanabe, M Arisawa. Biological characterization of cyclothialidine, a new DNA gyrase inhibitor. Antimicrob Agents Chemother 37:2656–2661, 1993.
40. MT Record Jr, WS Reznikoff, ML Craig, KL McQuade, PJ Schlax. *Escherichia coli* RNA polymerase ($E\sigma^{70}$), promoters, and the kinetics of the steps of transcription initiation. In: FC Neidhardt, ed. *Escherichia coli* and *Salmonella*: Cellular and Molecular Biology. Washington, DC: ASM Press, 1996, pp 792–821.
41. CA Gross, C Chan, A Dombroski, T Gruber, M Sharp, J Tupy, B Young. The func-

tional and regulatory roles of sigma factors in transcription. Cold Spring Harb Symp Quant Biol 63:141–155, 1998.
42. C Coynault, V Robbe-Saule, F Norel. Virulence and vaccine potential of *Salmonella typhimurium* mutants deficient in the expression of the RpoS (σ^S) regulon. Mol Microbiol 22:149–160, 1996.
43. H Yu, JC Boucher, NS Hibler, V Deretic. Virulence properties of *Pseudomonas aeruginosa* lacking the extreme-stress sigma factor AlgU (σ^E). Infect Immun 64: 2774–2781, 1996.
44. G Zhao, TI Meier, WK Yeh. Penicillin-binding proteins as antimicrobial targets: expression, purification, and assay technologies. In: HA Kirst, WK Yeh, MJ Zmijewski, Jr., eds. Enzyme Technologies for Pharmaceutical and Biotechnological Applications. New York: Marcel Dekker, 2001, pp. 263–287.
45. A Banerjee, E Dubnau, A Quemard, V Balasubramanian, KS Um, T Wilson, D Collins, G de Lisle, WR Jacobs Jr. *inhA*, a gene encoding a target for isoniazid and ethionamide in *Mycobacterium tuberculosis*. Science 263:227–230, 1994.
46. RJ Heath, JR Rubin, DR Holland, E Zhang, ME Snow, CO Rock. Mechanism of triclosan inhibition of bacterial fatty acid synthesis. J Biol Chem 274:11110–11114, 1999.
47. DTW Chu, JJ Plattner, L Katz. New directions in antibacterial research. J Med Chem 39:3853–3874, 1996.
48. SN Maiti, OA Phillips, RG Micetich, DM Livermore. B-Lactamase inhibitors: agents to overcome bacterial resistance. Curr Med Chem 56:441–456, 1998.
49. P Ball, A Geddes, G Rolinson. Amoxycillin clavulanate: an assessment after 15 years of clinical application. J Chemother 9:167–198, 1997.
50. CM Perry, A Markham. Piperacillin/tazobactam: an updated review of its use in the treatment of bacterial infections. Drugs 57:805–843, 1999.
51. C Lai, B Weisblum. Altered methylation of ribosomal RNA in an erythromycin-resistant strain of *Staphylococcus aureus*. Proc Natl Acad Sci (USA) 68:856–860, 1971.
52. H Westh, DM Hougaard, J Vuust, VT Rosdahl. Prevalence of *erm* gene classes in erythromycin-resistant *Staphylococcus aureus* strains isolated between 1959 and 1988. Antimicrob Agents Chemother 39:369–373, 1995.
53. J Clancy, BJ Schmieder, JW Petitpas, M Manousos, JA Williams, JA Faiella, AE Girard, PF McGuirk. Assays to detect and characterize synthetic agents that inhibit the ErmC methyltransferase. J Antibiot 48:1273–1279, 1995.
54. PJ Hajduk, J Dinges, JM Schkeryantz, D Janowick, M Kaminski, M Tufano, DJ Augeri, A Petros, V Nienaber, P Zhong, R Hammond, M Coen, B Beutel, L Katz, SW Fesik. Novel inhibitors of Erm methyltransferases from NMR and parallel synthesis. J Med Chem 42:3852–3859, 1999.
55. CT Walsh, SL Fisher, IS Park, M Prahalad, Z Wu. Bacterial resistance to vancomycin: five genes and one missing hydrogen bond tell the story. Curr Biol 3:21–28, 1996.
56. TM Perl. The threat of vancomycin resistance. Am J Med 106:26S–37S, 1999.
57. M Arthur, P Courvalin. Genetics and mechanism of glycopeptide resistance in enterococci. Antimicrob Agents Chemother 37:1563–1571, 1993.
58. Z Wu, CT Walsh. Phosphinate analogs of D-, D-dipeptides: slow-binding inhibition

and proteolysis protection of VanX, a D-, D-dipeptidase required for vancomycin resistance in *Enterococcus faecium*. Proc Natl Acad Sci (USA) 92:11603–11607, 1995.

59. SD Pratt, X Xuei, AC Mackinnon, AM Nilius, DM Hensey-Rudolf, P Zhong, L Katz. Development of a coupled VanA/VanX assay: screening for inhibitors of glycopeptide resistance. J Biomol Screening 2:241–247, 1997.
60. M Crouvoisier, D Mengin-Lecreulx, J van Heijenoort. UDP-N-acetylglucosamine: N-acetylmuramoyl-(pentapeptide) pyrophosphoryl undecaprenol N-acetylglucosamine transferase from *Escherichia coli*: overproduction, solubilization, and purification. FEBS Lett 449:289–292, 1999.
61. PE Reynolds, F Depardieu, S Dutka-Malen, M Arthur, P Courvalin. Glycopeptide resistance mediated by enterococcal transposon Tn1546 requires production of VanX for hydrolysis of D-alanyl-D-alanine. Mol Microbiol 13, 1065–1070, 1994.
62. TDH Bugg, S Dutka-Malen, M Arthur, P Courvalin, CT Walsh. Identification of vancomycin resistance protein VanA as a D-alanine:D-alanine ligase of altered substrate specificity. Biochemistry 30, 2017–2021, 1991.
63. M Arthur, C Molinas, S Dutka-Malen, P Courvalin. Structural relationship between the vancomycin resistance protein VanH and 2-hydroxycarboxylic acid dehydrogenases. Gene 103(1):133–134, 1991.
64. A Matagne, J Lamotte-Brasseur, JM Frere. Catalytic properties of class A β-lactamases: efficiency and diversity. Biochem J 330:581–598, 1998.
65. KC Usher, LC Blaszczak, GS Weston, BK Shoichet, SJ Remington. Three-dimensional structure of AmpC β-lactamase from *Escherichia coli* bound to a transition-state analogue: possible implications for the oxyanion hypothesis and for inhibitor design. Biochemistry 37:16082–16092, 1998.
66. EB Chaibi, D Sirot, G Paul, R Labia. Inhibitor-resistant TEM β-lactamases: phenotypic, genetic, and biochemical characteristics. J Antimicrob Chemother 43:447–458, 1999.
67. P Swaren, D Golemi, S Cabantous, A Bulychev, L Maveyraud, S Mobashery, JP Samana. X-ray structure of the Asn276Asp variant of the *Escherichia coli* TEM-1 β-lactamase: direct observation of electrostatic modulation in resistance to inactivation by clavulanic acid. Biochemistry 38:9570–9576, 1999.
68. GS Weston, J Blazquez, F Baquero, BK Shoichet. Structure-based enhancement of boronic acid-based inhibitors of AmpC β-lactamase. J Med Chem 41:4577–4586, 1998.
69. M Paetzel, RE Dalbey, NC Strynadhka. Crystal structure of a bacterial signal peptidase in complex with a β-lactam inhibitor. Nature 396:186–190, 1998.
70. G Wu, S Daniel-Issakani, K LaMarco, B Strulovici. Automated high throughput filtration assay: application to polymerase inhibitor identification. Anal Biochem 245:226–230, 1997.
71. JE Casnellie. Assay for protein kinases using peptides with basic residues for phosphocellulose binding. Meth Enzymol 200:115–120, 1991.
72. S Roychoudhury, SM Collins, BA Hynd, CN Parker. High throughput autophosphorylation assay for bacterial histidine kinases. J Biomol Screening 2:85–90, 1997.
73. S Roychoudhury, SE Blondelle, SM Collins, MC Davis, HD McKeever, RA Houghten, CN Parker. Use of combinatorial library screening to identify inhibitors of a bacterial two-component kinase. Molecular Diversity 4:173–182, 1998.

74. JJ Hillard, RM Goldschmidt, L Licata, EZ Baum, K Bush. Multiple mechanisms of action for inhibitors of histidine protein kinases from bacterial two-component systems. Antimicrob Agents Chemother 43:1693–1699, 1999.
75. XY Yu, JM Hill, G Yu, W Wang, AF Kluge, P Wendler, P Gallant. Synthesis and structure-activity relationships of a series of novel thiazoles as inhibitors of aminoacyl-tRNA synthetases. Boorg Med Chem Lett 9:375–380, 1999.
76. G Bahador, A Bowman, S Davis, VJ Hernandezy, M Cashel, J Watson. Screening for antibacterials that affect ppGpp metabolism. Abstracts of the 37th Interscience Conference on Antimicrobial Agents and Chemotherapy, American Society for Microbiology, Washington, DC, 1997, p 186.
77. M Hennig, A D'arcy, I Hampele, D Stueber, G Dale, HJ Schoenfeld, B Poeschl, B Wipf, PJ Hartman. Dihydroneopterin aldolase: a new target in the fight against multiresistant staphylococci. Abstracts of the 37th Interscience Conference on Antimicrobial Agents and Chemotherapy, American Society for Microbiology, Washington, DC, 1997, p 186.
78. J Zhang, D Christensen, JE Kranz. RNase P as a novel antibacterial target. Abstracts of the 37th Interscience Conference on Antimicrobial Agents and Chemotherapy, American Society for Microbiology, Washington, DC, 1997, p 186.
79. H Yuan, P Gallant, X Shen, J Tao. Bacterial SecA as an antimicrobial target. Abstracts of the 38th Interscience Conference on Antimicrobial Agents and Chemotherapy, American Society for Microbiology, Washington, DC, 1998, p 273.
80. D Chen, C Rosenow, J Trias, RJ White, Z Yuan. Pathway screening: novel technology for identifying inhibitors of MurA-F in a single incubation. Abstracts of the 38th Interscience Conference on Antimicrobial Agents and Chemotherapy, American Society for Microbiology, Washington, DC, 1998, p 273.
81. C Rosenow, J Trias. Mur ligases in *S. pneumoniae* targets for novel antibiotic development. Abstracts of the 38th Interscience Conference on Antimicrobial Agents and Chemotherapy, American Society for Microbiology, Washington, DC, 1998, p 274.
82. MK Pope, RL Asano, SB Delcardayre, JE Davies. Coenzyme A reductase: a novel drug target for staphylococci and enterococci. Abstracts of the 38th Interscience Conference on Antimicrobial Agents and Chemotherapy, American Society for Microbiology, Washington, DC, 1998, p 274.

11

Penicillin-Binding Proteins as Antimicrobial Targets: Expression, Purification, and Assay Technologies

Genshi Zhao, Timothy I. Meier, and Wu-Kuang Yeh
Eli Lilly and Company, Indianapolis, Indiana

I. INTRODUCTION

Penicillin-binding proteins (PBPs), the target enzymes of β-lactam antibiotics, play an essential role in bacterial cell wall biosynthesis, cell division, and cell elongation. PBPs constitute a family of essential membrane-bound transpeptidase and transglycosylase enzymes that catalyze the final steps of cell wall peptidoglycan cross-linking and polymerization, respectively [1–10]. Two classes of PBPs are recognized mainly on the basis of their molecular sizes [1,2]. Low-molecular-mass PBPs exhibit DD-carboxypeptidase or DD-endopeptidase activity which appears to be important but not essential for cell growth [11–14]. High-molecular-mass PBPs have been further divided into class A and class B, which differ in part by the sequences of their N-terminal regions [15]. The C-terminal domains of both classes of PBPs appear to possess transpeptidase activity. The functions of the N-terminal domains of these PBPs are not well established, but for the class A PBPs, evidence suggests that they possess transglycosylase activity [1–6,8,10,15]. The N-terminal domains of the class A PBPs contain four conserved motifs that are also present in monofunctional transglycosylases and appear to be responsible for their transglycosylase activity [1,2,9,15,16]. Several members of the class A PBPs catalyze the polymerization of the polysaccharide backbone of peptidoglycan *in vitro* [1–6,8,10,15].

The peptidoglycan cross-linking activity of PBPs, otherwise known as transpeptidase activity, is inhibited by β-lactam antibiotics because of their structural similarity to the natural substrates (stem peptides) and their ability to form a covalent penicilloyl complex via the active-site serine residues [1,2,17,18]. Formation of the penicilloyl-PBP enzyme complex leads to the functional inactivation of PBPs. The peptidoglycan polymerization activity, which is also known as transglycosylase activity of class A high-molecular-mass PBPs, is inhibited directly by moenomycin and indirectly by vancomycin and glycopeptides [4,19–22].

The transpeptidase activity of PBPs is well characterized, mainly due to its important role in bacterial resistance to β-lactam antibiotics [1–3,15,23–27]. The transpeptidase activity can be characterized by its ability to bind to different β-lactams and to hydrolyze analogs of bacterial cell wall stem peptides [1,2,15,28–35]. A crystal structure has been described for the transpeptidase domain of the *Streptococcus pneumoniae* PBP2x protein, a member of class B PBPs [36].

In contrast, the transglycosylase activity of the class A PBPs is poorly characterized, mainly because of the unavailability and complexity of its substrates, lipid intermediates of stem peptides [1–6,8,10,15]. The transglycosylase activity is routinely assayed by a complicated procedure involving the use of bacterial membrane preparations that contain radioisotope-labeled native substrates for the enzyme [3,6,8,10]. Unlike the transpeptidase domain of PBPs, a crystal structure has not yet been described for a transglycosylase domain of any PBPs.

II. ROLE OF PBPs IN BACTERIAL RESISTANCE TO β-LACTAM ANTIBIOTICS

The emerging resistance of bacteria to penicillin is a frightening problem in a number of common pathogenic bacterial species such as *S. pneumoniae*, *Staphylococcus aureus*, *Enterococcus*, *Shigella dysenteriae*, and *Mycobacterium tuberculosis* [24–26,37,38]. For example, *S. pneumoniae*, a major human pathogen of the upper respiratory tract, has developed resistance to many β-lactams [39–44]. *S. pneumoniae* contains five high-molecular mass PBPs, 1a, 1b, 2a, 2b, and 2x, that were detected using sodium dodecyl sulfate-polyacrylamide gel electrophoresis (SDS-PAGE) by their binding of radiolabeled penicillin [23,45,46]. Extensive studies have shown that the intrinsic resistance of the organism to β-lactam antibiotics has occurred by development of altered high-molecular-mass PBPs that have reduced affinity for the antibiotics [24–26,39–44]. Studies have also established that some of the penicillin-resistant *pbp* genes of *S. pneumoniae* were evolved by interspecies recombinational events that replaced portions of the sensitive *pbp* genes with parts of the *pbp* genes of closely related streptococcal species. In addition, biochemical and cell wall composition studies have suggested that the penicillin-resistant PBPs, unlike the penicillin-sensitive (wild-type) PBPs, utilize branched stem peptides as their physiological substrates in the cell. First, the cell wall peptidoglycan of penicillin-sensitive strains of *S. pneumoniae* contains a majority of monomeric and oligomeric forms of linear stem peptides [47–50]. The cell wall peptidoglycan of penicillin-resistant clinical isolates of *S. pneumoniae*, on the other hand, contains abnormal branched stem peptides. These results suggest that the abnormal stem peptides might be the preferred substrates for the penicillin-resistant PBPs *in vivo*. Second, biochemical studies demonstrated that the penicillin-sensitive PBP2x protein, a major resistance determinant, of *S. pneumoniae* differed significantly from the penicillin-resistant PBP2x protein with respect to kinetic properties [30,31,35]. Together, the biochemical and cell wall composition studies suggest that mutations in the penicillin-resistant *pbp* genes might have resulted in the change of the substrate specificity of their encoded enzymes *in vivo* as well as the enzymes involved in the synthesis of the substrates for PBPs.

III. PBPs AS ANTIBACTERIAL TARGETS

Despite the widespread resistance of bacteria to all classes of β-lactam antibiotics and vancomycin [20,22,24–26,37,38], PBPs are still attractive targets for discovering novel antimicrobials for several reasons. PBPs are present in all major human pathogens of Gram-negative and Gram-positive bacteria [2,15,51–55]. Their transpeptidase and transglycosylase activities, responsible for peptidoglycan cross-linking and polymerization, are required for the growth of all cell wall-containing bacteria [1–4,7–9,15]. Probably due to their essential roles in the bacterial cell wall biosynthesis, there exist multiple PBPs in the cell that perform essentially the same functions. This functional and physical redundancy of PBPs has been a determining factor that has undoubtedly delayed the course of penicillin-resistance for decades. PBPs are localized on the surface of the cell [1,2,15], which can therefore minimize potential transport problems that drugs might encounter. PBPs are also unique in that there do not exist homologs of PBPs in humans. β-Lactam antibiotics are very safe. As a result, these antibiotics have been widely prescribed for pediatric and also prenatal use. Therefore, PBPs are valuable targets that might offer a great potential for finding novel antimicrobial agents alternative to the β-lactam antibiotics.

In this chapter, we summarize the progress made in the expression, purification, and assay technologies of PBPs, and propose new assay formats amenable to high-throughput screening operations which can be used to identify inhibitors that specifically inhibit the transpeptidase activity of PBPs. We also outline potential impacts of new advances in assay technologies, molecular biology, and structural biology on future antibiotic discovery.

IV. STRATEGIES AND TECHNOLOGIES FOR EXPRESSION OF PBPs

PBPs are membrane-bound enzymes. In general, PBPs, especially high-molecular-mass PBPs, consist of a short cytoplasmic domain in their N-terminus, followed by a hydrophobic transmembrane domain, then a transglycosylase domain (for class A PBPs) or a non-penicillin-binding domain (also called an unknown functional domain for class B PBPs), and finally a transpeptidase domain located in their C-terminus (Fig. 1; [1,2,15,56]). Thus, purification of PBPs traditionally involved the isolation of membranes and solubilization of PBPs from the membrane preparation by detergents. Since PBPs are present at a very low level (approximately 100–300 molecules per cell) [57–59] and also in multiple forms in the cell, this classical approach was not particularly efficient for producing a significant amount of a purified PBP protein of particular interest.

To facilitate the purification of PBPs, two specific strategies for expression of PBPs have been designed and extensively applied to large-scale production of high-quality PBP proteins that have been used for biochemical characterizations and structural determinations [30,31,33,35,36,60–75]. The most effective strategy involves the removal of the hydrophobic transmembrane domain coding region along with the short cytoplasmic domain coding region of a *pbp* gene of

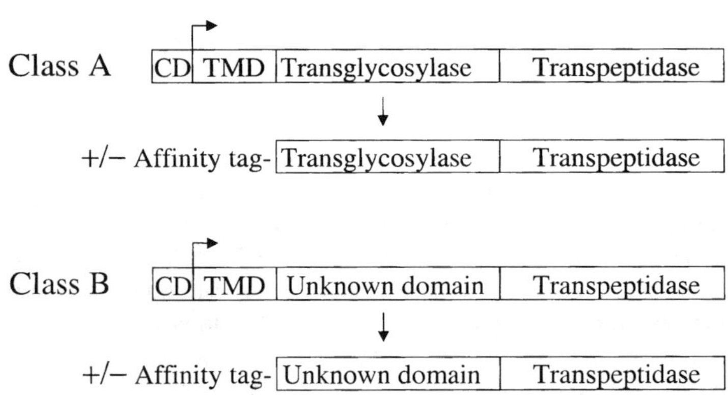

Figure 1 Strategies for generation and overexpression of soluble PBPs in *E. coli*. The CD (cytoplasmic domain) and TMD (transmembrane domain) coding regions of the genes that encode the class A and B PBPs are removed by PCR or other molecular approaches. The resulting truncated *pbp* genes that contain the coding regions for the transglycosylase and transpeptidase domains (for class A PBPs) or the non-penicillin-binding and transpeptidase domains (for class B PBPs) can be cloned into different expression systems that lead to the production of PBPs with (+) or without (−) an affinity tag. Affinity tags can be also fused to the C-termini of PBPs.

interest. It is then followed by cloning of the remaining gene into different expression systems such as pET (Novagen, Inc.), pGEX-2T (Pharmacia LKB Biotechnology), and others (Fig. 1). *Escherichia coli* cells, transformed with the resulting expression plasmids, were used to express PBPs. One distinct advantage of this approach is that in most cases, once expressed, PBPs are soluble and retain their native enzymatic properties in the cell or in the periplasm, although they are not physically associated with the membrane. The genes encoding the penicillin-sensitive and -resistant PBP2x proteins of *S. pneumoniae* [35,62,76], PBP2a (MecA) of *S. aureus* [67,69,74,75], PBP5 of *E. coli* [70], and PBP2 of penicillin-sensitive and -resistant *Neisseria gonorrhoeae* [73] were highly expressed in *E. coli* by using this approach. The above PBPs were expressed as nontagged or polyhistidine(His)-tagged proteins in *E. coli*, and purification of these highly expressed PBPs led to substantial amounts of enzymatically active enzymes (see below). The PBP1a, PBP2a, and PBP2x of *S. pneumoniae* were also expressed as glutathione-S-transferase (GST)-tagged proteins in *E. coli* [33,63,64,77]. The expression levels of the His-tagged PBP2x protein (His-PBP2x), GST-PBP2a of *S. pneumoniae*, and the His-PBP2a of *S. aureus* were estimated to be 30%, 40%, and 30%, respectively, of total cellular proteins [33,35,71; Zhao, unpublished]. For instance, in order to produce soluble forms of the penicillin-sensitive and -resistant PBP2x proteins of *S. pneumoniae*, the short cytoplasmic and hydrophobic transmembrane domain coding regions of both *pbp2x* genes were removed by PCR manipulations. The resulting truncated *pbp2x* genes were then cloned into the pET expression vectors, pET5a and pET16. When pET5a was used, the PBP2x proteins were expressed as nontagged proteins in *E. coli*, but when pET16b was used, both PBPs were expressed as His-tagged proteins in *E. coli* [35]. In other cases, PBPs were produced in *E. coli* as a fusion protein with a cleavable OmpA signal peptide or maltose-binding protein that is attached to the N-terminus of the protein and subsequently secreted into the periplasm of *E. coli* as a nontagged or maltose-binding protein-tagged protein [66,71,78].

The other strategy that has also been often used for expression of PBPs involves a direct addition of an affinity tag to the C- or N-terminal part of a full-length PBP of interest. For example, the PBP1 of *Mycobacterium leprae* [68,79], PBP7 of *Pseudomonas aeruginosa* [80], and PBP1b of *E. coli* [81] were produced in *E. coli* as His-tagged proteins that were properly incorporated into the membrane of *E. coli*. Finally, PBPs can also be overexpressed simply by cloning these PBP genes into a high-copy-number plasmid such as pUC and pBR derivatives or other plasmids. For example, PBP3 and PBP5 of *E. coli* [70,78], PBP5 of *Enterococcus hirae* [82], and PBP3 of *P. aeruginosa* [59] were expressed using these plasmids for different purposes of study.

Various expression systems have been successfully used to express many different PBPs in *E. coli*. Obviously, each expression system offers different advantages, depending on the nature of studies. For screening operations, biochemical characterizations, and structural studies of PBPs, the pET and pGEX-2T expression systems in conjunction with the removal of the PBP hydrophobic transmembrane coding regions have been proven to be very effective [33,35,62–64,69,70,72,74,75]. In our studies, we found that a given expression system might not work for a particular gene [33]. Therefore, it is necessary to examine different expression systems for their suitability of expressing a particular PBP of interest. For physiological and topological studies of PBPs, expression of PBPs using low-copy-number plasmids, such as pACYC or pBR derivatives, clearly offers advantages over other expression systems that rely on high-copy-number plasmids. The amounts of PBPs produced in the cell using low-copy-number plasmids probably would not drastically exceed their *in vivo* levels, thereby being close to their physiological state in the cell. Addition of an affinity tag to a PBP protein with or without its transmembrane domain can significantly facilitate its purification (see below) and recovery. A number of studies suggested that addition of an affinity tag (His, GST, or other tag) to PBPs with or without their transmembrane domains did not alter their enzymatic properties [33,35,36,61–68,71,73–75]. Therefore, if a large amount of a purified PBP protein is required, addition of an affinity tag to the PBP protein of interest is recommended.

V. STRATEGIES AND TECHNOLOGIES FOR PURIFICATION OF PBPs

Purification of PBPs in the past was a difficult task for several reasons. Multiple PBPs are present in all species of bacteria, PBPs share similar biochemical and physical properties, and their cellular amounts are very low [57,58,60]. However, PBPs are well known to bind covalently to different β-lactam antibiotics. Thus, PBPs were traditionally purified mainly on the basis of this property, by using β-lactam affinity column chromatography followed by other types of column chromatography [58,83–92]. The inclusion of affinity column chromatography is a very critical step for enriching PBPs, which otherwise would be very difficult to purify. The covalently bound PBPs were released from the column by hydroxylamine. The enriched PBPs were then further purified by other column chromatographic methods. Thus, a typical purification scheme consists of isolation of membranes, solubilization of PBPs from the membrane preparation by detergents, and affinity and other column chromatographic methods for the solubilized PBPs.

It is also known that a given β-lactam antibiotic binds to only a certain number of PBPs, and affinities of PBPs for β-lactam antibiotics vary greatly [1,2,25,58]. Therefore, a PBP that binds specifically to a given β-lactam can in principle be purified using an affinity column containing this β-lactam. Alternatively, a PBP can be purified by varying elution conditions or by differential binding to different β-lactams to remove the unwanted PBPs before affinity column chromatography [58,92]. By using these affinity column chromatographic methods, many PBPs from different organisms were purified. These include PBP1a, PBP1b, PBP2, PBP3, PBP5, and PBP6 of *E. coli* [58,65,83,84,89]; PBP1, PBP2b, and PBP4 of *Bacillus subtilis* [90]; PBP5 of *Streptococcus faecium* [93]; PBP1 of *Bacillus licheniformis* [88]; and a PBP with a molecular mass of 49.5 kDa from *Mycobacterium smegmatis* [85].

Some members of class A high-molecular-mass PBPs were found to bind to moenomycin [64,94]. *E. coli* PBP1a and 1b were purified by using moenomycin-affinity column chromatography in combination with other column chromatographic methods.

Some PBPs were also found to bind to different dyes, such as Cibacron, Dylon, Procion, and Reactive Blue [62,66,72,76,95–97]. Several PBPs, including PBP3, PBP4, and PBP5 of *E. coli*, PBP2a of *S. aureus*, and PBP2x of *S. pneumoniae*, were purified using dye-affinity column chromatography in combination with other column chromatographic methods. There was one report that PBP2a from a methicillin-resistant *S. aureus* strain was purified by using a monoclonal antibody prepared against the PBP protein [98].

Advances in molecular biology have made it possible to express PBPs in *E. coli* at very high levels. As a result, many PBPs have been overexpressed in *E. coli*. Overexpression of PBPs has made their purification considerably easier, especially when affinity tags are used. By using different expression systems along with affinity tags, large quantities of soluble PBPs were purified in one step by nickel or glutathione affinity column chromatography [35,63,68,71,74,77,79]. These include PBP2a from a methicillin-resistant *S. aureus* isolate, PBP1 from *M. leprae*, PBP1a from *S. pneumoniae*, and PBP2x from penicillin-sensitive and -resistant *S. pneumoniae*. For instance, substantial amounts of purified His-PBP2a of *S. aureus* (6 mg protein/liter of culture) and GST-PBP1a of *S. pneumoniae* (8–10 mg protein/liter of culture) were obtained [63,67]. As shown in Figure 2, the His-tagged penicillin-sensitive and -resistant PBP2x proteins of *S. pneumoniae* were purified by using nickel column chromatography [35]. Biochemical and spectrometric analyses established that the purified proteins were enzymatically active and exhibited proper kinetic properties. In some cases, PBPs, when overproduced in the cell, are present in inclusion bodies [33,61,64,67,73]. In all cases reported, the inclusion bodies produced were isolated and the proteins were refolded into a soluble, enzymatically active form of PBPs by using 3-[(3-cholamidopropyl)dimethylammonio]-1-propanesulfonate (CHAPS), urea, or guanidine HCl [33,61,64,67,73]. The solubilized proteins were then further purified by using affinity column chromatography. In one case, however, the isolated inclusion bodies of the *S. pneumoniae* PBP2a protein were apparently pure and therefore

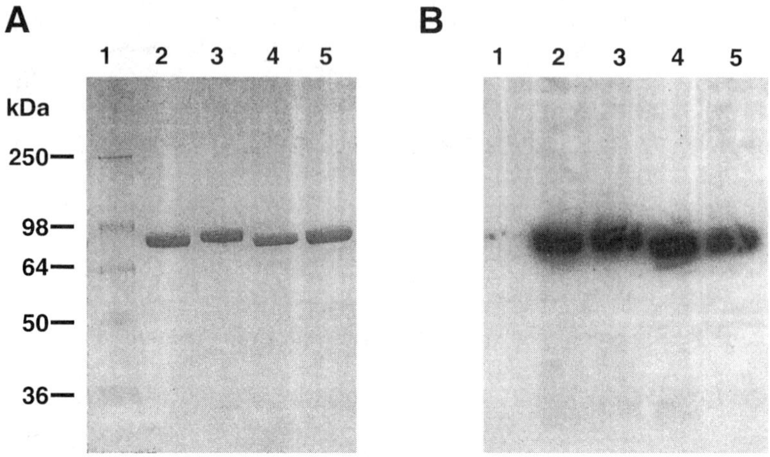

Figure 2 SDS-PAGE analysis of the purified penicillin-sensitive and -resistant PBP2x proteins of *S. pneumoniae* and their ability to bind ^{125}I-labeled penicillin V. The PBP2x proteins were purified, labeled with ^{125}I-labeled penicillin V, and separated on a 10% gel as described [35]. The gel was stained with Coomassie blue (A), dried, and exposed to an X-ray film (B). (A and B), Lane 1, prestained molecular mass standards (from top to bottom: myosin, 250 kDa; BSA, 98 kDa; glutamic dehydrogenase, 64 kDa; alcohol dehydrogenase, 50 kDa; and carbonic anhydrase, 36 kDa). Lane 2, His-tagged penicillin-sensitive PBP2x. Lane 3, His-tagged penicillin-resistant PBP2x. Lane 4, nontagged penicillin-sensitive PBP2x. Lane 5, nontagged penicillin-resistant PBP2x. (From Ref. 33.)

did not require further purification after solubilization and refolding. In this study, 37 mg of a purified, refolded, and soluble GST-PBP2a protein preparation was obtained from 1 liter of *E. coli* culture [33].

For purification of PBPs that are expressed as nontagged proteins in *E. coli*, a number of methods, including dye affinity, anion exchange, adsorption (hydroxyapatite), cation exchange, hydrophobic interaction, and gel filtration have been successfully used. The following PBPs were purified using these methods: penicillin-sensitive and -resistant PBP2x of *S. pneumoniae* [30,31,35,62,76], PBP3 and PBP5 of *E. coli* [65,66], and PBP2a of *S. aureus* [69,72]. For instance, the following procedures were used to purify the penicillin-sensitive and -resistant PBP2x proteins of *S. pneumoniae*. First, both PBPs were expressed as nontagged proteins in *E. coli* using the pET expression system. The penicillin-sensitive and -resistant PBP2x proteins were purified by Q-Sepharose, hydroxyapatite, and Source-S column chromatographic methods. After the three steps of purification, the penicillin-sensitive or -resistant PBP2x protein was homogenous as judged by SDS-PAGE (Fig. 2). This three-step purification scheme yielded 6–8 mg of the purified protein per liter of culture [35]. The purified PBP2x proteins exhibited identical kinetic properties to those of the purified His-PBP2x proteins. In another case, PBP2a of *S. aureus* was expressed as a nontagged protein in *E. coli* using the expression vector pOW200 [72,75]. The PBP2a protein of *S. aureus* was purified by Q-Sepharose, CM-Sepharose, and dye-ligand (Reactive Blue/4-agarose) column chromatographic methods. After the three steps of purification, the PBP2a protein preparation was found to be nearly homogenous as judged by SDS-PAGE. Using these expression and purification schemes, 100 mg of the PBP2a protein was obtained from 200 g wet weight of *E. coli* cells [72].

Obviously, modern molecular biology has drastically changed the way PBPs are purified. The power of various bacterial expression systems has made it possible to obtain large quantities of soluble PBPs for extensive biochemical and structural studies, and also screening operations of these proteins. Although β-lactam-affinity column chromatography has been widely used for the purification of PBPs, obtaining a substantial amount of a homogenous PBP preparation for biochemical and in particular structural studies as well as screening operations remains a challenge. Many bacterial genomes, including *E. coli*, *B. subtilis*, *H. influenzae*, *S. pneumoniae*, and *M. tuberculosis*, have been sequenced, which span from Gram-negative to Gram-positive bacteria [51–55]. Therefore, purification of a particular PBP for biochemical and structural studies, and screening operations through overexpression of the protein, can be more readily achieved than purification of the PBP directly from its native source. Generating a soluble form of PBP is critical to the success of structural studies. The strategies that have been often used for expression of soluble PBPs have been proven to be very effective in generating substantial amounts of PBPs suitable for various biochemical and structural studies [30–33,35,36,62–64,69–70,72].

In summary, purification of overexpressed PBPs can be achieved by a variety of methods as described above. Nickel, glutathione, and maltose affinity column chromatographic methods have been used for the purification of His-, GST- and maltose-binding protein-tagged PBPs, respectively. For convenience, purification of PBPs can also be easily carried out in a batch mode rather than by column chromatography. For purification of nontagged PBPs, anion and cation exchange, adsorption, and gel-filtration column chromatographic methods all have been used [30,31,35,62,69,72,76]. Dye-affinity column chromatography was useful in some cases, but appeared to be limited because affinities of these dyes for other PBPs are not known and have to be determined [95,96]. Although β-lactam-affinity column chromatography has been widely used for the purification of PBPs without overexpression, it has not been used much for the purification of overexpressed PBPs, for several reasons. The high levels of expression of PBPs did not appear to require any specific affinity column chromatography.

β-Lactam-affinity resins are not commercially available. Finally, the PBPs covalently bound to β-lactam-affinity columns are routinely eluted using hydroxylamine, which may result in inactivation of PBPs [86,87].

VI. STRATEGIES AND TECHNOLOGIES FOR ASSAY OF PBPs

PBPs were initially discovered by their ability to bind to penicillin [17,18,60,99–101]. Since then, PBPs have been routinely detected by penicillin-binding assays [60,99–101]. In the past 30 years, ^{3}H- and ^{14}C-labeled penicillins have often been used for the detection of PBPs. A typical procedure for the detection of the bacterial membrane-bound PBPs consists of the preparation of membranes, incubation of the membrane preparations with ^{3}H- or ^{14}C-labeled penicillin, separation of the labeled PBPs by SDS-PAGE, and detection by fluorography [60,99–103]. Due to the low energy levels of these radioisotopes, the use of ^{3}H- or ^{14}C-labeled penicillin for the detection of PBPs routinely required days to weeks [29,34,60,102,104]. To shorten the time required for the detection of PBPs, ^{125}I-labeled penicillins were synthesized [102,105–107]. The use of ^{125}I-labeled β-lactams as labeling reagents for the detection of PBPs has significantly reduced the time to, typically, 1 day. Although the ^{125}I-labeled β-lactam method offers some distinct advantages over the ^{3}H- and ^{14}C-labeled penicillin methods with regard to its time-saving feature, ^{125}I-labeled β-lactams have not been widely used as labeling reagents for the detection and characterization of PBPs for several reasons. ^{125}I-labeled β-lactams are not commercially available [102–107]. The use of ^{125}I-labeled materials in general imposes many restrictions due to their extremely hazardous nature. Finally, the high specificity of ^{125}I-labeled β-lactams may have limited their use in kinetic studies of PBPs.

To explore alternative nonradioisotopic approaches for the detection of PBPs, β-lactams with an amino group, such as ampicillin, 6-aminopenicillanic acid, and 7-aminocephalosporanic acid, have been conjugated to biotin, fluoresceins, or digoxigenin [29,104,108–112]. Using β-lactam-biotin or -digoxigenin conjugates, the detection of PBPs is achieved by immunoblotting and chemiluminescence. A typical procedure consists of the following steps: labeling bacterial membrane preparations or whole cells with the β-lactam conjugates, separating the labeled proteins by SDS-PAGE, transferring the labeled proteins to a membrane (nitrocellulose or nylon), blotting the membrane with streptavidin-conjugated peroxidase or antibodies prepared against digoxigenin, and detecting the labeled PBPs by chemiluminescence or color change [108,111,112]. Using fluorescein-labeled β-lactams as labeling reagents, PBPs can be directly detected with the aid of an ultraviolet transilluminator, an automatic DNA sequencer, or a fluorescence microscope [29,104,109,110,113]. PBPs from several different bacterial species, including *Borrelia burgdorferi*, *E. coli*, *H. influenzae*, *S. pneumoniae*, and *S. aureus*, were detected and characterized using these β-lactam conjugates as labeling reagents [108,111,112]. Compared with the ^{3}H-, ^{14}C-, and ^{125}I-labeled β-lactam methods, one of the major advantages of the β-lactam conjugate methods for the detection of PBPs is their nonradioisotopic nature. The sensitivity of these methods is similar to those of the traditional radioactive methods [29,104,108,110–112]. The disadvantages of these methods utilizing β-lactam conjugates as labeling reagents are as follows. They require synthesis and preparation because of the lack of commercial availability [29,104,108–112]. These β-lactam conjugates are relatively unstable and last for only about 2 weeks [108,111,112]. In addition, using digoxigenin-conjugated β-lactams requires antibodies prepared against digoxigenin.

Recently, BOCILLIN FL, a fluorescent penicillin, was synthesized for the detection and characterization of PBPs [34]. BOCILLIN FL, a derivative of penicillin V, is an orange solid with extinction coefficient of 68,000 M^{-1} cm^{-1} and a maximal absorption at 504 nm (Molecular Probes, Inc.). It fluoresces at 511 nm upon excitation at 504 nm. BOCILLIN FL has been used to detect PBPs from the membrane preparations of several bacterial species, including *S. pneumoniae*, *E. coli*, and *P. aeruginosa* (Fig. 3). A typical procedure of using BOCILLIN FL for the detection of PBPs involves preparation of the bacterial cytoplasmic membranes, incubation of the membrane preparations with BOCILLIN FL, and visualization of PBPs by a UV transilluminator or with the aid of a

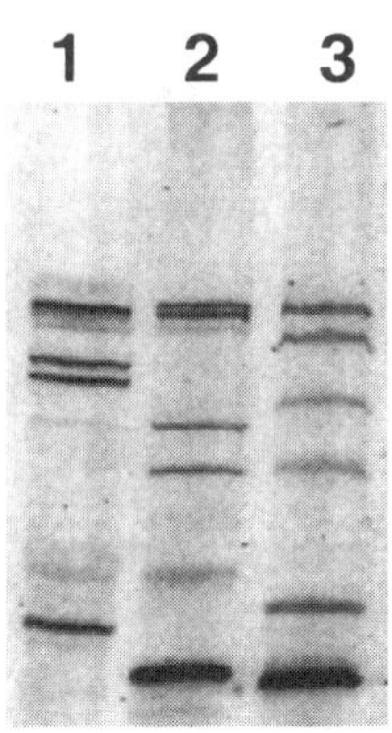

Figure 3 Detection of PBPs of *S. pneumoniae*, *E. coli*, and *P. aeruginosa* using BOCILLIN FL as a labeling reagent. The membrane fractions of *S. pneumoniae*, *E. coli*, and *P. aeruginosa* were prepared, labeled with BOCILLIN FL, separated by SDS-PAGE, and visualized using a FluorImager as described [34]. Lanes 1–3, the BOCILLIN FL-labeled membranes preparations of *S. pneumoniae*, *E. coli*, and *P. aeruginosa*, respectively. (From Ref. 34.)

FluorImager [34]. This method allowed rapid detection of 30 ng of a purified PBP protein under UV light and of 2–4 ng of the protein with the aid of a FluorImager. The sensitivity of this method is similar to those of the methods using the fluorescein- and radioisotope-labeled β-lactams [29,34,104,110]. The use of BOCILLIN FL as a labeling reagent for the detection of PBPs offers several advantages over the radioisotopic methods [60,99–103]. The BOCILLIN FL method, like the fluorescein-labeled and the biotin- and digoxigenin-conjugated β-lactam methods, does not involve the use of radioactive materials. Thus, no hazardous materials are produced. Like the fluorescein-labeled β-lactam method, results can be obtained immediately after the completion of SDS-PAGE, since no gel manipulations are required for detection by using this method. The ^{3}H-, ^{14}C-, and ^{125}I-labeled penicillin methods, as discussed earlier, normally require days to weeks. Fluorescein-labeled, and biotin- and digoxigenin-conjugated β-lactams offer similar advantages, yet they are not commercially available and do not appear to be stable for more than 2 weeks. BOCILLIN FL, however, appeared to be stable and resistant to freeze and thaw [34].

Because PBPs are required for multiple cellular processes and represent a unique class of attractive antimicrobial targets and also because bacteria have become resistant to virtually all antibiotics available on the market, the development of high-throughput assays for the identification of potential inhibitors of PBPs is critical to discovery of novel antibiotics that can be used to combat the current crisis of antibiotic resistance. A standard industrial high-throughput assay format, in general, must meet the following criteria. The assay format must be robust, being able to screen against a large number of compounds in a very short period of time, typically screening a library of $0.5–1 \times 10^{6}$ compounds in 2–4 weeks. The assay should be a homogenous format, i.e., no separation or washing during the assay. Although under certain conditions only a nonhomogenous assay format is available, which requires a step of separation or washing during the assay, a homogenous assay format is always preferred. All reagents used in the assay must be stable. Variations of the assay must be minimal ($<15\%$). The high-throughput assay formats that have been most often used in industry are scintillation proximity assay (SPA), fluorescent resonance energy transfer (FRET), fluorescence polarization (FP), and various colorimetric assays [114–117]. These assay formats meet the above-described criteria. Recently, we have developed a FP assay for the PBP2x protein of *S. pneumoniae* [34], a primary penicillin-resistance determinant in the organism. This assay possesses all necessary features and can be used directly for the identification of inhibitors of the PBP2x protein. This FP assay is a homogenous assay that utilizes the commercially available BOCILLIN FL as a labeling reagent. The assay was carried out in 96-well microtiter plates. By using this assay, we determined the kinetic parameters of the PBP2x protein and IC_{50} values of the protein for different β-lactam antibiotics (Fig. 4). The kinetic parameters and IC_{50} values determined by using

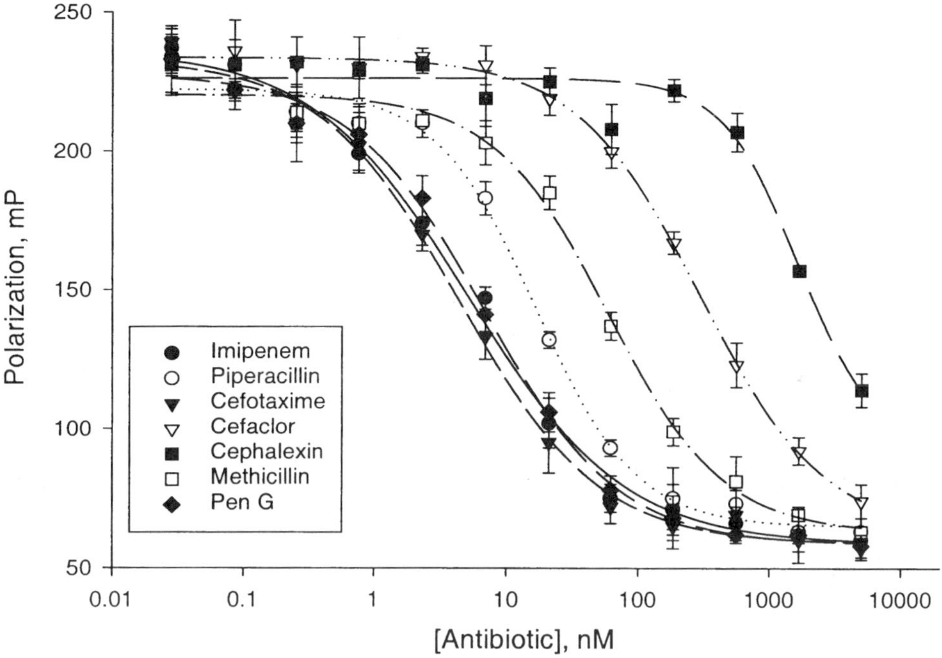

Figure 4 Determination of the IC_{50} values of the penicillin-sensitive PBP2x protein of *S. pneumoniae (hex)* R6 for different β-lactam antibiotics by fluorescence polarization assays using BOCILLIN FL as a labeling reagent. The following antibiotics were used: imipenem, piperacillin, cefotaxime, cefaclor, cephalexin, methicillin, and penicillin G (PenG). The fluorescence polarization for the penicillin-sensitive PBP2x protein (1.3 nM) in a competitive interaction with BOCILLIN FL (2 nM) and increasing concentrations of unlabeled antibiotics was measured as described [34]. Basal polarization for the BOCILLIN FL probe was 60 mP and the maximum polarization determined in the absence of competitor was 240 mP. Data points represent the average of four replicates (±SD), and the curve is the predicted nonlinear regression result.

the FP assay were similar to those reported previously by using other methodologies [34]. This FP assay is very sensitive, which allows the detection of nanogram quantities of the PBP2x protein. A similar FP assay can also be developed for other PBPs by using BOCILLIN FL as a labeling reagent. Since the fluorescein-labeled β-lactams share similar properties to those of BOCILLIN FL, a FP assay using the fluorescein-labeled β-lactams can be also envisioned if these reagents are proven to be stable. One disadvantage associated with the FP assays is the requirement that target proteins (enzymes) must have very high binding affinities for ligands used in the assay.

There is another reported high-throughput assay format for the PBP2a of *S. aureus*. This assay format is a membrane-based binding assay that utilized ^{3}H-penicillin as a labeling reagent and required TCA precipitation of the PBP2a-^{3}H-penicillin complex, separation of the complex from free PBP2a, drying in 96-well microtiter plates in which the assay was carried out, and quantifying radioactivity associated with the plates. By using this method, the kinetic properties such as K_m and k_{cat} of the PBP2a protein were determined and the inhibition of the PBP2a protein by non-β-lactam compounds was evaluated [118]. The disadvantage of this assay format is clearly its discontinuous nature, which requires three steps of precipitation, separation, and drying before counting.

With the availability of a variety of scintillant-coated plates or beads from commercial suppliers such as Amersham Pharmacia Biotechnology and New England Nuclear Science, a number of high-throughput assay formats for PBPs can be now envisioned. For instance, a SPA-based penicillin-binding activity assay can be easily designed simply by biotinylation of a purified PBP of particular interest (Fig. 5). Penicillin-binding assays using a ^{3}H- or ^{14}C-labeled penicillin as a labeling reagent can be performed in 96-well microtiter plates. The PBP-^{3}H- or ^{14}C-penicillin complexes are then captured by mixing with streptavidin-SPA beads (containing scintillant) and the penicillin-binding activity of PBP is finally measured by counting the plates directly (Fig. 5). In the presence of an inhibitor, the penicillin-binding activity of PBP should be reduced according to its potency. Thus, this assay format can be used to identify potential inhibitors of any given PBP in a high-throughput mode. This assay is also a homogenous format that does not require any separation, washing, or filtration.

Other types of high-throughput assay formats can be also designed based on the penicillin-binding activity of a particular PBP protein. Several types of 96-well microtiter plates are commercially available, such as Flashplates from NEN, that contain scintillant. Certain types of the plates are also coated with glutathione or streptavidin for capturing GST-tagged or biotinylated proteins (Amersham Pharmacia Biotechnology). These plates can be coated with a PBP either through its direct binding to the plates or through the binding of its GST tag to the glutathione on the plates or its biotin group to the streptavidin on the plates. The penicillin-binding activity of a PBP, therefore, can be measured directly by addition of ^{3}H- or ^{14}C-labeled β-lactam to plates that have already been coated with PBP. In the presence of a potential inhibitor of PBP, the penicillin-binding activity of PBP should be decreased, depending on its potency. This assay is also a homogenous format that does not require any separation, washing, or filtration once microtiter plates are coated with a PBP. Thus, the Flashplates-based assays should be also applicable to the large-scale identification of inhibitors of the penicillin-binding activity of all PBPs.

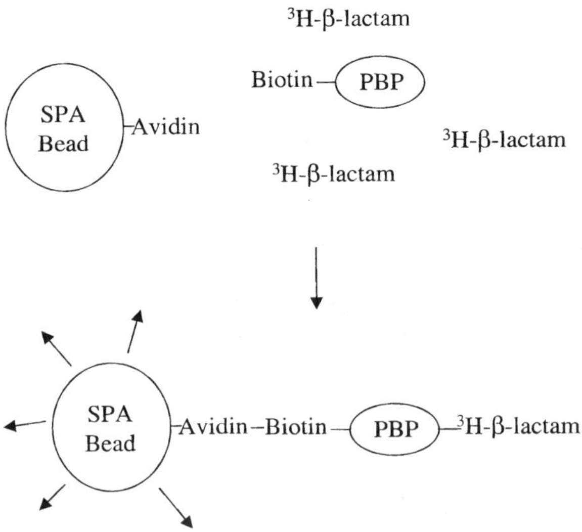

Figure 5 A penicillin-binding activity-based high-throughput SPA assay for the identification of potential inhibitors of the transpeptidase activity of PBPs. A particular PBP of interest is first labeled with biotin. The biotinylated PBP is then mixed with a ^{3}H-labeled β-lactam in 96-well microtiter plates and the reaction mixtures are incubated at room temperature for a period of time. After incubation, the biotionylated PBP-^{3}H-β-lactam complexes are captured by mixing with streptavidin coated SPA beads. The β-lactam-binding activity of the PBP is measured by counting the microtiter plates after mixing with the SPA beads.

The interaction between various β-lactams and PBP2a of *S. aureus* has been explored as a possible approach for the identification of inhibitors of PBP2a [119]. It was observed that the binding of β-lactams to the PBP2a protein induced precipitation of the protein. In addition, the binding of nitrocefin to the PBP2a protein was found to cause a red shift in the maximum absorption of nitrocefin. Based on this observed absorption shift, a chromogenic assay for the PBP2a protein was developed [119]. This assay involves incubating purified PBP2a with nitrocefin, collecting PBP2a precipitates, dissolving the precipitates in urea, and measuring the absorbance increase at 510 nm. This binding assay can be developed into a 96-well format for screening purposes. However, this assay format has several limitations. This format is not a homogenous assay, since it requires collection of precipitated PBP2a protein. The amount of protein required for a screen may not be attainable, although it is possible that this might be overcome by improving expression of the protein. Finally, this assay format may not be sensitive enough to be used in a high-throughput mode.

Various ester and thioester derivatives of hippuric acid have been synthesized and used as chromogenic substrates for the transpeptidases of PBPs. These substrates have been used for the characterization of the transpeptidase activity of a number of PBPs [27,28,30–32,35,63,64,120]. These chromogenic depsipeptide substrates, upon cleavage by the transpeptidase of PBPs, exhibit an increase in absorbance at around 250 nm, but with relatively low extinction coefficients ranging mainly from 500 to 1000 M^{-1} cm^{-1} [27,28,30–32,120]. However, the sensitivity of these transpeptidase activity assays using thioester substrates is increased significantly when a thiol-group capture agent is used to capture the released thiol group in the reaction [35,121]. The hydrolysis activity of these substrates by PBPs has been shown to be inhibited by various β-lactam antibiotics, and the penicillin-binding activity of some PBPs tested is also inhibited by these substrates [27,35]. Thus, it appears to be possible that a colorimetric assay using one of these thioester substrates can be used to identify inhibitors of the transpeptidase activity of PBPs. However, there are several major limitations to this assay format. First, these substrates all undergo a significant amount of spontaneous hydrolysis [27,28,30–32,35,120], which leads to a very high background. Second, these substrates do not appear to have good affinities for PBPs [30–33,62,63,120]. Thus, a very high concentration of the substrates must be used in the assay, which, in conjunction with their high levels of spontaneous hydrolysis, makes it almost impossible to implement this assay in a high-throughput mode.

In summary, BOCILLIN FL and fluorescein-labeled penicillins should be used for routine detection of PBPs, and ^{3}H-penicillins are better suited for kinetic characterizations of PBPs due to their low energy level. For the development of high-throughput assays for PBPs, BOCILLIN FL and ^{3}H-penicillins are clearly superior to other labeled penicillins.

VII. PERSPECTIVES

PBPs, the targets of β-lactam antibiotics, are attractive antibacterial targets for the discovery of novel antibiotics. A considerable amount of progress has been made in the last 30 years in the expression and purification of PBPs, and in particular, in assay development. Advances in molecular biology have made it quite feasible to carry out large-scale purification of PBPs for screening operations and structural determinations. Leveraging molecular biology and structural biology to better understand the structure–function relationship of PBPs and using high-throughput assays to systematically screen various PBPs against a diverse set of chemical and natural product libraries will be an extremely important part of the future of drug discovery efforts. In addition, how to utilize the information obtained from structural studies and screening operations of PBPs will be crucial to future success in the discovery of novel antibiotics.

ACKNOWLEDGMENTS

We thank Q. M. Wang, A. Kraft, and K. A. McAllister for critical reading of the manuscript.

REFERENCES

1. JM Ghuysen. Serine β-lactamases and penicillin-binding proteins. Annu Rev Microbiol 45:37–67, 1991.
2. JM Ghuysen, G Dive. Biochemistry of the penicilloyl serine transferases. In: JM Ghuysen, R Hakenbeck, eds. Bacterial Cell Wall. Amsterdam, The Netherlands: Elsevier Science, 1994, pp 103–129.
3. F Ishino, K Mitsui, S Tamaki, M Matsuhashi. Dual enzyme activities of cell wall peptidoglycan synthesis: peptidoglycan transglycosylase and penicillin-sensitive transpeptidase in purified preparations of *Escherichia coli* penicillin-binding protein 1a. Biochem Biophys Res Commun 97:287–293, 1980.
4. M Matsuhashi. Utilization of lipid-linked precursors and the formation of peptidoglycan in the process of cell growth and division: membrane enzymes involved in the final steps of peptidoglycan synthesis and the mechanism of their regulation. In: JM Ghuysen, R Hakenbeck, eds. Bacterial Cell Wall. Amsterdam, The Netherlands: Elsevier Science, 1994, pp 55–72.
5. J Nakagawa, S Tamaki, S Tomioka, M Matsuhashi. Functional biosynthesis of cell wall peptidoglycan by polymorphic bifunctional polypeptides. J Biol Chem 259: 13937–13946, 1984.
6. H Suzuki, Y van Heijenoort, T Tamura, J Mizoguchi, Y Hirota, J van Heijenoort. *In vitro* peptidoglycan polymerization catalyzed by penicillin binding protein 1b of *Escherichia coli* K-12. FEMS Microbiol Lett 110:245–249, 1980.
7. S Tamaki, S Nakajima, M Matsuhashi. Thermosensitive mutation in *Escherichia coli* simultaneously causing defects in penicillin-binding protein-1b and in enzyme activity for peptidoglycan synthesis *in vitro*. Proc Natl Acad Sci (USA) 74:5472–5476, 1977.
8. Y van Heijenoort, M Gomez, M Derrien, J Ayala, J van Heijenoort. Membrane intermediates in the peptidoglycan metabolism of *Escherichia coli*: possible roles of PBP1b and PBP3. J Bacteriol 174:3549–3557, 1992.
9. SY Yousif, JK Broome-Smith, BG Spratt. Lysis of *Escherichia coli* by β-lactam antibiotics: deletion analysis of the role of penicillin-binding proteins 1a and 1b. J Gen Microbiol 131:2839–2845, 1985.
10. CAL Zijderveld, Q Waisfisz, MEG Aarsman, N Nanninga. Hybrid proteins of the transglycosylase and the transpeptidase domains of PBP1b and PBP3 of *Escherichia coli*. J Bacteriol 177:6290–6293, 1995.

11. C Schuster, B Dobrinski, R Hakenbeck. Unusual septum formation in *Streptococcus pneumoniae* mutants with an alternation in the DD-carboxypeptidase penicillin-binding protein 3. J Bacteriol 172:6499–6505, 1990.
12. A Severin, C Schuster, R Hakenbeck, A Tomasz. Altered murein composition in a DD-carboxypeptidase mutant of *Streptococcus pneumoniae*. J Bacteriol 174:5152–5155, 1992.
13. B Korat, H Mottl, W Keck. Penicillin-binding protein 4 of *Escherichia coli*: molecular cloning of the *dacB* gene, controlled overexpression, and alterations in murein composition. Mol Microbiol 5:675–684, 1991.
14. T Romaeis, JV Holtje. Penicillin-binding protein 7/8 of *Escherichia coli* is a DD-endopeptidase. Eur J Biochem 224:597–604, 1994.
15. JM Ghuysen, P Charlier, J Coyette, C Duez, E Fonze, C Fraipont, C Goffin, B Joris, M Nguyen-Disteche. Penicillin and beyond: evolution, protein fold, multimodular polypeptides, and multiprotein complexes. Microb Drug Resist 2:163–175, 1996.
16. MD Berardino, A Dijkstra, D Stuber, W Keck, M Gubler. The monofunctional glycosyltransferase of *Escherichia coli* is a member of a new class of peptidoglycan-synthesizing enzymes. FEBS Lett 392:184–188, 1996.
17. D Tipper, JL Strominger. Mechanism of action of penicillins: a proposal based on their structural similarity to acyl-D-alanyl-D-alanine. Proc Natl Acad Sci (USA) 54:1133–1141, 1965.
18. EM Wise, JT Park. Penicillin: its basic site of action as an inhibitor of a peptide cross-linking reaction in cell wall muropeptide synthesis. Proc Natl Acad Sci (USA) 54:75–81, 1965.
19. NE Allen, JN Hobbs Jr, TI Nicas. Inhibition of peptidoglycan biosynthesis in vancomycin-susceptible and -resistant bacteria by a semisynthetic glycopeptide antibiotic. Antimicrob Agents Chemother 40:2356–2362, 1996.
20. TD Bugg, CT Walsh. Intracellular steps of bacterial cell wall peptidoglycan biosynthesis: enzymology, antibiotics, and antibiotic resistance. Nat Prod Rep 9:199–215, 1992.
21. F Huber, F Nesemann. Moenomycin: an inhibitor of cell wall synthesis. Biochem Biophys Res Commun 30:7–13, 1968.
22. TI Nicas, ML Zekel, DK Braun. Beyond vancomycin: new therapies to meet the challenge of glycopeptide resistance. Trends Microbiol 5:240–249, 1997.
23. R Hakenbeck, A Konig, I Kern, M van der Linden, W Keck, D Billot-Klein, R Legrand, B Schoot, L Gutmann. Acquisition of five high-Mr penicillin-binding protein variants during transfer of high-level β-lactam resistance from *Streptococcus mitis* to *Streptococcus pneumoniae*. J Bacteriol 180:1831–1840, 1998.
24. HC Neu. The crisis in antibiotic resistance. Science 257:1064–1073, 1992.
25. BG Spratt. Resistance to β-lactam antibiotics. In: JM Ghuysen, R Hakenbeck, eds. Bacterial Cell Wall. Amsterdam, The Netherlands: Elsevier Science, 1994, pp 517–534.
26. A Tomasz, R Munoz. β-lactam antibiotic resistance in Gram-positive bacterial pathogens of the upper respiratory tract: a brief overview of mechanisms. Microb Drug Resist 1:103–109, 1995.
27. M Adam, C Damblon, M Jamin, W Zorzi, V Dusart, M Galleni, A El Eharroubi, G Piras, BG Spratt, W Keck, J Coyote, JM Ghuysen, M Nguyen-Disteche, JM

Frere. Acyltransferase activities of the high-molecular mass essential penicillin-binding proteins. Biochem J 279:610–614, 1991.
28. M Adam, C Damblon, B Plaitin, L Christiaens, JM Frere. Chromogenic depsipeptide substrates for β-lactamases and penicillin-sensitive DD-peptidases. Biochem J 270:525–529, 1990.
29. B Granier, M Jamin, M Adam, M Galleni, B Lakaye, W Zorzi. J Grandchamps, JM Wilkin, C Fraipont, B Joris, C Duez, M Nguyen-Disteche, J Coyette, M Leyh-Bouille, J Dusart, L Christiaens, JM Frere, JM Ghuysen. Serine-type D-ala-D-ala peptidases and penicillin-binding proteins. Meth Enzymol 244:249–266, 1994.
30. M Jamin, C Damblon, S Millier, R Hakenbeck, JM Frere. Penicillin-binding protein 2x of *Streptococcus pneumoniae*: enzymic activities and interactions with β-lactams. Biochem J 292:735–741, 1993.
31. M Jamin, R Hakenbeck, JM Frere. Penicillin-binding protein 2x as a major contributor to intrinsic β-lactam resistance of *Streptococcus pneumoniae*. FEBS Lett 331: 101–104, 1993.
32. M Jamin, M Adam, C Damblon, L Christiaens, JM Frere. Accumulation of acyl-enzyme in DD-peptidase-catalysed reactions with analogues of peptide substrates. Biochem J 280:499–506, 1991.
33. G Zhao, TI Meier, J Hoskins, SR Jaskunas. Penicillin-binding protein 2a of *Streptococcus pneumoniae*: expression in *Escherichia coli* and purification and refolding of inclusion bodies into a soluble and enzymatically active enzyme. Prot Expr Purif 16:331–339, 1999.
34. G Zhao, TI Meier, SD Kahl, KR Gee, LC Blaszczak. BOCILLIN FL, a sensitive and commercially available reagent for detection of penicillin-binding proteins. Antimicrob Agents Chemother 43:1124–1128, 1999.
35. G Zhao, WK Yeh, RH Carnahan, J Flowkowitsch, TI Meier, WE Alborn, GW Becker, SR Jaskunas. Biochemical characterization of penicillin-resistant and -sensitive penicillin-binding protein 2x transpeptidase activities of *Streptococcus pneumoniae* and mechanistic implications in bacterial resistance to β-lactam antibiotics. J Bacteriol 179:4901–4908, 1997.
36. S Pares, N Mouz, Y Petillot, R Hakenbeck, O Dideberg. X-ray structure of *Streptococcus pneumoniae* PBP2x, a primary penicillin target enzyme. Nat Struct Biol 3: 284–289, 1996.
37. ML Cohen. Epidemiology of drug resistance: implications for a post-antimicrobial era. Science 257:1050–1055, 1992.
38. J Davies. Inactivation of antibiotics and the dissemination of resistance genes. Science 264:375–382, 1994.
39. CG Dowson, A Hutchison, JA Brannigan, RC George, D Hansman, J Linares, A Tomasz, J Maynard-Smith, BG Spratt. Horizontal transfer of penicillin-binding protein genes in penicillin-resistant clinical isolates of *Streptococcus pneumoniae*. Proc Natl Acad Sci (USA) 86:8842–8846, 1989.
40. CG Dowson, A Hutchison, BG Spratt. Extensive remodeling of the transpeptidase domain of penicillin-binding protein 2B of a penicillin-resistant South African isolate of *Streptococcus pneumoniae*. Mol Microbiol 3:95–102, 1989.
41. CG Dowson, A Hutchison, N Woodford, AP Johnson, RC George, BG Spratt. Penicillin-resistant viridans streptococci have obtained altered penicillin-binding protein

genes from penicillin-resistant strains of *Streptococcus pneumoniae*. Proc Natl Acad Sci (USA) 87:5858–5862, 1990.
42. CG Dowson, TJ Coffey, BG Spratt. Origin and molecular epidemiology of penicillin-binding protein-mediated resistance to β-lactam antibiotics. Trends Microbiol 2:361–366, 1994.
43. R Hakenbeck, M Tarpay, A Tomasz. Multiple changes of penicillin-binding proteins in penicillin-resistant clinical isolates of *Streptococcus pneumoniae*. Antimicrob Agents Chemother 17:364–371, 1980.
44. G Laible, BG Spratt, R Hakenbeck. Interspecies recombinational events during the evolution of altered PBP2x genes in penicillin-resistant clinical isolates of *Streptococcus pneumoniae*. Mol Microbiol 5:1993–2002, 1991.
45. J Hoskins, P Matsushima, DL Mullen, J Tang, G Zhao, TI Meier, T Nicas, SR Jaskunas. Gene disruption studies of penicillin-binding proteins 1a, 1b, and 2a in *Streptococcus pneumoniae*. J Bacteriol 181:6552–6555, 1999.
46. G Laible, R Hakenbeck, MA Sicard, B Joris, JM Ghuysen. Nucleotide sequences of the *pbpX* genes encoding the penicillin-binding proteins 2x from *Streptococcus pneumoniae* R6 and a cefotaxime-resistant mutant, C506. Mol Microbiol 3:1337–1348, 1989.
47. J Garcia-Bustos, A Tomasz. A biological price of antibiotic resistance: major changes in the peptidoglycan structure of penicillin-resistant strains of *Streptococcus pneumoniae*. Proc Natl Acad Sci (USA) 87:5414–5419, 1990.
48. A Severin, AM Figueiredo, A Tomasz. Separation of abnormal cell wall composition from penicillin-resistance through genetic transformation of *Streptococcus pneumoniae*. J Bacteriol 178:1788–1792, 1996.
49. A Severin, A Tomasz. Naturally occurring peptidoglycan variants of *Streptococcus pneumoniae*. J Bacteriol 178:168–174, 1996.
50. A Severin, MV Vaz Pato, AM Figureiredo, A Tomasz. Drastic changes in the peptidoglycan composition of penicillin-resistant laboratory mutants of *Streptococcus pneumoniae*. FEMS Microbiol Lett 130:31–35, 1995.
51. FR Blattner, G Plunkett III, CA Bloch, NT Perna, V Burland, M Riley, J Collado-Vides, JD Glasner, CK Rode, GF Mayhew, J Gregor, NW Davis, HA Kirkpatrick, MA Goeden, DJ Rose, B Mau, Y Shao. The complete genome sequence of *Escherichia coli* K-12. Science 277:1453–1474, 1997.
52. RH Baltz, FH Norris, P Matsushima, BS DeHoff, P Rockey, G Porter, S Burgett, R Peery, J Hoskins, L Braverman, I Jenkins, P Solenberg, M Young, MA McHenney, PL Skatrud, PR Rosteck Jr. DNA sequence sampling of the *Streptococcus pneumoniae* genome to identify novel targets for antibiotic development. Microb Drug Resist 4:1–9, 1998.
53. ST Cole, R Brosch, J Parkhill, T Garnier, C Churcher, D Harris, et al. Deciphering the biology of *Mycobacterium tuberculosis* from the complete genome sequence. Nature 393:537–544, 1998.
54. RD Fleischmann, MD Adams, O White, RA Clayton, EF Kirkness, AR Kerlavage, et al. Whole-genome random sequencing and assembly of *Haemophilus influenzae* Rd. Science 269:496–512, 1995.
55. F Kunst, N Ogasawara, I Moszer, AM Albertini, G Alloni, V Azevedo, et al. The

complete genome sequence of the gram-positive bacterium *Bacillus subtilis*. Nature 390:249–256, 1997.
56. C Goffin, JM Ghuysen. Multimodular penicillin-binding proteins: an enigmatic family of orthologs and paralogs. Microbiol Mol Biol Rev 62:1079–1093, 1998.
57. TJ Dougherty, K Kennedy, RE Kessler, MJ Pucci. Direct quantitation of the number of individual penicillin-binding proteins per cell in *Escherichia coli*. J Bacteriol 178:6110–6115, 1996.
58. T Tamura, H Suzuki, Y Nishimura, J Mizoguchi, Y Hirota. On the process of cellular division in *Escherichia coli*: isolation and characterization of penicillin-binding proteins 1a, 1b, and 3. Proc Natl Acad Sci (USA) 77:4499–4503, 1980.
59. X Liao, REW Hancock. Cloning and characterization of the *Pseudomonas aeruginosa pbpB* gene encoding penicillin-binding protein 3. Antimicrob Agents Chemother 39:1871–1874, 1995.
60. BG Spratt. Properties of the penicillin binding proteins of *Escherichia coli* K12. Eur J Biochem 72:341–352, 1977.
61. J Bartholome-De Belder, M Nguysen-Disteche, N Houba-Herin, JM Ghuysen, IN Maruyama, H Hara, Y Hirota, M Inouye. Overexpression, solubilization, and refolding of a genetically engineered derivative of the penicillin-binding protein 3 of *Escherichia coli* K12. Mol Microbiol 2:519–526, 1988.
62. P Charlier, G Buisson, O Dideberg, J Wierenga, W Keck, G Laible, R Hakenbeck. Crystallization of a genetically engineered water-soluble primary penicillin target enzyme. J Mol Biol 232:1007–1009, 1993.
63. AM Di Guilmi, N Mouz, JP Andrieu, J Hoskins, SR Jaskunas, J Gagnon, O Dideberg, T Vernet. Identification, purification, and characterization of transpeptidase and glycosyltransferase domains of *Streptococcus pneumoniae* penicillin-binding protein 1a. J Bacteriol 180:5652–5659, 1998.
64. AM Di Guilmi, N Mouz, L Martin, J Hoskins, SR Jaskunas, O Dideberg, T Vernet. Glycosyltransferase domain of penicillin-binding protein 2a from *Streptococcus pneumoniae* is membrane associated. J Bacteriol 181:2773–2781, 1999.
65. LCS Ferreira, U Schwarz, W Keck, P Charlier, O Dideberg, JM Ghuysen. Properties and crystallization of genetically engineered, water-soluble derivative of penicillin-binding protein 5 of *Escherichia coli* K12. Eur J Biochem 171:11–16, 1988.
66. C Fraipont, M Adam, M Nguyen-Disteche, W Keck, J Van Beeumen, JA Ayala, B Dranier, H Hara, JM Ghuysen. Engineering and overexpression of periplasmic forms of the penicillin-binding protein 3 of *Escherichia coli*. Biochem J 298:189–195, 1994.
67. LJ Frank, D Wisniewski, GG Hammond, J Hermes, A Marcy, PM Cameron. High-yield expression, refolding, and purification of penicillin-binding protein 2a from methicillin-resistant *Staphylococcus aureus* strain 27R. Prot Expr Purif 6:671–678, 1995.
68. S Lepage, P Dubois, TK Ghosh, B Joris, S Mahapatra, M Kundu, J Basu, P Chakrabarti, ST Cole, M Nguyen-Disteche, JM Ghuysen. Dual multimodular class A penicillin-binding proteins in *Mycobacterium leprae*. J Bacteriol 179:4627–4630, 1997.
69. WP Lu, Y Sun, MD Bauer, S Paule, PM Koenigs, WG Kraft. Penicillin-binding protein 2a from methicillin-resistant *Staphylococcus aureus*: kinetic characteriza-

tion of its interactions with β-lactams using electrospray mass spectrometry. Biochemistry 38:6537–6546, 1999.
70. RA Nicholas, JL Strominger. Site-directed mutants of a soluble form of penicillin-binding protein 5 from *Escherichia coli* and their catalytic properties. J Biol Chem 263:2034–2040, 1988.
71. KD Pryor, B Leiting. High-level expression of soluble protein in *Escherichia coli* using a His_6-tag and maltose-binding protein double-affinity fusion system. Prot Expr Purif 10:309–319, 1997.
72. S Roychoudhury, JE Dotzlaf, S Ghag, WK Yeh. Purification, properties, and kinetics of enzymatic acylation with β-lactams of soluble penicillin-binding protein 2a: a major factor in methicillin-resistant *Staphylococcus aureus*. J Biol Chem 269: 12067–12073, 1994.
73. DE Schultz, BG Spratt, RA Nicholas. Expression and purification of a soluble form of penicillin-binding protein 2 from both penicillin-susceptible and penicillin-resistant *Neisseria gonorrhoeae*. Prot Expr Purif 2:339–349, 1991.
74. CYE Wu, J Hoskins, LC Blaszczak, DA Preston, PL Skatrud. Construction of a water-soluble form of penicillin-binding protein 2a from a methicillin-resistant *Staphylococcus aureus* isolate. Antimicrob Agents Chemother 36:533–539, 1992.
75. CYE Wu, LC Blaszczak, MC Smith, PL Skatrud. Construction of a modified penicillin-binding protein 2a from methicillin-resistant *Staphylococcus aureus* and purification by immobilized metal affinity chromatography. J Bacteriol 176:1539–1541, 1994.
76. G Laible, W Keck, R Lurz, H Mottl, JM Frere, M Jamin, R Hakenbeck. Penicillin-binding protein 2x of *Streptococcus pneumoniae*: expression in *Escherichia coli* and purification of a soluble enzymatically active derivative. Eur J Biochem 207: 943–949, 1992.
77. N Mouz, AM Di Guilmi, E Gordon, R Hakenbeck, O Dideberg, T Vernet. Mutations in the active site of penicillin-binding protein PBP2x from *Streptococcus pneumoniae*. J Biol Chem 274:19175–19180, 1999.
78. C Goffin, C Fraipont, J Ayala, M Terrak, M Nguyen-Disteche, JM Ghuysen. The non-penicillin-binding module of the tripartite penicillin-binding protein 3 of *Escherichia coli* is required for folding and/or stability of the penicillin-binding module and the membrane-anchoring module confers cell separation activity on the folded structure. J Bacteriol 178:5402–5409, 1996.
79. J Basu, S Mahapatra, M Kundu, S Mukhopadhyay, M Nguyen-Disteche, P Dubois, B Joris, J Van Beeumen, ST Cole, P Chakrabarti, JM Ghuysen. Identification and overexpression in *Escherichia coli* of a *Mycobacterium leprae* gene, *pon1*, encoding a high-molecular-mass class A penicillin-binding protein, PBP1. J Bacteriol 178:1701–1711, 1996.
80. J Song, G Xie, PK Elf, KD Young, RA Jensen. Comparative analysis of *Pseudomonas aeruginosa* penicillin-binding protein 7 in the context of its membership in the family of low-molecular-mass PBPs. Microbiology 144:975–983, 1998.
81. RA Nicholas, DR Lamson, DE Schiltz. Penicillin-binding protein 1B from *Escherichia coli* contains a membrane associated site in addition to its transmembrane anchor. J Biol Chem 268:5632–5641, 1993.
82. ME Mollerach, P Partoune, J Coyette, JM Ghuysen. Importance of the E-46 D-

160 polypeptide segment of the non-penicillin-binding module for the folding of the low-affinity, multimodular class B penicillin-binding protein 5 of *Enterococcus hirae*. J Bacteriol 178:1774–1175, 1996.
83. H Aranuma, JL Strominger. Purification and properties of penicillin-binding proteins 5 and 6 from *Escherichia coli* membranes. J Biol Chem 255:11173–11180, 1980.
84. H Aranuma, JL Strominger. Purification and properties of penicillin-binding proteins 5 and 6 from the *dacA* mutant strain of *Escherichia coli* (JE11191). J Biol Chem 259:1294–1298, 1984.
85. J Basu, R Chattopadhyay, M Kundu, P Chakrabarti. Purification and partial characterization of a penicillin-binding protein from *Mycobacterium smegmatis*. J Bacteriol 174:4829–4832, 1992.
86. PM Blumberg, JL Strominger. Isolation by covalent affinity chromatography of the penicillin-binding components from membranes of *Bacillus subtilis*. Proc Natl Acad Sci (USA) 69:3751–3755, 1972.
87. PM Blumberg, JL Strominger. Covalent affinity chromatography of penicillin-binding components from bacterial membranes. Meth Enzymol 22:401–405, 1974.
88. HA Chase, PE Reynolds, JB Ward. Purification and characterization of the penicillin-binding protein that is the lethal target of penicillin in *Bacillus megaterium* and *Bacillus licheniformis*: protein exchange and complex stability. Eur J Biochem 88: 275–285, 1978.
89. SJ Curtiss, JL Strominger. Purification of penicillin-binding protein 2 of *Escherichia coli*. J Bacteriol 145:398–403, 1981.
90. G Kleppe, JL Strominger. Studies of the high molecular weight penicillin-binding proteins of *Bacillus subtilis*. J Biol Chem 254:4856–4862, 1979.
91. ST Shepherd, HA Chase, PE Reynolds. The separation and properties of two penicillin-binding proteins from *Salmonella typhimurium*. Eur J Biochem 78:521–532, 1977.
92. DJ Waxman, DM Lindgren, JL Strominger. High-molecular-weight penicillin-binding proteins from membranes of *Bacilli*. J Bacteriol 148:950–955, 1981.
93. A Grossato, YR Cheng, E Tonin, P Jacques, R Fontana. Purification of *Streptococcus faecium* penicillin-binding protein 5, a multifunctional penicillin-binding protein. Microbiologica 9:21–28, 1986.
94. W Vollmer, M von Rechenberg, JV Holtje. Demonstration of molecular interactions between the murein polymerase PBP1B, the lytic transglycosylase MltA, and the scaffolding protein MipA of *Escherichia coli*. J Biol Chem 274:6726–6734, 1999.
95. H Mottl, and W Keck. Purification of penicillin-binding protein 4 of *Escherichia coli* as a soluble protein by dye-affinity chromatography. Eur J Biochem 200:767–773, 1991.
96. H Mottl, and W Keck. Rapid screening of a large number of immobilized textile dyes for the purification of proteins by use of penicillin-binding protein 4 of *Escherichia coli* as a model enzyme. Prot Expr Purif 3:403–409, 1992.
97. MPG Van Der Linden, LD Haan, W Keck. Domain organization of penicillin-binding protein 5 from *Escherichia coli* analyzed by C-terminal truncation. Biochem J 289:593–598, 1993.

98. CR Harrington, DM O'Hara, PE Reynolds. Characterization of a monoclonal antibody and its use in the immunoaffinity purification of penicillin-binding protein 2′ of methicillin-resistant *Staphylococcus aureus*. FEMS Microbiol Lett 53:143–147, 1989.
99. PM Blumberg, JL Strominger. Interaction of penicillin with the bacterial cell: penicillin-binding proteins and penicillin-sensitive enzymes. Bacteriol Rev 38:291–335, 1974.
100. BG Spratt, AB Pardee. Penicillin-binding proteins and cell shape in *E. coli*. Nature (London) 254:516–517, 1975.
101. H Suginaka, PM Blumberg, JL Strominger. Multiple penicillin-binding components in *Bacillus subtilis*, *Bacillus cereus*, *Staphylococcus aureus*, and *Escherichia coli*. J Biol Chem 247:5279–5288, 1972.
102. JM Masson, P Baron, D Michelot, R Labia. Evaluation of ^{125}I-radiolabeled penicillin-X in penicillin-binding protein studies. Drugs Exp Clin Res 9:821–825, 1983.
103. DA Preston, CYE Wu, LC Blaszczak, DE Seitz, NG Halligan. Biological characterization of a new radioactive labeling reagent for bacterial penicillin-binding proteins. Antimicrob Agents Chemother 34:718–721, 1990.
104. M Galleni, B Lakaye, S Lepage, M Jamin, I Thamn, B Joris, JM Frere. A new, highly sensitive method for the detection and quantification of penicillin-binding proteins. Biochem J 291:19–21, 1993.
105. LC Blaszczak, NG Halligan, DE Seitz. Radioiododestannylation: convenient synthesis of a stable penicillin derivative for rapid penicillin binding protein (PBP) assay. J Labelled Compd Radiopharmacol 27:401–406, 1989.
106. JM Masson, R Labia. Synthesis of a ^{125}I-radiolabeled penicillin for penicillin-binding proteins studies. Anal Biochem 128:164–168, 1983.
107. FB Wientjes, TJ Olijhoek, U Schwarz, N Nanninga. Labelling pattern of major penicillin-binding proteins of *Escherichia* coli during the division cycle. J Bacteriol 153:1287–1293, 1983.
108. M Dargis, F Malouin. Use of biotinylated β-lactams and chemiluminescence for study and purification of penicillin-binding proteins in bacteria. Antimicrob Agents Chemother 38:973–980, 1994.
109. J Hao, KE Kendrick. Visualization of penicillin-binding proteins during sporulation of *Streptomyces griseus*. J Bacteriol 180:2125–2132, 1998.
110. B Lakaye, C Damblon, M Jamin, M Galleni, S Lepage, B Joris, J Macrchand-Brynaert, C Frydrych, JM Frere. Synthesis, purification, and kinetic properties of fluorescein-labeled penicillins. Biochem J 300:141–145, 1994.
111. MV Norgard, SI Baker, JD Radolf. Chemiluminescent analysis of *Borrelia burgdorferi* penicillin-binding proteins using ampicillin conjugated to digoxigenin. Microb Pathogen 19:257–272, 1995.
112. LM Weigel, JT Belisle, JD Radolf, MV Norgard. Digoxigenin-ampicillin conjugate for detection of penicillin-binding proteins by chemiluminescence. Antimicrob Agents Chemother 38:330–336, 1994.
113. S Lepage, M Galleni, B Lakaye, B Joris, I Thamm, JM Frere. Kinetic properties of the *Bacillus licheniformis* penicillin-binding proteins. Biochem J 309:49–53, 1995.

114. JJ Burbaum, NH Sigal. New technologies for high-throughput screening. Curr Opin Chem Biol 1:72–78, 1997.
115. PB Fernandes. Technological advances in high-throughput screening. Curr Opin Chem Biol 2:597–603, 1998.
116. L Silverman, R Campbell, JR Broach. New assay technologies for high-throughput screening. Curr Opin Chem Biol 2:397–403, 1998.
117. GS Sittampalam, SD Kahl, WP Janzen. High-throughput screening: advances in assay technologies. Curr Opin Chem Biol 1:384–391, 1997.
118. JH Toney, GG Hammond, B Leiting, KD Pryor, JK Wu, GC Cuca, DL Pompliano. Soluble penicillin-binding protein 2a: β-lactam binding and inhibition by non-β-lactams using a 96-well format. Anal Biochem 255:113–119, 1998.
119. S Roychoudhury, RE Kaiser, DN Brems, WK Yeh. Specific interaction between β-lactams and soluble penicillin-binding protein 2a from methicillin-resistant *Staphylococcus aureus*: development of a chromogenic assay. Antimicrob Agent Chemother 40:2075–2079, 1996.
120. J Grandchamps, M Nguyen-Disteche, C Damblon, JM Frere. *Streptomyces* K15 active-site serine DD-transpeptidase: specificity profile for peptide, thiol ester carbonyl donors and pathways of the transfer reactions. Biochem J 307:335–339, 1995.
121. JA Kelly, SG Waley, M Adam, JM Frere. Crystalline enzyme kinetics: activity of the *Streptomyces* R61 D-alanyl-D-alanine peptidase. Biochem Biophys Acta 1119: 256–260, 1992.

12

Development of a High-Throughput Screen for *Streptococcus pneumoniae* UDP-N-Acetylmuramoyl-Alanine: D-Glutamate Ligase (MurD) for the Identification of MurD Inhibitors

Michele C. Smith, James A. Cook, Gary M. Birch, Stephen A. Hitchcock, Robert B. Peery, Joann Hoskins, Paul L. Skatrud, Raymond C. Yao, and Karen L. Cox
Eli Lilly and Company, Indianapolis, Indiana

I. INTRODUCTION

The need for new antibiotics is driven by the recent rise in the incidence of resistance to commonly used antibiotics. The emergence of multiple-drug resistance to community-acquired infections, such as those caused by *Streptococcus pneumoniae*, is particularly alarming due to the ease of transmission [1–4]. Recent reports show that methicillin-resistant *Staphylococcus aureus*, the common cause of hospital-acquired infections, has also moved into the community [5].

An approach to combat this situation is to discover and develop novel antibiotics with new mechanisms of action. Analysis of the recent genomic data from bacteria estimates that, of the thousands of genes in any given genome, only 250–300 genes are essential for bacterial life [6–8]. This minimal gene set includes approximately 25 genes whose products are inhibited by the current arsenal of antibiotics. The potential for the discovery and development of new antibiotics is staggering, since only 10% of the possible targets have been exploited.

Cell wall biosynthesis is a target of many commonly used antibiotics. The unique nature of the cell wall allows specific inhibitors of this process to be used therapeutically with few side effects. Penicillins, the most commonly used antibiotics, inhibit cell wall biosynthesis by virtue of their interaction with the enzymes responsible for cross-linking peptidoglycan. Many enzymes in the cell wall biosynthetic pathway remain unexploited, including the cytosolic ligases responsible for the synthesis of the stem peptide precursor of peptidoglycan as shown in Figure 1. Four enzymes, MurC, MurD, MurE, and MurF, in this pathway carry out nonribosomal peptide synthesis and are essential for cell wall-containing bacteria [9–15].

The MurD enzyme is a particularly attractive target for drug development since one of its substrates, D-glutamic acid, is not found in mammalian cells. The D-glutamic acid-binding site may be utilized to generate an active-site inhibitor specific for MurD. A competitive inhibitor with respect to D-glutamic acid would be expected to have less toxicity because of the absence of the D configuration in the host. With this rationale in mind, we set out to develop a high-throughput screen to search for *S. pneumoniae* MurD inhibitors.

The published assays for the Mur enzymes are not amenable to high-throughput screening. Radiolabeled substrates are commonly used to follow ligase reactions where the product is separated either by HPLC [16], activated-charcoal adsorption, or some other appropriate adsorbent [9,17,18]. The separation of radiolabeled substrates from products is cumbersome and necessitates disposal of large amounts of radioactive waste from a high-throughput screen. A spectrophotometric assay, which follows the equimolar generation of ADP, has also been described [19]. ADP is a substrate for pyruvate kinase (PK), and the products from this reaction are substrates for lactate dehydrogenase (LDH). PK and LDH can be coupled to an ADP-generating reaction, such as the one catalyzed by MurD. The reaction progress can be monitored spectrophotometrically at 340 nm since LDH generates NAD from NADH, leading to a decrease in absorbance as ADP is produced. In spite of its attractiveness as a continuous assay, the PK/LDH coupled spectrophotometric assay suffers from low sensitivity and interference from natural products and some organic compounds.

We developed an alternative assay format using a coupled enzyme assay with MurD as the rate-limiting enzyme reaction and MurE and MurF as the coupling enzymes. The final product formed, stem peptide or UDP-MurNAc-pentapeptide, contained a pentapeptide that was found to be immunogenic, thus allowing the product to be detected in an ELISA. The rationale for using a coupled enzyme assay to generate stem peptide arose from the fact that the minimum epitope size for antibody recognition in proteins is six amino acids [19]. We reasoned that the pentapeptide might be immunogenic, since it was close in length to the minimum of six amino acids and it contained three amino acids in the D configuration as well as an unusual isoglutamic acid linkage to lysine. We present here the results of developing a robust high-throughput MurD coupled enzyme assay with ELISA detection. The methodology presented here can be adapted to

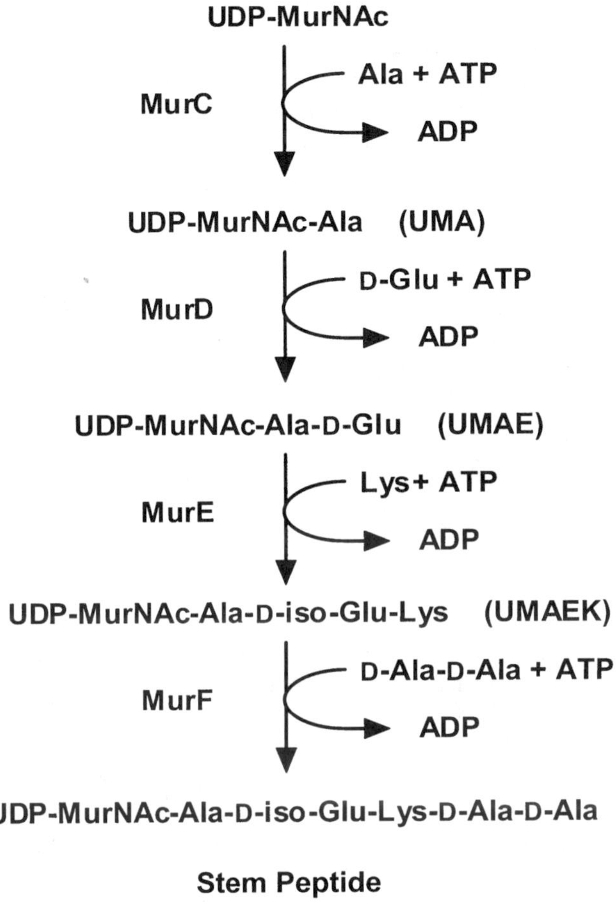

Figure 1 Stem peptide biosynthetic pathway.

other enzyme systems to produce small-molecular-weight products that can be detected using ELISA.

II. ENZYME ASSAY REAGENTS

The substrate for MurD (UDP-MurNAc-L-Ala) and purified Mur enzymes were required to develop a coupled enzyme assay. None of the Mur enzymes or their substrates were available commercially, although purification procedures for the enzymes are reported as well as the enzymatic preparation of the MurD substrate, UDP-MurNAc-L-Ala (UMA) [9,18,20–22]. The other reagents for the coupled enzyme assay were all commercially available. Since our ultimate goal was to develop an antibacterial agent to combat *S. pneumoniae*, we cloned recombinant MurD and MurE from *S. pneumoniae*. Both enzymes contain a D-glutamic acid-binding site, the target we hoped to exploit. MurF, required as the last step in the coupling assay, was cloned from *Escherichia coli*. All enzymes were expressed and purified from *E. coli*.

Biosynthetic methods to produce the MurD substrate, UMA, are not suitable for making the large quantities required for a high-throughput screen. Methods for chemically synthesizing the substrate were therefore developed and scaled up so as to provide the multigram quantities required for the screen and its development.

A. Cloning of MurD, MurE, and MurF

The *mur*D and *mur*E genes from *S. pneumoniae (hex)* R6, as well as the *mur*F gene from *E. coli*, were amplified by PCR with primers containing convenient restriction sites for cloning into *E. coli* expression vectors. Upstream PCR primers to amplify these genes were designed around the ATG initiation codon, with *Bam*HI or *Eco*RI and *Nde*I restriction sites included for cloning purposes. These primers also contained sequences encoding a His_8 chelating peptide or purification tag, followed by a Factor Xa cleavage site (IEGR) for removal of the His tag following purification. The downstream primers for all three genes spanned the stop codon and incorporated a *Bam*HI restriction site into the resulting fragment.

Primer pairs used for the amplification of the *S. pneumoniae mur*D gene, PS225 and PS226 [23], were designed using the patented *mur*D gene sequence [24,25]. Primers SPMUREEXP-1 and SPMUREEXP-2 [26] used for the amplification of the *S. pneumoniae mur*E gene, were designed using the patented *mur*E gene sequence [27,28]. The primer pair PS214 and PS215 [29] was designed from the published *E. coli mur*F sequence (Accession X15432, GenBank) using the ATG of the presynthetase enzyme.

Standard PCR reaction conditions were used with either *S. pneumoniae (hex)* R6 genomic DNA or genomic DNA isolated from *E. coli* K-12 as the template and AmpliTaq DNA polymerase (Perkin Elmer) for the enzyme. Resulting PCR products were washed, digested with *Nde*I and *Bam*HI and ligated to pET11a (Novagen). Clones containing an insert of the correct size were sequenced in their entirety to verify that no errors were introduced by PCR. DNA sequencing was performed using PE-ABI Prism Dye Terminator Cycle Sequencing Ready Reaction fluorescent-based chemistry. Sequence data were collected on an ABI377 and analyzed using PE-ABI Sequence Analysis v. 3.0 software. Data were edited using Sequencher v. 3.0.

B. Expression of *E. coli* MurF and *S. pneumoniae* MurD and MurE

All proteins were expressed in *E. coli* using identical conditions. The expression plasmids were transformed into BL21DE3 cells (Novagen). An isolated colony was used to inoculate 200 ml of TY broth with 100 μg/ml ampicillin. This culture was grown overnight at 30°C, with aeration. In the morning, the overnight culture was diluted 1/50 into 2 liters of fresh TY broth containing 100 μg/ml ampicillin. The culture was shifted to 37°C and the OD_{600} was monitored. At $OD_{600} = 0.5–1.0$, expression from the T7 promoter was induced by adding IPTG to a final concentration of 0.4 mM. After 3 hr, the cells were harvested by centrifugation at 12,000 × *g* for 10 min.

C. Purification of MurD, MurE, and MurF

High expression levels of soluble and active Mur enzymes in *E. coli* enabled the rapid purification of large amounts of enzyme for the high-throughput screen. Approximately 120 mg of purified Mur enzymes could be purified from each liter of expression culture.

E. coli expressing His-tagged Mur enzymes were lysed with lysozyme and sonication and the proteins purified by Ni^{2+}-IMAC. Typically, 4.0 ml (32 mg total protein) of *E. coli* supernatant was injected onto a 1.0 × 10 cm column of ToyoPearl AF-Chelate 650M equilibrated with 50 mM sodium phosphate, 1.0 M NaCl, pH 7.2 (buffer A). The column was eluted with a linear gradient of buffer A to 100% 50 mM sodium phosphate, 1.0 M NaCl, and 500 mM imidazole, pH 7.2 (buffer B) over 12 column volumes. Each of the Mur enzymes eluted from the column with 35% buffer B and was pooled and dialyzed against 50 mM Tris, pH 8.1, at 4°C. The dialyzed protein was collected and stored at −20°C in 10% glycerol. MurE and MurF also contained 1 mM DTT to keep the protein reduced. All three enzymes were characterized by amino acid analysis (Beckman 6300), N-terminal sequencing (Perkin-Elmer Procise N-terminal amino acid sequencer), and mass spectrometry (Perkin-Elmer API III Biomolecular mass analyzer). A gel of the purified proteins is shown in Figure 2.

D. MurD Substrate

UDP-MurNAc-L-Ala (UMA) and UDP-MurNAc-L-Ala-D-Glu (UMAE) were synthesized as described for the stem peptide [30]. The synthesis of UMA was

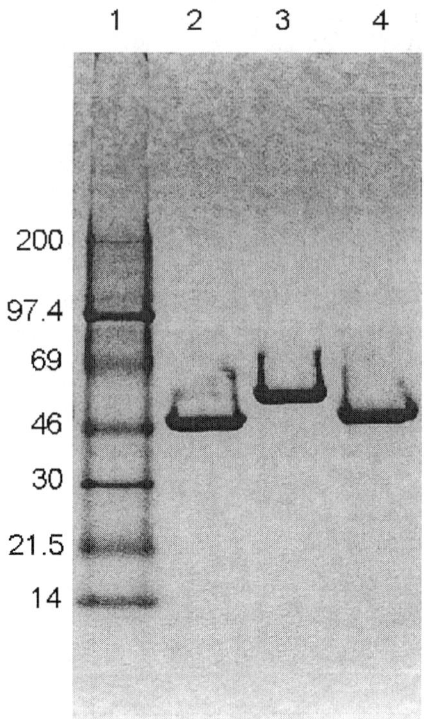

Figure 2 SDS-PAGE of purified Mur enzymes used in the coupled enzyme assay. A Novex 10–20% Tricine gel was used and stained with Coomassie blue. Lane 1 contains the molecular-weight standards (kDa); lanes 2–4 contain 3.5 μg of MurD, MurE, and MurF, respectively.

conducted on a multigram scale to afford enough substrate to run a high-throughput screen or 600,000 assays.

III. ENZYME ASSAYS

Since MurD was the primary target, it was made the rate-limiting step in the coupled enzyme assay and conditions were optimized for that catalytic reaction. The kinetic parameters were determined for MurD. The K_m for UMA and D-Glu guided the choice of substrate concentrations for the MurD reaction so the enzyme was optimally efficient [31]. The kinetic parameters for the amino acid or dipeptide and tripeptide substrates for MurE and MurF were then determined under the optimal MurD conditions. We developed a sensitive HPLC assay system to follow each separate reaction and measure the amount of each intermediate in the coupled enzyme reactions. Measuring the amount of each intermediate verified that MurD was controlling the rate of product formation during the optimization of each coupled enzyme reaction. This empirical approach made efficient use of the enzymes and substrates while maintaining a kinetically robust assay for MurD.

A. HPLC Assay

The progression of the reaction catalyzed by MurD was monitored by HPLC. The reaction was quenched at various time points by adding an equal volume of 2% acetic acid and UMA and UMAE were separated by HPLC using a MetaChem Inertsil ODS-3 5 μm 4.6 × 250 mm column at a flow rate of 1.5 ml/min with detection at 263 nm. Buffer A was 50 mM NaH_2PO_4, pH 4.0, and buffer B was 9 volumes of buffer A and 1 volume of acetonitrile. The gradient used to separate all the UDP-MurNAc peptides consisted of a 3.3-min wash with 10% B after the sample was injected, followed by a 10-min gradient to 30% B, and then a sharper 13-min gradient to 90% B. The column was washed for 14 min with 90% B, followed by a 15-min equilibration with 10% B before injecting the next sample. UMA eluted at ~9 min, UMAE eluted at ~11 min, UDP-MurNAc-Ala-D-iso-Glu-Lys (UMAEK) at ~8 min, and stem peptide (UM-PP) at ~12 min. The HPLC assay was also used to assess the stability of the enzymes, substrates, and products as a function of time and storage conditions.

B. Optimization of the MurD Reaction

The MurD reaction conditions were optimized at 22°C with 50 μM UMA by varying the other reaction components using experimental design software by JMP (SAS Institute, Cary, NC). The pH and concentrations of the following components were varied within the ranges given: pH (7.0–8.8), Tris HCl (25–100 mM), $MgCl_2$ (3–50 μM), D-glutamate (50–510 μM), ATP (0.5–1.5 mM), BSA (0–100 μg/ml), and MurD (2.5–1000 ng/ml). Only $MgCl_2$, D-glutamate, and enzyme concentrations affected the reaction rate. The initial velocities (10–20% product formation calculated by integrating both substrate and product peaks) were measured at various enzyme concentrations and found to be linear up to 5 ng/ml MurD. Initial velocity conditions were achieved with 3 ng/ml MurD for a 2-hr duration. A fixed-time assay of 2-hr duration accommodated the plate rotation times necessary for a high-throughput automated screen. The optimized conditions for the assay were 50 mM Tris, pH 7.7, 20 mM $MgCl_2$, 1.0 mM ATP, 400 μM D-glutamic acid, 50 μM UMA, 3 ng/ml MurD, and 100 μg/ml BSA. The HPLC system and an ADP coupling assay [19] were used to measure the K_m of MurD for UMA and D-Glu, which were found to be 30 and 105 μM, respectively (Fig. 3).

C. Coupled Enzyme Assay

The next step was to couple the MurD reaction to MurE and then MurF so as to produce stem peptide with a rate of formation dependent on MurD concentra-

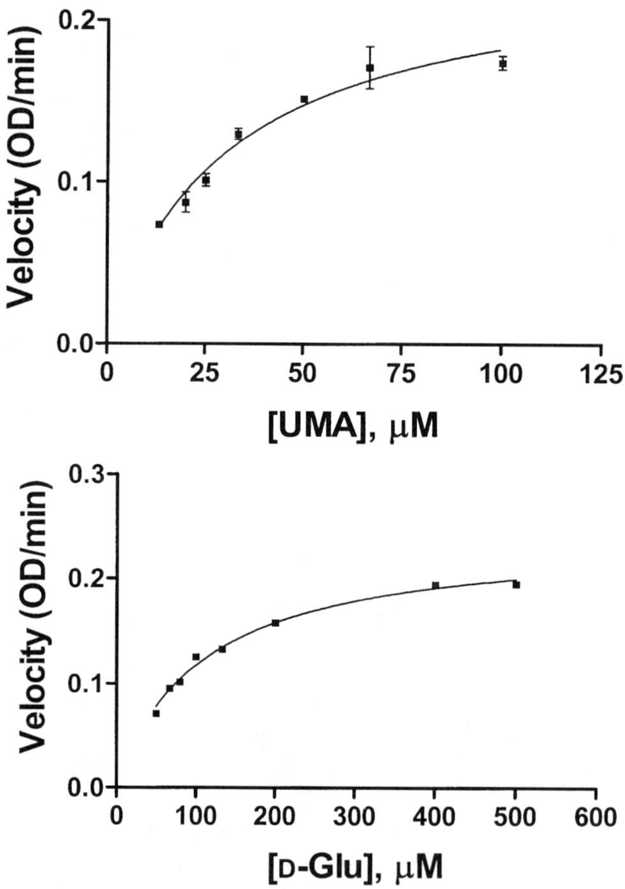

Figure 3 Determination of *S. pneumoniae* MurD kinetic parameters. Initial velocities as a function of substrate concentration were measured using the ADP coupled enzyme assay with PK and LDH. Data were fit with a nonlinear regression analysis using GraphPad Prism. Top panel is for UMA and bottom panel is for D-Glu.

tion. The reactions were followed using the HPLC system described above. MurD conditions were used to determine the saturating lysine concentration for the MurE reaction under initial velocity conditions for a 2-hr period. The presence of a reducing agent, such as DTT, was also examined, since MurE contains two cysteines, but no effect was observed on the rate of the reaction. With the optimal lysine concentration determined for the MurE reaction (2 mM), a coupled assay with MurD and MurE was then performed. The MurD conditions with 2 mM lysine were used and the MurE concentration was varied to give an initial velocity of UMAEK production equivalent to that of generating UMAE when MurD was used alone. The intermediate MurD product was absent in the HPLC, proving that MurE was in excess.

The MurF concentration required to drive the production of UM-PP, at the same initial velocities as UMAE when MurD was used alone, was determined in the presence of varied D-Ala-D-Ala concentrations, with MurD and E under the optimized conditions. The minimal D-Ala-D-Ala concentration (50 μM) required to saturate the system was chosen so as to give the same initial velocities.

The optimized MurD coupled enzyme reaction contained 50 mM Tris, pH 7.7, 20 mM $MgCl_2$, 1 mM ATP, 100 μg/ml BSA, 400 μM D-glutamic acid, 2 mM lysine, 50 μM D-Ala-D-Ala, 50 μM UMA, 3 ng/ml MurD, 15 ng/ml MurE, and 100 ng/ml MurF in 100 μL at 22°C. The reaction was started with the addition of UMA.

IV. ELISA DETECTION OF THE COUPLED ENZYME REACTION

A. ELISA Reagents

The first step in developing an ELISA to detect the stem peptide product of the coupled enzyme reaction was to determine if antibodies to the stem peptide could be raised in animals. Only small quantities of stem peptide could be isolated or prepared enzymatically, so we used the commercially available pentapeptide portion of the stem peptide as the hapten in the hopes of generating high-affinity antibodies that would cross-react with the stem peptide.

Pentapeptide (Ala-D-iso-Glu-Lys-D-Ala-D-Ala) was conjugated to Keyhole Limpet hemocyanin (both from Sigma) at a 220:1 ratio using glutaraldehyde and was used to immunize rabbits (Hazelton Labs). A second conjugate, BSA-pentapeptide, was prepared for use in measuring the antibody titer as well as in the ELISA, so as to eliminate any cross-reactivity with anti-hemocyanin antibodies. BSA-pentapeptide was prepared in two steps by reacting BSA (Intergen Bovuminar Cohn Fraction V) with glutaraldehyde followed by dialysis to remove residual glutaraldehyde. Glutaraldehyde-activated BSA was reacted with 15 equivalents of a blocked form of pentapeptide (Ala-D-iso-Glu-Lys(N-ε-OCF3)-

D-Ala-D-Ala) [32] followed by removal of the lysine-protecting group with three equivalents of LiOH in THF/water at pH 11 for 2 hr at room temperature, followed by neutralization to pH 7.0. This method ensured that pentapeptide was linked via the N-terminus, thus exposing the maximum amount of peptide for antibody recognition. The antibody produced by the rabbits was affinity purified using pentapeptide as a ligand, and then stored at 1 mg/ml in PBS with 10% glycerol and sodium azide at −20°C.

B. ELISA

A competitive ELISA was developed to detect stem peptide generated from the coupled enzyme reaction as well as to measure the cross-reactivity of the antibody with small peptides. The ELISA format is pictured in Figure 4. BSA-pentapeptide (BSA-PP) was immobilized onto the wall of a microtiter plate. Samples or standards containing pentapeptide were then added, followed by rabbit anti-pentapeptide antibody. The amount of anti-pentapeptide antibody bound to immobilized BSA-PP was visualized by adding a second anti-rabbit IgG antibody

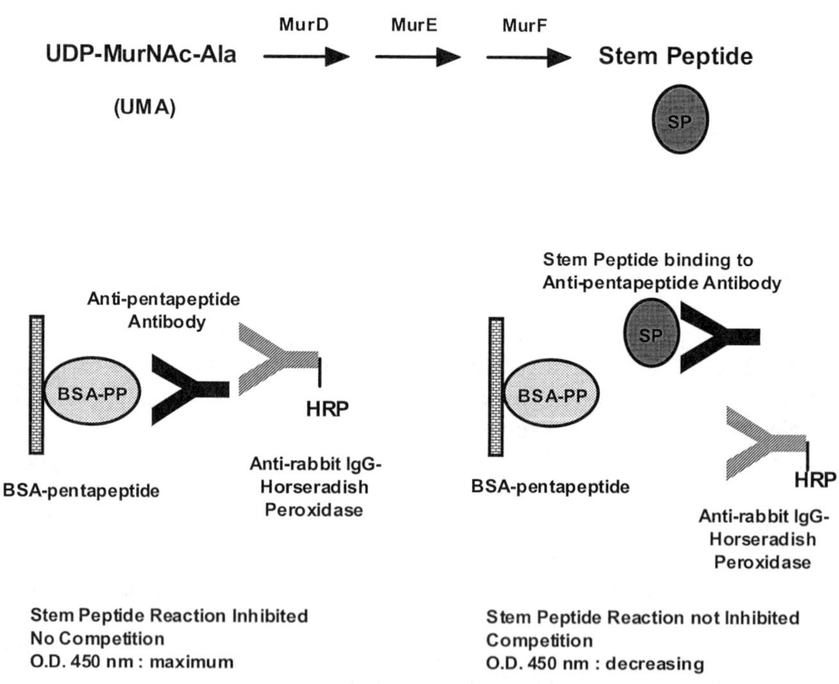

Figure 4 Schematic of ELISA for stem peptide detection.

labeled with horseradish peroxidase. In the absence of free pentapeptide, maximum binding of the antibody was observed and gave a maximum OD reading after adding the peroxidase substrate. Molecules reactive with the antibody, such as stem peptide or pentapeptide fragments, bound to the anti-pentapeptide antibody and prevented it from binding to the immobilized BSA-PP, which resulted in a lower absorbance. The decrease in absorbance was proportional to the increase in the concentration of the competing species.

Figure 5 shows the results of a competitive ELISA with Ala-D-iso-Glu-Lys-D-Ala-D-Ala (pentapeptide), Lys-D-Ala-D-Ala, Ala-D-Ala-D-Ala, Ala-D-iso-Glu, Ala-D-iso-Glu-Lys, D-Ala-D-Ala, and Lys-D-Ala-D-Lac. The antibody reacted strongly with pentapeptide and Lys-D-Ala-D-Ala but weakly with Ala-D-Ala-D-Ala, suggesting that the immunogenic determinant of the pentapeptide to this antibody resides in the last three residues, Lys-D-Ala-D-Ala. Figure 6 shows that the cross-reactivities of the pentapeptide and stem peptide in the competitive ELISA were virtually identical.

The MurD, E, and F coupled reaction was performed with and without enzymes and the amount of stem peptide generated was measured by HPLC and ELISA. The amount of product measured by both methods was the same.

These results validated the ELISA detection method for stem peptide. In addition, the other components of the enzyme reaction mixture were tested in the ELISA, and none cross-reacted with the antibody at concentrations used in the assay.

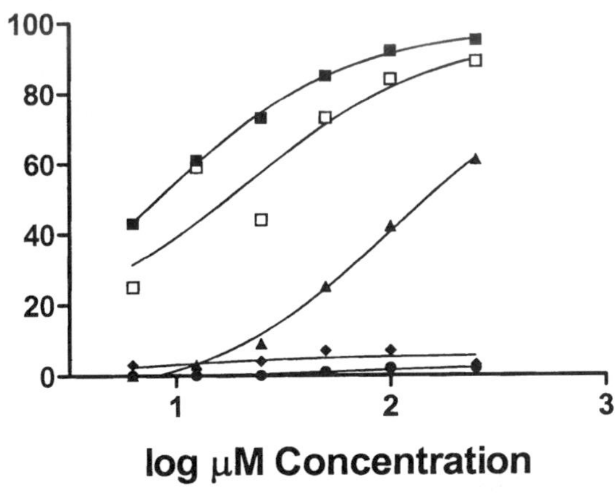

Figure 5 Reactivity of the anti-pentapeptide antibody. A competitive ELISA was carried out as described in the text, and the percent displacement by Lys-D-Ala-D-Ala (■), D-Ala-D-Ala (●), pentapeptide (□), D-Ala-D-Ala-D-Ala (▲), and Lys-D-Ala-D-Lac (◆) is shown.

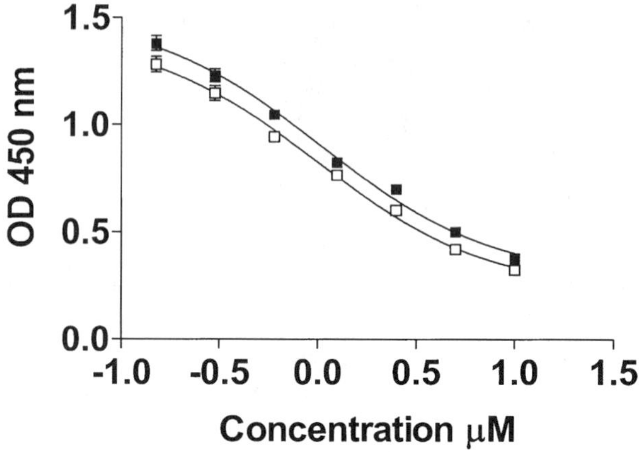

Figure 6 Reactivity of the anti-pentapeptide antibody with stem peptide and pentapeptide. A competitive ELISA was carried out as described in the text, and the OD at 450 nm is shown for stem peptide (■) and pentapeptide (□).

V. HIGH-THROUGHPUT SCREEN

The two-part protocol for the high-throughput screen is depicted in Figure 7. The first part involved the use of Titertek's Multidrop liquid dispenser to automate the addition of reagents for the coupled enzyme assay and quenching the reaction. Samples from the coupled enzyme assay were then transferred to a second liquid-dispensing system (MRD8) for running the ELISA to detect the product of the coupled enzyme assay. To ensure a high-quality run through the screen, stability studies were performed on the reagents and products used in the screen. Variability studies were performed using robots.

A. High-Throughput Screen

Compounds for testing were dissolved in 10% DMSO and diluted 1/20 in water to give a final stock concentration of 50 μM. A 20-μl aliquot of the compound stock was added to each well in a 96-well plate, followed by 50 μl of the enzyme solution. The reaction was started by adding 30 μl of 166.67 μM UMA to all wells. The final reaction well contained 50 mM Tris, pH 7.7, 20 mM $MgCl_2$, 1.0 mM ATP, 0.4 mM D-Glu, 2.0 mM Lys, 0.05 mM D-Ala-D-Ala, 3 ng/ml MurD, 15 ng/ml MurE, 100 ng/ml MurF, and 100 μg/ml BSA. The 96-well plates were either stacked or covered and incubated for 2 hr at room temperature. The reaction was stopped by adding 20 μl of 250 mM EDTA followed by mixing.

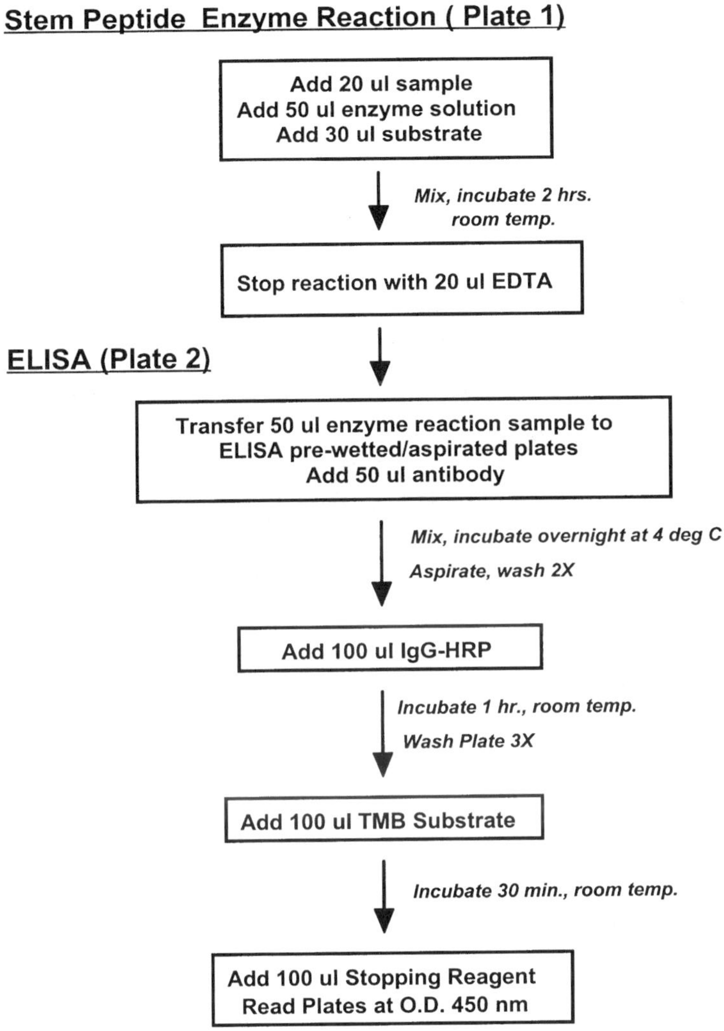

Figure 7 Flow chart of the MurD ELISA-based high-throughput screen.

A 50-μl aliquot of the reaction was transferred to the ELISA plate, which had been prepared earlier using the following protocol. Each well of a plate (# 3590, Costar) was coated with 100 μl of 250 ng/ml BSA-pentapeptide in PBS overnight at 4°C. The plates were then washed three times with PBS-0.05% Tween 20 using 120 μl/well. The plates were blocked with 120 μl/well of Pierce

Superblock for 3 hr at room temperature and then aspirated and dried in a vacuum oven at room temperature for 3 hr. The plates were stored in plastic bags containing desiccant at 4°C. Plates were soaked with 120 μl/well of PBS-0.05% Tween 20, 1% BSA in 0.02% Thimerosal, and then aspirated before use.

After adding the quenched enzyme reaction sample to the ELISA plate, 50 μl of a 25-ng/ml solution of affinity-purified pentapeptide antibody was added to each well, and the plates were allowed to incubate overnight at 4°C. Liquid was aspirated from the wells, followed by two 120-μl washes with PBS-0.05% Tween 20. After washing, each well received 100 μl of anti-rabbit IgG conjugated with horseradish peroxidase (American Qualex) at a dilution, usually 1/10,000, sufficient to give an OD reading between 1.5 and 2.0 on plates reacted with antibody alone. The second antibody was incubated on the plate for 1 hr, washed three times with 120 μl of PBS-0.05% Tween 20, followed by TMB (3,3′,5,5′-tetramethylbenzidine) peroxidase substrate addition (100 μl of a 1:1 solution of TMB and peroxidase according to the protocol of Kirkegaard & Perry Laboratories, Gaithersburg, Maryland). After a 30-min incubation, 100 μl of 1 M phosphoric acid was added to stop the peroxidase reaction. The plates were read at 450 nm.

B. Reagent and Product Stability Studies

Stability studies on all the enzyme reaction components showed that the reagents were stable over an 8-hr period at room temperature and for months when stored at −20°C. Since the stem peptide enzyme reaction was performed one day and the ELISA the following day, the stability of the stem peptide product was measured using both HPLC and ELISA. There was no detectable change in the amount of stem peptide when measured immediately or when the enzyme reaction was stored overnight at 4°C. The anti-rabbit IgG-HRP conjugate, purchased from American Qualex, showed some instability after a week of using material from an opened vial. To circumvent this stability issue, a new vial was used each week after titering it to an OD of 1.5–2.0. The antibody was stable, as measured by its reactivity to pentapeptide, for at least 5 months when stored at 4°C in PBS with 0.02% sodium azide.

The effect of solvents such as DMSO, ethanol, and methanol, on the MurD coupled enzyme reaction and the ELISA was studied. The MurD enzyme reaction was run in the presence and absence of Mur enzymes to generate minimum and maximum signals. Final solvent concentrations up to 2% had no effect on the reaction when measured by either HPLC or in the ELISA.

C. Performance Validation

Multidrop (Titertek Instruments; Huntsville, Alabama) liquid dispensers were used to add the enzyme solution to the compound plate, followed by the substrate,

UMA, to start the reaction. The reaction was stopped by adding EDTA using Titertek's Multidrop liquid dispenser. The ELISA was automated using an eight-head MRD8 (Titertek Instruments, Huntsville, AL). The instrument is designed specifically for ELISAs and allows washes, additions, and programmed timing of the various steps. Both the MRD8 and the Multidrop were calibrated each day prior to running the assay. This was done by dispensing 100 μl of wash containing a yellow dye and reading at 450 nm. If the coefficient of variation on either instrument was greater than 1.5%, then the instrument was calibrated and the validation repeated.

A two-plate, 3-day study was performed to measure the inter- and intraplate variation at the minimum, maximum, and 50% signals. The minimum signal was generated by letting the enzyme reaction proceed without inhibition so as to produce free stem peptide, which would compete with immobilized BSA-PP for antibody binding. The maximum signal was produced by omitting the enzymes from the MurD reaction so no stem peptide was produced, so as to give maximum antibody binding to immobilized BSA-PP. The 50% signal was made by adding 20 μl/well of 3 μM pentapeptide in the absence of Mur enzymes, to provide enough competing peptide to decrease antibody binding to the plate. Each plate had CVs of less than 5% with one or two exceptions where a plate contained a single high-value outlier. The CVs for the same signal on all plates on a single day were less than 10%, with the higher values arising from the outliers. The CVs for the means over the 3-day period were 10.05% for the minimum signal, 7.78% for the middle signal, and 4.45% for the maximum signal, which satisfied the criteria of less than 20% variation between days. No significant edge effects or drifts in signal across the plate were observed.

VI. EVALUATION OF MurD INHIBITORS

Hits from the screen were used to select related compounds for further testing from the Lilly collection of compounds [33]. The HPLC assay was used to measure MurD inhibition and standard antimicrobial assays were carried out with *S. pneumoniae*, *Haemophilus influenzae* 76, *Staphylococcus aureus* 027, and *Moraxella catarrhalis* BC-1. A series of indole and phenoxypropyl amine derivatives with activity against MurD, were discovered that exhibited broad-spectrum antibacterial activity.

VII. CONCLUSION

A high-throughput screen for *S. pneumoniae* MurD was developed that did not require radioactivity. The high expression levels of soluble active Mur enzymes in *E. coli* allowed large quantities of enzymes to be purified rapidly. The chemical synthesis of multigram quantities of the MurD substrate made the development and implementation of the screen possible. Assay conditions were optimized for the detection of MurD inhibitors, and the optimized assay was validated in high-throughput format. Stem peptide produced in the assay was dependent on MurD activity because MurE and MurF were present in excess. The immunogenicity of the pentapeptide portion of stem peptide allowed the stem peptide product of the coupled enzyme reaction to be detected using a competitive ELISA. A high-throughput screen employing this assay was used to screen a large library of compounds and natural products. The screen was reproducible and efficiently identified inhibitors of MurD. Inhibitors were verified with the HPLC assay and then tested for antibacterial activity with *S. pneumoniae*. Many of the inhibitors identified by the screen also exhibited antibacterial activity.

REFERENCES

1. DWM Crook, BG Spratt. Multiple antibiotic resistance in *Streptococcus pneumoniae*. Br Med Bull 54:595–610, 1998.
2. GV Doern, MA Pfaller, K Kugler, J Freeman, RN Jones. Prevalence of antimicrobial resistance among respiratory tract isolates of *Streptococcus pneumoniae* in North America: 1997 results from the SENTRY antimicrobial surveillance program. Clin Infect Dis 27:764–770, 1998.
3. AE Simor, M Louie, TCBS Network, DE Low. Canadian national survey of prevalence of antimicrobial resistance among clinical isolates of *Streptococcus pneumoniae*. Antimicrob Agents Chemother 40:2190–2193, 1996.
4. C Thornsberry, PT Ogilvie, HP Holley Jr, DF Sahm. Survey of susceptibilities of *Streptococcus pneumoniae*, *Haemophilus influenzae*, and *Moraxella catarrhalis* isolates to 26 antimicrobial agents: a prospective U.S. study. Antimicrobial Agents Chemother 43:2612–2623, 1999.
5. Centers for Disease Control and Prevention. Four pediatric deaths from community-acquired methicillin-resistant *Staphylococcus aureus*—Minnesota and North Dakota, 1997–1999. Morb Mortal Wkly Rep 48:707–710, 1999.
6. EV Koonin, AR Mushegian, KE Rudd. Sequencing and analysis of bacterial genomes. Curr Biol 6:404–416, 1996.
7. AR Mushegian, EV Koonin. A minimal gene set for cellular life derived by comparison of complete bacterial genomes. Proc Natl Acad Sci (USA) 93:10268–10273, 1996.
8. CM Fraser, JD Gocayne, O White, MD Adams, RA Clayton, RD Fleishmann, CJ Bult, AR Kerlavage, G Sutton, JM Kelley, JL Fritchman, JF Weidman, KV Small, M Sandusky, J Fuhrmann, D Nguyen, TR Utterback, DM Saudek, CA Phillips, JM Merrick, J-F Tomb, BA Dougherty, KF Bott, P-C Hu, TS Lucier, SN Peterson, HO Smith, CA Hutchison, JC Venter. The minimal gene complement of *Mycoplasma genitalium*. Science 270:397–403, 1995.
9. J van Heijenoort. Assembly of the monomer unit of bacterial peptidoglycan. Cell Mol Life Sci 54:300–304, 1998.

10. TD Bugg, CT Walsh. Intracellular steps of bacterial cell wall peptidoglycan biosynthesis: enzymology, antibiotics, and antibiotic resistance. Nat Prod Rep 9:199–215, 1992.
11. RH Baltz, FH Norris, P Matsushima, BS DeHoff, P Rockey, G Porter, S Burgett, R Peery, J Hoskins, L Braverman, I Jenkins, P Solenberg, M Young, MA McHenney, PL Skatrud, JPR Rosteck. DNA sequence sampling of the *Streptococcus pneumoniae* genome to identify novel targets for antibiotic development. Microb Drug Resist 4: 1–9, 1998.
12. EJJ Lugtenberg, L de Haas-Menger, WHM Ruyters. Murein synthesis and identification of cell wall precursors of temperature-sensitive lysis mutants of *Escherichia coli*. J Bacteriol 109:326–335, 1972.
13. EJJ Lugtenberg, A van Schijndel-van Dam. Temperature-sensitive mutants of *Escherichia coli* K-12 with low activities of the L-alanine adding enzyme and the D-alanyl-D-alanine adding enzyme. J Bacteriol 110:35–40, 1972.
14. EJJ Lugtenberg, A van Schijndel-van Dam. Temperature-sensitive mutant of *Escherichia coli* K-12 with an impaired D-alanine: D-alanine ligase. J Bacteriol 113:96–104, 1973.
15. D Mengin-Lecreulx, B Flouret, J van Heijenoort. Cytoplasmic steps of peptidoglycan synthesis in *Escherichia coli*. J Bacteriol 151:1109–1117, 1982.
16. The abbreviations used are as follows: UDP-MurNAc, uridine diphosphate N-acetylmuramic acid; MurC, UDP-N-acetyl-muramoyl:L-alanine ligase; MurD, UDP-N-acetylmuramoyl-alanine:D-glutamate ligase; MurE, UDP-N-acetylmuramoyl-alanine-D-glutamate:lysine ligase; MurF, UDP-N-acetylmuramoyl-alanine-D-iso-glutamate-lysine:D-alanine-D-alanine ligase; UMA, UDP-N-acetylmuramoyl-alanine; UMAE, UDP-N-acetylmuramoyl-alanine-D-glutamate; UMAEK, UDP-N-acetylmuramoyl-alanine-D-iso-glutamate-lysine; HPLC, high-performance liquid chromatography; ATP, adenosine triphosphate; ADP, adenosine diphosphate; PK, pyruvate kinase; LDH, lactate dehydrogenase; NAD, nicotinamide adenine dinucleotide; NADH, reduced nicotinamide adenine dinucleotide; ELISA, enzyme-linked immunosorbent assay; PCR, polymerase chain reaction; DNA, deoxyribonucleic acid; TY, trypton yeast extract; OD, optical density; IPTG, isopropyl thiogalactopyranoside; IMAC, immobilized metal-ion affinity chromatography; Tris, tris(hydroxymethyl) aminomethane; DTT, dithiothreitol; BSA, bovine serum albumin; PBS, phosphate-buffered saline; IgG immunoglobulin G; DMSO, dimethylsulfoxide; EDTA, ethylenediaminetetraacetic acid; TMB, 3,3′,5,5′-tetramethylbenzidine; HRP, horseradish peroxidase.
17. B Flouret, D Mengin-Lecreulx, J van Heijenoort. Reverse-phase high-pressure liquid chromatography of uridine diphosphate N-acetylmuramyl peptide precursors of bacterial cell wall peptidoglycan. Anal Biochem 114:59–63, 1981.
18. KK Wong, DW Kuo, RM Chabin, C Fournier, LD Gegnas, ST Waddell, F Marsilio, B Leiting, DL Pompliano. Engineering a cell-free murein biosynthetic pathway: combinatorial enzymology in drug discovery. J Am Chem Soc 120:13527–13528, 1998.
19. E Daub, LE Zawadzke, D Botstein, CT Walsh. Isolation, cloning, and sequencing of the *Salmonella typhimurium* dd1A gene with purification and characterization of its product, D-alanine:D-alanine ligase (ADP forming). Biochemistry 27:3701–3708, 1988.

20. F Pratviel-Sosa, D Mengin-Lecreulx, J van Heijenoort. Over-production, purification and properties of the uridine diphosphate N-acetylmuramoyl-L-alanine:D-glutamate ligase from *Escherichia coli*. Eur J Biochem 202:1169–1176, 1991.
21. G Auger, L Martin, J Bertrand, P Ferrari, E Fanchon, S Vaganay, Y Petillot, J van Heijenoort, D Blanot, O Dideberg. Large-scale preparation, purification, and crystallization of UDP-*N*-acetylmuramoyl-L-alanine: D-glutamate ligase from *Escherichia coli*. Protein Expr Purif 13:23–29, 1998.
22. K Duncan, J van Heijenoort, CT Walsh. Purification and characterization of the D-alanyl-D-alanine-adding enzyme from *Escherichia coli*. Biochemistry 29:2379–2386, 1990.
23. The sequence for primer PS225 is 5′-AATAGTCGGAGTAGAAATGGATCCCGA CATATGCATCATCATCATCATCATCATCACATCGAAGGTCGTAAAGTAA TAGATCAATTTAAA-3′. The sequence for primer PS226 is 5′-CCTGTAAAGAC AATTTTTTGGATCCTTTATTCTTTTAACTCCGCT-3′.
24. JA Hoskins, RB Peery, PL Skatrud, C-yE Wu. Gene murD stem peptide biosynthesis enzyme of *Streptococcus* and method and kit for identification of inhibitors of MurD. U.S. Patent 5,681,694, Eli Lilly and Company, 1997.
25. J Hoskins, RB Peery, PL Skatrud. Peptidoglycan biosynthetic *murD* gene from *Strept ococcus pneumoniae*. U.S. Patent 5,834,270, Eli Lilly and Company, 1998.
26. The sequence for primer SPMUREEXP-1 is 5′-TACGACTTTCGGTTTACAATAG AACATATGCATCATCATCATCATCATCATCACATCGAAGGTCGTATTAA GATTGAAACCGTATTAG-3′. The sequence for primer SPMUREEXP-2 is 5′-AT CGGAATAAACTAAAGGATCCAATCAGATGTCACTTGAAAACAAATTGG AAC-3′.
27. RB Peery, PL Skatrud. Peptidoglycan biosynthetic *murE* protein from *Streptococcus pneumoniae*. U.S. Patent 5,712,108, Eli Lilly and Company, 1998.
28. RB Peery, PL Skatrud. Peptidoglycan biosynthetic *murE* gene from *Streptococcus pneumoniae*. U.S. Patent 5,821,096, Eli Lilly and Company, 1998.
29. The sequence for primer PS214 is 5′- CGCGTCTGCTGGGGGAATTCATATGCA TCATCATCATCATCATCATCACATCGAAGGTCGTATTAGCGTAACCCTT-3′. The sequence for primer PS215 is 5′-ACCAAATGTTCGGGGATCCAAACTA ACATG-3′.
30. SA Hitchcock, CN Eid, JA Aikins, M Zia-Ebrahimi, LC Blaszczak. The first total synthesis of bacterial cell wall precursor UDP-N-acetylmuramyl-pentapeptide (park nucleotide). J Am Chem Soc 120:1916–1917, 1998.
31. IH Segel, Enzyme kinetics. Behavior and Analysis of Rapid Equilibrium and Steady-State Enzyme Systems. New York: John Wiley, & Sons, 1975.
32. CN Eid, MJ Nesler, M Zia-Ebrahimi, CYE Wu, R Yao, K Cox, J Richardson. Synthesis of a radioiodinated park nucleotide analog: a new tool for antibacterial screen development. Labelled Compounds Radiopharmaceut XLI:705–716, 1998.
33. GM Birch, CYE Wu, DL Letourneau, SK Sigmund, JH Wikel, RF Bruns, NG Halligan, F Victor, MA Tebbe, CJ Boylan, JW Foley, TR Parr Jr, JA Cook, MC Smith. Antibacterial activity and *Streptococcus pneumoniae (hex)* R6 MurD enzyme inhibition by indole and phenoxy propyl amine derivatives (abstr. #1277). ICAAC, San Francisco, 1999.

13
Purification and Assay Development for Human Rhinovirus Proteases

Q. May Wang and Robert B. Johnson
Eli Lilly and Company, Indianapolis, Indiana

I. INTRODUCTION

As the major cause of the common cold in humans, human rhinoviruses (HRV) are a group of plus-strand RNA viruses belonging to the picornavirus family. More than a hundred HRV serotypes have been identified to date, which may be responsible for over 50% of common colds in adults and children [1–4]. The genome of HRVs is around 7.2 kb in length and contains a single open reading frame which can be translated into a ~250-kDa viral polyprotein precursor in infected cells [2,3] (Fig. 1). Subsequent processing of this precursor protein is achieved by two virally encoded proteases, designated 2A and 3C [1,2]. The first cleavage of the precursor protein is performed by the viral 2A protease as a co-translational event [2,3]. This cleavage, taking place at the junction of the capsid protein and the N-terminus of 2A itself as seen in Fig. 1, separates the structural proteins from the nonstructural ones [2,3]. Most of the remaining cleavages are governed by the 3C protease or its precursor 3CD protein [2,3]. Although the order of the 3C cleavage cascade has not been clearly delineated, the activity of this protease eventually results in the generation of mature viral proteins and enzymes required for the completion of the viral replication cycle (Fig. 1). Since the generation of mature viral proteins and functional enzymes is essential for viral replication and life cycle, both HRV 2A and 3C protease are viewed as important targets for antiviral intervention.

On the basis of sequence alignment and their response to classic protease inhibitors [5–11], both HRV 2A and 3C proteases are classified as cysteine prote-

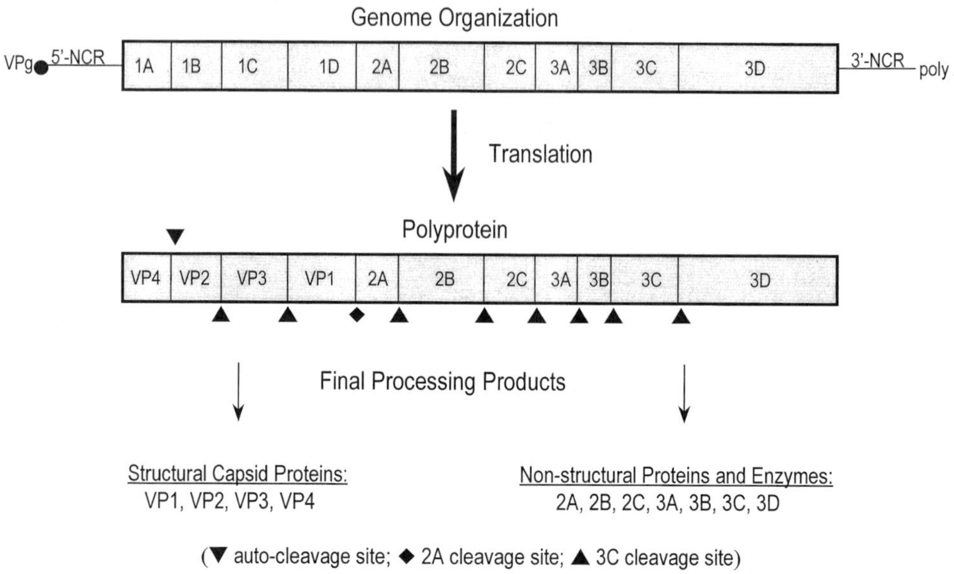

Figure 1 Schematic representation of HRV genome organization, polyprotein translation, and maturation cleavage. The cleavage sites on the viral polyprotein for HRV 2A and 3C proteases are indicated. The amino acid sequence flanking the processing site VP1/2A for HRV14 2A protease is -RKGDIKSY//GLGPRYGG-. The 3C protease cleaves the polyprotein at multiple sites. The amino acid sequence at the 2C/3A site for HRV14 3C protease is -DSLETLFQ//GPVYKDLE-.

ases. For HRV 2A protease, the catalytic triad has been defined as His-Asp-Cys through mutational analysis [5,12,13]. For the 3C protease, its catalytic triad is composed of His-Glu-Cys [2,14,15]. However, the overall architecture of 2A and 3C proteases was originally predicted to be similar to small trypsin-type serine proteases rather than to a typical cysteine protease [6,11,12]. This hypothesis has been confirmed by the X-ray crystal structure of the 3C protein from HRV14 [14]. The active site cysteine in a serine protease-like structure places 2A and 3C in a new class of proteases. Their unique protein structures have been viewed as a great advantage for development of antiviral agents against HRV infections because of expected inhibitor selectivity against human cysteine and serine proteases.

In addition to their protease activity, both 2A and 3C proteases are believed to participate in the regulation of host cell metabolism. The two viral proteases have been found to cleave several important host cell enzymes and transcription factors. For example, the eukaryotic initiation factor eIF-4γ, also termed p220, is cleaved by the 2A protease in infected cells [13,16–19]. Similarly, HRV 3C protease is able to cleave a number of cellular proteins including histone H3 protein, TATA-binding protein, a subunit of transcription factor IIIC, and the cAMP-responsible element-binding protein [20–25]. As most of these proteins are cellular transcription factors, their degradation by 2A and 3C protease causes the inhibition of host cell transcription [16–27].

The active 2A and 3C proteases from different HRV serotypes have been overproduced in bacterial cells and purified by several groups [18,28–42]. The availability of large quantities of the recombinant viral proteases has greatly facilitated their biochemical characterization, assay development, and efforts to identify specific inhibitors against these enzymes. We have worked on the purification and assay development for the two proteases from HRV serotype-14 (HRV14) as part of our efforts in antiviral development. The experimental procedures described below may provide a general approach toward the purification and development of high-throughput assays for the 2A and 3C proteases of other HRV serotypes as well as other members of the picornaviral family.

II. PREPARATION OF ACTIVE HRV14 2A PROTEASE

HRV14 2A protease was purified from bacterial cells expressing the 2A gene coding region. For expression, the 2A gene was cloned into the *Escherichia coli* expression vector, pH10, and the resulting construct pH10/2A was transformed into the *E. coli* strain RV308 cells as described previously [43]. Transformants that contained pH10/2A were selected and incubated at 30°C in 2X TY growth medium supplemented with 10 mg/ml tetracycline until the OD_{600} reached approximately 0.7. Production of the recombinant HRV14 2A protease was achieved by shifting the cell culture temperature to 42°C for 3 hr [43]. The cells were then harvested, lysed, and analyzed for 2A expression by sodium dodecyl sulfate-polyacrylamide gel electrophoresis (SDS-PAGE).

Harvested bacterial cells expressing HRV14 2A protein were collected from 2-liter cultures and resuspended in buffer A (25 mM N-[2-hydroxyethyl] piperazine-N′-[2-ethanesulphonic acid] (HEPES), pH 8.0, 5 mM dithiothreitol (DTT), and 5% glycerol). The cells were treated with DNase (1 unit/ml) and lysozyme (0.5 μg/ml) for 60 min, followed by the addition of NaCl to a final concentration of 1 M. Cells were then lysed by sonication, and inclusion bodies containing HRV14 2A were collected by centrifugation at 10,000*g* for 20 min. The pellet was first washed with 100 ml of 1 M NaCl, then 1 M urea, and followed by deionized water.

The isolated inclusion bodies were solubilized overnight with 7 M urea in buffer A, and insoluble materials were removed by centrifugation at 10,000*g* for 30 min. The supernatant, containing the denatured 2A protein, was collected and diluted with buffer A to a concentration of 0.1 mg/ml. To refold 2A protein, the diluted sample preparation was dialyzed overnight at 4°C against buffer B (25 mM HEPES, pH 8.0, 5% glycerol, and 150 mM NaCl) in the presence of 0.1 mM $ZnCl_2$. After dialysis, preparation was centrifuged at 10,000*g* for 30 min and then was loaded onto a Mono Q 5/5 column (Pharmacia). The bound proteins were eluted with a linear gradient of 0.15–1 M NaCl in buffer A. Fractions containing 2A protein were collected after SDS-PAGE (Fig. 2). These fractions were loaded onto a Superdex-75 Hiload 26/60 column (Pharmacia) and then eluted with buffer B. The peak containing the 2A protein was identified, pooled, and stored at −20°C.

The purification of HCV14 2A protein after each column is summarized in Fig. 2. As seen in Fig. 2, the protein had a molecular weight of ~16 kDa, matching its estimated molecular mass of 15,976 Da. The identity of this purified protein as HRV14 2A was further confirmed by mass spectrometric analysis (data not shown). Amino acid sequence analysis was also performed on the purified HRV14 2A protein. Its N-terminal amino acids were identified as GLGPRYGGIYTSN-, which was identical to the expected sequence.

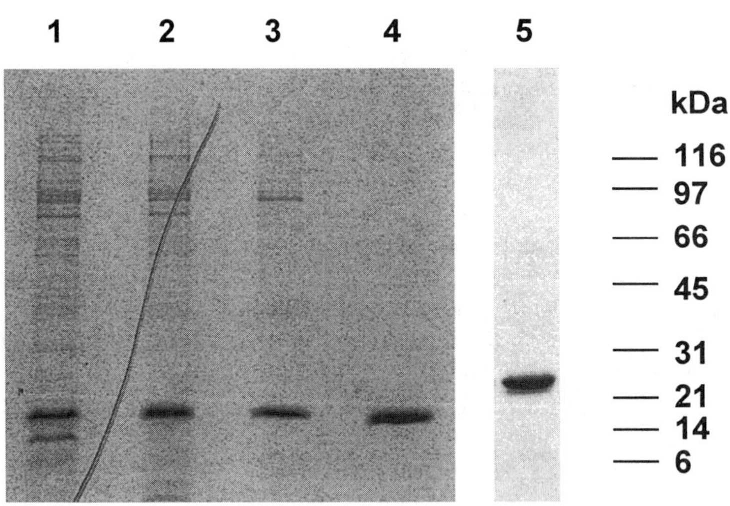

Figure 2 SDS-PAGE analysis of purified HRV14 2A and 3C proteases. Protein samples (~1–2 µg) were separated by electrophoresis on 16% gels and then stained with Coomassie blue. Lanes 1–4 represent the purification of HRV14 2A protein. Lanes: 1, transformed cell lysate; 2, urea-solubilized inclusion bodies; 3 and 4, 2A protease preparations after Mono Q and Superdex-75 columns, respectively. Lane 5 represents the purified HRV14 3C protease.

III. PREPARATION OF ACTIVE HRV14 3C PROTEASE

Expression of HRV14 3C protease has been described by several groups [31–42]. To overproduce 3C protein in a bacterial system, we used the expression and purification protocols described by Birch et al. with slight modifications [41]. Briefly, the 3C coding region was cloned into the same expression vector pH10 as described above for the 2A protease [41,43]. The constructed pH10/3C expression vector was then transformed into *E. coli* BL21-DE3 cells. Similar to HRV14 2A overexpression, production of 3C protease in transformed cells was induced by incubation at 42°C for 3 hr [41]. Cells were collected for purification.

Similar to the purification of the 2A protease, cytoplasmic inclusion bodies were collected and prepared as described above. After solubilized in 7 M urea in the buffer containing 50 mM Tris, pH 7.5, 1 mM EDTA, and 20 mM cysteine [41], the supernatant fraction was collected and fractionated through two ion exchange columns, Q-sepharose and SP-sepharose. Fractions containing 3C protein were identified by SDS-PAGE and pooled. To refold the protein, purified 3C protease was dialyzed against 20 mM 2-[N-morpholine]ethane sulfonic acid (MES), pH 6.5, 1 mM ethylenediaminetetraacetic acid (EDTA), and 1 mM DTT overnight at 4°C. After centrifugation at 30,000*g* for 30 min, refolded 3C protease, present in the supernatant, was collected and loaded onto a size-exclusion column (Superdex-75). The 20-kDa 3C protease was eluted at the expected position using a buffer containing 20 mM MES, pH 6.5, 1 mM EDTA, and 10% glycerol. The final preparation was shown to contain only one protein band around 20 kDa as judged by SDS-PAGE analysis (Fig. 2, lane 5). Mass spectrometric analysis demonstrated that the molecular weight of the purified protein was 19,997 ± 1 Da, identical to the calculated mass of HRV14 3C protease. In addition, its catalytic activity was demonstrated through several assays using peptide substrates derived from the authentic 3C processing site on the viral polyprotein precursor (see below).

IV. HIGH-THROUGHPUT ASSAY DEVELOPMENT FOR 2A AND 3C PROTEASES

Because development of a prophylactic vaccine against HRV and picornavirus infections has been found to be problematic due to the presence of large numbers of serotypes and genotypes, searching for efficient treatment for diseases associated with these viruses has been the focus of these years. These efforts include development of antiviral agents specifically targeting viral enzymes. To develop novel inhibitors against 2A and 3C proteases, accurate and sensitive enzyme assays suitable for both high-capacity screening and enzyme-inhibition kinetic studies are required. When the highly purified 2A and 3C enzyme preparations became available, we explored several different approaches to develop high-throughput assays for these enzymes.

A. A Continuous Colorimetric Assay Using *p*-Nitroanilide Peptides

The most common method used for measurement of 2A and 3C protease activity *in vitro* was high-performance liquid chromatography (HPLC)-based assays utilizing various synthetic peptides [44–48]. Although HPLC-based assays are quantitative and feasible for enzyme kinetic studies [44–48], analysis of the cleavage products by HPLC is usually time-consuming and thus is not suitable for large numbers of samples. Several groups have used the HPLC-based assays to study the cleavage specificity of 2A and 3C proteases [44–48]. Based on these studies, it has been proposed that both 2A and 3C protease cleaved synthetic peptides containing their native processing sites *in vitro* and had a specific amino acid requirement at position P1′ [44–48].

However, the crystal structure of HRV14 3C protease, solved recently, suggested that this protease might not have an absolute requirement for glycine at position P1′ [14]. On the basis of the 3C protease model, we conducted a rational design of chromogenic peptide substrates containing *p*-nitroaniline (pNA) as the only moiety at the prime side to see if these peptides could be cleaved by the viral 3C protease. In these peptides, amino acids downstream from the original cleavage site have all been replaced with a chromophoric *p*-nitroaniline moiety which is linked directly to the bond undergoing enzymatic cleavage, thereby generating a new cleavage site Gln-pNA for the enzyme (Table 1). Hydrolysis of the pNA peptides by 3C at the newly formed scissile bond was expected to release free *p*-nitroaniline, a yellow-colored product that could be continuously monitored by its absorbance at 405 nm [49–50]. As the extinction coefficient of free *p*-nitroaniline and pNA peptides were determined to be 10,360 M^{-1} cm^{-1}, and ~70 M^{-1} cm^{-1}, respectively, this large difference in molar extinction coefficient would lead to a very sensitive assay [50].

Several peptides mimicking the native 3C cleavage site were synthesized with an N-acylated *p*-nitroaniline at position P1′. These peptides, with four to eight amino acids upstream of the cleavage bond, were all derived from the original 2C/3A processing site on the HRV14 polyprotein (Table 1 and Fig. 1). To test if they were substrates for 3C protease, cleavage reactions were performed directly in microtiter plates and monitored at a wavelength of 405 nm using a temperature-controlled microplate reader (Molecular Devices). Cleavage efficiency of these pNA peptides by HRV14 3C protease, expressed as k_{cat}/K_m, is shown in Table 1. The 3C enzyme cleavage of peptide pNA4 was enzyme-concentration and time-dependent (Fig. 3A), with K_m and V_{max} values of 1.2 mM and 12.21 μM/min, respectively. Analyses of the reaction products by both HPLC and

Table 1 Cleavage Efficiency of pNA Peptides by HRV14 2A and 3C Protease

		Catalytic efficiency k_{cat}/K_m (M^{-1} s^{-1})	
pNA peptide	Amino acid sequence	HRV14 2A	HRV14 3C
pNA1	T-L-F-Q--pNA	ND	40.0
pNA2	E-T-L-F-Q--pNA	ND	230
pNA3	D-S-L-E-T-L-F-Q--pNA	NC	465
pNA4	E-A-L-F-Q--pNA	NC	840 ± 14
pNA5	R-K-G-D-I-K-S-Y--pNA	335 ± 20	NC
pNA6	T-R-P-I-I-T-T-A--pNA	58 ± 8	NC

Note: The k_{cat}/K_m values were determined by plotting the initial reaction rates versus substrate concentrations using the equation of $v = (k_{cat}/K_m)[E][S]$, where [E] and [S] are the enzyme and substrate concentrations, respectively. The enzyme concentration was used at 0.1 μM for 3C and 0.4 μM for 2A protease. The reactions were performed at 30°C in a 200-μl reaction mixture containing 25 mM HEPES, pH 7.5, 150 mM NaCl, 1 mM EDTA, and 250 μM pNA peptide substrate and monitored by absorbance at 405 nm. Amino acid sequences of the pNA peptides are given in single-letter amino acid code. NC, no cleavage; ND, not determined.

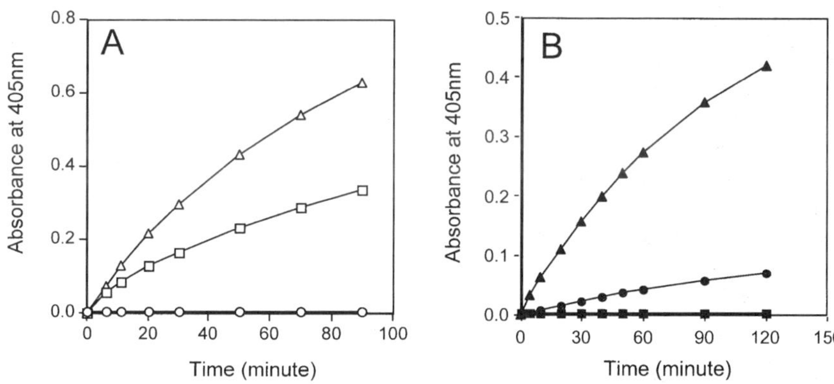

Figure 3 Cleavage activity of HRV14 2A and 3C proteases toward colorimetric peptides. Each pNA peptide was used at 250 μM in these reactions and absorbance at 405 nm was measured at the indicated time points. The reactions were performed at 30°C for the time indicated in a 200-μl reaction mixture containing 25 mM HEPES, pH 7.5, 150 mM NaCl, 1 mM EDTA, and 250 μM pNA peptide substrate. (A) Cleavage of peptide pNA4 by purified HRV14 3C protease. Reactions were performed under the conditions described in the text using 0 (-○-), 0.2 μM (-□-), and 0.4 μM (-△-) purified 3C protease, respectively. (B) Time-dependent cleavage of pNA peptides by purified HRV14 2A protease. Shown is the 2A cleavage of peptide pNA4 (-■-), pNA5 (-▲-), and pNA6 (-●-) as described in the text.

mass spectrometry confirmed that the 3C protease cleaved these pNA peptides at the expected Gln-pNA bond (data not shown). These results strongly suggested that the minimal structure requirement for an efficient 3C protease cleavage was located at the N-terminal side of the scissile bond.

Since the 2A protease has a structural folding similar to that of the 3C protease, as mentioned above, we decided to investigate if the 2A enzyme could also recognize and hydrolyze peptides containing pNA as the only prime-side moiety. An octapeptide pNA5, representing the original HRV14 2A processing site VP1/2A, was designed and synthesized (Table 1 and Fig. 1). Similar to the pNA peptide for the 3C protease, the pNA moiety was also linked to the bond undergoing enzymatic cleavage to create a new recognition site of Tyr-pNA for the 2A protease. The 2A protease reactions were conducted in the same buffer system and monitored in the same way as the 3C protease described above.

When peptide pNA5 was incubated with the purified HRV14 2A protease, an absorbance increase at 405 nm was detected, indicating that the 2A enzyme did not have a stringent amino acid requirement at position P1′. The 2A protease was able to cleave the amide bond newly formed between tyrosine and pNA, as revealed by HPLC and mass spectrometric analysis (not shown). The peptide hydrolysis reactions were dependent on both time and 2A enzyme concentration (Fig. 3B). The kinetic values for the 2A enzyme toward peptide pNA5 were also determined, with K_m and k_{cat}/K_m values of 0.96 mM and $\sim$335 M^{-1} s^{-1}, respectively. Interestingly, peptide pNA6, representing the HRV2 2A recognition site, could be cleaved by the purified HRV14 2A enzyme, but with approximately sixfold less efficiency (Table 1). However, the pNA peptides designed for 3C enzyme could not be hydrolyzed by the 2A protease, implying that amino acids upstream of the scissile bond were critical for the 2A enzyme as seen with the 3C protease. In addition, we had observed similar recognition features for the 2A protease from HRV2 [49]. Therefore, it is possible that this characteristic is common for the 2A and 3C proteases from different HRV serotypes or other members of the picornavirus family.

B. Fluorescence Assays Based on Resonance Energy Transfer

Fluorescence assays have been widely used to detect cellular and viral protease activity [51,52], especially for the proteases requiring specific amino acids on the prime side of the cleavage bond. In general, fluorescence assays for proteolytic enzymes employ peptide substrates containing a fluorescence donor and quencher flanking the scissile bond [51]. Cleavage of these internally quenched peptides by proteases results in a fluorescence signal increase due to the release of the quenching effect of the quencher. Reactions using fluorescent peptides could be continuously monitored by a fluorometer with preferred excitation and emission wavelengths. The major advantage of fluorescence assays over the colorimetric methods described above is their high sensitivity. For example, the cleavage of a fluorescence peptide containing a resonance energy transfer donor/quencher pair by human immunodeficiency virus protease caused over 34-fold increase in fluorescence quantum yields [53]. In order to develop fluorescence assays for the two viral proteases, we designed several internally quenched peptides using different fluorescence donor and quencher pairs and examined the cleavage efficiency of HRV14 2A and 3C protease toward these peptides.

Fluorescence assays using Edans [5-[(2-aminoethyl)amino]naphthalene-1-sulphonic acid] as a fluorescent donor and Dabcyl [4-(4-dimethylaminophenylazo)benzoic acid] as a quenching acceptor have been widely described for the human immunodeficiency virus protease and several other viral and cellular proteases [53–56]. Cleavage of internally quenched peptides containing the Edans/Dabcyl pair by a protease could cause 10–34-fold fluorescence increase [53–56]. For HRV14 3C protease, we synthesized several peptides containing the Edans and Dabcyl moieties. As seen in Table 2, the amino acid sequences of these peptides were all derived from the 2C/3A site of the HRV14 polyprotein with the Edans and Dabcyl moieties attached to the side chain of a glutamic acid and a lysine residue, respectively. To prepare these peptides, the two modified amino acids, Edans-modified glutamic acid and Dabcyl-modified lysine, were first prepared through solution chemistry synthesis [57]. These two modified amino acids could be incorporated into a peptide at any desired position through a traditional solid-phase peptide synthesis method [57].

Of the three Edans/Dabcyl peptides synthesized, peptide F3 exhibited much better solubility in aqueous solution than the first two, and thus was tested for 3C

Table 2 Cleavage of Fluorogenic Peptides by HRV14 2A and 3C Protease

		Catalytic Efficiency k_{cat}/K_m (M^{-1} s^{-1})	
Peptides	Amino acid sequence	HRV14 2A	HRV14 3C
F1	E(Edans)VLFQ//GPK(Dabcyl)YR	ND	ND
F2	LE(Edans)VLFQ//GPK(Dabcyl)YRD	ND	ND
F3	DSE(Edans)EVLFQ//GPVK(Dabcyl)RD	NC	13,275
F4	RKGDIKTY(NO_2)//GLGPK(Abz)Y	<10	ND
F5	RKGDIKTY(NO_2)//GPGPK(Abz)Y	126 ± 9	ND
F6	GRTTLSTY(NO_2)//GPPRK(Abz)Y	418 ± 35	NC

Note: The protease reactions were performed under the conditions described in the text and monitored using a Perkin-Elmer LS50B fluorometer. The k_{cat}/K_m values were determined as described in Table 1 legend. Peptide sequences are given in single-letter amino acid code. NC, no cleavage; ND, not detected.

cleavage efficiency. We found that a complete hydrolysis of peptide F3 resulted in ~24-fold increase in fluorescence signal. The peptide F3 cleavage progress curve is shown in Fig. 4A. Under the typical 3C protease assay conditions described above, purified 3C protease cleaved peptide F3 with a k_{cat}/K_m value of 13,275 M^{-1} s^{-1}. By examining 3C inhibition by a panel of classic protease inhibitors, we demonstrated that both colorimetric and fluorogenic assays were suitable for inhibitor evaluation. Shown in Table 3 is the comparison of the 3C inhibition profiles generated using the two assays described in this report.

Development of a more sensitive fluorescence assay for the purified 2A protease was also undertaken. Although Edans/Dabcyl is an excellent donor/quencher pair for making fluorogenic peptides, attachment of Edans and Dabcyl moieties to both glutamic acid and lysine requires several-step solution chemistry synthesis [57]. To avoid this complicated chemistry procedure, we used a different donor and quencher pair: aminobenzoic acid (Abz) as a fluorescence donor and 3-nitrotyrosine as an accepting quencher. Similar to the Edans/Dabcyl pair, the efficient intramolecular quenching of Abz fluorescence by 3-nitrotyrosine acts through resonance energy transfer [58]. Therefore, cleavage of fluorogenic pep-

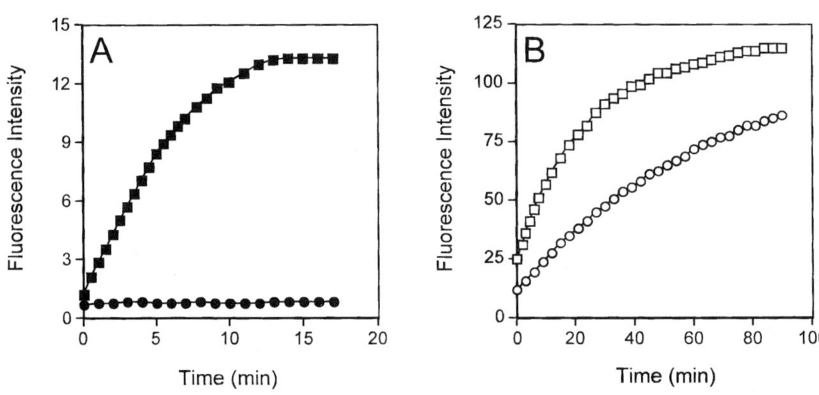

Figure 4 Fluorogenic peptide cleavage by purified HRV14 2A and 3C proteases. (A) Cleavage of Edans/Dabcyl peptide F3 by the 3C protease. Reactions were performed under the conditions described in the text using 35 μM peptide F3 in the absence (-●-) or presence of 0.1 μM purified 3C protease (-■-). Reaction progress was monitored continuously by a fluorometer with excitation at 340 nm and emission at 510 nm at bandwidths of 2.5 nm. (B) Cleavage of fluorogenic peptides by purified HRV14 2A protease. Shown is the time-dependent 2A (0.1 μM) cleavage of peptides F5 (-○-) and F6 (-□-). Reactions were monitored by a fluorometer with emission and excitation wavelengths of 415 and 340 nm, respectively.

Table 3 Inhibition of HRV14 3C Protease by Classic Protease Inhibitors

			HRV14 3C inhibition (IC_{50})	
Inhibitor	Target	Conc. tested	pNA assay	Fluoresc. assay
Control		No addition	NI	NI
EDTA	Metalloprotease	50 mM	NI	NI
EGTA	Metalloprotease	50 mM	NI	NI
E-64	Cys Protease	100 μM	NI	NI
Iodoacetamide	Cys Protease	Varied	1.1 mM	1.7 mM
Pepstatin	Asp Protease	20 μM	NI	NI
Aprotinin	Ser Protease	15 μM	NI	NI
Leupeptin	Ser/Cys Protease	Varied	0.75 mM	1.8 mM
PMSF	Ser/Cys Protease	Varied	8.0 mM	7.5 mM
TLCK	Ser/Cys Protease	1 mM	>1.0 mM	>1.0 mM

Note: Reactions were performed under the conditions described in the text with either colorimetric peptide pNA4 (250 μM) or fluorogenic peptide F3 (35 μM) as a substrate. The IC_{50} values represent the inhibitor concentration required to reduce the protease activity by 50% of the control containing no inhibitor. NI, no inhibition was observed at the concentrations indicated. E64, *trans*-epoxysuccinyl-L-leucylamide-(4-guanidino)-butane; PMSF, phenylmethylsulphonyl fluoride; TLCK, tosyl-L-lysine-chloromethyl ketone.

tides by a protease is expected to generate an increased fluorescent signal as seen with the Edans/Dabcyl peptides. More important, Abz-modified lysine and 3-nitrotyrosine are commercially available and readily used for solid-phase peptide synthesis.

Several peptides were designed and synthesized for the 2A protease. Peptide F4 and F5 contained the amino acids derived from the HRV14 2A processing site with a leucine and a proline at P2′, respectively, while peptide F6 was derived from the 2A recognition site of elF4γ protein, a substrate identified for HRV 2A protease [59]. Our preliminary data showed that the catalytic efficiency was extremely low for peptide F4, containing the original P2′ leucine (Table 2). As the 2A protease is known to prefer a proline at P2′ [48], peptide F5 was then synthesized and tested for 2A cleavage. With such a single amino acid change, the 2A catalytic efficiency toward peptide F5 was found to increase over 13-fold (Table 2). Encouraged by these results, peptide F6, containing a P2′ proline with other amino acids derived from the 2A cleavage site on elF4A protein, was designed and synthesized. The cleavage of this peptide by 2A was found to be more efficient than that of F4 and F5 peptides, indicating that the amino acid sequence of peptide F6 might fit the 2A active site better [60]. Nevertheless, the cleavage of peptide F6 by HRV14 2A caused over 10-fold fluorescence intensity increase and was time- and enzyme-concentration-dependent (Fig. 4B). The k_{cat}/K_m value was determined as 418 M^{-1} s^{-1} (Table 2), similar to that determined using the chromogenic peptide pNA5.

V. CONCLUSIONS

The 2A and 3C proteases encoded by HRV represent attractive targets for antiviral intervention. In this report, we describe production of active recombinant HRV14 2A and 3C proteases through refolding and sequential chromatographic purification. The availability of these highly purified viral enzymes permitted us to develop more convenient, sensitive, and quantitative assays which can be used for evaluation of viral protease inhibitors [61–63]. For the two continuous protease assays described here, the major advantage of the colorimetric assay is convenience; cleavage reactions can be monitored with either a conventional spectrophotometer or a microplate reader. Compared to the colorimetric assay, the fluorescence assay based on efficient resonance energy transfer is more sensitive, in addition to its excellent quantitative and time-saving features. Fluorescence signal could be detected even when substrate was used at low micromolar concentrations [53,57,60]. It is noteworthy that every viral protease has different substrate specificity requirements, and even a single amino acid change may affect the protease catalytic efficiency dramatically. Therefore, the substrate specificity of a given protease must be considered during development of colorimetric or fluorogenic peptide substrates. Nevertheless, it is hoped that the availability of these two types of assays for HRV proteases will not only aid in biochemical characterization of the two virus proteases but also facilitate antiviral chemotherapeutic efforts.

REFERENCES

1. RB Couch. Rhinoviruses. In: BN Fields, ed. Virology. Philadelphia: Raven, 1996, pp 713–734.
2. AG Porter. Picornavirus nonstructural proteins: emerging roles in virus replication and inhibition of host cell functions. J Virol 67:6917–6921, 1993.
3. AC Palmenberg. Proteolytic procession of picornaviral polyprotein. Annu Rev Microbiol 44:603–623, 1990.
4. G Stanway. Structure, function, and evolution of picornaviruses. J Gen Virol 71: 2483–2501, 1990.
5. H Toyoda, MJ Nicklin, MG Murray, CW Anderson, JJ Dunn, FW Studier, E Wimmer. A second virus-encoded proteinase involved in proteolytic processing of poliovirus polyprotein. Cell 45:761–770, 1986.

6. JF Bazan, RJ Fletterick. Viral cysteine proteases are homologous to the trypsin-like family of serine proteases: structural and functional implications. Proc Natl Acad Sci (USA) 85:7872–7876, 1988.
7. SF Yu, RE Llyod. Identification of essential amino acid residues in the functional activity of poliovirus 2A protease. Virology 182:615–625, 1991.
8. W Sommergruber, J Seipelt, F Fessl, T Skern, HD Liebig, G Casari. Mutational analyses support a model for the HRV2 2A proteinase. Virology 234:203–214, 1991.
9. KC Cheah, LE Leong, AG Porter. Site-directed mutagenesis suggests close functional relationship between a human rhinovirus 3C cysteine protease and cellular trypsin-like serine proteases. J Biol Chem 265:7180–7187, 1990.
10. AE Gorbalenya, VM Blinov, AP Donchenko. Poliovirus-encoded proteinase 3C: a possible evolutionary link between cellular serine and cysteine proteinase families. FEBS Lett 194:253–257, 1986.
11. MA Lawson, BL Semler. Poliovirus thiol proteinase 3C can utilize a serine nucleophile within the putative catalytic triad. Proc Natl Acad Sci (USA) 88:9919–9923, 1991.
12. W Sommergruber, M Zorn, D Blaas, F Fessl, P Volkmann, I Maurer-Fogy, P Pallai, V Merluzzi, M Matteo, T Skern, E Kuechler. Polypeptide 2A of human rhinovirus type 2: identification as a protease and characterization by mutational analysis. Virology 169:68–77, 1989.
13. CUT Hellen, M Fäcke, HG Kräusslich, CK Lee, E Wimmer. Characterization of poliovirus 2A proteinase by mutational analysis: residues required for autocatalytic activity are essential for induction of cleavage of eukaryotic initiation factor 4F polypeptide p220. J Virol 65:4226–4231, 1991.
14. DA Matthews, WA Smith, RA Ferre, B Codon, G Budahazi, JE Villafranca, W Sisson, CA Janson, HE McElroy, CL Gribskov, S Worland. Structure of human rhinovirus 3C protease reveals a trypsin-like polypeptide fold, RNA-binding site, and means for cleaving precursor polyprotein. Cell 77:761–771, 1994.
15. KM Kean, NL Teterina, D Marc, M Girard. Analysis of putative active site residues of the poliovirus 3C protease. Virology 181:609–619, 1991.
16. HG Krausslich, MJ Nicklin, H Toyoda, D Etchison, E Wimmer. Poliovirus proteinase 2A induces cleavage of eucaryotic initiation factor 4F polypeptide p220. J Virol 61:2711–2718, 1987.
17. RE Lloyd, MJ Grubman, E Ehrenfeld. Relationship of p220 cleavage during picornavirus infection to 2A proteinase sequencing. J Virol 62:4216–4223, 1988.
18. HD Liebig, E Ziegler, R Yan, K Hartmuth, H Klump, H Kowalski, D Blaas, W Sommergruber, L Frasel, B Lamphear, R Rhoads, E Kuechler, T Skern. Purification of two picornaviral 2A proteinases: interaction with elF-4γ and influence on in vitro translation. Biochemistry 32:7581–7588, 1993.
19. W Sommergruber, H Ahorn, H Klump, J Seipelt, A Zoephel, F Fessl, E Krystek, D Blaas, E Kuechler, HD Liebig, and T Skern. 2A Proteinases of coxsackie- and rhinovirus cleave peptides derived from elF-4γ via a common recognition motif. Virology 198:741–745, 1994.
20. MM Falk, PR Grigera, IE Bergmann, A Zibert, G Multhaup, E Beck. Foot and mouth disease virus protease 3C induces specific proteolytic cleavage of host cell histone H3. J Virol 64:748–756, 1990.

21. ME Clark, T Hämmerle, E Wimmer, A Dasgupta. Poliovirus proteinase 3C converts an active form of transcription factor IIIC to an inactive form: a mechanism for inhibition of host cell polymerase III transcription by poliovirus. EMBO J 10:2941–2947, 1991.
22. ME Clark, PM Lieberman, AJ Berk, A Dasgupta. Direct cleavage of human TATA-binding protein by poliovirus protease 3C in vivo and in vitro. Mol Cell Biol 13: 1232–1237, 1993.
23. M Joachims, KS Harris, D Etchison. Poliovirus protease 3C mediates cleavage of microtubule-associated protein 4. Virology 211:451–461, 1995.
24. P Yalamanchili, U Datta, A Dasgupta. Inhibition of host cell transcription by poliovirus: cleavage of transcription factor CREB by poliovirus-encoded protease 3Cpro. J Virol 71:1220–1226, 1997.
25. P Yalamanchili, K Weidman, A Dasgupta. Cleavage of transcriptional activator Oct-1 by poliovirus encoded protease 3Cpro. Virology 239:176–185, 1997.
26. MV Davies, J Pelletier, K Meerovitch, N Sonenberg, RJ Kaufman. The effect of poliovirus proteinase 2Apro expression on cellular metabolism. Inhibition of DNA replication, RNA polymerase II transcription, and translation. J Biol Chem 266: 14714–14720, 1991.
27. I Novoa, F Martinez-Abarca, P Fortes, J Ortin, L Carrasco. Cleavage of p220 by purified poliovirus 2A(pro) in cell-free systems: effects on translation of capped and uncapped mRNAs. Biochemistry 36:7802–7809, 1997.
28. H König, B Rosenwirth. Purification and partial characterization of poliovirus protease 2A by means of a functional assay. J Virol 62:1243–1250, 1988.
29. R Aldabe, E Feduchi, I Novoa, L Carrasco. Expression of poliovirus 2Apro in mammalian cells: effects on translation. FEBS Lett 377:1–5, 1995.
30. JC Alvey, EE Wyckoff, SF Yu, R Lloyd, E Ehrenfeld. Cis- and trans-cleavage activities of poliovirus 2A protease expressed in *Escherichia coli*. J Virol 65:6077–6083, 1991.
31. LA Ivanoff, T Towatari, J Ray, BD Korant, SR Petteway Jr. Expression and site-specific mutagenesis of the poliovirus 3C protease in *Escherichia coli*. Proc Natl Acad Sci (USA) 83:5392–5396, 1986.
32. MJ Nicklin, KS Harris, PV Pallai, E Wimmer. Poliovirus proteinase 3C: large-scale expression, purification, and specific cleavage activity on natural and synthetic substrates in vitro. J Virol 62:4586–4593, 1988.
33. RT Libby, D Cosman, MK Cooney, JE Merriam, CJ March, TP Hopp. Human rhinovirus 3C protease: cloning and expression of an active form in *Escherichia coli*. Biochemistry 27:6262–6268, 1988.
34. Y Takahara, N Ando, M Kohara, K Hagino-Yamagishi, A Nomoto, H Itoh, N Numao, K Kondo. Purification of enzymatically active poliovirus proteinase 3C produced in *Escherichia coli*. Gene 79:249–258, 1989.
35. JA Knott, DC Orr, DS Montgomery, CA Sullivan, A Weston. The expression and purification of human rhinovirus protease 3C. Eur J Biochem 182:547–555, 1989.
36. B Aschauer, G Werner, J McCray, B Rosenwirth, H Bachmayer. Biologically active protease 3C of human rhinovirus 1A is expressed from a cloned cDNA segment in *Escherichia coli*. Virology 184:587–594, 1991.

37. EZ Baum, GA Bebernitz, O Palant, T Mueller, SJ Iotch. Purification, properties, and mutagenesis of poliovirus 3C protease. Virology 185:140–150, 1991.
38. BA Malcolm, SM Chin, DA Jewell, JR Stratton-Thomas, KB Thudium, R Ralston, S Rosenberg. Expression and characterization of recombinant hepatitis A virus 3C proteinase. Biochemistry 31:3358–3363, 1992.
39. K Miyashita, M Kusumi, R Utsumi, T Komano, N Satoh. Expression and purification of recombinant 3C proteinase of Coxsackievirus B3. Biosci Biotechnol Biochem 56:746–750, 1992.
40. SA Harmon, W Updike, XY Jia, DF Summers, E Ehrenfeld. Polyprotein processing in cis and in trans by hepatitis A virus 3C protease cloned and expressed in *Escherichia coli*. J Virol 66:5242–5247, 1992.
41. GM Birch, T Black, SK Malcolm, MT Lai, RE Zimmerman, SR Jaskunas. Purification of recombinant human rhinovirus 14 3C protease expressed in *Escherichia coli*. Protein Expr Purif 6:609–618, 1995.
42. GJ Davis, QM Wang, GA Cox, RB Johnson, M Wakulchik, CA Dotson, EC Villarreal. Expression and purification of recombinant rhinovirus 14 3CD proteinase and its comparison to the 3C proteinase. Arch Biochem Biophys 346:125–130, 1997.
43. QM Wang, RB Johnson, GA Cox, EC Villarreal, LM Churgay, JE Hale. Enzymatic characterization of refolded human rhinovirus type 14 2A protease expressed in *Escherichia coli*. J Virol 72:1683–1687, 1998.
44. MG Cordingley, RB Register, PL Callahan, VM Garsky, RJ Colonno. Cleavage of small peptides in vitro by human rhinovirus 14 3C protease expressed in *Escherichia coli*. J Virol 63:5037–5045, 1989.
45. DC Orr, AC Long, J Kay, BM Dunn, JM Cameron. Hydrolysis of a series of synthetic peptide substrates by the human rhinovirus 14 3C proteinase, cloned and expressed in *Escherichia coli*. J Gen Virol 70:2931–2942, 1989.
46. MG Cordingley, PL Callahan, VV Sardana, VM Garsky, RJ Colonno. Substrate requirements of human rhinovirus 3C protease for peptide cleavage in vitro. J Biol Chem 265:9062–9065, 1990.
47. T Skern, W Sommergruber, H Auer, P Volkmann, M Zorn, HD Liebig, F Fessl, D Blaas, E Kuechler. Substrate requirements of a human rhinoviral 2A proteinase. Virology 181:46–54, 1991.
48. W Sommergruber, H Ahorn, A Zöphel, I Maurer-Fogy, F Fessl, G Schnorrenberg, HD Liebig, D Blaas, E Kuechler, T Skern. Cleavage specificity on synthetic peptide substrates of human rhinovirus 2 proteinase 2A. J Biol Chem 267:22639–22644, 1992.
49. QM Wang, W Sommergruber, RB Johnson. Cleavage specificity of human rhinovirus-2 2A protease for peptide substrates. Biochem Biophys Res Commun 235:562–566, 1997.
50. QM Wang, RB Johnson, GA Cox, EC Villarreal, RJ Loncharich. A continuous colorimetric assay for rhinovirus-14 3C protease using peptide *p*-nitroanilides as substrates. Anal Biochem 252:238–245, 1997.
51. CG Knight. Fluorometric assays of proteolytic enzymes. Meth Enzymol 248:18–35, 1995.
52. P Wu, L Brand. Resonance energy transfer: methods and applications. Anal Biochem 218:1–13, 1994.

53. ED Matayoshi, GT Wang, GA Krafft, J Erickson. Novel fluorogenic substrates for assaying retroviral proteases by resonance energy transfer. Science 247:954–958, 1990.
54. GT Wang, CC Chung, TF Holzman, GA Krafft. A continuous fluorescence assay of renin activity. Anal Biochem 201:351–359, 1993.
55. GA Krafft, GT Wang. Synthetic approaches to continuous assays of retroviral proteases. Meth Enzymol 241:70–86, 1994.
56. BP Holskin, M Bukhtiyarova, BM Dunn, P Baur, J de Chastonay, MW Pennington. A continuous fluorescence-based assay of human cytomegalovirus protease using a peptide substrate. Anal Biochem 226:148–155, 1995.
57. QM Wang, RB Johnson, JD Cohen, GT Voy, JM Richardson, LN Jungheim. Development of a continuous fluorescence assay for rhinovirus-14 3C protease using synthetic peptides. Antiviral Chem Chemother 8:303–310, 1997.
58. M Meldal, K Breddam. Anthranilamide and nitrotyrosine as a donor-acceptor pair in internally quenched fluorescent substrates for endopeptidases: multicolumn peptide synthesis of enzyme substrates for subtilisin Carlsberg and pepsin. Anal Biochem 195:141–147, 1991.
59. BJ Lamphear, R Yan, F Yang, D Waters, HD Liebig, H Klump, E Kuechler, T Skern, RE Rhoads. Mapping the cleavage site in protein synthesis initiation factor elF-4 gamma of the 2A proteases from human Coxsackievirus and rhinovirus. J Biol Chem 268:19200–19203, 1993.
60. QM Wang, RB Johnson, W Sommergruber, TA Shepherd. Development of *in vitro* peptide substrates for human rhinovirus-14 2A protease. Arch Biochem Biophys 356:12–18, 1998.
61. JS Kong, S Venkatraman, K Furness, S Nimkar, TA Shepherd, QM Wang, J Aube, RP Hanzlik. Synthesis and evaluation of peptidyl Michael acceptors that inactivate human rhinovirus 3C protease and inhibit virus replication. J Med Chem 41:2579–2587, 1998.
62. LN Jungheim, JD Cohen, RB Johnson, EC Villarreal, M Wakulchik, RJ Loncharich, QM Wang. Inhibition of human rhinovirus 3C protease by homophthalimides. Bioorg Med Chem Lett 7:1589–1594, 1997.
63. QM Wang, RB Johnson, LN Jungheim, JD Cohen, EC Villarreal. Dual inhibition of human rhinovirus 2A and 3C proteases by homophthalimides. Antimicrob Agents Chemother 42:916–920, 1998.

14

Screening for Parasiticides Using Recombinant Microorganisms

Timothy G. Geary
Pharmacia Animal Health, Kalamazoo, Michigan

I. INTRODUCTION: CURRENT PROSPECTS FOR ANTIPARASITIC DRUG DISCOVERY—MORE OBSTACLES THAN SOLUTIONS

A. Therapeutic Challenges of Parasites

All animals and plants are plagued by parasites. Although it is not commonly appreciated, more animal species can be called ''parasite'' than any other classification [1]. Pathogens targeted for control are found in multiple protozoan, helminth (Table 1), and insect (Table 2) phyla. This extraordinary diversity complicates antiparasitic chemotherapy, in which breadth of spectrum is a key attribute. The demand for broad spectrum provides one of the most difficult challenges for the discovery of commercially viable antiparasitic drugs. The challenge is made more difficult by the fact that parasites inhabit virtually every tissue of the host, requiring delivery of antiparasitic drugs to multiple target compartments. Evolution of resistance to these drugs by parasites is the rule rather than the exception and obviously makes matters worse. Resistance has been used to justify the search for new drugs and for nonchemotherapeutic methods of control. In agriculture, control of nematode and arthropod pests poses huge challenges, including resistance and environmental concerns.

Although parasites are among the most prevalent pathogens of humans [2–4] and their domesticated animals [5], relatively little attention is paid to them in either basic research or drug discovery programs [5–8]. Reasons for this unfortunate status can be summarized as follows.

Table 1 Examples of Protozoan and Helminth Parasites

Phylum	Genus	Diseases
	Protozoa	
Apicomplexa	*Plasmodium*, *Toxoplasma*, *Eimeria*	Malaria, coccidiosis, abortion, birth defects
Sarcodina	*Entamoeba*	Amebiasis
Zoomastigina	*Trypanosoma*, *Leishmania*	Sleeping sickness, Chagas' disease, kala-azar
	Helminths	
Platyhelminthes	*Schistosoma*, *Fasciola*, *Echinococcus*, *Taenia*	Schistosomiasis, cysticercosis, hydatid disease
Nematoda	*Ascaris*, *Ancylostoma*, *Onchocerca*, *Brugia*, *Trichinella*, *Ostertagia*, *Meloidogyne*, *Globodera*	River blindness, elephantiasis, anemia, trichinellosis, root-knot gall disease

1. Human parasitoses occur primarily among the poor; there is little incentive for wealthy governments (which support most research) or pharmaceutical industries (which prosper by selling drugs at a profit) to invest in diseases of the poor. Although the enormous rise in opportunistic protozoal infections in AIDS patients stimulated research on parasites such as *Toxoplasma gondii* and *Cryptosporidium parvum*, the remarkable success of combination antiviral chemotherapy subsequently reduced the incidence of these infections. Not surprisingly, the impetus to discover new antiprotozoal compounds lessened. Recent efforts to stimulate public and private investment in malaria research

Table 2 Examples of Insect Parasites

Order	Examples
	Phylum Arthropoda
Phthiraptera	Chewing and sucking lice
Hemiptera	Whiteflies, cinch bugs (plant pests)
Heteroptera	Kissing bugs (reduviids), bedbugs
Siphonoptera	Fleas
Diptera	Mosquitoes, biting flies
Arachnida	Ticks, mites

involve a very small fraction of the funds devoted to other (Western) diseases, despite the fact that *Plasmodium falciparum* causes $>10^6$ deaths/year [4].

2. Helminth infections of humans are rarely immediately life-threatening. Competing health care demands in poor countries, including basic nutrition, sanitation, HIV infection and tuberculosis, dwarf the demand for better control of worms.
3. Veterinary diseases attract little government support for research. Animal health applications are typically cost-constrained by the economics of animal production. Consequently, the profit found in veterinary products is usually lower than in human pharmaceuticals, which draws industrial research investment away from animal health, including parasitology.

B. Current Status of Antiparasitic Drug Discovery

The relatively primitive condition of parasitology research threatens the survival of antiparasitic drug discovery in the continually evolving screening strategies of the pharmaceutical industry (Table 3). All available antiparasitic drugs were initially discovered in screens that employed whole organisms, both target and host, in protocols pioneered by Ehrlich and perfected during the second era of discovery. Subsequent developments, including the paradigm known as drug design (era 3), required increasingly intensive investment in personnel training and

Table 3 Eras of Drug Discovery

1. *Gleaning* from herbal medicine: the genesis of pharmacology (digitalis; salicylic acid—aspirin; morphine; quinine; artemisinin; reserpine; taxol). Ancient history–present.
2. *Screening* in whole animals or against whole organisms or cells in culture (Trypan red; arsphenamine; prontosil and sulfa drugs; penicillin and all other antibiotic classes; almost all antiparasitic drugs; most anticancer, antiviral, and antifungal agents). 1910–present.
3. *Drug design*, based on crystal structures of target incorporated with substrate or inhibitor (primarily for enzymes). More useful for lead optimization than for primary discovery. Can be configured to serve as electronic (''virtual'') random screen. 1980s–present.
4. *Mechanism-based, high-throughput random screening*, utilizing isolated proteins (enzymes, receptors), cloned regulatory domains, or recombinant cells. Trend toward miniaturization and increasing throughput and automation. Mid-1980s–present.

equipment as well as in-depth knowledge of the target. Parasites compete poorly in terms of return on investment and state-of-the-art target knowledge.

The fourth era of discovery has seen a return to the paradigm of random screening that provided antiparasitic drugs in the preceding 80 years or so. However, instead of using whole animals, tissues, or cells, the new approach focuses on the discovery of compounds that interact specifically with a defined target. The development of a plethora of elegant systems for rapidly screening hundreds of thousands of compounds has revolutionized drug discovery in all therapeutic areas. Unfortunately, these systems are expensive to implement and operate and are typically located in a core facility of a company. Highly trained personnel are needed to design, implement, and run the screens. Different formats are required for different targets, so a considerable investment may be required before screening can begin. Consequently, access to the screening facility and the associated expertise is limited by the perceived quality of the drug target and by the size of the market for which the drug is targeted. Parasitological drug targets are often difficult to validate (due to the scarcity of research), and animal health is not very competitive in terms of market size. Since it is not generally possible to routinely obtain large amounts of parasite tissue for target purification to support high-throughput screening (HTS), recombinant systems are absolutely required [6,7]. Additional up-front investment is required to fit parasite targets into systems that can process tens of thousands of samples per week. Facing a future in which antiparasitic drug discovery could become extinct in the pharmaceutical industry, we sought to develop an alternative platform for parasiticide screening that would enable the field to keep pace with changing technology.

II. BASIS FOR THE USE OF RECOMBINANT MICROBES IN HTS

A. The First Screen Employed Microbes

The onset of the second era of discovery can be traced to the decision by Ehrlich to screen a collection of industrial dyes for activity in mice infected with trypanosomes; the first screen was thus a search for an antiparasitic drug [9,10]. This research eventually led to the discovery of arsenical treponemacides and then to the sulfa antibiotics, the first of which was Prontosil, based on industrial dyes. Later, recognition of the antibiotic activity of penicillin led to the widespread adoption of whole-bacteria screens to test both natural products and synthetic compounds for antibiotic activity [10]. Consequently, the first HTS technology was based on simple assays of bacterial viability. The extension of this paradigm to other pathogens became the standard. It is worth noting that all antiparasitic drugs were either isolated from herbal remedies or discovered in relatively simple *in vivo* or *in vitro* screens.

B. Microbial Screens for Nonantibiotic Discovery

Further development of microbial-based screening platforms for nonantibiotic discovery programs was exemplified by the work of Hitchings and Elion (see [11–13] for review). Focusing initially on the discovery of compounds that interrupted specific steps in purine metabolism, they developed a screening system based simply on the survival of *Lactobacillus casei*. Initial screening was done in medium that lacked purines. Compounds that inhibited growth of *L. casei* under this condition were retested in medium containing purines or purine precursors to determine if any could reverse the toxicity. Observation of reversal with a particular supplement pinpointed the site of inhibition. This represents perhaps the first example of what we have termed a *nutrient-dependent viability screen* [14]. The approach was extended to screens for inhibitors of folate metabolism using *Enterococcus faecium* [15], among many other examples [16,17], primarily by Japanese workers.

The key conceptual development in the paradigm established by Hitchings and Elion is that screens based on bacterial viability can discover compounds for nonantibiotic indications; their purine antimetabolites found clinical employment as anticancer drugs and immunosuppressive and antiviral agents. However, in this case, the drug targets were bacterial enzymes, which bear an unknown resemblance to the homologous enzyme in the target species. The extension of nutrient-dependent viability screening to specifically constructed recombinant microorganisms represents the next logical evolution of the paradigm. The principle of complementation of a bacterial mutation by functional expression of a homologous enzyme has been well established for many microorganisms [13]. It is possible to construct a system in which viability of the recombinant microbe is dependent on the function of the foreign (drug target) protein under specific nutritional conditions.

C. Antiparasitic Drug Discovery Using Recombinant Microbes

1. *Screens Employing* Escherichia coli

a. A Screen Using Multiple Complemented Strains. An initial approach toward the use of recombinant microorganisms for antiparasitic drug discovery is found in the example of a key enzyme in glycolysis in the parasitic protozoan *Trypanosoma brucei* [18]. A *T. brucei* gene encoding 6-phosphogluconate dehydrogenase (6-PGD) was cloned by complementation of a strain of *Escherichia coli* in which the homologous gene had been inactivated by mutation. The mutant could not utilize glucose as a sole carbon source, a trait that was reversed by expression of the parasite enzyme. A screen for inhibitors of parasite 6-PGD was proposed in which compounds would be screened for activity against the recombinant bacterial strain in the presence of glucose as the sole carbon source. Specificity of inhibition could be determined by testing active compounds against either wild-type bacteria or against the mutant complemented with the homologous human gene.

b. Nutrient-Dependent Viability Screens for Antiparasitic Drugs. The approach proposed for 6-PGD relies on species-dependent differences in potency to identify active compounds. However, as a primary screen, it is often preferable to identify specific enzyme inhibitors that can be refined to achieve selectivity through medicinal chemistry. Thus, the recombinant microbe paradigm was further refined by the addition of nutrient-dependent viability as a screening variable. Practical application of this principle was first demonstrated by development of a screen for inhibitors of the rate-limiting step in glycolysis, phosphofructokinase (PFK; EC 2.7.1.11), from the parasitic nematode *Haemonchus contortus* [14,19]. A very similar strategy was simultaneously developed for another nematode enzyme that plays a crucial role in energy generation, phosphoenolpyruvate carboxykinase (PEPCK; EC 4.1.1.32) [14,20].

The screening strategy for nematode PFK inhibitors is diagrammed in Figure 1. A strain of *E. coli* that carries mutations in the two loci that encode PFK (*pfk A*, *pfk B*) was used to clone by complementation a cDNA encoding PFK from *H. contortus* [19]. The parent strain is unable to grow on media in which a hexose (conveniently, mannitol) is the sole carbon source. The transformed, complemented strain expresses fairly high levels of PFK enzyme activity compared to wild-type *E. coli*, and grows well (but slightly slower than wild-type *E. coli*) on mannitol (doubling times 1.5 versus 1.1 hr, ref. 19). To be consistent with modern screening formats and automated pipetting platforms, we configured a 96-well microtiter plate screen using the complemented strain of *E. coli*. Each well received an aliquot of minimal medium + mannitol and an inoculum of the complemented strain that provided linear growth for at least 24 hr. After overnight growth at 37°C, an aliquot of a solution of a viability dye, Alamar blue, was added to each well. Alamar blue is a redox-sensitive dye that has been widely used in viability assays for mammalian cells and microbes [21,22]. Wells in which cellular replication has occurred turn from blue to pink, depending on the extent of growth. The dye has both a visible and fluorescent spectrum, so that a correlation between absorbance or emission and viability can be drawn. However, it has been our experience that simple visual inspection of 96-well plates identifies wells that differ from control wells in color. In the PFK assay, inoculum size and time of incubation are chosen to allow a minimum of 10 doubling times. Wells are read within 1 hr of Alamar blue addition. Wells in which bacterial replication was minimal appear blue or purple and are easy to distinguish from the vast majority of pink wells, which contain innocuous compounds.

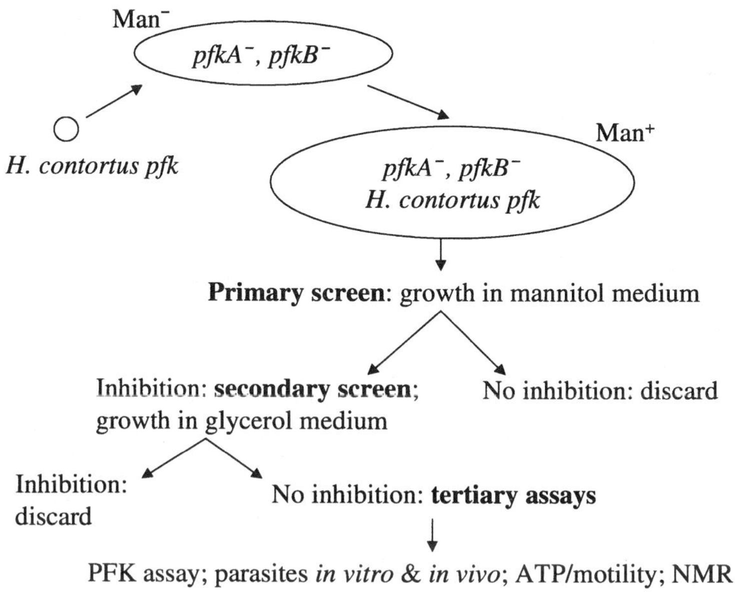

Figure 1 Diagrammatic representation of screening stream for inhibitors of nematode phosphofructokinase (PFK). A strain of *E. coli* that lacks PFK activity due to mutations in the *pfkA* and *pfkB* genes cannot survive on mannitol (Man^-) as a sole carbon source. This strain was transformed with a cDNA encoding *Haemonchus contortus* PFK. The complemented strain grew well on mannitol (Man^+). Collections of fermentation extracts and synthetic chemicals were tested for toxicity to this strain in mannitol medium (*primary screen*). Inactive agents were discarded, while agents that inhibited growth under these conditions were retested against the same recombinant strain in glycerol-containing medium (*secondary screen*). Compounds that inhibited growth in both mannitol and glycerol medium cannot act via inhibition of PFK and were discarded. Compounds that are toxic in mannitol medium, but not in glycerol medium, are candidate PFK inhibitors and are further characterized in a battery of additional tests using whole organisms or semipurified PFK (*tertiary assays*). These assays include a comparison of candidate-induced reductions in motility and ATP levels in *H. contortus* and an NMR assay in which the distribution of metabolites of [^{13}C]mannitol is determined in the presence of the candidate inhibitor.

Active compounds were retested in medium containing glycerol as a carbon source. Glycerol does not require PFK for introduction into bacterial energy metabolism pathways. Compounds that were as toxic to the recombinant bacteria in the presence of glycerol as they were in mannitol could not be acting via inhibition of PFK. Conversely, compounds that are significantly more toxic in mannitol medium than in glycerol medium are candidate PFK inhibitors and are further characterized in tertiary assays (see sec. III.B.3).

c. Other Examples. A number of additional parasiticide screens have been developed in *E. coli*. Much work has been done on hypoxanthine phosphoribosyl transferase (HPRT; EC 2.4.2.8). The gene encoding this enzyme in *P. falciparum* was shown to complement purine metabolism mutants of *E. coli* [23] and *Salmonella typhimurium* [24]. Subsequent studies used HPRT-encoding genes from this parasite as well as those from the protozoan parasites, *Tritrichomonas foetus* and *Trypanosoma cruzi*, and the helminth *Schistosoma mansoni* to construct an assay for enzyme inhibitors [25,26]. Inclusion of a recombinant strain of *E. coli* expressing human HPRT permitted the early determination of host–parasite selectivity. Later work converted the assay to a liquid format in 96-well microtiter plates, which is more compatible with modern HTS platforms [27]. The assay is performed in a semidefined medium in which HPRT function is required for replication. The authors measured optical density after exposure to test compounds as an index of bacterial replication.

Another example of a recombinant, nutrient-dependent viability screen for a parasite enzyme is built on a bifunctional aldehyde dehydrogenase-alcohol dehydrogenase from *Entameoba histolytica* (EhADH2). A gene encoding this enzyme (*EhADH2*) was cloned by complementation of a strain of *E. coli* with a mutation in aldehyde dehydrogenase (*adhE*) gene [28]. *E. histolytica* is an obligate anaerobe and requires EhADH2 for fermentation of glucose to ethanol. AdhE is necessary for anaerobic growth in *E. coli*, and *EhADH2* was identified by its ability to support growth of the *adhE*$^-$ strain of *E. coli* in the absence of oxygen. Using O_2 as the discriminating nutrient, a screen was proposed for inhibitors of EhADH2 in which compounds are tested against the recombinant bacterium under anaerobiosis. Specific inhibition of this enzyme by active compounds can be shown by retesting them in the presence of O_2, as EhADH2 inhibitors will not be active in this case, while actives that do not target this enzyme should still be toxic.

Many other parasite enzymes have been functionally expressed in strains of *E. coli* in complementation formats. Most examples are protozoal enzymes that function in glycolysis, including glucose phosphate isomerase from *Plasmodium falciparum* [29]; phosphoglycerate kinase from *T. brucei* [30]; triosephosphate isomerase from *Giardia lamblia* [31]; and glucose 6-phosphate isomerase and enolase from *T. gondii* [32]. In addition, inosine monophosphate dehydrogenase (IMPDH) from *T. foetus* has been expressed in a strain of *E. coli* in which the *IMPDH* gene was deleted [33]. A cDNA encoding superoxide dismutase from *Leishmania donovani* was expressed in a strain of *E. coli* lacking superoxide dismutase (SOD) activity. Complementation was demonstrated by protection from paraquat toxicity in the recombinant strain [34]; paraquat is a herbicide that is metabolized to a strongly oxidizing derivative. An additional example is malic enzyme from the nematode *H. contortus*, which has also been functionally expressed in *E. coli* [35].

2. *Screens Using* Saccharomyces cerevisiae

The yeast *Saccharomyces cerevisae* is well known as a host for the expression of heterologous proteins. It is also exceptionally useful for nutrient-dependent viability screening, and is even more versatile than *E. coli*. Examples of the myriad ways in which genes encoding heterologous proteins can be functionally expressed in yeast for HTS are readily available [13,36–39]. However, few applications for antiparasitic drug discovery have been described.

Perhaps the first yeast-based, high-throughput, nutrient-dependent viability screen for antiparasitic drugs was designed to discover inhibitors of nematode (*H. contortus*) ornithine decarboxylase (ODC; E.C. 4.1.1.17) [40]. A strain of *S. cerevisiae* in which the ODC gene was deleted did not grow in the absence of exogenous polyamines, a defect corrected by expression of the nematode ODC gene. The *H. contortus* ODC gene was transcribed at a low level in this system, but still provided robust growth in polyamine-free medium [40]. The screen was performed in 96-well microtiter plates. Each well was inoculated with 100 recombinant yeast in the presence of minimal medium and test compounds. After 48 hr at 30°C, plates received an aliquot of Alamar blue solution and were read as described above. Active compounds were retested in the presence of putrescine. Compounds that were candidate ODC inhibitors would not be toxic in the presence of putrescine, while compounds that acted against any other target were expected to demonstrate polyamine-independent toxicity.

Further modifications using the same strain of ODC^- *S. cerevisiae* reconstituted a bacterial/plant polyamine synthesis pathway in yeast [41]. The ODC^- strain was transformed with plasmids encoding arginine decarboxylase and agmatine ureohydrolase, which conferred polyamine-independent growth on the recombinant microbe. A similar construction could be used to screen for inhibitors of the homologous enzymes from Apicomplexan protozoa, which synthesize polyamines through this pathway [42].

Of interest is a recently described yeast-based, nutrient-dependent viability screen for inhibitors of protozoal dihydrofolate reductase (DHFR) [43,44]. Antiprotozoal activity of DHFR inhibitors is well known, and $DHFR^-$ yeast complemented with the DHFR gene derived from the malaria parasite *P. falciparum* have been used to characterize the molecular pharmacology of resistance to the antimalarial DHFR inhibitors pyrimethamine and cycloguanil [45,46]. The subsequent development of a screen was based on the demonstration that the protozoal enzymes could complement the mutation in yeast. The screen includes multiple strains of recombinant yeast, each transformed with and dependent on a different heterologous DHFR gene, including those from *P. falciparum*, *T. gondii*, *C. parvum*, *Pneumocystis carinii*, and *Homo sapiens*. As designed, all these recombinant strains are simultaneously exposed to test compounds in order to identify any with selective activity. The screen can be run in 96-well microtiter plates, with yeast replication measured by changes in optical density [43].

Other parasite proteins of interest have been functionally expressed in *S. cerevisiae*, including a transporter, pgh, thought to be at least partially responsible for resistance to certain quinoline-containing antimalarial drugs in *P. falciparum* [47]. The gene encoding this transporter, *pfmdr1*, complements a pheromone transporter in *S. cerevisiae*, which has allowed preliminary analysis of its function in a heterologous system [48]. Two other examples illustrate the variety of targets that can be studied in yeast. A *Schistosoma mansoni* gene encoding a sarcoplasmic endoreticulum Ca^{2+}-ATPase complemented a mutant strain of *S. cerevisiae* [49]. The mutant strain lacked both yeast genes encoding homologs of this Ca^{2+} sequestration pump and could not grow on normal (low-Ca^{2+}) medium. In another example, the gene encoding an important trypanosomal glycosyltransferase was cloned by complementation in *S. cerevisiae* [50]. In trypanosomes, many proteins are anchored to the cell membrane through glycosylphosphatidylinositol (GPI) anchors. The enzyme that donates a mannose residue for the synthesis of GPI and other purposes was cloned by complementation of a temperature-sensitive mutation in an essential yeast gene that is involved in protein glycosylation.

III. CONSIDERATIONS OF ASSAY DESIGN

A. Control of Level of Target Expression

Overexpression of a target enzyme in a recombinant screen would obviously make lead identification more difficult, simply because more of a compound would be needed to cause a detectable decrease in enzyme activity [43]. In the PFK assay [14], enzyme activity was abundant in recombinant bacteria, but the assay was configured to make the rate of growth tightly dependent on available mannitol. Compounds that reduced mannitol catabolism were thus easily detectable. Conversely, the nematode ODC construct expressed in the yeast screen was present at very low levels [40]. As relatively small amounts of polyamines are needed for cell growth, read-through transcription of the ODC gene behind a galactose-responsive (GAL) promoter provided sufficient amounts of the enzyme for robust yeast growth. Expression levels of different *HPRT* genes in *E. coli* had to be modified in order to permit comparative screening, and sulfanilide had to be added to the system to improve the sensitivity of detection of known DHRF inhibitors in some of the recombinant strains [44].

B. Assay Design

1. Endpoint

There are many ways to estimate microbial growth. The simplest is visual inspection of colonies growing on agar plates, though this method is difficult to adapt for HTS. There is a well-known correlation between cell density and optical density, which can be exploited in a 96-well microtiter plate format (e.g., [43]). Measurement of the incorporation of radioactive nutrients is an excellent quantitative method, but has fallen from favor due to concerns about spills and contamination. Finally, both spectrophotometric and fluorimetric assays are conveniently adapted to HTS formats, and Alamar blue is only one example of the tools available for this purpose. As mentioned, we have even found it convenient to use simple visual inspection of Alamar blue plates to identify wells of interest. However, quantification obtainable with a microplate reader is attractive in many settings.

2. Secondary Testing

Many screens for antiparasitic drugs based on recombinant microorganisms are constructed to allow simultaneous evaluation of test compounds against the target enzyme from parasite and host. While this format is useful for small numbers of compounds, it makes screening large compound collections much more difficult. Instead, such collections should be initially screened against the recombinant microbe under nutrient conditions that isolate the parasite enzyme (e.g., in mannitol medium for PFK or polyamine-free medium for ODC). Active compounds in the primary screen can then be tested for specificity in one of two ways. One can ask for host:parasite selectivity at an early stage by testing against a recombinant microbe that is complemented with the host enzyme. Conversely, one can first ask for specificity of inhibition against the target enzyme by retesting against the parasite-complemented strain under nutrient conditions that do not require function of the parasite enzyme (glycerol medium for PFK or putrescine-supplemented medium for ODC). In this case, actives that do not interfere with the target protein will retain toxicity. Specific compounds will not be toxic under conditions in which the function of the target protein is not required.

The availability of medicinal chemistry resources can decide which approach to take; in the absence of chemistry, a specific enzyme inhibitor that does not distinguish between host and parasite has little value. If chemistry is available, however, identifying a specific enzyme inhibitor can lead to the discovery of parasite-selective compounds. Random screens that search for novel structures are based on the premise that discovery of a specific inhibitor is an extremely rare event, and so the nutrient-dependent viability format is more desirable for HTS.

3. Tertiary Tests

Further testing of confirmed positives in the microbial screening phases, using nutrient dependence of toxicity as a variable, is required to determine if the candidate inhibits the target enzyme, and thus merits a medicinal chemistry program. In the PFK screen, for example, several glycolytic enzymes lie between the target (PFK) and the enzymatic entry of glycerol into energy metabolism. Candidate compounds could inhibit any of them. It is also possible that compounds could appear to be active by inhibiting the uptake of mannitol, but not glycerol. Therefore, actives were tested for inhibition of *H. contortus* PFK in enzyme assays employing whole-worm homogenates. In addition, we measured the response of *H. contortus* motility to exposure to actives, and correlated reductions in motility with reductions in worm ATP levels (e.g., [51]). In this secondary assay, neuroactive substances reduce motility without affecting ATP levels, while the two variables diminish in concert in the presence of metabolic poisons, such as a PFK inhibitor. Finally, whole, live *H. contortus* were exposed to candidate inhibitors in the presence of [^{13}C]glucose, with the distribution of glycolytic intermediates determined by NMR (as described in [51]). One can predict that a PFK inhibitor will cause an accumulation of specific metabolic intermediates and a depletion of those formed downstream from PFK. The same approach can be used for candidate inhibitors of other glycolytic enzymes. Used together, these tests both confirm if specific inhibition of PFK has been attained, and document the consequences of that inhibition for parasite viability.

Similarly, candidate leads discovered in the ODC assay could act by inhibiting *H. contortus* ODC or yeast S-adenosylmethionine decarboxylase [40]. A single confirmed positive was tested for inhibition of both enzymes in assays using extracts of the recombinant *S. cerevisiae*. The compound, stilbamidine isethionate, was shown in these experiments to inhibit S-adenosylmethionine decarboxylase, but not ODC [40], and no further work was done with it.

4. Test Substances

Collections of synthetic compounds are easily screened in recombinant microbes [6,13]. Fermentation extracts present more problems, since they may contain nutrients that reduce dependence on the target protein. It proved difficult to screen fermentation extracts in the ODC assay [40], since even trace amounts of polyamines in the extract eliminate dependence on the parasite enzyme. Conversely, the PFK assay was compatible with fermentations. The suitability of fermentation extracts must be empirically determined in each case.

5. Risk/Benefit Analysis of Recombinant Microbe-Based Screens

Summarized in Table 4, the advantages of an HTS platform based on recombinant microbes are obvious for relatively small screening operations (e.g., parasiticide

Table 4 Risk/Benefit Analysis of Screens Based on Recombinant Microorganisms

Advantages	Disadvantages
Minimize equipment costs	Microbial permeability limits may give false negatives
Minimize personnel training costs	
Simple, robust assay endpoint	Not suitable for all targets (no useful phenotype; expression problems)
Focus attention on interesting compounds at early stage	
Complete genome sequences available for *E. coli*, *S. cerevisiae*, *C. elegans*	
Can accommodate targets not suitable for other formats	

discovery). Two profound benefits for resource consumption are evident. A distinct financial and time advantage arises from the use of a single format for multiple, diverse targets. This limits the amount of training required for personnel in the screening laboratory, and also greatly reduces equipment expense, since the endpoint (microbial viability) is so easily measured. The second advantage is a reduction in downstream resources spent on nonspecific actives. The inclusion of the nutrient-dependent viability test (or testing against host and parasite targets in different recombinant strains) drastically limits the number of positives that require follow-up. For example, we found a single confirmed positive fermentation extract in the nematode PFK screen out of 40,000 tested; this extract reduced motility and ATP levels in *H. contortus* simultaneously, caused an accumulation of [^{13}C]glucose 6-phosphate in *H. contortus*, and inhibited nematode PFK in enzyme assays (unpublished observations). Similarly, we found one confirmed candidate ODC inhibitor in a collection of 80,000 synthetic chemicals [40].

The unique benefit of recombinant microbes for screening is that target specificity is evaluated early in the process. Compounds that nonspecifically affect enzyme function (by sulfhydryl modification, cation chelation, alkylation, etc.) will not survive the comparative testing phase. The hit rate in our laboratories in these types of screens has been under 0.01% (unpublished observations), which means that additional attention can be focused only on the most promising candidates. In contrast, screens that employ purified recombinant enzymes for HTS often have hit rates ≥2.5% (unpublished observations). Further characterization of so many candidates requires a considerable effort in both biology and chemistry laboratories.

The biggest concern over the use of recombinant microbes is that microbial cell walls constitute a permeability barrier for test compounds. Enzyme inhibitors that cannot accumulate in bacterial or yeast cytoplasm will appear as false negatives in a screen. Permeability characteristics of *E. coli* and *S. cerevisiae* have not been rigorously described, so it is not possible to arrange for a different testing scheme for compounds with particularly unfavorable physical chemical properties. We have argued elsewhere [13] that this concern is overstated. Missing some active compounds is of less concern than missing all of them by not running a screen. Furthermore, compounds that have difficulty entering microbial cells may also have difficulty crossing lipid membranes in the parasite or host [13]. Screens that use purified proteins demand no accounting for pharmaceutical properties in active compounds, and often identify ''leads'' with poor behavior in animals. Such leads can consume considerable amounts of chemistry and biology resources in follow-up work. Therefore, whether permeability concerns constitute a problem or an advantage for screens based on recombinant microbes is moot.

Undeniably, not all targets are suited for HTS in recombinant microbes. In our own experience, we found that we were unable to complement a PFK$^-$ strain of *S. cerevisiae* with the cDNA encoding *H. contortus* PFK, despite the fact that it was functionally expressed in *E. coli* (unpublished observations). Conversely, we initially attempted to construct an ODC screen using a strain of *E. coli* that was unable to synthesize polyamines [40]. However, this strain required mutations in multiple genes in order to achieve the polyamine$^-$ phenotype, and proved to be too leaky for routine screening [40]. The ODC$^-$ strain of *S. cerevisiae* was much better suited for HTS.

It must also be acknowledged that some targets are not suitable for screens that employ purified proteins. The most convenient assay for PFK activity requires the presence of several additional enzymes that link to generate a product that can be quantified by spectrophotometry. ODC enzyme activity is most commonly measured by trapping [^{14}C]CO_2 for quantification by scintillation spectrometry. Neither assay is well suited for current HTS formats. It would have been technically difficult to screen for inhibitors of either enzyme without the use of recombinant microorganisms.

IV. FUTURE DIRECTIONS

The advantages of screens based on this paradigm are most evident for smaller-scale screening operations. However, their robust nature and ease of operation could make them attractive options for any screening laboratory. Expected developments in bioinformatics will only accentuate the benefits of the format.

A. Sequenced Genomes

The availability of complete genome sequences for *E. coli* [52], *S. cerevisiae* [53], and *C. elegans* ([54]; see below) provides an enormous advantage for the construction of specifically mutated strains. Furthermore, as noted above, HTS formats in *S. cerevisiae* have been developed for a variety of drug targets that have no yeast homolog [13,36–39]. Screens can be devised for inhibitors of protein:protein interactions, protein:DNA interactions, and protein:RNA interactions; alternative screening strategies for such purposes are not often apparent. These developments extend the utility of yeast as a screening tool far beyond the search for enzyme inhibitors, to include a variety of neurotransmitter receptors, ion channels, and transcriptional regulators. To date, antiparasitic drug discovery programs have not employed such novel strategies (at least publicly). It is not unreasonable to expect developments in this area in the future.

B. *C. elegans* as a Screening Tool

The free-living nematode *C. elegans* was the first metazoan organism to have its genome sequence defined [54]. The utility of this organism for anthelmintic screening has been recognized for 20 years (see [6–8] for review). Specific screens for anthelmintic as well as nonanthelmintic leads can be based on changes in behavior or survival of *C. elegans* [55]. It is also well known that phenotypes caused by mutations in *C. elegans* genes can be complemented by homologous genes from other organisms, including humans.

However, there are as yet no public examples of the use of recombinant *C. elegans* for the discovery of anthelmintics. The potential of this system is illustrated by the use of *C. elegans* to confirm that a specific mutation in a gene encoding β-tubulin from the parasitic species *H. contortus* underlies the phenotype of benzimidazole resistance [56]. It is to be expected that the enormous potential of *C. elegans* will be exploited in a variety of creative ways for anthelmintic discovery. Developments in genetic technology for the protozoan parasite *T. gondii* [57] may likewise portend future benefits for the discovery of antiprotozoal drugs.

REFERENCES

1. DA Windsor. Most of the species on Earth are parasites. Int J Parasitol 28:1939–1941, 1998.
2. M-S Chan. The global burden of intestinal nematode infection—fifty years on. Parasitol Today 13:438–444, 1997.
3. DWT Crompton. How much human helminthiasis is there in the world? J Parasitol 85:397–403, 1999.
4. SI Hirst, LA Stapley. Parasitology: the dawn of a new millenium. Parasitol Today 16:1–3, 2000.
5. AM Zajac, NC Sangster, TG Geary. Why veterinarians should care more about parasitology. Parasitol Today 16:504–506. 2000.

6. DP Thompson, RD Klein, TG Geary. Prospects for rational approaches to anthelmintic discovery. Parasitology 113:S217–S238, 1996.
7. TG Geary, DP Thompson, RD Klein. Mechanism-based screening: discovery of the next generation of anthelmintics depends upon more basic research. Int J Parasitol 29:105–112, 1999.
8. TG Geary, NC Sangster, DP Thompson. Frontiers in anthelmintic pharmacology. Vet Parasitol 84:275–295, 1999.
9. P Ehrlich, K Shiga. Farbentherapeutische Versuche bei Trypanosomenerkrankung. Berlin klin Wochscher 41:329–332, 362–365, 1904.
10. F Hawking. History of chemotherapy. In: RJ Schitzer and F Hawking, eds. Experimental Chemotherapy, Vol. 1. New York: Academic Press, 1963, pp 1–24.
11. GB Elion. The purine path to chemotherapy. Science 244:41–47, 1989.
12. GH Hitchings. The Third Paul Ehrlich lecture: basic research as a forerunner of new medications. Med Res Rev 11:1–15, 1991.
13. RD Klein, TG Geary. Recombinant microorganisms as tools for high throughput screening for nonantibiotic compounds. J Biomol Screen 2:41–49, 1997.
14. RD Klein, TG Geary. Method of identifying compounds useful as antiparasitic drugs. U.S. Patent 5,079,143, 1992.
15. S Ōmura, M Murata, K Kimura, S Matsukura, T Nishihara, H Tanaka. Screening for new antifolates of microbial origin and a new antifolate AM-8402. J Antibiot 38:1016–1024, 1985.
16. S Ōmura. Philosophy of new drug discovery. Microbiol Rev 50:259–279, 1986.
17. H Tomoda, S Ōmura. New strategy for discovery of enzyme inhibitors: screening with intact mammalian cells or intact microorganisms having special functions. J Antibiot 43:1207–1222, 1990.
18. RW Le Page, MP Barrett. Cloning of the 6-phosphoglucanate dehydrogenase gene from *Trypanosoma brucei* by complementation in *Escherichia coli*. Biochem Soc Trans 18:724–727, 1990.
19. RD Klein, ER Olson, MA Favreau, CA Winterrowd, NT Hatzenbuhler, MH Shea, SC Nulf, TG Geary. Cloning of a cDNA encoding phosphofructokinase from *Haemonchus contortus*. Mol Biochem Parasitol 48:17–26, 1991.
20. RD Klein, CA Winterrowd, NT Hatzenbuhler, MH Shea, MA Favreau, SC Nulf, TG Geary. Cloning of a cDNA encoding phosphoenolpyruvate carboxykinase from *Haemonchus contortus*. Mol Biochem Parasitol 50:285–294, 1992.
21. B Page, C Page, M Noel. A new fluorometric assay for cytotoxicity measurements in vitro. Int J Oncol 3:473–476, 1993.
22. LA Collins, SG Franzblau. Microplate Alamar blue assay versus BACTEC 460 system for high-throughput screening of compounds against *Mycobacterium tuberculosis* and *Mycobacterium avium*. Antimicrob Agents Chemother 41:1004–1009, 1997.
23. G Vasanthakumar, RL Davis Jr, MA Sullivan, JP Donahue. Cloning and expression in *Escherichia coli* of a hypoxanthine-guanine phosphoribosyl transferase-encoding cDNA from *Plasmodium falciparum*. Gene 91:63–69, 1990.
24. M Shahabuddin, J Scaife. The gene for hypoxanthine phosphoribosyl transferase of *Plasmodium falciparum* complements a bacterial *HPT* mutation. Mol Biochem Parasitol 41:281–288, 1990.
25. AE Eakin, R Nieves-Alicea, R Tosado-Acevedo, MS Chin, CC Wang, SP Craig III.

Comparative complement selection in bacteria enables screening for lead compounds targeted to a purine salvage enzyme of parasites. Antimicrob Agents Chemother 39:620–625, 1995.
26. AE Eakin, A Guerra, PJ Focia, J Torres-Martinez, SP Craig III. The hypoxanthine phosphoribosyltransferase (HPRT) from *Trypanosoma cruzi* as a target for structure based drug design: crystallization and inhibition studies with purine analogs. Antimicrob Agents Chemother 41:1686–1692, 1997.
27. B Canyuk, SP Craig III, AE Eakin. Bacterial complementation as a means to test enzyme-ligand interactions. Appl Microbiol Biotechnol 50:181–186, 1998.
28. T-S Yong, E Li, D Clark, SL Stanley Jr. Complementation of an *Escherichia coli adhE* mutant by the *Entamoeba histolytica EhADH2* gene provides a method for the identification of new antiamebic drugs. Proc Natl Acad Sci (USA) 93:6464–6469, 1996.
29. DC Kaslow, S Hill. Cloning metabolic pathway genes by complementation in *Escherichia coli*. Isolation and expression of *Plasmodium falciparum* glucose phosphate isomerase. J Biol Chem 265:12337–12341, 1990.
30. K Alexander, T Hill, J Schilling, M Parsons. Microbody phosphoglycerate kinase of *Trypanosoma brucei*: expression and complementation in *Escherichia coli*. Gene 90:215–220, 1990.
31. MR Mowatt, EC Weinbach, TC Howard, TE Nash. Complementation of an *Escherichia coli* glycolysis mutant by *Giardia lamblia* triosephosphate isomerase. Exp Parasitol 78:85–92, 1994.
32. F Dzierszinski, O Popescu, C Toursel, C Slomianny, B Yahiaoui, S Tomavo. The protozoan parasite *Toxoplasma gondii* expresses two functional plant-like glycolytic enzymes. Implications for evolutionary origin of apicomplexans. J Biol Chem 274: 24888–24895, 1999.
33. JT Beck, S Zhao, CC Wang. Cloning, sequencing, and structural analysis of the DNA encoding inosine monophosphate dehydrogenase (EC 1.1.1.205) from *Tritrichomonas foetus*. Exp Parasitol 78:101–112, 1994.
34. SO Ismail, YAW Skeiky, A Bhatia, LA Omara-Opyene, L Gedamu. Molecular cloning, characterization, and expression in *Escherichia coli* of iron superoxide dismutase cDNA from *Leishmania donovani chagasi*. Infect Immun 62:657–664, 1994.
35. L Stols, G Kulkarni, BG Harris, MI Donnelly. Expression of *Ascaris suum* malic enzyme in a mutant *Escherichia coli* allows production of succinic acid from glucose. Appl Biochem Biotechnol 63–65:153–158, 1997.
36. DR Kirsch. Development of improved cell-based assays and screens in *Saccharomyces* through the combination of molecular and classical genetics. Curr Opin Biotech 4:543–552, 1993.
37. MH Pausch. G protein-coupled receptors in *Saccharomyces cerevisiae*: high-throughput screening assays for drug discovery. Trends Biotechnol 15:487–494, 1997.
38. SG Oliver. From gene to screen in yeast. Curr Opin Gen Develop 7:405–409, 1997.
39. T Munder, A Hinnen. Yeast cells as tools for target-oriented screening. Appl Microbiol Biotechnol 52:311–320, 1999.
40. RD Klein, MA Favreau, SJ Alexander-Bowman, SC Nulf, L Vanover, CA Winterrowd, N Yarlett, M Martinez, JS Keithly, MA Zantello, EM Thomas, TG Geary.

Haemonchus contortus: cloning and functional expression of a cDNA encoding ornithine decarboxylase and development of a screen for inhibitors. Exp Parasitol 87: 171–184, 1997.
41. RD Klein, TG Geary, AS Gibson, MA Favreau, CA Winterrowd, SJ Upton, JS Keithly, G Zhu, RL Malmberg, MP Martinez, N Yarlett. Reconstitution of a bacterial/plant polyamine biosynthesis pathway in *Saccharomyces cerevisiae*. Microbiology 145:301–307, 1999.
42. JS Keithly, G Zhu, SJ Upton, KM Woods, MP Martinez, N Yarlett. Polyamine biosynthesis in *Cryptosporidium parvum* and its implications for chemotherapy. Mol Biochem Parasitol 88:35–42, 1997.
43. CH Sibley, VH Brophy, S Cheesman, KL Hamilton, EG Hankins, JM Wooden, B Kilbey. Yeast as a model system to study drugs effective against apicomplexan proteins. Methods 13:190–207, 1997.
44. VH Brophy, J Vasquez, RG Nelson, JR Forney, A Rosowsky, CH Sibley. Identification of *Cryptosporidium parvum* dihydrofolate reductase inhibitors by complementation in *Saccharomyces cerevisiae*. Antimicrob Agents Chemother 44:1019–1028, 2000.
45. JM Wooden, LH Hartwell, B Vasquez, CH Sibley. Analysis in yeast of antimalaria drugs that target the dihydrofolate reductase of *Plasmodium falciparum*. Mol Biochem Parasitol 85:25–40, 1997.
46. JF Cortese, CV Plowe. Antifolate resistance due to new and known *Plasmodium falciparum* dihydrofolate reductase mutations expressed in yeast. Mol Biochem Parasitol 94:205–214, 1998.
47. MB Reed, KJ Saliba, SR Caruana, K Kirk, AF Cowman. Pgh1 modulates sensitivity and resistance to multiple antimalarials in *Plasmodium falciparum*. Nature (London) 403:906–909, 2000.
48. S Volkman, D Wirth. Functional analysis of *pfmdr1* gene of *Plasmodium falciparum*. Meth Enzymol 292:174–181, 1998.
49. E Talla, RL de Mendonça, I Degand, A Goffeau, M Ghislain. *Schistosoma mansoni* Ca^{2+}-ATPase SMA2 restores viability to yeast Ca^{2+}-ATPase-deficient strains and functions in calcineurin-mediated Ca^{2+} tolerance. J Biol Chem 273:27831–27840, 1998.
50. R Mazhari-Tabrizi, V Eckert, M Blank, R Müller, D Mumberg, M Funk, RT Schwarz. Cloning and functional expression of glycosyltransferases from parasitic protozoans by heterologous complementation in yeast: the dolichol phosphate mannose synthase from *Trypanosoma brucei*. Biochem J 316:853–858, 1996.
51. TG Geary, SM Sims, EM Thomas, L Vanover, JP Davis, CA Winterrowd, RD Klein, NFH Ho, DP Thompson. *Haemonchus contortus*: ivermectin-induced paralysis of the pharynx. Exp Parasitol 77:88–96, 1993.
52. FR Blattner, G Plunkett III, CA Bloch, NT Perna, V Burland, M Riley, J Collado-Vides, JD Glasner, CK Rode, GF Mayhew, J Gregor, NW Davis, HA Kirkpatrick, MA Goeden, DJ Rose, B Mau, Y Shao. The complete genome sequence of *Escherichia coli* K-12. Science 277:1453–1474, 1997.
53. A Goffeau, BG Barrell, H Bussey, RW Davis, B Dujon, H Feldmann, F Galibert, JD Holheisel, C Jacq, M Johnston, EJ Louis, HW Mewes, Y Murakami, P Philippsen, H Tettelin, SG Oliver. Life with 6000 genes. Science 274:563–567, 1996.

54. *C. elegans* Sequencing Consortium. Genome sequence of the nematode *C. elegans*: a platform for investigating biology. Science 282:2012–2018, 1998.
55. JB Rand, CD Johnson. Genetic pharmacology: interactions between drugs and gene products in *Caenorhabditis elegans*. Meth Cell Biol 48:187–204, 1995.
56. MSG Kwa, JG Veenstra, M Van Dijk, MH Roos. β-Tubulin genes from the parasitic nematode *Haemonchus contortus* modulate drug resistance in *Caenorhabditis elegans*. J Mol Biol 246:500–510, 1995.
57. M Soete, C Hettman, D Soldati. The importance of reverse genetics in determining gene function in apicomplexan parasites. Parasitology 118:S53–S61, 1999.

15

Screening for Inhibitors of Lipid Metabolism

Hiroshi Tomoda and Satoshi Ōmura
The Kitasato Institute and Kitasato University, Tokyo, Japan

I. INTRODUCTION

In lipid metabolism, there is elegant balance in the levels of end-product lipids, and the enzymes and genes involved in their biosynthesis, as well as close cooperation with other metabolisms to maintain homeostasis. When the balance is lost, obesity or hyperlipidemia will develop, leading to a variety of serious diseases including atherosclerosis, hypertension, diabetes, functional depression of certain organs, and so on. Therefore, the control of lipid metabolism by drugs could lead to the prevention or treatment of these diseases.

Hypercholesterolemia involves heterogeneous disorders of lipid metabolism characterized by elevated levels of plasma total cholesterol and low-density lipoprotein (LDL)-derived cholesterol. It is definitively linked to increased morbidity and mortality due to myocardial infarction. Cholesterol enters the body in two ways; absorption from diet or endogenous biosynthesis. The interference of either process would provide an effective means of lowering plasma total cholesterol. 3-Hydroxy-3-methylglutaryl coenzyme A (HMG-CoA) reductase, a rate-limiting enzyme in the cholesterol biosynthetic pathway, was considered a promising target of inhibition, and in the 1970s, screening for inhibitors of this enzyme was carried out extensively. Eventually, the structurally related compounds compactin (ML236B) and mevinolin (monacolin K) were discovered from fungal culture broths as potent and specific inhibitors of HMG-CoA reductase [1–4]. Since then, the derivatives, lovastatin and pravastatin, and mevinolin, have been used clinically for the treatment of hypercholesterolemia [5]. Now, novel synthetic inhibitors of this enzyme, such as cerivastatin [6], are acknowledged as more effective agents.

Recent reports on clinical trials of pravastatin and simvastatin have shown a significant reduction in patient mortality rates for both hypercholesterolemic patients without known coronary heart disease and for those with existing coronary heart disease [7,8]. These trials have established cholesterol-lowering agents as an effective treatment for coronary disease and have stimulated the search for new cholesterol-lowering agents with other mechanisms of action.

For the last 25 years, we have searched for new biologically active compounds of microbial origin, including a number of enzyme inhibitors [9]. The enzyme inhibitors discovered by our group are summarized in Table 1. They are categorized into three classes on the basis of how they were discovered. The first class of compounds, cerulenin [10], herbimycin [11,12], lactacystin [13,14], and staurosporine [15], were originally discovered in assay systems nonspecified as enzyme inhibitors. Later studies on the mechanism of action revealed that they are enzyme inhibitors. The second class of compounds, arisugacins [16], pyripyropenes [17,18], and pepticinnamins [19], were discovered in direct assays using target enzymes themselves. So, it is necessary to test their specificity since they might inhibit other enzymes. The third class, diazaquinomycins [20,21], hymeglusin (1233A or F-244) [21,22], phosalacine [23], and triacsins [21], were discovered in mechanism-based assays using intact mammalian cells or micoorganisms with specific functions [21,24]. This kind of assay is expected to facilitate the discovery of highly specific enzyme inhibitors.

In this chapter, our recent experiences in screening for inhibitors of lipid metabolism are reviewed. We have concentrated on novel enzymes or proteins

Table 1 Enzyme Inhibitors of Microbial Origin Discovered by Ōmura and Co-workers

Compound	Discovered as/in	Target enzyme	Ref.
Cerulenin	Antibiotic	Fatty acid synthase	10
Herbimycin	Herbicide	Tyrosine kinase	11,12
Pyripyropenes	Enzyme inhibitor	ACAT	17,18
Pepticinnamins	Enzyme inhibitor	Protein farnesyltransferase	19
Lactacystin	Inducer of neurite outgrowth	Proteasome	13,14
Hymeglusin	Mechanism-based screen	HMG-CoA synthase	21,22
Arisugacins	Enzyme inhibitor	Acetylcholine esterase	16
Staurosporine	Chemical screen	Protein kinases	15
Phosalacine	Mechanism-based screen	Glutamate synthase	23
Diazaquinomycins	Mechanism-based screen	Thymidylate synthase	20,21
Triacsins	Mechanism-based screen	Acyl-CoA synthetase	21

as a target of inhibition for drug discovery, and in some cases, established intricate assay systems even in the primary screens for drug discovery. The practical assay methods are also described.

II. NEW TARGET ENZYMES AND PROTEINS OF LIPID METABOLISM

We established a screening system for inhibitors of mevalonate biosynthesis from microbial metabolites by utilizing intact animal cells as a test organism [22]. Microbial culture broths were screened whose cytotoxic effects were rescued by the addition of mevalonate to the cell culture medium, resulting in the discovery of fungal beta-lactone hymeglusin (1233A or F-244). Studying the mechanism of action revealed that hymeglusin inhibits HMG-CoA synthase specifically by modifying the active-site cysteine of the enzyme covalently [25,26]. Second, although HMG-CoA reductase inhibitors have proved effective and clinically well tolerated, the enzyme steps after formation of farnesyl pyrophosphate in the cholesterol biosynthetic pathway are considered preferable as targets of inhibition, because a potentially dose-limiting toxicity might arise from the consequent reduced levels of essential isoprenoid precursors, the antioxidant ubiquinone, or the dolichols. Similarly, inhibiting enzymes after the formation of lanosterol has the potential to result in toxicity arising from the utilization of intermediates having the complete sterol ring system. Therefore, squalene synthase, squalene epoxidase, and lanosterol synthase (oxidosqualene cyclase) remain for further consideration as enzyme inhibition targets. Now, considerable effort has been directed toward screening for inhibitors of these enzymes. Glaxo and Merck researchers independently discovered a series of similar fungal metabolites, squalestatins and zaragozic acids, as potent inhibitors of squalene synthase, respectively [27–29]. However, the inhibitors also have an ability to inhibit protein farnesyltransferase because both enzymes recruit farnesyl pyrophosphate as a common substrate [30], so specific inhibitors have been sought or chemically synthesized on the basis of squalestatin or zaragozic acid as a lead [31].

Third, acyl-CoA:cholesterol acyltransferase (ACAT) [EC 2.3.1.26], an enzyme that works after the formation of cholesterol, was considered a unique target of inhibition [32]. ACAT catalyzes the synthesis of cholesteryl esters from cholesterol and long-chain fatty acyl-CoA. ACAT plays important roles in the body, for example, in the absorption of dietary cholesterol from the intestines, production of lipoprotein in liver and formation of foam cells from macrophages in arterial walls. Therefore, ACAT inhibition is expected not only to lower plasma cholesterol levels but also to have a direct effect at the arterial wall. A number of synthetic ACAT inhibitors such as ureas, imidazoles, and acyl amides have been developed [33]. Several groups have searched for novel ACAT inhibitors of natural origin with structures different from the synthetic ones. We discovered purpactins [34], glisoprenins [35], terpendoles [36,37], and pyripyropenes [17,18] from fungal culture broths. Among the natural inhibitors so far reported, pyripyropenes show the most potent inhibitory activity against ACAT, and proved to be orally active in hamsters, reducing cholesterol absorption from intestines [38]. Issues surrounding the toxic effect of ACAT inhibitors on the adrenal gland are pervasive [39]. There has been no conclusive data as to whether the toxic effects on the adrenal gland are due to the mechanism of action of these drugs. However, certain synthetic inhibitors proved effective *in vivo* but had no effect on the adrenal gland [40,41]. Furthermore, recent molecular biological studies have afforded new information about ACAT. Chang and co-workers succeeded in cloning cDNA for an ACAT gene from macrophages [42], now known as ACAT-1. ACAT-1 is ubiquitously expressed, and high-level expression is observed in sebaceous glands, steroidogenic tissues, and macrophages. ACAT-1-deficient transgenic mice [43,44] were viable, fertile, and ostensibly healthy, with no evidence of adrenal dysfunction, indicating that the observed toxicity is not related to ACAT inhibition and that the loss of ACAT-1 is not incompatible with life. Furthermore, the knockout mice had a reduced ability to synthesize cholesteryl esters, but only in specific tissues, suggesting the presence of another ACAT. These findings have led to the discovery of ACAT-2, responsible for cholesterol esterification in the liver and intestine [45–47]. Therefore, ACAT inhibitors need to be reevaluated for their specificity for ACAT-1 or -2 inhibition and toxicity to the adrenal gland.

More attention is now being paid to lipid metabolic pathways as promising targets of inhibition for the treatment of infectious diseases. An alternative mevalonate-independent (nonmevalonate) pathway has been defined specifically in some bacteria [48] and malaria [49]. *Mycobacterium tuberculosis*, the organism that causes tuberculosis, has unique cell wall lipids. An enzyme in lipid biosynthesis was found to be a target of isoniazid, an important first-line antituberculosis drug [50]. Therefore, enzymes involved in these pathways are expected to make novel targets for antibacterial, antimalarial, and antituberculosis drugs.

III. SCREENING FOR INHIBITORS OF DIACYLGLYCEROL ACYLTRANSFERASE

Triacylglycerol (TG) synthesis in mammals is important in many processes, including lactation, energy storage in fat and muscle, fat absorption in the intestine, and the assembly of lipoprotein particles in the liver and small intestine. Too much accumulation of TG in certain organs and tissues of the body causes fatty liver, obesity, and hypertriglyceridemia. Diacylglycerol acyltransferase (acyl-CoA:1,2-diacyl-*sn*-glycerol *O*-acyltransferase, abbreviated as DGAT) [EC 2.3.1.20] catalyzes the reaction of acyl residue transfer from acyl-CoA to diacylglycerol to form TG. The reaction is the final step of *de novo* triacylglycerol biosynthesis, and is the only pathway which is exclusively involved in triacylglycerol formation (Fig. 1) [51]. Therefore, DGAT is considered a potential target of inhibition for control of such disorders. However, few DGAT inhibitors have been reported [52].

A. Screening Method

The enzyme DGAT has not been purified to date, probably because it is a hydrophobic and integral membrane protein. Therefore, DGAT activity was measured using rat liver microsomes as an enzyme source and radiolabeled palmitate as a substrate by the method of Mayorek and Bar-Tana [52] with some modifications [53]. The reaction mixture contains microsomal protein, BSA, [^{14}C]palmitoyl-CoA, $MgCl_2$, diisopropyl fluorophosphate, 1,2-dioleoyl-*sn*-glycerol, and a test sample in a total volume of 0.2 ml. After a 15-min incubation at 23°C, lipids are extracted and separated by thin-layer chromatography (TLC). The distribution of radioactivity on TLC is analyzed with a radioscanner to determine the amount of [^{14}C]TG.

To investigate the specificity of DGAT inhibitors, synthesis of lipids including TG, phosphatidylcholine (PC), and phosphatidylethanolamine (PE) was measured in an intact cell assay using Raji cells by our established method [54]. Raji cells are incubated in the presence of [^{14}C]oleic acid with or without a test sample in a total volume of 0.2 ml. After a 20-min incubation at 37°C, [^{14}C]oleic acid is incorporated mainly into PC, PE, and TG, which are extracted and separated by TLC. The distribution of radioactivity in these lipids on TLC is analyzed.

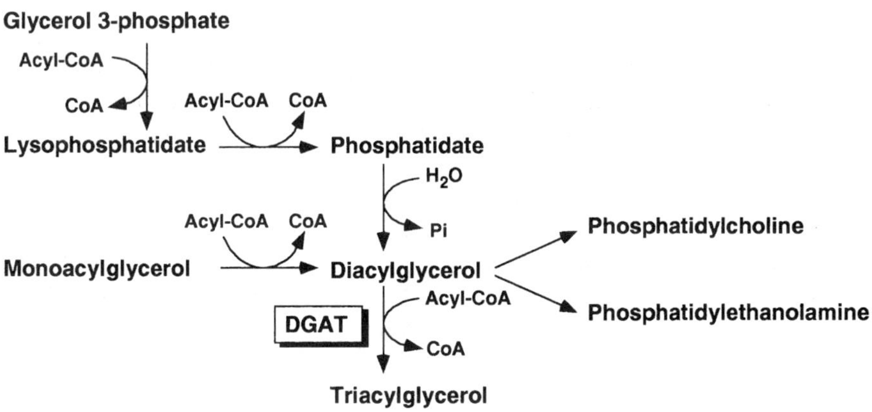

Figure 1 Triacylglycerol biosynthetic pathway.

If the sample inhibits DGAT activity specifically in Raji cells, incorporation of [^{14}C]oleic acid only into the TG fraction should be inhibited.

B. Inhibitors

About 15,000 samples, mostly microbial culture broths, were subjected to our screening program for DGAT inhibitors. Finally, we obtained amidepsines and roselipins from fungal strains, and xanthohumols from a plant extract as novel DGAT inhibitors (Fig. 2).

Five amidepsines, A to E, were isolated from the culture broths of fungal strains FO-2942 and FO-5969, which were considered to belong to the genus *Humicola* [53,55,56]. They are structurally related, with the common skeleton of tridepside gyrophoric acid. Many gyrophoric acid derivatives were isolated from lichens and amidepsine D was identical with 2,4-di-*O*-methylgyrophoric acid [57]. However, the other amidepsines are a new type of compound having an amino acid (alanine or valine) attached to the skeleton. Two chalcones were isolated from extracts obtained by treatment of hop of *Humulus lupulus* (L.) with 70% methanol [58]. One was identical with xanthohumol and the other, named xanthohumol B, was found to be a new compound. Very recently, four novel glycolipids, named roselipins, were isolated from the culture broth of *Gliocladium roseum* KF-1040, a marine isolate [59–61]. They are composed of highly methylated fatty acid, hexose, and alditol moieties. The different terminal hydroxy residue of the alditol is bound to the carboxylic acid of the fatty acid to form the diastereoisomers of roselipins 1A and 1B, or roselipins 2A and 2B.

The effects of these compounds on triacylglycerol synthesis are summarized in Table 2. In the assay using rat liver microsomes, the compounds all inhibited DGAT activity in a dose-dependent manner, and amidepsines A, B, and D, and roselipins are potent, with IC_{50} values of 10–22 μM. In the intact Raji cell assay, amidepsine B showed inhibition of TG synthesis, but almost no effects on phosphatidylcholine and phosphatidylethanolamine syntheses, indicating that the drug inhibits DGAT activity in Raji cells specifically. Amidepsine D and xanthohumols also showed specific inhibition of DGAT to some extent. However, roselipins did not show selective TG inhibition, probably due to lack of permeability through the cell membrane.

The DGAT enzyme has been known to exist for years, and a number of researchers have contributed to characterizing its biochemical properties. Recently, Farese and co-workers identified a gene encoding a mammalian DGAT [62]. Interestingly, the translation of a full-length cDNA predicts an open reading frame encoding a 498-amino acid protein that is about 20% identical to mouse ACAT, with the most highly conserved region in the C terminus. From the predicted protein sequence, a potential *N*-linked glycosylation site, a putative tyrosine phophorylation site, and an active site serine residue also found in ACAT are shown. The protein has multiple hydrophobic domains and 6–12 possible

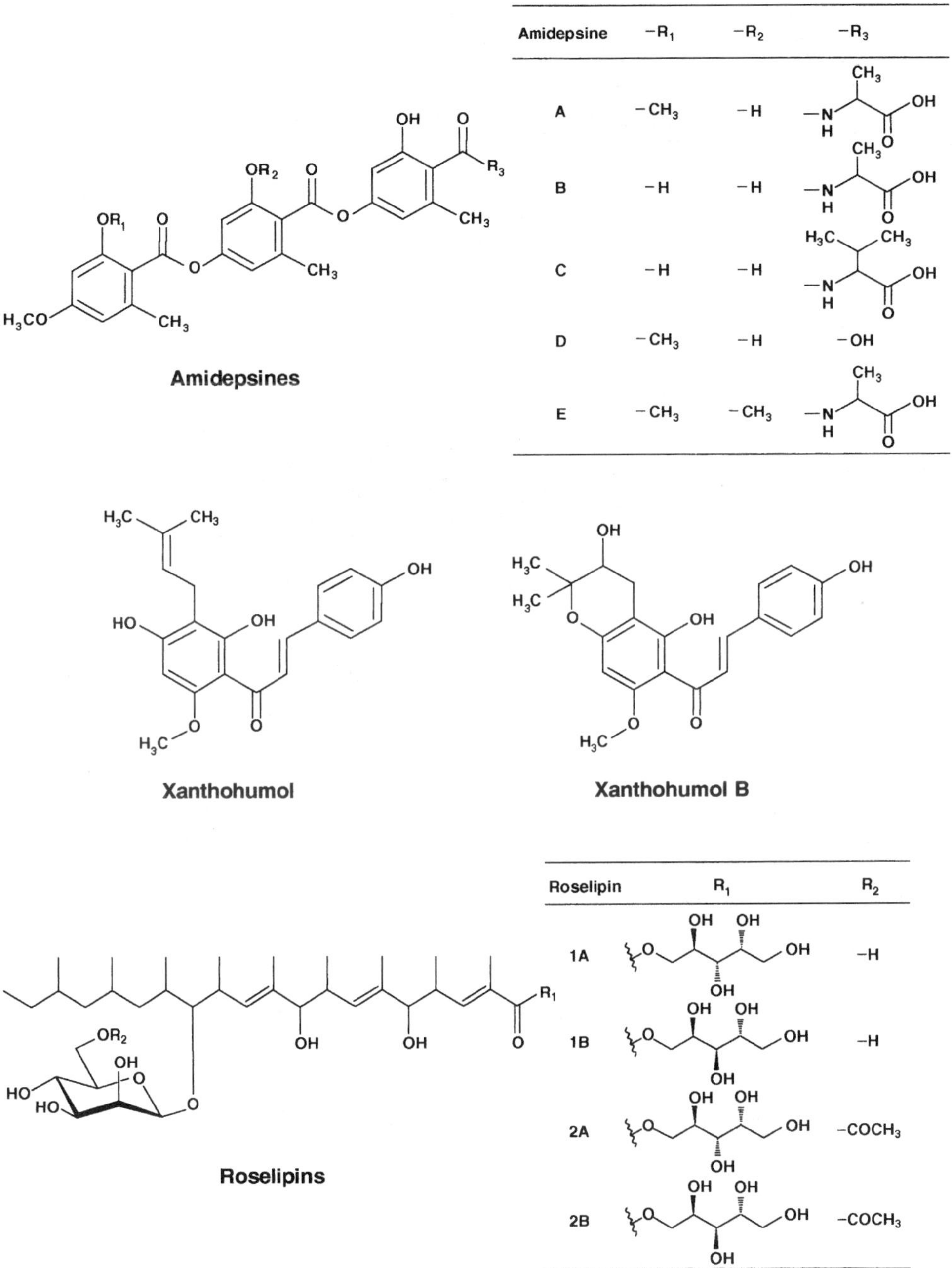

Figure 2 Structures of DGAT inhibitors.

Table 2 Inhibition of Triacylglycerol Synthesis by DGAT Inhibitors in Assays Using Rat Liver Microsomes and Intact Raji Cells

	Rat microsomes	Raji cells		
DGAT inhibitor IC_{50} (μM)	TG IC_{50} (μM)	TG IC_{50} (μM)	PC IC_{50} (μM)	PE IC_{50} (μM)
Amidepsine A	10	16	>100	>100
Amidepsine B	19	3.4	>100	>100
Amidepsine C	52	17	52	>100
Amidepsine D	18	2.8	10	>100
Amidepsine E	124	91	100	100
Xanthohumol	50	21	>100	>100
Xanthohumol B	190	34	>100	>100
Roselipin 1A	17	30	30	>30
Roselipin 1B	15	24	24	28
Roselipin 2A	22	20	20	20
Roselipin 2B	18	20	20	20

TG, triacylglycerol; PC, phosphatidylcholine; PE, phosphatidylethanolamine.

transmembrane domains. Its mRNA expression was detected in every mammalian tissue, and the highest expression levels were found in the small intestine. These findings are consistent with a proposed role for DGAT in intestinal fat absorption. However, mRNA expression was relatively low in the livers of humans regardless of high DGAT activity, suggesting the existence of a second DGAT in livers. They established a DGAT expression system, and samples including our DGAT inhibitors are to be evaluated. Thus, further understanding of DGAT at a molecular level will help us search for DGAT inhibitors, leading to potential approaches for treating hypertriglyceridemia or obesity in humans.

IV. SCREENING FOR INHIBITORS OF CHOLESTERYL ESTER TRANSFER PROTEIN

Cholesteryl ester transfer protein (CETP) promotes exchange and transfer of neutral lipids such as cholesteryl ester (CE) and TG between plasma lipoproteins [63–65]. The function of CETP is illustrated in Fig. 3. CETP is a very hydrophobic and heat-stable glycoprotein with an apparent molecular weight of 74 kDa as determined by SDS-PAGE analysis [66,67]. The cDNA from human liver was cloned and sequenced [68]. It encodes for a 476-amino acid protein (53 kDa), suggesting that the apparent higher molecular weight is due to the addition of carbohydrate residues by posttranslational modification.

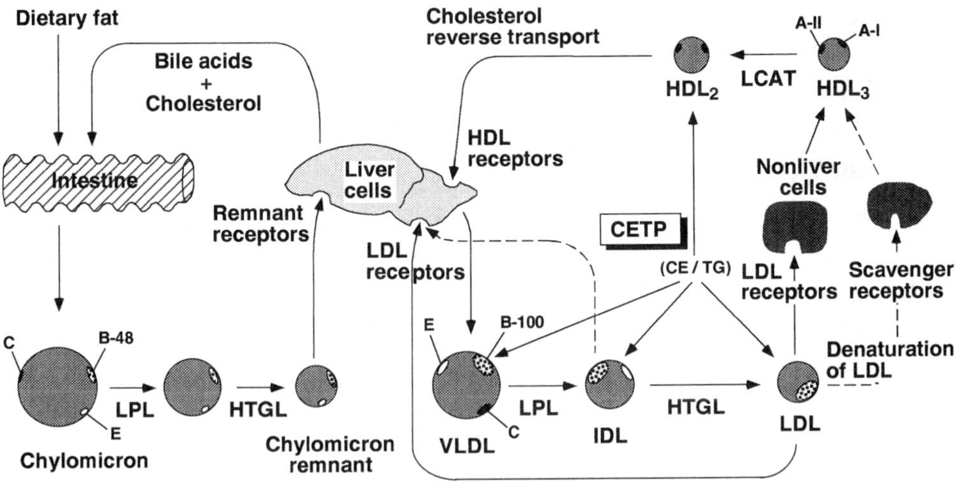

Figure 3 Function of CETP, LPL (lipoprotein lipase) and HTGL (hepatic triglyceride lipase).

Evidence has been accumulating of the importance of CETP in atherosclerosis: (1) CETP decreased the cholesterol concentration in high density lipoprotein (HDL) *in vitro* [69] and *in vivo* [70]; (2) rats and mice deficient in CETP activity have high plasma HDL and are resistant to atherosclerosis; (3) human subjects with a genetic deficiency of CETP have very high HDL and low LDL cholesterol levels and are resistant to atherosclerosis [71,72]; (4) human CETP gene-introduced transgenic mice have a redistribution of cholesterol from HDL to LDL [70,73,74] and exhibit a marked increase in susceptibility to diet-induced atherosclerosis [74]; (5) antisense oligonucleotides and antibody against CETP inhibited the development of atherosclerosis in cholesterol-fed rabbits [75]; and (6) antibody against CETP inhibited the development of atherosclerosis in cholesterol-fed rabbits [76]. Therefore, inhibition of CETP is proposed as a novel target for anti-atherosclerotic drugs. Interestingly, the mechanism of CETP-mediated lipid transfer is still unclear. In fact, small molecules modulating the CETP activity have been searched for extensively for therapeutic and biochemical purposes.

A. Screening Methods

1. In Vitro *Assay*

Three methods have been reported for CETP assay. The rationale of each method is illustrated in Figure 4. Methods B and C are good for a high-throughput screening (HTS) format.

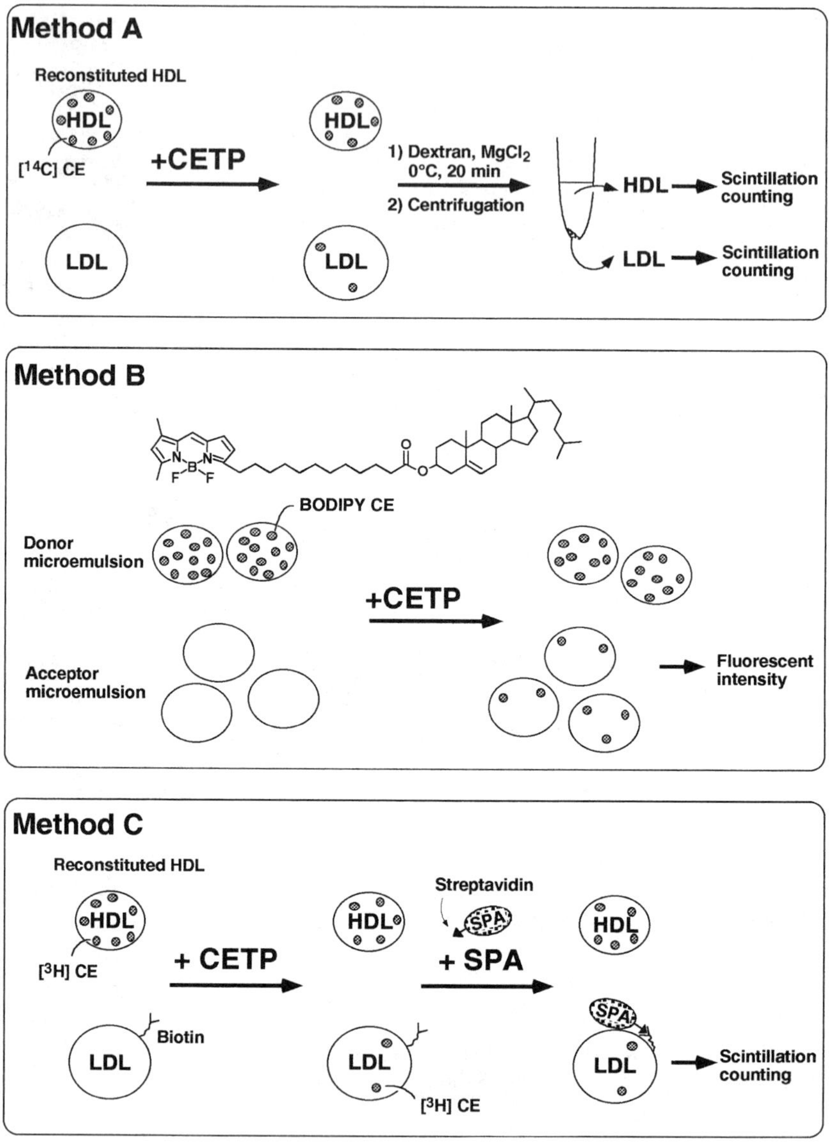

Figure 4 Rationales of three CETP assays.

Method A. The rationale of method A is that HDL and LDL are separated by selective precipitation of LDL by dextran sulfate and Mg^{2+} after the reaction between LDL and reconstituted HDL containing radiolabeled CE by CETP. The method was originally described by Kato et al. [77]. The assay mixtures consist of reconstituted [^{14}C]CE-HDL as the donor for CE, LDL as the acceptor, 5,5′-dithiobis-2-nitrobenzoic acid, bovine serum albumin (BSA), partially purified CETP, and a test sample in Eppendorf tubes (1.5 ml). After a 30-min incubation at 37°C, the reaction is terminated by the addition of an LDL-precipitation solution. After standing for 20 min in an ice bath, the assay mixtures are centrifuged, and the supernatant solution containing [^{14}C]CE-HDL is analyzed for radioactivity. Furthermore, the [^{14}C]CE-LDL precipitate is also analyzed for radioactivity if necessary. Usually the blank and control transfer values are about 6% and 34% of initial [^{14}C]CE-HDL added under the assay conditions, respectively.

Method B. Bisgaier et al. [78] reported a CETP assay using fluorescent cholesteryl 4,4-difluoro-5,7-dimethyl-4-bora-3a,4a-diaza-3-indacenedodecanoate (BODIPY-CE, Fig. 4) in microemulsions. Microemulsions for donor/acceptor contain triolein, BODIPY-CE/cholesteryl oleate, and 1-hexadecanoyl-2-[*cis*-9-octadecenoyl]-*sn*-glycero-3-phosphocholine. The assay mixtures consist of acceptor microemulsions, donor microemulsions, and a test sample in each well of flat-bottom 96-well plates. After preincubation at 37°C for 10 min, the reaction is initiated by addition of CETP solution. The fluorescent intensity in each well of the plates is periodically (every 10 sec) detected at 37°C with a fluorescent 96-well plate reader equipped with 485- and 538-nm bandpass filters in the excitation and emission paths, respectively.

Method C. Most researchers have carried out CETP assay using a Scintillation Proximity Assay (SPA) kit [79]. The reaction mixtures containing a test sample, [^{3}H]CE-HDL, and biotin-LDL, are thoroughly mixed. The reaction is initiated by the addition of CETP. After a 4-hr incubation at 37°C, the reaction is terminated by the addition of streptavidin SPA beads, and allowed to stand for 1 hr at room temperature. Finally, the transfer of [^{3}H]CE from [^{3}H]CE-HDL to LDL is measured by scintillation counter.

2. Ex Vivo *Assay*

Ex vivo tests for CETP inhibitors were reported using transgenic mice expressing human CETP and human apo A-I [80] or hamsters [81]. For example, a test sample dissolved in Cremophor EL solution (4 μl, final 10 mg/kg) is administered to male mice (fasted overnight). Blood is taken at 4 and 24 hr after dosing, and is centrifuged immediately to obtain plasma. The plasma (25 μl) is used as a CETP source to determine the CETP activity by method A.

B. Inhibitors

Over 20,000 samples of microbial culture broths were subjected to our screening program for CETP inhibitors by method A. At first no BSA was added to the assay mixture, but many false-positive compounds such as fatty acids were isolated. To prevent this, the optimal concentration of BSA was tested and set up as 200 μM, resulting in a low hit rate in the primary screen. The serum albumin concentration in the assay is similar to that in human plasma. Finally, we discovered erabulenols from a fungal strain, and ferroverdins from an actinomycete strain, as novel CETP inhibitors (Fig. 5).

Erabulenols A and B were isolated from the culture broths of fungal strain FO-5637, considered to belong to the genus *Penicillium* [82,83]. Scleroderolide, previously reported as a fungal metabolite, was also isolated from culture broth. They have a tetracyclic ring, and erabulenols consist of a phenalenone skeleton and a trimethyltetrahydrofuran moiety. Very recently, three ferroverdins were isolated as very potent CETP inhibitors from the culture broth of *Streptomyces* sp. WK-5344 [84,85]. Ferroverdin A was known as a green pigment, but ferroverdins B and C were new compounds. They are a complex of one iron ion and three ligands with structures of *para*-vinylphenyl-nitrosohydroxybenzoate derivatives.

Novel compounds of synthetic or natural origin (Fig. 5) were reported to inhibit CETP activity. PD 140195 [86], cholesterol derivatives of U-617 and U-95594 [87], and an isoflavan CGS 25159 [81] are synthetic inhibitors. Recently, synthetic pyridine derivatives [88,89], and bis-(2-aminophenyl) disulfides and 2-aminophenyl thio derivatives [90,91] were disclosed. Wiedendiols [92] and suberitenones [93] were isolated from marine sponges, and U-106305 was isolated as a novel CETP inhibitor from a *Streptomyces* sp. [94]. A peptide from hog plasma was also reported to inhibit the activity [95]. Additionally, known compounds such as strongylin A, puupehenone, chloropuupehenone, aureol, and avarol were reported to show CETP inhibitory activity [92]. They were all previously isolated from marine sponges and are structurally similar to wiedendiols. We also observed that known fungal metabolites, sclerotiorin originally isolated as a yellow pigment [96], and L681512 compounds isolated as elastase inhibitors [97] inhibit CETP activity [80,98].

The IC_{50} values of these compounds for CETP activity are summarized in Table 3. Although it is difficult to compare their inhibitory activity due to the different assay methods, chloropuupehenone and ferroverdin B show very potent CETP inhibition, with nanomolar IC_{50} values. Some compounds gave quite different IC_{50} values depending on the assay conditions. For example, the IC_{50} value of sclerotiorin is 19 μM when assayed in the absence of BSA, and 67 μM in the presence of BSA. Therefore, the addition of a high concentration of serum albumin or whole plasma to the assay mixture is recommended to exclude a nonspecific-binding type of inhibitor. Accordingly, the CETP inhibitory activity of cer-

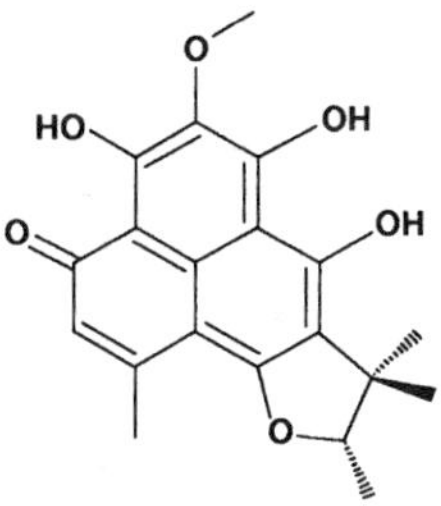

Erabulenol A

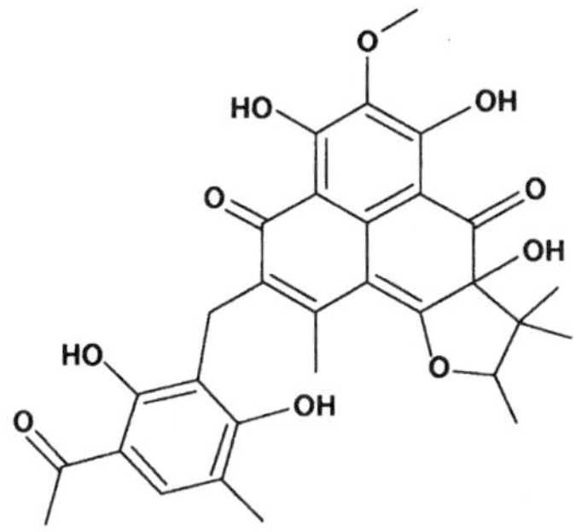

Erabulenol B

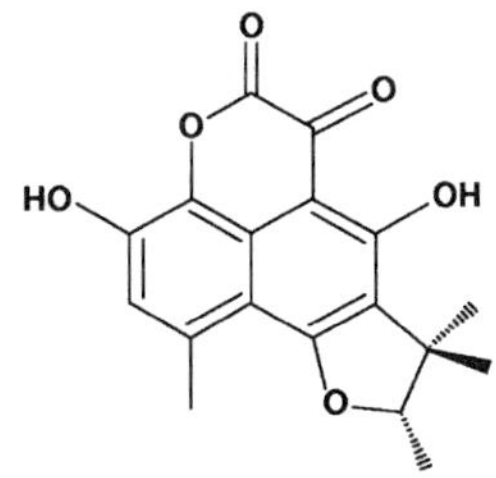

Scleroderolide

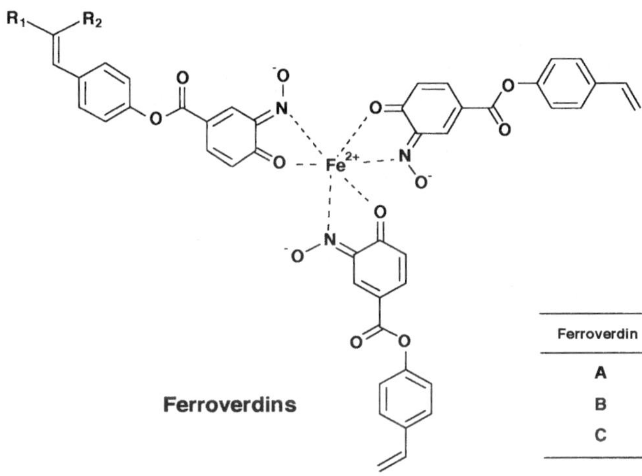

Ferroverdins

Ferroverdin	R_1	R_2
A	H	H
B	OH	H
C	H	COOH

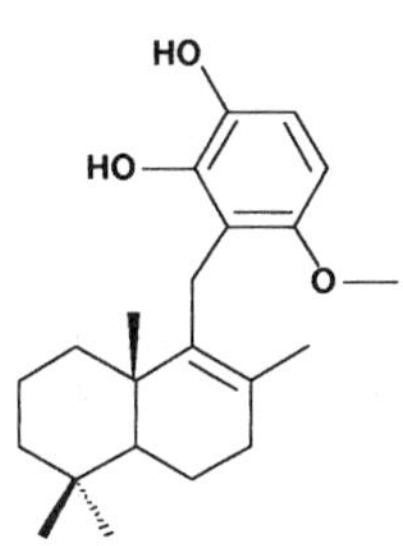

Wiedendiol-A

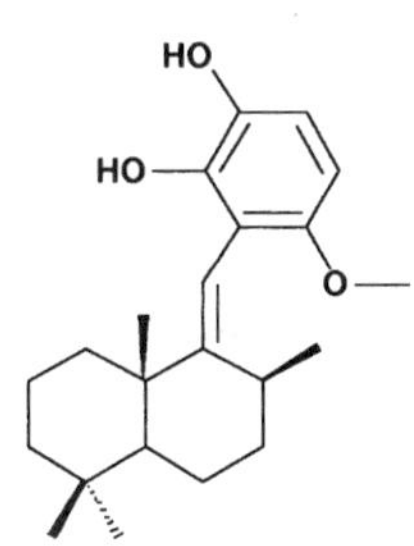

Wiedendiol-B

(A)

Figure 5 Structures of CETP inhibitors of natural (A) and synthetic (B) origin.

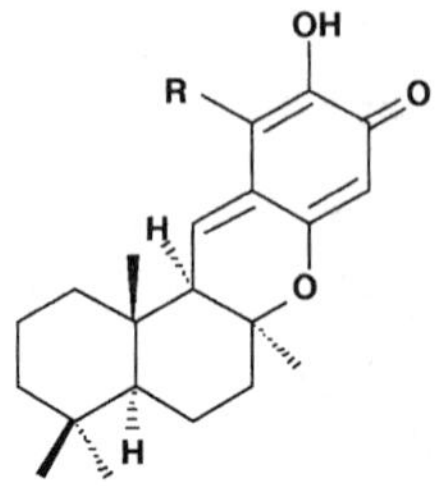

	R
Chloropuupehenone	Cl
Puupehenone	H

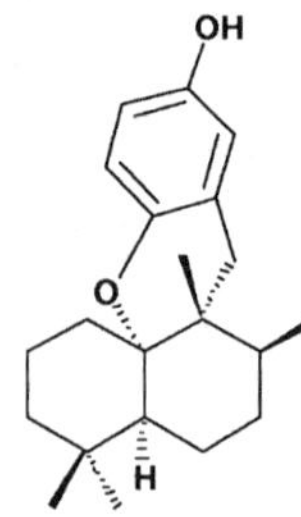

Aureol

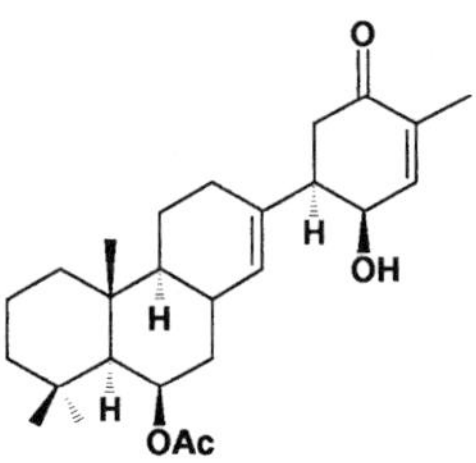

Suberitenone A

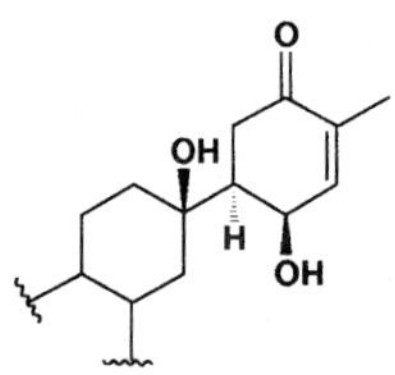

Suberitenone B

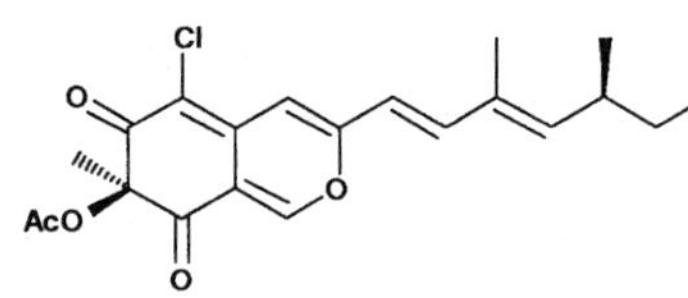

Sclerotiorin

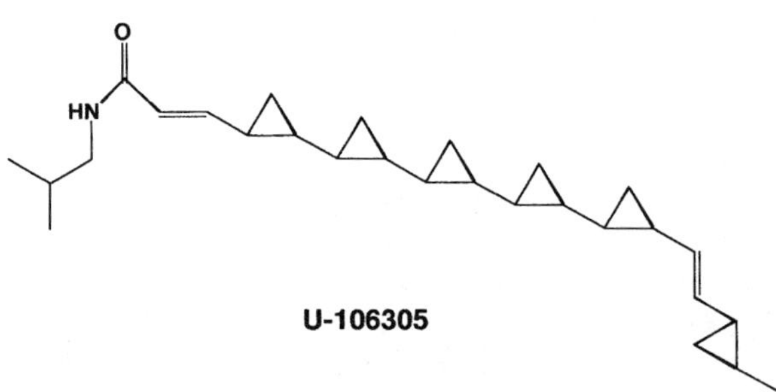

U-106305

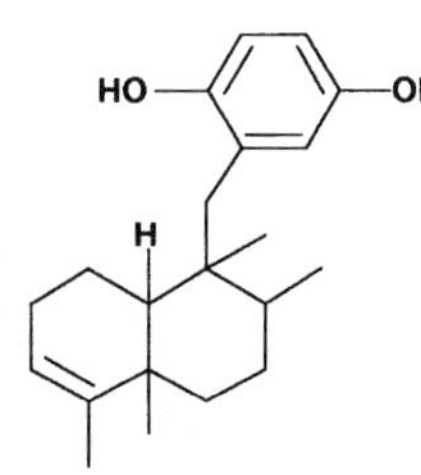

Avarol

(A Continued)

Figure 5 Continued.

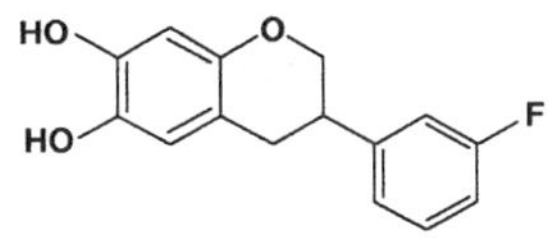

CGS 25159

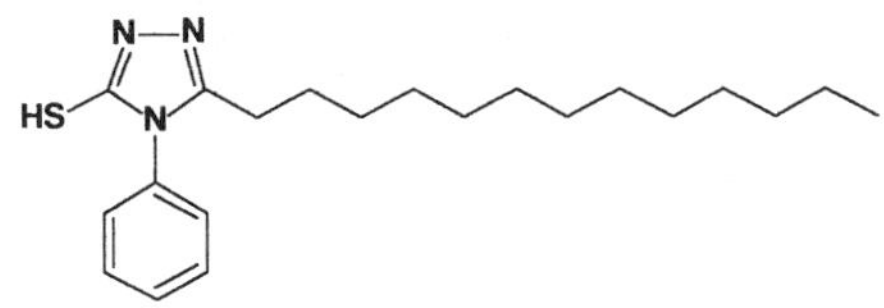

PD 140195

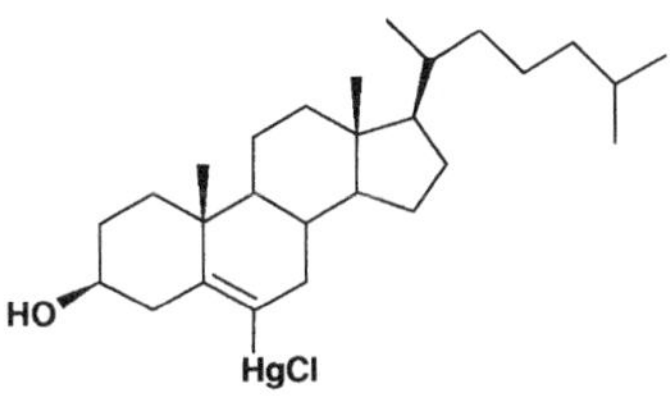

U-617

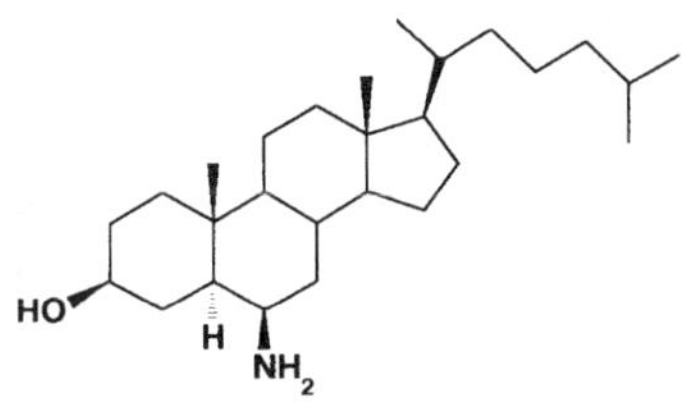

U-95594

(B)

Figure 5 Continued.

tain compounds should be reevaluated in the presence of serum albumin or plasma. In this sense, the highly potent CETP inhibition by ferroverdins B and C even in the presence of BSA is promising and further *ex vivo* and *in vivo* evaluation of the compounds is warranted.

Data have been presented for the *ex vivo* and *in vivo* evaluation of CETP inhibitors. *Ex vivo* efficacy was shown using hamsters for CGS 25159 (oral administration at 10 mg/kg) [81] and using transgenic mice for sclerotiorin [80] and L681512 (oral administration at 10 mg/kg) [98]. Furthermore, analysis of plasma lipoproteins from CGS 25159-treated (10 and 30 mg/kg, p.o.) hypercholesterolemic hamsters showed *in vivo* efficacy with an increase in HDL cholesterol [81].

C. Application of Inhibitors for the Biochemical Study of CETP

Since sclerotiorin, a member of the azaphilone family, was found to inhibit CETP activity, the effect of azaphilones on CETP activity was tested. A structure-specific CETP inhibition by azaphilones was shown. Electrophilic ketone(s) and/or enone(s) at both the C-6 and C-8 positions of the isochromane-like ring are neces-

Table 3 CETP Inhibitors

Compound	Origin	Assay method	IC_{50} (μM)	Ref.
Aureol	Sponge	C	22	92
Avarol	Sponge	C	25	92
Chloropuupehenone	Sponge	C	0.30	92
CGS25159	Synthetic	A	10	81
Erabulenol A	Fungus	A	48	82,83
Erabulenol B	Fungus	A	58	82,83
Ferroverdin A	Actinomycete	A	21	84,85
Ferroverdin B	Actinomycete	A	0.62	84,85
Ferroverdin C	Actinomycete	A	2.2	84,85
L-681,512-1∼-4	Fungus	A	1.6∼2.5	97,98
PD 140195	Synthetic	B/B(+BSA)	2.0/50	86
Puupehenone	Sponge	C	6.0	92
Sclerotiorin	Fungus	A (−BSA)	19	80
Scleroderolide	Fungus	A	95	82,83
Suberitenone A	Sponge	C		93
Suberitenone B	Sponge	C	10	93
U-106305	Actinomycete	B	25	94
U-617	Synthetic	B		87
U-95594	Synthetic	B		87
Wiedendiol A	Sponge	C	5.0	92
Wiedendiol B	Sponge	C	5.0	92

sary for eliciting CETP inhibitory activity (Fig. 6A). Several experiments strongly suggested that the drug inhibits CETP irreversibly. Sclerotiorin was reported to react with a primary amine such as ammonia and methylamine to yield sclerotioramine and *N*-methylsclerotioramine, respectively [99]. Similarly, sclerotiorin reacts with α- and ε-amino residues of lysine to form the sclerotiorin adducts in our model reaction (Fig. 6B), which seems to explain the structure-specific CETP inhibition by azaphilones. Therefore, it is plausible that the drug modifies primary amines such as lysine or the *N*-terminal amino acid in the CETP molecule covalently. The amino acid sequence of human CETP revealed 25 lysine residues in the molecule [68], some of which are responsible for CETP activity [100]. Furthermore, the *N*-terminal cysteine of both human and rabbit plasma CETP was modified by *para*-chloromercuriphenylsulfonate, resulting in inhibition of TG transfer activity [101]. Therefore, covalent modification of such residues by sclerotiorin might impair the CETP activity.

Swenson et al. observed the transfer of CE from vesicles to CETP by separating the vesicles and CETP through a gel-filtration column [102]. Under similar conditions, 24% of [^{14}C]CE is transferred from [^{14}C]CE/PC vesicles to CETP. A

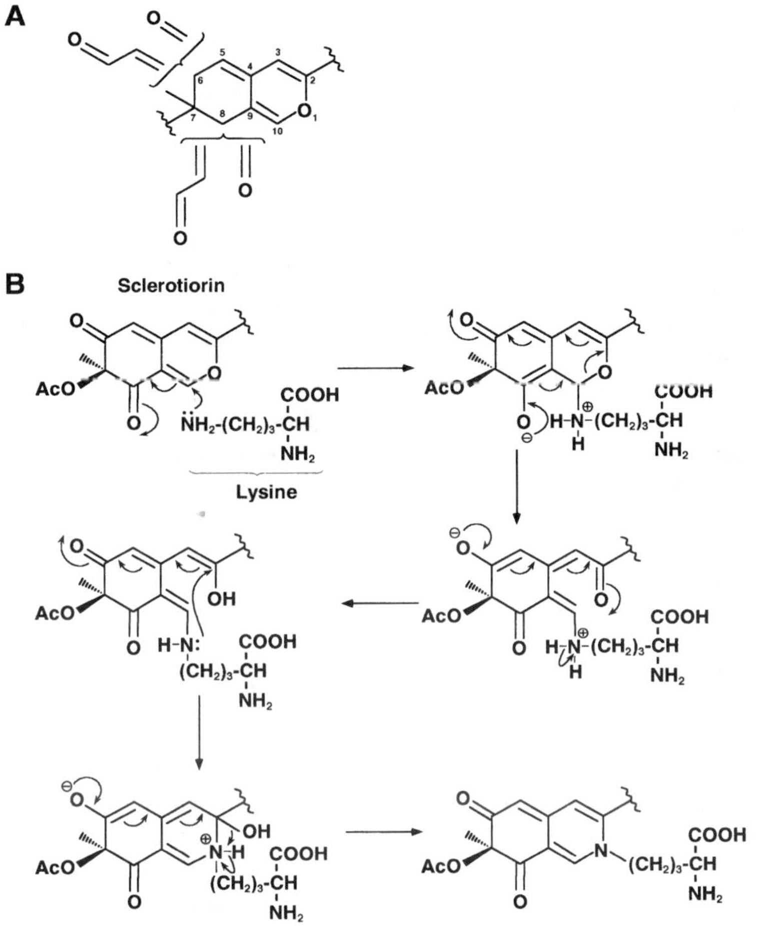

Figure 6 Consensus structure of azaphilones for CETP inhibition (A) and hypothetical mechanism of reaction of sclerotiorin with ε-amino residue of lysine (B).

thiol-modifying reagent, *para*-chloromercuriphenylsulfonate, inhibits this process [102], while sclerotiorin does not affect it essentially, but CETP lost the ability to transfer bound [^{14}C]CE to LDL [80]. These findings indicate that the inhibitory mechanism of sclerotiorin is different from that of *para*-chloromercuriphenylsulfonate. Previous studies showed that the binding of CETP to lipoproteins involves mainly ionic interactions, that is, negatively charged lipoproteins and positively charged CETP [103,104]. Jiang et al. [100] demonstrated, in fact, that point mutagenesis of positively charged amino acids including protonated ^{233}Lys within the conserved region of CETP reduced the HDL binding and CETP activities markedly. Therefore, it is plausible that sclerotiorin can modify such important lysine residues especially on the surface of the molecule to form a covalent bond as predicted from the model reaction.

The biological properties of sclerotiorin have not been studied in detail. We have investigated the effects of the drug on several enzymatic and biochemical reactions such as acyl-CoA:cholesterol acyltransferase, DGAT, phospholipase A2, PAF acetyl hydrolase, and gp 120-CD4 binding activities. Sclerotiorin did not show any inhibitory effects on the reactions at 100 μM except on type II phospholipase A2 (IC_{50} 1.5 μM) [105]. If our model of the reaction mechanism is valid, several lysine residues in the protein molecules might be modified by sclerotiorin. However, most enzymes appear to maintain their activity even after such modification, while CETP and type II phospholipase A2 lost their activity, suggesting differences in functional importance of lysine residues among proteins. Therefore, sclerotiorin is a useful tool with which to investigate whether lysine residues in proteins play a critical role in their biological functions. The binding site(s) of sclerotiorin on CETP remains to be defined.

V. SCREENING FOR INHIBITORS OF MACROPHAGE-DERIVED FOAM CELL FORMATION

From recent advanced research on ACAT genes, two isozymes, ACAT-1 and -2, are known to be present and to play different roles in the body [45–47]. ACAT-1 is ubiquitous and expressed most in sebaceous gland, steroidogenic tissues, and macrophages, while ACAT-2 is expressed in the liver and intestine. The findings indicate that ACAT-1 contributes deeply to macrophage-derived foam cell formation in atherosclerosis and that ACAT-2 is responsible for absorption of dietary cholesterol from the intestine and lipoprotein production in the liver. A number of ACAT inhibitors including synthetic and natural inhibitors were reported. Researchers carried out assays using rat liver microsomes as the enzyme source in most cases, and inhibitors were evaluated in an animal model by measuring cholesterol absorption from intestines. The course of experiments is now recognized as an assay for the ACAT-2 isozyme.

We are interested in ACAT-1 inhibitors, which are expected to affect macrophages directly. In the early stages of atherosclerogenesis, macrophages penetrate the intima, efficiently take up modified LDL, store cholesterol and fatty acids as a form of neutral lipids such as CE and TG in the cytosolic lipid droplets, and are converted into foam cells, leading to the development of atherosclerosis in the arterial wall. We established an assay system of lipid droplet formation using intact mouse macrophages and searched for microbial inhibitors of the formation in macrophages. Once a compound is discovered, the inhibition site should be defined.

A. Screening Method

Nishikawa et al. [106] reported that when mouse peritoneal macrophages are cultured in the presence of negatively charged liposomes, they take up the liposomes via the scavenger receptors and metabolize their components such as phospholipids and cholesterol to form lipid droplets containing neutral lipids in the cytosol (Fig. 7). On the basis of their observation, we have developed cell-based assays [107] by microscopic observation of oil red O-stained lipid droplets (morphological assay) and by measurement of [^{14}C]neutral lipids (CE and TG) synthesized from [^{14}C]oleic acid (biochemical assay). When macrophages are cultured in the presence (Fig. 8A) of liposomes, a number of lipid droplets are observed in the cytosols in the morphological assay. For the primary screen, culture broths were selected that caused a reduction of the size and/or the number of lipid droplets without cytotoxic effect on macrophages. Then, the inhibition was confirmed in the biochemical assay. In control experiments with liposomes, the incorporation of [^{14}C]oleic acid into CE, TG, and PL is about 23%, 20%, and 10% of the total radioactivity added, respectively.

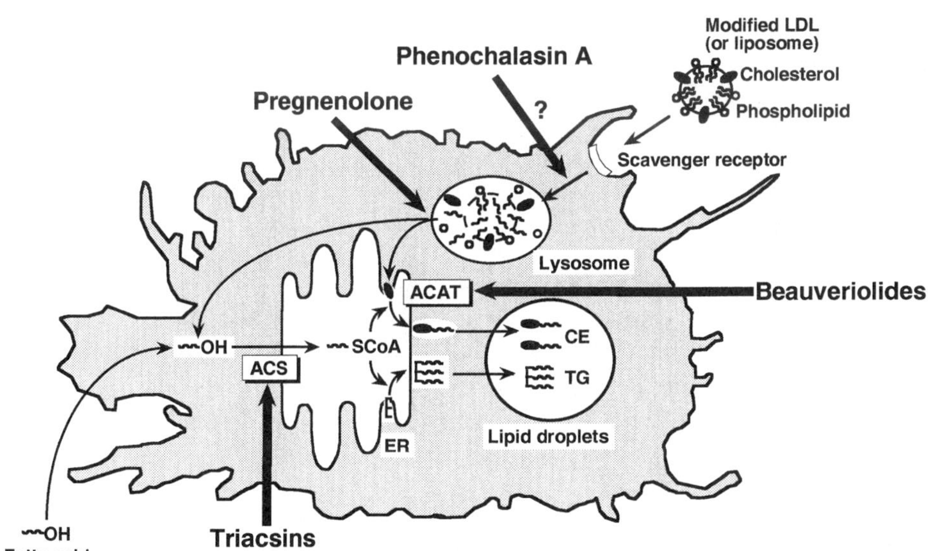

Figure 7 Process of lipid droplet formation in macrophages and inhibition sites of inhibitors. ACAT (acyl-CoA:cholesterol acyltransferase) and ACS (acyl-CoA synthetase).

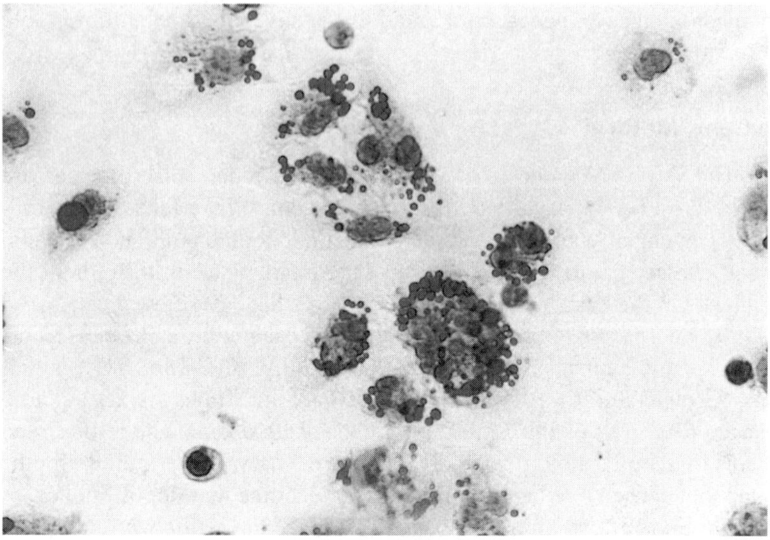

(A)

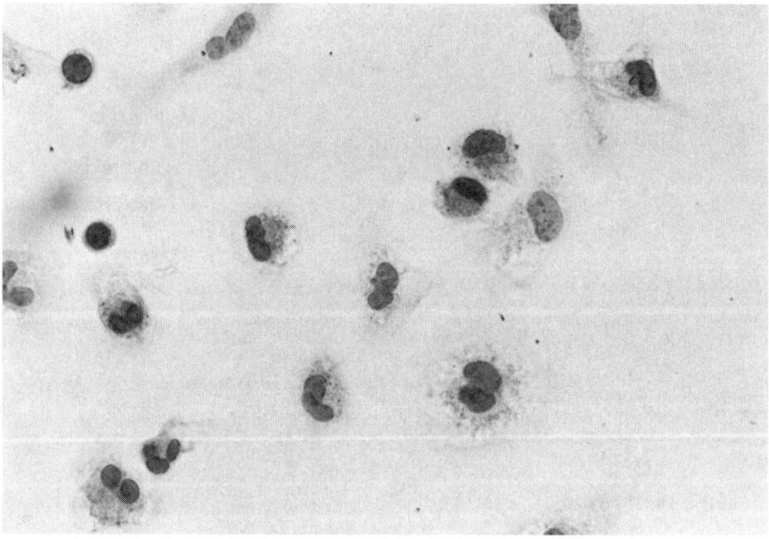

(B)

Figure 8 Lipid droplet formation in macrophages. A number of lipid droplets are formed in the cytosols of mouse peritoneal macrophages when macrophages are incubated at 37°C for 14 hr in the presence of liposomes (A). No lipid droplets are formed in the absence of liposomes (B).

B. Effects of Known Lipid Metabolism Inhibitors on the Assays

First, effects of known inhibitors of lipid metabolism on the lipid droplet formation and the neutral lipid synthesis in macrophages were examined. Cerulenin (fatty acid synthase inhibitor) [10], compactin (HMG-CoA reductase inhibitor) [1,2], and hymeglusin (HMG-CoA synthase inhibitor) [25] showed no effects, indicating that *de novo* biosynthesis of fatty acid and cholesterol is not involved in the process of lipid droplet formation. Pregnenolone (inhibitor of cellular cholesterol transport) [108] and CL-283,546 (ACAT inhibitor) [109] inhibited [^{14}C]CE synthesis specifically, with IC_{50} values of 5.0 and 0.038 μM, respectively. The inhibitors decreased cytosolic lipid droplets only partially even at the highest doses, which caused almost complete inhibition of [^{14}C]CE synthesis. In contrast, triacsin C (inhibitor of long-chain acyl-CoA synthetase) [110] inhibited both [^{14}C]CE and [^{14}C]TG syntheses, with similar IC_{50} values of 0.19 and 0.10 μM, respectively. Furthermore, the triacsin C dose-dependent inhibition of lipid droplet formation was almost complete at 0.59 μM, a concentration that showed about 90% inhibition of [^{14}C]CE and [^{14}C]TG syntheses. These results show that inhibition of acyl-CoA synthetase by triacsin C causes a decrease in the cellular levels of acyl-CoA, the common substrate for CE and TG syntheses, leading to inhibition of the synthesis of neutral lipids and eventually to the complete disappearance of cytosolic lipid droplets in the macrophages. These findings imply that TG synthesis as well as CE synthesis is responsible for lipid droplet formation in mouse macrophages.

C. Inhibitors

Over 10,000 samples of microbial culture broths were subjected to the screening program in the morphological assay. We discovered beauveriolides and phenochalasins from fungal culture broths as novel inhibitors of lipid droplet formation in macrophages (Fig. 9).

Two structurally related cyclodepsipeptides, beauveriolide I [111] and a novel compound named beauveriolide III, were isolated from the culture broth of *Beauveria* sp. FO-6979 [112,113]. Beauveriolides I and III caused a reduction in the number and size of cytosolic lipid droplets in macrophages at 10 μM without any cytotoxic effect on macrophages. They inhibited [^{14}C]CE synthesis specifically, with IC_{50} values of 0.8 and 0.4 μM, respectively. Studies on the mode of action revealed that they inhibit ACAT activity in microsomes prepared from both mouse macrophages and mouse liver, suggesting that beauveriolides inhibit ACAT-1 and -2 to similar extents. Among cyclodepsipeptides tested, beauveriolides I and III are the only ACAT inhibitors effective in morphological and bio-

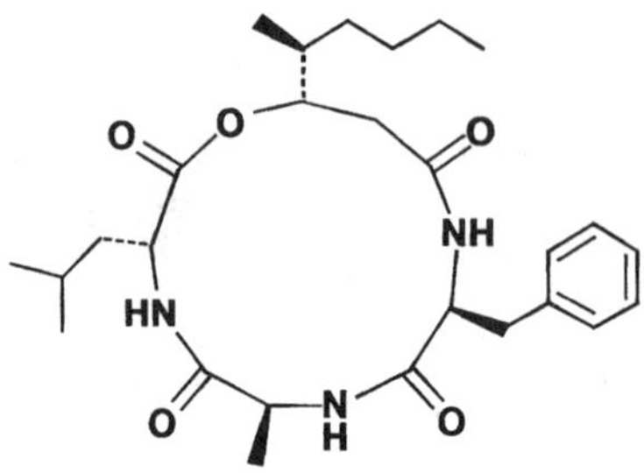

Beauveriolide I

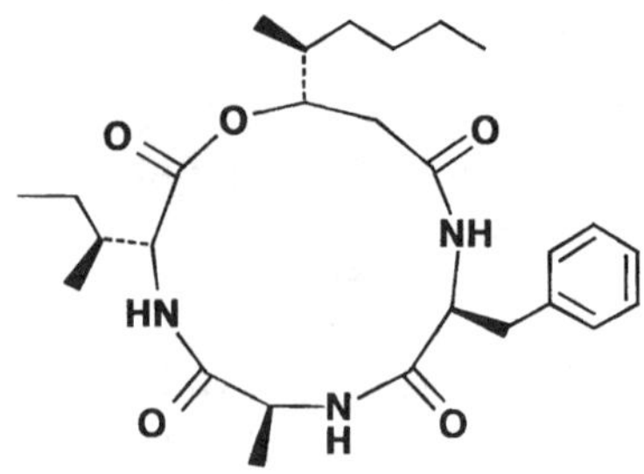

Beauveriolide III

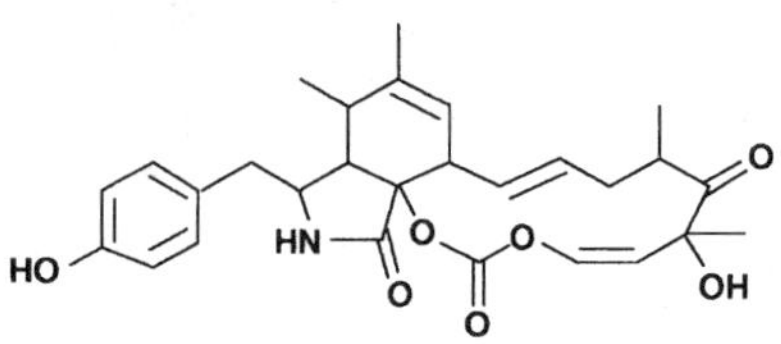

Phenochalasin A

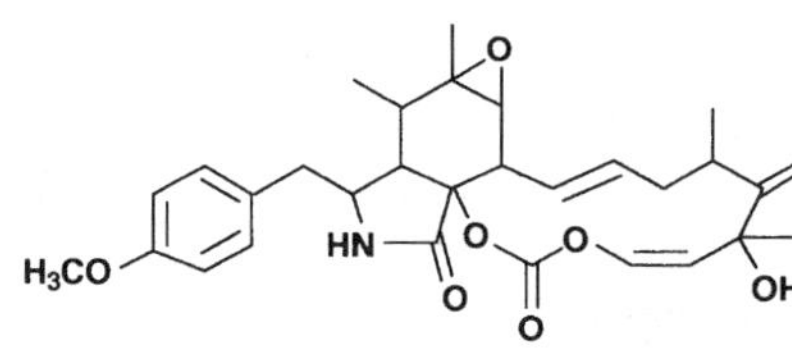

Phenochalasin B

Figure 9 Structures of inhibitors of lipid droplet formation in macrophages.

chemical assays using intact macrophages. We are planning to test beauveriolides in an *in vivo* model of atherosclerosis using LDL receptor-deficient mice.

Recently, phenochalasins A and B, belonging to the cytochalasan family, were isolated from the fungal culture broth of *Phomopsis* sp. FT-0211 [114,115]. Phenochalasin A has a unique phenol residue. Therefore, the effects of phenochalasins A and B and cytochalasin D on lipid droplet formation in macrophages were tested in morphological and biochemical assays [116]. Phenochalasin A inhibits the formation in a dose-dependent manner at 0.5–20 μM without any toxic effect on macrophages. Cytochalasin D also inhibits lipid droplet formation but only in a narrow range of concentrations (0.5–1.5 μM), and with cytotoxic effects at higher concentrations (>2.0 μM). Phenochalasin B has a toxic effect on macrophages at concentrations tested (0.2–20 μM). From the biochemical assay, phenochalasin A and cytochalasin D inhibit CE synthesis specifically, with IC_{50} values of 0.64 and 2.4 μM, respectively, while phenochalasin B inhibits both CE and TG syntheses. Thus, these members of the cytochalasan family have different effects on lipid droplet formation in macrophages. Tabas et al. reported that cytochalasin D inhibits lipid droplet formation by specifically blocking CE synthesis in mouse macrophages, indicating that actin cytoskeleton is responsible for CE synthesis [117]. However, our experiments suggested that phenochalasin A does not inhibit actin cytoskeleton, but the mechanism of action remains to be defined. Anyway, phenochalasin A is the best inhibitor of lipid droplet formation in macrophages among 10 cytochalasans tested.

From this screening work, it is expected that novel microbial products which affect the ACAT-1, TG synthetic pathway or unknown sites responsible for macrophage-derived foam cell formation will be discovered, leading to a new type of anti-atherosclerotic agent and providing a novel target for pharmaceutical intervention.

VI. THE FUTURE OF SCREENING

Recent advances in biotechnology and molecular biology have provided researchers with a new strategy and methodology for drug discovery screens. Mammalian cells or microorganisms are easily manipulated to express or disrupt target genes, and swap similar genes, and such artificial organisms or expressed enzymes are introduced to automatic robotic or high-throughput screening systems.

Enzymes involved in biosynthetic or metabolic pathways have been targets in the development of new drugs. Squalene epoxidase in the sterol biosynthetic pathway, for example, is important not only as a promising target for cholesterol-lowering agents as described first, but also as a potential target for antifungal drugs. A screening system for squalene epoxidase inhibitors was constructed by using artificial microorganisms. Allylamine derivatives such as terbinafine and tolnaftate were originally synthesized as antifungal agents inhibiting fungal (or yeast) squalene epoxidase [118]. Later, Banyu researchers found a new allylamine, NB-598, to be a potent inhibitor of mammalian squalene epoxidase [119], with the important implication that mammalian and fungal squalene epoxidases show different sensitivity to certain drugs. Meanwhile, the genes encoding squalene epoxidase were cloned from yeast, fungus and rat. Sakakibara et al. succeeded in cloning the rat squalene epoxidase gene by utilizing these inhibitors [120]. They introduced the rat gene to yeast (*Schizosaccharomyces pombe* JY266) to construct a transformant *S. pombe* RSE, which possesses squalene epoxidases of both yeast and mammalian origins. The wild JY266 and transformant RSE strains have been utilized for screening for specific inhibitors of yeast or mammalian squalene epoxidase. As shown in Figure 10, growth of the wild strain is inhibited almost completely by terbinafine even at 0.2 μg/ml and dose-dependently by tolnaftate (0.2–2.0 μg/ml), but growth is not inhibited by NB-598 (2.0 μg/ml). Conversely, the growth of the transformant in the presence of terbinafine or tolnaftate (1.0 μg/ml) to block yeast squalene epoxidase is inhibited by NB-598 at 0.2 μg/ml, but not by a further addition of terbinafine nor tolnaftate at 2.0 μg/ml. Ideally, a yeast strain swapping the yeast squalene epoxidase gene for the

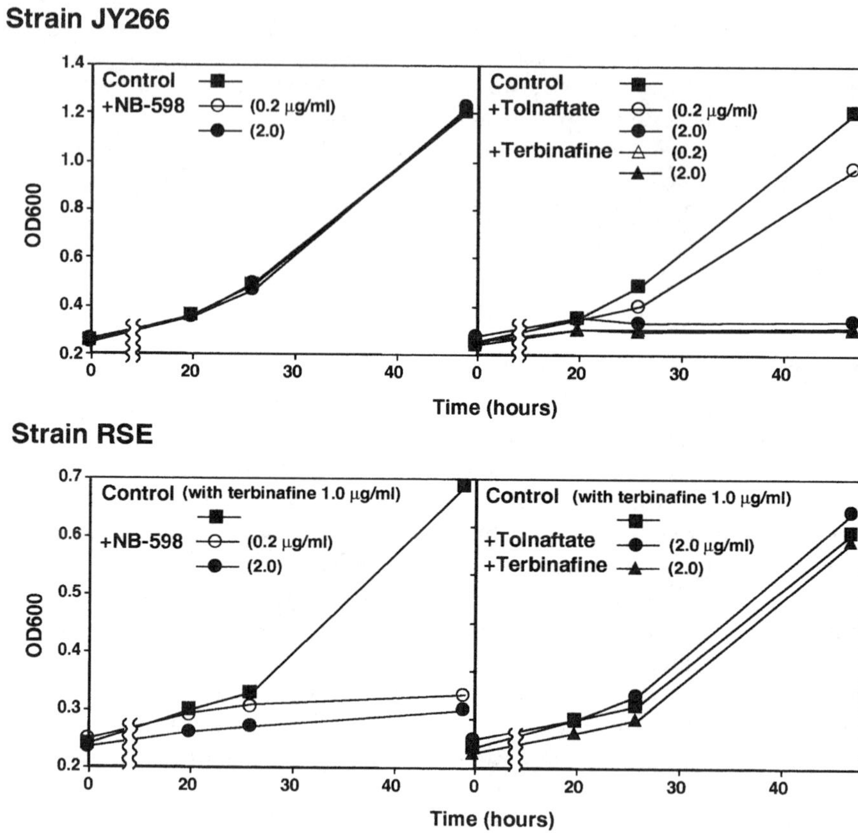

Figure 10 Selective growth inhibition of *S. pombe* JY266 and RSE strains by terbinafine and tolnaftate (inhibitor of fungal squalene epoxidase) and by NB-598 (inhibitor of mammalian squalene epoxidase). Strain JY-266 possesses the fungal squalene epoxidase gene and transformant strain RSE possesses fungal and rat squalene epoxidase genes.

rat one is better than the transformant RES as a test organism for the screening because terbinafine is not needed for the assay. This strategy of using intact artificial organisms and manipulating desired genes is expected to spread in the near future. For example, even though host and pathogenic organisms possess the same enzymes, but with different characteristics like squalene epoxidase, this strategy will be useful in screening for specific inhibitors of the target enzymes.

Target enzymes expressed biotechnologically are also introduced to HTS systems. Merck researchers reported a new approach to drug screening by combinatorial enzymology [121]. They have engineered a cell-free bacterial cell wall biosynthetic pathway to assay simultaneously six of the enzymes in the biosynthetic sequence. More than 40 compounds including penicillin and other beta-lactams and glycopeptides are used as antibiotics, but most of them inhibit the enzymes responsible for the last few steps in the cell-wall biosynthetic pathway. They aimed at the murein biosynthetic enzymes (MurA through MurF, Fig. 11) involved in the earlier steps in this pathway as valid and promising targets for drug discovery. Since the pathway enzymes have evolved conserved binding motifs to bind structurally related pathway metabolites, an inhibitor that recognizes homologous binding motifs will likely bind to more than one enzyme in the pathway. They expect that modest inhibition of several enzymes will reduce flux through the pathway more effectively than potent inhibition of a single enzyme, and further that the frequency of target-mediated resistance to such a compound would be negligible. This approach can be applied to any therapeutically relevant metabolic pathway.

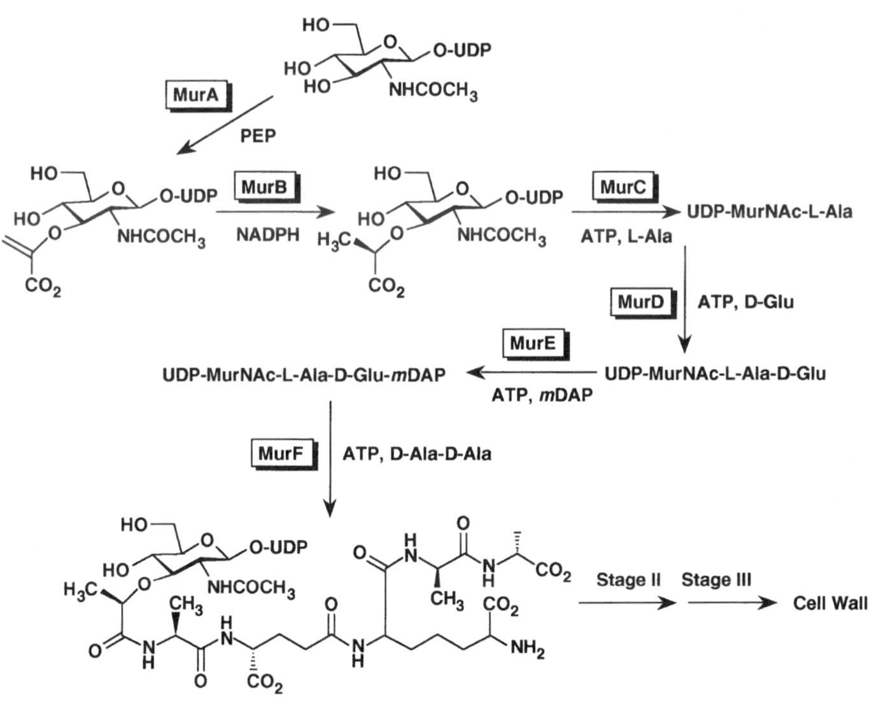

Figure 11 Sequential murein biosynthetic pathway, the early steps of the bacterial cell wall biosynthesis.

A number of projects on genome sequencing including that of human and pathogenic organisms are in progress worldwide. The achievements promise to be of great benefit to drug discovery. In 1998, the genome sequencing of *Mycobacterium tuberculosis* was completed [122]. It was believed that tuberculosis caused by *M. tuberculosis* would eventually be eliminated, but the microorganism has proved to be very resilient and the disease continues to be a public health threat. *M. tuberculosis* has unique cell wall lipids called mycolic acids. These extremely long fatty acids form a broad family of more than 500 closely related structures and comprise about 30% of the dry weight of *M. tuberculosis*. Recently, Barry and co-workers identified an enzyme in mycolic acid biosynthesis that is targeted by isoniazid, the most widely used antituberculosis drug [50]. Moreover, the complete genome sequence predicts that the microorganism produces about 250 distinct enzymes involved in fatty acid metabolism, while there are only 50 for *E. coli*. These findings imply that the enzymes are promising targets for new antituberculosis drugs.

New enzymes for drug discovery have been identified through biosynthetic studies on microbial metabolites. It has been long accepted that isopentenyl diphosphate, an intermediate of sterols and terpenoids, is synthesized only through

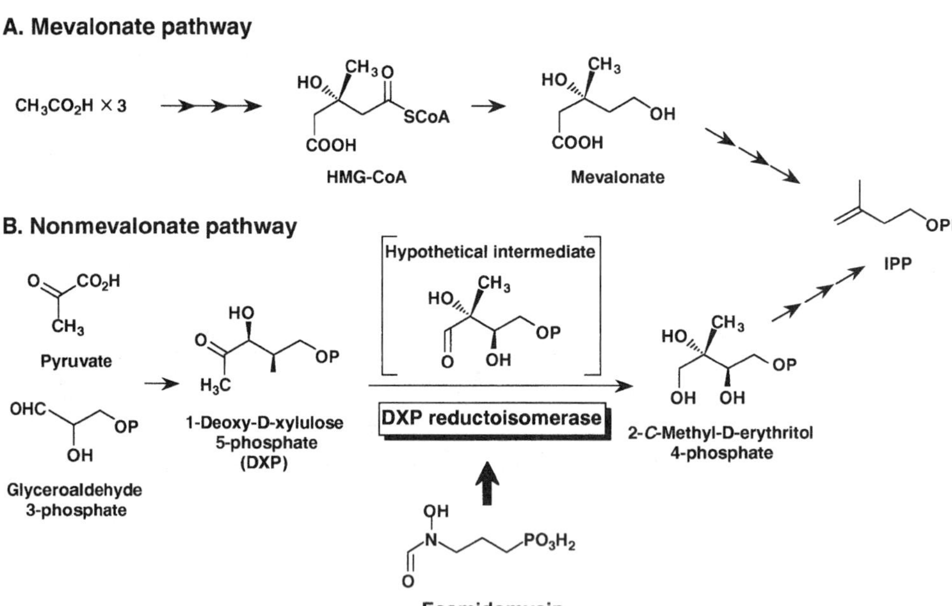

Figure 12 Mevalonate pathway and nonmevalonate pathway. Antibiotic fosmidomycin inhibits 1-deoxy-D-xylulose 5-phosphate (DXP) reductoisomerase.

the mevalonate pathway. However, it has been shown that a mevalonate-independent (nonmevalonate) pathway of isopentenyl diphosphate biosynthesis is present and essential for some bacteria [123,124], algae [125], plants [126,127], and malaria [49]. Recent studies [128–131] revealed that the initial step of this pathway is the formation of 1-deoxy-D-xylulose 5-phosphate (DXP) by condensation of pyruvate and glyceraldehyde 3-phosphate and the second step is the intramolecular rearrangement of DXP to 2-*C*-methyl-D-erythritol 4-phosphate via hypothetical intermediate 2-*C*-methylerythrose 4-phosphate (Fig. 12). Seto and co-workers succeeded in the cloning and overexpression of the gene encoding the second reaction (DXP reductoisomerase), and established an assay system, leading to the finding that a known antibiotic fosmidomycin is a potent and specific inhibitor of DXP reductoisomerase [48]. Moreover, *Plasmodium falciparum* strains possess the nonmevalonate pathway, and fosmidomycin was shown to be active against the malarial enzyme and to inhibit the *in vitro* growth of multidrug-resistant *P. falciparum* strains [49]. Thus, enzymes involved in this nonmevalonate pathway are effective targets for new antibacterial and antimalarial drugs.

ACKNOWLEDGMENTS

This work was supported by grants from the ''Research for the Future'' Program of The Japan Society for the Promotion of Science (JSPS-RFTF96I00304), a Grant-in-Aid for Scientific Research (B) from the Ministry of Education, Science, Sports and Culture of Japan (09480147), and the Japan Keirin Association.

REFERENCES

1. A Endo, M Kuroda, Y Tsujita. ML-236A, ML-236B, and ML-236C, new inhibitors of cholesterogenesis produced by *Penicillium citrinum*. J Antibiot 29:1336–1348, 1976.
2. AG Brown, TC Smale, TJ King, R Hassenkamp, RH Thompson. Crystal and molecular structure of compactin, a new antifungal metabolite from *Penicillium brevicompactum*. J Chem Soc Perkin Trans 1976:1165–1169.
3. A Endo. Monacolin K, a new hypocholesterolemic agent produced by a *Monascus* species. J Antibiot 32:852–854, 1979.
4. AW Albert, J Chen, G Kuron, V Hunt, J Huff, C Hoffman, J Rothrock, M Lopez, H Joshua, E Harris, A Patchett, R Monaghan, S Currie, E Stapley, G Albers-Schonberg, O Hensens, J Hirshfield, K Hoogsteen, J Liesch, J Springer. Mevinolin, a highly potent competitive inhibitor of hydroxymethylglutaryl-coenzyme A reductase and cholesterol-lowering agent. Proc Natl Acad Sci (USA) 77:3957–3961, 1980.

5. A Endo. The discovery and development of HMG-CoA reductase inhibitors. J Lipid Res 33:1569–1582, 1992.
6. M Hanefeld, JP Deslypere, L Ose, PN Durrington, M Farnier, N Schmage. Efficacy and safety of 300 micrograms and 400 micrograms cerivastatin once daily in patients with primary hypercholesterolaemia: a multicentre, randomized, double-blind, placebo-controlled study. J Int Med Res 27:115–129, 1999.
7. J Shepherd, SM Cobbe, I Ford, CG Isles, AR Lorimer, PW MacFarlane, JH McKillop, CJ Packard. Prevention of coronary heart disease with pravastatin in men with hypercholesterolemia. West of Scotland Coronary Prevention Study Group. N Engl J Med 33:1301–1307, 1995.
8. TR Pedersen, J Kjekshus, K Berg, T Haghfelt, O Faergeman, G Thorgeirsson, K Pyorala, T Miettinen, AG Olsson, H Wedel, L Wilhelmsen. Randomised trial of cholesterol lowering in 4444 patients with coronary heart disease: the Scandinavian Simvastatin Survival Study (4S). Lancet 344:1383–1389, 1994.
9. S Ōmura. Splendid Gifts from Microorganisms. 2nd ed. Tokyo: Research Center for Biological Function, The Kitasato Institute, 1998.
10. S Ōmura. The antibiotic cerulenin, a novel tool for biochemistry as an inhibitor of fatty acid synthesis. Bacteriol Rev 40:681–697, 1976.
11. S Ōmura, Y Iwai, Y Takahashi, N Sadakane, A Nakagawa, H Oiwa, Y Hasegawa, T Ikai. Herbimycin, a new antibiotic produced by a strain of *Streptomyces*. J Antibiot 32:255–261, 1979.
12. Y Uehara, H Fukazawa. Use and selectivity of herbimycin A as inhibitor of protein-tyrosine kinases. Meth Enzymol 201:370–379, 1991.
13. S Ōmura, T Fujimoto, K Otoguro, K Matsuzaki, R Moriguchi, H Tanaka, Y Sasaki. Lactacystin, a novel microbial metabolite, induces neuritogenesis of neuroblastoma cells. J Antibiot 44:113–116, 1991.
14. G Fenteany, RF Standaert, WS Lane, S Choi, EJ Corey, SL Schreiber. Inhibition of proteasome activities and subunit-specific amino-terminal threonine modification by lactacystin. Science 268:726–731, 1995.
15. S Ōmura, Y Sasaki, Y Iwai, H Takeshima. Staurosporine, a potentially important gift from a microorganism. J Antibiot 48:535–548, 1995.
16. K Otoguro, F Kuno, S Ōmura. Arisugacins, selective acetylcholinesterase inhibitors of microbial origin. Pharmacol Ther 76:45–54, 1997.
17. H Tomoda, YK Kim, H Nishida, R Masuma, S Ōmura. Pyripyropenes, novel inhibitors of acyl-CoA:cholesterol acyltransferase produced by *Aspergillus fumigatus*. I. Production, isolation, and biological properties. J Antibiot 47:148–153, 1994.
18. H Tomoda, H Nishida, YK Kim, R Obata, T Sunazuka, S Ōmura, J Bordner, M Guadllana, PG Dormer, AB Smith III. Relative and absolute stereochemistry of pyripyropene A, a potent, bioavailable inhibitor of acyl-CoA: cholesterol acyltransferase. J Am Chem Soc 116:12097–12098, 1994.
19. S Ōmura, D Van der Pyl, J Inokoshi, Y Takahashi, H Takeshima. Pepticinnamins, new farnesyl-protein transferase inhibitors produced by an actinomycete. I. Producing strain, fermentation, isolation and biological activity. J Antibiot 46:222–228, 1993.
20. M Murata, T Miyasaka, H Tanaka, S Ōmura. Diazaquinomycin A, a new antifolate antibiotic, inhibits thymidylate synthase. J Antibiot 38:1025–1033, 1985.

21. H Tomoda, S Ōmura. New strategy for discovery of enzyme inhibitors: screening with intact mammalian cells or intact microorganisms having special functions. J Antibiot 43:1207–1222, 1990.
22. H Kumagai, H Tomoda, S Ōmura. Method of search for microbial inhibitors of mevalonate biosynthesis using animal cells. J Antibiot 43:397–402, 1990.
23. S Ōmura, M Murata, H Hanak, K Hinotozawa, R Oiwa, H Tanaka. Phosalacine, a new herbicidal antibiotic containing phosphinothricin. Fermentation, isolation, biological activity and mechanism of action. J Antibiot 37:829–835, 1984.
24. H Tomoda, S Ōmura. New strategy for search of enzyme inhibitors. In: S Ōmura, ed. The Search for Bioactive Compounds from Microorganisms. New York: Springer-Verlag, 1992, pp 161–170.
25. S Ōmura, H Tomoda, H Kumagai, MD Greenspan, JB Yodkovitz, JS Chen, AW Albert, I Martin, S Mochales, RL Monaghan, JC Chabala, RE Schwartz, AA Patchett. Potent inhibitory effect of antibiotic 1233A on cholesterol biosynthesis which specifically blocks 3-hydroxy-3-methylglutaryl coenzyme A synthase. J Antibiot 40:1356–1357, 1987.
26. H Tomoda, H Kumagai, H Tanaka, S Ōmura. F-244 specifically inhibits 3-hydroxy-3-methylglutaryl coenzyme A synthase. Biochim Biophys Acta 922:351–356, 1987.
27. MJ Dawson, JE Farthing, PS Marshall, RF Middleton, MJ O'Neill, A Shuttleworth, C Stylli, RM Tait, PM Taylor, HG Wildman, AD Buss, D Langley, MV Hayes. The squalestatins, novel inhibitors of squalene synthase produced by a species of *Phoma*. I. Taxonomy, fermentation, isolation, physico-chemical properties and biological activity. J Antibiot 45:639–647, 1992.
28. A Baxter, BJ Fitzgerald, JL Hutson, AD McCarthy, JM Motteram, BC Ross, M Sapra, MA Snowden, NS Watson, RJ Williams, C Wright. Squalestatin 1, a potent inhibitor of squalene synthase, which lowers serum cholesterol *in vivo*. J Biol Chem 267:11705–11708, 1992.
29. JD Bergstrom, MM Kurtz, DJ Rew, AM Amend, JD Karkas, RG Bostedor, VS Bansal, C Dufresne, FL VanMiddlesworth, OD Hensens, JM Liesch, DL Zink, KE Wilson, J Onishi, JA Milligan, G Bills, L Kaplan, MN Omstead, RG Jenkins, L Huang, MS Meinz, L Quinn, RW Burg, YL Kong, S Mochales, M Mojena, I Martin, F Palaez, MT Diez, AW Alberts. Zaragozic acids: a family of fungal metabolites that are picomolar competitive inhibitors of squalene synthase. Proc Natl Acad Sci (USA) 90:80–84, 1993.
30. C Dufresne, KE Wilson, SB Singh, DL Zink, JD Bergstrom, D Rew, JD Polishook, M Meinz, L Huang, KC Silverman, RB Lingham, M Mojena, C Cascales, F Pelaez, JB Gibbs. Zaragozic acids D and D2: potent inhibitors of squalene synthase and of Ras farnesyl-protein transferase. J Nat Prod 56:1923–1929, 1993.
31. M Yonemoto, T Satoh, H Arakawa, I Suzuki-Takahashi, Y Monden, T Kodera, K Tanaka, T Aoyama, Y Iwasawa, T Kamei, S Nishimura, K Tomimoto. J-104,871, a novel farnesyltransferase inhibitor, blocks Ras farnesylation *in vivo* in a farnesyl pyrophosphate-competitive manner. Mol Pharmacol 54:1–7, 1998.
32. DR Sliskovic, AD White. Therapeutic potential of ACAT inhibitors as lipid lowering and anti-atherosclerotic agents. Trends Pharmacol Sci 12:194–199, 1991.

33. FG Kathawala, JC Heider. Acyl CoA:cholesterol acyltransferase inhibitors and lipid lipoprotein metabolism. In: DT Witiak, HAI Newman, DR Feller, eds. Antilipidemic Drugs. Amsterdam: Elsevier, 1992, pp 159–195.
34. H Tomoda, H Nishida, R Masuma, J Cao, S Okuda, S Ōmura. Purpactins, new inhibitors of acyl-CoA:cholesterol acyltransferase produced by *Penicillium purpurogenum*. I. Production, isolation and physico-chemical and biological properties. J Antibiot 44:136–143, 1991.
35. H Tomoda, XH Huang, H Nishida, R Masuma, YK Kim, S Ōmura. Glisoprenins, new inhibitors of acyl-CoA:cholesterol acyltransferase produced by *Gliocladium* sp. FO-1513. I. Production, isolation and physico-chemical and biological properties. J Antibiot 45:1202–1206, 1992.
36. XH Huang, H Tomoda, H Nishida, R Masuma, S Ōmura. Terpendoles, novel ACAT inhibitors produced by *Albophoma yamanashiensis*. I. Production, isolation and biological properties. J Antibiot 48:1–4, 1995.
37. H Tomoda, N Tabata, DJ Yang, H Takayanagi, S Ōmura. Terpendoles, novel ACAT inhibitors produced by *Albophoma yamanashiensis*. III. Production, isolation and structure elucidation of new components. J Antibiot 48:793–804, 1995.
38. S Ōmura, H Tomoda, YK Kim, H Nishida. Pyripyropenes, highly potent inhibitors of acyl-CoA:cholesterol acyltransferase produced by *Aspergillus fumigatus*. J Antibiot 46:1168–1169, 1993.
39. JF Reindel, MA Dominick, TM Bocan, AW Gough, EJ McGuire. Toxicologic effects of a novel acyl-CoA:cholesterol acyltransferase inhibitor in cynomolgus monkeys. Toxicol Pathol 22:510–518, 1994.
40. A Tanaka, T Terasawa, H Hagihara, Y Sakuma, N Ishibe, M Sawada, H Takasugi, H Tanaka. Inhibitors of acyl-CoA:cholesterol O-acyltransferase. 2. Identification and structure-activity relationships of a novel series of N-alkyl-N-(heteroaryl-substituted benzyl)-N′-arylureas. J Med Chem 41:2390–2410, 1998.
41. DR Sliskovic, JA Picard, PM O'Brien, P Liao, WH Roark, BD Roth, MA Anderson, SB Mueller, TM Bocan, RF Bousley, KL Hamelehle, R Homan, JF Reindel, RL Stanfield, D Turluck, RB Krause. alpha-Substituted malonester amides: tools to define the relationship between ACAT inhibition and adrenal toxicity. J Med Chem 41:682–690, 1998.
42. CC Chang, HY Huh, KM Cadigan, TY Chang. Molecular cloning and functional expression of human acyl-coenzyme A:cholesterol acyltransferase cDNA in mutant Chinese hamster ovary cells. J Biol Chem 268:20747–20755, 1993.
43. VL Meiner, S Cases, HM Myers, ER Sande, S Bellosta, M Schambelan, RE Pitas, J McGuire, J Herz, RV Farese Jr. Disruption of the acyl-CoA:cholesterol acyltransferase gene in mice: evidence suggesting multiple cholesterol esterification enzymes in mammals. Proc Natl Acad Sci (USA) 93:14041–14046, 1996.
44. RV Farese Jr. Acyl CoA:cholesterol acyltransferase genes and knockout mice. Curr Opin Lipidol 9:119–123, 1998.
45. RA Anderson, C Joyce, M Davis, JW Reagan, M Clark, GS Shelness, LL Rudel. Identification of a form of acyl-CoA:cholesterol acyltransferase specific to liver and intestine in nonhuman primates. J Biol Chem 273:26747–26754, 1998.
46. S Cases, S Novak, YW Zheng, HM Myers, SR Lear, E Sande, CB Welch, AJ Lusis, TA Spencer, BR Krause, SK Erickson, RV Farese Jr. ACAT-2, a second

mammalian acyl-CoA:cholesterol acyltransferase. Its cloning, expression, and characterization. J Biol Chem 273:26755–26764, 1998.

47. P Oelkers, A Behari, D Cromley, JT Billheimer, SL Sturley. Characterization of two human genes encoding acyl coenzyme A:cholesterol acyltransferase-related enzymes. J Biol Chem 273:26765–26771, 1998.
48. T Kuzuyama, T Shimizu, S Takahashi, H Seto. Fosmidomycin, a specific inhibitor of 1-deoxy-D-xylulose 5-phosphate reductoisomerase in the nonmevalonate pathway for terpenoid biosynthesis. Tetrahedron Lett 39:7913–7916, 1998.
49. H Jomaa, J Wiester, S Sanderbrand, B Altincicek, C Weidemeyer, M Hintz, I Turbachova, M Eberl, J Zeidler, HK Lichtenthaler, D Soldati, E Beck. Inhibitors of nonmevalonate pathway of isoprenoid biosynthesis as antimalarial drugs. Science 285:1573–1576, 1999.
50. K Mdluli, RA Slayden, Y Zhu, S Ramaswamy, X Pan, D Mead, DD Crane, JM Musser, CE Barry III. Inhibition of a *Mycobacterium tuberculosis* beta-ketoacyl ACP synthase by isoniazid. Science 280:1607–1610, 1998.
51. RM Bell, RA Coleman. Enzymes of triacylglycerol formation in mammals. The Enzymes 16:87–111, 1983.
52. N Mayorek, J Bar-Tana. Inhibition of diacylglycerol acyltransferase by 2-bromooctanoate in cultured rat hepatocytes. J Biol Chem 260:6528–6532, 1985.
53. H Tomoda, M Ito, N Tabata, R Masuma, Y Yamaguchi, S Ōmura. Amidepsines, inhibitors of diacylglycerol acyltransferase produced by *Humicola* sp. FO-2942. I. Production, isolation and biological properties. J Antibiot 48:937–941, 1995.
54. H Tomoda, K Igarashi, JC Cyong, S Ōmura. Evidence for an essential role of long chain acyl-CoA synthetase in animal cell proliferation. Inhibition of long chain acyl-CoA synthetase by triacsins caused inhibition of Raji cell proliferation. J Biol Chem 266:4214–4219, 1991.
55. H Tomoda, N Tabata, M Ito, S Ōmura. Amidepsines, inhibitors of diacylglycerol acyltransferase produced by *Humicola* sp. FO-2942. II. Structure elucidation of amidepsines A, B and C. J Antibiot 48:942–947, 1995.
56. H Tomoda, Y Yamaguchi, N Tabata, T Kobayashi, R Masuma, H Tanaka, S Ōmura. Amidepsine E, an inhibitor of diacylglycerol acyltransferase produced by *Humicola* sp. FO-5969. J Antibiot 49:929–931, 1996.
57. JA Elix, VK Jayanthi, CC Leznoff. 2,4-Di-*O*-methylgyrophoric acid and 2,4,5-tri-*O*-methylhiascic acid. New tridepsides from *Parmelia damaziana*. Austral J Chem 34:1757–1761, 1981.
58. N Tabata, M Ito, H Tomoda, S Ōmura. Xanthohumols, diacylglycerol acyltransferase inhibitors, from *Humulus lupulus*. Phytochemistry 46:683–687, 1997.
59. S Ōmura, H Tomoda, N Tabata, Y Ohyama, A Abe, M Namikoshi. Roselipins, novel fungal metabolites having a highly methylated fatty acid modified with a mannose and an arabinitol. J Antibiot 52:586–589, 1999.
60. H Tomoda, Y Ohyama, A Abe, N Tabata, M Namikoshi, Y Yamaguchi, R Masuma, S Ōmura. Roselipins, inhibitors of diacylglycerol acyltransferase, produced by *Gliocladium roseum* KF-1040. J Antibiot 52:689–694, 1999.
61. N Tabata, Y Ohyama, H Tomoda, T Abe, M Namikoshi, S Ōmura. Structure elucidation of roselipins, inhibitors of diacylglycerol acyltransferase, produced by *Gliocladium roseum* KF-1040. J Antibiot 52:815–826, 1999.

62. S Cases, SJ Smith, Y-W Zheng, HM Myers, SR Lear, E Sande, S Novak, C Collins, CB Welch, AJ Lusis, SK Erickson, RV Farese Jr. Identification of a gene encoding an acyl CoA:diacylglycerol acyltransferase, a key enzyme in triacylglycerol synthesis. Proc Natl Acad Sci (USA) 95:13018–13023, 1998.
63. AR Tall. Plasma cholesteryl ester transfer protein. J Lipid Res 34:1255–1274, 1993.
64. P Barter, K-A Rye. Cholesteryl ester transfer protein: its role in plasma lipid transport. Clin Exp Pharmacol Physiol 21:663–672, 1994.
65. L Lagrost. Regulation of cholesteryl ester transfer protein (CETP) activity: review of *in vitro* and *in vivo* studies. Biochim Biophys Acta 1215:209–236, 1994.
66. CB Hesler, TL Swenson, AR Tall. Purification and characterization of a human plasma cholesteryl ester transfer protein. J Biol Chem 262:2275–2282, 1987.
67. AS Jarnagin, W Kohr, C Fielding. Isolation and specificity of a Mr 74,000 cholesteryl ester transfer protein from human plasma. Proc Natl Acad Sci (USA) 84:1854–1857, 1987.
68. D Drayna, AS Jarnagin, J McLean, W Henzel, W Kohn, C Fielding, R Lawn. Cloning and sequencing of human cholesteryl ester transfer protein cDNA. Nature (London) 327:632–634, 1987.
69. GJ Hopkins, LBF Chang, PJ Barter. Role of lipid transfers in the formation of a subpopulation of small high density lipoproteins. J Lipid Res 26:218–229, 1985.
70. LB Agellon, A Walsh, T Hayek, P Moulin, XC Jiang, SA Shelanski, JL Breslow, AR Tall. Reduced high density lipoprotein cholesterol in human cholesteryl ester transfer protein transgenic mice. J Biol Chem 266:10796–10801, 1991.
71. J Koizumi, H Mabuchi, A Yoshimura, I Michishita, M Takeda, H Itoh, Y Sakai, T Sakai, K Ueda, R Takeda. Deficiency of serum cholesteryl-ester transfer activity in patients with familial hyperalphalipoproteinaemia. Atherosclerosis 58:175–186, 1985.
72. ML Brown, A Inazu, CB Hesler, LB Agellon, C Mann, ME Whitelock, YL Marcel, RW Milne, J Koizumi, H Mabuchi, T Takeda, AR Tall. Molecular basis of lipid transfer protein deficiency in a family with increased high-density lipoproteins. Nature (London) 342:448–451, 1989.
73. KR Marotti, CK Castle, RW Murray, EF Rehberg, HG Polites, GW Melchior. The role of cholesteryl ester transfer protein in primate apolipoprotein A-I metabolism. Insights from studies with transgenic mice. Arterioscler Thromb 12:736–744, 1992.
74. K Marotti, CK Castle, TP Boyle, AH Lin, RW Murray, GW Melchior. Severe atherosclerosis in transgenic mice expressing simian cholesteryl ester transfer protein. Nature (London) 364:73–75, 1993.
75. M Sugano, N Makino, S Sawada, S Otsuka, M Watanabe, H Okamoto, M Kamada, A Mizushima. Effect of antisense oligonucleotides against cholesteryl ester transfer protein on the development of atherosclerosis in cholesterol-fed rabbits. J Biol Chem 273:5033–5036, 1998.
76. LJ Thomas, MD Picard, DP Miller, CM Honan, H Adari, CD Emmett, HC Marsh, US Ryan, CL Petty, CW Rittershaus. A vaccine to produce anti-cholesteryl ester transfer protein (CETP) antibodies for the prevention/treatment of atherosclerosis. FASEB J 13(Part II):A693, 1999.
77. H Kato, T Nakanishi, H Arai, I Nishida, T Nishida. Purification, microheterogen-

eity, and stability of human lipid transfer protein. J Biol Chem 264:4082–4087, 1989.
78. CL Bisgaier, LL Minton, AD Essenburg, A White, R Homan. Use of fluorescent cholesteryl ester microemulsions in cholesteryl ester transfer protein assays. J Lipid Res 34:1625–1634, 1993.
79. L Lagrost, N Loreau, P Gambert, C Lallemant. Immunospecific scintillation proximity assay of cholesteryl ester transfer protein activity. Clin Chem 41:914–919, 1995.
80. H Tomoda, C Matsushima, N Tabata, I Namatame, H Tanaka, H Arai, M Fukazawa, K Inoue, S Ōmura. Structure-specific inhibition of cholesteryl ester transfer protein by azaphilones. J Antibiot 52:160–170, 1999.
81. HV Kothari, KJ Poirier, WH Lee, Y Satoh. Inhibition of cholesterol ester transfer protein CGS 25159 and changes in lipoproteins in hamsters. Atherosclerosis 128: 59–66, 1997.
82. H Tomoda, N Tabata, R Masuma, SY Si, S Ōmura. Erabulenols A and B, inhibitors of cholesteryl ester transfer protein, produced by *Penicillium* sp. FO-5637. I. Production, isolation and biological properties. J Antibiot 51:618–623, 1998.
83. N Tabata, H Tomoda, S Ōmura. Erabulenols A and B, inhibitors of cholesteryl ester transfer protein, produced by *Penicillium* sp. FO-5637. II. Structure elucidation of erabulenols A and B. J Antibiot 51:624–628, 1998.
84. H Tomoda, N Tabata, M Shinose, Y Takahashi, B Woodruff, S Ōmura. Ferroverdins, potent inhibitors of cholesteryl ester transfer protein, produced by *Streptomyces* sp. WK-5344. I. J Antibiot 52:1101–1107, 1999.
85. N Tabata, H Tomoda, S Ōmura. Ferroverdins, potent inhibitors of cholesteryl ester transfer protein, produced by *Streptomyces* sp. WK-5344. II. Structure elucidation. J Antibiot 52:1108–1113, 1999.
86. CL Bisgaier, AD Essenburg, LL Minton, R Homan, CJ Blankley, A White. Cholesteryl ester transfer protein inhibition by PD 140195. Lipids 29:811–818, 1994.
87. DE Epps, KA Greenlee, JS Harris, EW Thomas, CK Castle, JF Fisher, RR Hozak, CK Marschke, GW Melchior, FL Kezdy. Kinetics and inhibition of lipid exchange catalyzed by plasma cholesteryl ester transfer protein (lipid transfer protein). Biochemistry 34:12560–12569, 1995.
88. G Schmidt, R Angerbauer, A Brandes, M Logers, M Muller-Gliemann, H Bischoff, D Schmidt, S Wohlfeil. 2-Aryl-substituted pyridines. U.S. patent 5,925,645, 1999.
89. C Schmeck, M Muller-Gliemann, G Schmidt, A Brandes, R Angerbauer, M Logers, KD Bremm, H Bischoff, D Schmidt, J Schuhmacher. Heterocyclic-fused pyridines. U.S. patent 5,932,587, 1999.
90. H Shinkai, K Maeda, H Okamoto. Compounds effective as CETP inhibitors. Japanese Patent 11,049,743, 1999.
91. H Shinkai, K Maeda, H Okamoto. CETP inhibitors. Japanese Patent 11,222,428, 1999.
92. SJ Coval, MA Conover, R Mierzwa, A King, MS Puar. Wiedendiol-A and B, cholesteryl ester transfer protein inhibitors from the marine sponge *Xestesponogia wiedenmayeni*. Bioorg Med Chem Lett 5:605–610, 1995.
93. J Shin, Y Seo, JR Rho, E Baek, BM Kwon, TS Jeong, SH Bok. Suberitenones A

and B: sesterterpenoids of an unprecedented skeletal class from the antarctic sponge *Suberites* sp. J Org Chem 60:7582–7588, 1995.
94. MZ Kuo, RJ Zielinski, JI Cialdella, CK Marschke, MJ Dupuis, GP Li, DA Kloosterman, CH Spilman, VP Marshall. Discovery, isolation, structure elucidation and biosynthesis of U-106305, a cholesteryl ester transfer protein inhibitor from UC11136. J Am Chem Soc 117:10629–10634, 1995.
95. KH Cho, JY Lee, MS Choi, JM Cho, JS Lim, YB Park. A peptide from hog plasma that inhibits human cholesteryl ester transfer protein. Biochim Biophys Acta 1391: 133–144, 1998.
96. TP Curtin, J Reilly. Sclerotiorin, $C_{20}H_{20}O_5Cl$, a chlorine-containing metabolic product of *Penicillium sclerotiorum* van Beyma. Biochem J 34:1419–1421, 1940.
97. BM Ashe, DS Fletcher. Anti-inflammatory and antidegenerative compounds isolated from L 681512. U.S. Patent 4,874,755, 1989.
98. N Tabata, H Tomoda, Y Yamaguchi, R Masuma, MJ Bamberger, S Ōmura. Inhibition of cholesteryl ester transfer protein by fungal L681,512 compounds. J Antibiot 52:1042–1045, 1999.
99. RA Eade, H Page, A Alexander, K Turner, WB Whalley. The chemistry of fungi. Part XXVIII. Sclerotiorin and its hydrogenation products. J Chem Soc 1957:4913–4924.
100. XC Jiang, C Bruce, T Cocke, S Wang, M Boguski, AR Tall. Point mutagenesis of positively charged amino acids of cholesteryl ester transfer protein: conserved residues within the lipid transfer/lipopolysaccharide binding protein gene family essential for function. Biochemistry 34:7258–7263, 1995.
101. H Kotake, LB Agellon, S Yokoyama. Modification of the N-terminal cysteine of plasma cholesteryl ester transfer protein selectively inhibits triglyceride transfer activity. Biochim Biophys Acta 1347:69–74, 1997.
102. TL Swenson, RW Brocia, AR Tall. Plasma cholesteryl ester transfer protein has binding sites for neutral lipids and phospholipids. J Biol Chem 263:5150–5157, 1988.
103. D Sammett, AR Tall. Mechanisms of enhancement of cholesteryl ester transfer protein activity by lipolysis. J Biol Chem 260:6687–6697, 1985.
104. HI Nishida, H Arai, T Nishida. Cholesterol ester transfer mediated by lipid transfer protein as influenced by changes in the charge characteristics of plasma lipoproteins. J Biol Chem 268:16352–16360, 1993.
105. M Nakamura, T Kino, K Niko, S Kiyoto, M Okuhara: Phospholipase A2 inhibitor. Japanese Patent 2,255,615, 1990.
106. K Nishikawa, H Arai, K Inoue. Scavenger receptor-mediated uptake and metabolism of lipid vesicles containing acidic phospholipids by mouse peritoneal macrophages. J Biol Chem 265:5226–5231, 1990.
107. I Namatame, H Tomoda, H Arai, K Inoue, S Ōmura. Complete inhibition of mouse macrophage-derived foam cell formation by triacsin C. J Biochem (Tokyo) 125: 317–327, 1999.
108. K Aikawa, T Furuchi, Y Fujimoto, H Arai, K Inoue. Structure-specific inhibition of lysosomal cholesterol transport in macrophages by various steroids. Biochim Biophys Acta 1213:127–134, 1994.

109. ASJ Katocs, CH Wang, EE Largis. The hypercholesterolemic activity of the ACAT inhibitor CL 283,546 in rat, rabbit and monkey. FASEB J 2:A1219, 1988.
110. H Tomoda, K Igarashi, S Ōmura. Inhibition of acyl-CoA synthetase by triacsins. Biochim Biophys Acta 921:595–598, 1987.
111. K Mochizuki, K Oomori, H Tamura, Y Shizuri, S Nishiyama, E Miyoshi, S Yamamura. The structures of bioactive cyclodepsipeptides, beauveriolide I and II, metabolites of entomopathogenic fungi *Beauveria* sp. Bull Chem Soc Jpn 66:3041–3046, 1992.
112. I Namatame, H Tomoda, S Si, Y Yamaguchi, R Masuma, S Ōmura. Beauveriolides, specific inhibitors of lipid droplet formation in mouse macrophages, produced by *Beauveria* sp. FO-6979. J Antibiot 52:1–6, 1999.
113. I Namatame, H Tomoda, N Tabata, S Si, S Ōmura. Structure elucidation of fungal beauveriolide III, a novel inhibitor of lipid droplet formation in mouse macrophages. J Antibiot 52:7–12, 1999.
114. H Tomoda, I Namatame, S Si, K Kawaguchi, R Masuma, M Namikoshi, S Ōmura. Phenochalasins, inhibitors of lipid droplet formation in mouse macrophages, produced by *Phomopsis* sp. FO-0211. J Antibiot 52:851–856, 1999.
115. H Tomoda, I Namatame, N Tabata, K Kawaguchi, S Si, S Ōmura. Structure elucidation of fungal phenochalasins, inhibitors of lipid droplet formation in mouse macrophages. J Antibiot 52:857–861, 1999.
116. I Namatame, H Tomoda, M Arai, S Ōmura. Effect of fungal metabolites cytochalasans on lipid droplet formation in mouse macrophages. J Antibiot 53:12–18, 2000.
117. I Tabas, X Zha, N Beatini, JN Myers, FR Maxfield. The actin cytoskeleton is important for the stimulation of cholesterol esterification by atherogenic lipoproteins in macrophages. J Biol Chem 269:22547–22556, 1994.
118. G Petranyi, NS Ryder, A Stutz. Allylamine derivatives: new class of synthetic antifungal agents inhibiting fungal squalene epoxidase. Science 224:1239–1241, 1984.
119. Y Hidaka, H Hotta, Y Nagata, Y Iwasawa, M Horie, T Kamei. Effect of a novel squalene epoxidase inhibitor, NB-598, on the regulation of cholesterol metabolism in Hep G2 cells. J Biol Chem 266:13171–13177, 1991.
120. J Sakakibara, R Watanabe, Y Kanai, T Ono. Molecular cloning and expression of rat squalene epoxidase. J Biol Chem 270:17–20, 1995.
121. KK Wong, DW Kuo, RM Chabin, C Fournier, LD Gegnas, ST Waddell, F Marsilio, B Leiting, DL Pompliano. Engineering a cell-free murein biosynthetic pathway: combinatorial enzymology in drug discovery. J Am Chem Soc 120:13527–13528, 1998.
122. ST Cole, R Brosch, J Parkhill, T Garnier, C Churcher, D Harris, SV Gordon, K Eiglmeier, S Gas, CE Barry III, F Tekaia, K Badcock, D Basham, D Brown, T Chillingworth, R Connor, R Davies, K Devlin, T Feltwell, S Gentles, N Hamlin, S Holroyd, T Hornsby, K Jagels, A Krogh, J McLean, S Moule, L Murphy, K Oliver, J Osborne, MA Quail, MA Rajandream, J Rogers, S Rutter, K Seeger, J Skelton, R Squares, JE Sulston, K Taylor, S Whitehead, BG Barrell. Deciphering the biology of *Mycobacterium tuberculosis* from the complete genome sequence. Nature (London) 393:537–544, 1998.
123. M Rohmer, M Knani, P Simonin, B Sutter, H Sahm. Isoprenoid biosynthesis in

bacteria: a novel pathway for the early steps leading to isopentenyl diphosphate. Biochem J 295:517–524, 1993.

124. M Rohmer, M Seemann, S Horbach, S Bringer-Meyer, H Sahm. Glyceraldehyde 3-phosphate and pyruvate as precursors of isoprenic units in an alternative non-mevalonate pathway for terpenoid biosynthesis. J Am Chem Soc 118:2564–2566, 1996.
125. J Schwender, M Seemann, HK Lichtenthaler, M Rohmer. Biosynthesis of isoprenoids (carotenoids, sterols, prenyl side-chains of chlorophylls and plastoquinone) via a novel pyruvate/glyceraldehyde 3-phosphate non-mevalonate pathway in the green alga *Scenedesmus obliquus*. Biochem J 316:73–80, 1996.
126. W Eisenreich, B Menhard, PJ Hylands, MH Zenk, A Bacher. Studies on the biosynthesis of taxol: the taxane carbon skeleton is not of mevalonoid origin. Proc Natl Acad Sci (USA) 93:6431–6436, 1996.
127. HK Lichtenthaler, J Schwender, A Disch, M Rohmer. Biosynthesis of isoprenoids in higher plant chloroplasts proceeds via a mevalonate-independent pathway. FEBS Lett 400:271–274, 1997.
128. GA Sprenger, U Schorken, T Wiegert, S Grolle, AA de Graaf, SV Taylor, TP Begley, S Bringer-Meyer, H Sahm. Identification of a thiamin-dependent synthase in *Escherichia coli* required for the formation of the 1-deoxy-D-xylulose 5-phosphate precursor to isoprenoids, thiamin, and pyridoxol. Proc Natl Acad Sci (USA) 94:12857–12862, 1997.
129. BM Lange, MR Wildung, D McCaskill, R Croteau. A family of transketolases that directs isoprenoid biosynthesis via a mevalonate-independent pathway. Proc Natl Acad Sci (USA) 95:2100–2104, 1998.
130. LM Lois, N Campos, SR Putra, K Danielsen, M Rohmer, A Boronat. Cloning and characterization of a gene from *Escherichia coli* encoding a transketolase-like enzyme that catalyzes the synthesis of D-1-deoxyxylulose 5-phosphate, a common precursor for isoprenoid, thiamin, and pyridoxol biosynthesis. Proc Natl Acad Sci (USA) 95:2105–2110, 1998.
131. S Takahashi, T Kuzuyama, H Watanabe, H Seto. A 1-deoxy-D-xylulose 5-phosphate reductoisomerase catalyzing the formation of 2-C-methyl-D-erythritol 4-phosphate in an alternative nonmevalonate pathway for terpenoid biosynthesis. Proc Natl Acad Sci (USA) 95:9879–9884, 1998.

16

Design and Development of a Selective Assay System for the Phospholipase A_2 Superfamily

Hsiu-Chiung Yang and Marian Mosior
Eli Lilly and Company, Indianapolis, Indiana

Edward A. Dennis
University of California, San Diego, California

I. INTRODUCTION

Bokay provided the first evidence for a phospholipid-hydrolyzing enzyme in 1877 [1]. Since then, this enzyme became a subject of scrutiny by many investigators, due to its ubiquitous presence and multifunctionality as well as its involvement in a wide variety of physiological and pathological responses in many tissue types and organisms (for review, see Refs. 2–6). To date, 10 distinct groups of phospholipase A_2 (PLA_2) have been identified and cloned from various sources (Table 1 and Refs. 2–4). Multiple forms of PLA_2 are often found in the same tissue or cell type. This molecular diversity makes it very difficult to delineate the role of each individual PLA_2 form in certain physiological and pathological processes. This problem becomes especially apparent when selective inhibitors for each group are not available. Therefore, an assay system, in which the activity of each of these PLA_2 can be measured sensitively and selectively in biological samples, is crucial to the advance of the understanding of their respective roles in physiological and pathophysiological processes. The purpose of this chapter is to outline general strategies for the rational design of such an assay system.

Table 1 Classification of Mammalian Phospholipase A_2

Localization	Group	Major sources	Molecular mass (kDa)	Chromosome localization	Physiological roles	No. of disulfide bonds	Characteristics
Secreted	IB	Human pancreas, lung, spleen, kidney, ovary	13.2[a]	12 (human)	Digestion, cell proliferation, cell contraction.	7	His–Asp pair, elapid (pancreatic) loop; characteristic disulfide bond Cys-11-77
	IIA	Human synovial fluid, hematopoetic cells, intestine	13.2[a]	1 (human) and distal part of chromosome 4 (mouse)	Inflammation	7	His–Asp pair; carboxyl extension; characteristic disulfide bond Cys-50-137
	IIC	Rat/mouse testis	14.6[a]	1 (human) and distal part of chromosome 4 (mouse)	Unknown	8	His–Asp pair; carboxyl extension; 4 amino acids insertion at positions 90–94; characteristic disulfide bonds Cys-50-137 and Cys 86-92; nonfunctional pseudogene in human
	IID	Rat spleen and thymus	14.5[a]	4 (mouse)	Unknown	7	His–Asp pair; carboxyl extension; characteristic disulfide bond Cys-50-137
	V	Human heart and lung, rat hippocampus	13.6[a]	1 (human) and distal part of chromosome 4 (mouse)	Production of lipid mediator	6	His–Asp pair, surface Trp 31 & 79
	VIIA	Human plasma	45	Unknown	Anti-inflammatory		GXSXG lipase consensus sequence; Ca^{2+}-independent catalytic activity

	X	Human leukocyte, spleen	13.6[a]	16 (human)	Inflammation and/or immunity	8	His–Asp pair; carboxyl extension, prepropeptide sequence; characteristic disulfide bonds Cys-11-77 and Cys-50-137
Intracellular	IVA	Human kidney, platelet, placenta, leukocytes, brain.	85[a]	1 (human; near COX-2 gene) 13 (rat)	Inflammation, parturition	0	Ser-228 in GXSXG lipase consensus sequence, CaLB domain, responsive to PIP_2 for catalytic activity and membrane association, multiple serine phosphorylation sites
	IVB	Human cerebellum, pancreas	114[a]	15 (human, near phosphoinositol bisphosphate phosphatase gene)	Unknown	0	N-terminal extension, CaLB domain, Ser-228 in GXSXG lipase consensus sequence
	IVC	Human heart, skeletal muscle	61[a]	19 (human, near calmodulin gene)	Unknown	0	Prenyl group-binding site motif, Ser-228 in GXSXG lipase consensus sequence; Ca^{2+}-independent catalytic activity

Table 1 Continued

Localization	Group	Major sources	Molecular mass (kDa)	Chromosome localization	Physiological roles	No. of disulfide bonds	Characteristics
	VI	Human pancreas, heart, brain, skeletal muscle, prostate, testis, thyroid, spinal cord	80–88	22 (human)	Phospholipid remodeling	0	GXSXG lipase consensus sequence; 8 ankyrin repeats; Ca^{2+}-independent catalytic activity, proline-rich consensus motif ($PX_5PX_8HHPX_{12}NX_4Q$), LH-iPLA$_2$ (long isoform, 88 kDa) but not SH-iPLA$_2$ (short isoform, 85 kDa) is activated by ATP
	VIIB	Human brain	29–30	Unknown	1. Modulation of PAF level 2. Regulation of neuronal migration during development	0	GDSXV modified consensus sequence of serine esterase family; catalytic triad Ser-His-Asp; β and γ subunits form a heterotrimer with a 45-kDa noncatalytic subunit; Ca^{2+}-independent catalytic activity
	VIII	Bovine brain	29	Unknown	Unknown	0	Ca^{2+}-independent catalytic activity

[a] Molecular mass was calculated based on the primary sequence.

IIE & IIF sPLA$_2$ were recently cloned (E Valentin, RS Koduri, JC Scimeca, G Carle, MH Gelb, M Lazdunski, G Lambeau. Cloning and recombinant expression of a novel mouse-secreted phospholipase A$_2$. J Biol Chem 274:19152–19160, 1999). Little is known about their properties and distribution. The sequence alignment of all murine sPLA$_2$s is given in Fig. 1. Groups III and IX PLA$_2$ are not of mammalian origin. CaLB domain is the Ca^{2+}- and lipid-binding domain, designated also as C2 domain.

II. CLASSIFICATION AND DISTRIBUTION

Phospholipase A_2 (E.C. 3.1.1.4) constitutes a growing family of enzymes. They are found in many cell types and tissues. Based on their cellular localization and their primary and secondary structures, including characteristic domains, they are divided into 10 major groups, often with a few subgroups within each group [2–4]. Table 1 summarizes the cellular localization, sources, chromosomal localization, physiological roles, and molecular characteristics of each mammalian PLA_2 group. Notably, there is no significant homology between secreted ($sPLA_2$) and intracellular PLA_2s. Among $sPLA_2$s, there is between 30% and 50% identity in their primary sequences. The major conserved regions among all $sPLA_2$s are the Ca^{2+} loop and the active-site regions. Figure 1 shows the sequence alignment of eight murine $sPLA_2$s (group IB, IIA, IIC, IID, IIE, IIF, V, and X). In general, the intracellular PLA_2s share very little sequence homology. The only recognizable similarity is the GXSXG lipase consensus motif.

Based on their cellular localization, mammalian PLA_2s can be divided into two general types: extracellular (secreted groups I, II, V, VIIA, and X) and intracellular (groups IV, VI, VIIB, and VIII) PLA_2s [2–4]. Generally, the secreted PLA_2s ($sPLA_2$s) have low molecular masses, between 13 and 16 kDa, and require millimolar calcium concentrations for catalytic activity, with the exception of group VIIA $sPLA_2$, which has a molecular mass of 45 kDa and does not require Ca^{2+} for catalytic activity. The general distinguishing structural features of $sPLA_2$s include the His–Asp pair, elapid loop, carboxyl extension, and multiple disulfide bonds. Based on those distinct molecular structural features of each enzyme, $sPLA_2$s are divided into groups I, II, III, V, VIIA, IX, and X. Among them, groups III and IX $sPLA_2$s are not found in mammals. The intracellular PLA_2s are divided into Ca^{2+}- dependent and -independent ones, based on the Ca^{2+} requirements needed for membrane association. These enzymes have larger molecular masses of about 29–120 kDa and are insensitive to thio-reducing agents such as dithiothrietol (DTT).

III. INVOLVEMENT OF PLA_2 IN DISEASES

PLA_2 catalyzes the hydrolysis of phospholipids in the *sn-2* position and generates free fatty acids, e.g., arachidonic acid as the major product, with the concomitant release of lysophospholipids. These products serve different functions within the cell. Arachidonic acid (AA) may act as an intracellular second messenger or can be further metabolized to pro-inflammatory mediators, e.g., prostaglandins, leukotrienes, thromboxanes, or hydroxyeicosatetraenoic acids [7–8]. The generation of AA by PLA_2 is the rate-limiting step for the production of eicosanoids [9], through cyclooxygenases (COX-1 and COX-2), lipoxygenase, and cyto-

```
                       1 [                                                    50
group_IB     29.7%     MKLLLLAALLTAGAAAHSISPRAVWQFRNMIKCT----IPGSDPLKDYNN
group_IIA    42.4%     -----------------------------MIRLK-----TGKRAELSYAF
group_IIC    32.7%     --MKGIAIFLVFIFYWTTSTLSSFWQFQRMVKHV-----TGRSAFFSYYG
group_IID    41.4%     --MR-LALLCGLLLAGITATQGGLLNLNKMVTHM-----TGKKAFFSYWP
group_IIE    39.6%     --MK-PPIALACLCLLVPLAGGNLVQFGVMIERM-----TGKPALQ-YND
group_IIF    31.8%     --MKKFFAIAVLAGSVVTTAHSSLLNLKSMVEAI-----THRNSILSFVG
group_V     100.0%     --MKGLLTLAWFLACSVPAVPGGLLELKSMIEKV-----TRKNAFKNYGF
group_X      27.5%     MLLLLLLLLLGPGPGFSEATRRSHVYKRGLLELAGTLDCVGPRSPMAYMN
                       prepropeptide sequence

                       51                                                    100
group_IB     29.7%     YGCYCGLGGWGTPVDDLDRCCQTHDHCYSQAKKLESCKFLIDNPYTNTYS
group_IIA    42.4%     YGCHCGLGGKGSPKDATDRCCVTHDCCYKSLEK-SGCG-----TKLLKYK
group_IIC    32.7%     YGCYCGLGGKGLPVDATDRCCWAHDCCYHKLKE-YGCQ-----PILNAYQ
group_IID    41.4%     YGCHCGLGGKGQPKDATDWCCQKHDCCYAHLKI-DGCK-----SLTDNYK
group_IIE    39.6%     YGCYCGVGGSHWPVDETDWCCHAHDCCYGRLEK-LGCD-----PKLEKYL
group_IIF    31.8%     YGCYCGLGGRGHPMDEVDWCCHAHDCCYEKLFE-QGCR-----PYVDHYD
group_V     100.0%     YGCYCGWGGRGTPKDGTDWCCQMHDRCYGQLEE-KDCA-----IRTQSYD
group_X      27.5%     YGCYCGLGGHGEPRDAIDWCCYHHDCCYSRAQD-AGCS-----PKLDRYP
                       Ca2+ loop        active site          pancreatic
                                                             loop

                       101                                                   150
group_IB     29.7%     YSCSGS-EITCSAKN---NKCEDFICNCDREAAICFSK--VPYNKEYKNL
group_IIA    42.4%     YSHQGG-QITCSANQ---NSCQKRLCQCDKAAAECFARNKKTYSLKYQFY
group_IIC    32.7%     FTIVNG-TVTCGCTVASSCPCGQKACECDKQSVYCFKENLATYEKAFKQL
group_IID    41.4%     YSISQG-TIQCSDNG---SWCERQLCACDKEVALCLKQNLDSYNKRLRYY
group_IIE    39.6%     FSITRD-NIFCAGR----TACQRHTCECDKRAALCFRHNLNTYNRKYAHY
group_IIF    31.8%     HRIENGTMIVCTELN--ETECDKQTCECDKSLTLCLKDH--PYRNKYRGY
group_V     100.0%     YRYTNG-LVICEHD----SFCPMRLCACDRKLVYCLRRNLWTYNPLYQYY
group_X      27.5%     WKCMDH-HILCGPAE---NKCQELLCRCDEELAYCLAG--TEYHLKYLFF

                       151                                  ] 187
group_IB     29.7%     DT-GKFC-----------------------------
group_IIA    42.4%     P--NMFCKGKKPKC----------------------
group_IIC    32.7%     FPTRPQCGRDKLQC----------------------
group_IID    41.4%     W--RPRCKGKTPAC----------------------
group_IIE    39.6%     P--NKLCTGPTPPC----------------------
group_IIF    31.8%     F--NVYCQGPTPNCSIYDPYPEEVTCGHGLPATPVST
group_V     100.0%     P--NFLC-----------------------------
group_X      27.5%     P--SILCEKDSPKCN---------------------
                            group II
                            COOH-terminal extension
```

Figure 1 Multiple alignment of murine secreted phospholipases A_2 ($sPLA_2$). Sequences were aligned using ClustalW algorithm with group V $sPLA_2$ as the reference sequence. The number following the designation of each isoform represents the percentage of identity with respect to group V $sPLA_2$. Structural elements are underlined and annotated.

chrome P450 pathways. Lysophospholipids may perturb membrane homeostasis and increase membrane fluidity and permeability, and they can be converted into a bioactive molecule, platelet-activating factor [10,11].

Numerous studies have suggested that PLA_2 activity changes in neural disorders [12–17], cancer [18–21], inflammatory disorders [8,22–24], and parturition [25,26]. In addition, lysophosphatidylcholine (LPC) causes demyelination [27]. Recent studies suggested that LPC, converted from oxidized low-density lipoprotein (ox-LDL), induces endothelium-dependent vasoconstriction [28] and abolishes the formation of the receptor G-protein complex [29].

IV. PROPERTIES OF PLA_2s

Table 2 summarizes properties of mammalian PLA_2s that are potentially involved in the generation of arachidonic acid. The $sPLA_2$s are stable under a wide range of conditions, due to the presence of multiple disulfide bonds within the molecule. However, the presence of thio-reducing agents, such as DTT, causes the loss of enzymatic activity [2,4]. In general, they all require high concentrations of Ca^{2+} (millimolar range) for catalytic activity. Among $sPLA_2$s, only the pancreatic $sPLA_2$-IB was activated by a micromolar concentration of Zn^{2+} and a millimolar concentration of Cd^{2+} [30]. In contrast, those cations had little effect on the group II $sPLA_2$. Secreted $sPLA_2$s display a differential preference toward phospholipid head groups, e.g., choline versus ethanolamine. In general, $sPLA_2$s do not have preference for fatty acids in the *sn-2* position of phospholipids, with the exception of the $sPLA_2$-V [31,32]. All mammalian $sPLA_2$s have distinct pI and pH optima for phospholipid hydrolysis, and are activated by the presence of anionic phospholipids, e.g., phosphatidylserine and phosphatidylglycerol. These anionic phospholipids regulate the catalytic activity of $sPLA_2$s in *in vitro* models by alteration of their membrane association, presumably through the interaction with the patches of positively charged residues on the surface [33].

Properties of the two predominantly intracellular PLA_2s are listed in Table 2 as well. The major difference between these two groups of PLA_2 is the requirement of Ca^{2+} for membrane translocation [34]. In addition, the activity of $iPLA_2$ appears to be affected by ATP [35,36], whereas that of $cPLA_2$ is increased by PIP_2, through the enhancement of its membrane association [37,38]. The activation of $cPLA_2$ by PIP_2 cannot be replaced by other anionic phospholipids, e.g., phosphotidylserine [38].

V. DESIGN OF A GROUP-SPECIFIC ASSAY SYSTEM

To date, there are numerous reports on the assay procedures for PLA_2 (for review, see Ref. 39). They range from simple procedures (e.g., titrametric and acidimet-

Table 2 Properties of Mammalian PLA_2s

Group	$sPLA_2$-IB	$sPLA_2$-IIA	$sPLA_2$-V	$sPLA_2$-X	$cPLA_2$-IVA	$iPLA_2$-VI
pH optimum	Acidic	Alkaline	Neutral	Unknown	Neutral	Neutral
Ca^{2+} requirement	mM	mM	mM	mM	μM	None
Calculated pI	7.6	9.3	8.5	5.3	5.2	6.9
Substrate specificity	PE = PC	PE > PC	PC > PE	Unknown	PC = PE	PC > PE
Fatty acid preference	No	No	LA > OA > PA > AA	Unknown	AA > LA > PA	LA > PA > OA > AA
Regulation	1. Zymogen processing 2. Bile salts 3. Anionic phospholipids 4. μM concentration of Zn^{2+}	1. Induced by proinflammatory stimuli 2. Glycosaminoglycans. 3. Activated by anionic phospholipids	1. Induced by proinflammatory stimuli 2. Activated by anionic phospholipids	Unknown	1. Ca^{2+}-dependent translocation 2. Phosphorylation 3. Induced by cytokines 4. Activated by PIP_2	ATP
Catalytic mechanism	His–Asp pair	His–Asp pair	His–Asp pair	His–Asp pair	Ser 228 and Asp-549	GXSXG motif

Abbreviations: AA, arachidonic acid; LA, linoleic acid; OA, oleic acid; PA, palmitic acid; PC, phosphatidylcholine; PE, phosphatidylethanolamine; PIP_2, phosphatidylinositol 4,5-bisphosphate.

ric) to those involving expensive instrumentation (e.g., nuclear magnetic resonance and electron spin resonance), and from low sensitivity (10 to 100 nmol), e.g., spectrophotometric or polarographic procedures, to very sensitive procedures (fentomoles), e.g., radiometric methods. Several enzyme-coupled assays to detect PLA_2 activity are also available. Among them, the acyl-CoA synthase-coupled assay is the most sensitive one; as low as 1–3 nmol of fatty acid can be detected. However, this type of assay is not suitable for detecting PLA_2 activities in crude cell and tissue extracts. Continuous assays, using fluorescent probe-labeled phospholipids, demonstrate a medium sensitivity (detecting limits at pmole range) and convenience for inhibitor kinetics analysis. However, they suffer from two major disadvantages. First, fluorescent probes attached at either the head group or the *sn-2* fatty acyl chain of phospholipids modify the chemical structure of those molecules. Those structural changes may alter the interaction of the modified substrate with the enzyme either directly or through a change in the physical properties of the phospholipid aggregates. Second, some of them involve the displacement of the fluorescent fatty acid probes from fatty acid-binding protein or BSA (bovine serum albumin) by fatty acids released as a result of PLA_2-catalyzed hydrolysis of phospholipids. The affinity of those proteins for fatty acids depends strongly on the length of the acyl chain. Thereby, the comparison of PLA_2s with different substrate specificity is difficult. Collectively, due to the sensitivity and the use of a native substrate, radiometric assays are the most commonly used for the detection of phospholipase A_2 activity. However, the direct measurement of products of PLA_2-catalyzed hydrolysis does not provide sufficient selectivity to differentiate between the different enzyme forms of PLA_2.

As demonstrated by many studies, all cells and tissues contain multiple forms of PLA_2 involved in phospholipid metabolism. Since these enzymes catalyze the same enzymatic reaction with only a modest difference in the specificity toward phospholipid substrates and phospholipid aggregates (e.g., detergent mixed micelles, small unilamellar vesicles, and large unilamellar vesicles), it is difficult to distinguish them in biological samples that contain mixtures of several PLA_2 types. Thus, the incorporation of their enzymatic and kinetic properties (as illustrated in Table 2) in the assay for each PLA_2 group should add a sufficient number of variables to design a selective assay system.

Table 3 shows an example of the assay system, which differentiates among PLA_2 groups [40]. Unique properties of $iPLA_2$, e.g., Ca^{2+}-independent catalytic activity, DTT insensitivity, and ATP activation, were used to design the assay to detect the activity of $iPLA_2$ selectively. As suggested by Mosior et al. [38], PIP_2 enhances the membrane affinity of $cPLA_2$ and reduces its Ca^{2+} requirement for the catalytic activity. In addition, due to the lack of disulfide bonds, $cPLA_2$ is also insensitive to the presence of thio-reducing agents. Therefore, the addition of a millimolar concentration of DTT in the $cPLA_2$ assay reduces the activities of $sPLA_2$s to a minimal level. Thus, those properties were included in the design

Table 3 Selectivity of PLA_2 Assays

	$cPLA_2$-IV ASSAY	$iPLA_2$-IV ASSAY	$sPLA_2$-V ASSAY	$sPLA_2$-IIA ASSAY
$cPLA_2$-IV	1 (16.7 ± 0.2)	<1/10,000	1/700	1/85
$iPLA_2$-VI	1/30 (0.56 ± 0.01)	1 (4.5 ± 0.2)	<1/7,000	1/720
$sPLA_2$-V	<1/5,000	1/10,000	1 (4.7 ± 0.04)	1/4 (6.3 ± 0.4)
$sPLA_2$-IIA	1/600	<1/10,000	1/4 (1.3 ± 0.1)	1 (25.1 ± 0.1)

Abbreviations: $cPLA_2$-IV, group IV cytosolic Ca^{2+}-dependent phospholipase A_2; $iPLA_2$-VI, group VI Ca^{2+}-independent phospholipase A_2; sPLA-V, group V secreted phospholipase A_2; $sPLA_2$-IIA, group IIA secreted phospholipase A_2. Each column illustrates the activities of all PLA_2S relative to the enzyme with the highest activity in a given assay. Numbers in parentheses represent specific activities (mean ± S.D., $n > 3$, in μmol min^{-1} mg^{-1}) of the enzyme in that assay. Detailed conditions for each group-selective assay were described in [40]. Briefly, the $cPLA_2$-IV activity was determined at 100 μM 1-palmitoyl-2-arachidonyl-*sn*-glycero-3-phosphorylcholine (PAPC)/phosphatidylinositol 4,5-biophosphate (PIP_2) 97/3, in Triton X-100 (400 μM) mixed micelles in 100 mM HEPES, pH 7.5, 80 μM Ca^{2+}, 2 mM DTT, and 0.1 mg/ml bovine serum albumin (BSA). The $iPLA_2$-VI selective assay was performed in 100 μM 1-palmitoyl-2-palmitoyl-*sn*-phosphorylcadine (DPPC) in Triton X-100 (400 μM) mixed micelles in 100 mM HEPES, pH 7.5, 5 mM ethylenediaminetetraacetic acid (EDTA), 2 mM DTT, and 1 mM ATP. The conditions for the $sPLA_2$-IIA selective assay were as follows: 100 μM 1-palmitoyl-2-linoleoyl-*sn*-glycero-3-phosphorylethanolamine (PLPE)/1-palmitoyl-2-oleoyl-*sn*-glycero-3-phosphorylserine (POPS), 1/1, sonicated vesicles in 100 mM HEPES, pH 7.5, 1 mM Ca^{2+}, and 1 mg/ml BSA. The activity for $sPLA_2$-V was measured at 100 μM DPPC/POPS, 3/1, as sonicated vesicles in 100 mM HEPES, pH 7.5, 5 mM Ca^{2+}, and 1 mg/ml BSA.

of the assay conditions for $cPLA_2$. Both $sPLA_2$-IIA and -V require a millimolar concentration of Ca^{2+} for the catalytic activity, are sensitive to thio-reducing agents, and are activated by anionic phospholipids, e.g., phosphotidylserine. These factors pose a major difficulty in the design of selective assays for these two enzymes. Fortunately, these enzymes have distinct preferences toward head groups and fatty acids in the *sn-2* position of phospholipids. Therefore, these properties allow us to design an assay system that differentiates among these $sPLA_2$s. As demonstrated in Table 2, $sPLA_2$-IIA and -V show different pH optima for the substrate hydrolysis. This property (pH 7.4 for $sPLA_2$-V and pH 9.0 for $sPLA_2$-IIA) may be used to further differentiate between their activities. Another property used to differentiate between secreted and intracellular PLA_2s is their dependence on the substrate presentation. The assay for both $iPLA_2$ and $cPLA_2$ used Triton X-100/phospholipid mixed micelles, whereas those for $sPLA_2$s used substrates in the form of small unilamellar vesicles (SUVs).

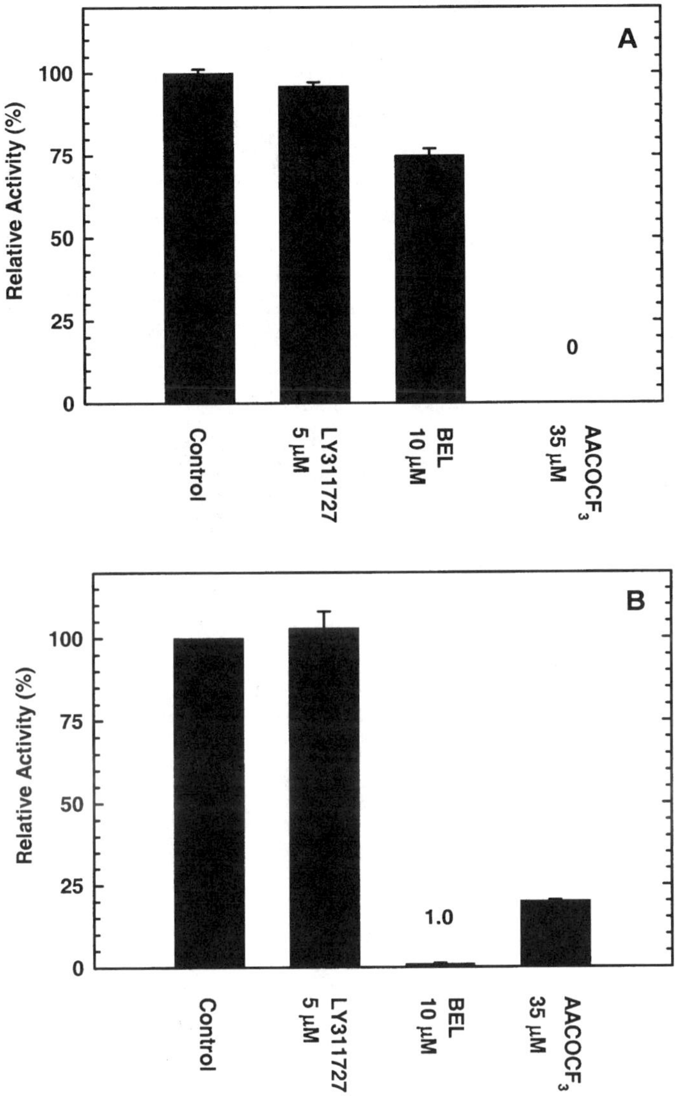

Figure 2 Analyses of $cPLA_2$-IV and $iPLA_2$-VI inhibitors. The cytosolic fraction of $P388D_1$ macrophage-like cells was obtained from the supernatant after centrifugation at $100{,}000 \times g$ for 1 hr. The (A) $cPLA_2$-IV and (B) $iPLA_2$-VI activities of this sample in the presence and absence of various PLA_2 inhibitors ($AACOCF_3$, BEL, and LY311727) were measured in their selective assays. The detailed conditions are described in Table 3 and Ref. 40.

As illustrated in Table 3, the specificity of this assay system toward three major subgroups of mammalian PLA_2s, that is, $sPLA_2$, $cPLA_2$, and $iPLA_2$, is high. In particular, both cytosolic groups (i.e., groups IV and VI) of PLA_2, display two or more orders of magnitude specificity in their respective assays. Particularly, the remarkable selectivity of the $iPLA_2$-VI assay, based on its lack of a requirement for Ca^{2+}, is greater than 10,000-fold over other PLA_2 isoforms. Distinguishing between closely related secreted PLA_2s is particularly challenging, owing to their extensive structural homology. Nevertheless, as illustrated in Table 3, one can at least identify them using their unique profiles in this assay system. The high specificity of those assays allows detection of these major PLS_2s in excess of other ones present in the same biological sample [40].

An example is given in Figure 2A and 2B. The cytosolic fraction of $P388D_1$ macrophage-like cells was tested in both $cPLA_2$-IV and $iPLA_2$-VI specific assays. A selective inhibitor for $iPLA_2$, BEL (bromoenol lactone), extinguished the activity measured in the $iPLA_2$-VI specific assay (Fig. 2B). In contrast, the activity measured in the $cPLA_2$-IV specific assay declined by only 25% (Fig. 2A). These results were consistent with the inhibition profile of this compound toward pure rat brain $iPLA_2$-VI and recombinant $cPLA_2$-IV, 100% and 20% inhibitions, respectively. Arachidonyl trifluoromethyl ketone ($AACOCF_3$), an inhibitor with some preference for pure $cPLA_2$-IV over pure $iPLA_2$-VI, obliterated the activity measured in the $cPLA_2$-IV specific assay and reduced by 80% that measured in the $iPLA_2$-VI specific assay. This is consistent with the inhibitory pattern of pure $iPLA_2$ from rat brain, which is 90% inhibited by 35 μM (7 mol%) $AACOCF_3$. In contrast, an inhibitor selective for $sPLA_2$-IIA, LY311727, had no effect in either of those assays (Fig. 2A and 2B).

As illustrated in Figure 2A and 2B, not all PLA_2 inhibitors are truly specific. Thus, a system of group-selective assays provides an additional tool in the identification of individual members of the PLA_2 superfamily in biological samples. In addition, this assay system will be helpful in defining the targets of compounds inhibiting PLA_2 activity in such samples.

VI. PURIFICATION OF $iPLA_2$ FROM RAT BRAIN USING THE GROUP-SPECIFIC ASSAY SYSTEM

Using the specific assay for the Ca^{2+}-independent PLA_2, ($iPLA_2$), an $iPLA_2$ was purified from $P388D_1$ macrophage-like cells [35] and rat brain [41]. The purification procedure for rat brain $iPLA_2$ uses ammonium sulfate precipitation, octyl-Sepharose, ATP-agarose, and calmodulin-agarose column chromatographic steps. SDS-PAGE analysis of fractions from this purification procedure is shown in Figure 3. In this purification procedure, one of the column chromatographic steps utilizes a specific property of this Ca^{2+}-independent PLA_2, namely, its inter-

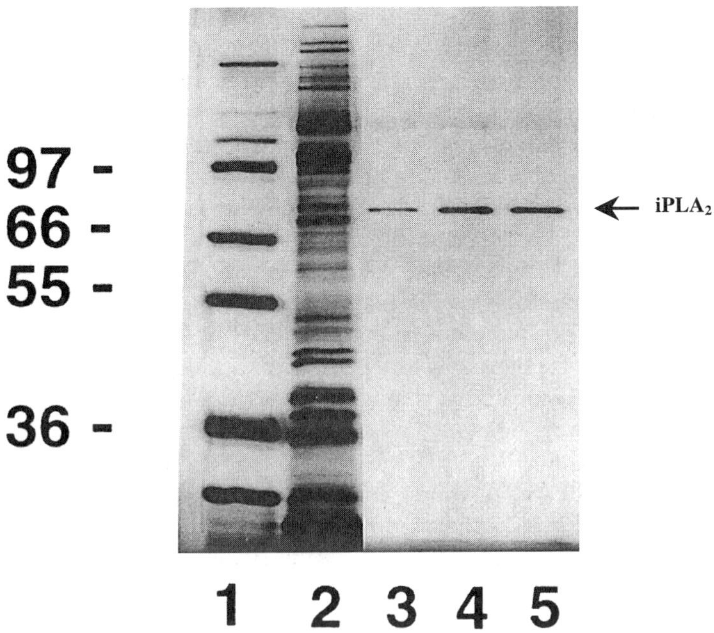

Figure 3 Silver stained SDS-PAGE gel of iPLA$_2$-VI purification. Lane 1, molecular mass markers; lane 2, 20 μl of pool from octyl-Sepharose column; lane 3, 20 μl of fraction #28 from the Calmodulin-agarose column; lane 4, 20 μl of fraction #29 from the Calmodulin-agarose column; and lane 5, 20 μl of fraction #30 from the Calmodulin-agarose column. (From Ref. 41.)

action with ATP. The ATP-agarose column chromatographic step enriched the iPLA$_2$ activity 1200-fold from the previous step [41].

VII. PLA$_2$ INHIBITORS

Specific inhibitors for a given enzyme, especially for those that are members of a superfamily of enzymes, offer the best way for the direct assessment of its role in a complex biological process. Some inhibitors may also have a therapeutic potential.

PLA$_2$ inhibitors were the subjects of two recent review articles [4,42]. Non-specific irreversible sPLA$_2$ inhibitors, *p*-bromophenacyl bromide (covalently modifying the active histidine residue) [43,44] and a natural product manoalide (covalently modifying lysine residue) [45,46], are often used to study physiological functions of $sPLA_2$s in various systems. These compounds inhibit $sPLA_2$, but not $cPLA_2$ or $iPLA_2$, in *in vitro* assays [4]. Therefore, they are often referred to as "specific $sPLA_2$ inhibitors." However, due to the irreversible nature of their interaction with amino acids, their use in *in vivo* or in cell-based assays contains an inherent risk of covalent modification of other proteins and they are known to be nonspecific. Structure based-design of inhibitors, based on the crystal structure of $sPLA_2$-IIA, resulted in several specific $sPLA_2$ inhibitors, e.g., indole compounds, LY311727 [47] and LY311299 [48], and a indolizine compound, 120-1032 [48]. Among them, LY311727 has been widely used in many cellular studies. This compound inhibits $sPLA_2$-IIA and $sPLA_2$-V in *in vitro* assay [32]. The indole compound is 1500-fold more potent toward $sPLA_2$-IIA than toward $sPLA_2$-IB [48]. It is also less potent toward $sPLA_2$-X in a transient expression system in HEK 293 cells compared to that toward $sPLA_2$-IIA and $sPLA_2$-V [49,50].

Less information is available on specific inhibitors of intracellular PLA_2s, $cPLA_2$ and $iPLA_2$. Two $cPLA_2$ inhibitors, arachidonyl trifluoromethyl ketone (AATFMK, also $AACOCF_3$) [51] and methyl arachidonyl fluorophosphonate (MAFP), have been reported [52]. $AACOCF_3$ is a reversible, tight- and slow-binding inhibitor [51]. MAFP is an irreversible inhibitor of $cPLA_2$ *in vitro*. In addition, both MAFP and $AACOCF_3$ also inhibit $iPLA_2$ with similar potency as that toward $cPLA_2$ [53].

Bromoenol lactone [BEL, known also as haloenol lactone suicide substrate (HELSS)] is a selective suicide inhibitor for $iPLA_2$ *in vitro* and in intact cells [54]. BEL is reported to inhibit a 40-kDa $iPLA_2$ activity from canine myocardium and the 80-kDa $iPLA_2$ from $P388D_1$ macrophages to a similar degree [35]. At high concentrations, it also inhibits Mg^{2+}-dependent phosphatidic acid phosphohydrolase and $cPLA_2$ [55–57].

This group-specific assay system will allow the evaluation of the pharmacokinetics and pharmacodynamics of compounds with PLA_2 inhibitory activity in complex systems, as exemplified in Figure 2.

VIII. CONCLUSION AND PERSPECTIVES

PLA_2s play important roles in phospholipid metabolism under normal physiological conditions and during pathological processes of many disorders (for review, see Refs. 2–4, 8). However, due to the lack of truly selective inhibitors for the PLA_2s, it is difficult to understand the roles played by each type of PLA_2 in different disease processes. Therefore, the assays with even a modest degree of selectivity for specific PLA_2 forms will be useful tools to further advance the elucidation of the role of each individual type of PLA_2 in physiological and pathological processes and to develop specific inhibitors.

In mammalian cells, there are multiple forms of enzymes within the same family, e.g., protein kinase C, phospholipase C, and caspase. A selective assay system would be instrumental in such cases to advance understanding of the respective role for each form within the same enzyme family. An example for design of a selective assay system for the superfamily of phospholipase A_2 is provided in this chapter. Such a selective assay system may play a significant enabling role for PLA_2 and other enzyme families in the discovery of inhibitors relevant to the treatment of pathologies involving those enzymes.

REFERENCES

1. A Bokay. Ueber die Verdaulichkeit des Nucleins und Lecithins. Z Physiol Chem 1: 157–164, 1877/78.
2. EA Dennis. Diversity of group types, regulation, and function of phospholipase A_2. J Biol Chem 269:13057–13060, 1994.
3. EA Dennis. The growing phospholipase A_2 superfamily of signal transduction enzymes. Trends Biochem Sci 22:1–2, 1997.
4. J Balsinde, MA Balboa, PA Insel, EA Dennis. Regulation and inhibition of phospholipase A_2. Annu Rev Pharmacol Toxicol 39:175–189, 1999.
5. B Chaminade, F Le Balle, O Fourcade, M Nauze, C Delagebeaudeuf, A Gassama-Diagne, MF Simon, J Fauvel, H Chap. New developments in phospholipase A_2. Lipids 34 (Suppl):S49–S55, 1999.
6. E Kaiser. Phospholipase A_2: its usefulness in laboratory diagnostics. Crit Rev Clin Lab Sci 36:65–163, 1999.
7. WL Smith. Eicosanoid biosynthesis and mechanism of action. Am J Physiol 263: F181–F191, 1992.
8. G Cirino. Multiple controls in inflammation. Extracellular and intracellular phospholipase A_2, inducible and constitutive cyclooxygenase, and inducible nitric oxide synthase. Biochem Pharmacol 55:105–111, 1998.
9. AA Farooqui, HC Yang, LA Horrocks. Involvement of phospholipase A_2 in neurodegeneration. Neurochem Int 30:517–522, 1997.
10. LP Vernon, JD Bell. Membrane structure, toxins and phospholipase A_2 activity. Pharmacol Ther 54:269–295, 1992.
11. H Naraba, Y Imai, I Kudo, Y Nakagawa, S Oh-ishi. Activation of phospholipase A_2 and acylation of lysophospholipids: the major regulators for platelet activating factor production in rat neutrophils. J Biochem (Tokyo) 118:442–447, 1995.
12. AA Farooqui, HC Yang, TA Rosenberger, LA Horrocks. Phospholipase A_2 and its role in brain tissue. J Neurochem 69:889–901, 1997.
13. E Shohami, Y Shapira, G Yadid, N Reisfeld, S Yedgar. Brain phospholipase A_2 is activated after experimental closed head injury in the rat. J Neurochem 53:1541–1546, 1989.
14. Y Owada, T Tominaga, T Yoshimoto, H Kondo. Molecular cloning of rat cDNA for cytosolic phospholipase A_2 and the increased gene expression in the dentate gyrus following transient forebrain ischemia. Brain Res 25:364–368, 1994.

15. G Rordorf, Y Uemura, JV Bonventre. Characterization of phospholipase A_2 (PLA_2) activity in gerbil brain: enhanced activities of cytosolic, mitochondrial, and microsomal forms after ischemia and reperfusion. J Neurosci 11:1829–1836, 1991.
16. JA Clemens, DT Stephenson, EB Smalstig, EF Roberts, EM Johnstone, JD Sharp, SP Little, RM Kramer. Reactive glia express cytosolic phospholipase A_2 after transient global forebrain ischemia in the rat. Stroke 27:527–535, 1996.
17. AD Edgar, J Strosznajder, LA Horrocks. Activation of ethanolamine phospholipase A_2 in brain during ischemia. J Neurochem 39:1111–1116, 1982.
18. LE Heasley, S Thaler, M Nicks, B Price, K Skorecki, RA Nemenoff. Induction of cytosolic phospholipase A_2 by oncogenic Ras in human non-small cell lung cancer. J Biol Chem 272:14501–14504, 1997.
19. CW Hendrickse, S Radley, IA Donovan, MR Keighley, JP Neoptolemos. Activities of phospholipase A_2 and diacylglycerol lipase are increased in human colorectal cancer. Br J Surg 82:475–478, 1995.
20. YL Wu, XR Jiang, DM Lillington, PD Allen, AC Newland, SM Kelsey. 1,25-Dihydroxyvitamin D3 protects human leukemic cells from tumor necrosis factor-induced apoptosis via inactivation of cytosolic phospholipase A_2. Cancer Res 58:633–640, 1998.
21. BP Kennedy, C Soravia, J Moffat, L Xia, T Hiruki, S Collins, S Gallinger, B Bapat. Overexpression of the nonpancreatic secretory group II PLA_2 messenger RNA and protein in colorectal adenomas from familial adenomatous polyposis patients. Cancer Res 58:500–503, 1998.
22. A Hietaranta, E Kemppainen, P Puolakkainen, V Sainio, R Haapiainen, H Peuravuori, E Kivilaakso, T Nevalainen. Extracellular phospholipases A_2 in relation to systemic inflammatory response syndrome (SIRS) and systemic complications in severe acute pancreatitis. Pancreas 18:385–391, 1999.
23. BO Anderson, EE Moore, A Banerjee. Phospholipase A_2 regulates critical inflammatory mediators of multiple organ failure. J Surg Res 56:199–205, 1994.
24. K Kim, SY Jung, DK Lee, JK Jung, JK Park, DK Kim, CH Lee. Suppression of inflammatory responses by surfactin, a selective inhibitor of platelet cytosolic phospholipase A_2. Biochem Pharmacol 55:975–985, 1998.
25. DG Skannal, DE Brockman, AL Eis, S Xue, TA Siddiqi, L Myatt. Changes in activity of cytosolic phospholipase A_2 in human amnion at parturition. Am J Obstet Gynecol 177:179–184, 1997.
26. JV Bonventre, Z Huang, MR Taheri, E O'Leary, E Li, MA Moskowitz, A Sapirstein. Reduced fertility and postischaemic brain injury in mice deficient in cytosolic phospholipase A_2. Nature 390:622–625, 1997.
27. V Dousset, B Brochet, A Vital, C Gross, A Benazzouz, A Boullerne, AM Bidabe, AM Gin, JM Caille. Lysolecithin-induced demyelination in primates: preliminary *in vivo* study with MR and magnetization transfer. Am J Neuroradiol 16:225–231, 1995.
28. T Murohara, K Kugiyama, M Ohgushi, S Sugiyama, Y Ohta, H Yasue. LPC in oxidized LDL elicits vasocontraction and inhibits endothelium-dependent relaxation. Am J Physiol 267:H2441–H2449, 1994.
29. NV Prokazova, ND Zvezdina, AA Korotaeva. Effect of lysophosphatidylcholine on transmembrane signal transduction. Biochemistry (Mosc.) 63:31–37, 1998.

30. M Lindahl, C Tagesson. Zinc (Zn^{2+}) binds to and stimulates the activity of group I but not group II phospholipase A_2. Inflammation 20:599–611, 1996.
31. J Chen, SJ Engle, JJ Seilhamer, JA Tischfield. Cloning and recombinant expression of a novel human low molecular weight Ca(2+)-dependent phospholipase A_2. J Biol Chem 269:2365–2368, 1994.
32. Y Chen, EA Dennis. Expression and characterization of human group V phospholipase A_2. Biochim Biophys Acta 1394:57–64, 1998.
33. DL Scott, SP White, JL Browning, JJ Rosa, MH Gelb, PB Sigler. Structures of free and inhibited human secretory phospholipase A_2 from inflammatory exudate. Science 254:1007–1010, 1991.
34. GY Xu, T McDonagh, HA Yu, EA Nalefski, JD Clark, DA Cumming. Solution structure and membrane interactions of the C2 domain of cytosolic phospholipase A_2. J Mol Biol 280:485–500, 1998.
35. EJ Ackermann, ES Kempner, EA Dennis. Ca^{2+}-independent cytosolic phospholipase A_2 from macrophage-like P388D_1 cells. Isolation and characterization. J Biol Chem 269:9227–9233, 1994.
36. Z Ma, X Wang, W Nowatzke, S Ramanadham, J Turk. Human pancreatic islets express mRNA species encoding two distinct catalytically active isoforms of group VI phospholipase A_2 (iPLA_2) that arise from an exon-skipping mechanism of alternative splicing of the transcript from the iPLA_2 gene on chromosome 22q13.1. J Biol Chem 274:9607–9616, 1999.
37. CC Leslie, JY Channon. Anionic phospholipids stimulate an arachidonoyl-hydrolyzing phospholipase A_2 from macrophages and reduce the calcium requirement for activity. Biochim Biophys Acta 1045:261–270, 1990.
38. M Mosior, DA Six, EA Dennis. Group IV cytosolic phospholipase A_2 binds with high affinity and specificity to phosphatidylinositol 4,5-bisphosphate resulting in dramatic increases in activity. J Biol Chem 273:2184–2191, 1998.
39. LJ Reynolds, WN Washburn, RA Deems, EA Dennis. Assay strategies and methods for phospholipases. In: EA Dennis, ed. Methods in Enzymology 197. Orlando; FL: Academic Press, 1991, pp 3–23.
40. HC Yang, M Mosior, CA Johnson, Y Chen, EA Dennis. Group-specific assays that distinguish between the four major types of mammalian phospholipase A_2. Anal Biochem 269:278–288, 1999.
41. HC Yang, M Mosior, B Ni, EA Dennis. Regional distribution, ontogeny, purification, and characterization of the Ca^{2+}-independent phospholipase A_2 from rat brain. J Neurochem 73:1278–1287, 1999.
42. AA Farooqui, ML Litsky, T Farooqui, LA Horrocks. Inhibitors of intracellular phospholipase A_2 activity: their neurochemical effects and therapeutical importance for neurological disorders. Brain Res Bulletin 49:139–153, 1999.
43. MF Roberts, RA Deems, TC Mincey, EA Dennis. Chemical modification of the histidine residue in phospholipase A_2 (Naja naja naja). A case of half-site reactivity. J Biol Chem 252:2405–2411, 1977.
44. JJ Volwerk, WA Pieterson, GH de Haas. Histidine at the active site of phospholipase A_2. Biochemistry 13:1446–1454, 1974.
45. RA Deems, D Lombardo, BP Morgan, ED Mihelich, EA Dennis. The inhibition of

phospholipase A_2 by manoalide and manoalide analogues. Biochim Biophys Acta 917:258–268, 1987.

46. KB Glaser, RS Jacobs. Inactivation of bee venom phospholipase A_2 by manoalide. A model based on the reactivity of manoalide with amino acids and peptide sequences. Biochem Pharmacol 36:2079–2086, 1987.
47. RW Schevitz, NJ Bach, DG Carlson, NY Chirgadze, DK Clawson, RD Dillard, SE Draheim, LW Hartley, ND Jones, ED Mihelich, JL Olkowski, DW Snyder, C Sommers, JP Wery. Structure-based design of the first potent and selective inhibitor of human non-pancreatic secretory phospholipase A_2. Nat Struct Biol 2:458–465, 1995.
48. K Kitadokoro, S Hagishita, T Sato, M Ohtani, K Miki. Crystal structure of human secretory phospholipase A_2-IIA complex with the potent indolizine inhibitor 120-1032. J Biochem (Tokyo) 123:619–623, 1998.
49. M Murakami, T Kambe, S Shimbara, I Kudo. Functional coupling between various phospholipase A_2s and cyclooxygenases in immediate and delayed prostanoid biosynthetic pathways. J Biol Chem 274:3103–3115, 1999.
50. M Murakami, S Shimbara, T Kambe, H Kuwata, MV Winstead, JA Tischfield, I Kudo. The functions of five distinct mammalian phospholipase A_2s in regulating arachidonic acid release. Type IIA and type V secretory phospholipase A_2s are functionally redundant and act in concert with cytosolic phospholipase A_2. J Biol Chem 273:14411–14423, 1998.
51. IP Street, HK Lin, F Laliberté, F Ghomashchi, Z Wang, H Perrier, NM Tremblay, Z Huang, PK Weech, MH Gelb. Slow- and tight-binding inhibitors of the 85-kDa human phospholipase A_2. Biochemistry 32:5935–5940, 1993.
52. Z Huang, P Payette, K Abdullah, WA Cromlish, B Kennedy. Functional identification of the active-site nucleophile of the human 85-kDa cytosolic phospholipase A_2. Biochemistry 35:3712–3721, 1996.
53. YC Lio, LJ Reynolds, J Balsinde, EA Dennis. Irreversible inhibition of Ca(2+)-independent phospholipase A_2 by methyl arachidonyl fluorophosphonate. Biochim Biophys Acta 1302:55–60, 1996.
54. L Hazen, LA Zupan, RH Weiss, DP Getman, RW Gross. Suicide inhibition of canine myocardial cytosolic calcium-independent phospholipase A_2. Mechanism-based discrimination between calcium-dependent and independent phospholipases A_2. J Biol Chem 266:7227–7232, 1991.
55. J Balsinde, EA Dennis. Bromoenol lactone inhibits magnesium-dependent phosphatidate phosphohydrolase and blocks triacylglycerol biosynthesis in mouse P388D1 macrophages. J Biol Chem 271:31937–31941, 1996.
56. MA Balboa, J Balsinde, EA Dennis. Involvement of phosphatidate phosphohydrolase in arachidonic acid mobilization in human amnionic WISH cells. J Biol Chem 273:7684–7690, 1998.
57. J Tang, RW Kriz, N Wolfman, M Shaffer, J Seehra, SS Jones. Novel cytosolic calcium-independent phospholipase A_2 contains eight ankyrin motifs. J Biol Chem 272: 8567–8575, 1997.

17

Understanding and Exploiting Bacterial Polyketide Synthases

Robert McDaniel
Kosan Biosciences, Inc., Hayward, California

Chaitan Khosla
Stanford University, Stanford, California

I. INTRODUCTION AND OVERVIEW

Although the structures of polyketide natural products have been (and continue to be) elucidated for over a century, the notion that microbial polyketides are synthesized through the action of modular enzyme assemblies called polyketide synthases (PKSs) is barely a decade old. As described below, PKSs catalyze the biosynthesis of the carbon-chain backbones of numerous bacterial polyketides using reactions that are mechanistically well understood but difficult to control (Fig. 1). Examples of resulting polyketides include well-known antibiotics, other pharmacologically active agents, and agricultural products (Fig. 2). Indeed, the spectacular structural diversity and complexity observed among the products of PKSs is probably unmatched in any other family of biosynthetic products, and has been the source of inspiration for the development of many aspects of modern organic chemistry.

The discovery of PKSs has led to the emergence of new horizons for the engineered biosynthesis of complex natural product-like molecules. The goals of this chapter are twofold. First, we will summarize our understanding of these systems at a molecular level, and the techniques that have emerged to dissect and manipulate these systems. Second, we will present selected examples of how PKSs can be engineered to synthesize novel biomolecules, and the pharmacological implications of such manipulation. Hopefully it will become clear that, al-

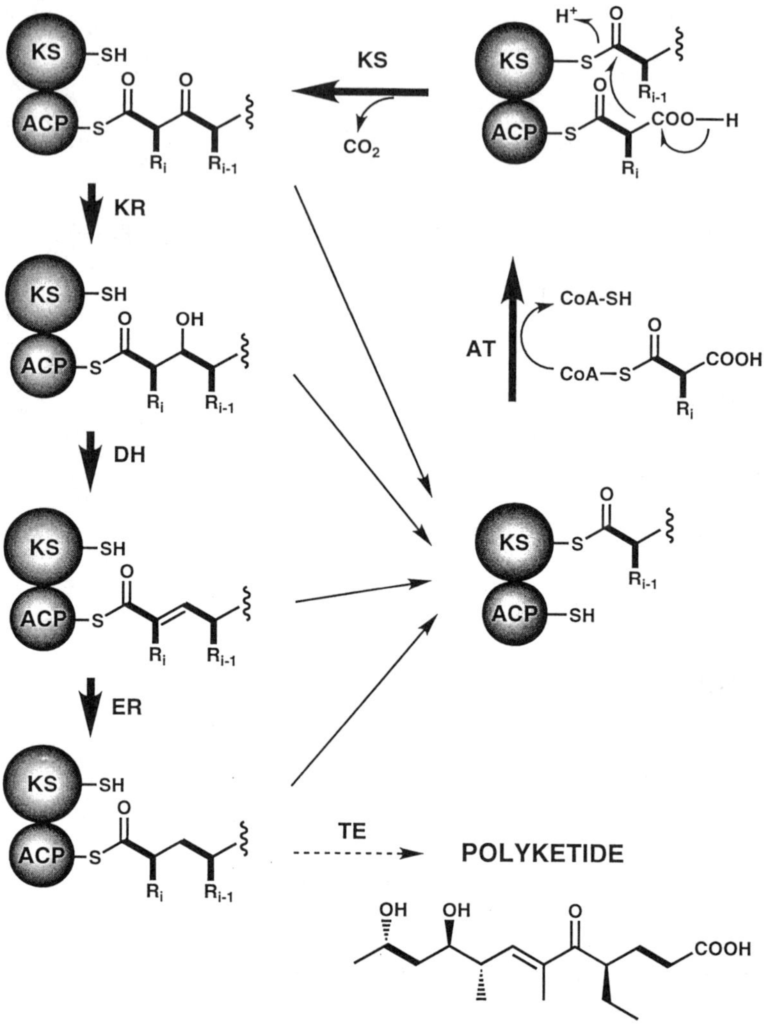

Figure 1 Polyketide biosynthesis. Polyketide backbones are formed via condensations from acyl-CoA thioesters of carboxylic acids. The β-ketone which results from each condensation can undergo a series of reductive steps analogous to fatty acid biosynthesis. However, either none or only some of the reductive activities may occur in a given cycle. This allows PKSs to generate diversity through selection of priming and extender units, variation of the reductive cycle, and stereoselectivity. (ACP, acyl carrier protein; AT, acyl transferase; KS, ketosynthase; DH, dehydratase; ER, enoylreductase; KR, ketoreductase; TE, thioesterase.) The structure depicted in the lower right-hand corner is representative of the possible structural variations that can arise during polyketide biosynthesis.

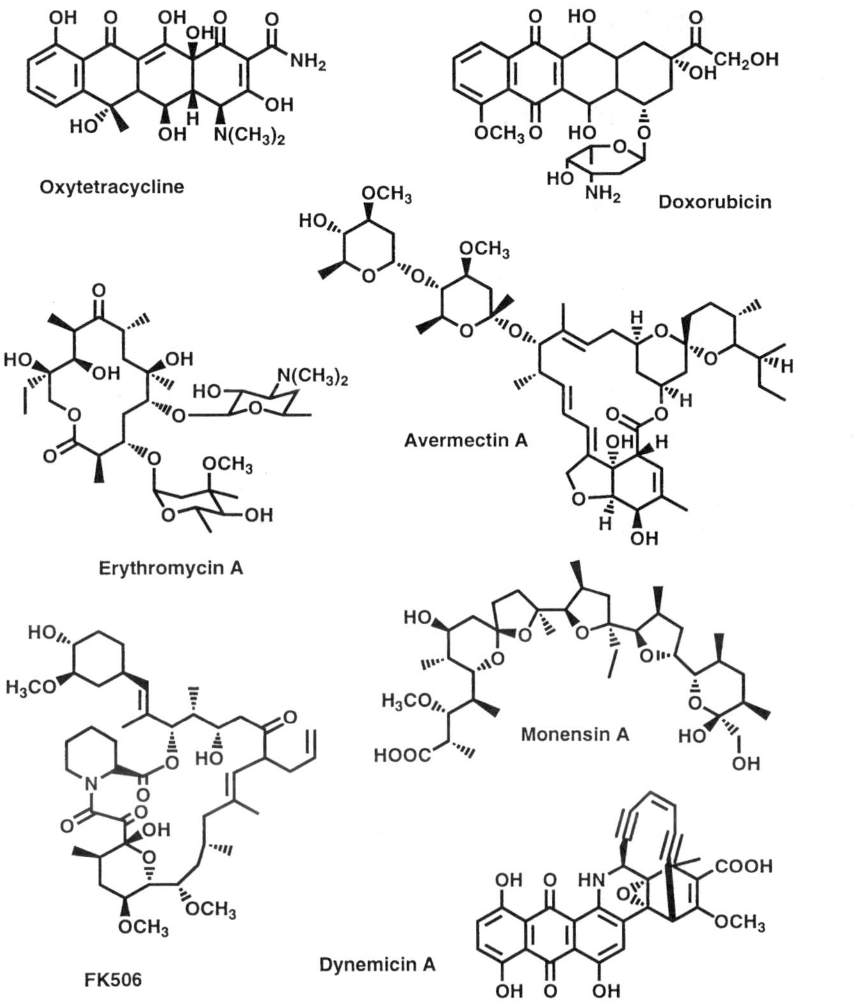

Figure 2 Examples of polyketides. Polyketides demonstrate a broad range of biological activities, including antibiotic (oxytetracycline and erythromycin), antitumor (doxorubicin and dynemicin), antiparasitic (avermectin), and immunosuppressive (FK506). Monensin is used as a bovine feed supplement and an anticoccidial agent.

though considerable progress has been made in both directions, our understanding of these multifunctional enzymatic systems and our ability to manipulate them productively are still in their infancy. In concluding, we speculate on future research directions that are likely to emerge from studies on PKSs.

II. GENES ENCODING POLYKETIDE SYNTHASES

Long before the discovery of PKSs themselves, a wealth of indirect information was gained regarding the biosynthetic properties of PKSs through incorporation experiments with [^{14}C]-, [^{13}C]-, [^{18}O]-, and [^{2}H]-labeled substrates and intermediate analogs [1]. However, the biochemical basis for these highly controlled synthetic processes remained virtually unknown until the advent of molecular genetic tools in this problem area.

In the 1980s, during the course of their genetic analysis of secondary metabolite biosynthesis in various *Streptomyces* species, Hopwood and his collaborators made the monumental discovery that the genes responsible for the biosynthesis, regulation, and self-resistance of bacterial natural products are clustered in the genomes of producer organisms [2]. Based on their work, a variety of genetic strategies have been developed to take advantage of nature's benevolence in order to clone complete biosynthetic gene clusters of interest. Below, the salient features that have emerged from DNA sequence analysis of polyketide synthase gene clusters are summarized.

A. Actinorhodin, Tetracenomycin, Doxorubicin, and Other Bacterial Aromatic Polyketide Synthases

A number of bacterial aromatic PKS gene clusters have been cloned and sequenced thus far. Three such examples are shown in Figure 3, those encoding the actinorhodin [3], the tetracenomycin [4], and the doxorubicin [5] PKS gene clusters. In the first two cases, biosynthesis involves the assembly of an elongated poly-β-ketone chain from malonyl-CoA units. Primer units are derived from decarboxylation of a malonyl unit. Chain assembly requires four polypeptides, collectively referred to as the minimal PKS. Three of these—the ketosynthase (KS), the chain length factor (CLF), and the acyl carrier protein (ACP)—are exclusively involved in biosynthesis of the natural product, whereas the fourth protein, the malonyl transferase (MAT), is shared between the fatty acid synthase and the bacterial aromatic PKSs in the bacterium, and is encoded outside the *act* gene cluster [6,7]. (Therefore it is not shown in Fig. 3.) The roles of the KS, the ACP, and the MAT are equivalent to those of their counterparts in fatty acid synthases and modular PKSs. The CLF has a high sequence similarity to the KS and associates tightly with it. While the CLF is believed to play an important role in chain length control, the actual biochemical mechanism for this control remains to be

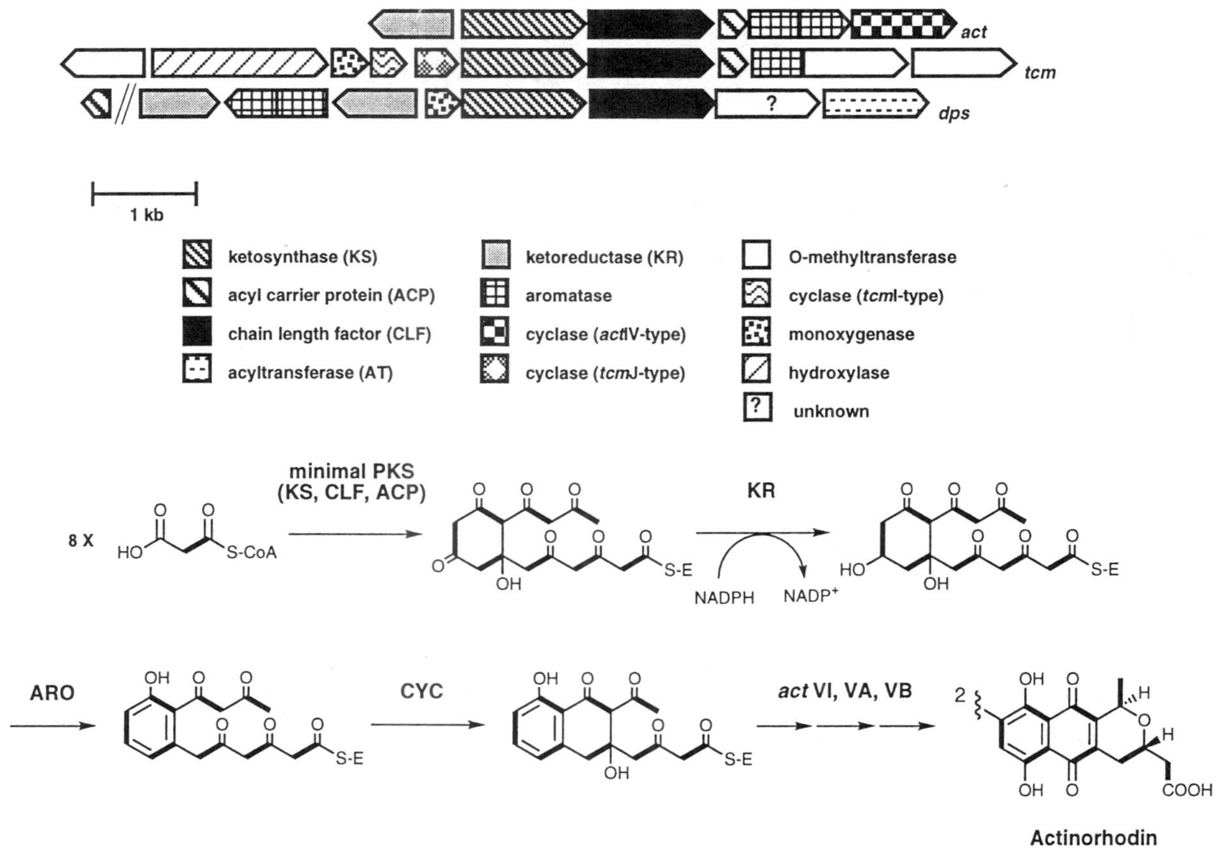

Figure 3 Aromatic PKS gene clusters and actinorhodin biosynthesis. Partial gene clusters for the polyketides actinorhodin (act), tetracenomycin (tcm), and doxorubicin (dps) are shown. Reconstitution of combinations of act genes in a PKS clean host have led to a proposed pathway by which the early stages of actinorhodin biosynthesis occur.

established. Certain bacterial aromatic PKSs use non-acetate primers. For example, the doxorubicin polyketide backbone is primed with a propionyl unit derived from propionyl-coenzyme A (propionyl-CoA). These gene clusters include two other genes that are relevant to this process—a gene encoding an acyl transferase and a ketosynthase homologous to the enzyme that primes fatty acid biosynthesis in bacteria. Lastly, the high lability of the poly-β-ketone products of these chain-building enzymes requires genes that encode for ketoreductases, cyclases, and aromatases.

B. Erythromycin, Rapamycin, and Other Modular Polyketide Synthases

Many macrocyclic as well as other polyketide natural products produced by bacteria are synthesized by modular PKSs, in which enzymes are organized into groups of active sites known as modules. As illustrated by the erythromycin [8,9] and rapamycin [10] PKSs (Fig. 4), each module is responsible for one cycle of polyketide chain extension and functional group modification. Within each module are catalytic domains of 100–400 amino acids each that are analogous in both function and sequence to the individual enzymes of fatty acid biosynthesis. All modules possess at a minimum β-keto acylthioester synthase (ketosynthase, KS), acyl transferase (AT), and acyl carrier protein (ACP) domains. In addition, specific combinations of ketoreductase (KR), dehydratase (DH), enoylreductase (ER), and thioesterase (TE) domains may be found in each module, according to the required degree of functional group modification taking place after each chain elongation cycle. From the structures of the resulting natural products, it can be deduced that certain enzymes are cryptic in function. At the N-terminus of module 1 in both synthases, one observes additional domains that play a role in priming. The erythromycin PKS, which is primed by a propionyl unit, includes an AT and an ACP, whereas the rapamycin PKS, which is primed by a cyclohexenoyl unit, possesses an acyl-CoA ligase (Lig). Likewise, sequence analysis also reveals the presence of specific enzymes responsible for chain termination. For example, at the C-terminus of the last module of the erythromycin PKS is a TE domain, which is responsible for formation of the 14-membered macrolactone, whereas the rapamycin PKS gene cluster includes a non-ribosomal peptide synthetase module (RapP) that presumably incorporates a pipecolyl moiety into the backbone, followed by macrocycle formation.

III. TOOLS FOR CLONING, EXPRESSING, AND MANIPULATING POLYKETIDE SYNTHASES

Within the past decade, the impact of molecular biology on natural-product biosynthetic research has been profoundly felt as a result of the rapidly increasing

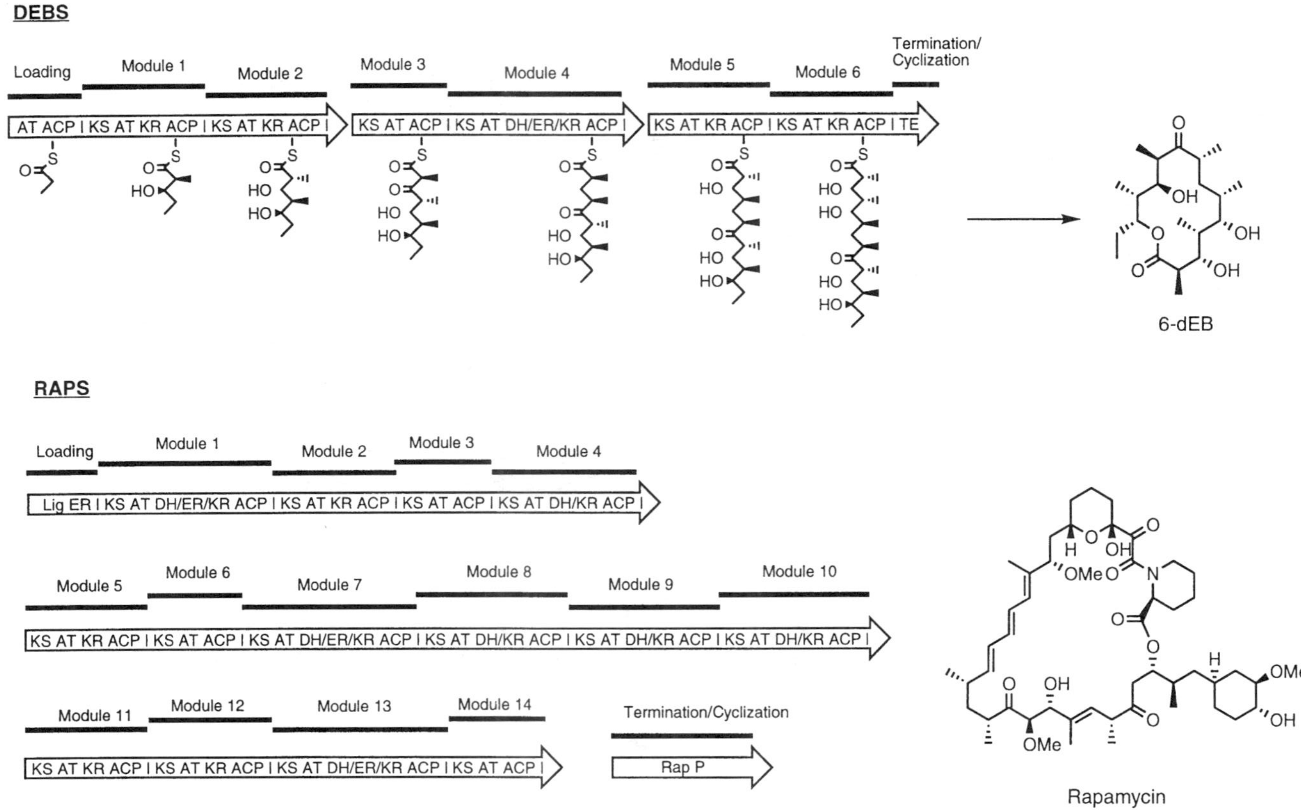

Figure 4 Modular PKS gene organization. The 6-deoxyerythronolide B (DEBS) and rapamycin (RAPS) PKSs are composed of 6 and 14 modules, respectively. DEBS catalyzes the formation of 6-deoxyerythronolide B (6-dEB) from a propionyl-CoA primer unit and six methylmalonyl-CoA extender units. Rapamycin is formed from a cyclohexenoyl starter unit, seven malonyl-CoA (modules 2, 5, 8, 9, 11, 12, and 14), and seven methylmalonyl-CoA (modules 1, 3, 4, 6, 7, 10, and 13) extender units, and pipecolic acid. For RAPS, only the domains which are believed to be functional in the PKS are shown.

capability for cloning new genes, constructing recombinant gene clusters, and analyzing the properties of their protein products, both *in vivo* and *in vitro.*

A. Cloning Strategies

The fact that all the genes responsible for polyketide biosynthesis and self-resistance are clustered in the genomes of bacteria and fungi makes it possible to clone complete biosynthetic gene clusters of interest. Frequently used strategies include complementation of blocked mutants, transfer of partial or complete pathways in a surrogate host, homology-based gene isolation, identification of resistance gene(s) through selection in a heterologous host, and reverse genetics based on limited amino acid sequence of a purified pathway enzyme. A variety of cloning vectors, selectable markers, and gene probes have been added to the genetic toolbox for studying secondary metabolism, especially in the actinomycetes [2].

B. Heterologous Expression and Mutagenesis of Polyketide Synthases

Two approaches have been exploited thus far for the generation and analysis of mutant PKSs. One strategy utilizes homologous recombination to replace or delete individual base pairs or even entire genetic segments in the chromosome of the native polyketide producer. All other native genes required for natural product biosynthesis, regulation, and precursor formation remain intact in the recombinant host. As a consequence, the reporter polyketide product is produced in the natural intracellular environment, and frequently undergoes some or all of the normal post-PKS transformations that are typically associated with polyketide natural product biosynthesis. For example, the early studies of Hopwood and co-workers on actinorhodin biosynthesis [11] and of Katz and co-workers on erythromycin biosynthesis [9] involved such methodologies. Homologous recombination is a particularly attractive strategy for manipulating extremely complex PKSs encoded by large gene clusters; however, it is often technically difficult and relatively slow, and therefore places serious constraints on the types of experiments that can be performed on a PKS of interest. An alternative strategy for mutagenesis involves heterologous expression of PKS genes in genetics-friendly hosts. In addition to enhancing the speed and convenience of manipulating PKS genes, heterologous expression offers the added advantage of relatively easy access to PKS proteins through overexpression compared to native producing hosts. However, metabolite production is often suboptimal (and sometimes nonexistent) in a heterologous host, presumably due to the poor supply of PKS substrates. For example, a bifunctional actinomyces-*Escherichia coli* vector (Fig. 5) with suitable control elements for the expression of PKS genes in the actinomycetes has successfully been used to functionally reconstitute the tetracenomycin [12],

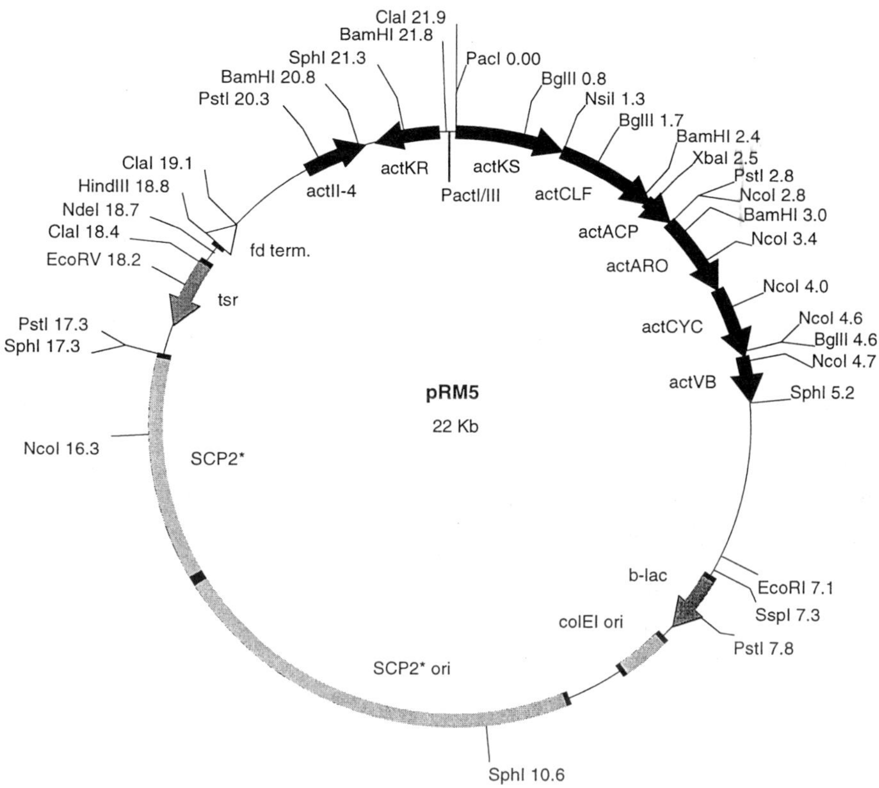

Figure 5 Plasmid pRM5. This shuttle vector contains the colE1 replication origin and β-lactamase gene for propagation and selection in *E. coli* (colE1). For use in *Streptomyces* it contains the low-copy SCP2* replication origin and thiostrepton (*tsr*) selectable marker. Expression of genes occurs from a bidirectional promoter (PactI/III) which is regulated by the transcriptional activator, actII-4, which is turned on during entry into stationary phase. A number of other PKS genes have been expressed in heterologous systems by replacing the *act* genes in pRM5.

erythromycin [13], picromycin/methymycin [14], and 6-methylsalicylic acid [15] PKS pathways in *Streptomyces coelicolor*, a model actinomycete with well-developed genetic tools. The resulting polyketide products are typically generated in yields that may range between 10 and 100 mg/liter culture. Moreover, PKS proteins are produced at 1–5% total cellular protein levels, which in turn has facilitated the development of convenient cell-free systems for polyketide synthesis. Indeed, with the recent cloning and analysis of the enzymes responsible for posttranslational modification of acyl carrier proteins, it has even become possible to functionally express PKSs in *E. coli* [16,17]. While this is proving to be an excellent source for active protein preparations, the absence of specialized precursors such as methylmalonyl-CoA in *E. coli* precludes the production of certain types of reporter metabolites *in vivo.*

C. Cell-Free Synthesis of Polyketides

An alternative to the cell-based production of natural and engineered polyketides would be to synthesize the same products *in vitro.* Although methylsalicylic acid synthase was reconstituted nearly 30 years ago [18], the ability to extend this principle to more complex PKSs had to wait until the application of recombinant DNA techniques to the field [19–21]. The power of cell-free synthesis stems from the fact that it is unencumbered by any metabolic constraints associated with qualitative or quantitative aspects of precursor availability; however, significantly reduced (microgram) quantities of polyketide products are ordinarily generated in such experiments. It should be noted that such limitations in product yield may be due not as much to lack of adequate protein as to the cost or availability of the necessary substrates, especially malonyl-CoA and methylmalonyl-CoA. Indeed, recent successes in the heterologous expression of active forms of bacterial and fungal PKS proteins in *E. coli* and yeast have paved the way for low-cost protein production, given well-established methodology for high-cell-density fermentation of these organisms. Moreover, the discovery that simple synthetic mimics of these substrates can efficiently substitute for CoA-thioesters in PKS-catalyzed reactions promises a convenient means of overcoming such difficulties [22].

IV. ARCHITECTURE OF POLYKETIDE SYNTHASES

A. Organization of Type II Systems

Type II PKSs such as the bacterial aromatic PKSs are comprised of several mono- or di-domain proteins. Although little is known about the interactions among these proteins, and the relevance of these interactions to enzyme function and selectivity, such interactions are presumably important, given the extremely high lability of the inferred biosynthetic intermediates in these pathways. More recently, protein chemical studies and kinetic analysis have provided evidence for interactions between the ACP and the rest of the minimal PKS, as well as between the auxiliary subunits and the minimal PKS [23,24].

B. Organization of a Type I Polyketide Synthase Module

Although high-resolution structures of type I PKS modules are not available, some insights into the topological organization of modules have been obtained. Electron microscopy on the vertebrate fatty acid synthase (whose domain layout is similar to that of modular PKSs as well as fungal PKSs) has provided evidence for the domain-like structure of these homodimeric proteins [25]. Analogous to vertebrate fatty acid synthases, gel filtration and sedimentation equilibrium studies have shown that the 6-deoxyerythronolide B synthase (DEBS) proteins exist as dimers [26]. Proteolytic analysis of a single module showed that the fragments containing KS/AT didomains are released as homodimers, whereas KR fragments do not form dimers [27]. Based on the sequential arrangement of domains in the linear polypeptide chain and the proteolysis data, a ''double helical'' model has been proposed [27]. Another approach for studying the functional topology and organization of various domains has involved *in vitro* complementation analysis, in which two inactive PKSs carrying mutations in different functional domains are reconstituted to a catalytically active heterodimer. Complementation studies with KS and ACP domains showed that KS2° (null mutation in KS domain of module 2) could complement either KS1° or ACP2° mutants, but KS1° and ACP2° did not complement each other [28]. These studies demonstrated that the KS and ACP domains of a single catalytic center are located on different dimer subunits. Moreover, the failure of KS1° and ACP2° mutants to complement suggested that the intermodular chain transfer takes place on the same polypeptide chain. A similar strategy was used to probe the orientation of the AT domain. Here, both the heterodimers generated between the inactive AT2° polypeptide and inactive KS1° or KS2° proteins supported polyketide synthesis, suggesting that AT2 could perform its function from either subunit [29]. These results illustrated that the AT domain can be shared between two clusters of active sites within the same dimeric module.

C. Intermodular Interactions

In multimodular PKSs such as DEBS, successive modules must be capable of interacting with each other to ensure accurate chain transfer. Two general types of models can be envisioned for these chain transfer events, which involve thioester exchange between the ACP domain of the upstream module and the KS domain of the downstream module. In one model, neighboring modules associate with each other through selective protein–protein interactions. Chain transfer might then occur simply as a result of the proximity of the ACP and KS domains. Alternatively, the transfer of chains between modules might depend on a ''switchlike'' mechanism in which specific conformational changes guide the growing polyketide chain into the downstream module once the upstream module has completed its assigned tasks. In both models, the selective association of modules is a necessary prerequisite for chain transfer; however, in the latter case the protein–protein interactions are not sufficient, and intermodular chain transfer is tightly coupled to the chemistry catalyzed by the donor module. Evidence for the existence of selective protein–protein interactions comes from protein chemical studies, in which DEBS has been isolated as an active $\alpha2\beta2\gamma2$ complex (see above), and from protein engineering studies, where intermodular linkers have been shown to play a crucial role in facilitating chain transfer between modules [17]. However, the question of whether ''switchlike'' mechanisms occur or not remains to be addressed.

V. MOLECULAR RECOGNITION FEATURES OF POLYKETIDE SYNTHASES

A. Priming of Polyketide Chains

As described above, the priming of PKSs is controlled by specific enzymes. The tolerance and specificity of these enzymes have been analyzed in the cases of various aromatic, modular, and fungal PKSs using *in vivo* and *in vitro* approaches. In DEBS, the AT-ACP loading domain has a distinct preference for propionyl primers, but will accept acetyl, butyryl, and possibly other alkyl primers as well [13,21,30]. Far greater tolerance, especially toward α-branched acyl groups, is observed in the case of the loading AT-ACP of the avermectin PKSs [31]. In both cases, the unnatural acyl groups are faithfully processed through the entire PKS, albeit with attenuated velocities in cases where kinetic analysis has been undertaken [32,33]. The DEBS system illustrates another kind of limitation to alternative primer unit incorporation, since the first module of this PKS can prime itself even in the absence of the loading domains, probably through decarboxylation of a methylmalonyl group [32,34]. This spontaneous priming mechanism competes with the incorporation of non-natural primers, especially in cases where the unnatural primer is substantially different from the (natural) propionyl primer [30,31]. More serious limitations in PKS priming are observed in other systems. For example, whereas the loading CoA-ligase-ACP domains of the rifamycin PKS appear to tolerate analogs of the normal primer, 3-amino-5-hydroxybenzoic acid (AHBA), extension of these anomalous primers was aborted at the tetraketide stage without any accumulation of full-length rifamycin analogs [35]. Thus there appears to be considerable diversity between PKSs regarding the ease with which naturally occurring priming mechanisms can be hijacked or altered to produce new natural-product analogs. Similar variability is also observed in the case of bacterial aromatic PKSs. For example, the doxorubicin PKS shows high specificity toward propionyl primer units (the minimal PKS will self-prime with a decarboxylated malonyl unit in the absence of the propionyl units or the priming enzymes) [36]. In contrast, the R1128 PKS, which synthesizes an anthracycline natural product, can tolerate a variety of alkyl primers [37]. Lastly, although extensive analysis of primer unit specificity has not been conducted, the 6-methylsalicylic acid synthase (MSAS) from *Penicillium patulum* can tolerate propionyl and butyryl units in addition to the natural acetyl primer unit [38].

B. Chain Elongation

In contrast to the apparently broad substrate specificity of priming enzymes in PKSs, the AT domains responsible for extender unit selection in aromatic, modular, and fungal PKSs are extremely fastidious. For example, the extender AT domains in DEBS exhibit a strict structural and stereochemical specificity for 2(S)-methylmalonyl-CoA [39]. Likewise, the fatty acid acyl transferase, that transfers extender units to the ACPs of aromatic PKSs, has a strict preference for malonyl groups [40]. Similar specificity toward malonyl extender units is also observed in the case of MSAS [38]. The AT domains therefore act as one of the key gatekeepers of polyketide biosynthesis, exercising tight control over the nature of the polyketide chain extension unit. The C-terminal end of AT domains appears to be important in controlling this substrate specificity [41]. Interestingly, the N-acetylcysteamine (NAC) analogs of malonyl- and methylmalonyl-CoA have been shown to be effective substitutes for the natural substrates of DEBS and MSAS, indicating that the extended pantetheinyl arm and the nucleotide moiety of CoA thioesters are not essential for substrate recognition by the AT domain [22,38].

In addition to choosing extender units, PKSs must also recognize the growing polyketide chains at each stage of the chain elongation process. In bacterial aromatic PKSs, chain extension occurs repetitively until the chain is ready for termination. In several cases, one or more carbonyls in the poly-β-ketone chain undergo reduction; these modifications are believed to occur after the full-length chain has been generated and not during chain growth [42]. Thus, the ketoreductase(s) presumably recognize full-length (but PKS-bound) substrates. However, since the absence of ketoreductases can lead to formation of prematurely truncated polyketide chains, these enzymes presumably associate tightly with the minimal PKS subunits [43]. More overt association between auxiliary enzymes and core PKSs is evident in the cases of fungal PKSs, where regioselective reductases and C-methyltransferases are responsible for chain modification during the elongation process [44,45]. These enzymes must be able to selectively recognize chains of specific lengths, although the structural basis for such recognition is unknown.

In modular PKSs the ability of a module to recognize the incoming chain is largely controlled by the KS domain. Until recently it was believed that KS domains had a high degree of specificity toward the length, functionality, and/or stereochemistry of incoming chains [33]. However more recent kinetic analysis has challenged this notion by demonstrating that different modules can turn over a given ketide substrate with surprisingly similar kinetic parameters [17]. Thus, while KS domains probably can discriminate between alternative incoming chains, this discrimination is modest, and is more pronounced when substrates are obtained from exogenous sources (such as NAC-thioesters) rather than from neighboring ACP domains (possibly via ''switchlike'' transfer mechanisms; see above). Modular PKSs also control the functionality and stereochemistry of the growing chain during every round of condensation. The control of functionality is hard-wired into modules, depending on the presence or absence of individual KR, DH, and ER domains. It is noteworthy, however, that certain KR, DH, and ER domains appear cryptic, and it remains to be established whether these domains are universally inactive or simply unable to recognize the substrate at hand. Our understanding of stereochemical control is even less clear. Genetic experiments suggest that KR domains are the primary determinants of alcohol stereochemistry [46], although biochemical experiments implicate a role for the KS domain as well [47]. Epimerase activity is believed to be associated with certain modules [39], and influences the stereochemistry of methyl-branched centers in these cases; however, the precise location of this epimerase activity remains a mystery.

C. Chain Termination

Fully extended polyketide chains of most PKSs are extremely labile, and termination of polyketide biosynthesis typically occurs with one or more concomitant regioselective cyclization events. The highly oxidized backbones of aromatic polyketides undergo successive cyclizations to form six-membered carbocyclic or heterocyclic rings. Indeed, these cyclization reactions are the source of much of the structural variety observed among naturally occurring bacterial aromatic polyketides. In some cases these six-membered rings undergo spontaneous aromatization (via the elimination of one or two water molecules); in others aromatization is enzyme-catalyzed. Cyclases and aromatases (Fig. 6A) are more diverse in sequence than the core PKS subunits, and their precise biochemical mechanisms remain unknown. However, genetic and chemical analysis on several PKSs has led to the identification of at least six different families of these enzymes, which in turn facilitates identification of newer members by homology alone. Since analogs of their cognate substrates are difficult to access, their molecular recognition features remain a mystery. However, genetic analysis on a family of didomain aromatases/cyclases has revealed a hierarchy in chain length specificity—proteins that naturally act upon longer chains can recognize shorter ones, but not vice versa [48]. Modular PKSs also possess specific enzymes for chain termination. The TE domain of DEBS can discriminate between substrates based on length, functionality, and stereochemistry [49], yet it has remarkable tolerance for generating macrocycles ranging from six-membered to at least 16-membered lactones (Fig. 6B) [50–52]. Likewise, the TE domain from the picromycin PKS can generate both 12- and 14-membered macrolactones [53].

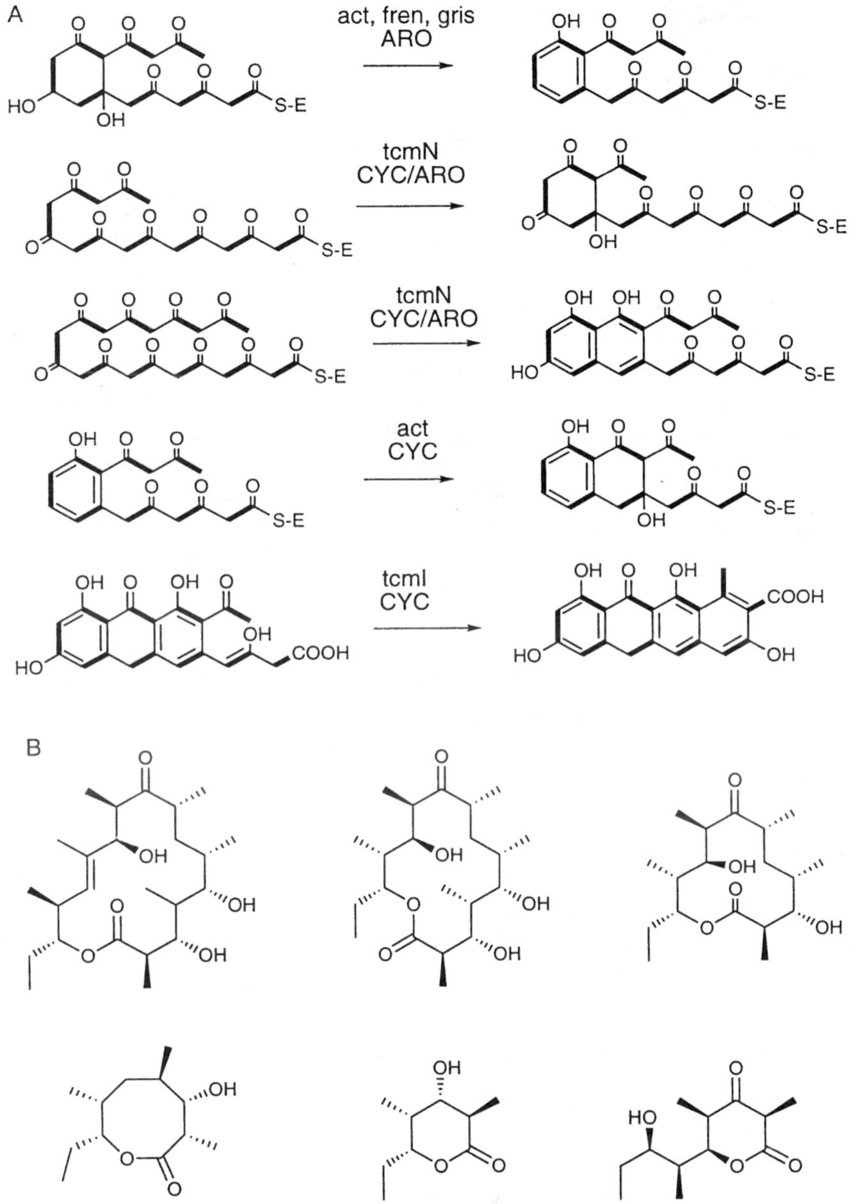

Figure 6 (A) Reactions catalyzed by aromatic cyclases and aromatases. These enzymes control the diverse cyclization patterns of aromatic polyketides and in general display high regioselectivity and substrate specificity. (B) Examples of known products with different cyclization patterns that are accessible via the thioesterase (TE) domain of DEBS.

VI. STRATEGIES FOR BIOSYNTHETIC MANIPULATION

Polyketide biosynthesis can be genetically manipulated by deletion, addition, or replacement of one or more catalytic activities in a pathway. Although these principles are applicable to both aromatic and modular systems, the architectural and biochemical differences discussed above have led to the development of different engineering approaches for each case. Most successful attempts at engineering aromatic polyketides result from "mixing and matching" of separate individual enzyme subunits. For example, KS/CLF pairs, ketoreductases, and cyclases/aromatases from different aromatic gene clusters have been combined to manipulate chain length, hydroxylation pattern, cyclization regiospecificity, and aromaticity. This has been especially fruitful in identifying unknown activities or specificities of PKS genes, leading to the rational design of some novel polyketides [48]. However, our incomplete understanding of how these enzymes control many features of polyketide biosynthesis currently leaves many potential manipulations out of the control of the molecular engineer.

In contrast, the assembly-line topography of modular PKSs invites more surgical protein engineering approaches. Each step of the biosynthetic pathway is catalyzed by a domain whose boundaries can be identified at the DNA level; similarly, the boundaries of entire modules can also be identified. Thus, PKS proteins can be readily engineered at the domain or module level using the genetic tools described above. Several laboratories have exploited this potential for protein engineering in recent years. For example, using DEBS and other modular PKSs, it has been demonstrated that domains and/or modules can be inactivated, deleted, substituted, and added while maintaining all upstream and downstream activities (Fig. 7 and 8). Although in many cases the yields of compound produced are diminished, these techniques have produced many compounds that would be difficult if not impossible to generate by other means. Collectively, they demonstrate that modular PKSs are highly tolerant toward structural manipulation as well as altered polyketide intermediates, both essential to the success of engineered and combinatorial biosynthesis.

A. Engineering Individual Catalytic Domains

The manipulation of a single catalytic domain or sets of domains is currently the most practiced technique and perhaps the best understood. β-Keto-processing domains offer perhaps the greatest degree of freedom for manipulating polyketide structures. Inactivation of an active site (Fig. 7, I) by site-directed mutagenesis is the most structurally conservative approach to bypassing any activity that is present in the naturally occurring PKSs [9,54]. Potential targets include the NADPH-binding sites of the KR and ER domains, and a similarly conserved motif in the DH domain. However, partial or complete deletion of domains (Fig.

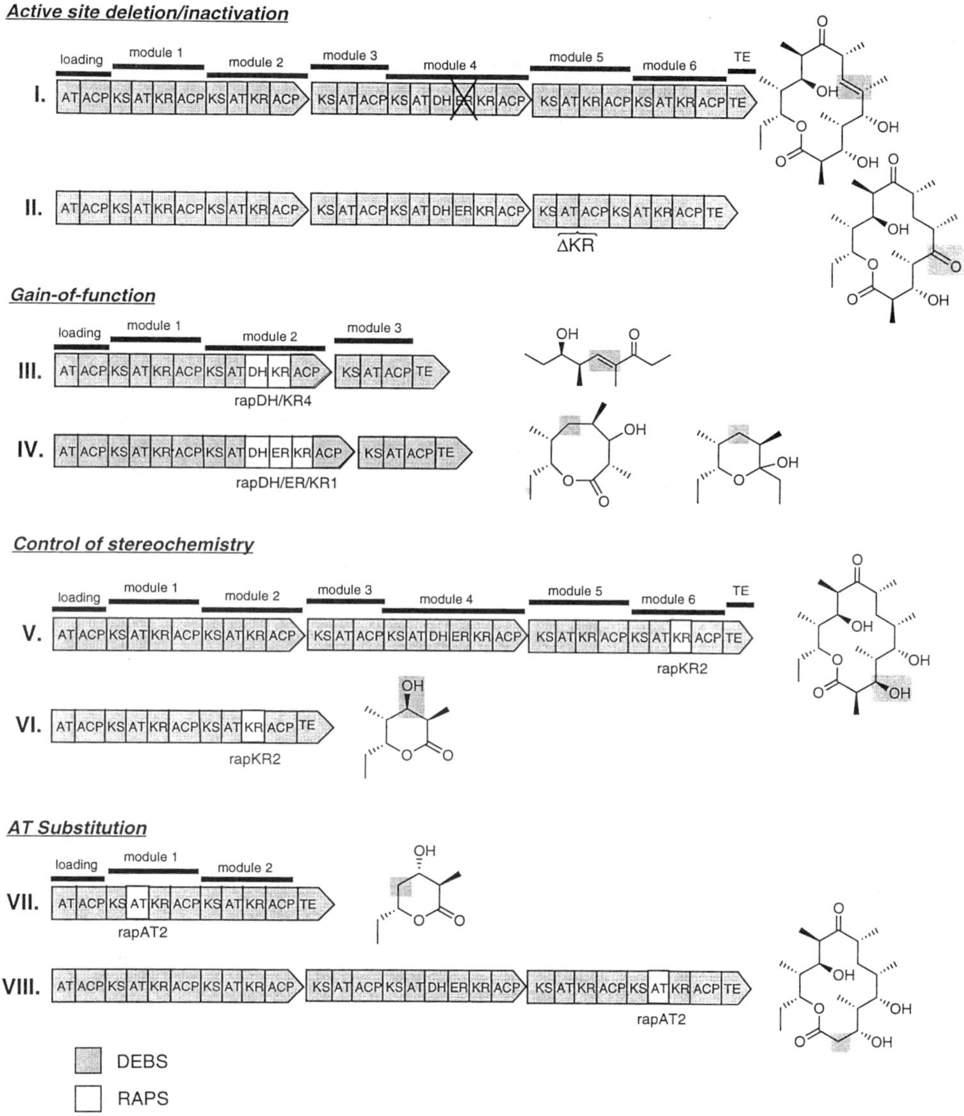

Figure 7 Mutagenesis of individual catalytic domains. These selected examples illustrate how individual catalytic domains can be manipulated to control reductive cycle, stereochemistry, and monomer incorporation in modular PKSs. See Sec. VI.A for details.

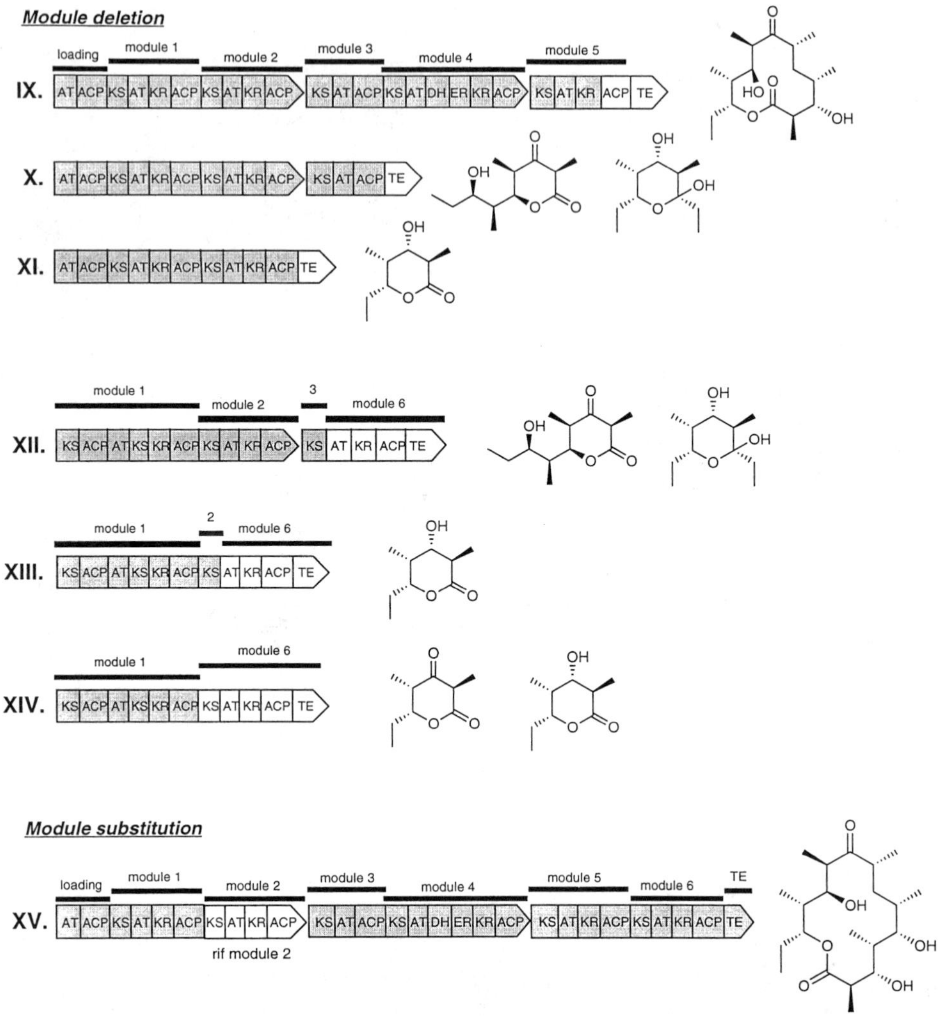

Figure 8 Module engineering. Entire modules within DEBS have been deleted or substituted. See Sec. VI.B for details.

7, II) has also proven to be viable and is more convenient when using cassette mutagenesis to perform many manipulations at once. For example, a short linker modeled after the last module of the rapamycin PKS has been used to delete many of the β-keto processing domains of DEBS [55]. Introduction of β-keto-processing activities, or gain-of-function mutagenesis, represents a more difficult challenge, since newly introduced catalytic domains must compete for a substrate that is the natural substrate for transfer to the next module. The most successful approach reported thus far has been the complete replacement of the native domain(s) with the entire set of desired activities from a heterologous module (Fig. 7, III–IV). For example, replacement of a KR with a DH + KR from another module is performed, rather than inserting a DH to engineer a double bond [56]. Thus, while modules appear to be somewhat tolerant of heterologous β-keto-processing domains, the domains themselves may prefer their cognate partners for optimal activity. This could be due to substrate specificities, interactions between the domains, or the necessity of a particular linker sequences between the domains. Finally, while it has been shown that KR domains can be used to alter the stereochemistry of hydroxyl groups within the terminal module of a PKS (Fig. 7, V–VI) [46,55], it has not been possible to achieve this effect for any internal modules of DEBS.

Substitution of AT (Fig. 7, VII–VIII) and loading domains to incorporate alternative starter and extender units has been presented in a number of studies. Replacement of methylmalonyl-specific AT domains with malonyl-specific domains has been performed successfully in DEBS modules 1, 2, 3, 5, and 6, leading to several desmethyl analogs of erythromycin [57–59]. DEBS has also been modified to incorporate an ethyl side chain in place of methyl [58]. Replacement of loading domains also appears to be robust and has been performed with DEBS and the spiramycin and avermectin PKSs [31,60]. The selection of a host for engineering loading domains and ATs is an important consideration as sufficient intracellular concentrations of substrate are required. While most actinomycetes are likely to have good levels of malonyl-CoA and methylmalonyl-CoA, other substrates, such as ethylmalonyl-CoA or hydroxymalonyl-CoA, are rarer. In the case of the ethyl side-chain replacement mentioned above, *S. erythraea* had to be engineered to increase the supply of ethylmalonyl-CoA [58]. Furthermore, many AT substitutions resulted in slightly ‘‘leaky’’ properties [41,59]. Recent mutagenesis studies suggest that the 3′ end of the AT domains plays a role in determining specificity and may be a good target for further engineering [41].

Finally, thioesterases and other chain-terminating and cyclizing domains can be targets of manipulation. For example, the properties of the picTE were used to improve the production profile of an engineered DEBS PKS which produced 3-keto compounds [14]. Although yet to be demonstrated, the attachment of non-ribosomal peptide synthetase domains, as in the rapamycin and FK506 PKSs, to heterologous modules would represent a significant advance.

B. Engineering Modules

Entire modules can also potentially be manipulated by deletion, substitution, and insertion like individual domains, although the latter has yet to be demonstrated.

The notion that modules can function independently of other modules was first demonstrated by a series of nested deletions in which the end modules of DEBS were successively deleted (Fig. 8, IX–XI). Truncated bi-, tri-, penta-, and hexamodular versions of DEBS have all been produced [50,61–63]. In most cases the amount of polyketide produced was similar to amounts produced by the complete PKS complex. Perhaps the most important factor to consider for module deletions *in vivo* is the ability of the intermediate to cyclize or undergo decarboxylation, as free carboxylic acids are difficult to isolate. Deletion of internal modules has also been examined. Early experiments in which module 6 of DEBS was fused to either module 2 or 3 (Fig. 8, XII–XIV) resulted in polyketide production only when the fusion boundary was chosen downstream of the module 2 or 3 KS domain, leading to the conclusion that KS domains were an important element of intermodular recognition [64]. More recent experiments have resulted in substitution of complete modules in DEBS with heterologous modules from the rifamycin PKS (Fig. 8, XV) [17]. A key element to success was the discovery that a specific type of linker region upstream from the KS domain is important for functional connectivity. Given the number of modules now available from sequenced PKS gene clusters, one can imagine designing a''toolbox'' of modules with desired activities and plugging them into an engineered scaffold, or designing modular PKSs *de novo*. Finally, it has also been shown that modules can be synthetically linked together, which may be useful for the latter as well as expression in eukaryotic hosts where a single transcript is desired [64].

C. Precursor-Directed Biosynthesis

Combining the advantages of modern synthetic chemistry with complex polyketide biochemistry provides a potentially powerful approach toward engineering polyketide chains with unique features. This has been achieved by constructing strains which are blocked at a particular step in the pathway or in the ability to produce a necessary precursor, and supplying synthetically derived precursors which mimic intermediates but contain different substituents. In the case of DEBS, a mutation was introduced in the active site of the KS domain in module 1 (KS1°) which inactivates the module, preventing priming of the PKS (Fig. 9). When this strain was fed N-acetylcysteamine thioesters of diketides, they were incorporated into module 2, and completely processed into 6-dEB analogs [52]. Since these diketides are apparently guided to module 2 by the functionality and stereochemical configuration of the α and β carbons, the functional group corresponding to the primer unit could be precisely controlled [52,65–67]. [This functional group can also be manipulated by replacing the loading domain of the DEBS PKS with that of the avermectin PKS (see above).] Precursor directed biosynthesis extends the capabilities of genetic engineering by allowing for the incorporation of intermediates at other modules. For example, when an α,β-unsat-

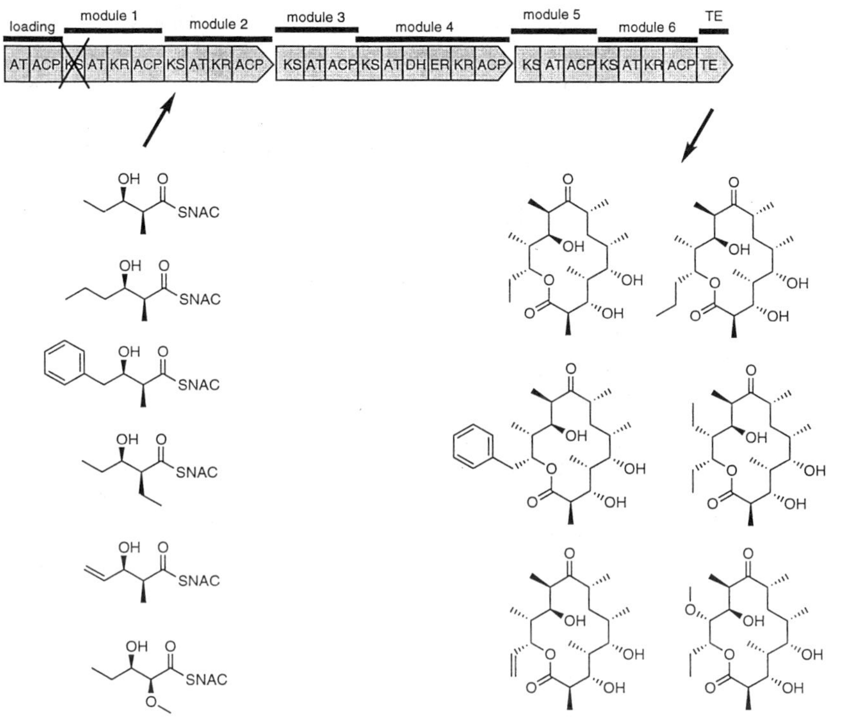

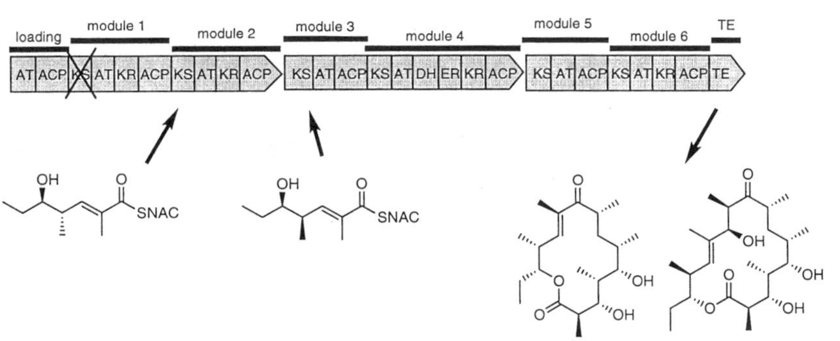

Figure 9 Precursor-directed biosynthesis. A block in the 6-dEB pathway was created by a Cys→Ala mutation in the active site of the KS of module 1. Feeding diketides with different α and γ substitutions resulted in the 6-dEB analogs in which the starter unit or first extender unit was modified. The stereochemistry of an unsaturated triketide dictated whether it was incorporated into module 2 or 3, leading to either a 14-membered or 16-membered macrolactone. See Sec. VI.C for details.

urated triketide intermediate was fed to the same KS1° DEBS strain, it was incorporated into module 3, leading to production of a 6-dEB analog with a double bond [65]. However, when the C-4 epimer of the triketide was fed, it was unexpectedly incorporated into module 2 rather than module 3 (Fig. 9) [52]. This resulted in production of a novel 16-membered macrolide similar to tylactone.

D. Cell-Free Enzymatic Synthesis

As discussed above, the purification and reconstitution of active PKSs from a variety of heterologous expression systems (including *E. coli*) are now feasible. Given the substantial tolerance of PKSs toward altered substrates and intermediates, it should therefore be possible to exploit this catalytic potential in a far more powerful way in cell-free systems than in intracellular systems. The primary limitations are with regard to the scale of synthesis. Attempts to stabilize and reuse the enzymes, in conjunction with the development of cheaper sources of natural and unnatural substrates and recycling systems for NADPH, should go a long way toward ameliorating this limitation.

VII. PHARMACOLOGICAL APPLICATIONS OF "UNNATURAL" NATURAL PRODUCTS

The appeal of manipulating polyketide synthases to create novel compounds stems from the wealth of polyketides that have been successfully developed into pharmaceuticals. Well over 30 polyketide natural products are currently in use for human therapeutics or agricultural applications, including anti-infectives, immunosuppressives, anticancer agents, cholesterol-lowering drugs, and insecticidal compounds. Why do polyketides possess such a useful and broad spectrum of pharmacological properties? In some cases the evolutionary advantage of producing the compound is suggested by its activity, for example, antibacterials or antifungals. However, in other cases, such as the anticholesterol or immunosuppressive activities, it is less clear. Whatever the reason, the ability to find such an abundance of useful activities within this class of compounds has spawned hope that libraries of unnatural natural products will aide the future development of natural products-based therapeutics.

A. Aromatic Polyketides

Until the biosynthesis of aromatic polyketides can be controlled as precisely as modular PKSs, the most likely utility for this class of PKS is in the generation of libraries to screen for new activities. Due to the highly reactive nature of the aromatic polyketide backbones, a variety of cyclization patterns have been created through combinatorial biosynthesis with aromatic PKS gene clusters

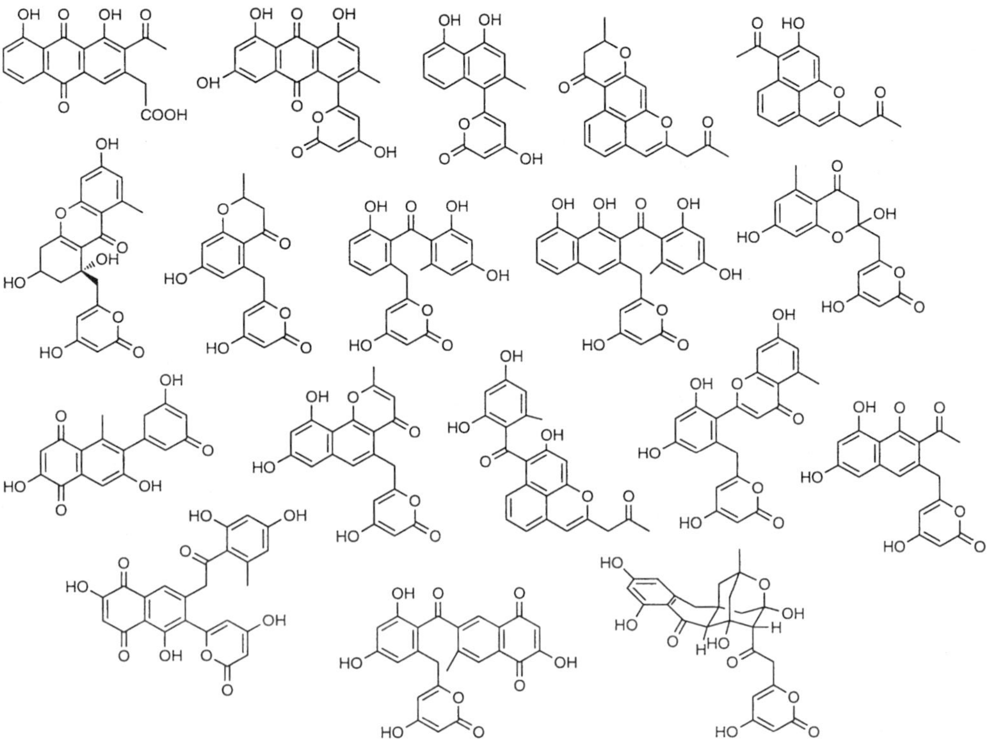

Figure 10 Examples of unnatural aromatic polyketides. Libraries of aromatic polyketides have been generated using combinations from the actinorhodin, tetracenomycin, frenolicin, griseusin, and *whiE* spore pigment gene clusters.

(Fig. 10) [48]. Strategies to rapidly increase the size and diversity of compounds are important, for example, by isolation of PKSs encoding new chain lengths, and ketoreductases and cyclases/aromatases with different regiospecificities. For example, the *whiE* minimal PKS which synthesizes an unknown spore pigment was recently shown to produce over 30 aromatic compounds with sizes ranging up to 24 carbons and an array of cyclization patterns [68]. This holds promise that even greater numbers and diversity could be achieved as PKSs encoding longer chain lengths are isolated.

B. Macrocyclic Products

Macrocyclic polyketides have so far been the primary focus of combinatorial biosynthesis technologies. The mutagenesis techniques outlined in Sec. VI suggest that the structures of some very complex molecules, such as erythromycin, FK506, rapamycin, or avermectin, can be manipulated in a number of ways to develop ''next generation'' derivatives. Furthermore, if the structure can be modified in a way that introduces a new reactive group, these chemical ''handles'' can be used to perform further synthetic derivatization. The precursor-directed biosynthesis approach highlighted above is a particularly attractive tool for analog construction in light of its general applicability. In particular, this may be a favorable method for constructing new molecules in organisms for which genetic manipulation is difficult. Relatively little sequence information is required to construct such mutations and, once the investment to construct the mutation has been made, many derivatives of the molecule could be produced by fermentation of a single strain with different feeds of synthetically produced substrate.

Finally, the potential number of compounds that could theoretically be produced by a modular PKS if all aspects of structural diversity could be controlled is enormous. For example, a six-module PKS, such as the erythromycin PKS, has a theoretical potential of tens of thousands of compounds. This number increases exponentially as the number of modules increases. While the theoretical number of polyketides may never be achieved, the possibility of generating even a small

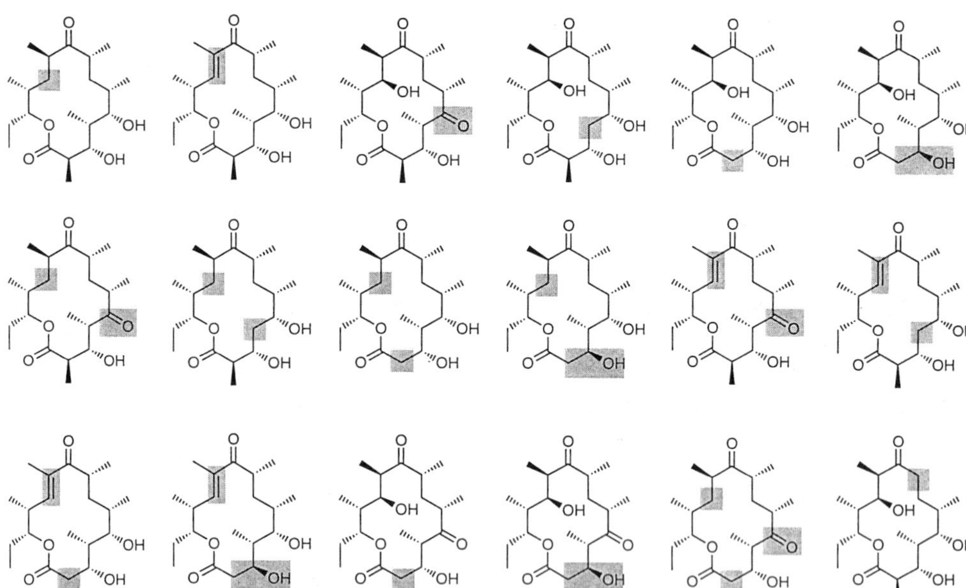

Figure 11 Examples of unnatural modular polyketides. These compounds are part of a combinatorial library that was engineered using DEBS and include single, double, and triple modifications in modules 2, 3, 5, and 6.

fraction of these justifies the development of combinatorial biosynthesis approaches. Although initial mutagenesis experiments performed with DEBS and other PKSs revealed the considerable tolerance of DEBS for the alteration of a single biosynthetic step, the ability to combinatorially produce large numbers of compounds with modular PKSs depends on the enzyme's flexibility toward multiple changes in the biosynthetic pathway. The recent macrolide library produced with DEBS (Fig. 11) addressed a key concern over the size of libraries and extent of diversity that can be expected from the combinatorial manipulation of modular PKSs. This library was produced by first engineering several single mutants and combining the successful ones to generate a library of over 100 compounds in which up to three modules could be manipulated simultaneously [55]. Although the production levels of many of the engineered PKSs was low, it demonstrated that the enzymes are capable of producing a large number of structures. The high degree of success in combining functional single mutants also suggested that development of general methods for increasing production levels of engineered PKSs could potentially increase the number of accessible structures.

VIII. FUTURE DIRECTIONS

Polyketide synthases represent a fundamentally interesting and pharmacologically attractive problem at the interface of chemistry and biology. As illustrated by the history of genetic engineering and protein engineering, the easier it gets to manipulate polyketide structure, the wider will be the applicability of these technologies. Therefore, future efforts aimed at developing improved insights into the enzymology of PKSs, in conjunction with more sophisticated tools for gene cloning and manipulation, are likely to remain the focus of such research in the next decade. A major boost to such protein engineering efforts is likely to come from acquisition of more detailed structural information on these systems. Likewise, for production of polyketide libraries, it will be important to develop technologies which permit more rapid mutant construction and screens for productive mutations. Lastly, the impact of genomic technologies is sure to be felt by the polyketide community. With the *S. coelicolor* genome over 50% sequenced at the time of this writing, the stage is set to explore how these organisms control secondary metabolite production and guide metabolic engineers toward the development of useful production hosts.

REFERENCES

1. D O'Hagan. The Polyketide Metabolites. Chichester, U.K.: Ellis Horwood, 1991.
2. DA Hopwood. Genetic contributions to understanding polyketide synthases. Chem Rev 97:2465–2498, 1997.

3. MA Fernández-Moreno, E Martínez, L Boto, DA Hopwood, F Malpartida. Nucleotide sequence and deduced functions of a set of cotranscribed genes of *Streptomyces coelicolor* A3(2) including the polyketide synthase for the antibiotic actinorhodin. J Biol Chem 267:19278–19290, 1992.
4. MJ Bibb, S Biró, H Motamedi, JF Collins, CR Hutchinson. Analysis of the nucleotide sequence of the *Streptomyces glaucescens* tcmI genes provides key information about the enzymology of polyketide biosynthesis. EMBO J 8:2727–2736, 1989.
5. A Grimm, K Madduri, A Ali, CR Hutchinson. Characterization of the *Streptomyces peucetius* ATCC29050 genes encoding doxorubicin polyketide synthase. Gene 151: 1–10, 1994.
6. CW Carreras, C Khosla. Purification and *in vitro* reconstitution of the essential protein components of an aromatic polyketide synthase. Biochemistry 37:2084–2088, 1998.
7. W Bao, E Wendt-Pienkowski, CR Hutchinson. Reconstitution of the iterative type II polyketide synthase for tetracenomycin F2 biosynthesis. Biochemistry 37:8132–8138, 1998.
8. J Cortes, SF Haydock, GA Roberts, DJ Bevitt, PF Leadlay. An unusually large multifunctional polypeptide in the erythromycin-producing polyketide synthase of *Saccharopolyspora erythraea*. Nature 348:176–178, 1990.
9. S Donadio, MJ Staver, JB McAlpine, SJ Swanson, L Katz. Modular organization of genes required for complex polyketide biosynthesis, Science 252:675–679, 1991.
10. T Schweke, JF Aparicio, I Molnar, A Konig, LE Khaw, SF Haydock, M Oliynyk, P Caffrey, J Cortes, JB Lester, G Bohm, J Staunton, PF Leadlay. The biosynthetic gene cluster for the polyketide immunosuppressant rapamycin. Proc Natl Acad Sci (USA) 92:7839–7843, 1995.
11. C Khosla, S Ebert-Khosla, DA Hopwood. Targeted gene replacements in a *Streptomyces* polyketide synthase gene cluster; role for the acyl carrier protein. Mol Microbiol 6:3237–3249, 1992.
12. R McDaniel, S Ebert-Khosla, DA Hopwood, C Khosla. Engineered biosynthesis of novel polyketides. Science 262:1546–1550, 1993.
13. CM Kao, L Katz, C Khosla. Engineered biosynthesis of a complete macrolactone in a heterologous host. Science 265:509–512, 1994.
14. L Tang, H Fu, M Betlach, R McDaniel. Elucidating the mechanism of chain termination switching in the picromycin/methymycin polyketide synthase. Chem Biol 6: 553–558, 1999.
15. DJ Bedford, E Schweizer, DA Hopwood, C Khosla. Expression of a functional fungal polyketide synthase in the bacterium *Streptomyces coelicolor* A3(2). J Bacteriol 177:4544–4548, 1995.
16. J Kealey, L Liu, DV Santi, M Betlach, PJ Barr. Production of a polyketide natural product in non-polyketide producing prokaryotic and eukaryotic hosts. Proc Natl Acad Sci (USA) 95:505–509, 1998.
17. RS Gokhale, SY Tsuji, DE Cane, C Khosla. Dissecting and exploiting intermodular communication in polyketide synthases. Science 284:482–485, 1999.
18. P Dimroth, H Walter, F Lynen. Biosynthese von 6-methylsalicysaure. Eur J Biochem 13:98–108, 1970.

19. B Shen, CR Hutchinson. Enzymatic synthesis of a bacterial polyketide from acetyl and malonyl coenzyme A. Science 262:1535–1540, 1993.
20. R Pieper, G Luo, DE Cane, C Khosla. Cell-free biosynthesis of polyketides by recombinant erythromycin polyketide synthases. Nature 378:263–266, 1995.
21. KEH Wiesmann, J Cortes, MJB Brown, AL Cutter, J Staunton, PF Leadlay. Polyketide synthesis *in vitro* on a modular polyketide synthase. Chem Biol 2:583–589, 1995.
22. NL Pohl, RS Gokhale, DE Cane, C Khosla. Synthesis and incorporation of an N-acetyl cysteamine analog of methylmalonyl-CoA by a modular polyketide synthase. J Am Chem Soc 120:11206–11207, 1998.
23. J Dreier, AN Shah, C Khosla. Kinetic analysis of the actinorhodin aromatic polyketide synthase. J Biol Chem 274:25108–25112, 1999.
24. RJX Zawada, C Khosla. Heterologous expression, purification, reconstitution and kinetic analysis of an extended type II polyketide synthase. Chem Biol 6:607–615, 1999.
25. S Smith. The animal fatty acid synthase: one gene, one polypeptide, seven enzymes, FASEB J 8:1248–1259, 1994.
26. P Caffrey, DJ Bevitt, J Staunton, PF Leadlay. Identification of DEBS 1, DEBS 2 and DEBS 3, the multienzyme polypeptides of the erythromycin-producing polyketide synthase from *Saccharopolyspora erythraea*. FEBS Lett 304:225–228, 1992.
27. J Staunton, P Caffrey, JF Aparicio, GA Roberts, SS Bethell, PF Leadlay. Evidence for a double-helical structure for modular polyketide synthases. Nature Struct Biol 3:188–192, 1996.
28. CM Kao, R Pieper, DE Cane, C Khosla. Evidence for two catalytically independent clusters of active sites in a functional modular polyketide synthase. Biochemistry 35:12363–12368, 1996.
29. RS Gokhale, J Lau, DE Cane, C Khosla. Functional orientation of the acyltransferase domain in a module of the erythromycin polyketide synthase. Biochemistry 37: 2524–2528, 1998.
30. R Pieper, G Luo, DE Cane, C Khosla. Remarkably broad substrate specificity of a modular polyketide synthase in a cell-free system. J Am Chem Soc 117:11373–11374, 1995.
31. AFA Marsden, B Wilkinson, J Cortes, NJ Dunster, J Staunton, PF Leadlay. Engineering broader specificity into an antibiotic-producing polyketide synthase. Science 279:199–203, 1998.
32. R Pieper, S Ebert-Khosla, DE Cane, C Khosla. Erythromycin biosynthesis: kinetic studies on a fully active modular polyketide synthase using natural and unnatural substrates. Biochemistry 35:2054–2060, 1996.
33. J Chuck, M McPherson, H Huang, JR Jacobsen, C Khosla, DE Cane. Molecular recognition of diketide substrates by a β-ketoacyl-ACP synthase domain within a bimodular polyketide synthase. Chem Biol 4:757–766, 1997.
34. R Pieper, RS Gokhale, G Luo, DE Cane, C Khosla. Purification and characterization of bimodular and trimodular derivatives of the erythromycin polyketide synthase. Biochemistry 36:1846–1851, 1997.
35. D Hunziker, TW Yu, CR Hutchinson, HG Floss, C Khosla. Primer unit specificity

in rifamycin biosynthesis principally resides in the later stages of the biosynthetic pathway. J Am Chem Soc 120:1092–1093, 1998.
36. W Bao, PJ Sheldon, E Wendt-Pienkowski, CR Hutchinson. The *Streptomyces peucetius* dpsC gene determines the choice of starter unit in biosynthesis of the daunorubicin polyketide. Biochemistry 38:9752–9757, 1999.
37. Y Hori, Y Abe, M Ezaki, T Goto, M Okuhara, M Kohsaka. R1128 substances, novel non-steroidal estrogen-receptor antagonists produced by a *Streptomyces*. I. Taxonomy, fermentation, isolation, and biological properties. J Antibiot 46:1055–1062, 1993.
38. MT Richardson, NL Pohl, JT Kealey, C Khosla. Tolerance and specificity of recombinant 6-methylsalicylic acid synthase. Metab Eng 1:180–187, 1999.
39. AFA Marsden, P Caffrey, JF Aparicio, MS Loughran, J Staunton, PF Leadlay. Stereospecific acyl transfers on the erythromycin-producing polyketide synthase. Science 263:378–380, 1994.
40. WP Revill, MJ Bibb, DA Hopwood. Purification of a malonyltransferase from *Streptomyces coelicolor* A3(2) and analysis of its genetic determinant. J Bacteriol 177: 3946–3952, 1995.
41. J Lau, H Fu, DE Cane, C Khosla. Dissecting the role of acyltransferase domains of modular polyketide synthases in the choice and stereochemical fate of extender units. Biochemistry 38:1643–1651, 1999.
42. PL Bartel, CB Zhu, JS Lampel, DC Dosch, NC Conners, WR Strohl, JM Beale, HG Floss. Biosynthesis of anthraquinones by interspecies cloning of actinorhodin biosynthesis genes in *Streptomyces*: clarification of actinorhodin gene functions. J Bacteriol 172:4816–4826, 1990.
43. TW Yu, Y Shen, R McDaniel, HG Floss, C Khosla, DA Hopwood, BS Moore. Engineered biosynthesis of novel polyketides from *Streptomyces* spore pigment polyketide synthases. J Am Chem Soc 120:7749–7759, 1998.
44. J Beck, S Ripka, A Siegner, E Schiltz, E Schweizer. The multifunctional 6-methylsalicylic acid synthase gene of *Penicillium patulum*: its gene structure relative to that of other polyketide synthases. Eur J Biochem 192:487–498, 1990.
45. M Offenzeller, G Santer, K Totschnig, Z Su, H Moser, R Traber, E Schneider-Scherzer. Biosynthesis of the unusual amino acid (4R)-4-[(E)-2-butenyl]-4-methyl-L-threonine of cyclosporin A: enzymatic analysis of the reaction sequence including identification of the methylation precursor in a polyketide pathway. Biochemistry 35:8401–8412, 1996.
46. CM Kao, M McPherson, R McDaniel, H Fu, DE Cane, C Khosla. Alcohol stereochemistry in polyketide backbones is controlled by the β-ketoreductase domains of modular polyketide synthases. J Am Chem Soc 120:2478–2479, 1998.
47. IE Holzbaur, RC Harris, M Bycroft, J Cortes, C Bisang, J Staunton, BAM Rudd, PF Leadlay. Molecular basis of Celmer's rules: the role of two ketoreductase domains in the control of chirality by the erythromycin modular polyketide synthase. Chem Biol 6:189–195, 1999.
48. R McDaniel, S Ebert-Khosla, DA Hopwood, C Khosla. Rational design of aromatic polyketide natural products by recombinant assembly of enzymatic subunits. Nature 375:549–554, 1995.
49. RS Gokhale, D Hunziker, DE Cane, C Khosla. Mechanism and specificity of the

terminal thioesterase domain from the erythromycin polyketide synthase. Chem Biol 6:117–125, 1999.
50. CM Kao, G Luo, L Katz, DE Cane, C Khosla. Manipulation of macrolide ring size by directed mutagenesis of a modular polyketide synthase. J Am Chem Soc 117: 9105–9106, 1995.
51. CM Kao, M McPherson, RN McDaniel, H Fu, DE Cane, C Khosla. Gain of function mutagenesis of a modular polyketide synthase II. Engineered biosynthesis of an eight-membered ring tetraketide lactone. J Am Chem Soc 119:11339–11340, 1997.
52. JR Jacobsen, CR Hutchinson, DE Cane, C Khosla. Precursor directed biosynthesis of novel erythromycin analogs by an engineered polyketide synthase. Science 277: 367–369, 1997.
53. Y Xue, L Zhao, HW Liu, DH Sherman. A gene cluster for macrolide antibiotic biosynthesis in *Streptomyces venezuelae*: Architecture of metabolic diversity. Proc Natl Acad Sci (USA) 95:12111–12116, 1998.
54. S Donadio, JB McAlpine, PJ Sheldon, M Jackson, L Katz. An erythromycin analog produced by reprogramming of polyketide synthesis. Proc Natl Acad Sci (USA) 90: 7119–7123, 1993.
55. R McDaniel, A Thamchaipenet, C Gustafsson, H Fu, M Betlach, G Ashley. Multiple genetic modifications of the erythromycin polyketide synthase to produce a library of novel ''unnatural'' natural products. Proc Natl Acad Sci (USA) 96:1846–1851, 1999.
56. R McDaniel, CM Kao, H Fu, P Hevezi, C Gustafsson, M Betlach, G Ashley, DE Cane, C Khosla. Gain-of-function mutagenesis of the erythromycin polyketide synthase. J Am Chem Soc 119:4309–4310, 1997.
57. X Ruan, A Pereda, DL Stassi, D Zeidner, RG Summers, M Jackson, A Shivakumar, S Kakavas, MJ Staver, S Donadio, L Katz. Acyltransferase domain substitutions in erythromycin polyketide synthase yield novel erythromycin derivatives. J Bacteriol 179:6416–6425, 1997.
58. DL Stassi, SJ Kakavas, KA Reynolds, G Gunawardana, S Swanson, D Zeidner, M Jackson, H Liu, A Buko, L Katz. Ethyl-substituted erythromycin derivatives produced by directed metabolic engineering. Proc Natl Acad Sci (USA) 95:7305–7309, 1998.
59. L Liu, A Thamchaipenet, H Fu, M Betlach, G Ashley. Biosynthesis of 2-nor-6-deoxyerythronolide B by rationally designed domain substitution. J Am Chem Soc 119:10553–10554, 1997.
60. S Kuhstoss, M Huber, JR Turner, JW Paschal, RN Rao. Production of a novel polyketide through the construction of a hybrid polyketide synthase. Gene 183:231–236, 1996.
61. J Cortes, KEH Wiesmann, GA Roberts, MJB Brown, J Staunton, PF Leadlay. Repositioning of a domain in a modular polyketide synthase to promote specific chain cleavage. Science 268:1487–1489, 1995.
62. CM Kao, G Luo, L Katz, DE Cane, C Khosla. Engineered biosynthesis of a triketide lactone from an incomplete modular polyketide synthase. J Am Chem Soc 116: 11612–11613, 1994.
63. CM Kao, G Luo, L Katz, DE Cane, C Khosla. Engineered biosynthesis of structur-

ally diverse tetraketides by a trimodular polyketide synthase. J Am Chem Soc 118: 9184–9185, 1996.
64. RN McDaniel, CM Kao, SJ Hwang, C Khosla. Engineered intra- and intermodular polyketide synthase fusions. Chem Biol 4:667–674, 1997.
65. JR Jacobsen, DE Cane, C Khosla. Dissecting the evolutionary relationships between 14-membered and 16-membered macrolides. J Am Chem Soc 120:9096–9097, 1998.
66. JR Jacobsen, A Keatinge-Clay, DE Cane, C Khosla. Precursor directed biosynthesis of 12-ethyl erythromycin. Bioorg Med Chem 6:1171–1177, 1998.
67. D Hunziker, N Wu, K Kinoshita, DE Cane, C Khosla. Precursor directed biosynthesis of novel 6-deoxyerythronolide B analogs containing non-natural oxygen substituents and reactive functionalities. Tetrahedron Lett 40:635–638, 1999.
68. Y Shen, P Yoon, TW Yu, HG Floss, D Hopwood, BS Moore. Ectopic expression of the minimal whiE polyketide synthase generates a library of aromatic polyketides of diverse sizes and shapes. Proc Natl Acad Sci (USA) 96:3622–3627, 1999.

18

Polyketide Synthases: Analysis and Use in Synthesis

Kira J. Weissman and James Staunton
University of Cambridge, Cambridge, United Kingdom

I. INTRODUCTION

Imagine going to the supermarket. You select a few items off the shelves, purchase them, and go home. In your kitchen, you fill a pot with water and add the ingredients. You heat the mixture gently and leave it stirring. A few hours later, your pot is brimming with medicine. This seemingly far-fetched scenario is the subject of this chapter.

As any graduate student organic chemist will attest, conventional synthesis of highly complex molecules is a painstaking and often laborious process—the stuff of Ph.D. programs. The synthesis of a natural product pharmaceutical, for example, may require tens of steps carried out on an ever-decreasing supply of increasingly precious compound. If these challenges weren't enough, the chemicals used in organic synthesis are frequently toxic, flammable, explosive, environmentally hazardous, or all of the above. The end result of thousands of man-hours, kilograms of starting materials, thousands of dollars of grant money, and thousands of gallons of harmful waste is usually a few milligrams of the target molecule. Although creative and intellectually rewarding, laboratory synthesis is generally an inefficient means to produce large quantities of pharmaceuticals.

As an example, consider the preparation of the antibiotic erythromycin A (Fig. 1). The total chemical synthesis of erythromycin was first completed in R. B. Woodward's laboratory in 1981. It took 49 people and 49 steps, and resulted in an overall yield of $<0.02\%$ relative to starting material [1–3]. In contrast, a lone bacterium can make a molecule of erythromycin A in a matter of seconds; a liter of such bacteria can provide up to 100 mg of erythromycin in a matter of

CHEMICAL SYNTHESIS

49 people and **49** steps:

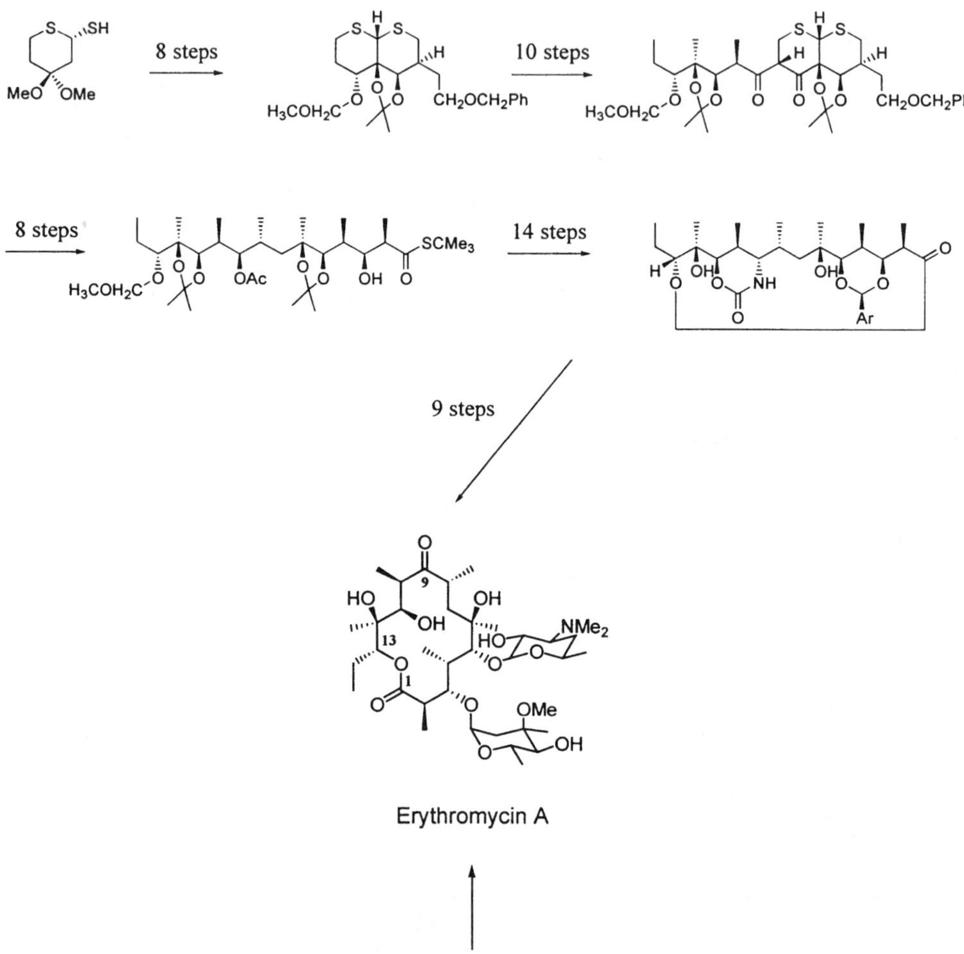

1 bacterium

BIOLOGICAL SYNTHESIS

Figure 1 A comparison of the chemical and biological synthesis of the antibiotic erythromycin A.

minutes. Removing the biosynthetic machinery—the protein catalysts—from the bacterial cells would allow a *single person* to carry out a ''one-pot'' total synthesis of this very complicated molecule starting from readily available precursors. The technology is also not limited to creating existing natural products. Genetic engineering has enabled the design of new enzymes capable of synthesising novel metabolites unknown in nature [4,5]. Such modified proteins could also be pulled from the cells and recruited to perform *in vitro.* (For a full discussion of the potential of genetic engineering, see Chap. 17.)

The ''grocery store'' approach to synthesis is not the stuff of fantasy. All that is required in principle are simple, commercially available precursors and a suitable set of purified biosynthetic enzymes. The use of purified enzymes has several obvious advantages over biosynthesis in the bacterial cell. *In vivo* assembly of complex molecules relies on the transient expression of biocatalysts that are often degraded after use; *in vitro* systems are free from such enzymatic proteolysis. Furthermore, products formed *in vivo* must be separated from all other undesired cellular metabolites by nontrivial methods, while the use of enzymes *in vitro* is a clean process—only one product results. Additionally, the reaction conditions *in vitro* can be precisely tuned for optimum turnover by manipulating the type and ionic strength of reaction buffer, pH, temperature, concentration of substrates, etc. Finally, *in vitro* systems provide a means for the controlled synthesis of compounds that might otherwise be inaccessible by genetic engineering. For example, precursors unavailable in the cell may be incorporated to yield analogs of the natural product.

Total synthesis *in vitro* is a major step forward from the conventional use of enzymes in organic synthesis. In traditional synthesis, an enzyme catalyst is employed to carry out only a *single* transformation in a multistep pathway. In this new technology, however, *all* of the processes are biological, including the key carbon–carbon bond-forming reactions.

This chapter describes the progress toward the one-pot total synthesis of the complex natural products called polyketides, molecules that hold a preeminent position in human medicine. It also details from this perspective the significant insights into polyketide biosynthesis that have been gained by studying the purified proteins *in vitro.*

II. THE CHEMISTRY OF POLYKETIDE BIOSYNTHESIS

The polyketides (Fig. 2) are a group of structurally diverse metabolites, synthesized primarily by soil microorganisms such as *Streptomyces* and related filamentous bacteria, but also by fungi and plants. The compounds are of great interest because they comprise a significant fraction not only of the number of microbial metabolites with physiological activity, but of the much smaller num-

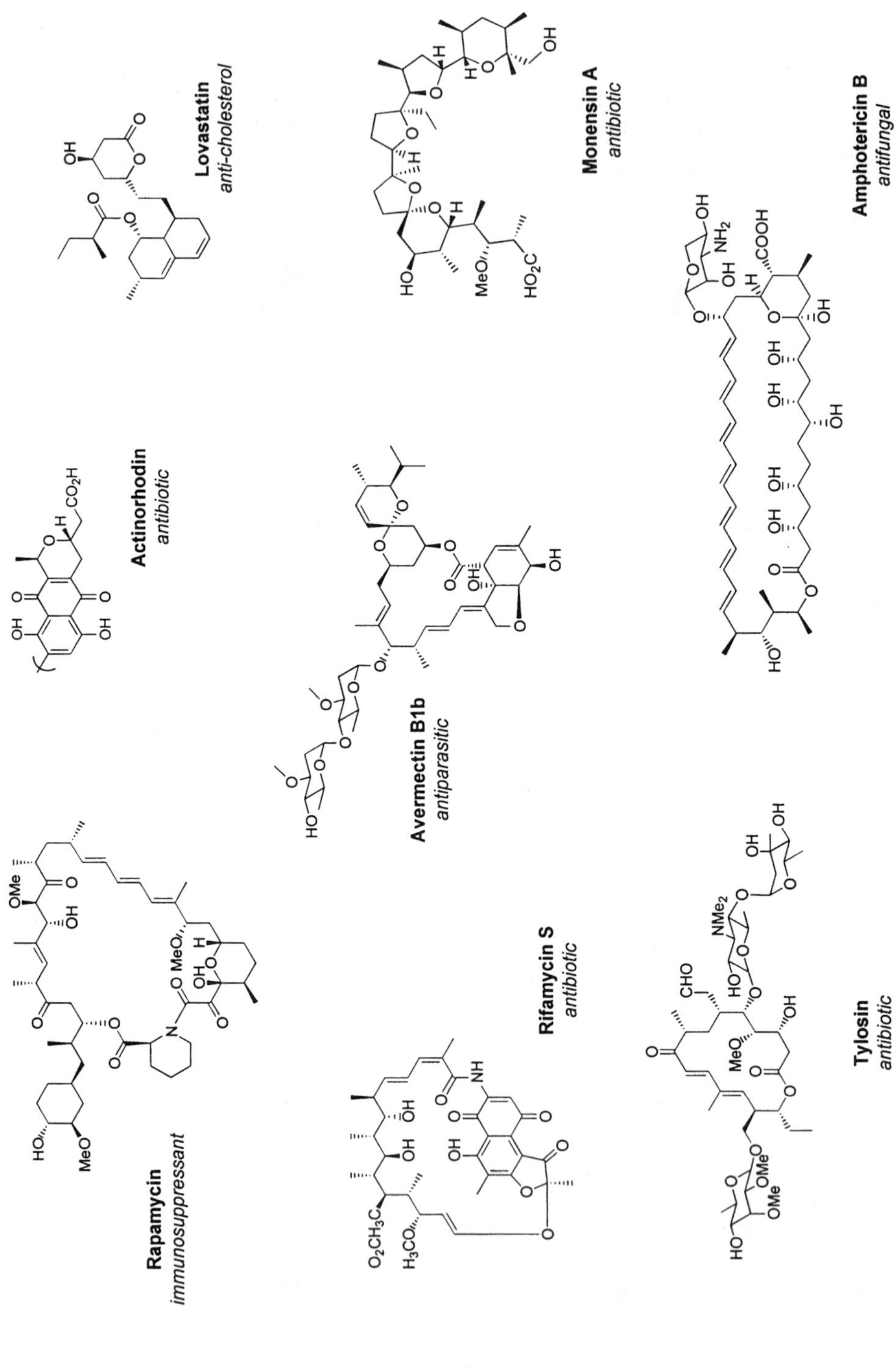

Figure 2 A selection of polyketide natural products and their biological activities.

ber which have found the greatest commercial application [6]; sales from antibiotic, anticholesterol, and immunosuppressant polyketides alone exceed $10 billion per annum worldwide (T. Buss, personal communication, 1997).

Like the related fatty acid synthases (FASs), polyketide synthases (PKSs) are multifunctional enzymes that catalyze the decarboxylative (Claisen) condensation of simple carboxylic acids, activated as their coenzyme A (CoA) thioesters. While FASs typically use acetyl-CoA as the starter unit and malonyl-CoA as the extender unit, PKSs often employ acetyl- or propionyl-CoA to initiate biosynthesis, and malonyl-, methylmalonyl-, and occasionally ethylmalonyl-CoA or propylmalonyl-CoA as a source of chain-extension units. After each condensation, FASs catalyze the full reduction of the β-ketothioester to a methylene by way of ketoreduction, dehydration, and enoyl reduction (Fig. 3). In contrast, PKSs shortcut the FAS pathway in one of two ways (Fig. 4). The aromatic PKSs (Fig. 4a) leave the β-keto groups substantially intact to produce aromatic products, while the modular PKSs (Fig. 4b) catalyze a variable extent of reduction to yield the so-called complex polyketides. In the latter case, reduction may not occur, or there may be formation of a β-hydroxy, double-bond, or fully saturated methylene; additionally, the outcome may vary between different cycles of chain extension (Fig. 4b). This inherent variability in keto reduction, the greater variety of

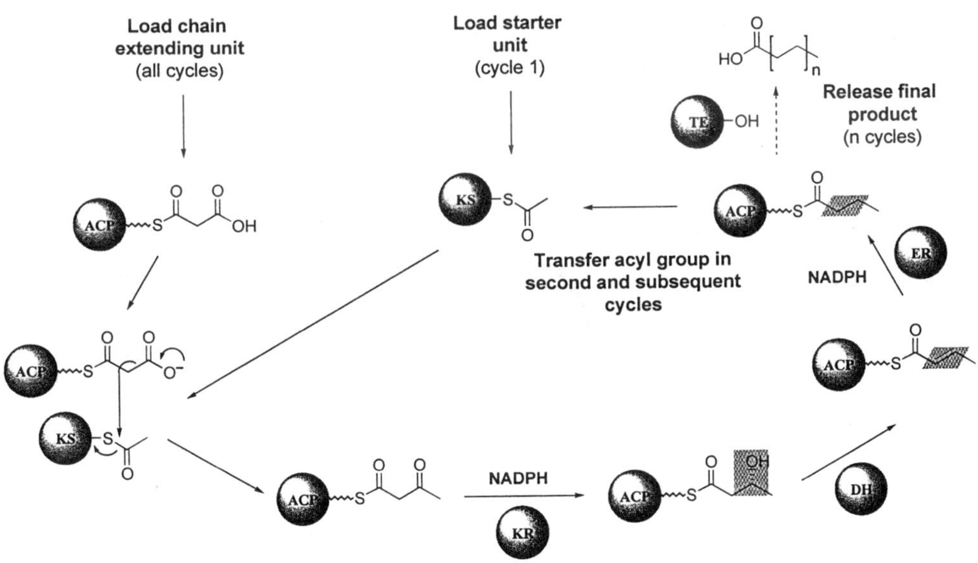

Figure 3 The fatty acid biosynthetic cycle (ACP, acyl carrier protein; KS, β-ketoacyl synthase; KR, β-ketoacyl reductase; DH, dehydratase; ER, enoyl reductase; TE, thioesterase).

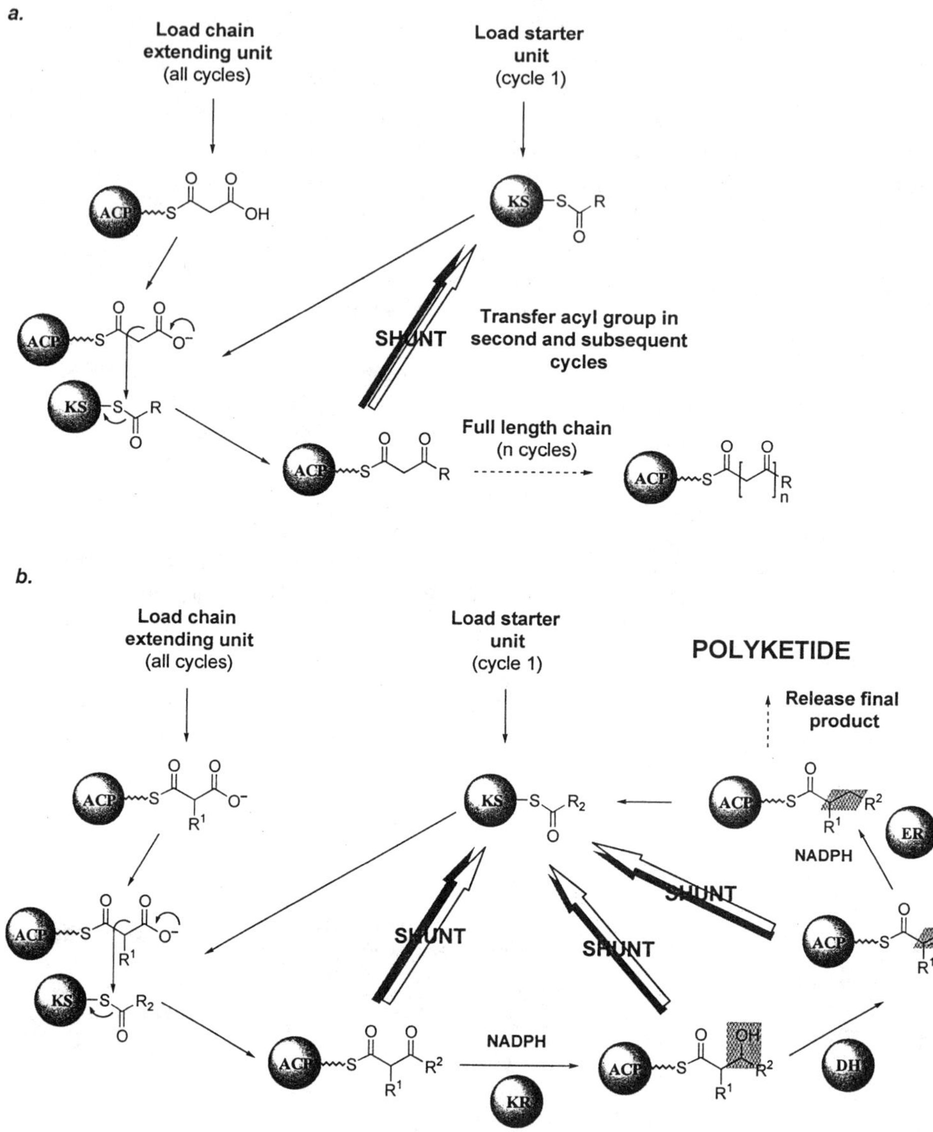

Figure 4 (a) The aromatic polyketide biosynthetic cycle (ACP, acyl carrier protein; KS, β-ketoacyl synthase; KR, β-ketoacyl reductase; DH, dehydratase; ER, enoyl reductase). (b) The complex polyketide biosynthetic cycle.

monomer units used, and the controlled variation in chain length account for the far greater diversity of polyketides compared to fatty acids. The variety of polyketide structures is further enhanced by alternative modes of chain cyclization, as well as additional modification by downstream enzymes, for example, through oxidation, reduction, glycosylation, and methylation [4]. It is these late ''tailoring'' steps that confer biological activity on the polyketide metabolites.

III. THE ENZYMOLOGY OF POLYKETIDE BIOSYNTHESIS

The cloning and sequence analysis of the clustered genes governing polyketide biosynthesis have revealed that aromatic PKS genes are organized very differently from those responsible for the modular PKSs (reviewed in Ref. 7). Typically, aromatic PKS genes encode a set of discrete proteins, each carrying a single active site, together with an acyl carrier protein (ACP) to which the growing polyketide chain is covalently tethered during chain extension. The aromatic PKS responsible for the biosynthesis of actinorhodin, for example, contains six activities: a ketosynthase (KS) to carry out the condensation reactions, a so-called chain length factor (CLF) (originally proposed as the determinant of chain length, but since also shown to initiate biosynthesis in many aromatic PKSs), an ACP, and three activities, a ketoreductase (KR), an aromatase (ARO), and a cyclase (CYC), involved in processing the fully formed chain. At least some of the constituent activities in these synthases act in every successive cycle, and so the aromatic PKSs have been called iterative.

In contrast to the aromatic PKSs, the PKSs for synthesis of the complex polyketides consist of gigantic multienzymes in which the activities are covalently linked and organized into ''modules.'' Each module contains the complement of enzymes required to carry out one particular cycle of chain extension and to catalyze the appropriate level of keto reduction; in contrast to the aromatic PKSs, each activity *normally* functions only once. The erythromycin PKS from *Saccharopolyspora erythraea* is the most extensively studied of the modular-type synthases. The PKS consists of three enormous protein subunits, 6-deoxyerythronolide B synthase (DEBS) 1, 2, and 3, which cooperate to assemble 6-deoxyerythronolide B (6-dEB), the aglycone core of the erythromycin molecule (Fig. 5). Each of the DEBS proteins contains two modules incorporating the three essential domains necessary for carrying out one cycle of chain extension [KS, acyltransferase (AT) and ACP], as well as the optional activities required to catalyze the reduction of the resulting intermediate [KR, dehydratase (DH) and enoyl reductase (ER)]. DEBS 1 is fronted by two domains (together termed a loading didomain) which accept the starter unit, while DEBS 3 terminates with a thioesterase (TE) which is thought to catalyze the off-loading and cyclizing of the full-length chain to form the enzyme-free intermediate 6-dEB. Again, it is the one-

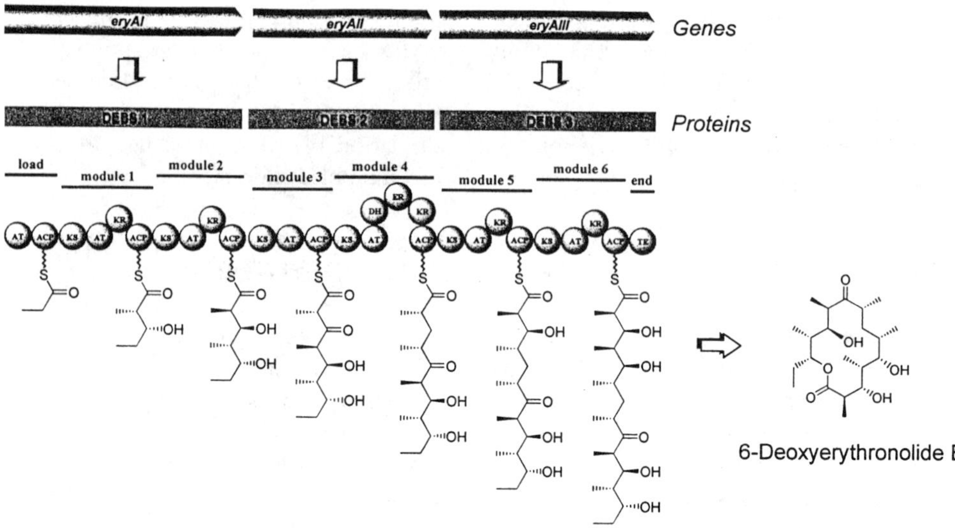

Figure 5 Domain organization of the erythromycin polyketide synthase. Putative domains are represented as circles and the structural residues are ignored. Each module incorporates the essential KS, AT, and ACP domains, while all but one include optional reductive activities. AT, acyltransferase; ACP, acyl carrier protein; KS, β-ketoacyl synthase; KR, β-ketoacyl reductase; DH, dehydratase; ER, enoyl reductase; TE, thioesterase.

to-one correspondence between catalytic activities and biosynthetic transformations that makes this type of PKS modular.

IV. BIOSYNTHESIS *IN VITRO* OF AROMATIC POLYKETIDES

The mechanisms by which aromatic PKSs synthesise a highly reactive poly-β-ketone intermediate of precise chain length and guide it toward a regiospecifically reduced and cyclized product are only poorly understood. Toward this end, mutagenesis and heterologous expression of recombinant bacterial aromatic PKSs have provided some information on the different protein components [8]. From the results of such experiments, a set of ''design rules'' has been deduced for the rational production of aromatic polyketides [9]. According to these rules, there is a ''minimal PKS'' which consists of the KS, CLF, and ACP. These activities are proposed to be sufficient to form a polyketone chain of an appropriate length and to carry out one cyclization to form an aromatic ring. The rules also describe the behavior of other PKS components, including the KR, ARO, and CYC [10]. Although these experiments have laid the groundwork for the biosynthesis of novel aromatic polyketides, they have not shed significant light on the mechanisms of catalysis, molecular recognition, and assembly of these multifunctional enzymes [8]. Such issues are best investigated by reconstituting biosynthesis using purified proteins *in vitro.*

Significant progress toward the development of fully controlled *in vitro* systems was made by the expression of components of aromatic PKSs in the non-native host *Escherichia coli*, whose biology is better understood [11–13]. As a result of these experiments, the first solution structure of a PKS component (ACP) was obtained [14,15]. Another important step was the biosynthesis *in vitro* of tetracenomycin F2, a precursor to the natural-product antibiotic tetracenomycin C. The synthesis was accomplished by adding acetyl- and malonyl-CoA to a *cell-free* extract of the natural producer of tetracenomycin C, *Streptomyces glaucescens*, in which the tetracenomycin minimal PKS genes and an ARO/CYC activity had been overexpressed (Fig. 6) [16]. A strain of *S. glaucescens* missing the crucial ACP activity was also created. When purified ACP from *E. coli* was added to a cell-free extract of the deficient strain, biosynthesis of tetracenomycin

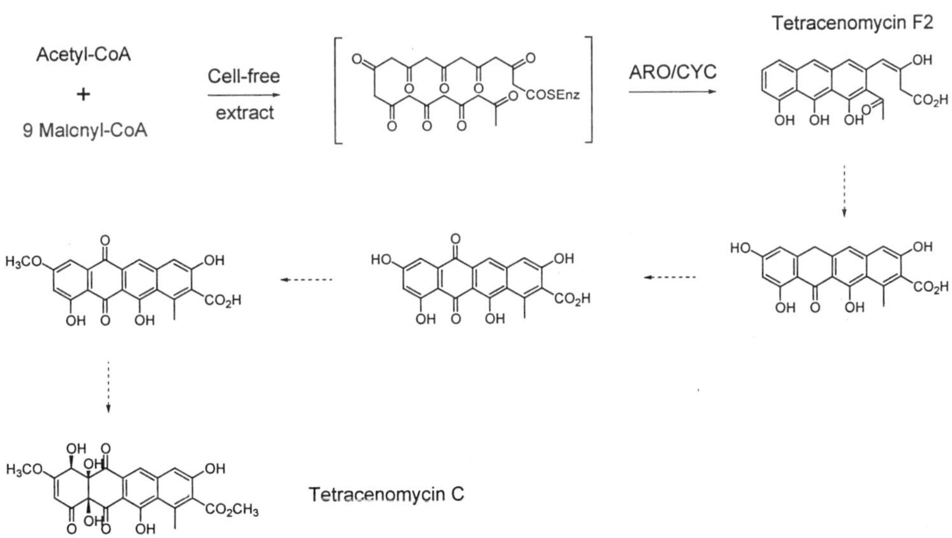

Figure 6 Cell-free biosynthesis of tetracenomycin F2. A strain of *Streptomyces glaucescens* in which the tetracenomycin-minimal PKS genes and an ARO/CYC activity had been overexpressed produced tetracenomycin F2, a precursor to the natural product tetracenomycin A (the remainder of the pathway is indicated by dashed arrows). The same result has been achieved using purified proteins *in vitro.*

F2 was restored. More recently, the biosynthesis of tetracenomycin F2 has been achieved with fully purified proteins [17].

In similar experiments, cell-free preparations of a *Streptomyces coelicolor* strain containing the minimal PKS involved in actinorhodin biosynthesis were shown to produce compounds called SEK4 and SEK4b, molecules generated as shunt products of the normal pathway (Fig. 7). When the strain also included genes for the KR, ARO, and CYC activities, a more elaborate structure, 3,8-dihydroxy-1-methylanthraquinone-2-carboxylic acid (DMAC), was obtained (Fig. 7) [8]. Biosynthesis of SEK4 and SEK4b has also been accomplished *in*

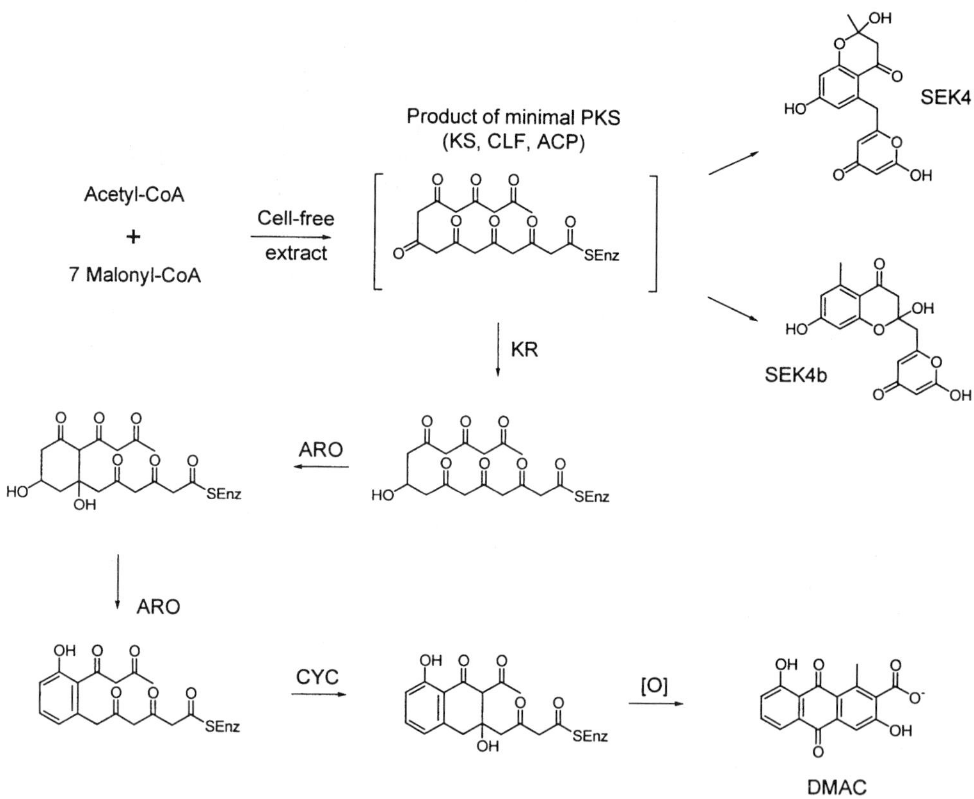

Figure 7 Cell-free biosynthesis of SEK4, SEK4b, and DMAC. A strain of *Streptomyces coelicolor* in which the actinorhodin-minimal PKS genes had been overexpressed, produced both SEK4 and SEK4b. Expression of the KR, ARO, and CYC genes in addition to the minimal PKS resulted in the biosynthesis of DMAC. Synthesis of SEK4 and SEK4b has also been achieved using purified proteins *in vitro*.

vitro using the purified components of the minimal PKS, although an additional activity from FAS was required (see below) [18]. The *in vitro* studies have recently been extended by incorporating various heterologous KR and ARO/CYC activities [19]. In all cases, the synthases generated the products observed *in vivo* from the same combination of activities.

Although *in vitro* biosynthesis of aromatic polyketides has been achieved, a substantially better understanding of the programming rules is required before such systems can be rationally controlled. The following crucial issues, among others, remain unresolved: (1) the structure of the active complex; (2) the activities involved in determining chain length and the control of cyclizations and aromatizations; and (3) the tolerance of various activities toward heterologous domains.

There is good evidence that if certain components of the natural pathway are missing, biosynthesis is at best inefficient. Shunt metabolites released from such incomplete PKSs are highly reactive and subject to spontaneous chemistry; such chemical events make it difficult to predict the structure of the product. Finally, a possibly serious limitation to the production of large numbers of novel aromatic polyketides *in vitro* by combining elements from different PKSs is whether the ''degrees of freedom'' represented by the domains are truly independent. The most recent results [20–23] strongly suggest that they are not independent and that optimistic estimates [24,25] of the potential diversity among aromatic polyketides should be regarded with caution [4].

V. MECHANISTIC INSIGHTS FROM BIOSYNTHESIS *IN VITRO* OF AROMATIC POLYKETIDES

Although *in vitro* synthesis has not yet yielded large libraries of novel compounds, it has nonetheless provided significant insight into the roles of various PKS domains.

The so-called chain length factor (CLF) of aromatic systems is similar in sequence to the KS domain responsible for condensation, but has a highly conserved glutamine (Q) residue in place of the active-site cysteine (C) of the KS. Early experiments involving the expression of heterologous domains *in vivo* led to the proposal that the CLF determines the chain length [9]. Subsequent studies, however, showed that the CLF might not alter the product chain length as predicted [21,23], that the CLF exerts its major influence only in the presence of its cognate ketosynthase, and that the ketoreductases, cyclases [21,23,26] and aromatases [27] may all have a decisive influence on the outcome.

Experiments both *in vivo* and *in vitro* have now pinpointed the most probable primary role for the CLF [28]. A purified KS/CLF complex, in which the KS had been inactivated by mutation, was shown to catalyze the efficient decar-

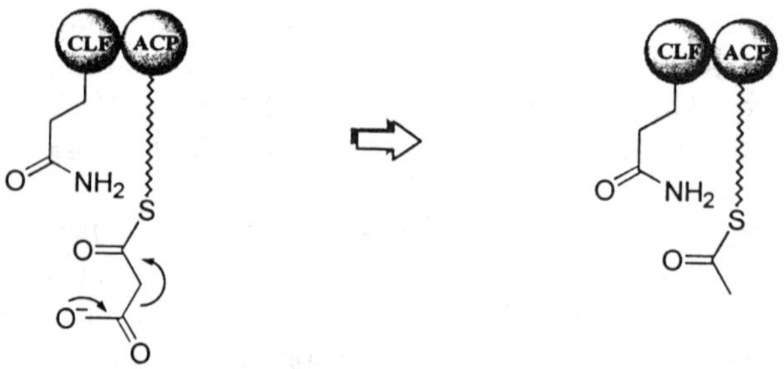

Figure 8 Decarboxylation of malonyl-ACP by the CLF. Decarboxylation of malonyl-ACP by the CLF results in acetyl-ACP, an enzyme-bound starter unit. Initiation of chain extension can then occur by transfer of the acetate starter unit to the KS domain.

boxylation of malonyl-CoA to acetyl-CoA (Fig. 8). This result unequivocally implicated the CLF as the activity responsible for the provision of acetate starter units in the biosynthesis of most aromatic polyketides. An analogous activity (designated KS_Q) has been identified in modular systems [28].

Another issue that has been investigated is whether polyketide synthesis *in vitro* absolutely requires the presence of an enzyme involved in fatty acid biosynthesis, the malonyl-CoA:ACP acyl transferase (MCAT). In the FAS, the MCAT enzyme is responsible for charging the ACP with malonate units. As there is no comparable acyltransferase activity in the aromatic PKS clusters, it has been suggested that the MCAT derived from the FAS cluster serves the same role in both systems [17,18]. However, *in vitro* experiments with rigorously purified ACP have provided strong evidence for its ability to self-malonylate, which would obviate the requirement for an MCAT [29–31]. Although the MCAT is not required *in vitro,* the relevance of self-malonylation to the living organism remains a matter of vigorous disagreement [32].

Two other findings from the MCAT experiments have relevance to achieving total biosynthesis *in vitro.* Investigations into the effects of varying protein concentrations on polyketide production demonstrated that the optimal composition of a minimal PKS is not stoichiometric in each of the components (KS, CLF, and ACP) [30]. Increasing the proportion of the ACP relative to the KS/CLF pair caused an enhancement in turnover; even at more than 100-fold excess of ACP, there appeared to be no indication of saturation. This result has important implications for the composition of efficient one-pot biosynthetic assemblies. The experiments also illustrated that domains can be usefully engineered to improve yields. In this case, the ACP was modified to prevent its inactivation by formation of an internal disulfide between its active thiol and a remote cysteine. Optimizing other activities by genetic engineering methods may be an important component of *in vitro* biosynthesis.

VI. BIOSYNTHESIS *IN VITRO* OF COMPLEX POLYKETIDES

Experiments *in vitro* with modular PKSs are considerably more straightforward than those with aromatic synthases. Whereas synthesis of aromatic polyketides requires the expression and purification of at minimum three activities, and as many as seven, a single multienzyme from a modular PKS can be competent for synthesis. Additionally, the design rules for the modular PKSs are considerably simpler, as there is a direct correlation between the number and nature of activities present and the resulting structure of the polyketide product.

As with the aromatic polyketides, the first *in vitro* biosynthesis of complex polyketides was achieved using a cell-free system. In this case, the three large multienzymes thought to be responsible for assembling 6-dEB were overexpressed in a heterologous host, *S. coelicolor* [33]. The recombinant strain was shown to produce 6-dEB as well as an analog of 6-dEB incorporating acetate in place of the normal starter unit, propionate. The proteins were partially purified from the extract and then repooled. When the natural substrates [propionyl-CoA, methylmalonyl-CoA, and reduced nicotinamide adenine dinucleotide phosphate (NADPH)] were added to the reconstituted complex, it showed comparable biosynthetic activity to the original cell-free extract.

Although cell-free synthesis of 6-dEB was achieved, kinetic studies demonstrated that the process was very inefficient [34]. The low rate of biosynthesis was likely due to the fact that the association of the three DEBS proteins *in vitro* was suboptimal. To simplify *in vitro* synthesis and to facilitate mechanistic analysis, a truncated version of the erythromycin PKS was created. The protein, DEBS 1-TE, was engineered by relocating the thioesterase (TE) from the end of DEBS 3 to the terminus of DEBS 1 (Fig. 9a,b) [35]. *In vivo,* this bimodular construct synthesizes two triketide lactone products, one derived from propionate as a starter unit, and the other from acetate.

The DEBS 1-TE mini-PKS offers several advantages over the complete three-protein DEBS system for studies *in vitro.* Most importantly, it is a single protein whose size (MW ~390 kDa) makes it amenable to purification by size-exclusion chromatography. And unlike the three DEBS proteins, purified DEBS 1-TE can function autonomously to catalyze two complete condensation cycles to give a lactone product. Finally, the lactone incorporates much of the functionality and stereochemistry present in the macrolide, but is significantly easier to analyze than 6-dEB by various techniques such as GC-MS and ^{1}H NMR. DEBS 1-TE can therefore serve as a convenient model system for the far more elaborate

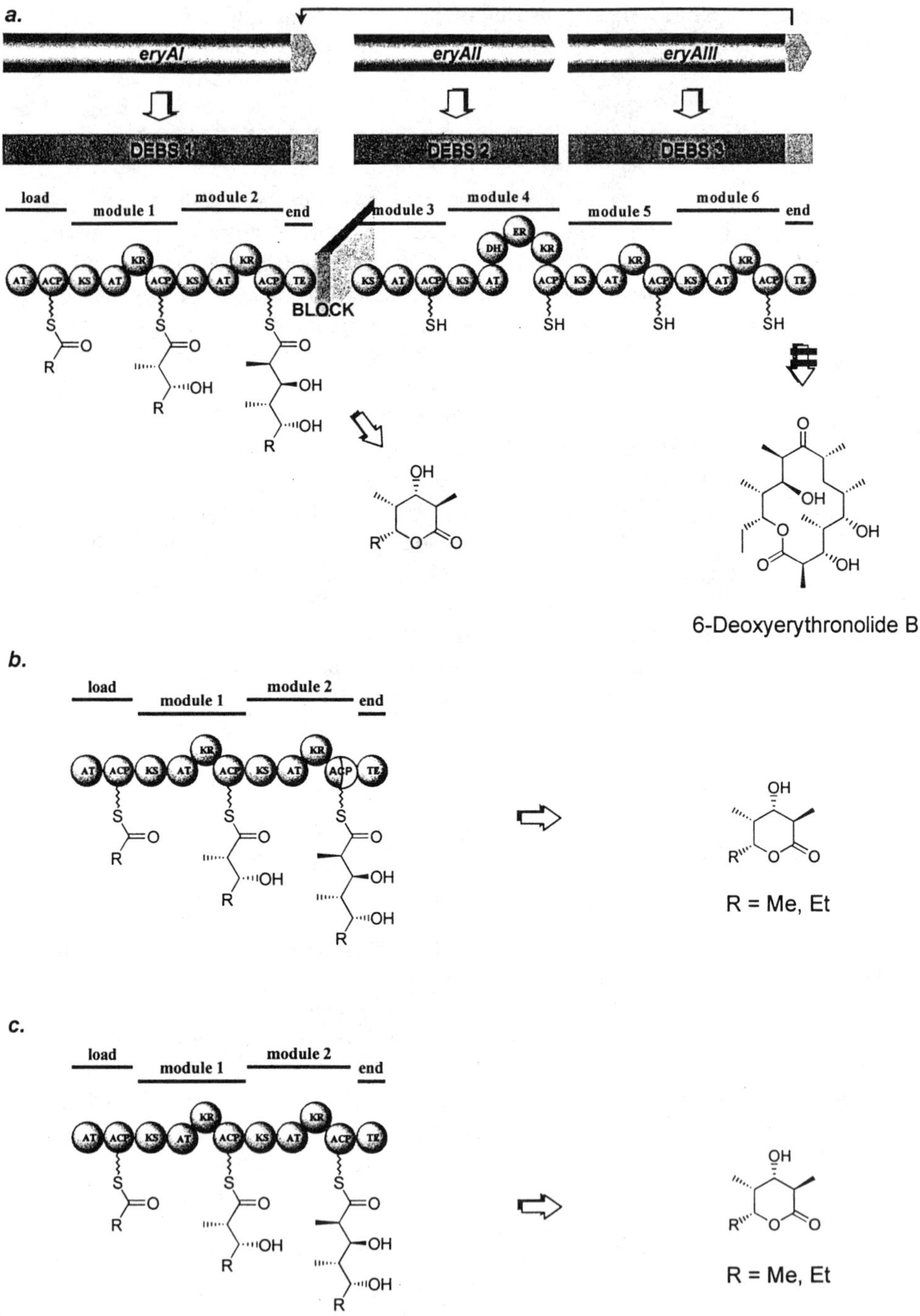
a.
eryAI
eryAII
eryAIII
DEBS 1
DEBS 2
DEBS 3
load
module 1
module 2
end
module 3
module 4
module 5
module 6
end
AT ACP KS AT KR ACP KS AT KR ACP TE
BLOCK
KS AT ACP KS DH ER KR AT ACP KS AT KR ACP KS AT KR ACP TE
SH
6-Deoxyerythronolide B
b.
load
module 1
module 2
end
R = Me, Et
c.
load
module 1
module 2
end
R = Me, Et

parent DEBS system, and has been used in the majority of *in vitro* studies of the mechanisms of erythromycin biosynthesis.

In the presence of its natural substrates, propionyl-CoA, methylmalonyl-CoA, and NADPH, DEBS 1-TE was initially shown to catalyze the formation of lactone in a cell-free system [36]. Concomitant work on a similar bimodular system called DEBS 1+TE (Fig. 9c) [37] in a cell-free extract and with partially purified protein, demonstrated that it too was competent for biosynthesis of the triketide lactone [33]. These experiments set the stage for more rigorous investigation of mechanistic aspects of erythromycin biosynthesis.

VII. MECHANISTIC INSIGHTS FROM BIOSYNTHESIS *IN VITRO* BY DEBS 1-TE AND DEBS 1+TE

The original DEBS 1-TE cell-free system allowed several other features of polyketide chain extension to be examined by using well-known inhibitors of specific enzyme activities. Incubation of DEBS 1-TE with the serine protease inhibitor phenylmethylsulfonyl fluoride (PMSF) resulted in a significant decrease in biosynthetic activity. This result emphasized the mechanistic similarity between the targeted TE domain and the serine protease enzymes. Addition of the fatty acid inhibitor cerulenin [38] also reduced production of the lactone, consistent with the evolutionary kinship between these two classes of enzymes. Inhibition of DEBS 1+TE by cerulenin was also reported [33].

The DEBS 1-TE multienzyme was purified to 90–95% homogeneity and then used in another series of experiments to establish the extent to which alternative starter units could be used by the polyketide synthase [36]. Substantial amounts of lactones were obtained in the presence of acetyl-, *n*-butyryl-, and isobutyryl-CoA, illustrating that the loading didomain exhibits a relaxed specificity for the starter unit (Fig. 10). The utilization of acetyl-CoA and *n*-butyryl-CoA by DEBS 1+TE was demonstrated in a cell-free system [39]. Additionally, in the absence of the reducing cofactor NADPH, cell-free DEBS 1+TE converted

Figure 9 Construction of bimodular polyketide synthases. (a) Chromosomal repositioning of the thioesterase domain from the C-terminus of module 6 to the end of module 2 in the erythromycin PKS leads to production of triketide lactones and the disruption of erythromycin biosynthesis. (b) DEBS 1-TE contains a fusion within the ACP domains of modules 2 and 6. In *Saccharopolyspora erythraea* and *Streptomyces coelicolor* the construct produced both propionate and acetate-derived lactones. (c) DEBS 1+TE contains a fusion between ACP_2 and the thioesterase domain. In *S. coelicolor*, the protein biosynthesized the same lactones.

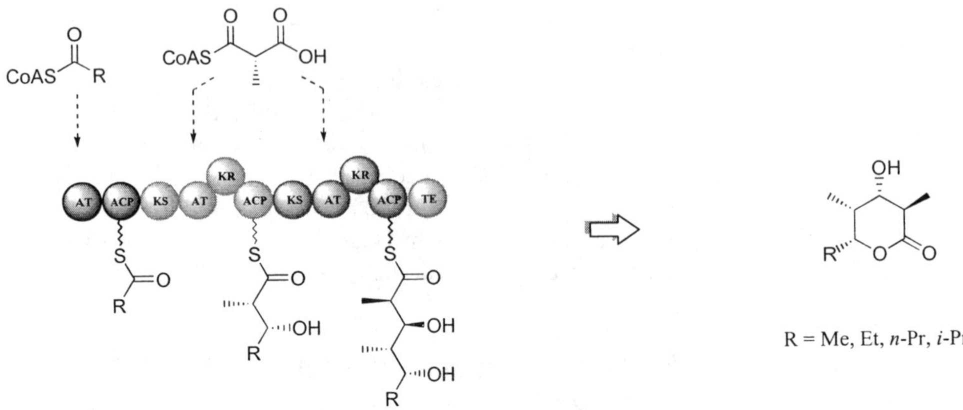

Figure 10 Biosynthesis of triketide lactones *in vitro* by DEBS 1-TE. Incubation of propionyl-CoA or acetyl-CoA with methylmalonyl-CoA and NADPH *in vitro* resulted in synthesis of the appropriate lactones. The mini-PKS also accepted the unnatural starter units *n*-butyryl-CoA and isobutyryl-CoA.

propionyl-CoA and methylmalonyl-CoA into a pyrone analog of the lactone (Fig. 11) [39]. This result demonstrated that the PKS tolerates the unreduced poly-β-keto intermediates, similar to those found in aromatic polyketide biosynthesis.

Purified DEBS 1-TE has also been used to investigate another interesting aspect of polyketide biosynthesis: the control of stereochemistry. As noted by W. D. Celmer in 1965, the macrolide polyketides (the class of complex polyketide to which erythromycin belongs) have the same absolute configuration at all comparable stereocenters (Fig. 12) [40,41]. These homologies suggest that there exists

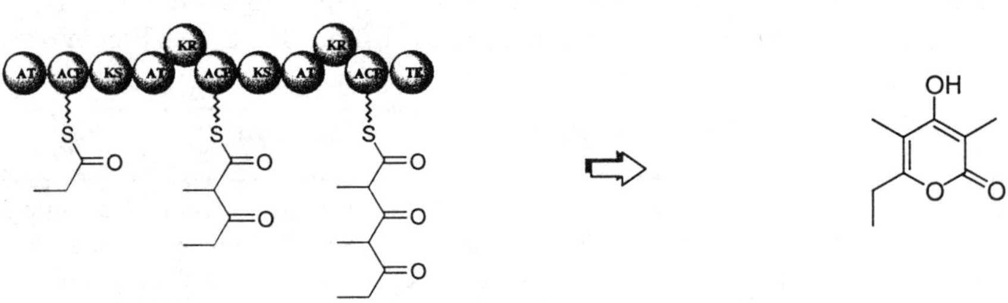

Figure 11 Incubations with DEBS 1+TE in the absence of NADPH. In the absence of reducing agent NADPH, biosynthesis from propionyl-CoA starter and methylmalonyl-CoA resulted in the pyrone analog of the lactone. This experiment showed that unreduced chains could be transferred between modules.

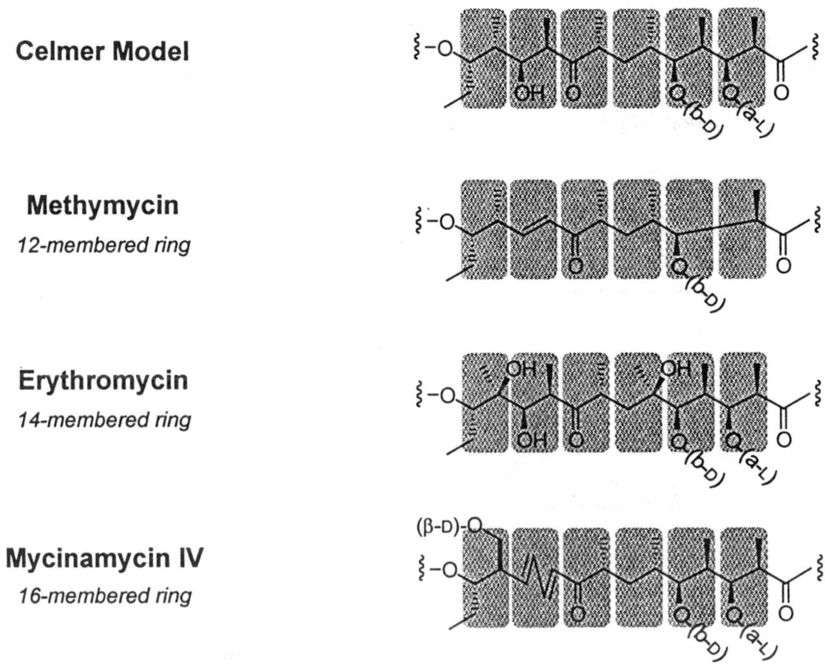

Figure 12 Stereochemical model for the complex polyketides. This figure shows the Celmer model for the structure and stereochemistry of the macrolides and its comparison to actual polyketide structures.

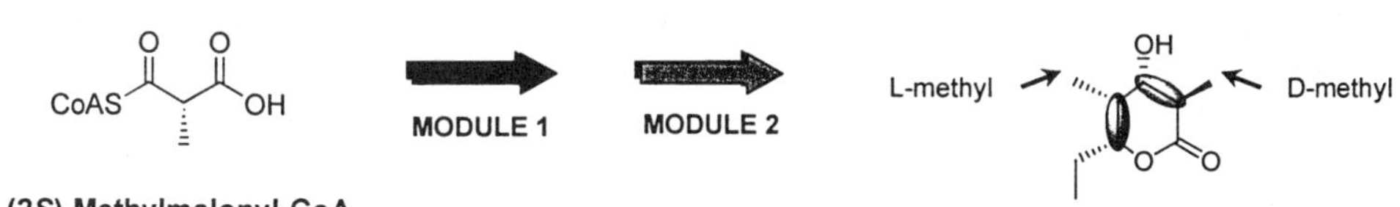

Figure 13 Stereochemistry of chain extension in erythromycin biosynthesis. DEBS 1-TE uses only the (2*S*)-isomer of methylmalonyl-CoA to generate both a D- and an L-center in its lactone product. The D-methyl stereochemistry corresponds to condensation of (2*S*)-methylmalonyl-CoA with retention of configuration, while the L-methyl corresponds to condensation with inversion.

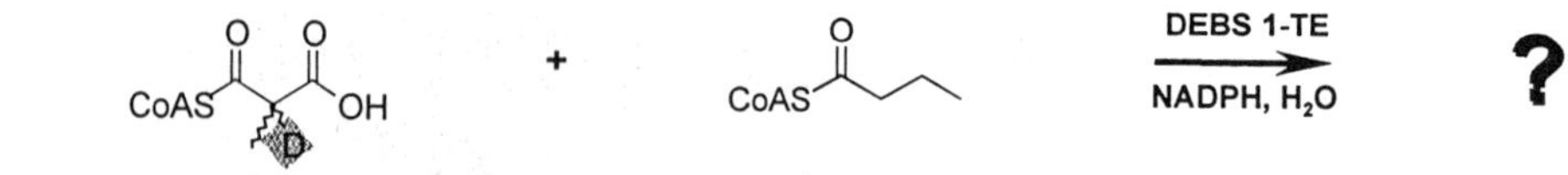

(2*RS*)-[2-^{2}H]Methylmalonyl-CoA

n-Butyryl-CoA

MECHANISM	CONDENSE WITH RETENTION	CONDENSE WITH INVERSION	EPIMERIZE	EXPECTED PRODUCT
I	Module 1	Module 2		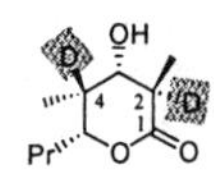
II	Modules 1 & 2		Module 2	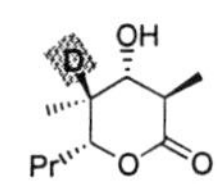
III		Modules 1 & 2	Module 1	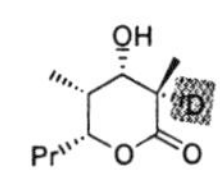
IV	Module 2	Module 1	Modules 1 & 2	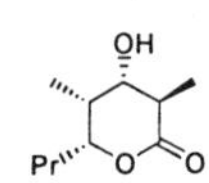

Figure 14 Four plausible mechanisms by which one configuration (*S*) in the chain-extending methylmalonyl-CoA can become two in the propionyl lactone product. The methyl stereochemistry generated by module 1 (C-4 center) corresponds to condensation with retention of configuration. Therefore, any mechanism incorporating inversion in that module (e.g., III) must necessarily include an essential epimerization at that center. Similarly, the methyl stereochemistry generated in module 2 (C-2 center) corresponds to condensation with inversion of configuration. Therefore, any mechanism including retention of configuration in that module (e.g., II) must include an epimerization at that center.

An experiment was designed to discriminate between the possible mechanisms. Biosynthesis of *n*-butyryl lactone was carried out in the presence of (2*RS*)-[2-^{2}H]methylmalonyl-CoA and NADPH. In the absence of spontaneous loss of the deuterium, the resulting labeling pattern in the *n*-butyryl lactone was diagnostic for a particular mechanism. Any epimerization step in a mechanism acted to remove the label from the corresponding center. Structural analysis of the lactone product demonstrated labeling consistent with mechanism III.

an intimate evolutionary relationship between the various PKSs. Understanding how the stereochemistry of chain extension is controlled is therefore crucial to determining the common mechanism of these enzymes, and will greatly facilitate attempts to design hybrid polyketides with particular stereochemistry.

Initial experiments with DEBS 1-TE showed that only the (2*S*)-isomer of methylmalonyl-CoA is used to generate both stereocenters (D and L) in the triketide lactone (Fig. 13) [36]. In follow-up experiments, DEBS 1-TE was incubated in the presence of starter unit, NADPH, and deuterium-labeled (2*S*)-methylmalonyl-CoA [42]. Use of the isotopically labeled substrate allowed discrimination between possible mechanisms by which the synthase could generate two methyl stereochemistries from a precursor with a single methyl configuration (Fig. 14), because the deuterium labeling pattern in the resulting lactone was characteristic

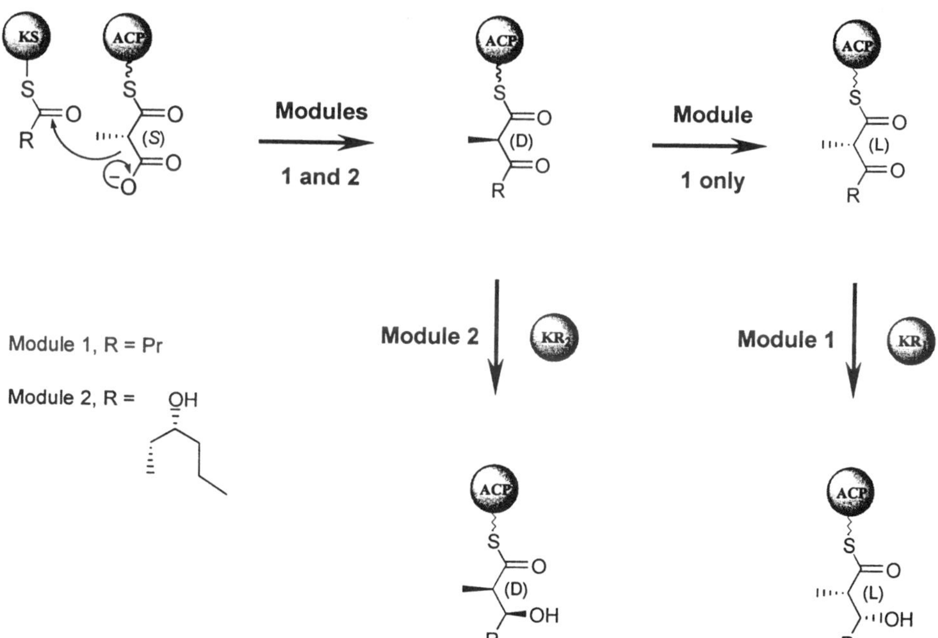

Figure 15 Stereochemistry of the condensation step in erythromycin biosynthesis. Experiments with bimodular DEBS 1-TE *in vitro* have demonstrated that condensation of (2*S*)-methylmalonyl-CoA in module 2 proceeds with decarboxylative inversion without cleavage of the C—H bond adjacent to the methyl group. The results are also consistent with the same stereochemistry of condensation, inversion, in module 1, followed by epimerization of the resulting D-methyl center to the required L-methyl configuration in the β-keto intermediate.

of a particular mechanism. These experiments demonstrated that condensation of (2*S*)-methylmalonyl-CoA in module 2 of DEBS 1-TE proceeds with inversion of configuration (as found in fatty acid biosynthesis) without cleavage of the C—H bond adjacent to the methyl group (Fig. 15). In contrast, in module 1, the chain extension process includes an obligatory epimerization with loss of the hydrogen attached to C-2 of methylmalonyl-CoA (Fig. 15). Investigations into the stereospecificity of ketoreduction by DEBS 1-TE *in vitro* provided further evidence for these conclusions [43]. Although the nature and timing of the epimerization remain to be established, this understanding of the control of stereochemistry in erythromycin biosynthesis should enable the creation of novel polyketides with designed stereochemical patterns.

Another aspect of erythromycin assembly that has been probed *in vitro* is the origin of the starter units [44]. It has been suggested that the KS domain of module 1 of DEBS 1+TE has the ability to prime its own biosynthesis by decarboxylation of methylmalonyl-CoA, generating starter-unit propionyl-CoA

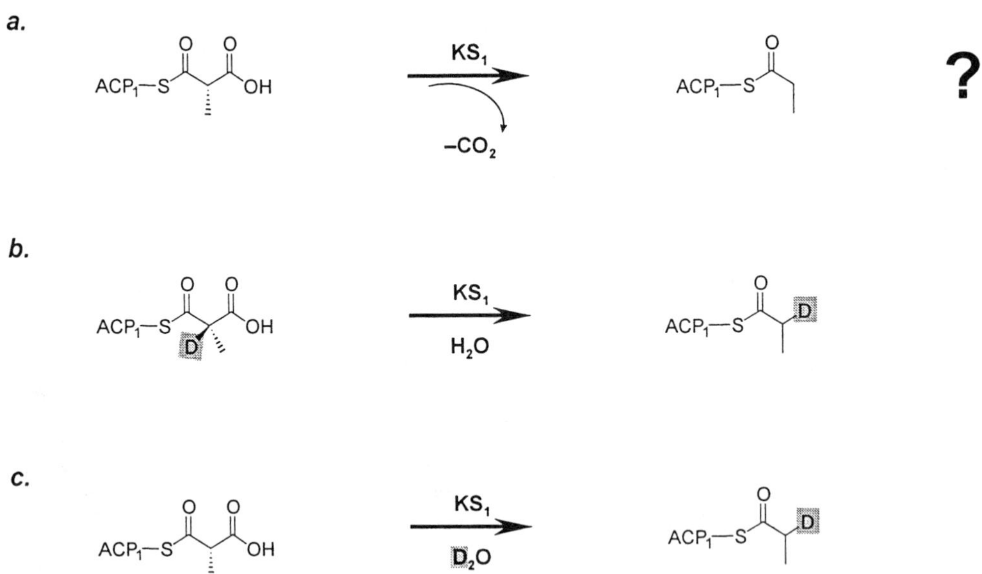

Figure 16 The putative KS-catalyzed decarboxylation of methylmalonyl-CoA. (a) In this proposed reaction, methylmalonyl-CoA bound to the ACP is decarboxylated by the KS domain to yield propionyl-ACP. The resulting propionate unit can then be transferred to the KS to initiate biosynthesis. (b) Decarboxylation of deuterium-labeled methylmalonyl-CoA in H_2O would result in labeled propionyl-ACP. (c) Decarboxylation of unlabeled methylmalonyl-CoA in D_2O would result in labeled propionyl-ACP.

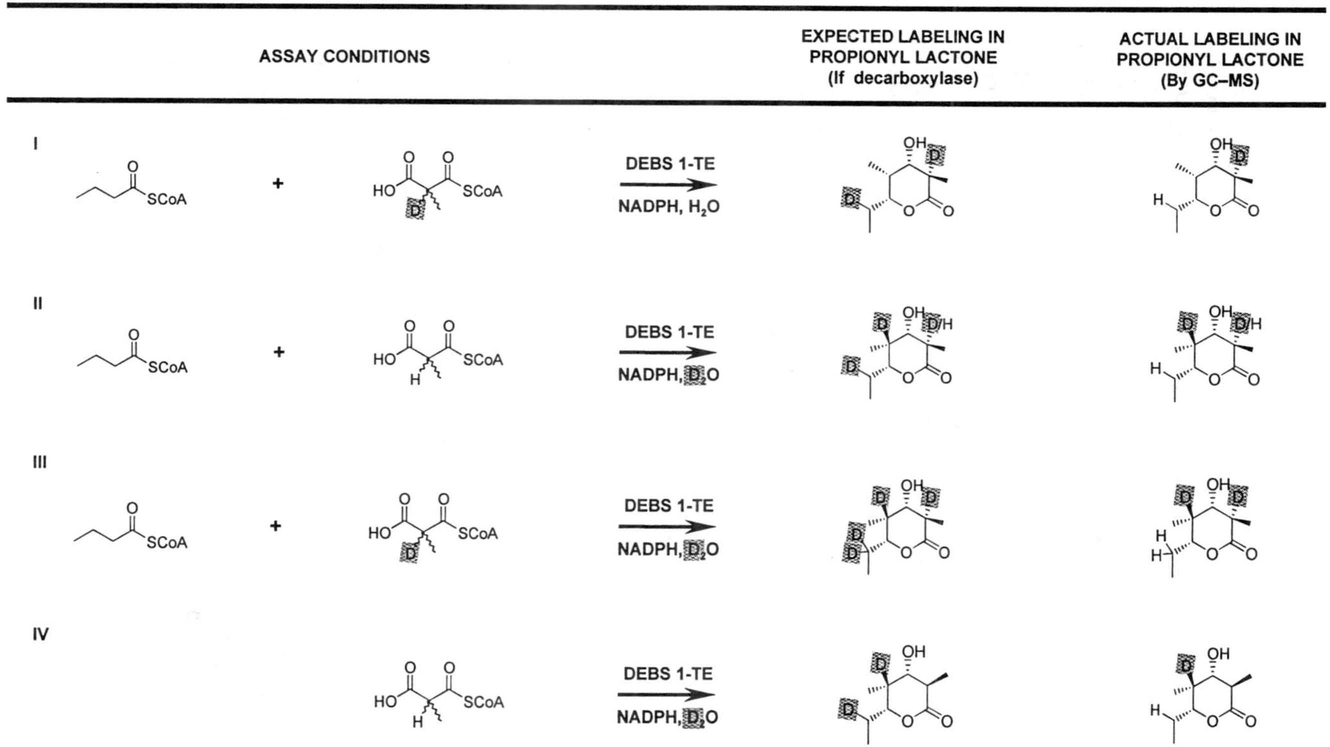

Figure 17 Design of assays to evaluate decarboxylation of methylmalonyl-CoA catalyzed by DEBS 1-TE. It was anticipated that a decarboxylase activity would incorporate deuterium into the starter unit of propionyl lactone and that such labeling would be visible by GC-MS analysis. As decarboxylation has been reported in the presence of primers other than propionyl-CoA, assays I–III included *n*-butyryl-CoA as a starter unit. Assay IV was designed to evaluate the possibility that *n*-butyryl-CoA suppresses decarboxylation. GC-MS analysis gave no evidence for labeling of the side chain in any assay. Therefore, decarboxylation is not a significant reaction of KS_1 under these conditions.

from the pool of extender units (Fig. 16) [34,45]. Such an activity would decrease the efficiency of incorporation of unnatural starter units in place of propionyl-CoA. To evaluate this putative function of the KS domain, studies were carried out *in vitro* with DEBS 1-TE, using deuterium-labeled (2*RS*)-methylmalonyl-CoA or D_2O (Figs. 16 and 17). Under these conditions, deuterium labeling in the starter unit was diagnostic for a decarboxylase activity in the KS. The experiments showed that, in contrast to earlier reports for DEBS 1+TE, DEBS 1−TE does not have a decarboxylase activity. This result bodes well for attempts to further exploit the relaxed specificity of the loading didomain toward analog production *in vitro.*

The use of DEBS 1−TE and DEBS 1+TE as model systems for erythromycin biosynthesis has enabled thorough investigation of the pathway *in vitro.* These studies have yielded significant insight into the complex workings of these multienzymes, and will most certainly aid in the design of novel proteins to produce desired targets both *in vivo* and *in vitro.*

VIII. ACCESSING DIVERSITY IN MODULAR POLYKETIDE BIOSYNTHESIS—THE *IN VIVO* MODEL

The future of *in vitro* ''grocery store'' biosynthesis lies not only in the generation of known natural products, but the production of analogs with potentially improved activities. Experiments aimed at engineering proteins capable of synthesizing novel metabolites have already been carried out with great success. As a comprehensive review is given in Chapter 17, the following account covers just some of the seminal advances in the field.

A. Altering the Starter Residue

An attractive target for manipulation is the starter unit. There are several potential methods for altering the starter acid of a particular polyketide. First, the inherent flexibility of the loading domains may be exploited to introduce ''non-natural'' starter units. As discussed earlier, although the loading didomain of DEBS accepts only propionyl-CoA and acetyl-CoA *in vivo,* it tolerates *n*-butyryl-CoA and the branched starter isobutyryl-CoA *in vitro* (Fig. 11) [36,39]. More remarkably, the loading didomain of the avermectin-producing polyketide synthase recognizes at least 44 different branched carboxylic acids or their derivatives when they are fed to *Streptomyces avermitilis* cultures [46], and this broad specificity has been exploited *in vivo* to produce an avermectin analog, Doramectin, which is more active than the parent macrolide (Fig. 18) [47].

Another strategy involves replacing the loading activities of one PKS with the comparable domains from another synthase. The viability of this approach

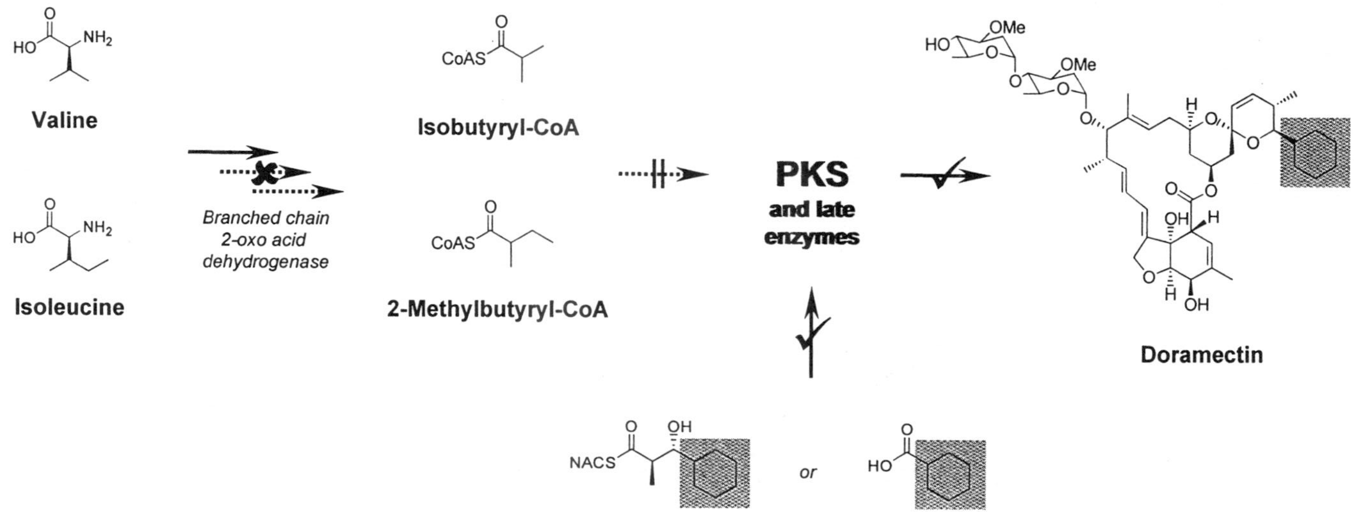

Figure 18 Incorporation of unnatural starter units into avermectins. The broad specificity of the avermectin PKS loading didomain was exploited by eliminating competition from the normal branched chain precursors of avermectin—isobutyryl-CoA and 2-methylbutyryl-CoA—by disabling branched-chain 2-oxo acid dehydrogenase activity. Feeding of cyclohexane carboxylic acid to this mutant avermectin producer resulted in the avermectin analog Doramectin incorporating a cyclohexane unit at the expected position. Administration of a synthetic diketide analog incorporating a cyclohexyl group also resulted in Doramectin, although in lower yields.

was demonstrated for the erythromycin PKS when its loading didomain was swapped with the comparable domains from the avermectin PKS, both in the context of DEBS 1−TE (Fig. 19) and the complete DEBS system (Fig. 20) [48]. The resulting hybrid PKSs produced the expected new compounds, demonstrating that the broad specificity of the avermectin loading didomain can be transferred to another modular polyketide synthase. More recently, the loading didomain of DEBS 1−TE was replaced with the comparable portion of the PKS responsible for biosynthesizing the polyether antibiotic monensin [28]. The hybrid synthase generated an acetate starter lactone, consistent with the change in specificity of the loading domains. Similar changes have been accomplished in other systems (e.g., Ref. [49]). Excitingly, some of the constructs have been improved to the stage where they can be used to produce novel metabolites on the metric ton scale by fermentation [50]. Attempts to replicate these results by modifying a traditional synthetic route would be doomed by the inherently low overall yield.

A third approach is precursor-directed biosynthesis in which altered starter unit functionality is introduced through the incorporation of a synthetic non-natural diketide intermediate into a polyketide product *in vivo*. This strategy was first

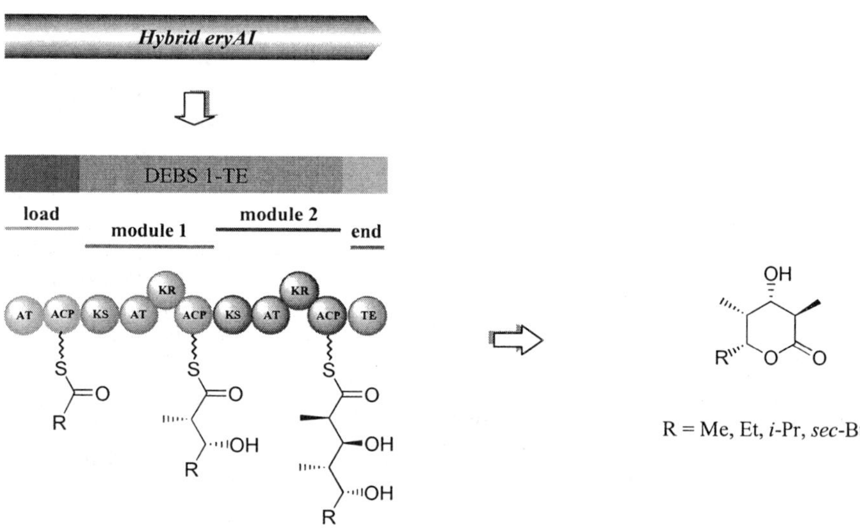

Figure 19 Altering the starter unit specificity of DEBS 1-TE. A hybrid PKS was constructed by replacing the erythromycin loading domains with those of the avermectin PKS. The resulting mini-PKS produced lactones incorporating the natural starter units of erythromycin biosynthesis, acetate and propionate, as well as those characteristic of avermectin, isobutyrate, and 2-methylbutyrate.

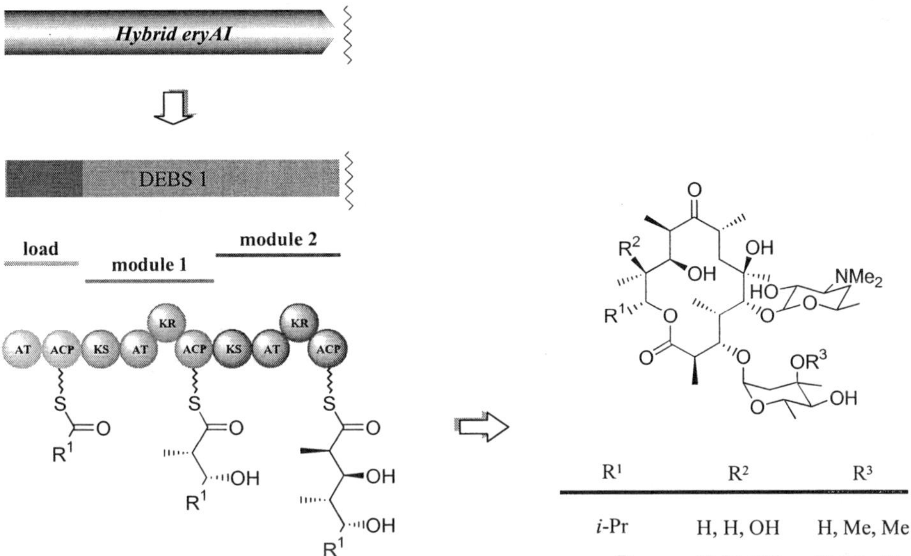

R¹	R²	R³
i-Pr	H, H, OH	H, Me, Me
sec-Bu	H, H, OH	H, Me, Me

Figure 20 Altering the starter unit specificity of DEBS. A hybrid PKS was constructed by introducing the avermectin loading domains in place of those of DEBS. Expression of the mutant strain in *Saccharopolyspora erythraea* resulted in erythromycin analogs incorporating both the expected isobutyrate and 2-methylbutyrate starter units characteristic of avermectin.

used to produce analogs of the avermectins altered in the region of the starter unit (Fig. 18) [51] and more recently has been employed successfully in the generation of similarly modified analogs of 6-dEB and erythromycin [52–54].

The feasibility of precursor-directed biosynthesis *in vitro* has been demonstrated for both the DEBS 1−TE [42,55] and DEBS 1+TE ($KS_1°$) [39,56] systems [DEBS 1+TE ($KS_1°$) is a DEBS 1+TE construct in which the KS domain of module 1 has been inactivated by deletion]. In both cases, the bimodular synthases processed analogs of the diketide intermediate into triketide lactones (Fig. 21). A significant limitation of this approach, however, is that it has so far failed to alter the stereochemistry of chain extension.

B. Altering Chain Extension—AT Swaps

Further targets for diversification are the nature and stereochemistry of chain extension. The choice of chain extension unit (whether malonyl-, methylmalonyl-, ethylmalonyl-, or propylmalonyl-CoA) is made by the acyltransferase (AT) do-

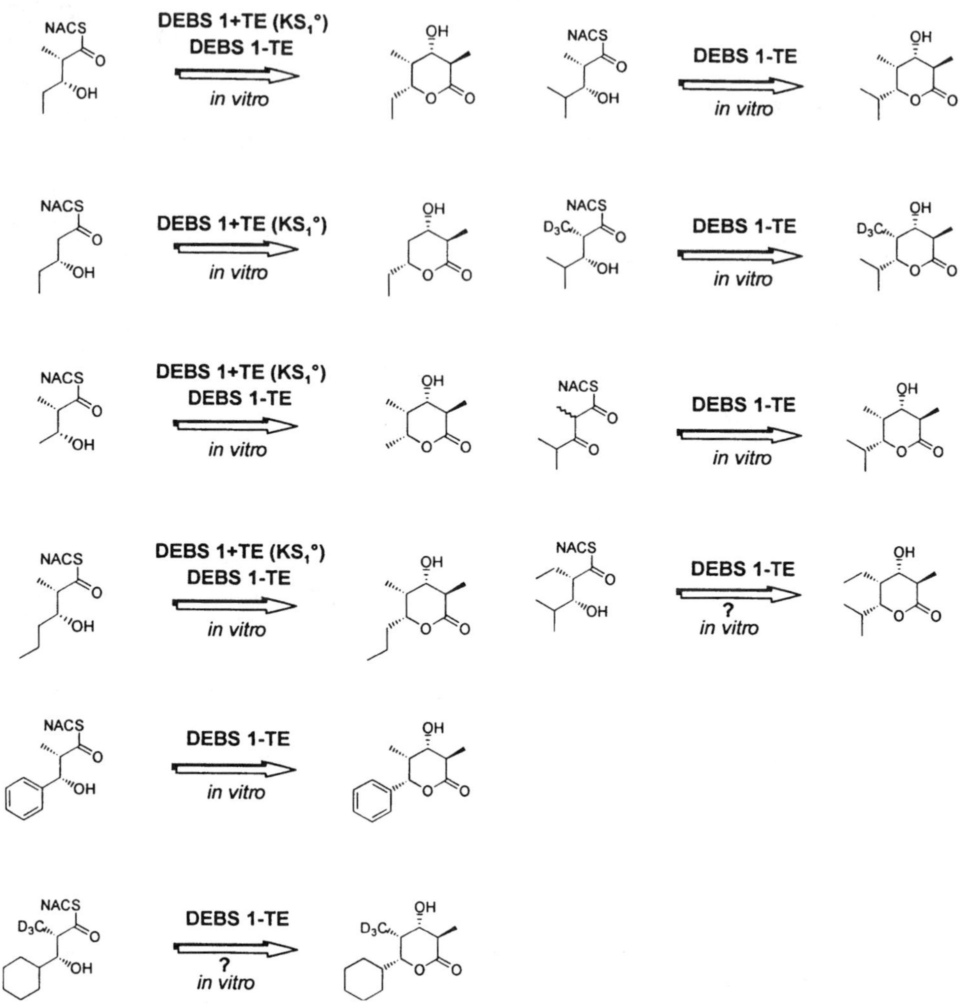

Figure 21 Precursor-directed biosynthesis using bimodular PKSs. Both DEBS 1-TE and DEBS 1+TE ($KS_1°$) were shown to process a range of diketide analogs to the corresponding triketide lactones *in vitro*.

Figure 22 Altering chain extension by DEBS 1. (a) A hybrid PKS was constructed by introducing a malonate-specific AT from RAPS in place of the methylmalonate-specific AT_1 in DEBS 1-TE. The resultant triketide lactones lacked a methyl group at C-4, consistent with the expected change in AT specificity. (b) A hybrid PKS was constructed by introducing a malonate-specific AT from RAPS in place of the methylmalonate-specific AT_2 in DEBS 1+TE. The resultant lactone lacked a methyl group at C-2, consistent with the expected change in AT specificity.

a.

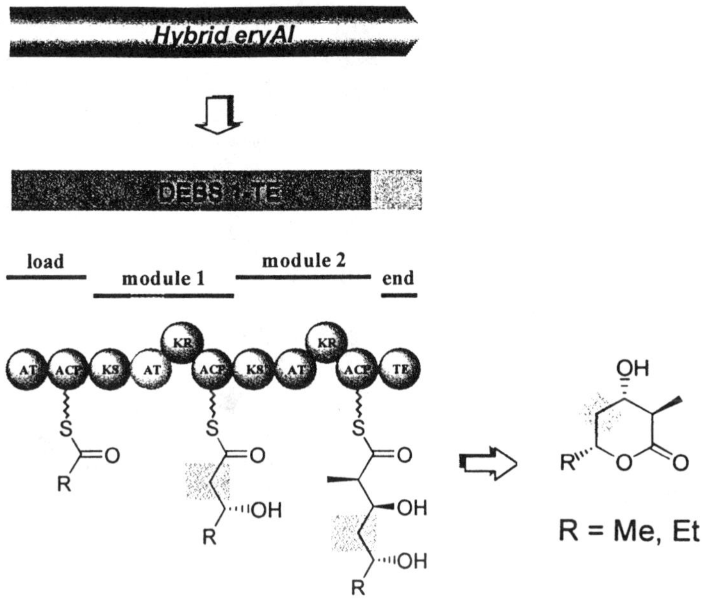
Hybrid eryAI
DEBS 1-TE
load
module 1
module 2
end
AT
ACP
KS
AT
KR
ACP
KS
AT
KR
ACP
TE
R = Me, Et

b.

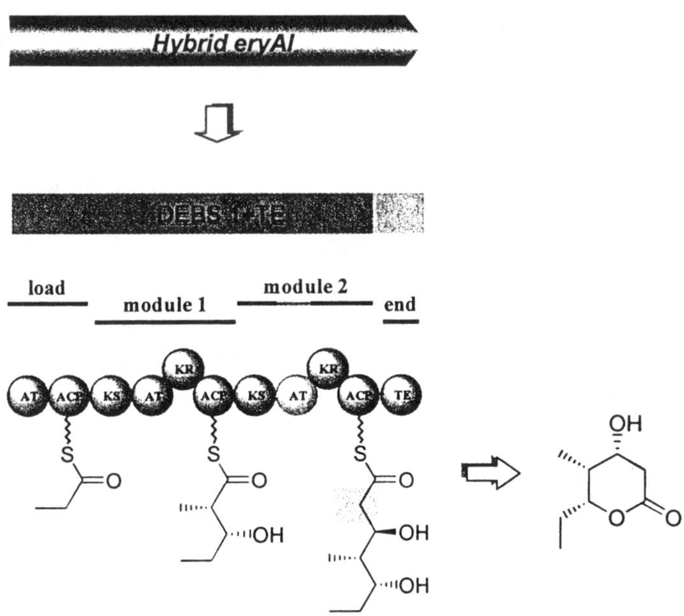
Hybrid eryAI
DEBS 1-TE
load
module 1
module 2
end
AT
ACP
KS
AT
KR
ACP
KS
AT
KR
ACP
TE

mains. Of these domains, malonyl-specific ATs can be distinguished from methylmalonyl-specific ATs on the basis of a particular sequence motif [57,58]. Therefore, chain extension can be altered rationally by selection of an appropriate AT domain. Replacement of an AT domain with one of alternative specificity should produce a modified polyketide, as long as the other activities within the PKS are unaffected and can tolerate the change in functionality.

The first successful demonstration of such specificity swapping was the replacement of the methylmalonate-specific AT of module 1 in DEBS 1−TE with a malonate-specific counterpart from module 2 of the rapamycin (RAPS) PKS (Fig. 22a) [59]. The resulting mutant strain of *S. coelicolor* produced acetate and propionate-derived lactones missing the methyl group at C-4, as expected. In a very similar experiment, the AT of module 2 in DEBS 1+TE was replaced with the same malonate-specific AT from RAPS (Fig. 22b) [58]. Again, the expected lactone was obtained. The clear conclusion from these experiments is that the domain swaps successfully and completely altered the nature of the extension unit recruited by the AT domains.

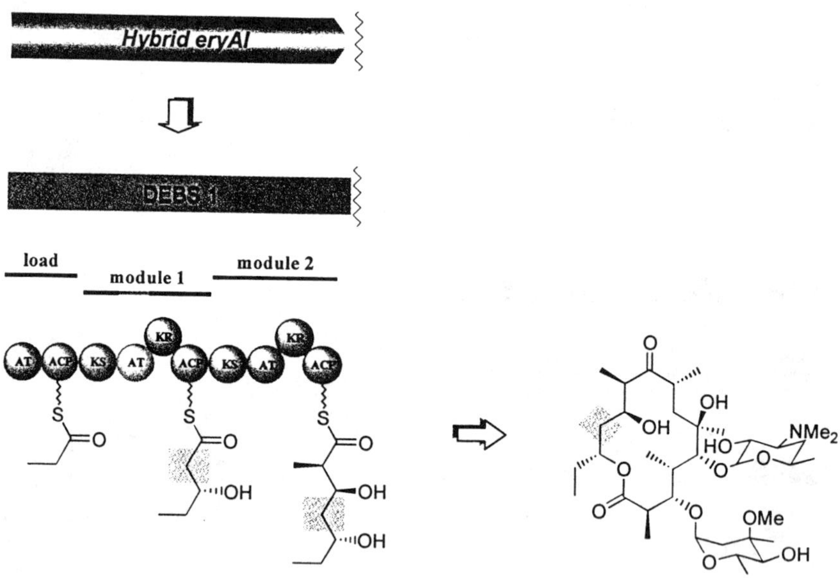

Figure 23 Altering chain extension by DEBS. A hybrid PKS was constructed by introducing malonate-specific ATs in place of the methylmalonate-specific AT_1 of DEBS. The resultant erythromycin analog lacked both the C-12 methyl and C-12 hydroxyl groups of the parent antibiotic.

Similar AT exchanges have also been achieved in the entire DEBS system. In one example, various heterologous AT domains were used to replace both AT_1 and AT_2 of DEBS in *S. erythraea* (Figs. 23 and 24) [60]. Both swap sites resulted in the production of erythromycin analogs lacking methyl groups at the anticipated positions. Finally, the methylmalonate-specific AT_6 of DEBS was replaced with the malonate-specific AT_2 from RAPS (Fig. 25) [61]. The mutant construct, when expressed in *S. coelicolor*, yielded analogs of 6-dEB missing the methyl group at C-2, as expected.

All of these examples involved replacing a methyl group with a sterically less demanding hydrogen. An *increase* in bulk at the C-6 position was achieved by introducing the ethylmalonate-specific AT from the niddamycin PKS [62] in place of DEBS AT_4 (Fig. 26) [63]. When sufficient intracellular pools of ethylmalonate were present, the resulting strain of *S. erythraea* produced a new metabolite, 6-ethylerythromycin A. Evidence for successful ethylmalonate swaps into modules 1–3 of DEBS was also obtained, but the corresponding AT replacements in modules 5 and 6 failed.

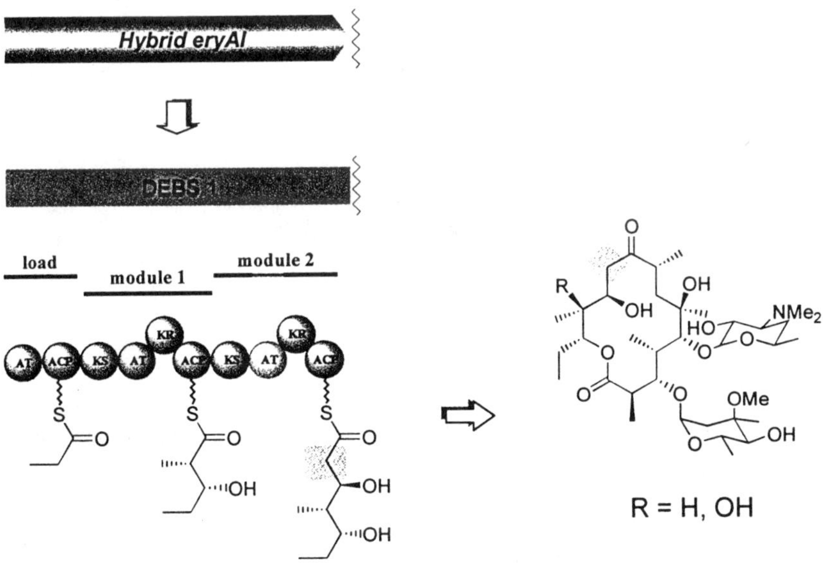

Figure 24 Altering chain extension by DEBS. A hybrid PKS was constructed by introducing malonate-specific ATs from the RAPS and ''Hyg'' clusters in place of the methylmalonate-specific AT_2 of DEBS. As expected, the resultant erythromycin analogs both lacked the C-10 methyl group.

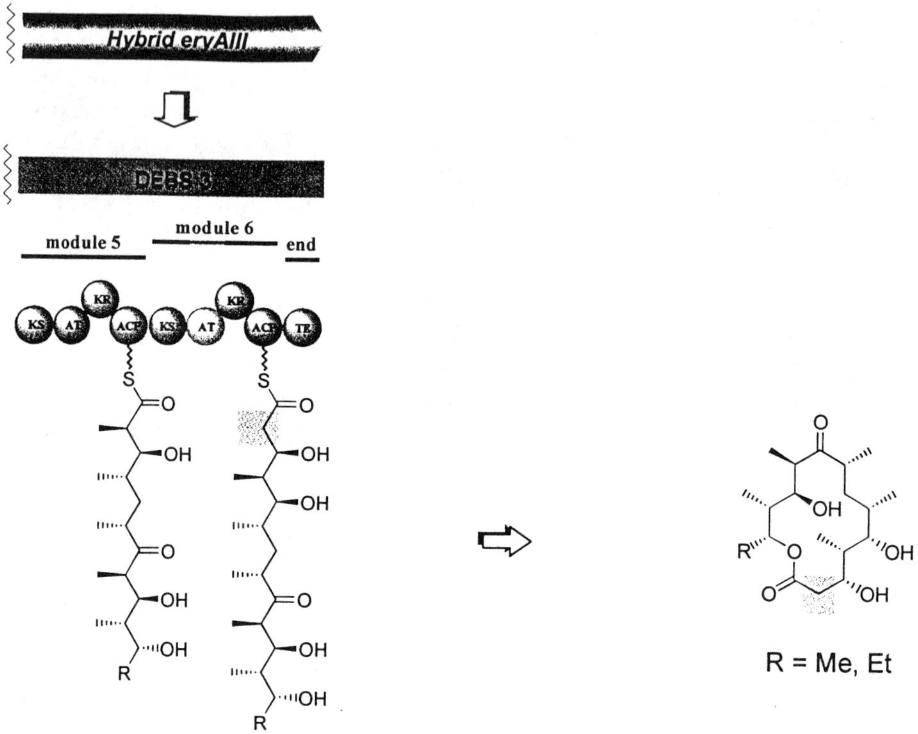

Figure 25 Altering chain extension by DEBS. A hybrid PKS was constructed by introducing a malonate-specific AT from RAPS in place of the methylmalonate-specific AT_6 of DEBS. The resultant 6-deoxyerythronolide B analogs had a C-2 methylene group, as anticipated.

C. Altering Chain Extension—KR Swaps

The ketoreductase domains catalyze the reduction of the β-keto groups established during chain extension. To study the control of the resulting hydroxyl stereochemistry, several KR replacements were engineered [64]. These experiments were carried in the context of another model system of the DEBS PKS, a tetraketide synthase called DEBS 1+module 3+TE, which generates two products (Fig. 27) [65]. The experiments involved replacing the KR of module 2 with KR_5 from DEBS and KR_2 and KR_4 from RAPS (Fig. 27). As expected, the KR_5 replacement yielded the normal products of the tetraketide synthase, as KR_5 in its natural context gives the same hydroxyl stereochemistry as KR_2. In contrast, the constructs containing the KR domains from RAPS both produced triketide

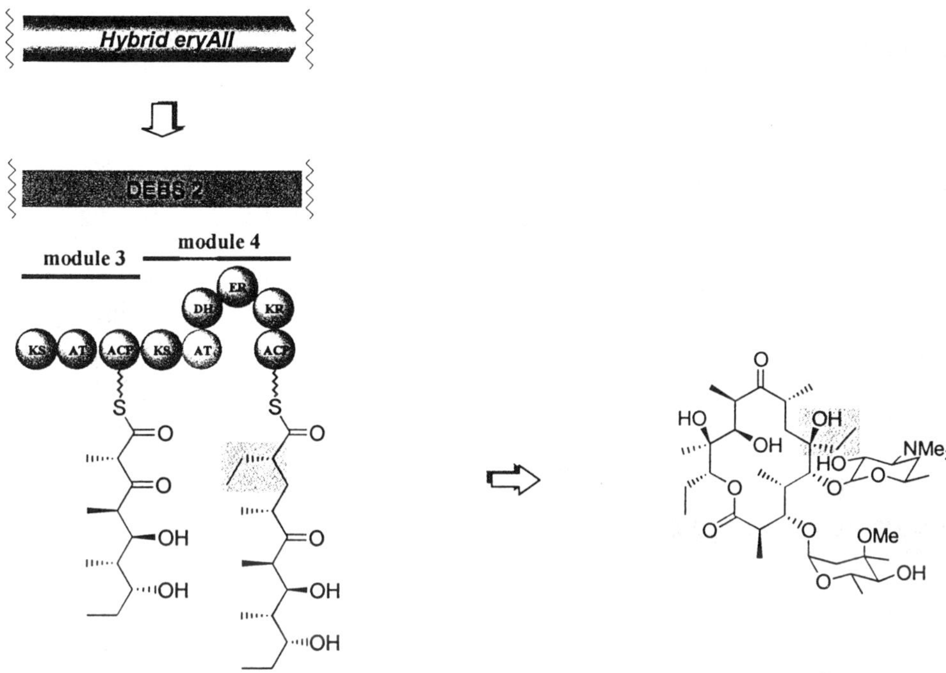

Figure 26 Altering chain extension by DEBS. A hybrid PKS was constructed by introducing an ethylmalonate-specific AT from the niddamycin PKS in place of the methylmalonate-specific AT_4 of DEBS. As expected, the resultant erythromycin analogue incorporated a C-6 ethyl group.

lactone incorporating the unnatural (3*R*)-hydroxyl (compare Figs. 9c and 27). In these experiments, at least, the KR domains were able to control the absolute stereochemistry of ketoreduction even in the presence of unnatural substitution and stereochemistry in their β-keto acyl substrates.

The experiments related here represent only some of the successful reengineering of the erythromycin PKS and other related synthases. In the future, this technology should provide many hybrid proteins for *in vitro* biosynthesis.

IX. CURRENT LIMITATIONS ON THE FUTURE OF BIOSYNTHESIS *IN VITRO*

Recent advances in molecular and structural biology have provided enzymes in large quantity for use in organic synthesis [66]. Thousands of these proteins are

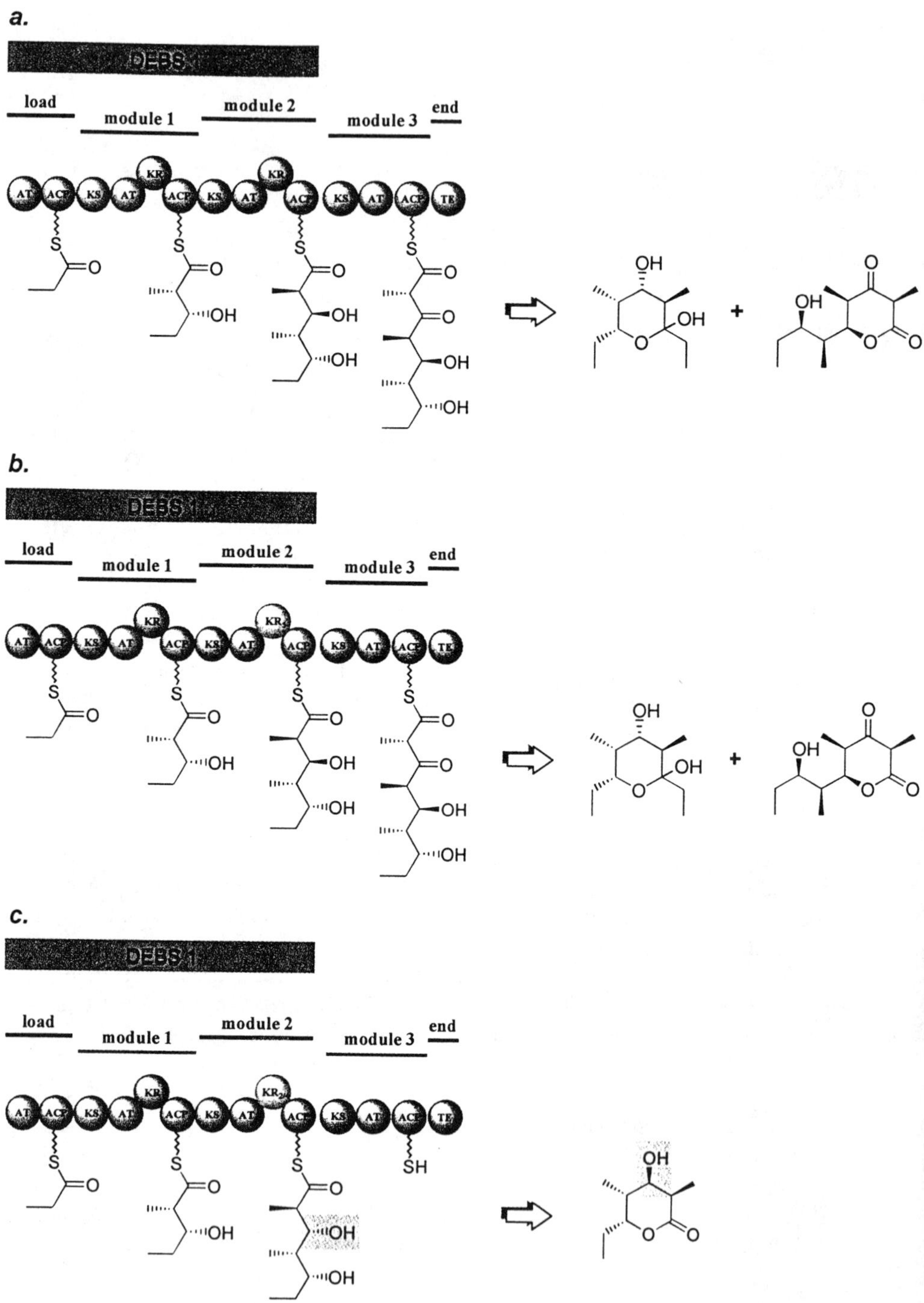
a.
load
module 1
module 2
module 3
end
AT
ACP
KS
KR
TE
OH
O
S
b.
load
module 1
module 2
module 3
end
c.
load
module 1
module 2
module 3
end
SH

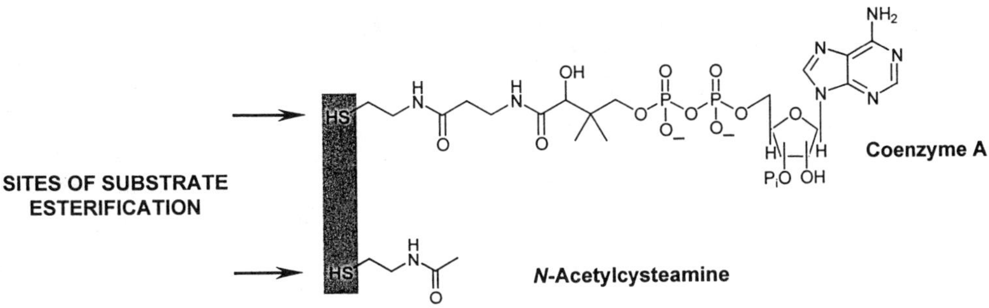

Figure 28 Comparison of the structures of coenzyme A with analog *N*-acetylcysteamine (NAC).

available commercially, and so it is often possible to select a catalyst for a specific transformation [67]. Some of the most generally useful enzymes are the lyases, the aldolases, the oxidoreductases, and the hydrolases [66,68]. The field has been further enhanced by advances in biotechnology such as techniques for the immobilization and stabilization of biocatalysts [69]. In contrast, the use of multi-enzyme polyketide synthases in organic synthesis remains a distant goal.

A. Expense of Reagents

One of the most significant limitations to large-scale *in vitro* synthesis is the expense of starting materials. To prepare 6-dEB, for example, requires NADPH (5 g/£480.50, Sigma), propionyl-CoA (25 mg/£194.40, Sigma), and methylmalonyl-CoA (25 mg/£394.30, Sigma). Clearly, a more economical source of substrates is crucial to the realization of preparative *in vitro* synthesis of polyketides. One strategy for making the process cheaper is to use *N*-acetylcysteamine (NAC) thioesters in place of CoA thioesters [70]. NAC is a good structural mimic of coenzyme A (Fig. 28), and can be purchased for considerably less than CoA (5 g NAC/£40.20 versus 1 g CoA/£665.70, Sigma). However, the use of NAC thioesters requires that they be synthesized by standard chemical methods. Addition-

---

Figure 27 Altering chain extension by DEBS 1+module 3+TE. (a) DEBS 1+module 3+TE makes two products *in vivo*. (b) KR_5 from DEBS was introduced in place of KR_2. The resultant hybrid PKS produced the expected tetraketide products. (c) The KR domains from RAPS modules 2 and 4 were introduced in place of DEBS KR_2. The resultant hybrid PKSs both produced a triketide lactone product incorporating novel (*R*)-hydroxyl stereochemistry at C-3.

ally, NAC compounds are incorporated into polyketides with less efficiency than their corresponding CoA thioesters [71].

A better alternative would be to augment an *in vitro* mixture with enzymes capable of biosynthesizing the required substrates *in situ* from simple, inexpensive precursors. For example, propionyl-CoA and methylmalonyl-CoA might be prepared from propionic acid (500 g/£52) by the enzymes acetyl-CoA synthetase and propionyl-CoA carboxylase [72]. The adenosine triphosphate (ATP) used in the synthesis of these precursors could be regenerated *in situ* by, for example, using phosphoenolpyruvate as the phosphorylating agent [73]. Likewise, the essential cofactor NADPH could be regenerated by a number of methods, including that based on formate dehydrogenase (FDH) [74] or that employing an alcohol dehydrogenase [75].

B. Availability of Proteins

Another barrier to the utility of *in vitro* biosynthesis is low yields of purified proteins. Standard methods of chromatography have typically produced 1–10 mg of pure DEBS 1−TE or DEBS 1+TE from 10 g of cells (1 liter of fermentation medium) [45,76]. However, for an enzyme to be commercially viable, purification must yield tens or hundreds of grams of protein [77]; production on this scale would require thousands of liters of bacteria! An additional challenge is the meager rate of turnover of the purified enzymes. For example, homogeneous DEBS 1−TE has a k_{cat} of ~1 min^{-1} under optimal conditions [42], and purified DEBS 1+TE gives a comparable level of synthesis (3–9 min^{-1}) [34,45]. It must be recognized, however, that growth and purification have been performed only in academic settings and on a small scale. Application of industrial growth conditions and high-producing strains, as well as optimization of purification, could have a revolutionary impact on the scale of synthesis. Until these limitations are addressed, the small quantities and low productivity of these synthases will be a significant impediment to the widespread application of this technique.

C. Structure and Immobilization of Synthases

A final crucial issue is the structure of the polyketide assembly lines. The aromatic PKSs apparently require association of their various subunits for proper functioning. In the absence of the appropriate PKS multidomain architecture, synthesis is generally inefficient and results in truncated metabolites that are subject to variable amounts of spontaneous chemistry [8,10,26]. It remains unclear what minimum complement of activities is required to achieve the natural PKS products formed *in vivo*. A consensus is growing, however, that chain length control is exerted by the PKS complex as a whole through some sort of cleft formed by the association of the domains [19]. As each of the PKS activities may play a crucial role in the structural integrity and functioning of the synthase, it will be essential to define the tolerance of the assembly to the introduction of heterologous domains. If interactions between non-native activities compromise the function of aromatic PKSs, the diversity accessible from these synthases will be significantly limited [4]. Additionally, the multicomponent nature of the synthases will greatly complicate efforts to immobilize them.

To the extent that they are covalently linked, the complex polyketide synthases are less reliant on association for function. In most systems, however, transient docking between multienzymes is required. Recently, it has been demonstrated that the regions of sequence upstream from N-terminal modules and downstream from C-terminal modules (referred to as *inter*polypeptide linkers) play a crucial role in the assembly of functional modules *in vivo* [78]. Clearly, the presence of such linkers will also be important for productive biosynthesis *in vitro* using multiprotein systems.

Alternatively, the requirement for efficient docking could be avoided by covalently linking all of the modules to form a giant synthase. Coupling the proteins together would have the added advantage of allowing the entire PKS to be immobilized on a matrix by a single covalent link. For example, conventional chemistry could be used to form a bond between a reactive residue in the protein and a complementary functionality on the matrix; obvious sites for such an amino acid would be the N- or C-terminal regions of the protein. Alternatively, affinity methods such as the addition to the protein of a six-histidine tag [79] or a maltose-binding protein [80] might be used to tether a mega-PKS to an appropriate matrix. To evaluate the feasibility of these approaches, it is useful to consider current models for the quaternary structure of a modular PKS.

There are currently two proposed structures for the erythromycin PKS, the "side-by-side" model and the helical model (Figs. 29 and 30). One undisputed feature of these models is that each of the DEBS polypeptides forms a homodimer (i.e., DEBS 1 associates with another copy of DEBS 1, and so on). In the side-by-side model (Fig. 29), the modules *within* each subunit are arranged in an *antiparallel*, side-by-side configuration. The resulting flat modules are then stacked one above the other and connected by linker regions running vertically at the outside of the stack. In contrast, in the helical model (Fig. 30), the modules are associated head-to-head and tail-to-tail and twisted together to form a ropelike structure [81]. Interestingly, the helical model predicts that even a single module should be catalytically active in the absence of other PKS components, which has been demonstrated [82], and that linkers or entire modules could be added onto or into the helical stack without significantly disrupting the overall topology of the PKS. Subunits with a helical structure would be amenable to covalent

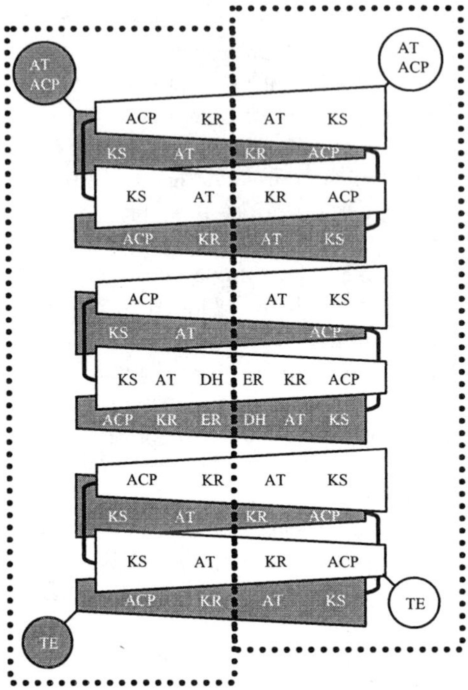

Figure 29 Proposed ''side-by-side'' structure for DEBS. In this model, the modules are arranged head-to-tail and side-by-side. The resulting flat modules are then stacked one on top of the other, and joined by linker regions running down the outside of the stack. Dashed boxes indicate the two catalytic centers—the domains that cooperate to form a polyketide chain.

Figure 30 Proposed ''double helical'' model for DEBS illustrated for modules 3 and 4 (the × on the KR indicates that it is inactive). (a) Mutant complementation studies and chemical cross-linking established two important facts about the operation of the PKS: (1) in going from one module to the next, the polyketide chain is passed from the ACP to the KS residing in the same chain; and (2) during the condensation reaction, the KS of one chain cooperates with the ACP on the complementary chain. Therefore, the growing chain is transferred both *within* the subunits and *between* them at different stages of its passage down the synthase. (b) In the side-by-side topology shown in (a), the KS and ACP domains that participate in chain extension are spatially separated. Twisting the chains together to form a helix brings the cooperating domains into close proximity. The double helical model is therefore in good agreement with the structural data relating to the DEBS PKS.

a.

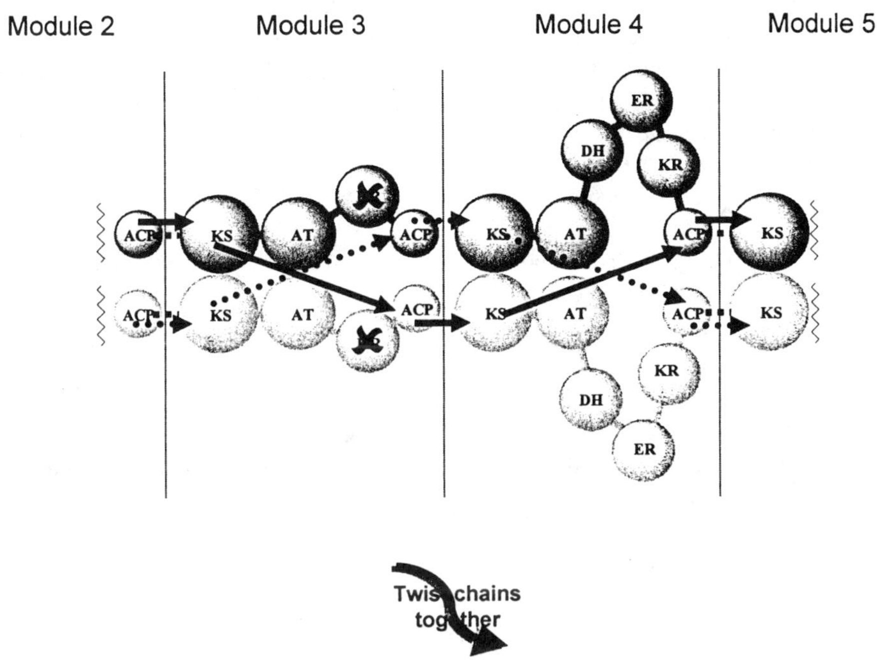
Module 2
Module 3
Module 4
Module 5
ACP
KS
AT
ACP
KS
AT
DH
ER
KR
ACP
KS
ACP
KS
AT
ACP
KS
AT
ACP
KS
KR
DH
ER
Twist chains together

b.

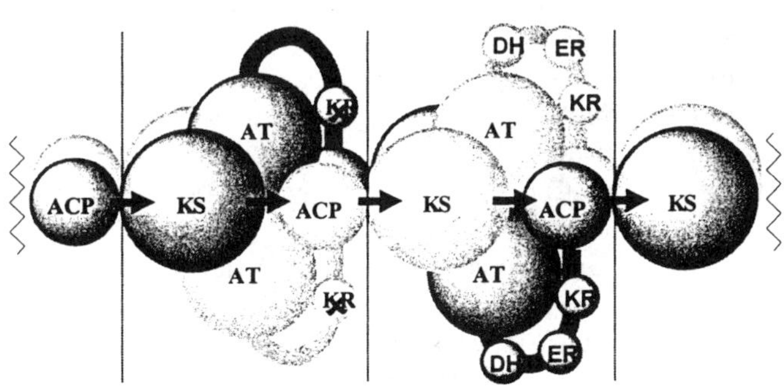
ACP
KS
AT
KR
ACP
AT
KR
KS
AT
DH
ER
KR
ACP
AT
KR
DH
ER
KS

coupling, as well as to the introduction of affinity peptides such as the six-histidine tag.

The most important distinction between the two structures is that only the helical model is compatible with all of the published structural data, including results obtained from limited proteolysis [81,83], chemical cross-linking [83], analytical ultracentrifugation [81], and mutant complementation [84,85]. Although the double helix provides a useful model with good predictive power, it will no doubt be subjected to further investigation by, for example, atomic force microscopy or X-ray crystallography. Such experiments should refine the structural information on PKSs, and point the way toward the productive modification and immobilization of the synthases.

X. OUTLOOK

This chapter has addressed only the generation of the core of polyketide metabolites, for example, the synthesis of 6-deoxyerythronolide B, the aglycone of erythromycin A (Fig. 5). In the living organism, these core molecules are further elaborated by hydroxylations, methylations, oxidations, reductions, and glycosylations to produce the active compounds. The total synthesis of a desired natural product, therefore, would necessitate that these accessory enzymes also be included *in vitro,* along with a suitable supply of their substrates (sugars, methyl group donors, etc.). In principle, there is no barrier to achieving such a mixture, but the current practice is to bypass this requirement by feeding the polyketide intermediates to whole organisms and relying on the enzymes in the cell to carry out the late-stage transformations. This approach is less efficient than carrying out the reactions *in vitro,* but is a satisfactory technique for the time being.

Despite this limitation, *in vitro* biosynthesis using purified biocatalysts promises to be a simple and inexpensive means to access a wealth of complex natural product structures under clean and controlled conditions. The versatility of this technology can be significantly enhanced by the application of genetic engineering to create novel proteins.

The field remains in its infancy, but academic researchers have already succeeded in producing molecules *in vitro* using both native and engineered proteins. While significant obstacles remain to the application of this approach to organic synthesis, the way forward is clear—it is time for researchers to take an active interest in preparing these proteins, not only for investigations into their structure and mechanism, but for use in one-pot total synthesis.

It may not be long before making complex medicines in the test tube is as easy as cooking chicken soup.

REFERENCES

1. RB Woodward, E Logusch, KP Nambiar, K Sakan, D Ward, B-W Au-Yeung, P Balaram, LJ Browne, PJ Card, CH Chen, RB Chênevert, A Fliri, K Frobel, H-J Gais, DG Garrat, K Hayakawa, W Heggie, DP Hesson, D Hoppe, I Hoppe et al. Asymmetric total synthesis of erythromycin. 1. Synthesis of erythronolide A seco acid derivative via asymmetric induction. J Am Chem Soc 103:3210–3213, 1981.
2. RB Woodward, E Logusch, KP Nambiar, K Sakan, D Ward, B-W Au-Yeung, P Balaram, LJ Browne, PJ Card, CH Chen, RB Chênevert, A Fliri, K Frobel, H-J Gais, DG Garrat, K Hayakawa, W Heggie, DP Hesson, D Hoppe, I Hoppe. Asymmetric total synthesis of erythromycin. 2. Synthesis of an erythronolide A lactone system. J Am Chem Soc 103:3213–3215, 1981.
3. RB Woodward, E Logusch, KP Nambiar, K Sakan, D Ward, B-W Au-Yeung, P Balaram, LJ Browne, PJ Card, CH Chen, RB Chênevert, A Fliri, K Frobel, H-J Gais, DG Garrat, K Hayakawa, W Heggie, DP Hesson, D Hoppe, I Hoppe. Asymmetric total synthesis of erythromycin. 3. Total synthesis of erythromycin. J Am Chem Soc 103:3215–3217, 1981.
4. PF Leadlay. Combinatorial approaches to polyketide biosynthesis. Curr Opin Chem Biol 1:162–168, 1997.
5. J Staunton. Combinatorial biosynthesis of erythromycin and complex polyketides. Curr Opin Chem Biol 2:339–345, 1998.
6. DE Cane. Introduction: polyketide and nonribosomal polypeptide biosynthesis. From Collie to *Coli*. Chem Rev 97:2463–2464, 1997.
7. L Katz, S Donadio. Polyketide synthesis: prospects for hybrid antibiotics. Annu Rev Microbiol 875–912, 1993.
8. CW Carreras, R Pieper, C Khosla. Efficient synthesis of aromatic polyketides *in vitro* by the actinorhodin polyketide synthase. J Am Chem Soc 118:5158–5159, 1996.
9. R McDaniel, S Ebert-Khosla, DA Hopwood, C Khosla, C. Rational design of aromatic polyketide natural products by recombinant assembly of enzymatic subunits. Nature 375:549–554, 1995.
10. DA Hopwood. Genetic contributions to understanding polyketide synthase. Chem Rev 97:2465–2497, 1997.
11. J Crosby, DH Sherman, MJ Bibb, WP Revill, DA Hopwood, TJ Simpson. Polyketide synthase acyl carrier proteins from *Streptomyces*. Expression in *Escherichia coli*, purification and partial characterization. Biochim Biophys Acta 1251:32–42, 1995.
12. HC Gramajo, J White, CR Hutchinson, MJ Bibb. Overproduction and localization of components of the polyketide synthase of *Streptomyces glaucescens* involved in the production of the antibiotic tetracenomycin C. J Bacteriol 173:6475–6483, 1991.
13. B Shen, RG Summers, H Gramajo, MJ Bibb, CR Hutchinson. Purification and characterization of the acyl carrier protein of the *Streptomyces glaucescens* tetracenomycin C polyketide synthase. J Bacteriol 174:3818–3821, 1992.
14. MP Crump, J Crosby, CE Dempsey, M Murray, DA Hopwood, TJ Simpson. Conserved secondary structure in the actinorhodin polyketide synthase acyl carrier pro-

tein from *Streptomyces coelicolor* A3(2) and the fatty acid synthase acyl carrier protein from *Escherichia coli*. FEBS Lett 391:302–306, 1996.
15. MP Crump, J Crosby, CE Dempsey, JA Parkinson, M Murray, DA Hopwood, TJ Simpson. Solution structure of the actinorhodin polyketide synthase acyl carrier protein from *Streptomyces coelicolor* A3(2). Biochemistry 36:6000–6008, 1997.
16. B Shen, CR Hutchinson. Enzymatic synthesis of a bacterial polyketide from acetyl and malonyl coenzyme A. Science 262:1535–1540, 1993.
17. W Bao, E Wendt-Pienkowski, CR Hutchinson. Reconstitution of the iterative type II polyketide synthase for tetracenomycin F2 biosynthesis. Biochemistry 37:8132–8138, 1998.
18. CW Carreras, C Khosla. Purification and *in vitro* reconstitution of the essential protein components of an aromatic polyketide synthase. Biochemistry 37:2084–2088, 1998.
19. RJX Zawada, C Khosla. Heterologous expression, purification, reconstitution and kinetic analysis of an extended type II polyketide synthase. Chem Biol 6:607–615, 1999.
20. VB Rajgarhia, WR Strohl. Minimal *Streptomyces* strain C5 daunorubicin polyketide biosynthetic genes required for aklanonic acid biosynthesis. J Bacteriol 179:2690–2696, 1997.
21. B Shen, RG Summers, E Wendt-Pienkowski, CR Hutchinson. The *Streptomyces glaucescens* TcmKL polyketide synthase and TcmN polyketide cyclase genes govern the size and shape of aromatic polyketides. J Am Chem Soc 117:6811–6821, 1995.
22. B Shen, CR Hutchinson. Deciphering the mechanism for the assembly of aromatic polyketides by a bacterial polyketide synthase. Proc Natl Acad Sci (USA) 93:6600–6604, 1996.
23. PJ Kramer, RJX Zawada, R McDaniel, CR Hutchinson, DA Hopwood, C Khosla. Rational design and engineered biosynthesis of a novel 18-carbon aromatic polyketide. J Am Chem Soc 119:635–639, 1997.
24. CJ Tsoi, C Khosla. Combinatorial biosynthesis of ''unnatural'' natural products: the polyketide example. Chem Biol 2:355–362, 1995.
25. C Khosla, RJX Zawada. Generation of polyketide libraries via combinatorial synthesis. Trends Biotechnol 14:335–341, 1996.
26. T-W Yu, Y Shen, R McDaniel, HG Floss, C Khosla, DA Hopwood, BS Moore. Engineered biosynthesis of novel polyketides from *Streptomyces* spore pigment polyketide synthases. J Am Chem Soc 120:7749–7759, 1998.
27. RJX Zawada, C Khosla. Domain analysis of the molecular recognition features of an aromatic polyketide synthase. J Biol Chem 272:16184–16188, 1997.
28. C Bisang, PF Long, J Cortés, J Westcott, J Crosby, A-L Matharu, RJ Cox, TJ Simpson, J Staunton, PF Leadlay. A chain initiation factor common to both modular and aromatic polyketide synthases. Nature 401:502–505, 1999.
29. TS Hitchman, J Crosby, KJ Byrom, RJ Cox, TJ Simpson. Catalytic self-acylation of type II polyketide synthase acyl carrier proteins. Chem Biol 5:35–47, 1998.
30. A-L Matharu, RJ Cox, J Crosby, KJ Byrom, TJ Simpson. MCAT is not required for *in vitro* polyketide synthesis in a minimal actinorhodin polyketide synthase from *Streptomyces coelicolor*. Chem Biol 5:699–712, 1998.

31. P Zhou, G Florova, KA Reynolds. Polyketide synthase acyl carrier protein (ACP) as a substrate and a catalyst for malonyl ACP biosynthesis. Chem Biol 6:577–584, 1999.
32. J Dreier, AN Shah, C Khosla. Kinetic analysis of the actinorhodin aromatic polyketide synthase. J Biol Chem 274:25108–25112, 1999.
33. R Pieper, G Luo, DE Cane, C Khosla. Cell-free synthesis of polyketides by recombinant erythromycin polyketide synthases. Nature 378:263–266, 1995.
34. R Pieper, S Ebert-Khosla, DE Cane, C Khosla. Erythromycin biosynthesis: kinetic studies on a fully active modular polyketide synthase using natural and unnatural substrates. Biochemistry 35:2054–2060, 1996.
35. J Cortés, KEH Wiesmann, GA Roberts, MJB Brown, J Staunton, PF Leadlay. Repositioning of a domain in a modular polyketide synthase to promote specific chain cleavage. Science 268:1487–1489, 1995.
36. KEH Wiesmann, J Cortés, MJB Brown, AL Cutter, J Staunton, PF Leadlay. Polyketide synthesis *in vitro* on a modular polyketide synthase. Chem Biol 2:583–589, 1995.
37. CM Kao, G Luo, L Katz, DE Cane, C Khosla. Manipulation of macrolide ring size by directed mutagenesis of a modular polyketide synthase. J Am Chem Soc 117: 9105–9106, 1995.
38. G D'Agnolo, IS Rosenfeld, J Awaya, S Omura, PR Vagelos. Inhibition of fatty acid synthesis by the antibiotic cerulenin; specific inactivation of β-ketoacyl-acyl carrier protein synthase. Biochem Biophys Acta 236:155–166, 1973.
39. R Pieper, G Luo, DE Cane, C Khosla. Remarkably broad substrate specificity of a modular polyketide synthase in a cell-free system. J Am Chem Soc 117:11373–11374, 1995.
40. WD Celmer. Macrolide stereochemistry. II. Configurational assignment at certain centers in various macrolide antibiotics. J Am Chem Soc 87:1799–1800, 1965.
41. WD Celmer. Macrolide stereochemistry. III. A configurational model for macrolide antibiotics. J Am Chem Soc 87:1801–1802, 1965.
42. KJ Weissman, M Timoney, M Bycroft, P Grice, U Hanefeld, J Staunton, PF Leadlay. The molecular basis of Celmer's rules: the stereochemistry of the condensation step in chain extension on the erythromycin polyketide synthase. Biochemistry 36: 13849–13855, 1997.
43. IE Holzbaur, RC Harris, M Bycroft, J Cortés, C Bisang, J Staunton, BAM Rudd, PF Leadlay. Molecular basis of Celmer's rules: the role of two ketoreductase domains in the control of chirality by the erythromycin modular polyketide synthase. Chem Biol 6:189–195, 1999.
44. KJ Weissman, M Bycroft, J Staunton, PF Leadlay. Origin of starter units for erythromycin biosynthesis. Biochemistry 37:11012–11017, 1998.
45. R Pieper, RS Gokhale, G Luo, DE Cane, C Khosla. Purification and characterization of bimodular and trimodular derivatives of the erythromycin polyketide synthase. Biochemistry 36:1846–1851, 1997.
46. CJ Dutton, SP Gibson, AC Goudie, KS Holdom, MS Pacey, JC Ruddock. Novel avermectins produced by mutational biosynthesis. J Antibiot 44:357–365, 1991.
47. W Traeder. Pharmacological characteristics of Doramectin—a new macrocyclic lactone derivative of the group of avermectins. Tierartzl Umsch 49:465–469, 1994.

48. AFA Marsden, B Wilkinson, J Cortés, NJ Dunster, J Staunton, PF Leadlay. Engineering broader specificity into an antibiotic-producing polyketide synthase. Science 279:199–202, 1998.
49. S Kuhstoss, M Huber, JR Turner, JW Paschal, N Rao. Production of a novel polyketide through the construction of a hybrid polyketide synthase. Gene 183:231–236, 1996.
50. MS Pacey, JP Dirlam, RW Geldart, PF Leadlay, HAI McArthur, EL McCormick, RA Monday, TN O'Connell, J Staunton, TJ Winchester. Novel erythromycins from a recombinant *Saccharopolyspora erythraea* strain NRRL 2338 pIG1. I. Fermentation, isolation and biological activity. J Antibiot 51:1029–1034, 1998.
51. CJ Dutton, AM Hooper, PF Leadlay, J Staunton. Avermectin biosynthesis. Intact incorporation of a diketide chain-assembly intermediate into the polyketide macrocylic ring. Tetrahedron Lett 35:327–330, 1994.
52. D Hunziker, N Wu, DE Cane, C Khosla. Precursor directed biosynthesis of novel 6-deoxyerythronolide B analogs. Tetrahedron Lett 40:635–638, 1999.
53. JR Jacobsen, CR Hutchinson, DE Cane, C Khosla. Precursor-directed biosynthesis of erythromycin analogs by an engineered polyketide synthase. Science 277:367–369, 1997.
54. JR Jacobsen, AT Keatinge-Clay, DE Cane, C Khosla. Precursor-directed biosynthesis of 12-ethyl erythromycin. Bioorg Med Chem 6:1171–1177, 1998.
55. KJ Weissman, M Bycroft, AL Cutter, U Hanefeld, EJ Frost, MC Timoney, R Harris, S Handa, M Roddis, J Staunton, PF Leadlay. Evaluating precursor-directed biosynthesis towards novel erythromycins through *in vitro* studies on a bimodular polyketide synthase. Chem Biol 5:743–754, 1998.
56. J-A Chuck, M McPherson, H Huang, JR Jacobsen, C Khosla, DE Cane. Molecular recognition of diketide substrates by a β-keto-acyl carrier protein synthase domain within a bimodular polyketide synthase. Chem Biol 4:757–766, 1997.
57. SF Haydock, JF Aparicio, I Molnár, T Schwecke, LE Khaw, A König, AFA Marsden, IS Galloway, J Staunton, PF Leadlay. Divergent sequence motifs correlated with the substrate specificity of (methyl)malonyl-CoA:acyl carrier protein transacylase domains in modular polyketide synthases. FEBS Lett 374:246–248, 1995.
58. J Lau, H Fu, DE Cane, C Khosla. Dissecting the role of acyltransferase domains of modular polyketide synthases in the choice and stereochemical fate of extender units. Biochemistry 38:1643–1651, 1999.
59. M Oliynyk, MJB Brown, J Cortés, J Staunton, PF Leadlay. A hybrid modular polyketide synthase obtained by domain swapping. Chem Biol 3:833–839, 1996.
60. X Ruan, A Pereda, DL Stassi, D Zeidner, RG Summers, M Jackson, A Shivakumar, S Kakavas, MJ Staver, S Donadio, L Katz. Acyltransferase domain substitutions in erythromycin polyketide synthase yield novel erythromycin derivatives. J Bacteriol 179:6416–6425, 1997.
61. L Liu, A Thamchaipenet, H Fu, M Betlach, G Ashley. Biosynthesis of 2-nor-6-deoxyerythronolide B by rationally designed domain substitution. J Am Chem Soc 119:10553–10554, 1997.
62. SJ Kakavas, L Katz, D Stassi. Identification and characterization of the niddamycin polyketide synthase genes from *Streptomyces caelestis* (NRRL-2821). J Bacteriol 179:7515–7522, 1997.

63. DL Stassi, SJ Kakavas, KA Reynolds, G Gunawardana, S Swanson, D Zeidner, M Jackson, H Liu, A Buko, L Katz. Ethyl-substituted erythromycin derivatives produced by directed metabolic engineering. Proc Natl Acad Sci (USA) 95:7305–7309, 1998.
64. CM Kao, M McPherson, R McDaniel, H Fu, DE Cane, C Khosla. Alcohol stereochemistry in polyketide backbones is controlled by the β-ketoreductase domains of modular polyketide synthases. J Am Chem Soc 120:2478–2479, 1998.
65. CM Kao, G Luo, L Katz, DE Cane, C Khosla. Engineered biosynthesis of structurally diverse tetraketides by a trimodular polyketide synthase. J Am Chem Soc 118: 9184–9185, 1996.
66. W-D Fessner. Enzyme mediated C-C bond formation. Curr Opin Chem Biol 2:85–97, 1998.
67. Mosbach, K. Immobilized enzymes in organic synthesis. In: Enzymes in Organic Synthesis. London: Pitman Publishing 1985, pp 57–70.
68. JB Jones. An illustrative example of a synthetically useful enzyme: horse liver alcohol dehydrogenase. In: R Porter, S Clark eds. Enzymes in Organic Synthesis. London: Pitman Publishing, 1985, pp 3–21.
69. CR Johnson, GW Wells. Organic synthesis using biocatalytically generated intermediates. Curr Opin Chem Biol 2:70–76, 1998.
70. NL Pohl, RS Gokhale, DE Cane, C Khosla. Synthesis and incorporation of an *N*-acetylcysteamine analogue of methylmalonyl-CoA by a modular polyketide synthase. J Am Chem Soc 120:11206–11207, 1998.
71. H Wong, JS Mattick, SJ Wakil. The architecture of the animal fatty acid synthetase. III. Isolation and characterisation of β-ketoacyl reductase. J Biol Chem 258:15305–15311, 1983.
72. S Donadio, MJ Staver, L Katz. Erythromycin production in *Saccharopolyspora erythraea* does not require a functional propionyl-CoA carboxylase. Mol Microbiol 19:977–984, 1996.
73. GM Whitesides. Applications of cell-free enzymes in organic synthesis. In: Enzymes in Organic Synthesis. London: Pitman Publishing, 1985, pp 76–96.
74. K Seelbach, B Riebel, W Hummel, MR Kula, VI Tishkov, AM Egorov, C Wandrey, U Kragl. A novel, efficient regenerating method of NADPH using a new formate dehydrogenase. Tetrahedron Lett 37:1377–1380, 1996.
75. CW Bradshaw, W Hummel, C-H Wong. *Lactobacillus kefir* alcohol dehydrogenase: a useful catalyst for synthesis. J Org Chem 57:1532–1536, 1992.
76. KJ Weissman. PhD dissertation, University of Cambridge, Cambridge, UK, 1999.
77. MD Scawen. Large scale purification of enzymes. In: Enzymes in Organic Synthesis. London: Pitman Publishing, 1985, pp 40–56.
78. RS Gokhale, S Tsuji, DE Cane, C Khosla. Dissecting and exploiting intermodular communications in polyketide synthases. Science 284:482–485, 1999.
79. MW Vandyke, M Sirito, M Sawadogo. Single-step purification of bacterially expressed polypeptides containing an oligo-histidine domain. Gene 111:99–104, 1992.
80. CV Maina, PD Riggs, AG Grandea, BE Slatko, LS Moran, JA Tagliamonte, LA McReynolds, C Diguan. An *Escherichia coli* vector to express and purify foreign proteins by fusion to and separation from maltose-binding protein. Gene 74:365–373, 1988.

81. J Staunton, P Caffrey, JF Aparicio, GA Roberts, SS Bethell, PF Leadlay. Evidence for a double-helical structure for modular polyketide synthases. Nature Struct Biol 3:188–192, 1996.
82. I Böhm, IE Holzbaur, U Hanefeld, J Cortés, J Staunton, PF Leadlay. Engineering of a minimal modular polyketide synthase, and targeted alteration of the stereospecificity of polyketide chain extension. Chem Biol 5:407–412, 1998.
83. JF Aparicio, P Caffrey, AFA Marsden, J Staunton, PF Leadlay. Limited proteolysis and active site studies of the first multienzyme component of the erythromycin-producing polyketide synthase. J Biol Chem 269:8524–8528, 1994.
84. CM Kao, R Pieper, DE Cane, C Khosla. Evidence for two catalytically independent clusters of active sites in a functional modular polyketide synthase. Biochemistry 35:12363–12368, 1996.
85. RS Gokhale, J Lau, DE Cane, C Khosla. Functional orientation of the acyltransferase domain in a module of the erythromycin polyketide synthase. Biochemistry 37: 2524–2528, 1998.

19

Enzymatic Synthesis of Fungal *N*-Methylated Cyclopeptides and Depsipeptides

Mirko Glinski, Till Hornbogen, and Rainer Zocher
Technical University Berlin, Berlin, Germany

I. INTRODUCTION

Bacteria and fungi produce numerous low-molecular-weight peptides, depsipeptides, peptidolactones, and lipopeptides, with linear, branched, or cyclic structures as secondary metabolites. Among these substances are many important pharmaceuticals, veterinary agents, and agrochemicals [1]. The enormous structural diversity and complexity of these biomolecules is impressive. Although the actual biological roles of each of these metabolites in the producing organisms are still unclear, it was speculated rather early that the mechanism of synthesis must differ from that of protein synthesis, since the vast majority of microbial peptides often contain unusual amino acids that are not found in proteins. Many of the nonproteinogenic amino acids have unique structures with respect to their carbon skeletons, presence of double bonds, unusual functional groups or D-configuration at the α-carbon atom. The amino acid constituents can also undergo extensive modifications, including *N*-methylation, acylation, glycosylation, and covalent linkage to other functional groups [2]. In general, the biosynthesis of these compounds is accomplished nonribosomally by large multifunctional enzymes called nonribosomal peptide synthetases, with molecular weights in the range from 100 to 1700 kDa [2]. Each of these peptide synthetases functions as a cellular factory representing the protein template for the construction of one defined peptide structure. Peptide synthetases are organized into coordinated groups of active sites termed modules, each encompassing a length of approximately 1000 amino acids and corresponding to a molecular mass equivalent of 120 kDa. Each module is responsible for catalysis of one cycle of polypeptide chain elongation and associated functional group modifications [3,4]. The modules show considerable sequence conservation to each other, and the number of these units equals the number of the amino acids serving as substrates of the corresponding enzymes [5,6].

N-methylated peptides such as cyclosporins and enniatins constitute a class of pharmacologically interesting compounds. They are synthesized by a special class of enzymes representing hybrid systems of peptide synthetases and integrated *N*-methyltransferases [3,4]. The latter constitute a new family of transmethylases sharing high homology within prokaryotes and eukaryotes [7].

Like other peptide synthetases, *N*-methylcyclopeptide synthetases follow a so-called thioltemplate mechanism, in which the substrate amino acids are activated by the corresponding modules as acyladenylates and then are thioesterified through covalent interaction with specific thiol groups of the multienzyme (thioltemplates) [8]. Thioester formation is mediated by enzyme-bound 4′-phosphopantetheine residues [9,10], which are covalently attached to the C-terminal region of each amino acid-activating module. At this stage the modification (*N*-methylation or epimerization) of the substrate amino acid occurs, followed by step-by-step linkage of the activated substrates (elongation reaction) to give a specific peptide chain with a defined sequence [3,11]. The completed peptide chain is covalently attached to the multienzyme as a thioester, and has to be removed by cyclization, amidation, or hydrolysis.

II. ENNIATINS AND BEAUVERICIN

Enniatins belong to the class of *N*-methylated cyclopeptides which are produced by various strains of the genus *Fusarium* [12]. As shown in Figure 1, enniatins consist of alternating residues of D-2-hydroxyisovaleric acid (D-Hiv) and a branched chain *N*-methyl-L-amino acid, linked by peptide and ester bonds. In the case of beauvericin the branched chain L-amino acid is substituted by L-phenylalanine (Fig. 1).

Enniatins exert antibiotic activities against various bacteria, exhibit immunomodulatory properties [13], and are potent inhibitors of mammalian cholesterol acyl transferase [14]. Besides, enniatins are well known for their behavior as ionophores with high specificity for sodium and potassium ions [15]. Toxin-producing *Fusaria* are non-host-specific plant pathogens [16]. Enniatins are postulated to play a role as wilt toxins during *Fusarium* infections of plants [17]. Enniatin production contributes to the virulence of *Fusarium avenaceum*, which contains an *esyn1* homologous gene [18,19]. Interestingly, enniatins, like the structurally related cyclodepsipeptides beauvericin and bassianolide, exhibit entomopathogenic properties [20].

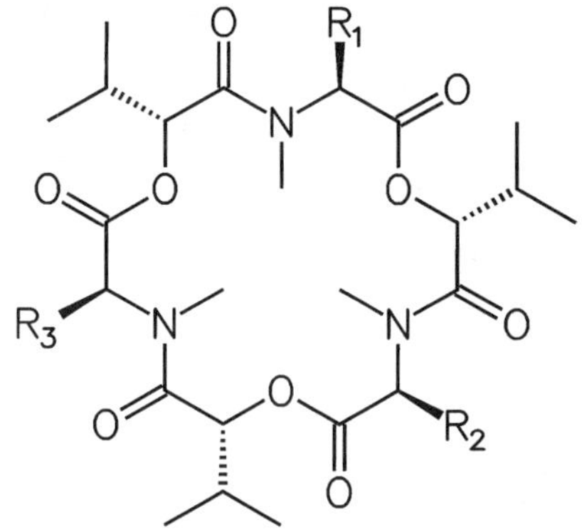

Figure 1 Structures of enniatins and beauvericin. Enniatin A, $R_1 = R_2 = R_3 =$ *sec*-butyl; enniatin A1, $R_1 =$ isopropyl, $R_2 = R_3 =$ *sec*-butyl; enniatin B, $R_1 = R_2 = R_3 =$ isopropyl; enniatin B1, $R_1 = R_2 =$ isopropyl, $R_3 =$ *sec*-butyl; beauvericin, $R_1 = R_2 = R_3$ = benzyl.

A. Enniatin Synthetase: Structure and Function

The biosynthesis of the depsipeptide enniatin proceeds via the thioltemplate mechanism on the multifunctional enzyme enniatin synthetase (Esyn) [21,22]. This enzyme is expressed constitutively during fermentative growth of the producer *Fusarium scirpi* [23]. Esyn was the first *N*-methylcyclopeptide synthetase to be characterized [21,22]. The Esyn corresponding gene (*esyn1*) from *F. scirpi* encoding an open reading frame of 9393 bp has been isolated, sequenced, and further characterized by heterologous expression in this laboratory (Fig. 2A) [7,24,25]. These studies revealed that the enzyme is a single polypeptide chain of 347 kDa [24]. After disruption of the *esyn1* homologous gene locus in *F. avenaceum*, no enniatin biosynthesis could be detected, indicating that indeed the *esyn1* gene is responsible for enniatin formation [18].

Esyn contains all catalytic functions necessary for the synthesis of enniatins from the primary precursors D-Hiv and branched-chain L-amino acid such as leucine, isoleucine, or valine, along with S-adenosyl-L-methionine (AdoMet) in the presence of adenosine 5′-triphosphate (ATP) [21,26]. AdoMet donates the methyl group in the *N*-methylated peptide bonds. 4′-Phosphopantetheine acts as the cofactor to facilitate the ordered shift of carboxy thioester-activated substrates between single modules. Sequence analysis of the *esyn1* gene indicated that the enzyme consists of two modules (EA and EB), which are each approximately 420 amino acid residues in size. Both modules show a high degree of similarity (39.5%) and identity (25%) to each other [24] and to other multifunctional peptide synthetases (e.g., cyclosporin synthetase (Cysyn [3]) and adenylate-forming en-

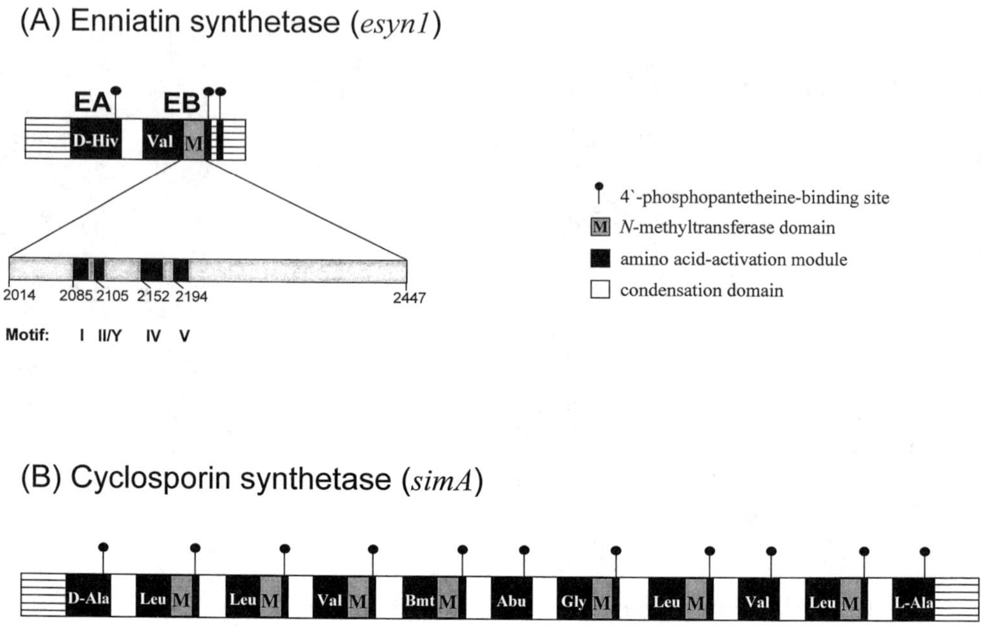

Figure 2 Order and organization of enniatin synthetase and cyclosporin synthetase as deduced from gene sequence and biochemical characterization. Symbols in the adenylate-forming modules (black boxes) indicate the corresponding activated amino acids. M stands for *N*-methyltransferase domain. Condensation domains are represented by white boxes. (A) Top: Structure of enniatin synthetase. EA represents the D-Hiv-activating module; EB represents the L-valine-activating module; D-Hiv is D-2-hydroxyisovaleric acid. Bottom: Structural features of the wild-type *N*-methyltransferase domain M of *esynl*. The black boxes indicate conserved motifs which can be found within methyltransferases and *N*-methyltransferase domains of peptide synthetases (see also Fig. 3). The numbers indicate the amino acid position in the sequence of Esyn. (B) Structure of cyclosporin synthetase. Abu = L-α-aminobutyric acid; Bmt = (4*R*)-4-[(E)-2-butenyl]-4-methyl-L-threonine.

zymes such as firefly luciferase [27] and 4-coumarate ligase [28], with the exception of an insertion of 434 amino acid residues representing the integrated *N*-methyltransferase domain M [24]. A putative condensation domain between EA and EB catalyzes the condensation of the covalently bound substrates to form the various intermediates.

Monoclonal antibodies directed to the multienzyme Esyn were used to map the catalytic sites of the enzyme [29]. The antibodies could be divided into three groups based on their influence on catalytic functions. Members of group one exclusively inhibited L-valine thioester formation, while members of group two interfered with D-Hiv thioester formation. Antibodies of group three inhibited both L-valine thioester and D-Hiv thioester formation as well as the *N*-methyltransferase. From these results it was concluded that the two modules of Esyn containing the two catalytic binding sites are situated close to each other in the three-dimensional structure of the enzyme. Module EA, located at the N-terminal end of the enzyme, is responsible for the activation and binding of D-Hiv, whereas the C-terminal module EB represents the L-amino acid-activating module [24,25,30].

The existence of two different sites in Esyn responsible for binding D-Hiv and the amino acid has also been deduced from inhibition studies using iodoacetamide and isovaleric acid [21,22]. These findings are in good agreement with results obtained from proteolytic digests of Esyn and expression studies of subclones derived from *esyn1* [24,30]. This result is also confirmed by the predicted Esyn protein sequence [25].

The sequential order of the amino acid-activating modules and their accompanying domains in peptide synthetase systems as deduced from DNA sequences of the corresponding genes is also shown in the case of the Cysyn corresponding gene (*simA*) (Fig. 2B).

B. Molecular Structure of the *N*-Methyltransferase Domain

Amino acid sequence comparison of the *N*-methyltransferase domain of enniatin synthetase with other methyltransferases showed no significant sequence similarities except for the conserved motif ^{2085}VLEIGTGSGMIL (motif I) (Fig. 3A) [7,24]. A conserved phenylalanine residue in the glycine-rich motif I was found to be crucial in positioning the adenine ring of AdoMet to the *Hha*I DNA methyltransferase [31]. The corresponding position of enniatin synthetases is occupied by the first glycine of motif I (position 2089) [24]. In the case of γ-DNA methyltransferases, the phenylalanine residue in motif I is also substituted [32]. A structure-guided sequence alignment using PSI-BLAST and the 3D-PSSM Web-based method [33,34] revealed additional conserved motifs (II, IV, and V) that were not obvious from simple alignments (Figs. 3A and 3B). This result permits tertiary structure prediction of the *N*-methyltransferase domain of enniatin synthetase and other AdoMet-dependent methyltransferases, e.g., M.*Taq*I methyltransferase and glycine *N*-methyltransferase [35,36], from their amino acid sequences. Both enniatin synthetases and γ-DNA methyltransferases contain a conserved phenylalanine in motif V downstream of motif IV. The phenylalanine in motif V of γ-DNA methyltransferases makes van der Waals contact to the AdoMet adenine [37]. Kagan and Clarke [38] identified two motifs (II, G/P T/Q Y/F D/A V/I I/F, corresponding to motif IV in DNA methyltransferase nomenclature, and III, L K/R PGGXL) conserved in the sequences of AdoMet-dependent methyltransferases that methylate small molecules.

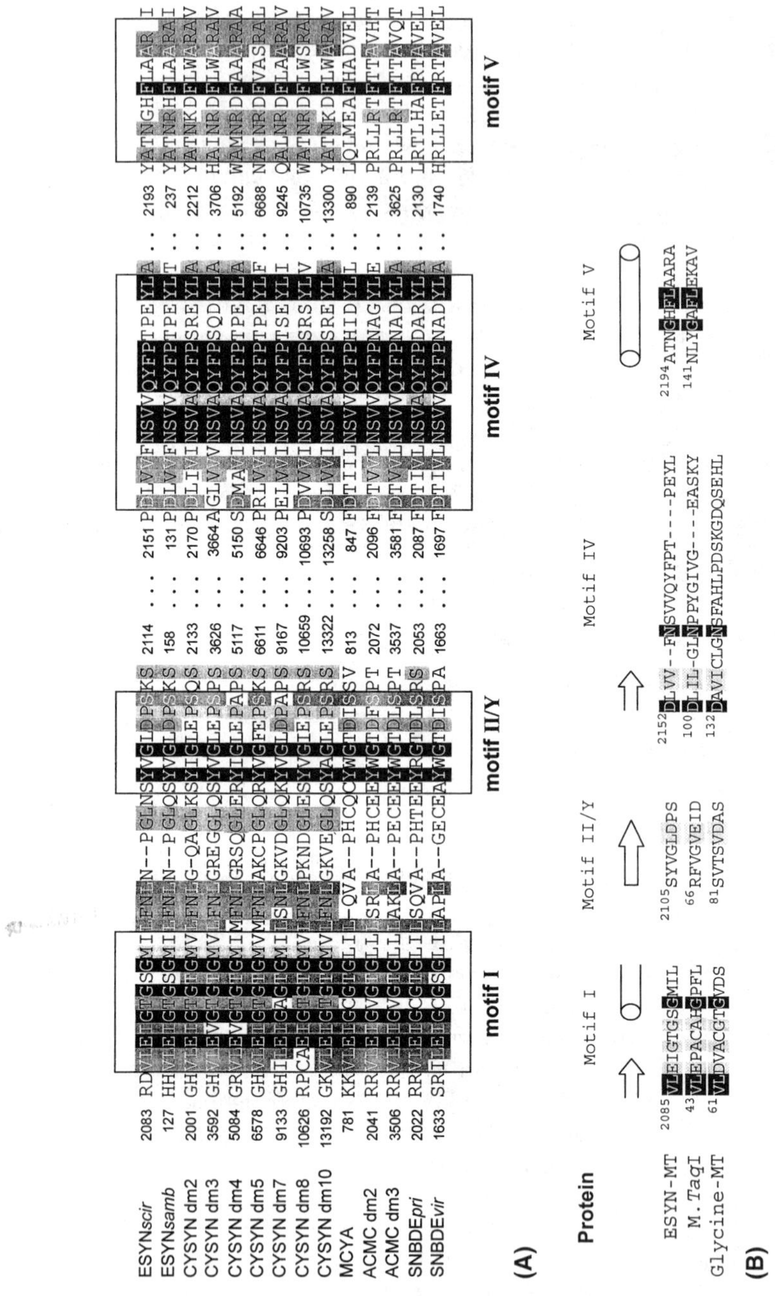
ESYNscir
ESYNsamb
CYSYN dm2
CYSYN dm3
CYSYN dm4
CYSYN dm5
CYSYN dm7
CYSYN dm8
CYSYN dm10
MCYA
ACMC dm2
ACMC dm3
SNBDEpri
SNBDEvir
motif I
motif II/Y
motif IV
motif V
(A)
Protein
ESYN-MT
M. TaqI
Glycine-MT
Motif I
Motif II/Y
Motif IV
Motif V
(B)

In a previous work on enniatin biosynthesis, Pieper et al. [30] found a photolabeled nonapeptide after UV irradiation of enniatin synthetase in the presence of radiolabeled AdoMet and subsequent chymotryptic digestion. This peptide is located C-terminal to motif I and contains a conserved tyrosine. The included motif was named motif Y (Fig. 3A) [39]. Motif Y shows high similarity to motif II in DNA methyltransferase nomenclature [32], both in sequence and in location relative to motif I. From photolabeling experiments, a conserved tyrosine residue (^{136}Tyr) was also suggested to play a role in AdoMet binding in rat guanidinoacetate methyltransferase [40]. Replacement of a conserved tyrosine residue (^{683}Tyr) in vaccinia virus mRNA (guanine-7-)methyltransferase by Ala and Ser abolished the methyltransferase AdoMet-binding activity, whereas the specific activity of the Phe mutant reached about 60% of that of wild type [41]. Previous investigations indicated that deletion of motif I in the *N*-methyltransferase portion of enniatin synthetase resulted in a loss of the ability to bind AdoMet as measured by photoaffinity labeling [25].

In Figure 3B an amino acid sequence alignment of the regions containing the conserved motifs of *N*-methyltransferase domains of peptide synthetases of eukaryotic and prokaryotic origin are shown.

Figure 3 Amino acid sequence alignment of conserved motifs within the methyltransferases. (A) Alignment of *N*-methyltransferase domains from peptide synthetases of procaryotic and eucaryotic origin. The alignment was made using the Clustal W alignment program in the GeneDoc package. The *N*-methyltransferase portions of the following sequences with their accession numbers were used: ESYN*scir* = Esyn from *Fusarium scirpi* Z18755 [24], ESYN*samb* = Esyn from *F. sambucinum* Z48743 [39], CYSYN = cyclosporin synthetase from *Tolypocladium niveum* Z28383 [42], MCYA = microcystin synthetase A from *Microcystis* spp. AB019578 [45], ACMC = actinomycin synthetase II from *Streptomyces crysomallus* [43], SNBDE*pri* = pristinamycin synthetase from *S. pristinaespiralis* Y11548, and SNBDE*vir* = virginiamycin synthetase from *S. virginiae* Y11547 [44]. Identical amino acid residues are black colored. More than 80% sequence similarities are dark shaded, more than 60% sequence similarities are shaded light gray. dm = *N*-methyltransferase domain. (B) Alignment of conserved motifs of M. *Taq*I, Glycine-MT, and ESYN-MT using PSI-BLAST and the 3D-PSSM Web-based method [33,34]. M.*Taq*I = methyltransferase from *Thermophilus aquaticus* [35], Glycine-MT = glycine *N*-methyltransferase from rat [36], ESYN-MT = *N*-methyltransferase domain of enniatin synthetases from *F. scirpi* [24]. Identical amino acids are shown as white letters against black background, conserved positions (described by Malone et al. [32]) are indicated by black letters on a gray-shaded background. The secondary structures of the methyltransferases derived from the 3D-PSSM method are indicated by cylinders (helices) and arrows (strands). The numbers indicate the positions of the first amino acid of the motif within the protein.

Deletion of the first 21 N-terminal amino acid residues of the *N*-methylation domain did not affect AdoMet-binding. Further shortening close to motif I resulted in loss of binding activity. Truncation of a short portion (38 amino acids) from the C-terminus and also deletions of internal sequences containing motifs I, II/Y, IV, and V [25] as well as only motif V, which contains an absolutely conserved Phe residue (Figs. 3A and 3B), led to complete loss of AdoMet-binding activity. Point mutations converting the conserved ^{2106}Tyr in motif Y (close to motif I) into valine, alanine, and serine strongly diminished AdoMet binding, whereas conversion of this residue to phenylalanine restored AdoMet-binding activity to about 70%, indicating that ^{2106}Tyr is important for AdoMet binding and that the aromatic residue of ^{2106}Tyr may be crucial for S-adenosyl-L-methionine binding in *N*-methyl peptide synthetases [7].

Today, sequence information about several *N*-methyltransferase domains of peptide synthetases of eukaryotic and prokaryotic origin is available. In Figure 4, an unrooted phylogenetic tree of *N*-methyltransferase domains of various peptide synthetases is shown. Sequence comparison revealed that the seven *N*-methyltransferase domains of the multifunctional enzyme Cysyn and the *N*-methyltransferase domains of Esyn from *F. scirpi* and *F. sambucinum* build one group [3,39,42] (the organization of the peptide synthetase modules and their integrated *N*-methyltransferase domains is shown in Fig. 2B). The domains of the different *Fusaria* share 92% identity. The Cysyn *N*-methyltransferase domains are 54–58% identical to each other. The identity of the Cysyn *N*-methyltransferase segments to the corresponding Esyn portions is 47–52%. The second group (procaryotic origin) includes the *N*-methyltransferase domains from various *Streptomyces* and *Microcystis* spp. [43–45]. The *N*-methyltransferases from the different *Streptomyces* exhibit 38–68% identity to each other. The *N*-methyltransferases from *Microcystis* spp. and the different *Streptomyces* share 27–30% sequence homology. The overall sequence homology between the two major groups is in the range of 18–25%.

C. Mechanism of Enniatin Biosynthesis

Esyn is a two-module enzyme but assembles three amino acids and three D-Hiv molecules in the final product enniatin. Studies on the mechanism of enniatin B formation revealed that the enniatin molecule is synthesized by three successive condensations of enzyme-bound dipeptidols with each other [22]. This result implies that, besides the two 4′-phosphopantetheine arms in the modules EA and EB, Esyn contains an additional 4′-phosphopantetheine at the C-terminus of module EB (waiting position), which picks up the intermediates of enniatin synthesis, i.e., the dipeptidol, tetrapeptidol, and hexapeptidol, to allow correct depsipeptide chain elongation. This proposal agrees with the finding of three putative 4′-phosphopantetheine-binding sites represented by the motif LGGXS [9,10,24] (Fig.

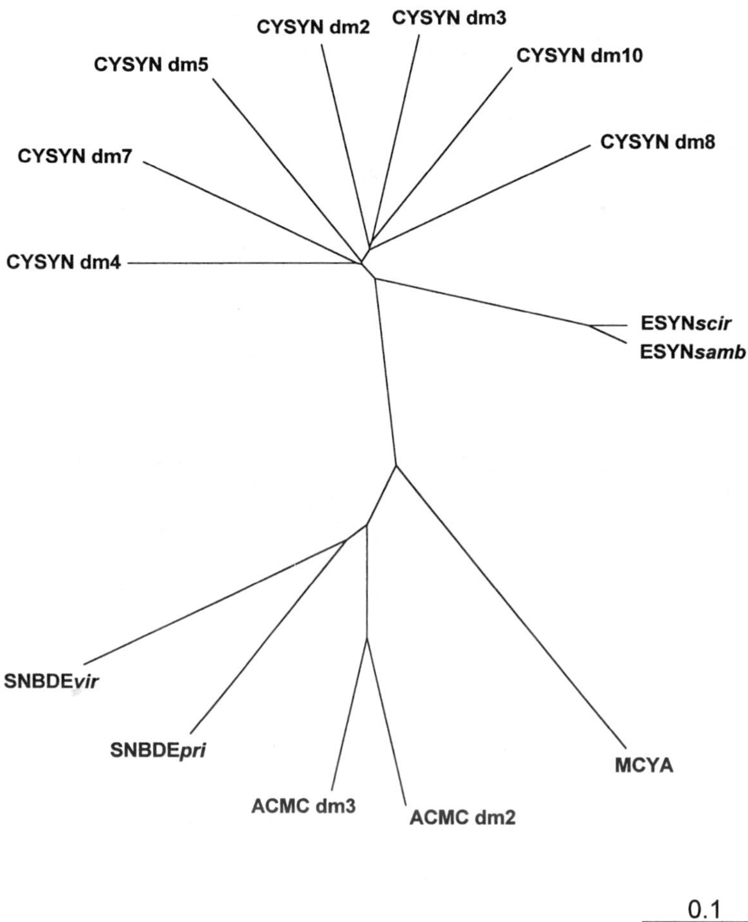

Figure 4 Unrooted phylogenetic tree of *N*-methyltransferase domains of peptide synthetases. The analysis was done by using a sequence alignment of *N*-methyltransferase domains of peptide synthetases by Clustal W (data not shown, [7]). The tree was built using the neighbor-joining method of PHYLIP. The branch length marker of 0.1 gives the distance in which an amino acid substitution is found in every tenth position (10% difference), e.g., ESYN*scir*/ESYN*samb* with a distance of 0.05 has a sequence difference of 5%. For sources of sequences, see legend to Figure 3A. CYSYN = cyclosporin synthetase, ESYN = enniatin synthetase, ACMC = actinomycin synthetase, SNBDE*pri* = pristinamycin synthetase, SNBDE*vir* = virginiamycin synthetase, MYCA = microcystin synthetase.

2A). Therefore, Esyn (together with beauvericin synthetase and PF1022A synthetase) belongs to the class of peptide synthetases which catalyze biosynthesis in an iterative manner. Dissecting the biosynthetic process in the individual steps catalyzed by the individual modules of Esyn has revealed the picture of reaction steps shown in Figure 5. The reaction sequence yielding the cyclohexadepsipeptide includes the activation of D-Hiv (module EA) and the L-amino acid (module EB) as covalently bound thioesters via the corresponding acyladenylates. Characteristically, the thioesterified L-amino acid is methylated with AdoMet, and thus *N*-methylation takes place prior to peptide bond formation and subsequent cyclization reactions [26]. After *N*-methylation, the amino acid and D-Hiv form a dipeptidol which is still covalently attached to the enzyme (module EB). The formed dipeptidol is then transferred to the waiting position in order to reinitiate a new reaction cycle. The hydroxyl group of the newly formed dipeptidol then attacks the carboxy group of the dipeptidol in the waiting position in a nucleophilic reaction yielding a tetrapeptidol, which is transferred to the waiting position. In a third condensation reaction the next new dipeptidol attacks the tetrapeptidol to form a hexapeptidol, which yields enniatin in the terminating cyclization reaction. Since the formation of cyclodepsipeptides larger than enniatin has not been observed, it is assumed that the size of a putative cyclization cavity apparently determines the length of the growing depsipeptide chain (Fig. 6).

D. Substrate Specificity of Enniatin Synthetases

Owing to the relatively broad substrate specificity of Esyn for amino and hydroxy acids, a variety of different enniatins can be synthesized by Esyn if appropriate concentrations of substrates (depending on the various K_m values) are used [46]. Studies on the substrate specificity revealed that the enzyme is capable of synthesizing enniatins A–C and also mixed-type enniatins containing more than one species of amino acid [47]. Addition of e.g., isoleucine to an ennaitin-producing culture of *F. oxysporum* results in a rather complex mixture of homologs [47]. The multienzyme also synthesizes different enniatins depending on the amino acid present in the *in vitro* reaction mixture [21].

Nevertheless, Esyns from different *Fusarium* strains may differ in their amino acid specificity as expressed by their V_{max}/K_m values. Esyn from the enniatin B-producing fungi *F. lateritium*, for instance, exhibits high affinity for L-valine, the constituent amino acid of enniatin B, and therefore strongly resembles the Esyn from *F. scirpi* [46]. In contrast, the multienzyme from the enniatin A producer *F. sambucinum* preferably incorporates L-leucine and L-isoleucine. This difference may be due to mutations in the amino acid-binding sites of the various Esyns. Sequence analysis of the amino acid activation domain of Esyns from *F. scirpi* and *F. sambucinum* showed a high degree of sequence identity with the exception of three oligopeptide regions [48]. Previous reports on selectivity-

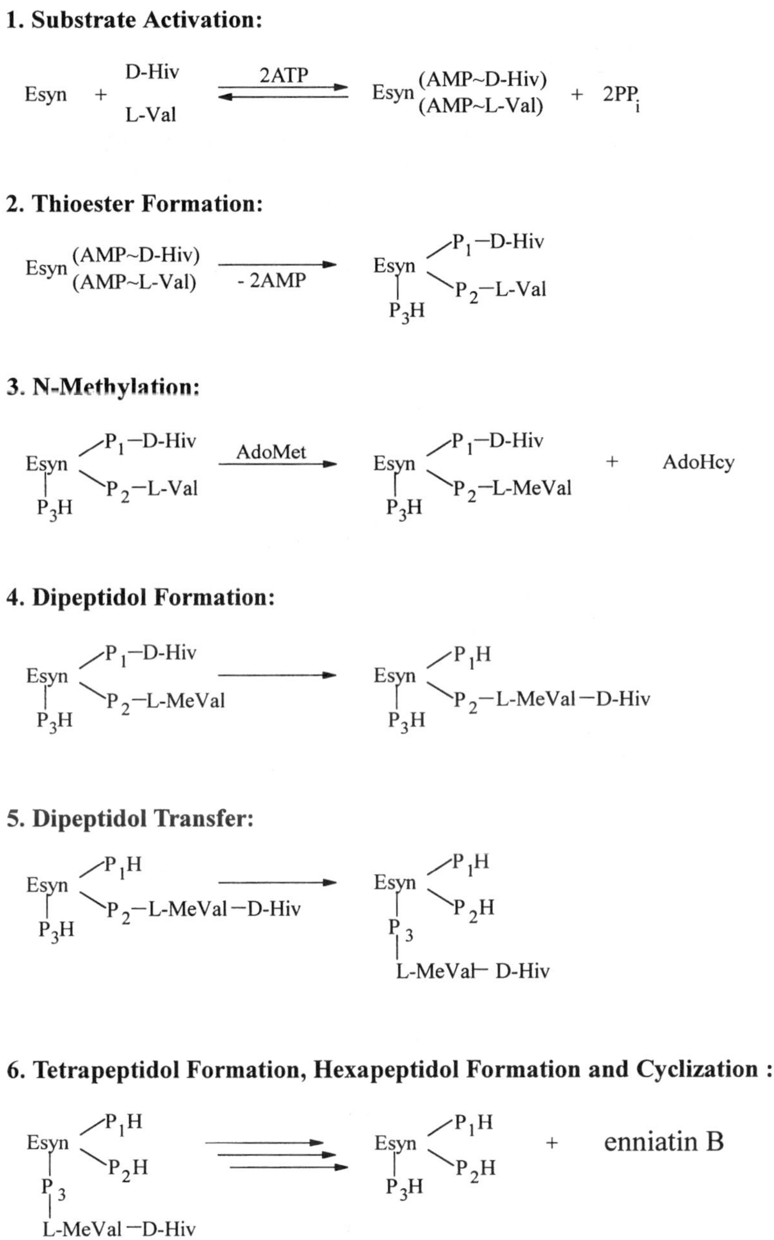

Figure 5 Scheme of partial reactions leading to enniatin B. P_1, P_2, P_3 = 4′-phosphopantetheine; P_3 represents the waiting position.

(A)

Cycle I
- **P1**-D-Hiv + **P2**-MeVal → **P1**-SH + **P2**-MeVal-D-Hiv
- **P2**-MeVal-D-Hiv + **P3**-SH → **P2**-SH + **P3**-MeVal-D-Hiv

Cycle II
- **P1**-D-Hiv + **P2**-MeVal → **P1**-SH + **P2**-MeVal-D-Hiv
- **P2**-MeVal-D-Hiv + **P3**-MeVal-D-Hiv → **P2**-$(\text{MeVal-D-Hiv})_2$ + **P3**-SH
- **P2**-$(\text{MeVal-D-Hiv})_2$ + **P3**-SH → **P2**-SH + **P3**-$(\text{MeVal-D-Hiv})_2$

Cycle III
- **P1**-D-Hiv + **P2**-MeVal → **P1**-SH + **P2**-MeVal-D-Hiv
- **P2**-MeVal-D-Hiv + **P3**-$(\text{MeVal-D-Hiv})_2$ → **P2**-$(\text{MeVal-D-Hiv})_3$ + **P3**-SH
- **P2**-$(\text{MeVal-D-Hiv})_3$ + **P3**-SH → **P2**-SH + **P3**-$(\text{MeVal-D-Hiv})_3$

Cyclization
- **P3**-$(\text{MeVal-D-Hiv})_3$ → **P3**-SH + **enniatin B**

(B)

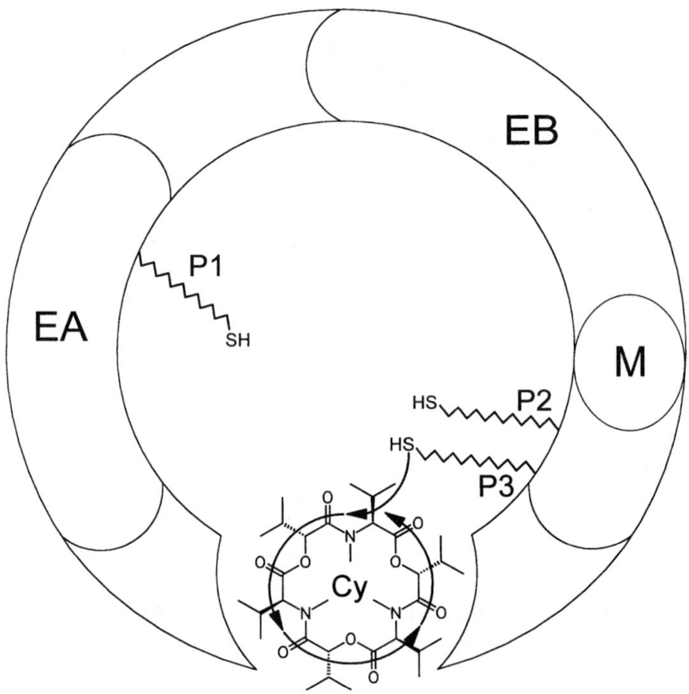

Figure 6 (A) Scheme of events of dipeptidol condensations on enniatin synthetase and the role of the additional 4′-phosphopantetheine-containing thioester domain as the waiting position. P_1, P_2, P_3 = 4′-phosphopantetheine. (B) Model of arrangement of catalytic sites of enniatin synthetase. Cy = cyclization cavity; EA represents the D-Hiv-activation module; EB represents the L-valine-activation module; M represents the *N*-methyltransferase domain.

conferring residues which mediate substrate specificity within gramicidin S synthetase I (PheA) and Cysyn are controversal with respect to the amino acid residues involved in substrate recognition in Esyn [48–52]. Our data agree only partially with these results. We suggest that substrate specificity is dictated from regions in the variable part of the adenylation domain [48; A. Doller, personal communication, 2000].

E. The *N*-Methylation Reaction

A characteristic property of almost all *N*-methyltransferases that have been studied so far is their sensitivity to inhibition by the reaction product S-adenosyl-L-homocysteine (AdoHcy), which is formed from the methyl group donor AdoMet in the *N*-methylation step [38]. In addition, sinefungin, another compound structurally related to AdoMet, is known to act as a competitive inhibitor of various *N*-methylating enzymes [53]. Billich and Zocher [26] found that, similar to other transmethylases, AdoHcy and sinefungin are potent inhibitors of the AdoMet-dependent reaction in enniatin synthesis. Kinetic analysis showed that AdoHcy exhibits an inhibition pattern typical for a partial competitive inhibitor, indicating that AdoHcy does not compete directly with AdoMet but binds to a discrete inhibitory site. In addition, AdoHcy inhibited the formation of the unmethylated depsipeptide formed in the absence of AdoMet. In contrast, sinefungin blocked enniatin (methylated product) formation competitively, indicating direct competition for the AdoMet-binding site, but does not exhibit an influence on the synthesis of desmethylenniatin even if present in excessive amounts. These results confirm the assumption that two different binding sites for the inhibitors must be present in the methyltransferase domain of Esyn. Thus, blocking the *N*-methyltransferase function by AdoHcy results in inhibition of the peptide bond formation (elongation reaction) and/or the cyclization ability of Esyn. This finding implies a dependence of the active sites for transacylation and *N*-methylation in the Esyn system.

F. Biosynthesis of Beauvericin

Beauvericin is a structural homolog of enniatins in which the branched-chain L-amino acid is substituted by the aromatic amino acid L-phenylalanine. Beauvericin synthetase, which has been isolated from the fungus *Beauveria bassiana* [54] and various strains of *Fusaria* [55], strongly resembles Esyn with respect to its molecular size and the reaction mechanism. In contrast to Esyn, which is only able to incorporate aliphatic amino acids, beauvericin synthetase exhibits high substrate specificity for aromatic amino acids such as phenylalanine. This capability is obviously caused by mutational alterations in the adenylation domain of this enzyme.

Recently, two new beauvericins were discovered [56]. These beauvericins are composed of L-phenylalanine, D-2-hydroxy-3-methyl-valerate, and D-2-hydroxyisovalerate, yielding mixed-type structures.

G. D-2-Hydroxyisovalerate Dehydrogenase

In some fungal species, namely, the enniatin producers of the genus *Fusarium*, there is a pathway leading from the primary precursor L-valine to D-Hiv via 2-ketoisovalerate [57,58]. The enzyme catalyzes the reversible reaction of 2-keto-isovalerate to D-Hiv, which is an intermediate in the biosynthetic pathway of enniatins in *Fusarium*. D-Hiv dehydrogenase consists of one polypeptide chain with a molecular mass of about 53 kDa. It is strictly dependent on NADPH and, in contrast to other NADPH-dependent oxidoreductases, exhibits a high substrate specificity with respect to 2-ketoisovalerate. This may explain the fact that D-Hiv is the exclusive hydroxy acid component in enniatins isolated from *Fusaria* [21]. D-Hiv dehydrogenase also has an important role in the biosynthesis of some other depsipeptides and peptolides, such as beauvericin [54], destruxin [59], valinomycin [60], bassianolide [61], and cyclosporin-like peptolide SDZ 214-103 [62], all of which contain D-Hiv.

III. PF 1022A AND RELATED CYCLOOCTADEPSIPEPTIDES

In the course of screening for new anthelmintic compounds, Sasaki et al. [63] isolated an *N*-methylated cyclodepsipeptide, PF1022A, from the fungus *Mycelia sterilia*. PF1022A exhibits strong anthelmintic properties in combination with low toxicity and is, therefore, one of the most outstanding anthelmintics [64,65]. PF1022A belongs to a new class of recently identified cyclooctadepsipeptides consisting of four alternating residues of *N*-methyl-L-leucine and four residues of D-lactate, D-phenyllactate, or D-hydroxyphenyllactate, as shown in Figure 7. Structurally, the PF1022 anthelmintics are related to bassianolide, which is an insecticidal metabolite from the fungus *Beauveria bassiana* [66], and to the enniatins.

A. Mechanism of Biosynthesis of PF1022-Type Cyclooctadepsipeptides

As in the case of enniatins, a 350-kDa depsipeptide synthetase, PF1022 synthetase (PFsyn), is responsible for PF1022A synthesis. The enzyme is capable of synthesizing all known natural cyclooctadepsipeptides of the PF1022 type (A, B, C, D, E, F, and 202) differing in the content of D-lactate, D-phenyllactate, and D-hydroxyphenyllactate [67,68]. In addition, the *in vitro* incubations pro-

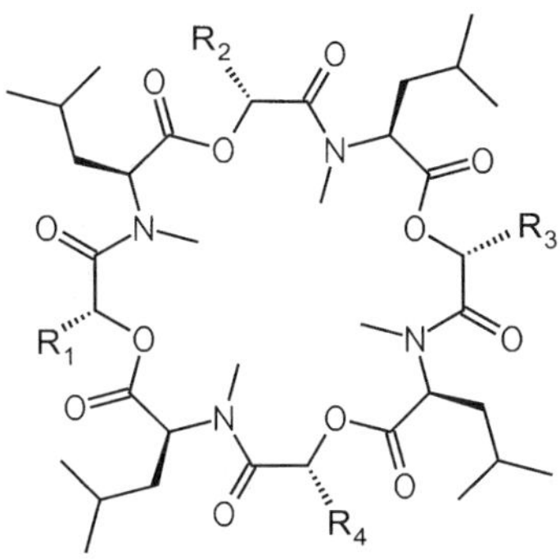

Figure 7 Structures of related cyclooctadepsipeptides from *Mycelia sterilia*. PF1022A is the main product in *Mycelia sterilia*. PF1022A, $R_1 = R_3 =$ benzyl, $R_2 = R_4 =$ methyl; PF1022B, $R_1 = R_2 = R_3 = R_4 =$ benzyl; PF1022C, $R_1 = R_2 = R_3 =$ benzyl, $R_4 =$ methyl; PF1022D, $R_1 =$ benzyl, $R_2 = R_3 = R_4 =$ methyl; PF1022E, $R_1 =$ benzyl, $R_2 = R_4 =$ methyl, $R_3 =$ *p*-hydroxybenzyl; PF1022F, $R_1 = R_2 = R_3 = R_4 =$ methyl; bassianolide, $R_1 = R_2 = R_3 = R_4 =$ isopropyl; PF1022-202, $R_1 = R_3 =$ methyl, $R_2 = R_4 =$ *p*-hydroxybenzyl.

duced di-, tetra-, and hexa-PF1022 homologs [68]. PF1022A, B, C, D, and F contain D-lactate and/or D-phenyllactate as the hydroxy acid constituent. PF1022E contains two D-lactate residues, one D-phenyllactate, and one D-hydroxyphenyllactate. The PF1022 homolog PF1022-202 consists of D-lactate, D-phenyllactate, and *N*-methyl-L-leucine. PFsyn strongly resembles the well-documented Esyn in size and mechanism. Interestingly, the molecular size of the purified enzyme in sodium dodecyl sulfate polyacrylamide gel electrophoresis (SDS-PAGE) is identical with that of Esyn, although the latter enzyme is responsible for the synthesis of a cyclohexadepsipeptide and therefore assembles only 6 residues instead of 8 to form the final product enniatin [69]. The molecular weight of 350 kDa estimated for PFsyn suggests that, like Esyn, PFsyn is also comprised of two peptide synthetase modules, each of more than 100 kDa, with an integrated methyltransferase portion of about 50 kDa [24,42].

Based on the assumption that PFsyn is a two-module enzyme like Esyn, catalyzing the biosynthesis of PF1022-related compounds via an iterative mechanism, we suggest that one module of PFsyn activates L-leucine and is responsible for *N*-methylation. The other module activates D-lactate as well as D-phenyllactate. The low substrate specificity of the latter module is also illustrated by the fact that the enzyme activates a wide range of different D-2-hydroxy acids [70]. The elongation reaction links the different combinations of the two possible dipeptidol units consisting either of D-lactate/*N*-methyl-L-leucine, D-phenyllactate/*N*-methyl-L-leucine, or D-hydroxyphenyllactate/*N*-methyl-L-leucine and therefore leads to the complete natural product spectrum of *M. sterilia*, including PF1022A, B, C, D, E, F, and PF1022-202.

B. D-Phenyllactate Dehydrogenase

The biosynthesis of cyclooctadepsipeptides of the PF1022 type *in vivo* is controlled by specific enzymes that are integrated in the biosynthesis pathway at key positions. They supply the precursors for the peptide synthetases [71,72] as in the case of D-Hiv in the biosynthetic pathway of enniatins [57,58]. D-Phenyllactate dehydrogenase (DPLDH) from the fungus *M. sterilia* stereospecifically catalyzes the reduction of phenylpyruvate to D-phenyllactate [70]. DPLDH exhibits a wide 2-ketoacid substrate spectrum. A preference for phenylpyruvate and *p*-hydroxyphenylpyruvate is reflected by low K_m and high k_{cat}/K_m values. The enzyme requires NADPH as a cofactor. Molecular mass determination of the denatured enzyme by SDS-PAGE gave an estimated value of 38 kDa, whereas gel-filtration calibration studies yielded a value of about 160 kDa, indicating a tetrameric structure. Due to substrate specificities and molecular properties of DPLDH, this enzyme can be classified as a D-isomer-specific 2-hydroxy acid dehydrogenase. Tryptic fragments of DPLDH show significant sequence homologies to several typical D-isomer-specific 2-hydroxyacid dehydrogenases [70].

IV. CYCLOSPORINS

Cyclosporins, produced by the filamentous fungus *Tolypocladium niveum* and by numerous strains of *Fusaria* and *Neocosmospora*, are a class of cyclic undecapeptides which are composed of hydrophobic aliphatic amino acids [73–75]. They exhibit antiinflammatory, immunosuppressive, antifungal, and antiparasitic properties [74]. The main metabolite, cyclosporin A, is in clinical use worldwide under the trade name SANDIMMUN® to prevent allograft rejection [77,78]. Besides cyclosporin A, there are 24 naturally occuring cyclosporins which have substitutions of amino acids in positions 1, 2, 4, 5, 7, and 11 and/or contain unmethylated peptide bonds in positions 1, 4, 6, 9, 10, or 11 [79–82]. Cyclosporin A contains three nonproteinogenic amino acids: D-alanine in position 8, L-α-aminobutyric acid in position 2, and, in position 1, the unusual amino acid 4(R)-4-[(E)-2-butenyl]-4-methyl-L-threonine (Bmt) (Fig. 8). All three amino acids have to be synthesized by a pathway independent of the primary metabolism. In addition, several peptide bonds of the cyclosporin molecule are *N*-methylated similar to the depsipeptides enniatin, beauvericin and PF1022A-related peptides.

A. Mechanism of Cyclosporin Biosynthesis

Studies made by feeding experiments with ^{14}C-labeled precursors showed that the biosynthesis of cyclosporins proceeds by a thioltemplate mechanism, which, owing to the *N*-methylating steps, has strong resemblance to that of enniatin

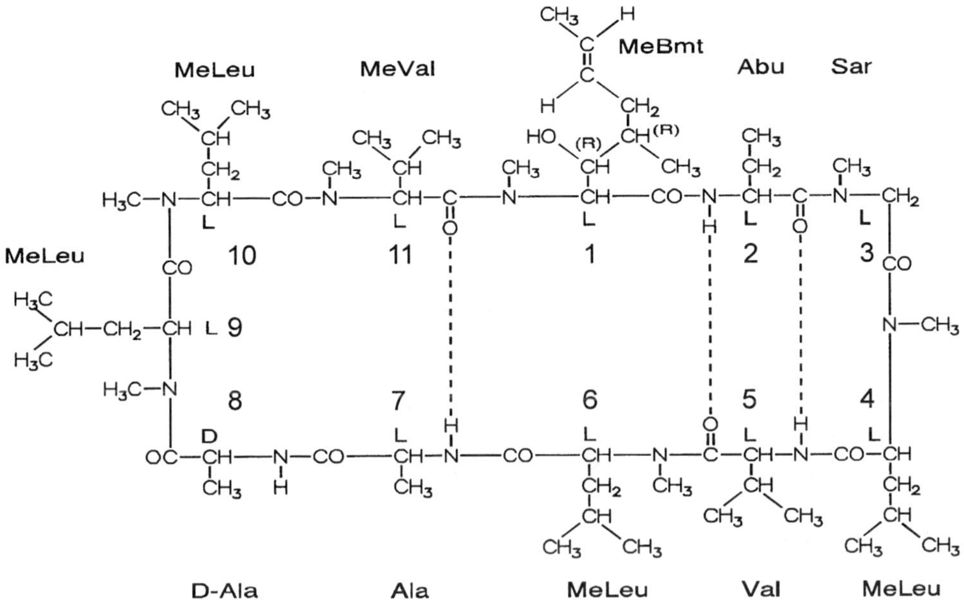

Figure 8 Structure of cyclosporin A. Abu = L-α-aminobutyric acid; Bmt = (4*R*)-4-[(E)-2-butenyl]-4-methyl-L-threonine; Sar = sarcosine (*N*-methylglycine).

synthesis [83]. The multifunctional enzyme Cysyn is responsible for cyclosporin biosynthesis [84]. Further experiments on the mechanism of cyclosporin formation were reported by Dittmann et al. [85]. The authors were able to isolate four enzyme-bound intermediate peptides of cyclosporin biosynthesis from a complex mixture of unidentified peptides. All four peptides carried alanine as the *N*-terminal amino acid. From these findings the authors conclude that D-alanine provides the ''starter amino acid'' of cyclosporin biosynthesis. The biosynthesis of cyclosporin A includes at least 40 reaction steps, which can be divided into 11 substrate activation reactions, 11 thioester formations, 7 *N*-methylations, 10 elongations, and the final cyclization reaction. Cysyn is therefore the most complex peptide synthetase so far described [84]. Besides the total synthesis of cyclosporins, the formation of several diketopiperazines can be observed. Diketopiperazines, which represent a partial sequence of the cyclosporin molecule, are cyclic dipeptides formed in a side reaction of Cysyn. This reaction is the cyclization of neighboring substrate amino acids on the enzyme under the consumption of ATP (and AdoMet).

Like Esyn, the multienzyme Cysyn is unable to carry out racemization reactions because of a lack of an integrated racemase function [18,42]. Enzyme-bound

L-alanine cannot be epimerized to D-alanine [72]. D-alanine is incorporated directly into the D-alanine position of cyclosporin. The D-alanine moiety is racemized from L-alanine by a separate specific alanine racemase, which plays a key role in cyclosporin biosynthesis [72]. The enzyme requires pyridoxal phosphate as the exclusive cofactor. Molecular mass determinations of the denatured racemase by SDS-PAGE gave a value of 37 kDa, whereas gel-filtration calibration studies yielded a value between 120 and 150 kDa, indicating an oligomeric native structure [72]. A similar situation is found in the case of the peptide synthetase that synthesizes HC-toxin [86]. The authors conclude that the *tox*G gene encodes an alanine racemase whose function is to synthesize D-alanine for incorporation into HC-toxin [87]. Detailed incorporation experiments as well as enzymatic studies revealed the origin of the unusual amino acid Bmt backbone. The basic assembly reaction, performed by Bmt polyketide synthase, ends at the stage of 3(*R*)-hydroxy-4(*R*)-methyl-6(*E*)-octenoyl-CoA [88], which then undergoes further methylation and reduction reactions to form the final precursor Bmt [89].

B. Molecular Structure of Cyclosporin Synthetase

Cyclosporin biosynthesis is catalyzed by a single multienzyme, the cyclosporin synthetase [90]. Cysyn is encoded by an open reading frame of 45.8 kb (Fig. 2B) [42]. Disruption of the Cysyn gene (*sim*A) resulted in loss of the ability to produce cyclosporins [91]. The enzyme has an estimated molecular mass of 1.69 MDa [42] containing 11 peptide synthetase modules, of which 7 are homologous to the module EB of Esyn carrying the *N*-methyltransferase domains [42,92]. From the arrangement of the modules and the fact that modules in peptide synthetases are collinear to the arrangement of amino acids in the peptides to be synthesized (with the exception of syringomycin synthetase from *Pseudomonas syringae* [93]), the authors conclude that the 5′-terminal module is responsible for D-alanine activation. The last domain at the 3′-end would represent the L-alanine-activating peptide synthetase module. This assumption was supported by the finding that a Cysyn fragment of 130 kDa could be isolated, which is capable of activating L-alanine. Edman degradation of this protein yielded a sequence with an N-terminus in the position of amino acid 13601 of Cysyn [42].

C. *N*-Methylation Reaction

Since the overall reaction of cyclosporin formation is a very complex process (at least 40 single reactions), for the investigation of the *N*-methylation reaction, the formation of the diketopiperazines cyclo(D-alanine-*N*-methylleucine) and cyclo (L-alanine-*N*-methylleucine) was used. In contrast to other *N*-methyltransferases such as guanidinoacetate methyltransferase [94] or indolethylamine methyltransferase [95], in which AdoHcy acts as a competitive inhibitor, in the case of the *N*-methyltransferase domain of Cysyn, AdoHcy interestingly exhibits a noncompetitive inhibition pattern [96]. Analogous to Esyn, the methyl group donor AdoMet and the inhibitor AdoHcy must therefore possess different binding sites on the *N*-methyltransferase domain. From the results that neither the adenylation reaction of the substrate amino acids nor the thioester formation is influenced by AdoHcy, it was concluded that, as in the case of Esyn, the inhibitor not only blocks the *N*-methylation but also the elongation and/or the cyclization reaction. Another similarity between Cysyn and Esyn is the effect of the antibiotic sinefungin on product formation. As in the case of enniatin formation, sinefungin inhibits the *in vitro* synthesis of cyclo(D-alanine-*N*-methylleucine) competitively, indicating a mutual binding site for the methyl group donor and the inhibitor, whereas the substrate activation reactions (adenylation and thioesterification) are not affected by sinefungin [96]. A model including the above results is shown in Figure 9. The organization of a peptide synthetase module (of Esyn and Cysyn) with an integrated *N*-methylation domain harbors a substrate amino acid-binding site and a separate ATP-binding site. The methyl group donor and the inhibitor

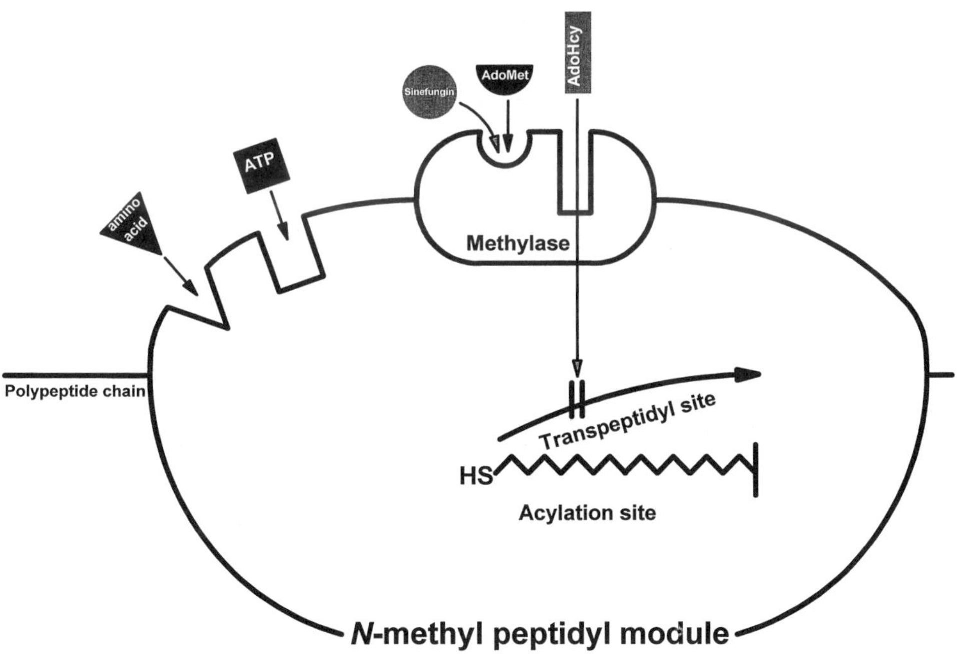

Figure 9 Organization of a peptide synthetase module containing an integrated *N*-methyltransferase domain.

sinefungin occupy the same binding site on the *N*-methyltransferase domain, whereas AdoHcy binds to a discrete inhibitory site.

D. Substrate Specificity of Cyclosporin Synthetase

From the spectrum of naturally occurring cyclosporins, it seems obvious that some of the peptide synthetase modules of Cysyn have a rather high substrate specificity and others, such as that responsible for Bmt, have a lower specificity, allowing incorporation of homolog substrates [79–82]. The only nonvariable positions in the natural cyclosporins are D-alanine (position 8) and sarcosine (*N*-methylglycine) in position 3 of the cyclosporin ring system (Fig. 8). Exchange of one or more of the unmethylated constitutive amino acids of cyclosporin A by various amino acids gives a picture of the substrate specificity of Cysyn *in vitro* [97]. The amino acid Bmt in position 1 and L-α-aminobutyric acid in position 2 of the cyclosporin molecule can be exchanged by a wide spectrum of amino acids, showing the great flexibility of these catalytic sites. Position 3 has a very high degree of substrate specificity, positions 4, 6, 7, 9, and 10 have marginally less. The variability of positions 5 and 11 is moderate, whereas position 8 shows only low substrate specificity *in vitro* [97].

E. Cyclosporin-Related Peptolide SDZ-214-103

Peptolide SDZ-214-103 is a cyclosporin-related undecapeptide lactone produced by the fungus *Cylindrotrichum oligospermum* [63]. It differs from the cyclosporins mainly at position 8. Because of the presence of a hydroxyacid (D-Hiv) instead of an amino acid (D-alanine in the natural cyclosporins [82]), one ester bond results in the ring system.

The peptolide is synthesized by the multifunctional enzyme SDZ 214-103 synthetase, in which the catalytic sites appear to be more specific than those of Cysyn. Interestingly, the D-2-hydroxyacid position (corresponding to position 8 of the cyclosporin ring) can be occupied by a wide range of substrates varying from D-lactic acid to D-2-hydroxyisocaproic acid [97]. The enzyme has been described to be a single polypeptide chain with a molecular mass of about 1400 kDa and strongly resembles Cysyn [98].

V. FUTURE DIRECTIONS

Future developments in peptide synthesis research will aim at clarifying the mechanisms of initiation, *N*-methylation, elongation, epimerization, and termination events on peptide synthetases. Parallel work will concentrate on the elucidation of the structures of peptide synthetase modules using biochemical techniques, such as cross-linking enzymes and mutational analysis of reaction sites. Understanding how peptide synthetases recognize and activate individual residues for incorporation into the final peptides is crucial to be able to engineer substrate-binding pockets of adenylation domains. The *N*-methylation reaction is still not fully understood [7]. The elucidation of the assembly of the AdoMet-binding site and the nature of the binding of AdoMet to the peptide synthetase is one major focus in the future.

It is of great interest to elucidate the three-dimensional structure of these multienzymes. For this purpose, X-ray crystallography and electron microscopy are appropriate methods. It will help to circumvent possible constraints in combining peptide synthetase modules into novel systems by recombinant DNA techniques, which will aid in the design of new bioactive compounds. In this context the biosynthesis of several novel polyketides, peptides, and *N*-methylated peptides by such manipulations has been reported [99–105]. The design and production of peptide antibiotics by targeted replacement of bacterial and fungal amino acid-activating modules demonstrate that these functional units are interchangeable between different organisms and produce peptides with modified amino acid sequences [105]. A novel peptide synthetase can be engineered by exchanging or recombining modules to achieve synthesis of new or altered peptides. This feat could be shown in the case of the bimodular actinomycin synthetase II by substituting the valine activation module with an *N*-methyl valine activation module, which contains an additional *N*-methylation activity. The resulting enzyme was able to catalyze synthesis of both the unmethylated and the *N*-methylated acyl dipeptides [43].

Combinatorial approaches in synthetic chemistry allow the high-throughput screening of millions of compounds. It is the aim to establish such approaches in biosynthetic systems by creating module assemblies, which provide a new tool to create peptide libraries. The large number of different modules in peptide synthetase systems is suitable for creating a pool of many different enzymes in order to detect a useful compound in the near future [2,51].

REFERENCES

1. H Kleinkauf, H von Döhren. Products of secondary metabolites. In: A Pühler, P Stadler, eds. Biotechnology. Weinheim: VCH, 1997, pp 1–55.
2. T Stein, J Vater. Amino acid activation and polymerization at modular multienzymes in nonribosomal peptide biosynthesis. Amino Acids 10:201–227, 1996.
3. R Zocher, U Keller. Thiol template peptide synthesis systems in bacteria and fungi. Adv Microb Physiol 38:85–131, 1997.

4. H von Döhren, U Keller, J Vater, R Zocher. Multifunctional peptide synthetases. Chem Rev 97:2651–2673, 1997.
5. T Stachelhaus, MA Marahiel. Modular structure of genes encoding multifunctional peptide synthetases required for non-ribosomal peptide synthesis. FEMS Microbiol Lett 125:3–14, 1995.
6. D Konz, MA Marahiel. How do peptide synthetases generate structural diversity? Chem Biol 6:39–48, 1999.
7. C Hacker, M Glinski, T Hornbogen, A Doller, R Zocher. Mutational analysis of the N-methyltransferase domain of the multifunctional enzyme enniatin synthetase. J Biol Chem 275:30826–30832, 2000.
8. MA Marahiel, T Stachelhaus, HD Mootz. Modular peptide synthetases involved in nonribosomal peptide synthesis. Chem Rev 97:2651–2673, 1997.
9. W Schlumbohm, T Stein, C Ullrich, J Vater, M Krause, MA Marahiel, V Kruft, B Wittmann-Liebold. An active site serine is involved in covalent substrate amino acid binding at each reaction center of gramicidin S synthetase. J Biol Chem 266: 23135–23141, 1991.
10. T Stein, J Vater, B Wittmann-Liebold, B Franke, M Panico, R McDowell, HR Morris. Detection of 4′-phosphopantetheine at the thioester binding site for L-valine of gramicidin S synthetase. FEBS Lett 340:39–44, 1994.
11. H Kleinkauf, H von Döhren. A nonribosomal system of peptide biosynthesis. Eur J Biochem 236:335–351, 1996.
12. PA Plattner, U Nager, A Boller. Welkstoffe und Antibiotika. Siebte Mitteilung über die Isolierung neuartiger Antibiotika aus *Fusarien*. Helv Chim Acta 31:594–602, 1948.
13. N Simon-Lavoine, M Forgeot. Novel biological substance from a fungus and the process for producing the same. German Patent DE2851629 (Science Union & Cie), 1980.
14. H Tomoda, XH Huang, J Cao, H Nishida, R Nagoa, S Okuda, H Tanaka, S Omura. Inhibition of acyl-CoA: cholesterol acyltransferase activity by cyclodepsipeptide antibiotics. J Antibiot 45:1626–1632, 1992.
15. HK Wipf, LAR Pieoda, Z Stefanac, W Simon. Komplexe von Enniatinen und anderen Antibiotika mit Alkalimetallionen. Helv Chim Acta 51:377–381, 1968.
16. RB Drysdale. The production and significance in phytopathology of toxins produced by species of *Fusarium*. In: MO Moss, JE Smith, eds. Applied Mycology of Fungi. Cambridge: Cambridge University Press, 1982, pp 95–105.
17. JD Walton. Peptide phytotoxins from plant-pathogenic fungi. In: H Kleinkauf, H von Döhren, eds. Biochemistry of Peptide Antibiotics. Berlin: W. de Gruyter, 1990, pp 179–203.
18. M Herrmann, A Haese, R Zocher. Effect of disruption of the enniatin synthetase gene on the virulence of *Fusarium avenaceum*. Mol Plant-Microbe Interact 9:226–232, 1996.
19. M Herrmann, R Zocher, A Haese. Enniatin production by *Fusarium* strains and its effect on potato tuber tissue. Appl Env Microbiol 62:393–398, 1996.
20. JF Grove, M Pople. The insecticidal activity of beauvericin and the enniatin complex. Mycopathologia 70:103–105, 1980.
21. R Zocher, U Keller, H Kleinkauf. Enniatin synthetase: a novel type of multifunc-

tional enzyme catalyzing depsipeptide synthesis in *Fusarium oxysporum*. Biochemistry 21:43–48, 1982.
22. R Zocher, U Keller, H Kleinkauf. Mechanism of depsipeptide formation catalyzed by enniatin synthetase. Biochem Biophys Res Commun 110:292–299, 1983.
23. A Billich, R Zocher. Constitutive expression of enniatin synthetase during fermentative growth of *Fusarium scirpi*. Appl Environ Microbiol 54:2504–2509, 1988.
24. A Haese, M Schubert, M Hermann, R Zocher. Molecular characterization of the enniatin synthetase gene encoding a multifunctional enzyme catalysing *N*-methyldepsipeptide formation in *Fusarium scirpi*. Mol Microbiol 7:905–914, 1993.
25. A Haese, R Pieper, T von Ostrowski, R Zocher. Bacterial expression of catalytically active fragments of the multifunctional enzyme enniatin synthetase. J Mol Biol 247:116–122, 1994.
26. A Billich, R Zocher. *N*-Methyltransferase function of the multifunctional enzyme enniatin synthetase. Biochemistry 26:8417–8423, 1987.
27. E Conti, NP Franks, P Brick. Crystal structure of firefly luciferase throws light on a superfamily of adenylate-forming enzymes. Structure 4:287–298, 1996.
28. K Turgay, M Krause, MA Marahiel. Four homologous domains in the primary structure of GrsB are related to domains in superfamily of adenylate-forming enzymes. Mol Microbiol 6:529–546, 1992.
29. A Billich, R Zocher, H Kleinkauf, DG Braun, D Lavanchy, HK Hochkeppel. Monoclonal antibodies to the multienzyme enniatin synthetase. Biol Chem Hoppe-Seyler 368:521–529, 1987.
30. R Pieper, A Haese, W Schröder, R Zocher. Arrangement of catalytic sites in the multifunctional enzyme enniatin synthetase. Eur J Biochem 230:119–126, 1995.
31. X Cheng, S Kumar, J Posfai, JW Pflugrath, RJ Roberts. Crystal structure of the *Hha*I DNA methyltransferase complexed with S-adenosyl-L-methionine. Cell 74: 299–307, 1993.
32. T Malone, RM Blumenthal, X Cheng. Structure-guided analysis reveals nine sequence motifs conserved among DNA amino-methyl-transferase, and suggests a catalytic mechanism for these enzymes. J Mol Biol 253:618–632, 1995.
33. LA Kelley, RM MacCallum, MJE Sternberg. Enhanced genome annotation using structural profiles in the program 3D-PSSM. J Mol Biol 299:501–522, 2000.
34. D Fischer, C Barret, K Bryson, A Elofsson, A Godzik, D Jones, KJ Karplus, LA Kelley, RM MacCallum, K Pawowski, B Rost, L Rychlewski, MJ Sternberg. CAFASP-1: critical assessment of fully automated structure prediction methods. Proteins: Structure, Function and Genetics (Suppl) 3:209–217, 1999.
35. G Schluckebier, M Kozak, N Bleimling, E Weinhold, W Saenger. Differential binding of S-adenosylmethionine, S-adenosylhomocysteine and sinefungin to the adenine-specific DNA methyltransferase M.*Taq*I. J Mol Biol 265:56–67, 1997.
36. H Ogawa, K Konishi, Y Takata, H Nakashima, M Fujioka. Rat glycine methyltransferase. Complete amino acid sequence deduced from a cDNA clone and characterization of the genomic DNA. Eur J Biochem 168:141–151, 1987.
37. G Schluckebier, M O'Gara, W Saenger, X Cheng. Universal catalytic domain structure of AdoMet-dependent methyltransferases. J Mol Biol 247:16–20, 1995.
38. RM Kagan, S Clarke. Widespread occurrence of three sequence motifs in diverse

S-adenosylmethionine-dependent methyltransferases suggest a common structure for these enzymes. Arch Biochem Biophys 310:417–427, 1994.
39. J Burmester, A Haese, R Zocher. Highly conserved N-methyltransferase as an integral part of peptide synthetases. Biochem Mol Biol Int 37:201–207, 1995.
40. Y Takata, M Fujioka. Identification of a tyrosine residue in rat guanidinoacetate methyltransferase that is photolabeled with S-adenosyl-L-methionine. Biochemistry 31:4369–4374, 1992.
41. X Mao, S Shuman. Vaccinia virus mRNA (guanine-7-)methyltransferase: mutational effects on cap methylation and AdoHcy-dependent photo-cross-linking of the cap to the methyl acceptor site. Biochemistry 35:6900–6910, 1996.
42. G Weber, K Schörgendörfer, E Schneider-Scherzer, E Leitner. The peptide synthetase catalysing cyclosporin production in *Tolypocladium niveum* is encoded by a giant 45.8-kilobase open reading frame. Curr Genet 26:120–125, 1994.
43. F Schauwecker, F Pfennig, N Grammel, U Keller. Construction and *in vitro* analysis of a new bi-modular polypeptide synthetase for the synthesis of *N*-methylated acylpeptides. Chem Biol 7:287–297, 2000.
44. V de Crécy-Lagard, W Saurin, D Thibaut, P Gil, L Naudin, J Crouzet, V Blanc. Streptogramin B biosynthesis in *Streptomyces pristinaespiralis* and *Streptomyces virginiae*: molecular characterization of the last structural peptide synthetase gene. Antimicrob Agents Chemother 41:1904–1909, 1997.
45. T Nishizawa, M Asayama, K Fujii, K Herada, M Shirai. Genetic analysis of the peptide synthetase genes for the cyclic heptapeptide microcystin in *Microcystis spp.* J Biochem (Tokyo) 126:520–529, 1999.
46. R Pieper, H Kleinkauf, R Zocher. Enniatin synthetases from different Fusaria exhibiting distinct amino acid specificities. J Antibiot 45:1273–1277, 1992.
47. N Madry, R Zocher, H Kleinkauf. Enniatin production by *Fusarium oxysporum* in chemically defined media. Eur J Appl Microbiol Biotechnol 17:75–79, 1983.
48. A Doller, A Haese, R Zocher. Molecular cloning of the amino acid activation domain of enniatin synthetase from *Fusarium sambucinum*. Symposium Enzymology of Biosynthesis of Natural Products, Abstract 66, Technische Universität Berlin, September 22–25, 1996.
49. GL Challis, J Ravel, CA Townsend. Predictive, structure-based model of amino acid recognition by nonribosomal peptide synthetase adenylation domains. Chem Biol 7:211–224, 2000.
50. H Husi, K Schörgendorfer, G Stempfer, P Taylor, MD Walkinshaw. Prediction of substrate-specific pockets in cyclosporin synthetase. FEBS Lett 414:532–536, 1997.
51. T Stachelhaus, HD Mootz, MA Marahiel. The specificity-conferring code of adenylation domains in nonribosomal peptide synthetases. Chem Biol 6:493–505, 1999.
52. E Conti, T Stachelhaus, MA Marahiel, P Brick. Structural basis for the activation of phenylalanine in the non-ribosomal biosynthesis of gramicidin S. EMBO J 16: 4174–4183, 1997.
53. GL Cantoni, HH Richards, PK Chiang. In: E Usdin, RT Borchardt, CR Creveling, eds. Transmethylation. Amsterdam: North Holland, 1979, pp 155–164.

54. H Peeters, R Zocher, N Madry, PB Oelrichs, H Kleinkauf, G Kraepelin. Cell-free synthesis of the depsipeptide beauvericin. J Antibiot 36:1762–1766, 1983.
55. A Logrieco, A Moretti, G Castella, M Kostecki, P Golinski, A Ritieni, J Chelkowski. Beauvericin production by *Fusarium* species. Appl Environ Microbiol 64: 3084–3088, 1998.
56. S Gupta, C Montilor, Y Hwang. Isolation of novel beauvericin analogues from the fungus *Beauveria bassiana*. J Natural Prod 58:733–738, 1995.
57. C Lee, H Görisch, H Kleinkauf, R Zocher. A highly specific D-hydroxyisovalerate dehydrogenase from the enniatin producer *Fusarium sambucinum*. J Biol Chem 267:11741–11744, 1992.
58. C Lee, R Zocher. The biochemical characterization of D-hydroxyisovalerate dehydrogenase, a key enzyme in the biosynthesis of enniatins. J Biochem Mol Biol 29: 493–499, 1996.
59. M Païs, BC Das, P Ferron. Depsipeptides from *Metarhizium anisopliae*. Phytochemistry 20:715–723, 1981.
60. GD Smith, WL Duax, DA Langs, GT Detitta, JW Edmonds, DC Rohrer, CM Weeks. The crystal and molecular structure of the triclinic and monoclinic forms of valinomycin. J Am Chem Soc 97:7242–7247, 1975.
61. Y Okumura. Peptidolactones. In: LC Vining, ed. Biochemistry and genetic regulation of commercially important antibiotics. Reading, MA: Addison-Wesley, 1983, pp 147–178.
62. MM Dreyfuss, MH Schreier, H Tscherter, R Wenger. Cyclic peptolides. European Patent EPO0296123 [Sandoz AG (CH); Sandoz AG (DE); Sandoz AG (AT)], 1988.
63. T Sasaki, M Takagi, T Yaguchi, S Miyadoh, T Okada, M Koyama. A new anthelmintic cyclodepsipeptide PF1022. J Antibiot 45:692–697, 1992.
64. M Terada, A Ishih, A Tungtrongchitr, M Sano, T Shomura. Effect of PF1022A on developing larvae of *Angiostrongylus costaricensis* in mice, with special reference to route, dose, and formulation. Jpn J Parasitol 42:199–210, 1993.
65. W Chen, M Terada, JT Cheng. Characterization of subtypes of gamma-aminobutyric acid receptors in an *Ascaris* muscle preparation by binding assay and binding of PF1022A, a new anthelmintic, on the receptors. Parasitol Res 82:97–101, 1996.
66. A Suzuki, M Kanaoka, A Isogai, S Murakoshi, M Ichinoe, S Tamura. Bassianolide, a new insecticidal cyclodepsipeptide from *Beauveria bassiana* and *Verticillium lecanii*. Tetrahedron Lett 25:2167–2170, 1977.
67. M Krause. Studies on the biosynthesis of the cyclooctadepsipeptide PF10022A. Ph.D. thesis, Technische Universität Berlin, Berlin, 1998.
68. W Weckwerth, K Miyamoto, K Iinuma, M Krause, M Glinski, T Storm, G Bonse, H Kleinkauf, R Zocher. Biosynthesis of PF1022 and related cyclooctadepsipeptides. J Biol Chem 275:17909–17915, 2000.
69. P Jeschke, G Bonse, G Thielking, W Etzel, A Harder, N Mencke, H Kleinkauf, R Zocher, K Iinuma, K Miyamoto. Process for the preparation of substituted aryl lactic acid containing cyclodepsipeptides with 24 ring atoms. WO9720945 (Bayer AG), 1998.
70. W Weckwerth. Studies on the biosynthesis of cyclooctadepsipeptides from *Mycelia sterilia*. Ph.D. thesis, Technische Universität Berlin, Berlin, 1998.

71. W Weckwerth, G Bonse, H Kleinkauf, R Zocher. Purification and characterization of D-2-hydroxyphenyllactate dehydrogenase from *Mycelia sterilia*. In preparation.
72. K Hoffmann, E Schneider-Scherzer, H Kleinkauf, R Zocher. Purification and characterization of eucaryotic alanine racemase acting as key enzyme in cyclosporin biosynthesis. J Biol Chem 17:12710–12714, 1994.
73. M Dreyfuss, E Härri, H Hoffmann, H Kobel, W Pache, H Tscherter. Cyclosporin A and C, new metabolites from *Trichoderma polysporum*. Eur J Appl Microbiol 3:125–133, 1976.
74. JF Borel. Ciclosporin and its future. Prog Allergy 38:9–18, 1986.
75. TH Aarino, SN Agathos. Production of extracellular enzymes and cyclosporin by *Tolypocladium inflatum* and morphologically related fungi. Biotechnol Lett 11: 759–764, 1989.
76. H Nakajima, T Hamasaki, K Tanaka, Y Kimura, SI Udagawa, Y Horie. Production of cyclosporin by fungi belonging to the genus *Neocosmospora*. Agric Biol Chem 53:2291–2292, 1988.
77. A von Wartburg, R Traber. Cyclosporins-fungal metabolites with immunsuppressive activities. Prog Med Chem 25:1–33, 1988.
78. G Feutren. Cyclosporin A: recent developments in the mechanism of action and clinical application. Curr Opin Immunol 2:239–245, 1989.
79. R Traber, M Kuhn, A Rüegger, H Lichti, HR Loosli, A von Wartburg. Die Struktur von Cyclosporin C. Helv Chim Acta 60:1247–1255, 1977.
80. R Traber, M Kuhn, HR Loosli, W Pache, A von Wartburg. Neue Cyclopeptide aus *Trychoderma polysporum*: die Cyclosporine B, D and E. Helv Chim Acta 60:1568–1578, 1977.
81. R Traber, HR Loosli, H Hofmann, M Kuhn, A von Wartburg. Isolierung und Struktermittlung der neuen Cyclosporine E, F, G, H und I. Helv Chim Acta 65:1655–1677, 1982.
82. R Traber, H Hoffmann, HR Loosli, M Ponelle, A von Wartburg. Neue Cyclosporine aus *Tolypocladium inflatum*. Die Cyclosporine K-Z. Helv Chim Acta 70:13–36, 1987.
83. R Zocher, N Madry, H Peeters, H Kleinkauf. Biosynthesis of cyclosporin A. Phytochemistry 23:549–551, 1984.
84. A Lawen, R Zocher. Cyclosporin synthetase-the most complex peptide synthesizing multienzyme polypeptide so far described. J Biol Chem 265:11355–11360, 1990.
85. J Dittmann, RM Wenger, H Kleinkauf, A Lawen. Mechanism of cyclosporin A biosynthesis. J Biol Chem 269:2841–2846, 1994.
86. JD Walton. Peptide phytotoxins from plant pathogenic fungi. In: H. Kleinkauf, H von Döhren, eds. Biochemistry of Peptide Antibiotics. Berlin: W de Gruyter, 1990, pp 179–203.
87. Y-Q Cheng, JD Walton. A eukaryotic alanine racemase gene involved in cyclic peptide biosynthesis. J Biol Chem 275:4906–4911, 2000.
88. M Offenzeller, Z Su, G Santer, H Moser, R Traber, K Memmert, E Schneider-Scherzer. Biosynthesis of the unusual amino acid (*4R*)-4-[*E*]-2-butenyl]-4-methyl-L-threonine of cyclosporin A. J Biol Chem 268:268-26127–26134, 1993.
89. M Offenzeller, G Santer, K Totschnig, Z Su, H Moser, R Traber, E Schneider-Scherzer. Biosynthesis of the unusual amino acid (4*R*)-4-[(*E*)-2-butenyl]-4-

methyl-L-threonine of cyclosporin A: enzymatic analysis of the reaction sequence including identification of the methylation precursor in a polyketide pathway. Biochemistry 35:8401–8412, 1996.
90. R Zocher, T Nihira, E Paul, N Madry, H Peeters, K Kleinkauf, U Keller. Biosynthesis of cyclosporin A: partial purification and properties of a multifunctional enzyme from *Tolypocladium inflatum*. Biochemistry 25:550–553, 1986.
91. G Weber, E Leitner. Disruption of the cyclosporin synthetase gene of *Tolypocladium niveum*. Curr Genet 26:461–467, 1994.
92. E Schneider, E Leitner, G Weber, K Schörgendorfer. Cyclosporin synthetase. European Patent 0578616 [Sandoz Ltd (CH); Sandoz AG (DE)], 1994.
93. E Guenzi, G Galli, I Grgurina, DC Gross, G Grandi. Characterization of the syringomycin synthetase gene cluster. A link between procaryotic and eucaryotic peptide synthetases. J Biol Chem 273:32857–32863, 1998.
94. YS Im, PK Chiang, GL Cantoni. Guanidoacetate methyltransferase. J Biol Chem 254:11047–11050, 1979.
95. RH Lin, N Narasimhachari, HE Himwich. Inhibition of indolethylamine-N-methyltransferase by S-adenosylhomocysteine. Biochem Biophys Res Commun 54:751–759, 1973.
96. M Glinski. Biosynthesis of cyclosporin A and enniatin B. Ph.D. thesis, Technische Universität Berlin, Berlin, 1999.
97. A Lawen, R Traber. Substrate specificities of cyclosporin synthetase and peptolide SDZ 214-103 synthetase. J Biol Chem 268:20452–20465, 1993.
98. A Lawen, R Traber, D. Geyl. *In vitro* biosynthesis of [Thr^2, Leu^5, D-Hiv^8, Leu^{10}]-cyclosporin, a cyclosporin-related peptolide, with immunosuppressive activities by a multienzyme polypeptide. J Biol Chem 266:15567–15570, 1991.
99. R McDaniel, S Ebert-Khosla, DA Hopwood, C Khosla. Engineered biosynthesis of novel polyketides. Science 262:1546–1550, 1993.
100. R McDaniel, S Ebert-Khosla, DA Hopwood, C Khosla. Rational design of aromatic polyketide natural products by recombinant assembly of enzymatic subunits. Nature 375:549–554, 1995.
101. J Cortes, KEH Wiesmann, GA Roberts, MJB Brown, J Staunton, PF Leadley. Repositioning of a domain in a modular polyketide synthase to promote specific chain cleavage. Science 268:1487–1489, 1995.
102. CR Hutchinson, I Fujii. Polyketide synthase gene manipulation: structure-function approach in engineering novel antibiotics. Annu Rev Microbiol 49:201–238, 1995.
103. CJ Tsoi, C Khosla. Combinatorial biosynthesis of ''unnatural'' natural products: the polyketide example. Chem Biol 2:355–362, 1995.
104. L Tang, H Fu, R McDaniel. Formation of functional heterologous complexes using subunits from the picromycin, erythromycin and oleandomycin polyketide synthases. Chem Biol 7:77–84, 2000.
105. T Stachelhaus, A Schneider, MA Marahiel. Rational design of peptide antibiotics by targeted replacement of bacterial and fungal domains. Science 269:69–72, 1995.

20

New Strategies for Target Identification, Validation, and Use of Enzymes in High-Throughput Screening

Joaquim Trias and Zhengyu Yuan
Versicor, Inc., Fremont, California

I. INTRODUCTION

The essentiality of any given gene in a microorganism depends on the genetic background of the strain and the outside environment. Thus, distinct environments can modify the essentiality of a gene; for example beta-lactamase genes are needed for bacterial survival in the presence of beta-lactams but not in their absence. By changing the environment, for instance, by adding an antibiotic to a medium or using *in vivo* models, one can expand the number of known essential genes of a given strain. All antibiotics in clinical use inhibit essential bacterial functions *in vivo*. Genes that are essential for growth, or required to develop an infection, or to confer resistance to antibiotics, are potential targets for drug discovery [1,2]. Despite the large amount of literature available on putative bacterial essential genes, only a few published reports provide conclusive evidence. Most experiments are based on negative data, such as the inability to knock out a specific gene [3–5] or use of conditional lethal mutants [6]. Both approaches suggest, but do not demonstrate, essentiality of a wild-type nonmutated target. Target-based antibiotic discovery programs are effort-intensive and require a solid foundation of the quality of the target. Methods that firmly establish the essential nature of a large number of genes are needed for target discovery and evaluation.

Enzymes are attractive targets for new antibacterial discovery. Some of the most successful antibiotics, such as beta-lactams or quinolones, inactivate the transpeptidation reaction of penicillin-binding proteins and DNA topoisomerases, respectively [7,8]; these enzymes catalyze essential reactions for bacterial growth. New enzyme inhibitors can be identified through mechanism-based drug design. This strategy has been successfully applied to the discovery of drugs such as HIV protease inhibitors [9] but is more difficult to implement in antibacterial discovery programs. Inhibition of bacterial growth is a multifactorial process; this complexity limits the application of rational drug design to antibiotic discovery. An active compound has to cross membranes, avoid being effluxed, face hydrolysis or modification by a myriad of enzymes, and reach its bacterial target at concentrations sufficient to have a significant impact on cell growth. A better drug discovery approach is to increase the probability of finding novel inhibitors by combining a mechanism-based pharmacophore, designed to inhibit a specific target, or set of targets sharing a common feature, with large structural diversity. Purified targets and bacterial strains are screened with compound collections or combinatorial chemical libraries that are made of transition-state analogs, or other rationally designed analogs, and are biased to inhibit these targets. Metallo-hydrolases or Mur ligases are well suited for this discovery strategy. Biased libraries can be set up even when there is little information on a specific target. For instance, a putative metallo-enzyme identified through motif or homology searches can be screened for inhibitors using highly diverse combinatorial chemical libraries synthesized around a chelator moiety, such as hydroxamic acid or sulfhydroxyl groups, even when the function of the target remains unknown. The chelator moiety adds a strong bias to the library toward metallo-enzymes that other random compound collections do not have.

Enzyme-based screens will identify hits that inactivate the target but do not necessarily stop cell growth. Compounds that inhibit both enzyme activity and cell growth are of obvious interest. However, these activities may not be related. It is necessary to establish whether there is a causative relation between inactivation of the target of interest and inhibition of bacterial growth. Otherwise the discovery program may be pursuing two unrelated activities that more often than not reflect underlying toxicity problems [10]. Thus, at least two kinds of assays are needed: one that measures the inhibition of the purified target and another that indicates whether inhibition of cell growth occurs through the expected mechanism of action.

This chapter focuses on a novel antibiotic discovery paradigm. Metallo-hydrolases and Mur ligases are used to illustrate this approach. New methods to identify and prioritize targets, develop screens, and evaluate new inhibitors are discussed. New developments in enzyme-based assays, such as pathway assays, are also presented. This new approach is opening new venues for screening targets that are difficult to screen because substrates are not easily available.

II. METALLO-HYDROLASES: A FAMILY OF TARGETS FOR NOVEL ANTIBIOTIC DISCOVERY

A. Evaluation of Candidate Metallo-Hydrolases and Construction of Screening Strains

A large number of metallo-hydrolases are present in bacteria. However, only a few are known to be essential for growth or involved in resistance (see examples in Table 1). None of them is the target of antibiotics in clinical use. Candidate metallo-hydrolases for new antibiotic discovery were identified from public genomes searching for homologs of known metal-enzymes or identifying open reading frames (ORFs) that have defined motifs of metallo-binding sites. Multiple alignment tools were used to identify orthologs or paralogs from the retrieved sequences. The distribution in key bacterial pathogens and eukaryotes was determined (see Table 1). Genetic evaluation of the candidate ORFs provides evidence of the essentiality of the target.

Targets are evaluated using gene-to-screen (GTS) technology. This technology provides conclusive data on the essentiality of a gene and screening strains. GTS strains have a specific chromosomal gene under regulatable promoter control [11]. If the gene is essential, then the strain is inducer-dependent for growth. Transcriptional control is chosen to modulate the level of a given target in *Escherichia coli* because of the extensive information available on promoter regulation. The positively regulated P_{BAD} promoter is an ideal regulatory system for modulating gene expression for several reasons: (1) the expression of an essential gene can be down-regulated to a level that does not support growth, an event similar to antibiotic action [11–13]; (2) P_{BAD} can be induced to high levels of expression [14]; and (3) the promoter is titratable over a range of arabinose concentrations [15]. GTS strains have only a single copy of the target gene in the chromosome in a monocistronic operon under regulatable promoter control. The original gene is deleted in all constructs to eliminate any potential polar effect on downstream genes [11].

GTS strains have been constructed for a number of genes that code for metallo-hydrolases (see Table 1) and used to evaluate these target candidates. GTS constructs have demonstrated essentiality of most of these genes with the exception of *ftsH*, coding for a zinc-containing protease. Temperature-sensitive mutants in the *ftsH* gene have been reported, and this gene was assumed to be essential [16,17]. However, the lack of arabinose dependence of the *ftsH*-GTS strain, and the ability to knock out this gene, show that this gene is not essential in *E. coli* (J. Trias, unpublished). These results illustrate the need for demonstrating essentiality with conclusive experiments for all candidate targets.

GTS constructs are also used for screening and establishing mechanism of action of enzyme inhibitors in bacterial cells. The intracellular concentration of the target under P_{BAD} control is modulated by the amount of inducer in the growth

Table 1 Candidate Metallo-Hydrolases for New Antibiotic Discovery in Bacteria

Enzyme (gene)	Distribution	Arabinose dependence of GTS strain[a]	Other methods to assess essentiality	Interest for drug discovery
Peptide deformylase (*def*)	Bacteria, chloroplasts	+ [18]	Inability to knock out gene [58]	Essential for growth
UDP-3-O-acyl-N-acetylglucosamine deacetylase (*lpxC*)	Gram-negative bacteria	+	Temperature-sensitive mutant [59]	Essential for growth
Methionine aminopeptidase (*pepM*)	Bacteria, eukaryotes	+	Gene under *lac* promoter control [60]	Essential for growth
FtsH protease (*ftsH*)	Bacteria, eukaryotes	−	Temperature-sensitive mutant [17]	Not suitable for drug discovery
N-succinyl-L-diaminopimelic acid desuccinylase (*dapE*)	Gram-negative and some Gram-positive bacteria	+	Temperature-sensitive mutant [23,61]	Essential for growth
D-Ala-D-Ala peptidase (*vanX*)	Vancomycin-resistant enterococci	NA		Potentiation of glycopeptides against vancomycin-resistant enterococci [27–29]
Beta-lactamase	*Aeromonas* spp., *Stenotrophomonas maltophilia, Bacteroides fragilis*	NA		Potentiation of beta-lactams against strains expressing a metallo-beta-lactamase [26]

[a] J. Trias, unpublished.
NA, not applicable.

medium [15]. At low arabinose concentrations, the strain becomes hypersusceptible to compounds that inhibit the down-regulated target (Fig. 1) (J. Trias unpublished, [11]). On the other hand, at high arabinose concentrations, the susceptibility to the specific inhibitor is much lower. This association does not exist for compounds that inhibit other targets (see Fig. 1). The selective hypersusceptible

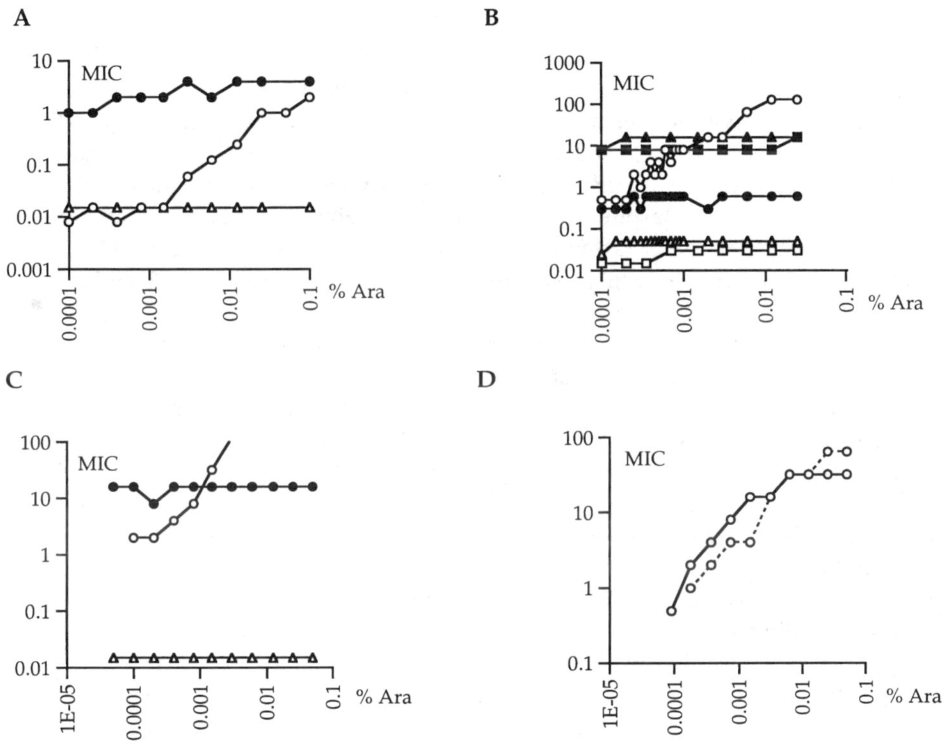

Figure 1 Susceptibility of strains with a target under arabinose promoter control to specific inhibitors of the regulated target (open circles), and to other antibiotics, at a range of arabinose concentrations (MICs in μg/ml). (A), P_{BAD}-*folA* strain; susceptibility to trimethoprim (open circles), fosfomycin (filled circles), and ciprofloxacin (open triangles). (B), P_{BAD}-*murA* strain; susceptibility to fosfomycin (open circles), tetracycline (filled circles), ciprofloxacin (open triangles), D-cycloserine (filled triangles), cefotaxime (open squares), and ampicillin (filled squares). (C), P_{BAD}-*def* strain; susceptibility to actinonin (open circles), fosfomycin (filled circles), and ciprofloxacin (open triangles); the susceptibility of P_{BAD}-*def* strain to actinonin at high arabinose concentrations was over 100 μg/ml. (D), P_{BAD}-*murA* strain; susceptibility to VRC474 (continuous line) and fosfomycin (broken line). (J. Trias, unpublished; [11].)

nature of these strains is used to screen for inhibitors of the down-regulated target and to determine the mechanism of action of inhibitors identified in screens that use purified enzyme. The method is independent of the function of the target, can be applied to any essential gene, and can be implemented with minimal effort to a large number of targets.

B. New Inhibitors of Metallo-Hydrolases

All the essential metallo-enzymes listed in Table 1 can be the starting point for a novel antibiotic discovery program. Peptide deformylase (PDF) and UDP-3-O-acyl-N-acetyl-glucosamine deacetylase are attractive because these enzymatic activities are absent in human cells [18–23]. Methionine aminopeptidase (MAP) is present in both bacteria and eukaryotes [24]. The latter have two isozymes, one of which is closely related to the bacterial enzyme, and the other resembling that of Archea [25]. Despite the high homology between the bacterial and human enzymes, it is conceivable to discover a selective inhibitor of bacterial MAP. D-Alanyl-D-alanine dipeptidase and metallo-beta-lactamases are involved in resistance to glycopeptides and beta-lactams [26–29]. Inhibition of these factors of resistance will expand the spectrum of activity of existing antimicrobials to include some resistant strains [26–29].

Onishi and co-workers [19] reported a series of oxazoline compounds with a hydroxamic acid moiety that inhibited UDP-3-O-acyl-N-acetyl-glucosamine deacetylase, a key enzyme of the lipopolysaccharide biosynthesis pathway in Gram-negative bacteria. The hydroxamic acid was required for enzyme inhibition, suggesting that inactivation took place through the chelation of the metal ion in the enzyme. The most potent analog, L-161,240 has an IC_{50} around 12 nM (see Fig. 2). This compound is active against enzymes from many different species, including *Pseudomonas aeruginosa*. However, L-161,240 is active only against *E. coli* and closely related species; *P. aeruginosa* and some enterobacterial species such as *Klebsiella pneumoniae* or *Enterobacter cloacae* are naturally resistant to this series of compounds. A few analogs were tested in an *E. coli* mouse septicemic model and protected mice from infection. Unfortunately, both bacteria and compound were given by the same route, intraperitoneal injection. A different experiment where the routes of infection and compound administration are different is needed to reach firm conclusions on the *in vivo* activity of this series.

Chromosomal metallo-beta-lactamases that hydrolyze carbapenem antibiotics, such as imipenem, meropenem, or biapenem, are present in some *Stenotrophomonas, Bacteroides*, and *Aeromonas* strains [26]. Some clinical *P. aeruginosa* and *Serratia marcescens* isolates have a plasmid that carries metallo-beta-lactamase genes [26]. These enzymes are not inactivated by inhibitors of serine-based beta-lactamases such as clavulanic acid or sulbactam analogs. Enzyme-

Compound	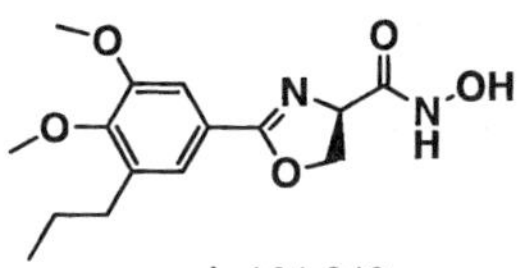	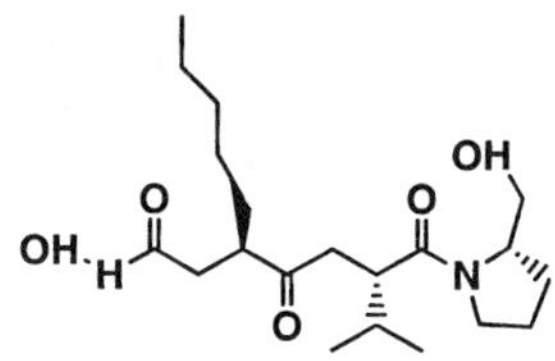	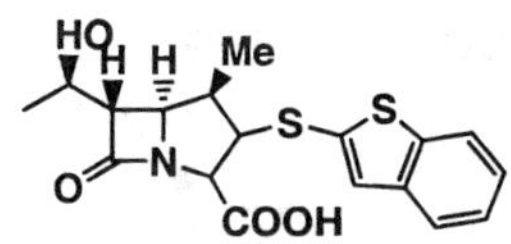
	L-161,240	Actinonin	J-110,441
Target	UDP-3-O-acyl-N-acetyl-glucosamine deacetylase	Peptide deformylase	Metallo-beta-lactamases
IC_{50} (nM)	12	0.8	<100
MIC (μg/ml)			
E. coli	1	>100	NA
S. aureus	>100	4	NA

Figure 2 Antimicrobial compounds that inhibit metallo-hydrolases. NA, not applicable.

based screens have identified several compounds (methylcarbapenems, biphenyl tetrazoles, and several thioester derivatives) that inhibit members of this class of beta-lactamases [30–32]. These compounds potentiate the activity of imipenem against carbapenem-resistant strains, and could be administered in combination with a carbapenem to extend the spectrum of activity.

Peptide deformylase is an iron-containing bacterial enzyme that removes the formyl moiety from nascent peptides. The gene that encodes for this activity, *def*, is essential in bacteria (see Table 1) and is present in all public bacterial genomes where the sequence is complete. The active center of the enzyme is similar to thermolysin and matrilysin, well-known metallo-hydrolases [18]. A hydroxamic acid-containing deformylase inhibitor, actinonin, was discovered by screening a biased compound collection [18]. This compound strongly inhibits deformylase from *E. coli* and *Staphylococcus aureus*, and is active against Gram-positive bacteria and fastidious Gram-negative organisms. The compound is actively effluxed in *E. coli*, as shown by high susceptibility of efflux pump *E. coli* mutants [18]; this is most likely the reason for the natural resistance of most Gram-negative bacteria.

A *def*-GTS construct was used to show that cell growth inhibition happens through the inhibition of PDF rather than another target [18]. The intracellular concentration of deformylase is modulated by the amount of inducer in the growth medium. At low arabinose concentrations the strain became hypersusceptible to compounds that inhibit the down-regulated target. On the other hand, at high arabinose concentrations the susceptibility to the specific inhibitor was much lower. This association does not exist for compounds that inhibit other targets (see Fig. 1, J. Trias unpublished, [11]). The susceptibility of a *def*-GTS strain to actinonin was strongly associated with the concentration of inducer, thus demonstrating that the major target of actinonin in *E. coli* is deformylase (Fig. 1 [18]).

III. PATHWAY ASSAYS: A NEW APPROACH FOR ENZYME-BASED SCREENING

Mur ligases catalyze the consecutive addition of amino acids and dipeptide to UDP-N-acetylmuramic acid, a building block of the bacterial peptidoglycan [33]. Arabinose-dependent GTS strains show that ligase genes are essential in *E. coli*. A fifth ligase, *mpl*, is not essential and is involved in the recycling of the peptidoglycan [34]. This enzyme is present only in enterobacteria and catalyses the addition of the tripeptide to UDP-N-acetylmuramic acid (Fig. 3). Mur ligase assays are cumbersome to run in high-throughput mode because of the lack of UDP-derived substrates. These compounds are not commercially available and have to be purified from cells [35], synthesized chemically [36] or enzymatically [37]. Two independent groups have recently developed ligase pathway assays [37–39]. These assays use six enzymes of the cytoplasmic steps of murein synthesis (MurA through MurF, see Fig. 1). MurA and MurB enzymes are involved in the synthesis of UDP-N-acetylmuramic acid; MurC, MurD, MurE, and MurF ligases catalyze the consecutive addition of amino acids and D-alanyl-D-alanine dipeptide to this substrate [33]. All six steps can be monitored in a single test vessel starting with the commercially available MurA substrate UDP-N-acetylglucosamine, required amino acids, and ATP, without the need to produce complex UDP-containing substrates for the intermediate steps.

Pathway screens require that the concentrations of enzymes and substrates are optimized so the rate of final product formation is linearly dependent on all enzyme concentrations. The concentration of each enzyme is chosen within the sensitive range at which any decrease of the enzyme concentration will cause a proportional signal decrease. In addition, the reaction signal is reduced to background if the concentration of any one of the enzymes or substrates is not included. High-throughput screening requires detection methods that are amenable to the large number of samples tested. Formation of product can be monitored by high-performance liquid chromatography (HPLC) or by direct absorption scintillation assay (DASA) (Z. Yuan, unpublished; [37–39]). The latter is based on the binding properties of a positively charged scintillant-containing solid surface that will selectively absorb the UDP-containing products, but not the amino acids, under proper binding conditions. Tritiated D-Ala-D-Ala, which itself does not bind to the plate, is used to monitor the production of the final reaction product. It is possible to follow the reaction progress of any individual ligase by using

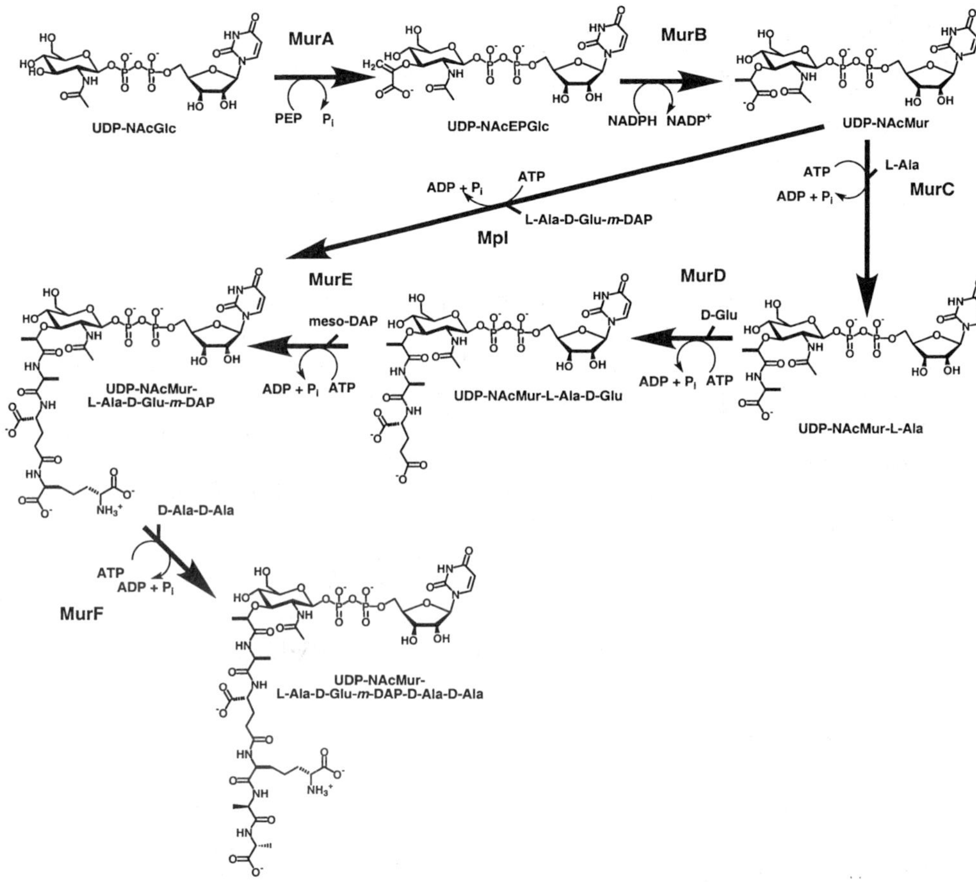

Figure 3 Schematic representation of the cytoplasmic steps of biosynthesis of UDP-N-acetylmuramic acid pentapeptide in *E. coli*. UDP-NAcGlc, UDP-N-acetylglucosamine; PEP, phosphoenolpyruvate; UDP-NAcEPGlc, UDP-N-acetylenolpyruvylglucosamine; UDP-NAcMur, UDP-N-acetylmuramic acid; *meso*-DAP, *meso*-diaminopimelic acid.

different radiolabeled amino acids. This detection method is significantly more efficient for high-throughput mode than the HPLC-based method.

Several hits with IC_{50} less than 10 μg/ml were identified using this DASA-pathway assay, including one compound with an IC_{50} of 0.08 μg/ml [38,40]. To deconvolute the effects of potential inhibitors in the pathway, inhibition of individual enzymes is tested for hits identified in the primary screening. For ex-

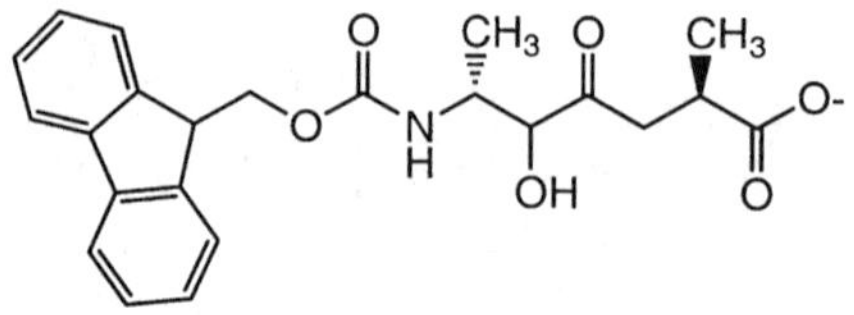

Figure 4 Chemical structure of MurA inhibitor VRC374 identified with DASA screen.

ample, VRC374 was identified as an inhibitor for the pathway assay with IC_{50} of 25 μg/ml (Fig. 4) [38,40]. It is apparent from these data that the inhibition observed in the pathway assay is due to the specific inhibition of MurA. This molecule inhibited a GTS strain carrying MurA under P_{BAD} control in an inducer concentration-dependent fashion, as expected for a MurA inhibitor (Fig. 1; J. Trias, unpublished; [11]).

Pathway assays have several advantages: (1) individual assays can be combined into one, saving labor and reagents; (2) the reaction product of a previous step is the substrate for the subsequent reaction and avoids the need to obtain the individual substrates; (3) a pathway assay can detect ''dead-end'' substrates, compounds that are capable of being incorporated into the product of previous reaction but prevent the further elongation of the pathway [37].

IV. CONCLUSIONS AND FUTURE DIRECTIONS

The ultimate goal of any target identification and evaluation program is to select targets that have higher chances for success in the discovery of novel antibiotics. Enzymes that perform functions essential for cell growth, virulence, or resistance to antibiotics are good choices for novel antibiotic drug-discovery projects. GTS technology demonstrates the essentiality of a target and provides screening strains. In addition, the same constructs are used to determine the mechanism of action of new enzyme inhibitors active against whole cells. This powerful technology has been developed in *E. coli* and is under development in other key pathogens. Evidence on the essentiality of a gene obtained in *E. coli* cannot always be extrapolated to other species. For example, lipoprotein leader peptidase is essential in *E. coli* [41]; on the other hand, the *lsp* homolog can be disrupted in *S. aureus* [42]. The major target of an antibiotic can also be different. The primary target of ciprofloxacin in *E. coli* is gyrase [7,43,44], while topoisomerase IV is a secondary target [7]. In *S. aureus* the order is inverted and topoisomerase IV is the major target [45,46]. GTS can be developed in other bacterial species. The key feature of GTS technology is the use of a regulatable promoter. P_{BAD} and other promoters regulated by members of the AraC/Xy1S family of positive transcriptional regulators have been successfully used in other Gram-negative bacteria [47,48], and are natural choices to develop GTS strains in this group of organisms. The use of this technology in Gram-positive bacterial pathogens is more difficult because of the limited number of known regulatable promoters. The *tet* regulon has been used in *S. aureus* [49] and a positively regulated promoter, P_A, has been reported in *S. pneumoniae* [50]. These promoters are good starting points for applying GTS technology in Gram-positive bacteria (J. Trias, unpublished).

Enzymes that are amenable to mechanism-based drug discovery are better choices for novel antibiotic discovery than targets that do not have any feature that can be used for designing an inhibitor. The combination of a family of targets, such as the metallo-enzyme family, and a biased compound collection based on molecules that have the potential to inactivate these targets, dramatically increases the chances of identifying novel inhibitors. The discovery of actinonin as a potent PDF inhibitor illustrates the success of this approach [18]. The number of members of the metallo-enzyme family will increase as new motifs that define metal-binding sites are identified. There are many other families of targets, for example, Mur ligases or transferases, where the same approach can be applied. The different classes of proteases are particularly promising; a few of them, such as serine- or aspartyl-protease families, have well-defined inhibitor pharmacophores [51]. A few bacterial proteases, such as leader peptidase I or HtrA protease, are essential for growth or play a major role in the processing of virulence factors and are the target of novel antimicrobial programs [52–54]. A large number of chemical entities derived from protease inhibitor projects of therapeutic areas other than bacterial infectious diseases are available for screening [51]. This universe of existing biased compounds makes protease targets very attractive for the discovery of novel antibacterials.

The large number of targets available requires the development of screens that save time, reagents, and test compounds. Pathway screens allow the testing of several targets at once, and use fewer resources than screens based on individual enzymes. In fact, Mur ligases can only be screened effectively in HTS format using this approach [37,38]. Other pathways, such as the postranslational modification steps of initiation of protein synthesis or fatty acid synthesis in bacteria, are also amenable to pathway screening. The increasing application of rational drug-design strategies, which are often target driven, emphasizes the need to develop assays that address other major issues involved in the susceptibility of bacteria to any given antibiotic. Screens that address the role of factors involved in natural resistance are needed. The role of efflux pumps and the permeability barrier of the outer membrane is well documented in *E. coli* and other Gram-negative bacteria, and hypersusceptible strains have been constructed [55]. Factors involved in natural resistance in Gram-positive bacteria remain largely unknown. Most of the research on this group of organisms is focused on acquired resistance [56] or on pathogens that are naturally resistant to most antibiotics in clinical use, such as *Mycobacterium* species [57]. A set of hypersusceptible mutants in Gram-negative and Gram-positive bacteria would increase the sensitivity of screens for HTS and help mechanism-based medicinal chemistry programs to improve the whole-cell activity of enzyme inhibitors.

REFERENCES

1. MY Galperin, EV Koonin. Searching for drug targets in microbial genomes. Curr Opin Biotechnol 10:571–578, 1999.
2. J Trias, EM Gordon. Innovative approaches to novel antibacterial drug discovery. Curr Opin Biotechnol 8:757–762, 1997.
3. BJ Akerley, EJ Rubin, A Camilli, DJ Lampe, HM Robertson, JJ Mekalanos. Systematic identification of essential genes by *in vitro* mariner mutagenesis. Proc Natl Acad Sci (USA) 95:8927–8932, 1998.
4. RH Baltz, FH Norris, P Matsushima, BS DeHoff, P Rockey, G Porter, S Burgett, R Peery, J Hoskins, L Braverman, I Jenkins, P Solenberg, M Young, MA McHenney, PL Skatrud, PR Rosteck, Jr. DNA sequence sampling of the *Streptococcus pneumoniae* genome to identify novel targets for antibiotic development. Microb Drug Resist 4:1–9, 1998.
5. CA Hutchison, SN Peterson, SR Gill, RT Cline, O White, CM Fraser, HO Smith, JC Venter. Global transposon mutagenesis and a minimal *Mycoplasma* genome. Science 286:2165–2169, 1999.
6. CG Sevastopoulos, CT Wehr, DA Glaser. Large-scale automated isolation of *Escherichia coli* mutants with thermosensitive DNA replication. Proc Natl Acad Sci (USA) 74:3485–3489, 1977.
7. K Drlica, X Zhao. DNA gyrase, topoisomerase IV, and the 4-quinolones. Microbiol Mol Biol Rev 61:377–392, 1997.
8. BG Spratt, KD Cromie. Penicillin-binding proteins of Gram-negative bacteria. Rev Infect Dis 10:699–711, 1988.
9. BD Dorsey, RB Levin, SL McDaniel, JP Vacca, JP Guare, PL Darke, JA Zugay, EA Emini, WA Schleif, JC Quintero, JH Lin, IW Chen, MK Holloway, PMD Fitzgerald, MG Axel, D Ostovic, PS Anderson, JR Huff. L-735,524: the design of a potent and orally bioavailable HIV protease inhibitor. J Med Chem 37:3443–3451, 1994.
10. JJ Hilliard, RM Goldschmidt, L Licata, EZ Baum, K Bush. Multiple mechanisms of action for inhibitors of histidine protein kinases from bacterial two-component systems. Antimicrob Agents Chemother 43:1693–1699, 1999.
11. D Young, NG Rafanan, CI Rosenow, D Chen, Z Yuan, R White, J Trias. A novel means of creating selective hypersusceptible screening strains by down regulation of essential genes. General Meeting of the American Society for Microbiology, Chicago, 1999, pp 23–24.
12. GJ Phillips, TJ Silhavy. The *E. coli ffh* gene is necessary for viability and efficient protein export. Nature 359:744–746, 1992.

13. RE Dalbey, W Wickner. Leader peptidase catalyzes the release of exported proteins from the outer surface of the *Escherichia coli* plasma membrane. J Biol Chem 260: 15925–15931, 1985.
14. S Johnston, JH Lee, DS Ray. High-level expression of M13 gene II protein from an inducible polycistronic messenger RNA. Gene 34:137–145, 1985.
15. LM Guzman, D Belin, MJ Carson, J Beckwith. Tight regulation, modulation, and high-level expression by vectors containing the arabinose P_{BAD} promoter. J Bacteriol 177:4121–4130, 1995.
16. Y Akiyama, T Ogura, K Ito. Involvement of FtsH in protein assembly into and through the membrane. I. Mutations that reduce retention efficiency of a cytoplasmic reporter. J Biol Chem 269:5218–5224, 1994.
17. JN Qu, SI Makino, H Adachi, Y Koyama, Y Akiyama, K Ito, T Tomoyasu, T Ogura, H Matsuzawa. The *tolZ* gene of *Escherichia coli* is identified as the *ftsH* gene. J Bacteriol 178:3457–3461, 1996.
18. DZ Chen, P Patel, CJ Hackbarth, W Wang, G Dreyer, DC Young, PS Margolis, C Wu, Z-J Ni, J Trias, RJ White, Z Yuan. Actinonin, a naturally occurring antibacterial agent, is a potent deformylase inhibitor. Biochemistry 39:1256–1262, 2000.
19. HR Onishi, BA Pelak, LS Gerckens, LL Silver, FM Kahan, MH Chen, AA Patchett, SM Galloway, SA Hyland, MS Anderson, CR Raetz. Antibacterial agents that inhibit lipid A biosynthesis. Science 274:980–982, 1996.
20. IM Fearnley, JE Walker. Two overlapping genes in bovine mitochondrial DNA encode membrane components of ATP synthase. EMBO J 5:2003–2008, 1986.
21. A Yoshida, M Lin. NH_2-terminal formylmethionine- and NH_2-terminal methionine-cleaving enzymes in rabbits. J Biol Chem 247:952–957, 1972.
22. T Yagi, Y Hatefi. Identification of the dicyclohexylcarbodiimide-binding subunit of NADH-ubiquinone oxidoreductase (Complex I). J Biol Chem 263:16150–16155, 1988.
23. B Wu, C Georgopoulos, D Ang. The essential *Escherichia coli msgB* gene, a multicopy suppressor of a temperature-sensitive allele of the heat shock gene *grpE*, is identical to *dapE*. J Bacteriol 174:5258–5264, 1992.
24. SM Arfin, RL Kendall, L Hall, LH Weaver, AE Stewart, BW Matthews, RA Bradshaw. Eukaryotic methionyl aminopeptidases: two classes of cobalt-dependent enzymes. Proc Natl Acad Sci (USA) 92:7714–7718, 1995.
25. PJ Keeling, WF Doolittle. Methionine aminopeptidase-1: the MAP of the mitochondrion? Trends Biochem Sci 21:285–286, 1996.
26. K Bush. Metallo-beta-lactamases: a class apart. Clin Infect Dis 27(Suppl 1):S48–S53, 1998.
27. Z Wu, CT Walsh. Phosphinate analogs of D-,D-dipeptides: slow-binding inhibition and proteolysis protection of VanX, a D-,D-dipeptidase required for vancomycin resistance in *Enterococcus faecium*. Proc Natl Acad Sci (USA) 92:11603–11607, 1995.
28. IA Lessard, CT Walsh. VanX, a bacterial D-alanyl-D-alanine dipeptidase: resistance, immunity, or survival function?. Proc Natl Acad Sci (USA) 96:11028–11032, 1999.
29. PE Reynolds, F Depardieu, S Dutka-Malen, M Arthur, P Courvalin. Glycopeptide resistance mediated by enterococcal transposon Tn1546 requires production of VanX for hydrolysis of D-alanyl-D-alanine. Mol Microbiol 13:1065–1070, 1994.

30. R Nagano, Y Adachi, H Imamura, K Yamada, T Hashizume, H Morishima. Carbapenem derivatives as potential inhibitors of various beta-lactamases, including class B metallo-beta-lactamases. Antimicrob Agents Chemother 43:2497–2503, 1999.
31. JH Toney, PM Fitzgerald, N Grover-Sharma, SH Olson, WJ May, JG Sundelof, DE Vanderwall, KA Cleary, SK Grant, JK Wu, JW Kozarich, DL Pompliano, GG Hammond. Antibiotic sensitization using biphenyl tetrazoles as potent inhibitors of *Bacteroides fragilis* metallo-beta-lactamase. Chem Biol 5:185–196, 1998.
32. GG Hammond, JL Huber, ML Greenlee, JB Laub, K Young, LL Silver, JM Balkovec, KD Pryor, JK Wu, B Leiting, DL Pompliano, JH Toney. Inhibition of IMP-1 metallo-beta-lactamase and sensitization of IMP-1-producing bacteria by thioester derivatives. FEMS Microbiol Lett 179:289–296, 1999.
33. WS Faraci. Cytosolic enzymes in peptidoglycan synthesis as potential antibacterial targets. In: J Sutcliffe, NH Georgopapadakou, eds. Emerging Targets in Antibacterial and Antifungal Chemotherapy. New York: Chapman & Hall, 1992, pp 176–204.
34. D Mengin-Lecreulx, J van Heijenoort, JT Park. Identification of the *mpl* gene encoding UDP-N-acetylmuramate: L-alanyl-gamma-D-glutamyl-meso-diaminopimelate ligase in *Escherichia coli* and its role in recycling of cell wall peptidoglycan. J Bacteriol 178:5347–5352, 1996.
35. B Flouret, D Mengin-Lecreulx, J van Heijenoort. Reverse-phase high-pressure liquid chromatography of uridine diphosphate N-acetylmuramyl peptide precursors of bacterial cell wall peptidoglycan. Anal Biochem 114:59–63, 1981.
36. SA Hitchcock, CN Eid, JA Aikins, M Zia-Ebrahimi, LC Blaszczak. The first total synthesis of bacterial cell wall precursor UDP-N-acetylmuramyl-pentapeptide (Park nucleotide). J Am Chem Soc 120:1916–1917, 1998.
37. J Trias, Z Yuan. Mining bacterial cell wall biosynthesis with new tools: multitarget screens. Drug Res Updates 2:358–362, 1999.
38. D Chen, C Rosenow, J Trias, RJ White, Z Yuan. Pathway screening: novel technology for identifying inhibitors of MurA-F in a single incubation. 38th Interscience Conference on Antimicrobial Agents and Chemotherapy, San Diego, CA, 1998, p 273.
39. KK Wong, DW Kuo, RM Chabin, C Fournier, LD Gegnas, ST Wadell, F Marsilio, B Leiting, DL Pompliano. Engineering a cell-free murein biosynthetic pathway: combinatorial enzymology in drug discovery. J Am Chem Soc 120:13527–13528, 1998.
40. A Bryskier. Novelties in the field of anti-infectives in 1998. Clin Infect Dis 29:632–658, 1999.
41. M Regue, J Remenick, M Tokunaga, GA Mackie, HC Wu. Mapping of the lipoprotein signal peptidase gene (*lsp*). J Bacteriol 158:632–635, 1984.
42. AM Lowe, DT Beattie, RL Deresiewicz. Identification of novel staphylococcal virulence genes by *in vivo* expression technology. Mol Microbiol 27:967–976, 1998.
43. A Sugino, CL Peebles, KN Kreuzer, NR Cozzarelli. Mechanism of action of nalidixic acid: purification of *Escherichia coli nalA* gene product and its relationship to DNA gyrase and a novel nicking-closing enzyme. Proc Natl Acad Sci (USA) 74: 4767–4771, 1977.
44. M Gellert, K Mizuuchi, MH O'Dea, T Itoh, JI Tomizawa. Nalidixic acid resistance: a second genetic character involved in DNA gyrase activity. Proc Natl Acad Sci (USA) 74:4772–4776, 1977.

45. L Ferrero, B Cameron, J Crouzet. Analysis of *gyrA* and *grlA* mutations in stepwise-selected ciprofloxacin-resistant mutants of *Staphylococcus aureus*. Antimicrob Agents Chemother 39:1554–1558, 1995.
46. L Ferrero, B Cameron, B Manse, D Lagneaux, J Crouzet, A Famechon, F Blanche. Cloning and primary structure of *Staphylococcus aureus* DNA topoisomerase IV: a primary target of fluoroquinolones. Mol Microbiol 13:641–653, 1994.
47. N Mermod, JL Ramos, PR Lehrbach, KN Timmis. Vector for regulated expression of cloned genes in a wide range of gram-negative bacteria. J Bacteriol 167:447–454, 1986.
48. R Sukchawalit, P Vattanaviboon, R Sallabhan, S Mongkolsuk. Construction and characterization of regulated L-arabinose-inducible broad host range expression vectors in *Xanthomonas*. FEMS Microbiol Lett 181:217–223, 1999.
49. Y Ji, A Marra, M Rosenberg, G Woodnutt. Regulated antisense RNA eliminates alpha-toxin virulence in *Staphylococcus aureus* infection. J Bacteriol 181:6585–6590, 1999.
50. C Rosenow, M Maniar, J Trias. Regulation of the alpha-galactosidase activity in *Streptococcus pneumoniae*: characterization of the raffinose utilization system. Genome Res 9:1189–1197, 1999.
51. R Goldman, W Kohlbrenner, P Lartey, A Pernet. Antibacterial agents specifically inhibiting lipopolysaccharide synthesis. Nature 329:162–164, 1987.
52. T Inada, DL Court, K Ito, Y Nakamura. Conditionally lethal amber mutations in the leader peptidase gene of *Escherichia coli*. J Bacteriol 171:585–587, 1989.
53. M Paetzel, RE Dalbey, NC Strynadka. Crystal structure of a bacterial signal peptidase in complex with a beta-lactam inhibitor. Nature 396:186–190, 1998.
54. MJ Pallen, BW Wren. The HtrA family of serine proteases. Mol Microbiol 26:209–221, 1997.
55. H Nikaido. Prevention of drug access to bacterial targets: permeability barriers and active efflux. Science 264:382–388, 1994.
56. PC Hsieh, SA Siegel, B Rogers, D Davis, K Lewis. Bacteria lacking a multidrug pump: a sensitive tool for drug discovery. Proc Natl Acad Sci (USA) 95:6602–6606, 1998.
57. J Trias. Antibiotic transport in mycobacteria. In: N Georgopapadakou, ed. Drug Transport in Antimicrobial and Anticancer Chemotherapy. New York: Marcel Dekker, 1995, pp 269–288.
58. D Mazel, S Pochet, P Marliere. Genetic characterization of polypeptide deformylase, a distinctive enzyme of eubacterial translation. EMBO J 13:914–923, 1994.
59. K Young, LL Silver, D Bramhill, P Cameron, SS Eveland, CR Raetz, SA Hyland, MS Anderson. The *envA* permeability/cell division gene of *Escherichia coli* encodes the second enzyme of lipid A biosynthesis. UDP-3-O-(R-3-hydroxymyristoyl)-N-acetylglucosamine deacetylase. J Biol Chem 270:30384–30391, 1995.
60. SY Chang, EC McGary, S Chang. Methionine aminopeptidase gene of *Escherichia coli* is essential for cell growth. J Bacteriol 171:4071–4072, 1989.
61. M Karita, ML Etterbeek, MH Forsyth, MK Tummuru, MJ Blaser. Characterization of *Helicobacter pylori dapE* and construction of a conditionally lethal *dapE* mutant. Infect Immun 65:4158–4164, 1997.

21
Use of Genomics for Enzyme-Based Drug Discovery

Molly B. Schmid
Microcide Pharmaceuticals, Inc., Mountain View, California

I. INTRODUCTION

Genomics has changed our view of the biological world in the past decade, providing both new information and new tools to characterize biological systems. Over the past few years, dozens of microbial genomes—including several of substantial clinical importance—have been completely sequenced, pushing the search for novel antimicrobial compounds into the postgenomic era.

Genomic information and associated new technologies have the potential to revolutionize the drug discovery process. Tremendous new opportunities exist to identify targets for therapeutic intervention, and to initiate new screens for therapeutic agents. The concurrent increase in high-throughput capacity allows the realization of the genomic potential for significant increases in drug discovery efforts. The new strategies used in antimicrobial drug discovery may have application in numerous therapeutic areas.

II. STATE OF BACTERIAL GENOMICS

The development of technologies and strategies for efficient genomic sequencing and contig assembly has resulted in the completion of dozens of microbial genome sequences within the past five years [1–5]. With both academic and com-

---

Author's current address: Genencor International, Inc., Palo Alto, California.

mercial organizations engaged in sequencing microbial genomes, prokaryotic genome information is increasing rapidly. In publicly available databases, there are currently 22 completed eubacterial genomes and at least 87 in progress (www.tigr.org, as of May 2000). Additional microbial sequences are available in commercial sequence databases, such as those offered by Incyte Genomics, Inc. (www.incyte.com) and Genome Therapeutics Corp. (www.genomecorp.com). Up-to-date information about microbial genome sequences and functional genomic analysis of the sequence information can be obtained from numerous web sites, a few of which are listed in Table 1.

Coupled with a widely recognized need for new antimicrobial agents due to the rapid emergence of drug resistance [6], microbial genomics has been embraced by the pharmaceutical industry. Underlying the pharmaceutical interest in microbial genome sequences is the belief that target-specific screening and new targets will allow identification of new structural classes of antimicrobial agents that will inhibit the growth of microbes that are resistant to currently available chemotherapeutic agents.

Several general features of microbes and microbial genomes have emerged through the analysis of individual genomes and comparisons between them, as discussed in the following paragraphs.

1. The sizes of bacterial genomes vary over 10-fold. The smallest known bacterial genomes are those of the symbiotic mycoplasmas. These microorganisms live in close association with eukaryotic cells and characteristically lack a cell wall and possess unique sterol-containing plasma membranes [7]. The sequenced genome of *Mycoplasma genitalium* is 580 kbp, and was predicted to encode 470 proteins [4], while *Mycoplasma pneumoniae* is 810 kbp and was predicted to encode 678 proteins [8]. The largest bacterial genome fully sequenced to date is *Escherichia coli* K-12 [1] at 4.6 Mbp, while the largest eubacterial genome sequence projects currently in progress are in *Pseudomonas aeruginosa* (5.9 Mbp) and *Pseudomonas putida* (6.0 Mbp).

2. Bacterial genomes are information-rich. Over 85% of most sequenced bacterial genomes encode protein or RNAs, with only a small fraction of the genome dedicated to control regions and noncoding features. The *E. coli* K-12 genome is relatively typical: 87.8% of the 4.6-Mbp genome is predicted to encode 4288 proteins, 0.8% encodes stable RNAs, 0.7% is noncoding repeat sequences, with the remaining 11% as regulatory sequences, undetermined, or nonfunctional DNA sequence [1].

3. Bacterial genes are relatively small. Due to the lack of introns, an average bacterial gene is roughly 1 kbp. Thus, the majority of information about encoded proteins and pathways can be understood without the assembly of the genome sequence into a single contig. Thus, for most pharmaceutical applications, including target identification and validation, "nearly complete" bacterial

Table 1 Sources of Genomic Information on the Internet

Organization	Web site	Primary information
TIGR	http://www.tigr.org	Listing of completed and in-progress microbial genome projects; DNA sequences
Swiss Institute for Bioinformatics, University of Geneva	http://www.ebi.ac.uk/genomes	Dedicated to analysis of protein sequences and structures
National Center for Biotechnology Information	http://www.ncbi.nlm.nih.gov	National resource for DNA sequence information
National Center for Genome Resources	http://www.ncgr.org	PathDB—database of biochemical pathways and metabolism
Argonne National Laboratory	http://www.wit.mcs.anl.gov	WIT site—system to support the curation of function assignments made to genes and the development of metabolic models
Institute for Chemical Research, Kyoto University	http://www.genome.ad.jp/KEGG/Kegg.html	KEGG—Kyoto Encyclopedia of Genes and Genomes—pathway maps, orthologs for sequenced genomes
Research Collaboratory Structural Bioinformatics	http://www.rcsb.org	Protein Data Bank (PDB) of 3D macromolecular structure data

genome sequences provide adequate information for target identification and pathway assessment [9].

4. The conservation of protein sequence information among eubacteria is sufficient to recognize proteins with similar structure and anticipated function (‘‘orthologs’’) among all groups of eubacteria [4]. Typically, orthologous pro-

teins between Gram-positive and Gram-negative species will have 30–40% amino acid identity, whereas in closely related bacterial species, such as the Gram-negative species *E. coli* and *Salmonella typhimurium*, orthologous proteins will have 75–95% amino acid identity. Currently, approximately 70% of the predicted proteins in a typical eubacterial genome have at least one ortholog in another sequenced microbial genome.

5. Related bacterial genomes differ by the addition and subtraction of genes and pathways. Numerous plasmids, bacteriophage, natural transformation, and transposable elements mediate horizontal transfer of genes between both closely related and distant microbial species [10], and potentially between prokaryotic species and eukaryotes [11–13]. In addition, these same transposable elements can mediate chromosome rearrangements—duplications, deletions, and inversions—through legitimate and illegitimate recombination events. Comparisons of closely related species, such as *S. typhimurium* and *E. coli*, or even different strains of the same species show that gains and losses of DNA segments are frequent events in bacterial evolution [14–16].

6. Many microbial genes are of unknown function. Even in the well-studied bacterium *E. coli*, over 38% of the predicted proteins have no experimental data to support an understanding of their functional role in the cell.

III. CHARACTERISTICS OF A GOOD ANTIMICROBIAL TARGET

The discovery of new chemotherapeutic agents is difficult because of the numerous safety and efficacy criteria that must be satisfied by a single compound. For antimicrobials, the hurdles are extremely high—patients are frequently in crisis, but the desired therapeutic outcome is complete cure. Thus, a good antimicrobial agent must possess a wide range of different properties: strong and rapid inhibition of microbial growth; little interaction with mammalian targets at therapeutic doses; and desirable physical, chemical, and pharmacological properties that allow appropriate formulation, sufficient residence time within the body to achieve the antimicrobial goals, and the lack of significant damage to mammalian cells, tissues, or organs. In addition, a good antimicrobial agent should have relatively low resistance development, so that the clinician can be assured of the agent's efficacy for each patient, and so that the developing pharmaceutical company can achieve a reasonable return on its research and development investment. Frequently, broad antimicrobial activity spectrum is also desirable because it allows empiric treatment and first-line therapy.

Microbial genomics has vastly increased the number of potential antimicrobial targets over the past few years, making it necessary to identify criteria to

Table 2 Characteristics of a Good Antimicrobial Agent

Property	Relationship to target
Bactericidal/fungicidal	Killing versus inhibition of growth is thought to be primarily a property of the target as is the ability to cause bacterial or fungal cell lysis [17,98].
Low resistance frequency	Certain essential genes can be bypassed by inhibition of another gene product. Such situations may result in high frequency of resistance emergence [43,99,100].
Microbial breadth of spectrum	Existence or nonexistence of the target can be assessed through bioinformatics; however, such studies do not guarantee essentiality in the new species. In addition, efflux-based properties can influence a compound's breadth of spectrum [101].
Selectivity of inhibitors	Presence or absence of the target in eukaryotes can be assessed through bioinformatics; however, certain prokaryotic enzymes with clear orthologs in eukaryotes are the targets of successful antibiotics: trimethoprim [102]; fluoroquinolones [59].
Potency of inhibitors	Intracellular amount of enzyme may influence amount of the inhibitor necessary for potency; protein concentrations in bacteria vary over at least 3 orders of magnitude [103].
Pharmacological properties (serum binding, distribution, clearance, half-life, etc.)	Many of these properties may be influenced by the nonspecific affinity of the compound for host proteins; however, these properties are not generally predictable from target choice at this time.
Synthetic accessibility, simplicity, patentability	Size, shape, and amino acid composition of the inhibitor-binding site may influence the chemical classes of inhibitors, as well as the achievable selectivity.

assess and prioritize the targets for subsequent screen development. A rationale can be made that certain characteristics of a chemical inhibitor will be influenced by the characteristics of the target, while other characteristics appear independent or only partially dependent on the target (see Table 2).

IV. USING GENOMIC INFORMATION IN TARGET ASSESSMENT

Genome sequences provide a significant source of new information for expanding and prioritizing the set of antimicrobial targets for drug discovery. With the current state of knowledge, experimental validation of the predictions is still necessary in most cases. Fortunately, new methods have been developed to allow large-scale experimental assessment.

A. Predicting Efficacy

The typically desired goal of antimicrobial chemotherapy is to kill microorganisms, or at least to rapidly inhibit their growth through the chemical inhibition of one or more essential cellular processes. Proteins that will provide good antimicrobial targets must cause loss of viability or cessation of growth when the target protein is inhibited. Experimental methods have shown that most genes in microbial genomes are not individually essential for growth [17,18]. Thus, many microbial enzymes will not make good therapeutic targets, and a selection process is necessary to identify those whose inhibition can provide therapeutic benefit.

In silico methods aim to identify the pharmaceutically relevant genes from genome sequence information [19–21]. One computational strategy assumed that bacteria accomplished their essential functions through common mechanisms that would be evolutionarily conserved. Comparison of the completed genomes of *Haemophilus influenza* and *M. genitalium* showed that approximately 250 genes existed in both of the bacterial genomes and these were hypothesized to comprise a ''minimal genome set'' necessary to support bacterial viability [20]. However, as additional genomes were sequenced, the set of completely conserved genes became very small [22], putting into question the initial assumption of the complete conservation of essential genes in bacteria.

An experimental test of the hypothesis that conserved genes were essential was made by comparing the genomes of *E. coli* and *M. genitalium*. Of 26 genes that were conserved between *E. coli* and the small *M. genitalium* genome, only six genes caused loss of viability when disrupted; 20 genes were nonessential for growth under standard laboratory conditions [21]. Thus, selecting genes that are evolutionarily conserved among microbial species is not sufficient to reliably predict whether the genes are necessary for microbial viability.

Experimental approaches for antimicrobial target validation rely on genetic inactivation of the gene product and assessment of the resulting growth properties. These approaches use genetic methods to predict the idealized antimicrobial properties of a chemical inhibitor of the target [23–25]. Experimentally, mutational inactivation of a potential target has been accomplished by gene disruption, conditional expression, or conditional-function mutations. Initial assessment is generally undertaken *in vitro*, using standard laboratory conditions; additional validation in relevant *in vivo* animal infection models can be determined in certain systems to assure that inhibition of a gene product will cause significant attenuation of an infection.

Conditional expression [26–29] and conditional lethal [17,30,31] mutants have a long history of use in identifying and characterizing essential genes in haploid organisms. This approach allows assessment of the physiological consequences of the loss of a gene's function, and often provides insight into the gene's normal physiological role, as well as its essential or nonessential character [23,32]. In addition, conditional lethal mutant strains have been used for target-directed whole-cell screening strategies [24].

Gene disruption has been applied to individual genes to test essentiality [32–35], or in high-throughput mode to provide a genome-wide assessment of the essential genes in an organism [18,36]. Most high-throughput methods use an efficient transposon system to generate a population of random insertions, then use molecular methods to display the insertion sites that remain in the population after a period of growth [18,37,38]. Strains that carry disruptions of genes that cause inviability or slow growth are lost from the population, so that the display identifies all nonessential genes that can be disrupted without loss of viability.

High-throughput gene disruption methods led to the estimate that 265–350 genes of *M. genitalium* were essential for growth of this microorganism under standard laboratory conditions [18]. Of these genes, approximately one-third were of unknown function. Genetic methods, based on the genetic redundancy in a collection of conditional lethal mutants, led to the estimate of 100–200 essential genes that could be identified by a conditional lethal approach in *S. typhimurium* [39]. Both of these estimates count genes that are individually essential; functions necessary for viability that can be provided by either of two gene products will not be counted as essential in these studies.

Disruption strategies have disadvantages and pitfalls that should be recognized when interpreting data and making predictions of the essentiality of bacterial genes. The methods frequently use negative evidence—the inability to isolate a viable strain—to demonstrate essentiality. In addition, many bacterial genes exist in operons, in which disruption of an upstream gene can decrease transcription of downstream genes in the operon [40]. Thus, disruption of an upstream gene can cause the erroneous conclusion of its essentiality, if the downstream genes are not independently assessed, or if the operon structure is not recognized. Disruption strategies designed to circumvent the problems of bacterial operons and polarity have been devised [41]. When using disruption methods for assessing gene essentiality, it is also useful to remain cognizant of the relatively frequent occurrence of tandem duplication. In *S. typhimurium*, the spontaneous tandem duplication frequency at some loci was as high as 1% of the cells in a population; duplication of a locus can allow inheritance of the gene disruption, while still providing gene function [42], which could confound some assessments of gene essentiality.

B. Predicting Resistance Frequencies

High resistance frequencies arise in targets and pathways that have high evolutionary flexibility. An unusually high frequency of reversion in conditional lethal mutants of a potential target gene may foreshadow high-frequency resistance to chemical inhibitors of the target. The peptidyl deformylase of eubacteria removes the N-terminal formyl group from the initiator methionine of most bacterial proteins after translation. The gene, *def*, is essential in *E. coli*—a strain carrying a disruption of the *def* gene is not viable. However, disruption of the *def* gene can be inherited if cells have also lost the function of the initiator methionyl-tRNA formyltransferase enzyme (*fmt*), which transfers a formyl group onto the methionine [43]. Inhibitors of the deformylase have been identified; as predicted by the genetic evidence, resistance to these inhibitors arises as a result of inactivation of the formyltransferase activity [44].

Low resistance emergence has been associated with antibiotics that inhibit more than one essential target, since it is expected that each gene will require mutation for the cell to become resistant. An excellent example supporting this contention is the highly successful beta-lactam antibiotic class, whose members inhibit bacterial penicillin-binding proteins (PBPs). In most eubacterial genomes, multiple PBPs exist [45], and in *E. coli*, several of the PBPs are known to be genetically essential [46]. Resistance due to target-based mutation does not arise frequently, and the only clinically important resistance mechanisms are not target based, but rather due to beta-lactamases and efflux mechanisms.

However, choosing multigene targets does not guarantee low resistance frequencies. The fluoroquinolones are highly successful antibiotics that inhibit the two bacterial Type II topoisomerases, DNA gyrase and topoisomerase IV [47]. These two topoisomerases are structurally related, each enzyme is genetically essential, and both of these paralogous topoisomerases are found in most eubacterial genomes [48]. However, while commercially successful, the fluoroquinolones are subject to both target-based and efflux-mediated resistance development. Analysis of resistant isolates shows that resistance mutations arise frequently, each of which provides modest increases in resistance, but whose additive effects cause substantial increases in resistance to the organism [49, 50].

Nonetheless, there are several unexploited gene families that may provide additional multigene targets for inhibitors, such as the two-component signal transduction systems [32,51], tRNA synthetases [52], RNA polymerase sigma subunits [53,54], and the aminoacyl ligases of peptidoglycan biosynthesis [55].

C. Predicting Breadth of Spectrum and Selectivity

Genomic information can provide a rapid assessment of the potential for broad-spectrum activity or high microbial selectivity of a target. It is relatively easy to identify orthologs in the eubacteria, because even distantly related eubacteria have orthologous proteins with over 30–40% amino acid identity. Certain publicly available tools, such as the WIT (http://www.wit.mcs.anl.gov) and KEGG (http://www.genome.ad.jp/KEGG/Kegg.html) sites, provide extensive databases of orthologous genes in sequenced genomes. These metabolic catalogs provide an invaluable tool for assessing and organizing lists of orthologous proteins in microbes.

Table 3 lists information on eubacterial DNA polymerases obtained from the WIT database. The eubacterial DNA polymerase III enzyme is a multiprotein complex, best studied in *E. coli*. As can be observed from the table, a number of the proteins that are necessary for efficient replication in *E. coli* are not found in the genomes of other prokaryotic species. This raises questions about whether structurally dissimilar proteins provide these missing functions, or whether certain microbial systems do not require these functions. In contrast, there are examples of species that have two or more paralogs of a protein that is found in single copy in *E. coli*, such as the multiple genes encoding orthologs of the DNA polymerase III alpha subunit that are found in Gram-positive bacteria, such as *Bacillus subtilis* and *Mycobacterium tuberculosis*. In addition, Table 3 shows the nomenclature difficulties that arise in describing these proteins, and the importance of database structures that make it easy to correlate information at the functional level.

Additional factors that will affect the breadth of spectrum of an antimicrobial inhibitor cannot be predicted at present from genome information alone. Among these factors, efflux transporters may modulate the species specificity of numerous antimicrobial agents [56,57]. While many classes of efflux pumps can be recognized from sequence homology [56], no prediction of the compounds that will be subject to efflux currently can be derived from sequence information.

Genomic information is being used to predict the potential of a target for broad antimicrobial activity spectrum and high selectivity. Within the bacterial kingdom, orthologs can be determined with relative certainty. However, the amount of sequence dissimilarity that is necessary to assure selectivity is unknown; similar three-dimensional protein structures can be determined by significantly different primary sequences [58], and the success of the fluoroquinolones shows that selectivity can be achieved even when the microbial target has a mammalian homolog [59]. These factors introduce uncertainty about the wisdom of hoping for selectivity based on simple sequence analysis, and of using sequence-based selectivity arguments to eliminate otherwise good antimicrobial targets.

Table 3 Prokaryotic DNA Polymerases (EC 2.7.7.7)

			Pol III								
Organism	Pol I	Pol II	Alpha	Epsilon	Theta	Gamma	Delta	Delta′	Chi	Psi	Beta
Escherichia coli	b3863 (polA)	b0060 (polB)	b0184 (dnaE)	b0215 (dnaQ)	b1842 (holE)	b0470 (dnaX)	b0640 (holA)	b1099 (holB)			b3701 (dnaN)
Haemophilus influenzae	HI0856		HI0739	HI0137		HI1229	HI0923	HI0455	HI1397	HI0011	HI0992
Helicobacter pylori	HP1470		HP1460	HP1387		HP0717		HP1231			HP0500
Mycoplasma genitalium			MG261 MG031			MG419 MG420					MG001
Mycoplasma pneumoniae			A19 orf 872 B01 orf 1443			C12 orf 681		D12 orf 253			K05 orf 380
Chlamydia trachomatis	CT493		CT545	CT261 CT536		CT187 CT334					CT075
Mycobacterium tuberculosis	Rv1629		Rv1547 Rv3370c	Rv3711c		Rv3721c		Rv3644c			Rv0002
Bacillus subtilis	polA		dnaE yorL polC			dnaX		holB			dnaN yshC

D. Are Some Proteins Better Targets Than Others?

Genomics has motivated a significantly increased effort in protein structure determination and structure prediction. The accumulation of three-dimensional protein structures has increased dramatically in the past few years. In 1999, approximately 2500 structures were added to the Protein Data Bank (PDB), a 20-fold increase over the 116 structures added to the PDB in 1989 (www.rcsb.org). Over 10,000 three-dimensional structures of proteins, peptides, and viruses now reside in the PDB. Of these, the number of prokaryotic proteins with structural information is significant: in *E. coli*, 771 of the 4288 predicted proteins have significant similarity to a protein with known three-dimensional structure [60].

In addition to increased structure determination, significant improvements have occurred in the ability to predict protein structures. The CASP3 competition to test current structure prediction methods concluded that alignment techniques—predicting structures for proteins with amino acid sequence similarity to a protein with known structure—worked very well for pairs of proteins with greater than 30% amino acid identity [61,62]. Since the homology among eubacterial orthologs is generally greater than 30%, the eubacterial proteins with a known three-dimensional structure can provide strong guidance for predicting the structures of orthologous and paralogous eubacterial proteins. Such guidance could be very important for structure-based design approaches to drug discovery.

It has been estimated that only 1,000–10,000 distinct protein folds exist in all proteins, based on the decreasing accumulation of novel structures as new proteins are added to the Protein Data Bank. Potentially, one strategy for prioritizing proteins as antibiotic targets would be to select proteins with unique folds (or to avoid those with ''popular'' folds) in an attempt to lessen the opportunities for toxicity based on mammalian proteins with similar binding sites.

Using newly available microbial genome sequences, the protein folds in each of eight microbial genomes were predicted from genome sequence information [63]. This analysis allowed the assembly and comparison of each organism's ''protein fold census.'' From this analysis, the distribution of occurrence of protein folds was determined for each species. The top-10 folds in each eubacterial genome accounted for 34–40% of the total ORFs in that genome. Certain folds are found frequently in prokaryotes, and less frequently in eukaryotes, such as the TIM-barrel, the NTP hydrolases with P-loop, and the Rossman fold [64]. In addition, there are numerous folds that are less common in prokaryotes, which are not predicted from the current analysis of eukaryotic genomes [64].

The quantity of a protein produced by a microbe may also be important in selecting a target. The amounts of proteins vary from less than one monomer per cell to over 10,000 monomers per cell. Thus, the necessary concentrations of an inhibitor could vary over several orders of magnitude, even for an inhibitor with nanomolar binding affinity.

The increasing number of co-crystals of protein with a bound ligand has improved and generalized our understanding of small-molecule binding to proteins. Such information may aid the development of additional tools for rational drug design, and increase the success of performing *in silico* screening and of biasing compound libraries for experimental screening. In addition, the development of these tools may allow the appropriate assessment of which proteins are more likely to be selectively inhibited by small molecules.

In general, small molecules bind to crevices and depressions in proteins, which can be active site locations, non-active site locations, or both [65,66]. Analysis of the binding sites of over 50 proteins revealed an overrepresentation of large bulky groups, such as tryptophan and tyrosine, as well as histidine and arginine in active sites [67]. Small molecules frequently displace well-ordered water molecules upon binding to a protein. The bulky amino acids found in active sites are thought to create ''rough'' surfaces that may increase binding affinity either directly, by increasing the available surface of interaction between the protein and ligand [68], or indirectly, through trapping and constraining the movement of water molecules that will be released upon ligand binding [69]. Strategies for identifying rough surfaces by fractal geometry have shown success in predicting active site locations [68].

V. GENOMIC METHODS FOR ENZYME-BASED LEAD IDENTIFICATION

For genomics to increase its value in the pharmaceutical setting, methods must be developed to improve and accelerate steps downstream of target selection. Several new biochemical and whole-cell screening strategies have emerged that allow homogenous assay formats on an expanded range of protein targets. Coupled with improvements in liquid-handling systems, robotics, and databases, these methods have significantly expanded the ability to perform high-throughput screening on a wide range of targets [70].

A. Traditional Biochemical Approaches for Lead Identification

Traditional target-directed screening methods have employed a biochemical screen against a particular target to assure high sensitivity and specificity for the target [71]. DNA sequence information, coupled with tagging methods and affinity purification, enable simpler methods for obtaining large quantities of a target protein from clinically relevant species.

Enzyme assay methods for screening have been augmented by new screening technologies that allow homogeneous assay formats. Scintillation proximity assays (SPA) rely on microbeads or plates that contain scintillant to provide a signal when a radioactive substance is physically nearby. Numerous binding assays have been developed, in which the receptor protein is bound to microbeads or plates, resulting in activation of the scintillant when a radioactively labeled ligand binds to the target protein [72,73]. Most methods have utilized biotin-streptavidin or antibody affinity to bind the target protein to the scintillant-containing solid support. Such methods require knowledge of a specific ligand that binds the target protein, as well as the ability to obtain radioactive ligand in sufficient quantity to perform the screening.

B. Ligand-Binding Approaches for Lead Identification

A traditional biochemical approach to screening limits the targets to those that have well-developed biochemical assays or known ligands. Currently, this approach is limited to a relatively small subset of the pharmaceutically relevant targets, even in well-studied microbial systems. Over 25% of the essential genes identified in *Staphylococcus aureus* have no functional characterization; many of the remaining essential genes do not have biochemical assays that could be readily implemented in high-throughput screens [39].

Recently, methods have been developed and implemented in high-throughput screens that can differentiate ligand-bound and unbound protein, based on the increased conformational stability of a protein when it is bound to a ligand [74]. Such methods have been applied to lead identification, using the assumption that ligands bound to the target provide a good starting point for the development of inhibitors. These methods enable screening to identify inhibitors of proteins for which there is little functional information. The ligand-binding approaches for lead identification enjoy the advantage of all biochemical assays, in their ability to provide quantitative and sensitive measures of the inhibitor's ability to bind the target protein. In addition, such assays do not require the inhibitor to overcome permeability barriers in order to be detected by the screen. Several experimental methods have been proposed to detect ligand-induced stabilization of a target protein [74–76].

Selection and enrichment methods have been used to identify peptide and oligonucleotide ligands that bind to target proteins [77–81]. Combinatorial peptide libraries are often produced using phage display methods, which incorporate random nucleotides into a site on the phage genome that allows transcription, translation, and incorporation of the encoded peptides onto the surface of the phage particle. Methods have also been described for identifying folded oligonucleotide molecules that bind tightly to a target protein. The Systematic Evolution of Ligands by Exponential enrichment technology (''SELEX'') is an enrichment strategy that can identify high-affinity oligonucleotides that bind to a protein target [82–85]. The peptide and oligonucleotide ligands identified by these methods can have binding affinities ranging from micromolar to subnanomolar after repeated rounds of enrichment.

Phage display and SELEX methods have been adapted to identify small-molecule inhibitors of target proteins. Peptide or nucleic acid inhibitors are not generally desirable as therapeutic agents, primarily due to their susceptibility to serum proteases or nucleases and lack of bioavailability. However, the peptides or oligomers can be used as reagents to identify small molecules that competitively inhibit binding of the peptide or oligomer ligand to the target [86,87].

C. Whole-Cell Approaches for Lead Identification

New genome-based whole-cell screening approaches have used genetically altered strains to create novel, target-specific phenotypes [23,24,88,89]. Several of the whole-cell screening approaches use hypomorphic phenotypes created by partial loss of function or underexpression of an essential gene. Cells carrying such mutations are hypersensitive to inhibitors of the target and can detect even weak inhibitors. These screens are the chemical equivalent of a genetic search for synthetic lethal mutations. Genetic synthetic lethality screens have shown that two mutations may be individually heritable but jointly noninheritable, either when the genes are functionally redundant or when the gene products interact physically or functionally with one another [90]. When one cellular component is weakened by mutation, it becomes more susceptible to inhibition by chemical inhibitors of that target, creating a synthetically lethal situation. This strategy has provided the basis for whole-cell target-directed screens [23,24,88,89].

VI. GENOMICS IMPACT ON LEAD COMPOUND CHARACTERIZATION

New approaches evolving from genomics may aid the process of determining mechanism of action for new classes of inhibitors. Rapid and high-throughput approaches for determining a compound's mechanism of action will allow the entry of this information earlier in the drug discovery process, at the stage of lead selection as well as in the functional characterization of diverse compound libraries.

Preliminary studies have shown the value that a combination of genetic and genomic techniques can have for providing high-throughput methods for determining mechanism of action. Manipulation of the gene dosage of an inhibitor's target by twofold was found to cause hypersensitivity or resistance to an inhibitor in *Saccharomyces cerevisiae* [88]. Similarly, mutants with weakened target function, obtained through semipermissive growth of conditional lethal mutants, were hypersensitive to target inhibitors [24]. Several groups have now recognized that a set of strains carrying alterations in possible target genes provides a tool for characterizing the mechanism of action of inhibitors [23,24, 91–93].

In addition, the ability of a transcriptional or proteomic profile to provide a biological ''fingerprint'' of the function of a compound has been shown recently [94–97]. Such studies should prove useful for identifying structure–activity relationships, identifying pharmacophores, and for toxicological assessments of new lead compounds.

VII. CONCLUSIONS

Over the past five years, genomics has significantly changed antimicrobial drug discovery. A much larger collection of targets is available for screening, and new methods for screening have taken advantage of the large number of targets and high-throughput robotics. Further advances are clearly necessary to reap the benefits of the postgenomic era. The ability to tailor chemical diversity libraries to be more likely to identify specific target inhibitors with high selectivity is still an important future goal. Advances in mammalian genomics will help in achieving this goal, as will the numerous emerging techniques and technologies for analysis of large-scale data sets that allow biological characterization of small-molecule lead compounds.

REFERENCES

1. FR Blattner, G Plunkett III, CA Bloch, NT Perna, V Burland, M Riley, J Collado-Vides, JD Glasner, CK Rode, GF Mayhgew, J Gregor, W Davis, HA Kirkpatrick, MA Goeden, DJ Rose, B Mau, Y Shao. The complete genome sequence of *Escherichia coli* K-12. Science 277:1453–1474, 1997.
2. RD Fleischmann, MD Adams, O White, RA Clayton, EF Kirkness, AR Kerlavage, CJ Bult, J-F Tomb, BA Dougherty, JM Merrick, K McKenney, G Sutton, W FitzHugh, C Fields, JD Gocayne, J Scott, R Shirley, L-I Liu, A Glodek, JM Kelley, et al. Whole-genome random sequencing and assembly of *Haemophilus influenzae* Rd. Science 269:496–512, 1995.
3. CM Fraser, S Casjens, WM Huang, GG Sutton, R Clayton, R Lathigra, O White, KA Ketchum, R Dodson, EK Hickey, M Gwinn, B Dougherty, J-F Tomb, RD Fleischmann, D Richardson, J Peterson, AR Kerlavage, J Quackenbush, S Salzberg, M Hanson, et al. Genomic sequence of a Lyme disease spirochaete, *Borrelia burgdorferi*. Nature 390:580–586, 1997.
4. CM Fraser, JD Gocayne, O White, MD Adams, RA Clayton, RD Fleishmann, CJ Bult, AR Kerlavage, G Sutton, JM Kelley, JL Fritchman, JF Weidman, KV Small, M Sandusky, J Fuhrmann, D Nguyen, TR Utterback, DM Saudek, CA Phillips, JM

Merrick, et al. The minimal gene complement of *Mycoplasma genitalium*. Science 270:397–403, 1995.

5. F Kunst, N Ogasawara, I Moszer, A Albertini, G Alloni, V Azevedo, M Bertero, P Bessieres, A Bolotin, S Borchert, R Borriss, L Boursier, A Brans, M Braun, S Brignell, S Bron, S Brouillet, C Bruschi, B Caldwell, V Capuano, et al. The complete genome sequence of the gram-positive bacterium *Bacillus subtilis*. Nature 390:249–256, 1997.
6. SB Levy. The challenge of antibiotic resistance. Sci Am 278:46–53, 1998.
7. JB Baseman, JG Tully. Mycoplasmas: sophisticated, reemerging, and burdened by their notoriety. Emerg Infect Dis 3:21–32, 1997.
8. R Himmelreich, H Hilbert, H Plagens, E Pirkl, B-C Li, R Herrmann. Complete sequence analysis of the genome of the bacterium *Mycoplasma pneumoniae*. Nucleic Acids Res 24:4420–4449, 1996.
9. RH Baltz, FH Norris, P Matsushima, BS DeHoff, P Rockney, G Porter, S Burgett, R Peery, L Hoskins, I Braverman, P Jenkins, P Solenberg, M Young, MA McHenney, PL Skatrud, PR Rosteck. DNA sequence sampling of the *Streptococcus pneumoniae* genome to identify novel targets for antibiotic development. Microbial Drug Resist 4:1–9, 1998.
10. AA Salyers, NB Shoemaker, AM Stevens, LY Li. Conjugative transposons: an unusual and diverse set of integrated gene transfer elements. Microbiol Rev 59: 579–590, 1995.
11. JA Heinemann, GF Sprague Jr. Transmission of plasmid DNA to yeast by conjugation with bacteria. Meth Enzymol 194:187–195, 1991.
12. GF Sprague Jr. Genetic exchange between kingdoms. Curr Opin Genet Dev 1:530–533, 1991.
13. C Grillot-Courvalin, S Goussard, F Huetz, DM Ojcius, P Courvalin. Functional gene transfer from intracellular bacteria to mammalian cells. Nature Biotechnol 16: 862–866, 1998.
14. W Martin. Mosaic bacterial chromosomes: a challenge en route to a tree of genomes. Bioessays 21:99–104, 1999.
15. JG Lawrence, H Ochman. Molecular archaeology of the *Escherichia coli* genome. Proc Natl Acad Sci (USA) 95:9413–9417, 1998.
16. JE LeClerc, B Li, WL Payne, TA Cebula. Promiscuous origin of a chimeric sequence in the *Escherichia coli* O157:H7 genome. J Bacteriol 181:7614–7617, 1999.
17. MB Schmid, N Kapur, DR Isaacson, P Lindroos, C Sharpe. Genetic analysis of temperature-sensitive lethal mutants of *Salmonella typhimurium*. Genetics 123: 625–633, 1989.
18. CA Hutchison, SN Peterson, SR Gill, RT Cline, O White, CM Fraser, HO Smith, JC Venter. Global transposon mutagenesis and a minimal Mycoplasma genome. Science 286:2165–2169, 1999.
19. J Maniloff. The minimal cell genome: on being the right size. Proc Natl Acad Sci (USA) 93:10004–10006, 1996.
20. A Mushegian, E Koonin. A minimal gene set for cellular life derived by comparison of complete bacterial genomes. Proc Natl Acad Sci (USA) 93:10268–10273, 1996.
21. F Arigoni, F Talabot, M Peitsch, MD Edgerton, E Meldrum, E Allet, R Fish, T

Jamotte, M Curchod, H Loferer. A genome-based approach for the identification of essential bacterial genes. Nature Biotechnol 16:851–856, 1998.
22. EV Koonin. Genome sequences: genome sequence of a model prokaryote. Curr Biol 7:656–659, 1997.
23. LH Hartwell, P Szankasi, CJ Roberts, AW Murray, SH Friend. Integrating genetic approaches into the discovery of anticancer drugs. Science 278:1064–1068, 1997.
24. KA Bostian, MB Schmid. New antibacterial targets and new approaches for drug discovery. In: WD Busse, HJ Zeiler, H Labischinski, eds. Antibacterial Therapy: Achievements, Problems and Future Perspectives. Berlin: Springer-Verlag, 1997.
25. RE Isaacson. Novel targets for antibiotics—an update. Exp Opin Invest Drugs 6: 1009–1017, 1997.
26. V Vagner, E Dervyn, SD Ehrlich. A vector for systematic gene inactivation in *Bacillus subtilis*. Microbiology 144:3097–3104, 1998.
27. T Parish, NG Stoker. Development and use of a conditional antisense mutagenesis system in mycobacteria. FEMS Microbiol Lett 154:151–157, 1997.
28. HE Takiff, T Baker, T Copeland, SM Chen, DL Court. Locating essential *Escherichia coli* genes by using mini-Tn10 transposons: the *pdxJ* operon. J Bacteriol 174:1544–1553, 1992.
29. WY Chow, DE Berg. Tn5*tac1*, a derivative of transposon Tn5 that generates conditional mutations. Proc Natl Acad Sci (USA) 85:6468–6472, 1988.
30. KP Lemon, I Kurtser, J Wu, AD Grossman. Control of initiation of sporulation by replication initiation genes in *Bacillus subtilis*. J Bacteriol 182:2989–2991, 2000.
31. M Ricard, Y Hirota. Process of cellular division in *Escherichia coli*: physiological study on thermosensitive mutants defective in cell division. J Bacteriol 116:314–322, 1973.
32. P Martin, T Li, D Sun, D Biek, M Schmid. Role in cell permeability of an essential two-component system in *Staphylococcus aureus*. J Bacteriol 181:3666–3673, 1999.
33. J Kato, S Fujisaki, K Nakajima, Y Nishimura, M Sato, A Nakano. The *Escherichia coli* homologue of yeast RER2, a key enzyme of dolichol synthesis is essential for carrier lipid formation in bacterial cell wall synthesis. J Bacteriol 181:2733–2738, 1999.
34. Y Nagano, R Matsuno, Y Sasaki. An essential gene of *Escherichia coli* that has sequence similarity to a chloroplast gene of unknown function. Mol Gen Genet 228:62–64, 1991.
35. E Bi, K Dai, S Subbarao, B Beall, J Lutkenhaus. FtsZ and cell division. Res Microbiol 142:249–252, 1991.
36. BJ Akerley, EJ Rubin, A Camilli, DJ Lampe, HM Robertson, JJ Mekalanos. Systematic identification of essential genes by *in vitro mariner* mutagenesis. Proc Natl Acad Sci (USA) 95:8927–8932, 1998.
37. DD Shoemaker, DA Lashkari, D Morris, M Mittmann, RW Davis. Quantitative phenotypic analysis of yeast deletion mutants using a highly parallel molecular bar-coding strategy. Nature Genet 14:450–456, 1996.
38. V Smith, K Chou, D Lashkari, D Botstein, PO Brown. Functional analysis of the genes of yeast chromosome V by genetic footprinting. Science 274:2069–2074, 1996.

39. MB Schmid, New targets and strategies for identification of novel classes of antibiotics. In: D Hughes, D Andersson, eds. Antibiotic Resistance and Antibiotic Development, London: Harwood, 2000.
40. JP Richardson. Transcription termination. Crit Rev Biochem Mol Biol 28:1–30, 1993.
41. A Link, D Phillips, G Church. Methods for generating precise deletions and insertions in the genome of wild-type *Escherichia coli*: application to open reading frame characterization. J Bacteriol 179:6228–6237, 1997.
42. RP Anderson, JR Roth. Tandem genetic duplications in phage and bacteria. Annu Rev Microbiol 31:473–505, 1977.
43. D Mazel, S Pochet, P Marliere. Genetic characterization of polypeptide deformylase, a distinctive enzyme of eubacterial translation. EMBO J 13:914–923, 1994.
44. PS Margolis, CJ Hackbarth, DC Young, W Wang, D Chen, Z Yuan, R White, J Trias. Peptide deformylase in *Staphylococcus aureus*: resistance to inhibition is mediated by mutations in the formyltransferase gene. Antimicrob Agents Chemother 44:1825–1831, 2000.
45. C Goffin, JM Ghuysen. Multimodular penicillin-binding proteins: an enigmatic family of orthologs and paralogs. Microbiol Mol Biol Rev 62:1079–1093, 1998.
46. SA Denome, PK Elf, TA Henderson, DE Nelson, KD Young. *Escherichia coli* mutants lacking all possible combinations of eight penicillin binding proteins: viability, characteristics, and implications for peptidoglycan synthesis. J Bacteriol 181: 3981–3993, 1999.
47. K Drlica, X Zhao. DNA gyrase, topoisomerase IV, and the 4-quinolones. Microbiol Mol Biol Rev 61:377–392, 1997.
48. WM Huang. Bacterial diversity based on type II DNA topoisomerase genes. Annu Rev Genet 30:79–107, 1996.
49. WV Kern, M Oethinger, AS Jellen-Ritter, SB Levy. Non-target gene mutations in the development of fluoroquinolone resistance in *Escherichia coli*. Antimicrob Agents Chemother 44:814–820, 2000.
50. LJ Piddock. Mechanisms of fluoroquinolone resistance: an update 1994–1998. Drugs 58 Suppl 2:11–18, 1999.
51. JF Barrett, RM Goldschmidt, LE Lawrence, B Foleno, R Chen, JP Demers, S Johnson, R Kanojia, J Fernandez, J Bernstein, L Libata, A Donetz, S Huang, DJ Hlasta, MJ Macielag, K Ohemeng, R Frechette, MB Frosco, DH Klaubert, JM Whiteley, et al. Antibacterial agents that inhibit two-component signal transduction systems. Proc Natl Acad Sci (USA) 95:5317–5322, 1998.
52. P Schimmel, J Tao, J Hill. Aminoacyl tRNA synthetases as targets for new anti-infectives. FASEB J 12:1599–1609, 1998.
53. R Losick. Summary: three decades after sigma. Cold Spring Harb Symp Quant Biol 63:653–666, 1998.
54. CA Gross, C Chan, A Dombroski, T Gruber, M Sharp, J Tupy, B Young. The functional and regulatory roles of sigma factors in transcription. Cold Spring Harb Symp Quant Biol 63:141–155, 1998.
55. SS Eveland, DL Pompliano, MS Anderson. Conditionally lethal *Escherichia coli* murein mutants contain point defects that map to regions conserved among murein

and folyl poly-gamma-glutamate ligases: identification of a ligase superfamily. Biochemistry 36:6223–6229, 1997.
56. MH Saier Jr, IT Paulsen, MK Sliwinski, SS Pao, RA Skurray, H Nikaido. Evolutionary origins of multidrug and drug-specific efflux pumps in bacteria. FASEB J 12:265–274, 1998.
57. A Lee, W Mao, MS Warren, A Mistry, K Hoshino, R Okumura, H Ishida, O Lomovskaya. Interplay between efflux pumps may provide either additive or multiplicative effects on drug resistance. J Bacteriol 182:3142–3150, 2000.
58. CA Orengo, DT Jones, JA Thornton. Protein superfamilies and domain superfolds. Nature 372:631–634, 1994.
59. Y Xia, ZY Yang, SL Morris-Natschke, KH Lee. Recent advances in the discovery and development of quinolones and analogs as antitumor agents. Curr Med Chem 6:179–194, 1999.
60. National Center for Biotechnology Information. http://www.ncbi.nlm.nih.gov, 2000.
61. MJ Sternberg, PA Bates, LA Kelley, RM MacCallum. Progress in protein structure prediction: assessment of CASP3. Curr Opin Struct Biol 9:368–373, 1999.
62. J Moult. Predicting protein three-dimensional structure. Curr Opin Biotechnol 10: 583–588, 1999.
63. M Gerstein. Patterns of protein-fold usage in eight microbial genomes: a comprehensive structural census. Proteins 33:518–534, 1998.
64. M Gerstein. http://bioinfo.mbb.yale.edu/genome, 2000.
65. D Ringe, C Mattos. Analysis of the binding surfaces of proteins. Med Res Rev 19: 321–331, 1999.
66. TJ Rydel, A Tulinsky, W Bode, R Huber. Refined structure of the hirudin-thrombin complex. J Mol Biol 221:583–601, 1991.
67. HO Villar, LM Kauvar. Amino acid preferences at protein binding sites. FEBS Lett 349:125–130, 1994.
68. FK Pettit, JU Bowie. Protein surface roughness and small molecular binding sites. J Mol Biol 285:1377–1382, 1999.
69. C Mattos, D Ringe. Locating and characterizing binding sites on proteins. Nature Biotechnol 14:595–599, 1996.
70. SA Sundberg. High-throughput and ultra-high-throughput screening: solution- and cell-based approaches. Curr Opin Biotechnol 11:47–53, 2000.
71. H Umezawa. Low-molecular-weight enzyme inhibitors of microbial origin. Annu Rev Microbiol 36:75–99, 1982.
72. PB Fernandes. Technological advances in high-throughput screening. Curr Opin Chem Biol 2:597–603, 1998.
73. L Silverman, R Campbell, JR Broach. New assay technologies for high-throughput screening. Curr Opin Chem Biol 2:397–403, 1998.
74. JU Bowie, AA Pakula. Screening method for identifying ligands for target proteins. U.S. Patent 5,585,277, Scriptgen Pharmaceuticals, 1996.
75. JU Bowie, A Pakula. Screening method for identifying ligands for target proteins. U.S. Patent 5,679,582, Scriptgen Pharmaceuticals, 1997.
76. MW Pantoliano, AW Rhind, FR Salemme. Microplate thermal shift assay for ligand development and multivariable protein chemistry optimization. U.S. Patent 6,020,141, 3-Dimensional Pharmaceuticals, 2000.

77. L Gold, B Polisky, O Uhlenbeck, M Yarus. Diversity of oligonucleotide functions. Annu Rev Biochem 64:763–797, 1995.
78. L Gold. The SELEX process: a surprising source of therapeutic and diagnostic compounds. Harvey Lect 91:47–57, 1995.
79. K Johnsson, L Ge. Phage display of combinatorial peptide and protein libraries and their applications in biology and chemistry. Curr Top Microbiol Immunol 243:87–105, 1999.
80. S Cabilly. The basic structure of filamentous phage and its use in the display of combinatorial peptide libraries. Mol Biotechnol 12:143–148, 1999.
81. KD Wittrup. Phage on display. Trends Biotechnol 17:423–424, 1999.
82. DW Drolet, RD Jenison, DE Smith, D Pratt, BJ Hicke. A high throughput platform for systematic evolution of ligands by exponential enrichment (SELEX). Comb Chem High Throughput Screen 2:271–278, 1999.
83. B Vant-Hull, A Payano-Baez, RH Davis, L Gold. The mathematics of SELEX against complex targets. J Mol Biol 278:579–597, 1998.
84. D Irvine, C Tuerk, L Gold. SELEXION. Systematic evolution of ligands by exponential enrichment with integrated optimization by non-linear analysis. J Mol Biol 222:739–761, 1991.
85. RD Jenison, SC Gill, A Pardi, B Polisky. High-resolution molecular discrimination by RNA. Science 263:1425–1429, 1994.
86. R Hyde-DeRuyscher, LA Paige, DJ Christensen, N Hyde-DeRuyscher, A Lim, ZL Fredericks, J Kranz, P Gallant, J Zhang, SM Rocklage, DM Fowlkes, PA Wendler, PT Hamilton. Detection of small-molecule enzyme inhibitors with peptides isolated from phage-displayed combinatorial peptide libraries. Chem Biol 7:17–25, 2000.
87. P Bridonneau, YF Chang, AV Buvoli, D O'Connell, D Parma. Site-directed selection of oligonucleotide antagonists by competitive elution. Antisense Nucleic Acid Drug Dev 9:1–11, 1999.
88. G Giaever, D Shoemaker, TW Jones, H Liang, EA Winzler, A Astromoff, RW Davis. Genomic profiling of drug sensitivities via induced haploinsufficiency. Nature Genet 21:278–283, 1999.
89. T Munder, A Hinnen. Yeast cells as tools for target-oriented screening. Appl Microbiol Biotechnol 52:311–320, 1999.
90. AG Davies, CA Spike, JE Shaw, RK Herman. Functional overlap between the *mec-8* gene and five *sym* genes in *Caenorhabditis elegans*. Genetics 153:117–134, 1999.
91. LM Shi, Y Fan, TG Myers, PM O'Connor, KD Paull, SH Friend, JN Weinstein. Mining the NCI anticancer drug discovery databases: genetic function approximation for the QSAR study of anticancer ellipticine analogues. J Chem Inf Comput Sci 38:189–199, 1998.
92. JN Weinstein, TG Myers, PM O'Connor, SH Friend, AJ Fornace Jr, KW Kohn, T Fojo, SE Bates, LV Rubinstein, NL Anderson, JK Buolamwini, WW van Osdol, AP Monks, DA Scudiero, EA Sausville, DW Zaharevitz, B Bunow, VN Viswanadhan, GS Johnson, RE Wittes, et al. An information-intensive approach to the molecular pharmacology of cancer. Science 275:343–349, 1997.
93. PM O'Connor, J Jackman, I Bae, TG Myers, S Fan, M Mutoh, DA Scudiero, A Monks, EA Sausville, JN Weinstein, S Friend, AJ Fornace Jr, KW Kohn. Characterization of the p53 tumor suppressor pathway in cell lines of the National Cancer

Institute anticancer drug screen and correlations with the growth-inhibitory potency of 123 anticancer agents. Cancer Res 57:4285–4300, 1997.

94. TR Hughes, MJ Marton, AR Jones, CJ Roberts, R Stoughton, CD Armour, HA Bennett, E Coffey, H Dai, YD He, MJ Kidd, AM King, MR Meyer, D Slade, PY Lum, SB Stepaniants, DD Shoemaker, D Gachotte, K Chakraburtty, J Simon, et al. Functional discovery via a compendium of expression profiles. Cell 102:109–126, 2000.
95. JA Simon, P Szankasi, DK Nguyen, C Ludlow, HM Dunstan, CJ Roberts, EL Jensen, LH Hartwell, SH Friend. Differential toxicities of anticancer agents among DNA repair and checkpoint mutants of *Saccharomyces cerevisiae*. Cancer Res 60: 328–333, 2000.
96. MJ Marton, JL DeRisi, HA Bennett, VR Iyer, MR Meyer, CJ Roberts, R Stoughton, J Burchard, D Slade, H Dai, DE Bassett Jr, LH Hartwell, PO Brown, SH Friend. Drug target validation and identification of secondary drug target effects using DNA microarrays. Nature Med 4:1293–1301, 1998.
97. GF Bammert, JM Fostel. Genome-wide expression patterns in *Saccharomyces cerevisiae*: comparison of drug treatments and genetic alterations affecting biosynthesis of ergosterol. Antimicrob Agents Chemother 44:1255–1265, 2000.
98. PM Dombrosky, MB Schmid, KD Young. Sequence divergence of the murB and rrfB genes from *Escherichia coli* and *Salmonella typhimurium*. Arch Microbiol 161:501–507, 1994.
99. J Wu, N Ohta, JL Zhao, A Newton. A novel bacterial tyrosine kinase essential for cell division and differentiation. Proc Natl Acad Sci (USA) 96:13068–13073, 1999.
100. S Mohan, TM Kelly, SS Eveland, CR Raetz, MS Anderson. An *Escherichia coli* gene (FabZ) encoding (3R)-hydroxymyristoyl acyl carrier protein dehydrase. Relation to fabA and suppression of mutations in lipid A biosynthesis. J Biol Chem 269:32896–32903, 1994.
101. H Nikaido. Antibiotic resistance caused by gram-negative multidrug efflux pumps. Clin Infect Dis 27 Suppl 1:32–41, 1998.
102. BI Schweitzer, AP Dicker, JR Bertino. Dihydrofolate reductase as a therapeutic target. FASEB J 4:2441–2452, 1990.
103. RA VanBogelen, KZ Abshire, B Moldover, ER Olson, FC Neidhardt. *Escherichia coli* proteome analysis using the gene-protein database. Electrophoresis 18:1243–1251, 1997.

22
Assigning Precise Function to Genes

Ridong Chen
University of Saskatchewan, Saskatoon, Saskatchewan, Canada

I. INTRODUCTION

The genomes of *Escherichia coli*, yeast, worm, and 18 microbes have been completely sequenced (http://www.ncbi.nlm.nih.gov/Entrez/Genome/org.html). The sequencing of the human genome is scheduled to be completed by the end of the year 2003, and the sequences of other genomes will become available in the coming decade. The emergence of genomics has changed the way that we discover and isolate genes. In the past we have proceeded from a peptide sequence or phenotype to the isolation of a new gene. Now, an investigator begins with the sequence of a key gene and searches for homologous genes in an organism of interest [1]. Certainly, genomics will have tremendous impact on enzyme industry. However, the extent to which using genomic approach for ''enzyme mining'' actually succeeds will depend on how accurately the function of a new protein can be defined without any *a-priori* functional knowledge. Here, a broadly applicable strategy is presented that allows us to assign precise function to genes.

II. ORTHOLOGS, PARALOGS, AND FUNCTIONAL PREDICTION

Over a period of more than 3 billion years, a large variety of protein molecules have evolved as biological catalysts (enzymes) to run the complex machinery of the present-day cells and organisms. The relationships between genes from different genomes are naturally represented as a system of homologous families that include both orthologs and paralogs. Orthologs are genes in different species that

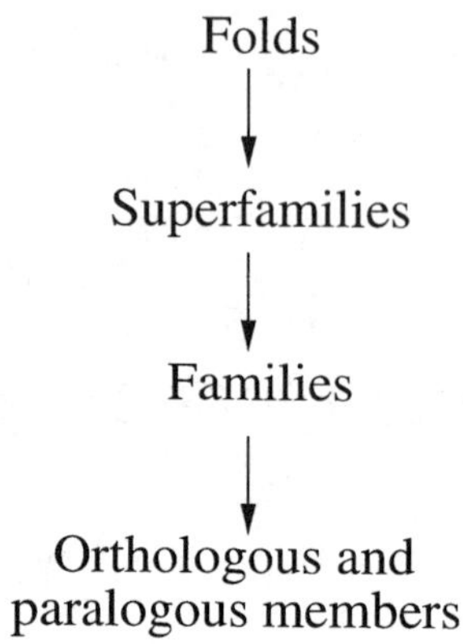

Figure 1 Hierarchy in protein taxonomy.

evolved from a common ancestral gene by speciation and retained the same function in the course of evolution. By contrast, paralogs are genes related by duplication within an organism and have evolved a related but different function [2]. Protein families are further connected into superfamilies, which are usually homologous, and folds which share common structural features but are not necessarily homologous (Fig. 1). Hence, a powerful approach to predicting the exact function of a new protein is to find its characterized orthologs [2,3]. Currently this is done by bioinformatics (biology with computers). Nevertheless, with incompletely sequenced genomes or large phylogenetic distances, there is always the chance that the real ortholog is not yet known and that the best match in a database hit is just a well-conserved paralog [2,3]. The analysis is further complicated by the fact that a significant fraction of functional annotations in databases is wrong or dubious. Using β-decarboxylating dehydrogenase family as a model, we demonstrate that, with insight into how distinct functions of orthologs and paralogs are conferred and evolved, it is feasible to identify orthologs with high confidence.

III. β-DECARBOXYLATING DEHYDROGENASES HAVE EVOLVED DIVERGENTLY FROM A COMMON ANCESTRAL GENE

β-Decarboxylating dehydrogenases are a family of bifunctional enzymes that catalyze the Mg^{2+}- and $NAD(P)^+$-dependent dehydrogenation at C2, followed by their Mg^{2+}-dependent decarboxylation at C3 of β-substituted malate:

$$\mathrm{HOOC{-}CH(R){-}CH(OH){-}COOH + NAD(P)^{+}} \xrightleftharpoons{\mathrm{Mg^{2+}}} \mathrm{HOOC{-}CH(R){-}CO{-}COOH\ NAD(P)H + H^{+}}$$

$$\mathrm{HOOC{-}CH(R){-}CO{-}COOH + H^{+}} \xrightleftharpoons{\mathrm{Mg^{2+}}} \mathrm{R{-}CH_2{-}CO{-}COOH + CO_2}$$

Three orthologs have been identified so far: NAD-dependent isocitrate dehydrogenase (NAD-IDH, EC 1.1.1.41), NADP-dependent isocitrate dehydrogenase (NADP-IDH, EC 1.1.1.42), and NAD-dependent isopropylmalate dehydrogenase (NAD-IMDH, EC 1.1.1.85). NAD-IDH is limited to eukaryotic organisms and participates in the supply of NADH used for respiratory ATP production in mitochondria, while NADP-IDH is present ubiquitously in both prokaryotes and eukaryotes and involved in the generation of both NADPH and α-ketoglutarate for biosynthetic pathways ([4], Fig. 2). NAD-IMDH is found in bacteria, fungi, and plants. This is the enzyme that catalyzes the third step of the pathway for leucine biosynthesis ([5], Fig. 3). NADP-IDH and NAD-IMDH are homodimers, while all the NAD-IDHs purified so far have a hetero-oligomeric structure [4]. The crystal structures of NADP-IDH from *E. coli* and NAD-IMDH from *Thermus thermophilus* have been solved [5,6]. Both enzymes share a common protein fold that lacks the βαβαβ motif characteristic of the nucleotide-binding Rossmann fold [7].

Phylogenic analyses indicate that NAD-IDH, NADP-IDH, and NAD-IMDH are homologous proteins that have evolved divergently from a common ancestral gene present in a progenitor of all extant organisms (Fig. 4) [8]. The gene might encode the primitive enzyme that possessed a very broad specificity, permitting it to react with a wide range of related substrates that share a common 2-R malate moiety. This would maximize the catalytic versatility of an ancestral cell that functioned with limited genetic information and enzyme resources [9]. The ancestral gene was then duplicated and acquired additional genetic information. Copies of the gene diverged via mutational modification, giving rise to contemporary

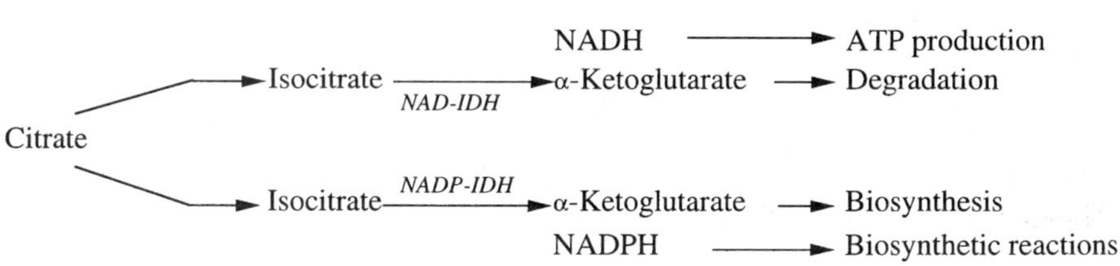

Figure 2 ''Division of labor'' between NAD-IDH and NADP-IDH in eukaryotic organisms.

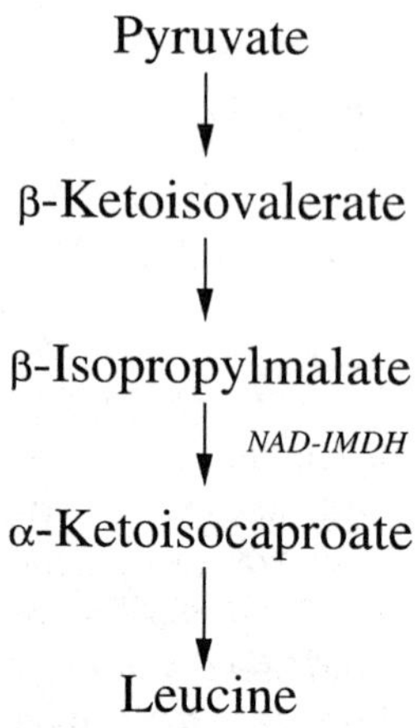

Figure 3 Leucine biosynthetic pathway.

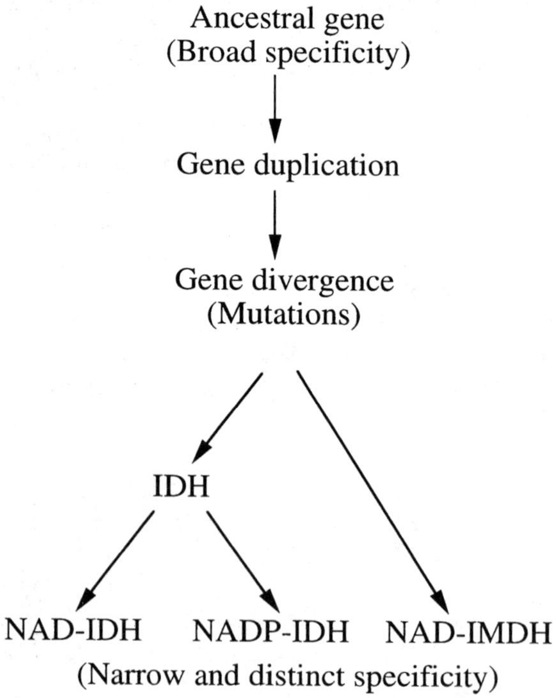

Figure 4 Evolution of the β-decarboxylating dehydrogenase gene family.

enzymes with strict specificity. It has been suggested that specificity toward isocitrate may have evolved before specificity toward NADP, and that the latter evolved around the time the eukaryotes first appeared [8]. Specialization of gene functions presumably allowed the improvement of metabolic efficiency and the evolution of new biochemical pathways [9].

IV. ONLY A FEW AMINO ACID REPLACEMENTS ARE RESPONSIBLE FOR DISTINCT FUNCTIONS OF ORTHOLOGS

Over evolutionary time, a large number of sequence differences has accumulated between the members of the β-decarboxylating dehydrogenase family. The present-day phylogenies are highly divergent, and the divergence between orthologous genes approaches and even exceeds the level of divergence between paralogs within a species (our unpublished results). Thus, it is not reliable to make assignment of new protein sequences to a particular ortholog based on the best BLAST hit or similarity searches. In fact, residues critical to substrate and coenzyme binding and catalysis can rarely be aligned properly using BLAST or CLUSTAL W. This is especially the case when the sequence of *E. coli* NADP-IDH is compared to eucaryotic NADP-IDHs which share no significant identities (<17%).

The substrates isocitrate and isopropylmalate are structurally similar, i.e., $^{-}OOC(HO)CHCH(X)COO^{-}$, where X represents the γ-moiety: the $—CH_2COO^{-}$ of isocitrate and the $-CH(CH_3)_2$ of isopropylmalate (Fig. 5). Coenzyme NADP differs from NAD only by a phosphate group esterified at the 2′C of the adenosine ribose. Presumably, distinct functions of NAD-IDH, NADP-IDH, and NAD-IMDH are conferred by their ability to recognize alternative substrates and coen-

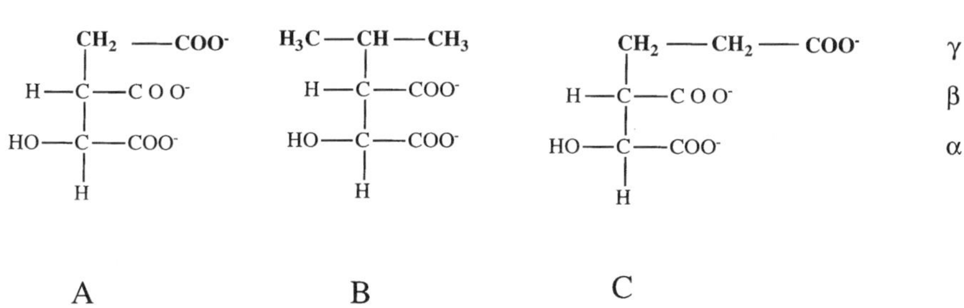

Figure 5 Structures of 2*R*,3*S*-isocitrate (A), 2*R*,3*S*-isopropylmalate (B), and 2*R*,3*S*-homoisocitrate (C). α, β refers to the α- and β-carboxyl groups, respectively. The unique γ-moieties are presented in bold.

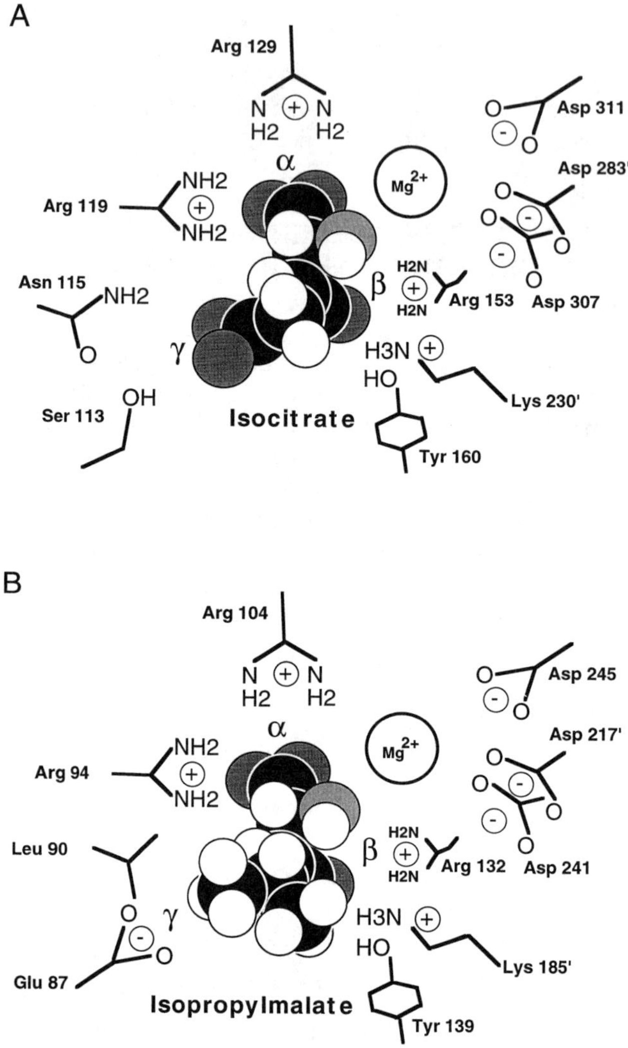

Figure 6 Schematic diagram of the active site of the *E. coli* NADP-IDH with bound 2*R*,3*S*-isocitrate (A) and the active site of *T. thermophilus* NAD-IMDH with bound 2*R*,3*S*-isopropylmalate (B). A prime indicates the residues donated from the second subunit. α, β refers to α- and β-carboxyl groups, respectively. The major determinants of substrate specificity are residues Ser113 and Asn115 in the NADP-IDH, and Glu87 and Leu90 in the NAD-IMDH, which interact with the unique γ-moieties of the substrates. (Modified from Ref. 15.)

zymes. Using protein engineering, we have demonstrated that, of hundreds of amino acid substitutions accumulated in these enzymes, only a few are involved in specificity determination [10–18].

A. Specificity Determinants in *E. coli* NADP-IDH

In *E. coli* NADP-IDH, all the residues involved in binding and catalysis have been identified by crystallographic analysis, site-directed mutagenesis, and protein engineering (Fig. 6A) [6,8,10–16,19–21]. The true substrate for the NADP-IDH is a Mg^{2+}-isocitrate complex which binds in a pocket, formed from residues donated from both monomers (Fig. 6A). Hydrogen bonds and some ion pairs are formed between the α-carboxylate of isocitrate and the side chains of Arg119, Arg129, and Arg153, and between the β-carboxylate of isocitrate and the side chains of Arg119, Arg 153, and Tyr160. The Mg^{2+} cation is coordinated to the α-carboxylate and α-hydroxyl groups of isocitrate, the side chains of Asp307, Asp311, and Asp283′ (the second subunit). The side chain of Asp283′ acts as a

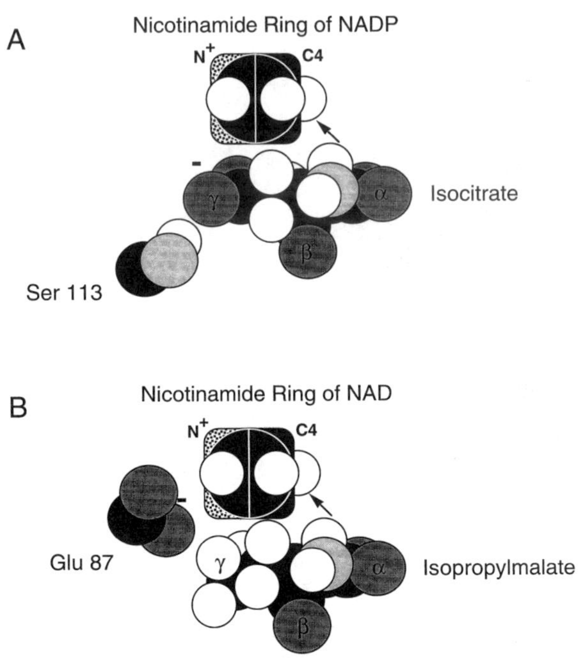

Figure 7 Model for the Michaelis complex of the *E. coli* NADP-IDH with isocitrate and NADP bound (A) and the *T. thermophilus* NAD-IMDH with isopropylmalate and NAD bound (B). The arrow shows how C_4 of the nicotinamide ring is positioned to receive the α-carbon hydride of substrates. (Modified from Ref. 15.)

base for the removal of a proton from the α-hydroxyl group during dehydrogenation, and the side chain of Lys230′ acts as an acid in the decarboxylation. The amino acid residues Ser113 and Asn115 on the helix d, and the γ-carboxylate of isocitrate, are the major determinants of substrate specificity (Fig. 6A) [11,14]. Ser113 forms a hydrogen bond with the γ-carboxylate of bound isocitrate, which, in turn, forms a salt bridge to the nicotinamide ring of the coenzyme (Fig. 7A). Asn115 also interacts with the γ-carboxylate of bound isocitrate. The interaction has a nonideal hydrogen bond angle and is probably purely electrostatic in nature. This residue also is coordinated to the amide of NADP in the Michaelis complex. The polar environment formed with Ser113 and Asn115 is not compatible with the hydrophobic γ-isopropyl group of bound isopropylmalate. More important, this γ-isopropyl group is unable to form the binding site for the nicotinamide

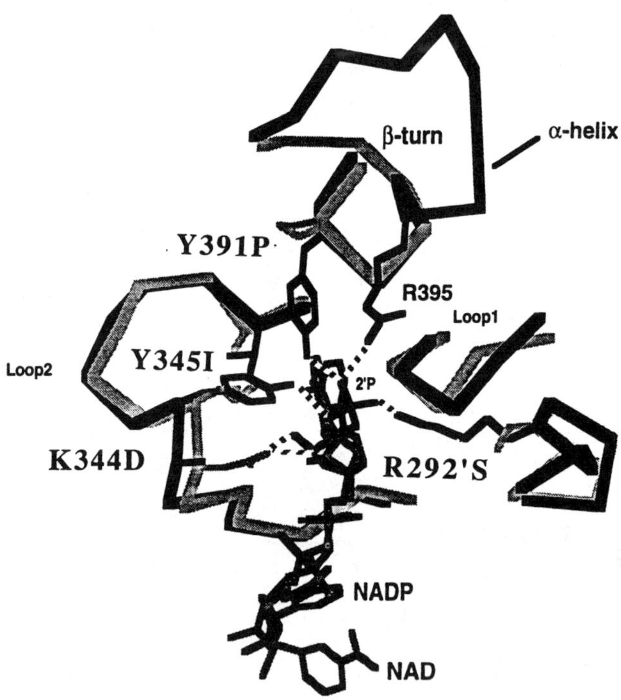

Figure 8 Superposition of the *T. thermophilus* NAD-IMDH binary complex with NAD (gray) on the *E. coli* NADP-IDH binary complex with NADP (black). NADP-IDH numbering is used throughout. It is noted that due to the difference of the local secondary structures, the NAD-IMDH does not have the counterpart of the Arg395 found in the NADP-IDH.

ring which is critical for hydride transfer. Consequently, in spite of the structural similarity of isocitrate and isopropylmalate, IDH does not catalyze a reaction with isopropylmalate (with reduced catalytic efficiency by a factor of at least greater than 10^8, our unpublished results). Coenzyme specificity in the NADP-IDH is conferred by hydrogen bonds between the side chains of Arg-395, Tyr-345, Tyr-391, and Arg-292′ and the 2′-phosphate of bound NADP (Fig. 8). The

Table 1 Structure-Based Alignment of β-Decarboxylating Dehydrogenase Sequences[a]

Enzymes	Sequences		
NADP-IDH			
	113 115 119 129	307 311	344 345
E. coli	R**S**L-**N**VAL**R**QELDLYICL**R**P	NLNG**D**YIS**D**A	HGTA-P**KY**AG
T. thermophilus	K**S**A-**N**VTL**R**KLFETYANV**R**P	NMNG**D**ILS**D**L	HGSA-P**KY**AG
A. fulgidus	R**S**L-**N**VTI**R**QVLDLYANV**R**P	NLNG**D**YLS**D**A	HGSA-P**KY**AG
S. cerevisiae	K**S**P-**N**GTI**R**NILGGTV-F**R**E	NLYG**D**ILS**D**I	HGTVTR**HY**FR
H. sapiens	K**S**P-**N**GTI**R**NILGGTV-F**R**E	NYDG**D**VQS**D**I	HGTVTR**HY**RE
S. tuberosum	R**S**P-**N**GTI**R**NILNGTV-F**R**E	NYDG**D**VQS**D**F	HGTVTR**HY**RV
NAD-IMDH			
	87 90 94 104	241 245	278 279
T. thermophilus	**E**TG-**L**LSL**R**KSQDLFANL**R**P	NIFG**D**ILS**D**L	HGSA-P**DI**AG
B. subtilis	**E**KG-**L**LSI**R**KELDLFVCL**R**P	NMFG**D**ILS**D**E	HGSA-P**DI**AG
S. cerevisiae	**E**QG-**L**LKI**R**KELQLYANL**R**P	NLYG**D**ILS**D**E	HGSA-P**DL**PK
B. napus	**E**TG-**L**LQL**R**AGLKVFANL**R**P	NIFG**D**ILS**D**E	HGSA-P**DI**AG
NAD-IDH			
Catalytic subunit			
S. cerevisiae IDH2	R**S**L-**N**LTL**R**KTFGLFANV**R**P	NLYG**D**ILS**D**L	HGSA-P**DI**AG
K. lactis IDH2	R**S**L-**N**LTL**R**KTFGLFANV**R**P	NLYG**D**ILS**D**L	HGSA-P**DI**AG
M. fascicularis α	R**S**M-**N**LLL**R**KTFDLYANV**R**P	NLYG**D**ILS**D**L	HGTA-P**DI**AG
Regulatory subunit			
S. cerevisiae IDH1	G**S**L-**N**VAL**R**KQLDIYANV**A**L	SMYG**T**ILG**N**I	HVGL--**DI**KG
K. lactis IDH1	G**S**L-**N**VAL**R**KQLDIFANV**A**L	NLYG**S**ILG**N**I	HVGL--**DI**KG
M. fascicularis β	A**S**Y-**D**MRL**R**RKLDLFANV**V**H	NLYG**N**IID**N**L	HPFA--**QA**VG
M. fascicularis γ	K**S**R-**N**NIL**R**TSLDLYANV**I**H	NLYG**N**IVN**N**V	RNT-GK**SI**AN
Incorrectly annotated sequence			
S. salivarius ''NADP-IDH''	R**S**L-**N**VAL**R**QELDLYACV**R**P	NLNG**D**YAS**D**A	HGTA-P**DI**AD
Tomato ''NADP-IDH''	S**S**L-**N**VQL**R**KELDLYASL**V**H	ISTE**I**WLT**N**T	AGNVGN**EK**IL
Arabidopsis ''NAD-IDH1''	S**S**L-**N**VQL**R**KELDLFASL**V**N	NLYG**N**LVA**N**T	AGNVGK**DK**IV
Arabidopsis ''NAD-IDH2''	S**S**L-**N**VNL**R**KELDLFASL**V**N	NLYG**N**LVA**N**T	AGNVGK**DT**TE
New gene: NAD-HDH			
S. cerevisiae ''IMDH''	**Y**SSP**I**VAL**R**REFDQYVNI**R**P	NLYG**D**ILS**D**L	HGSA-P**DI**FG

[a] Numbers are for the residues in bold for the first enzyme in each group. Alignment was performed as follows: residues involved in binding and catalysis were identified from the X-ray crystal structures of *E. coli* NADP-IDH and *T. thermophilus* NAD-IMDH. Other sequences were then aligned using active residues as key landmarks.

side chains of Lys344 may ion-pair with the 2′-phosphate. These interactions can not be formed with the 2′-hydroxyl of NAD. Hence, this enzyme is highly specific for NADP. Calculated as the ratio of k_{cat}/K_m, the enzyme displays a 7000-fold preference for NADP over NAD [10].

B. Specificity Determinants in *T. thermophilus* NAD-IMDH

Despite sharing only 25% sequence identity, *T. thermophilus* NAD-IMDH and *E. coli* NADP-IDH share a common protein fold, and their tertiary structures can be superimposed [5]. The amino acid residues involved in binding the 2*R*-malate core, common to 2*R*,3*S*-isocitrate and 2*R*,3*S*-isopropylmalate, are identical. The equivalent active residues in *T. thermophilus* NAD-IMDH are Arg94, Arg104, Arg132, Tyr139, Lys185′, Asp217′, Asp241, and Asp245 (Fig. 6B). However, the two enzymes differ in the amino acid residues involved in binding the γ-moieties of substrates. In NAD-IMDH, Asn115 is replaced by Leu90, which forms hydrophobic interactions with the γ-isopropyl group of bound isopropylmalate (Fig. 6B, Table 1). In addition, the helix d begins three amino acids earlier, with the consequence that the side chain of Glu87 occupies a position in close proximity to both Ser113 and the γ-carboxylate of isocitrate. The available structural data and the kinetic results suggest that the carboxyl group of Glu87 interacts with the nicotinamide ring and helps stabilize the Michaelis complex, mimicking the role played by the γ-carboxylate of the bound isocitrate in NADP-IDH (Fig. 7B) [15]. Meanwhile, this negatively charged residue lies close to an analogous position of the γ-moiety of bound isocitrate and thus would electrostatically repel this molecule as a substrate. Indeed, no activity is detectable with isocitrate [15]. Consequently, Glu87 and Leu90 are major determinants of substrate specificity in NAD-IMDH [15,17]. The coenzyme specificity toward NAD is mainly conferred by Asp278 (IMDH numbering), which forms a double hydrogen bond with the 2′- and 3′-hydroxyl groups of the adenosine ribose of NAD (Fig. 8) [13,22]. Meanwhile, this negatively charged residue repels the 2′-phosphate of NADP through electrostatic repulsion. As such, this enzyme is 100-fold more active with NAD than with NADP [13].

V. SPECIFICITY DETERMINANTS ARE RELIABLE MARKERS FOR ORTHOLOGS AND PARALOGS

A. NADP-IDHs and NAD-IMDHs from Other Species

The sequence of the NADP-IDHs and NAD-IMDHs from other species were compared with that of *E. coli* NADP-IDH and *T. thermophilus* NAD-IMDH (Table 1). It is noteworthy that the NADP-IDH/NAD-IMDH phylogenies are so divergent that correct alignment can be made only based on detailed knowledge of the X-ray crystallographic structures of *E. coli* NADP-IDH and *T. thermophilus* NAD-IMDH. It shows that all of the binding and catalytic residues identified in *E. coli* NADP-IDH are conserved in other NADP-IDHs except Lys344, which is either conserved or replaced by a positively charged residue His (Table 1). As expected. Ser113 and Asn115 are replaced by Glu and Leu, respectively, while Lys344 and Tyr345 are substituted by Asp and Ile or Leu in all known NAD-IMDHs (Table 1). Hence, these major specificity determinants can be used as landmarks for predicting the exact biochemical functions of new sequences belonging to the isocitrate and isopropylmalate dehydrogenases family.

B. Yeast and Mammalian NAD-IDHs

No crystal structure is available for NAD-IDHs. All the NAD-IDHs purified so far have a hetero-oligomeric structure which differs from the homodimeric form of NADP-IDHs and NAD-IMDHs [4]. Yeast NAD-IDH exists as a hetero-octamer consisting of four copies of the regulatory subunit (NAD-IDH1) and four copies of the catalytic subunit (NAD-IDH2) [23], while mammalian NAD-IDHs consist of three different subunits (α, β, and γ) having the tetrameric form of $\alpha_2\beta\gamma$ [24,25]. Catalytic residues are located in the α-subunit [26]. Site-directed mutagenesis analysis suggests that the yeast NAD-IDH has an active center formed from two of the multiple subunits which is essentially identical to that of *E. coli* NADP-IDH [27].

Structural knowledge-based alignment shows that the catalytic subunits of NAD-IDHs retain all the active residues involved in Mg^{2+}-isocitrate binding and catalysis (Table 1). Similar to NAD-IMDH, the residues Lys344 and Try345, which interact with NADP in *E. coli* NADP-IDH, are replaced by Asp and Ile in these NAD-IDH proteins (Table 1). These results are consistent with their coenzyme preference toward NAD. Most of the active residues, including the residues equivalent to Ser113 and Asn115 of *E. coli* NADP-IDH, are found in the regulatory subunits of both yeast and mammalian NAD-IDHs. However, both Asp307 and Asp311, which coordinate the Mg^{2+} of Mg^{2+}-isocitrate in the *E. coli* NADP-IDH, are substituted by Asn, Ser, or Thr in these sequences (Table 1). This replacement should abolish the binding with Mg^{2+}-isocitrate, the true substrate for IDHs. Indeed, only two Mg^{2+}-isocitrate sites per $\alpha_2\beta\gamma$ tetramer of yeast NAD-IDH have been observed [28]. Meanwhile, the Arg129 residue, which binds via ionic hydrogen bonds to the α-carboxylate of bound isocitrate in the *E. coli* NADP-IDH, is replaced by a hydrophobic branched-chain residue such as Val, Ile, or Ala (Table 1). This should significantly weaken the binding affinity with isocitrate. The substitutions of these important residues agree with the regulatory roles of the subunits. The residue Asp278 identified in the *T. thermophilus* NAD-IMDH is conserved in yeast IDH1 but is replaced by polar residues Gln or Ser in the mammalian β and γ subunits (Table 1). Thus, identifying the critical specificity determinants is sufficient to define the function of NAD-IDH proteins as well.

VI. FUNCTIONAL MISASSIGNMENT CAN BE CONFIDENTLY CORRECTED

Insight into the structural and functional determinants for enzyme specificity allowed us to access functional assignment of β-decarboxylating dehydrogenase sequences. Of the cDNA or genomic sequences that have been published or deposited in databases, at least 20 are incorrectly annotated. Case studies are presented here to demonstrate that correct annotation can be made with certainty.

A. Bacterial NAD-IDH Sequence

All the eubacterial IDHs purified so far are NADP-specific, while NAD-IDHs are only found in the mitochondria of eukaryotic organisms [4]. However, early enzymatic studies suggested that NAD-IDHs may exist in some bacteria [4]. This prompted our rigorous search for possible eubacterial NAD-IDH sequences. Of 17 nonredundant prokaryotic IDH sequences available in the databases, the sequence (Gene number Q59985) from *Streptococcus salivarius* was predicted to encode a NADP-IDH protein of 391 amino acid residues (Table 1). All of the substrate-binding and catalytic residues identified in *E. coli* NADP-IDH are conserved in this protein, including Ser113 and Asn115, which are the major determinants of specificity toward isocitrate. In contrast, the residues Lys344 and Tyr345 interacting with NADP in the *E. coli* NADP-IDH are replaced by Asp and Ile, as seen in NAD-IMDH. This observation allows us to assign the function of the *S. salivarius* protein as NAD-IDH. These results demonstrate that, similar to NADP-IDH, NAD-IDH is present in both prokaryotic and eukaryotic organisms.

B. Plant NAD-IDH Sequences

The tomato ''NADP-IDH'' cDNA sequence (accession number Y16126, Table 1) was isolated from tomato roots and was upregulated during arbuscular mycorrhiza colonization. This sequence has been placed in the database and annotated as an NADP-IDH-like protein. Both equivalent Ser113 and Asn115 residues are found in this sequence (Table 1), indicating that it is indeed an IDH protein. Nevertheless, the presence of the negatively charged Glu at residue 344 suggests that this IDH is NAD-dependent. Similar to the regulatory subunits of yeast and mammalian NAD-IDHs, the Arg129, Asp307, and Asp311 residues of the *E. coli* NADP-IDH involved in the binding of the Mg^{2+}-isocitrate complex are substituted by Val, Ile, and Asn (Table 1). This leads to the prediction that the current protein belongs to a regulatory subunit. It has been shown that plant NAD-IDHs are exclusively localized in mitochondria [4]. As such, this protein sequence can be annotated as a regulatory subunit of tomato mitochondrial NAD-IDH.

The two *Arabidopsis* cDNA clones, "NAD-IDH1" and "NAD-IDH2" (U81993 and U81994, Table 1), were identified by homology searches from the *Arabidopsis* EST database [29]. These sequences share 36–50% identity with yeast and mammalian NAD-IDHs and contain a typical mitochondrial import targeting peptide. However, the report concluded that it was impossible to make subunit assignments, comparable to those for mammalian and yeast NAD-IDHs, and suggested that a single-subunit form of *Arabidopsis* NAD-IDH may exist and these two clones may represent isozymes. With a careful examination of the active site residues in these two sequences, it is obvious that, like the tomato NAD-IDH protein, these gene products lack the Mg^{2+}-binding site and should correspond to different regulatory subunits of NAD-IDH (Table 1). Since the binding residues equivalent to Arg129, Asp307, and Asp311 found in *E. coli* NADP-IDH are missing from these sequences, neither subunit can form an active enzyme. This is consistent with the observation the cDNA clones failed to complement yeast NAD-IDH mutants [29]. The residues involved in the binding of Mg^{2+}-isocitrate complex must be present at the third subunit.

It is noteworthy that the N-terminal sequences of the four plant NAD-IDH proteins described here share common features of a mitochondrial transit peptide: rich in basic and hydroxylated amino acids but lacking acidic residues [30]. Furthermore, these sequences contain basic residues at the N-terminal extremity which are characteristic of mitochondrial transit peptides and are normally absent from chloroplastic transit peptides. In contrast, the motif of Val/Ile-Arg-Ala/Cys⊥Ala (⊥ cleavage site) is frequently found at the cleavage site in the end of chloroplastic transit peptides [31], including the one for the chloroplastic NAD-IMDH from rape [25], but is absent from all the NAD-IDH sequences. Taken together, these results allow us to predict that, like their mammalian counterparts, plant NAD-IDHs must consist of three different subunits. Recently, three cDNA clones encoding different NAD-IDH subunits have been isolated from tobacco plants [32]. Both biochemical and genetic studies of the enzyme indicated that indeed, the physiologically active form is composed of the three different subunits.

VII. NEW ORTHOLOGS CAN BE READILY DISCOVERED: NAD-HOMOISOCITRATE DEHYDROGENASE

It has been suggested that new enzyme functions are established most easily and most commonly by recruitment of proteins already catalyzing analogous reactions [9]. NAD-IDHs, NADP-IDHs, and NAD-IMDHs are enzymes of very ancient origin. Protein engineering demonstrates that a few amino acid residue substitutions are sufficient to alter their substrate and coenzyme specificities. Such a strategy may have been used to modify these enzymes into new orthologs that catalyze identical chemical reactions but have distinct substrate specificities. An example of this strategy is discussed below.

In fungi, lysine is synthesized via an α-aminoadipate pathway [33]. The conversion of homoisocitric acid to α-ketoadipic acid occurs due to enzyme homoisocitrate dehydrogenase (EC 1.1.1.87, NAD-HDH):

$$\text{Homoisocitric acid} + \text{NAD}^+ \xrightleftharpoons{Mg^{2+}} \alpha\text{-ketoadipic acid} + CO_2 + \text{NADH} + H^+$$

This reaction is clearly analogous to that of NAD-IDHs, NADP-IDHs, and NAD-IMDHs. Early studies showed that the enzyme of *S. cerevisiae* was separated from the NAD-IDH and has a different pH optimum [33]. The molecular mass of the enzyme is 48 kDa. These results suggest that NAD-HDH is a novel member of the β-decarboxylating dehydrogenase family. In spite of a number of biochemical and genetic studies, the gene encoding the enzyme has not been identified.

The substrate homoisocitrate is structurally similar to isocitrate and isopropylmalate, where the γ-moiety is $—CH_2CH_2COO^-$ (Fig. 5). With the principles governing the relationship between structure and function in NADP-IDHs and NAD-IMDHs, we predict that all of the substrate-binding and catalytic residues shared by *E. coli* NADP-IDH and *T. thermophilus* NAD-IMDH should be conserved in NAD-HDH. What makes this enzyme distinct would be the amino acid residues involved in binding the γ-moieties of substrate. Compared to isocitrate, the γ-carboxylate of homoisocitrate is one carbon farther away from C2. This difference prevents the formation of a salt bridge between the bound homoisocitrate and the nicotinamide ring of the coenzyme in the Michaelis complex that would otherwise pull the nicotinamide C4 out of the catalytic trajectory during hydride transfer. In this enzyme, the nicotinamide ring is probably aligned with the bound homoisocitrate by the protein itself, as seen in NAD-IMDH, where the carboxyl group of Glu87 interacts with the nicotinamide ring (Fig. 7B). However, the presence of a negatively charged residue such as Glu at this position would electrostatically repel the γ-moiety of bound substrate. Hence, the residue equivalent to Ser 113 of the NADP-IDH and Glu87 of the NAD-IMDH should be Tyr or Gln, which is able to interact with nicotinamide ring via hydrogen bonding. Meanwhile, the residue equivalent to Asn115 of NADP-IDH and Leu90 of NAD-IMDH is likely a nonpolar residue which interacts hydrophobically with the extra-carbon portion of the γ-moiety of the bound homoisocitrate (Fig. 6, Table 1).

The genome of *S. cerevisiae* has been completely sequenced. A search of the entire genomic sequence with *E. coli* NADP-IDH leads to the identification of a candidate sequence with functional annotation as NAD-IMDH (accession number P40495) (Table 1). This protein of 385 amino acid residues shares 32%, 34%, and 35% identity with the yeast NAD-IMDH and the regulatory and catalytic subunits of NAD-IDH, respectively. The molecular mass is estimated as 41 kDa. As expected, most of the substrate-binding and catalytic residues identified in *E. coli* NADP-IDH are found in this protein. Remarkably, the major specificity determinants in the NADP-IDH, Ser113 and Asn115, are replaced by Tyr and Ile, which makes this protein different from IDH and IMDH. Meanwhile, the residues Lys344 and Tyr345, which interact with NADP in *E. coli*, are replaced by Asp and Ile. This observation allows us to confidently assign the function of the protein as NAD-HDH.

VIII. CONCLUSION

Using the β-decarboxylating dehydrogenase family as a model, we demonstrate that rules for predicting the precise function of members of gene families can be established. The strategy is based on our recent findings that only a few amino acid residue substitutions in these enzymes are sufficient to change substrate and coenzyme specificities and thus alter enzyme functions. The few critical specificity determinants then serve as reliable markers for determining orthologous or paralogous relationships. The power of this approach has been demonstrated by correcting functional misassignments and discovering new genes. With the progress of genome-wide efforts to determine representative three-dimensional structures for all protein families [34], it is likely that the current approach will become broadly applicable. Extension of similar studies to other protein families would be much needed in order to take full advantage of the enormous wealth of biological information coming out of the EST and genome projects. In conclusion, such a powerful approach should allow us to significantly accelerate the discovery of new enzymes and related pathways in the coming years and thus herald an exciting new era for enzymologists.

ACKNOWLEDGMENTS

This work was supported by research grants from the Medical Research Council of Canada, the Natural Sciences and Engineering Research Council of Canada, and the Health Services and Utilization Research Commission.

REFERENCES

1. S Henikoff, EA Greene, S Pietrokovski, P Bork, TK Attwood, L Hood. Gene families: the taxonomy of protein paralogs and chimeras. Science 278:609–614, 1997.
2. RL Tatusov, KV Koonin, DJ Lipman. A genomic perspective on protein families. Science 278:613–617, 1997.
3. K Hofmann. Protein classification and functional assignment. Trends Guide to Bioinformatics (Trends suppl) 18–21, 1998.
4. RD Chen, P Gadal. Structure, function and regulation of NAD- and NADP-isocitrate dehydrogenases in higher plants and in other organisms. Plant Physiol Biochem 28: 411–427, 1990.
5. K Imada, M Sato, N Tanaka, Y Katsube, Y Matsuura, T Oshima. Three-dimensional structure of a highly thermostable enzyme, 3-isopropylmalate dehydrogenase of *Thermus thermophilus* at 2.2 Å resolution. J Mol Biol 222:725–738, 1991.
6. JH Hurley, P Thorsness, V Ramalingham, N Helmers, DE Koshland Jr, RM Stroud. Structure of a bacterial enzyme regulated by phosphorylation, isocitrate dehydrogenase. Proc Natl Acad Sci (USA) 86:8635–8639, 1989.
7. MG Rossmann, D Moras, KW Olsen. Chemical and biological evolution of a nucleotide-binding protein. Nature 250:194–199, 1974.
8. RD Chen, A Greer, JM Hurley, AM Dean. Engineering secondary structure to invert coenzyme specificity in isopropylmalate dehydrogenase. In: DR Marshak, ed. Techniques in Protein Chemistry VIII. New York: Academic Press, 1997, pp 809–816.
9. RA Jensen. Enzyme recruitment in evolution of new function. Annu Rev Microbiol 30:409–425, 1976.
10. RD Chen, AF Greer, AM Dean. A highly active decarboxylating dehydrogenase with rationally inverted coenzyme specificity. Proc Natl Acad Sci (USA) 92:11666–11670, 1995.
11. RD Chen, JA Grobler, JH Hurley, AM Dean (1996). Second-site suppression of regulatory phosphorylation in *Escherichia coli* isocitrate dehydrogenase. Protein Sci 5:287–295, 1996.
12. RD Chen, A Greer, AM Dean. Redesigning secondary structure to invert coenzyme specificity in isopropylmalate dehydrogenase. Proc Natl Acad Sci (USA) 93:12171–12176, 1996.
13. RD Chen, A Greer, AD Dean. Structural constraints in protein engineering: the coenzyme specificity of *Escherichia coli* isocitrate dehydrogenase. Eur J Biochem 250: 578–582, 1997.
14. AM Dean, AK Shiau, DE Koshland Jr. Determinants of performance in the isocitrate dehydrogenase of *Escherichia coli*. Protein Sci 5:341–347, 1995.
15. AM Dean, L Dvorak. The role of glutamate 87 in the kinetic mechanism of *Thermus thermophilus* isopropylmalate dehydrogenase. Protein Sci 4:2156–2167, 1995.
16. JH Hurley, RD Chen, AM Dean. Determinants of cofactor specificity in isocitrate dehydrogenase: structure of an engineered NADP → NAD specificity-reversal mutant. Biochemistry 35:5670–5678, 1996.
17. T Yaoi, K Miyazaki, T Oshima. Substrate recognition of isocitrate dehydrogenase and 3-isopropylmalate dehydrogenase from *Thermus thermophilus* HB8. J Biochem 121:78–81, 1997.

18. RD Chen. A general strategy for enzyme engineering. Trends Biotechnol 17:344–345, 1999.
19. JH Hurley, AM Dean, DE Koshland Jr, RM Stroud. Catalytic mechanism of $NADP^+$-dependent isocitrate dehydrogenase: implications from the structures of magnesium-isocitrate and $NADP^+$ complexes. Biochemistry 30:8671–8678, 1991.
20. BL Stoddard, AM Dean, DE Koshland Jr. Structure of isocitrate dehydrogenase with isocitrate, nicotinamide adenine dinucleotide phosphate, and calcium at 2.5-Å resolution: a pseudo-Michaelis ternary complex. Biochemistry 32:9310–9316, 1993.
21. JM Bolduc, DH Dyer, WG Scott, P Singer, RM Sweet, DE Koshland Jr, BL Stoddard. Mutagenesis and Laue structures of enzyme intermediates: isocitrate dehydrogenase. Science 268:1312–1318, 1995.
22. JH Hurley, AM Dean. Structure of 3-isopropylmalate dehydrogenase in complex with NAD^+: ligand-induced loop closing and mechanism for cofactor specificity. Structure 2:1007–1016, 1994.
23. DA Keys, L McAlister-Henn. Subunit structure, expression, and function of NAD(H)-specific isocitrate dehydrogenase in *Saccharomyces cerevisiae*. J Bacteriol 172:4280–4287, 1990.
24. N Ramachandran, RF Colman. Chemical characterization of distinct subunits of pig heart DPN-specific isocitrate dehydrogenase. J Biol Chem 255:8859–8864, 1980.
25. RS Ehrlich, RF Colman. Separation, recombination, and characterization of dissimilar subunits of the DNP-dependent isocitrate dehydrogenase from pig heart. J Biol Chem 258:7079–7086, 1983.
26. BJ Nichols, ACF Perry, L Hall, RM Denton. Molecular cloning and deduced amino acid sequences of the α- and β-subunits of mammalian NAD^+-isocitrate dehydrogenase. Biochem J 310:917–922, 1995.
27. JR Cupp, L McAlister-Henn. Kinetic analysis of NAD^+-isocitrate dehydrogenase with altered isocitrate binding site: contribution of IDH1 and IDH2 subunits to regulation and catalysis. Biochemistry 32:9323–9328, 1993.
28. GA Rutter, RM Denton. The binding of Ca^{2+} ions to pig heart NAD^+-isocitrate dehydrogenase and the 2-oxoglutarate dehydrogenase complex. Biochem J 263:453–462, 1989.
29. RH Behal, D Oliver. NAD^+-dependent isocitrate dehydrogenase from *Arabidopsis thaliana*. Characterization of two closely related subunits. Plant Mol Biol 36:691–698, 1998.
30. AL Moore, CK Wood, FZ Watts. Protein import into plant mitochondria. Annu Rev Plant Physiol Plant Mol Biol 45:545–575, 1994.
31. Y Gavel, GA von Hejne. A conserved cleavage-site motif in chloroplast transit peptides. FEBS Lett 261:455–458, 1990.
32. M Lancien, P Gadal, M Hodges. Molecular characterization of higher plant NAD-dependent isocitrate dehydrogenase: evidence for a heteromeric structure by the complementation of yeast mutants. Plant J 16:325–333, 1998.
33. JK Bhattacharjee. Evolution of α-aminoadipate pathway for the synthesis of lysine in fungi. In: RP Mortlock, ed. The Evolution of Metabolic Function. London: CRC Press, 1992, pp 47–80.
34. JM Thornton. The future of bioinformatics. Trends Guide to Bioinformatics (Trends suppl) 30–31, 1998.

23

Redesigning Binding and Catalytic Specificities of Enzymes

Ridong Chen
University of Saskatchewan, Saskatoon, Saskatchewan, Canada

I. INTRODUCTION

Enzyme technology is undergoing the most profound and exciting transformation in its history [1]. It promises unprecedented expansion in the scope and applications of new enzymes with desired physical and catalytic properties. Two themes have emerged that will dominate enzyme research in the coming years: discovering new enzymes with the powerful genomic approach and optimizing known enzymes by protein engineering. Two different complementary approaches to protein engineering are currently available: directed evolution involving random mutagenesis and *in vitro* recombination followed by screening to isolate mutants with the desired properties [2,3], and knowledge-based rational design using site-directed mutagenesis to engineer precise changes in amino acid sequences [4–6]. While both approaches have met with success, each suffers from drawbacks. Random mutagenesis requires a selectable phenotype in order to screen a large number of potential mutants, while rational design presumes a detailed understanding of the relations between structure and function. During the last few years, the directed evolution technique has received extensive reviews [7,8]. Here, emphasis is given on a rational design approach that has been successfully used for engineering specificities of isocitrate and isopropylmalate dehydrogenases [9–14].

II. IDEAL MODEL SYSTEM FOR ENZYME ENGINEERING: ISOCITRATE AND ISOPROPYLMALATE DEHYDROGENASES

Rational enzyme engineering refers to the creation of mutant proteins with predetermined enzymatic characteristics such as catalytic efficiency, pH optimum, or improved stability [14]. Precise changes in amino acid sequence are preconceived based on a detailed knowledge of protein structure, function, and mechanism and then introduced using oligonucleotide-directed mutagenesis. The power of this approach has been demonstrated by the generation of a faster superoxide dismutase, already one of the fastest known enzymes in nature [6], and conversion of lactate dehydrogenase into a highly specific malate dehydrogenase [5]. Despite these spectacular examples, numerous published or unpublished attempts have failed, illustrating the need for a model system and systematic approaches to establish general rules for rationally engineering proteins.

Isocitrate dehydrogenase (E.C. 1.1.1.42, IDH) catalyzes two sequential reactions: the oxidation of 2*R*,3*S*-isocitrate to oxalosuccinate, followed by the decarboxylation of this intermediate to α-ketoglutarate and CO_2 [15], viz.,

$$\text{Isocitrate}^{3-} + \text{NAD(P)}^+ \xrightleftharpoons{\text{Mg}^{2+}} \text{oxalosuccinate}^{3-} + \text{NAD(P)H} + \text{H}^+$$

$$\text{Oxalosuccinate}^{3-} + \text{H}^+ \xrightleftharpoons{\text{Mg}^{2+}} \alpha\text{-ketoglutarate}^{2-} + \text{CO}_2$$

Two distinct IDHs have been identified in living organisms: NAD-IDH (EC 1.1.1.41) is limited exclusively to eukaryotic organisms, and NADP-IDH (EC 1.1.1.42) is ubiquitously present in both prokaryotes and eukaryotes [15]. These enzymes play important biological roles: NAD-IDH is involved in the supply of NADH used for respiratory ATP production in mitochondria of eukaryotic organisms; the bacterial NADP-IDH is also part of the Krebs cycle machinery. However, the eukaryotic NADP-dependent enzyme is primarily involved in providing NADPH and α-ketoglutarate for biosynthesis [15].

The NADP-IDH from *Escherichia coli* has been thoroughly studied. It is a dimeric protein of two identical 40-kDa subunits. High-resolution X-ray crystal structures have been determined for the enzyme with and without substrate [16,17], and for the pseudo-Michaelis complex of the enzyme with isocitrate and NADP [18]. Structures of sequential intermediates formed during the catalytic action of IDH are also available [19]. Additionally, the kinetic and catalytic mechanisms have been determined in detail [20]. Amino acid residues which are involved in interactions with substrate, coenzyme, metal ions, and catalysis have been identified [10,21].

Both NADP-IDH and the homologous isopropylmalate dehydrogenase (IMDH, EC 1.1.1.85) belong to the non-Rossmann-fold β-decarboxylating dehydrogenase family. IMDH catalyzes chemically equivalent reactions, the oxidation of 3-isopropylmalate to isopropyloxaloacetate, followed by the decarboxylation of this intermediate to α-ketoisocaproate and CO_2 [22]:

$$\text{Isopropylmalate}^{2-} + \text{NAD}^{+} \underset{}{\overset{\text{Mg}^{2+}}{\rightleftharpoons}} \text{isopropyloxaloacetate}^{2-} + \text{NADH} + \text{H}^{+}$$

$$\text{Isopropyloxaloacetate}^{2-} + \text{H}^{+} \overset{\text{Mg}^{2+}}{\rightleftharpoons} \alpha\text{-ketoisocaproate}^{1-} + \text{CO}_2$$

This enzyme is involved in the biosynthetic pathway of leucine in bacteria, yeast, and higher plants. The IMDH from *Thermus thermophilus* has been extensively studied, and it is composed of two identical 35-kDa subunits. The kinetic and catalytic mechanisms have been determined, and X-ray crystal structures solved [22,23].

Despite sharing only 25% sequence identity, structural analysis indicates that both proteins of *E. coli* NADP-IDH and *T. thermophilus* NAD-IMDH are homodimers which share a common protein fold that lacks the βαβαβ motif characteristic of the nucleotide binding Rossmann fold [23]. The strict and distinct specificities of these enzymes provide an attractive model system for engineering specificity, while the extensive knowledge of substrate and coenzyme binding and catalysis provide the sound foundation critical for rational design.

III. SPECIFICITIES OF ISOCITRATE AND ISOPROPYLMALATE DEHYDROGENASES

Of central importance to biochemical systems is the remarkable ability of enzymes to discriminate between similar substrates. Both IDH and IMDH display striking coenzyme (recycled substrate) and substrate specificities.

A. Coenzyme Specificities

In biological reactions, the pyridine nucleotide coenzymes, NAD and NADP, provide the reducing equivalents necessary for energy transduction and biosynthesis in living cells. The reduced coenzymes are generated by the transfer of a hydride anion from a wide variety of substrates and catalyzed by an equally wide variety of dehydrogenases. Structurally, NADP differs from NAD only by a phosphate group esterified at the 2′-carbon of the adenosine ribose. However, this difference confers the two types of coenzymes with fundamentally distinct roles in metabolism: whereas the NADH system is specialized in collecting reducing equivalents to be oxidized for ATP production, the NADPH system mainly gathers reducing equivalents to be donated during synthetic processes. Hence, discrimination between these coenzymes by dehydrogenases represents a striking illustration of the importance of enzyme specificity in biochemical reactions and provides an excellent model system to elucidate specificity determinants by protein engineering.

E. coli NADP-IDH and the homologous *T. thermophilus* NAD-IMDH have similar nucleotide-binding pockets that are quite distinct from the Rossmann fold found in many other dehydrogenases [16,17,22,23]. The pocket is constructed from three loops and an α-helix in IDH, the latter being substituted by a β-turn in IMDH (Fig. 1). Calculated as the ratio of k_{cat}/K_m, the *E. coli* NADP-IDH is 7000-fold more active with NADP than with NAD [1], whereas IMDH exhibits a 100-fold preference for NAD.

Coenzyme specificity in IDH is conferred by interactions between Arg395, Tyr345, Tyr391, and Arg292′ (on the second subunit) with the 2′-phosphate of bound NADP (*E. coli* IDH numbering). These residues are conserved in prokaryotic NADP-dependent IDHs and substituted by a variety of residues in the NAD-dependent dehydrogenases (Table 1). The β-turn of IMDH has no site equivalent to Arg395 in the α-helix of IDH, while replacements by Ser292′, Ile345, and Pro391 eliminate all other favorable interactions with the 2′-phosphate. Coenzyme specificity in IMDH is conferred by the conserved Asp344, which forms a double hydrogen bond with the 2′- and 3′-hydroxyls of the adenosine ribose of NAD, shifting its position and perhaps changing the ribose pucker from C2′-endo to C3′-endo. Not only are these movements incompatible with the strong 2′-phosphate interactions seen in IDH, but the negative charge on Asp344 may repel NADP.

B. Substrate Specificities

E. coli NADP-IDH has shown detectable catalytic abilities toward various analogs of isocitrate, $^{-}OOC(HO)CHCH(X)COO^{-}$, where X represents the CH_2COO^{-} of isocitrate (Fig. 2). However, the enzyme displays spectacular substrate specificity: when X is changed to a methyl or ethyl group, the catalytic efficiency (measured by $k_{cat}/K_{i.NADP} \times K_m$) decreases by a factor of 10^6 and to a hydrogen by a factor of 10^8. The structures of the binary complex of IDH with bound isocitrate and the pseudo-Michaelis complex of IDH with NADP and isocitrate suggest that the amino acid residues Ser113 and Asn115 on helix d (Table 2), and the γ-carboxylate of isocitrate are involved in the binding and catalytic specificity of the substrate (Fig. 3). Ser113 forms a hydrogen bond with the γ-carboxylate of isocitrate. Asn115 is coordinated to the amide of NADP. The γ-carboxylate of isocitrate interacts electrostatically with the positively charged nicotinamide ring of NADP. The importance of these interactions has been confirmed by kinetic analysis of site-directed mutants in combination with synthetic substrate analogs [10,21].

An understanding of the determinants of substrate specificity has been significantly enhanced by comparative analysis of the homologous structures of

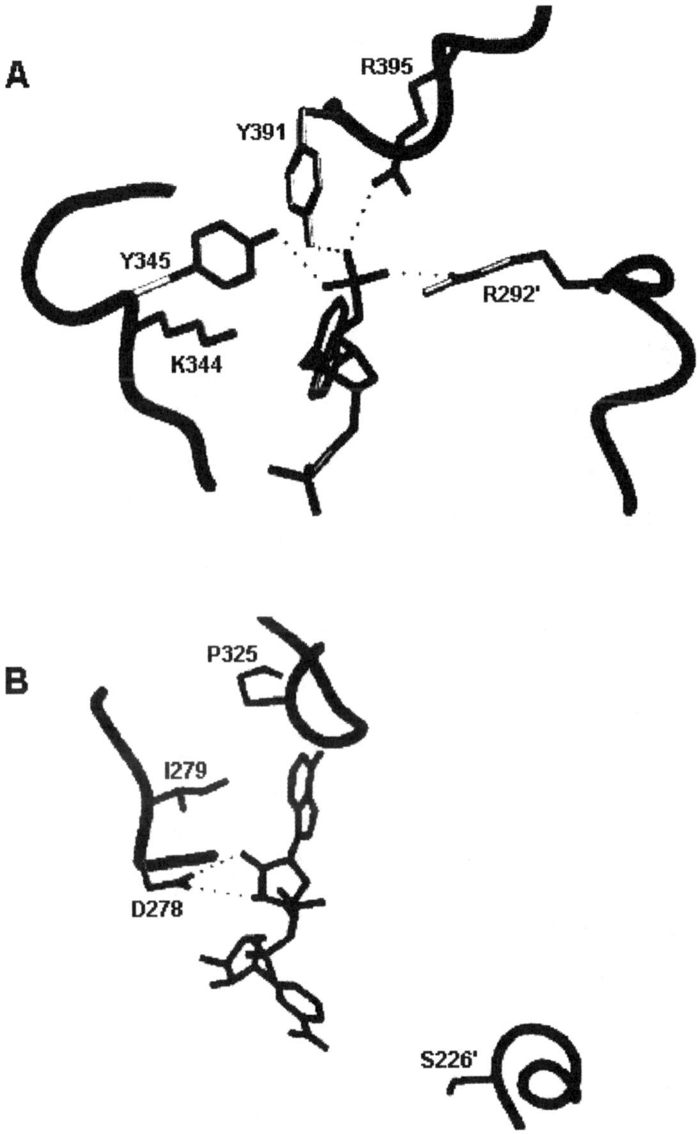

Figure 1 (A) Interaction of the nucleotide-binding pocket of the *E. coli* NADP-IDH with NADP and (B) that of *T. thermophilus* NAD-IMDH with NAD. It is noted that due to the difference of the local secondary structures, the NAD-IMDH does not have the counterpart of the Arg395 found in the NADP-IDH.

Table 1 Structure-Based Alignment of Representative NADP-IDH, NAD-IMDH, and NAD-IDH Sequences[a]

Enzymes	Sequences		
NADP-IDH	201	332 344 345 351	391 395
E. coli	IKP**C**SEEGTKR	DE**C**ALFEATHGTAP**KY**AGQDK**V**NP	T**Y**DFE**R**LM
Vibrio sp.	IKP**V**SKEGSQR	DE**V**AVFEATHGTAP**KY**AGKNK**V**NP	T**Y**DFE**R**LM
Anabaena sp.	IKP**I**SKTGSQR	DS**C**AVFEATHGTAP**KH**AGLDR**I**NP	T**Y**DLA**R**LL
Synechocystis sp.	IKP**I**SKTGSQR	DS**A**AIFEATHGTAP**KH**AGLDR**I**NP	T**Y**DLA**R**LM
NAD-IMDH			
T. thermophilus	TER**Y**SKPEVER	RG**T**PVFEPVHGSAP**DI**AGKGI**A**NP	P**P**DLG**G**SA
E. coli	TEV**Y**HRFEIER	EG**F**GLYEPAGGSAP**DI**AGKNI**A**NP	T**G**DLA**R**GA
S. cerevisiae	SEQ**Y**TVPEVQR	TA**F**GLYEPCHGSAP**DL**P-KNK**V**NP	T**G**DLG**G**SN
Y. lipolitica	TET**Y**SVPEVER	EA**F**GLYEPCHGSAP**DL**G-KQK**V**NP	T**A**DIG**G**SS
Engineered NAD-IDH			
E. coli	IKP**M**SEFKTKR	DE**Y**ALFESTHGTAP**DI**AGQDK**A**NP	T**K**DFE**S**LM

[a] Numbers are for the residues in bold found in *E. coli* NADP-IDH.

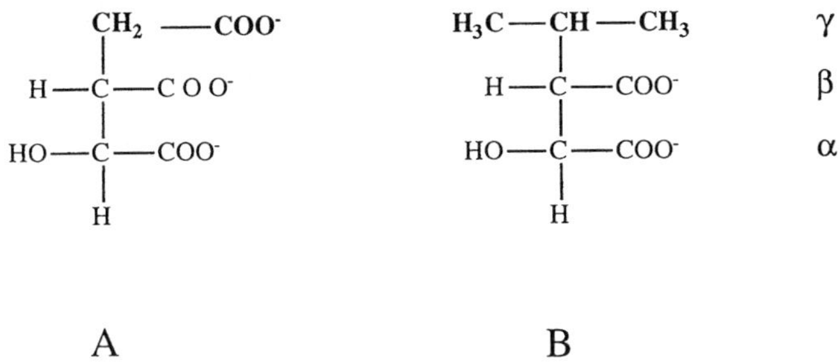

Figure 2 Structures of 2*R*,3*S*-isocitrate (A) and 2*R*,3*S*-isopropylmalate (B), where α, β refers to the α- and β-carboxyl groups, respectively. The unique γ-moieties are presented in bold.

IDH and IMDH. Isocitrate and isopropylmalate have the similar structure $^{-}OOC(HO)CHCH(X)COO^{-}$, where X represents the CH_2COO^{-} of isocitrate and the $CH(CH_3)_2$ of isopropylmalate (Fig. 2). In spite of the similarities of the substrates and the enzyme structures [16,23], IMDH does not catalyze a reaction with isocitrate and IDH does not catalyze a reaction with isopropylmalate (our

Table 2 Structure-Based Alignment of β-Decarboxylating Dehydrogenase Sequences[a]

Enzymes	Sequences
NADP-IDH	
	113 115 119 129
E. coli	R**S**L**N**VAL**R**QELDLYICL**R**P
T. thermophilus	K**S**A**N**VTL**R**KLFETYANV**R**P
A. fulgidus	R**S**L**N**VTI**R**QVLDLYANV**R**P
S. cerevisiae	K**S**P**N**GTI**R**NILGGTV-F**R**E
H. sapiens	K**S**P**N**GTI**R**NILGGTV-F**R**E
S. tuberosum	R**S**P**N**GTI**R**NILNGTV-F**R**E
NAD-IMDH	
	87 90 94 104
T. thermophilus	**E**TG**L**LSL**R**KSQDLFANL**R**P
B. subtilis	**E**KG**L**LSI**R**KELDLFVCL**R**P
S. cerevisiae	**E**QG**L**LKI**R**KELQLYANL**R**P
B. napus	**E**TG**L**LQL**R**AGLKVFANL**R**P

[a] Numbers are for the residues in bold for the first enzyme in each group.

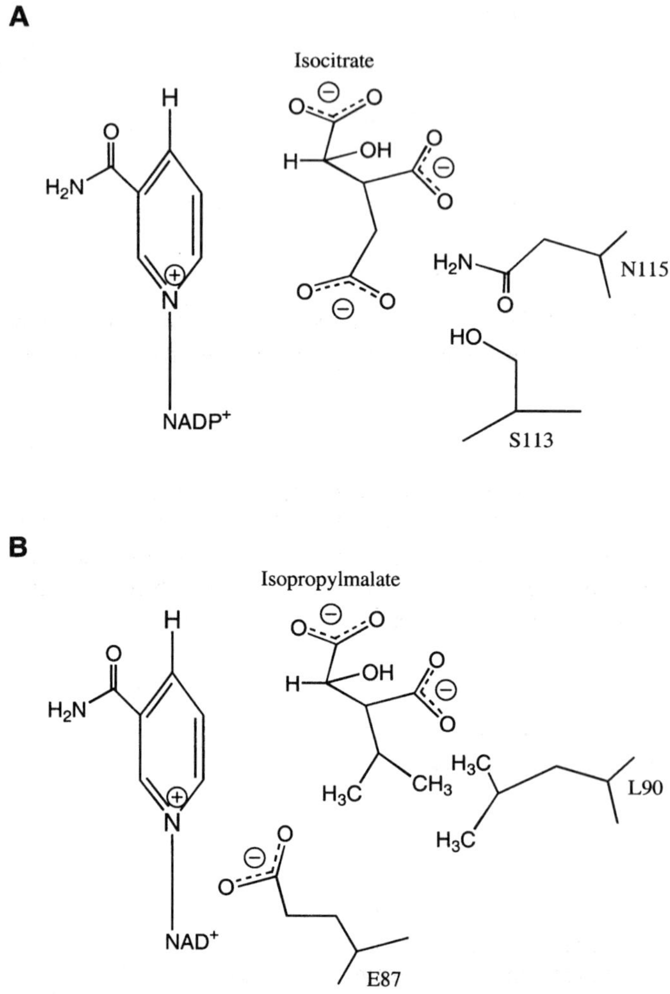

Figure 3 Schematic diagram of the active site of the *E. coli* NADP-IDH with bound isocitrate (A) and that of *T. thermophilus* NAD-IMDH with bound isopropylmalate (B).

unpublished results). All the amino acid residues involved in binding the 2R-malate core of both substrates are identical in both enzymes, indicating that the substrate specificity is determined by the interaction with γ-moieties attached to the 3*S*-position of the 2*R*-malate core. Indeed, the two enzymes differ in the amino acid sequences involved in binding the γ-moieties of substrates. In IMDH, Asn115 is replaced by leucine (Table 2, Fig. 3). In addition, helix d starts three amino acids earlier, with the consequence that a conserved glutamate occupied a position physically above, but nevertheless in close proximity to Ser113 and the γ-carboxylate of isocitrate (Table 2, Fig. 3). Site-directed mutagenesis and structural analysis have demonstrated that this glutamate may mimic the role of the γ-carboxylate of isocitrate for IDH, interacting with the nicotinamide ring of the coenzyme (Fig. 3 [24]).

IV. AN EFFICIENT STRATEGY FOR MUTANT SCREENING

In the rational design of enzymes, precise changes in amino acid sequence are preconceived based on the understanding of protein structure–function relationships and then introduced using oligonucleotide-directed mutagenesis. The conventional approach requires that mutations be identified by sequencing, the mutant enzymes purified and their kinetic/functional properties determined after each round of mutagenesis. This tedious approach is impractical for rational engineering of enzymes, where many rounds of mutagenesis are necessary to confer novel properties [14].

Using *E. coli* NADP-IDH and *T. thermophilus* NAD-IDH as models, a strategy for rapid screening of mutants has been developed. This strategy involves the optimization of protein expression at reproducible levels and the direct evaluation of enzyme performance in crude extracts of a suitable microbial host free from background enzyme activities (Fig. 4). This approach allows the rapid and reliable identification of mutant enzymes with effective combinations of beneficial mutations without DNA sequencing and protein purification [14]. As an example, the following procedure has been developed to screen mutants of NADP-IDH spectrophotometrically. Expression of recombinant proteins up to 50% of soluble cell protein is driven by the *icd* promotor. Following transformation of plasmids into *E. coli* strain SL4 (Δ*idh*), cells are incubated in 1 ml of 2XYT broth for 30 min at 37°C to allow expression of Amp^r, streaked on fresh 2XYT plates containing 60 μg/ml fresh ampicillin, and grown overnight at 37°C. Individual colonies are inoculated in 3 ml of fresh 2XYT medium containing 60 μg/ml fresh ampicillin and grown overnight at 37°C, and 1 ml of the overnight culture is centrifuged in a microfuge tube. The cells are resuspended in 1 ml of KAC buffer (25 M MOPS, 100 mM NaCl, 5 mM $MgCl_2$, and 1 mM DTT, pH 7.5), twice sonicated on ice for a total of 30 sec, and the debris is removed by centrifugation. Values for V_{max} are determined at saturating concentrations of both NADP and isocitrate, while values for K_m for coenzymes are estimated by varying the appropriate concentration of NAD or NADP. Extracts from each of three putative mutants displaying similar kinetics and three putative nonmutants are subjected to sodium dodecyl sulfate-polyacrylamide gel electrophoresis (SDS-PAGE) in order to compare IDH expression levels. Plasmid DNA isolated from

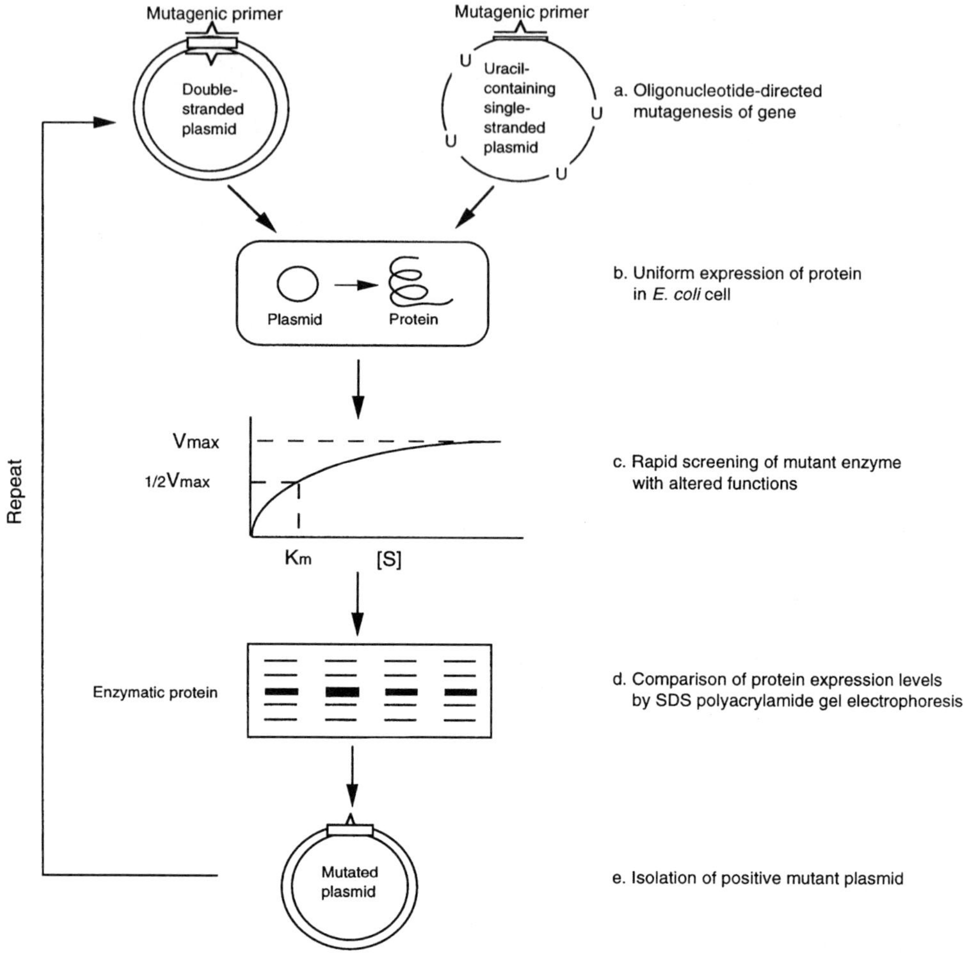

Figure 4 Schematic illustration of the efficient screening of mutant enzymes during multiple cycles of mutagenesis.

the remaining 2 ml of unsonicated culture from each of the positive mutants and nonmutants is sequenced, thus reducing the possibility of second-site mutations. The entire procedure for each round of mutagenesis takes no more than 7 days to complete.

Such a simple and efficient protocol greatly facilitates the rational design of enzymes when multiple cycles are necessary to engineer novel properties. The method also provides a rapid means to distinguish among multiple kinetic phenotypes generated by degenerate oligonucleotides and random *in vitro* mutagenic procedures. Moreover, this approach has broad applicability, since there is no prerequisite that the mutations confer any obvious phenotypic differences upon cells/colonies.

V. GENERATING A HIGHLY ACTIVE AND SPECIFIC NAD-IDH

In this study we combined two strategies common in protein engineering. Substitutions based on rational design within the nucleotide-binding pocket were used to convert *E. coli* IDH coenzyme specificity from NADP to NAD, while substitutions improving overall performance were identified by partial random mutagenesis at sites outside the nucleotide-binding pocket [5,12].

A. Rational Design

Structural analysis indicates that the specificity toward NADP is conferred by hydrogen bonds between Tyr345, Tyr391, Arg395, Arg292′, and the 2′-phosphate of bound NADP (Fig. 1A). Guided by the structures of IDH and IMDH with bound coenzymes and molecular modeling, five substitutions were introduced in the adenosine-2′-phosphate-binding pocket of the *E. coli* enzyme, K344D, Y345I, V351A, Y391K, and R395S, all of which were predicted to disrupt the favorable interaction with NADP (Fig. 1A, Table 1). Additionally, substituting Lys344 by Asp introduces a negative charge into the pocket that repels the 2′-phosphate of NADP. Formation of a hydrogen bond between the substituted Asp344 and the 2′-hydroxyl group on the NAD ribose requires that the latter move 3.5 Å closer to the carboxylate of Asp344 and that the ribose pucker change from 2′C to 3′C endo, as seen in the homologous NAD-IMDH from *T. thermophilus*. This requirement was accommodated by substituting Y345I and V351A. The engineered enzyme, DIAKS, displays a 100-fold preference for NAD over NADP, which is comparable to the similar 140-fold preference exhibited by the NAD-IMDH (Fig. 5A).

B. Partial Random Mutagenesis

The engineered DIAKS mutant, which was generated solely by rational means, displays a coenzyme preference comparable to its natural counterparts and exhibits a 30-fold increase in enzyme performance with NAD compared to wild-type IDH. Nevertheless, its performance is still 10-fold lower than other homologous NAD-dependent enzymes. Despite the availability of high-resolution structures, including Laue structures of sequential intermediates formed during catalysis [19], no other obviously favorable mutations could be predicted that might further improve enzyme performance. Hence, partial random mutagenesis was used to

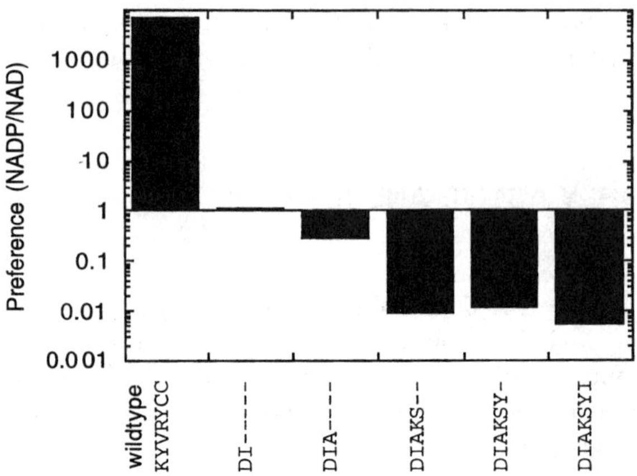

(A)

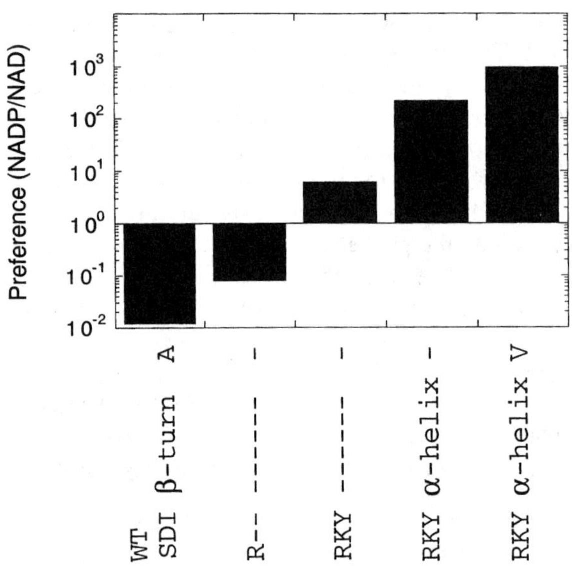

(B)

Figure 5 The systematic change in coenzyme preference generated by engineering mutants of IDH (A) and IMDH (B).

identify a mutant with two additional mutations outside the adenosine-binding pocket, C332Y and C201M, which lead to a net eightfold increase in performance [9]. Cys332 is adjacent to the catalytic center and is located at the amino terminus of a β-sheet that traverses the hydrophobic core and terminates in the nucleotide binding pocket just before His339. Substitution by Tyr increases k_{cat} by twofold and decreases K_m by threefold. Cys201 lies adjacent to a super-secondary structure containing the catalytic base K230′. Substitution by Met results in a twofold improvement in k_{cat}. ''Irrational'' random mutagenesis is usually employed when the relationship between structure and function is poorly understood, yet this study demonstrates that this approach can effectively augment a rational design approach.

C. Performance of the Final Mutant

The final mutant, DIAKSYM, contains seven amino acid substitutions generated through five rounds of mutagenesis. Its kinetics compare favorably with those from homologous NAD-dependent IDHs and IMDHs from various natural sources [9,15]. In fact, with an apparent dissociation constant for isocitrate ($K_{i.isocitrate}$) of 7 μM, the overall catalytic efficiency ($k_{cat}/K_{i.isocitrate} \times K_{m.NAD}$) of our engineered NAD-IDH exceeds, by a factor of 20, that of its natural counterpart from yeast (Table 3). The comparable enzyme performance and coenzyme specificities of the natural and engineered enzymes demonstrate that, with protein engineering, we may have identified all the major specificity determinants.

D. Crystal Structure of Engineered Enzyme

The X-ray crystallographic structure of the engineered *E. coli* NAD-IDH with bound NAD has been solved at 1.9-Å resolution [13]. As expected, and as seen

Table 3 Comparison of Overall Catalytic Efficiency of Engineered and Natural NAD-Dependent Isocitrate Dehydrogenases (IDH) from *E. coli* and *S. cerevisiae*, Respectively

Kinetic parameter	Engineered IDH at 21°C	Natural IDH at 24°C
$K_{m.NAD}$[a] (μM)	99	210
$K_{i.isocitrate}$[b] (μM)	7	150
k_{cat} (sec^{-1})	16.2	40
$k_{cat}/(K_{i.isocitrate} \times K_{m.NAD})$ (μM^{-2} sec^{-1})	23.3×10^{-3}	1.3×10^{-3}

[a] Michaelis constant for NAD.
[b] Apparent dissociation constant for isocitrate.

in IMDH, H-bonds between Asp344 and the 2′OH and 3′OH of the nicotinamide ring are established through a shift in the position of the NAD adenine ring and a change in the pucker of the adenosine ribose. The shift in the position and ribose pucker of NAD relative to NADP is crucial for hydrogen-bond formation with Asp344 and is accommodated by the adjacent Y345I replacement and the V351A mutation at the back of the pocket. In light of the structural information, we redesigned an alternative hydrogen-bonding scheme, one that does not require the adenine ring to shift. Modeling indicates that the extra methylene generated by a K344E replacement might allow a hydrogen bond to form with the 3′-OH of NAD, with the latter adopting the same position as NADP in the wild-type enzyme. A second Y345F substitution removes the hydrogen bond to the 2′-phosphate of NADP, and the bulky phenylalanine side chain should inhibit any shift in the adenine ring toward the left-hand side of the pocket. However, these attempts failed to generate NAD dependence [12]. These experiments and simulations suggest that the means by which specificity can be engineered toward NAD is constrained by the architecture of the NADP-IDH coenzyme-binding pocket. It also suggests that with the given structural scaffold, the engineered enzyme is probably as efficient an NAD-specific IDH as is currently possible to design.

VI. CREATING A NATURALLY NONEXISTENT NADP-IMDH

Structural analysis indicates that coenzyme specificity of IMDH is conferred by Asp278 (Fig. 1B), a conserved residue that forms a double hydrogen bond with the 2′- and 3′-hydroxyls of the adenosine ribose of NAD and which repels the negatively charged 2′-phosphate of NADP. Asp278 is replaced by Lys in IDH (Fig. 1), a rigidly conserved residue that probably interacts with the 2′-phosphate of NADP. Ser226′ (on the second subunit) and the conserved Ile279 of IMDH are replaced by Arg226′ and Tyr279 in all eubacterial IDHs (Fig. 1B), where they form hydrogen bonds to the 2′-phosphate of NADP. In addition, a β-turn in the coenzyme-binding pocket of IMDH is replaced by an α-helix and loop in IDH. Two additional interactions with the 2′-phosphate are found in IDH: Tyr325 (IMDH numbering) and Arg395 (IDH numbering) (Fig. 1A). These residues have no equivalent in IMDH, where the α-helix of IDH is replaced by a β-turn. Hence, converting the coenzyme specificity presents a new challenge requiring a change in the secondary structure of IMDH.

A. Engineering Individual Residues

Ser226′, Asp278, and Ile279 were replaced by Arg, Lys, and Tyr in sequential rounds of mutagenesis. These mutations should cause the loss of hydrogen bonds to the adenosine ribose hydroxyls of NAD and may lead to the formation of hydrogen bonds to the 2′-phosphate. Indeed, these substitutions result in a dramatic increase in K_m for NAD, from 31 to 1836 μM. In contrast, the K_m for NADP is improved from 722 to 14 μM, suggesting that hydrogen bonds between Lys278, Tyr279, and the 2′-phosphate have been successfully established. Nevertheless, due to a marked drop in k_{cat} with NADP, preference of the engineered enzyme now favors NADP over NAD only by a factor of 6.

B. Engineering a Secondary Structure Element

Further improvements in specificity require that the β-turn of IMDH be replaced by an IDH-like α-helix so that two additional residues, Tyr325 and Arg395, can interact with the 2′-phosphate of NADP. Molecular modeling indicated that engineering such a change in secondary structure would require additional substitutions to facilitate packing within the remaining IMDH structure. Phe327 was replaced by Leu to avoid steric crowding that would distort the helix and disrupt interactions with the 2′-phosphate. Met397 of IDH (Table 1, IDH numbering) is replaced by Ala, again to avoid steric crowding and to allow an Arg in IMDH to hydrogen bond to several main-chain carbonyls near the terminus of the helix (Table 1, IDH numbering). The two terminal amino acids of the IMDH β-turn, Ala332 and Gly333, were retained to avoid steric packing problems associated with substituting the Leu and Lys of IDH. However, at the proximal end, Pro324 was replaced by the more flexible Thr, found in some related sequences, to allow minor shifts in the peptide backbone that might facilitate hydrogen bonding between the Tyr which was substituted for Pro325 and the 2′-phosphate of NADP. With these substitutions, specificity of the enzyme for NADP increased from 6 to 208 (Fig. 5B). The K_m toward NAD increased 10-fold while the K_m toward NADP increased 2-fold, so k_{cat} improved 6-fold. The results demonstrate that the engineering of secondary structure has been successful.

C. Generating the Final Mutant

Substituting Ala285, which lies at the back of the nucleotide-binding pocket, by Val decreases performance with NAD by a factor of 5, while the activity toward NADP is largely maintained. Molecular modeling suggests that the bulkier side chain of Val forces the adenine ring to shift, disrupting a hydrogen bond between the adenine N2 and the main-chain amide at residue 286. The resulting loss in affinity might be compensated for in the case of NADP by improved interactions between the 2′-phosphate and the introduced D278K, I279Y, P325Y, and 395R. Either that, or the strong interactions with the 2′-phosphate had already forced NADP to shift out of the way. This final mutant favors NADP over NAD by a factor of 1000 (Fig. 5B). Like wild-type IDH, and unlike wild-type IMDH, this mutant binds tightly to Affi-Gel blue affinity columns. This provides additional evidence to support the notion that an IDH-like NADP-binding pocket has been successfully engineered in IMDH.

D. Kinetic Properties of the Final Mutant

Four amino acid substitutions, S226R, D278K, I279Y, and A285V, coupled with the replacement of a β-turn with a surface IDH-like α-helix and loop, converts the coenzyme specificity of IMDH from a 100-fold preference for NAD to a 1000-fold preference for NADP (Fig. 5B). Performance with NAD was impaired 630-fold, while that with NADP was improved 130-fold. Indeed, the performance of the redesigned enzyme toward NADP is 60% higher than that of the wild-type enzyme toward NAD. Inspection of the kinetic data reveals that throughout this engineering project the k_{cat}s toward NAD and NADP were generally similar and remained similar to the k_{cat} of the wild-type enzyme. The results suggest that the change in specificity arises primarily from discrimination in binding, rather than changes in catalysis.

VII. ENGINEERING SUBSTRATE SPECIFICITY OF NADP-IDH

Ser113 and Asn115 were replaced by the corresponding amino acids Glu and Leu in IMDH, respectively (Table 2), generating three mutants: S113E, N115L, and the double mutant. As expected, the performance of the mutant enzymes with isocitrate decreased dramatically, demonstrating that Ser113 and Asn115 on helix d are involved in the binding and catalytic specificity of the substrate. However, neither of the substitutions listed established a new substrate specificity. The wild-type enzyme prefers isocitrate to methylmalate by a factor of 10^6, whereas the best mutant, S113E, prefers methylmalate over isocitrate only five-fold [12].

The K_d of the wild-type IDH for methylmalate and isopropylmalate is around 100 μM, which is 10-fold higher than for isocitrate. In contrast, the k_{cat} is 10^3 lower, indicating that engineering substrate specificity from isocitrate to an alternative substrate is not merely a binding problem, as it is for coenzyme specificity. The most critical problem is the alignment of the nicotinamide ring for hydride transfer during catalysis. When isocitrate is replaced by isopropylmalate or other analogs, we also remove the ''substrate-assisted stabilization'' for the transition state of the Michaelis complex conferred by the negatively charged γ-carboxylate (Fig. 3). Replacement of Ser113 by glutamate was designed to mimic the role of Glu87 in IMDH, which interacts with the nicotinamide ring of the coenzyme. Indeed, the performance of mutant enzyme S113E toward the alternative substrate methylmalate has been improved by a factor of three- to fourfold [12]. However, Glu87 in IMDH occupies a position physically above Glu113 in IDH. Consequently, the introduced Glu113 in the dimeric IDH mutant would not be able to interact well with the nicotinamide ring, which may explain why improved performance toward methylmalate for this mutant is so modest. In order to install a more sterically equivalent Glu in IDH, helix d should be engineered to start three amino acids earlier, as in IMDH. This can be realized by secondary structure engineering: introducing the corresponding α-helix d of IMDH into IDH.

VIII. CONCLUSION

Naturally occurring proteins rarely prove optimal for technological purposes. Using isocitrate and isopropylmalate dehydrogenases as models, we demonstrated that enzyme properties can be rationally altered or optimized. In these enzymes, the kinetic and catalytic mechanisms have been thoroughly elucidated, and an extensive knowledge of three-dimensional structures is available. This knowledge has allowed us to engineer highly active and specific forms of NAD-IDH and NADP-IMDH. These enzymes were created by mutating individual amino acid residues as well as secondary structures. With the rapidly increasing numbers of structures solved by NMR or X-ray diffraction [25], this rational approach should become broadly applicable to the redesign and optimization of functions of many other enzymes.

Rational enzyme design is a powerful tool that also has tremendous potential for increasing our understanding of protein structure–function relations [9,11,26]. The successful redesign of coenzyme specificities in both IDH and IMDH demonstrates that coenzyme specificities in the β-decarboxylating dehydrogenases are determined primarily by interactions between the nucleotides and surface amino acid residues in the nucleotide-binding pockets. Nevertheless, additional residues not in contact with the coenzymes play key roles through modulating the direct interactions. Clearly, a deeper understanding of enzyme binding and catalytic specifities will in turn increase the success for enzyme engineering and lay the foundation for the functional prediction of new enzyme sequences (see Chap. 22).

ACKNOWLEDGMENTS

This work has been supported by research grants from the Medical Research Council of Canada, the Natural Sciences and Engineering Research Council of Canada, and the Health Services and Utilization Research Commission.

REFERENCES

1. G Georgiou, N Dewitt. Enzyme beauty. Nature Biotechnol 17:1161–1162, 1999.
2. BG Hall. Directed evolution of a bacterial operon. BioEssays 12:551–558, 1990.
3. WPC Stemmer. Rapid evolution of a protein *in vitro* by DNA shuffling. Nature 370: 389–391, 1994.
4. P Carter, JA Wells. Dissecting the catalytic triad of a serine protease. Nature 332: 564–568, 1988.
5. HM Wilks, KW Hart, R Feeney, CR Dunn, H Muirhead, WN Chia, DA Barstow, AR Atkinson, JJ Holbrook. A specific, highly active malate dehydrogenase by redesign of a lactate dehydrogenase framework. Science 242:1541–1544, 1988.
6. ED Getzoff, DE Cabelli, CL Fisher, HE Parge, MS Viezzoli, L Banci, RA Hallewell. Faster superoxide dismutase mutants designed by enhancing electrostatic guidance. Nature 358:347–351, 1992.
7. FH Arnold. Engineering proteins for nonnatural environments. FASEB J 7:744–749, 1993.
8. FH Arnold, AA Volkov. Directed evolution of biocatalysts. Curr Opin Chem Biol 3:54–59, 1999.
9. RD Chen, AF Greer, AM Dean. A highly active decarboxylating dehydrogenase with rationally inverted coenzyme specificity. Proc Natl Acad Sci (USA) 92:11666–11670, 1995.
10. RD Chen, JA Grobler, JH Hurley, AM Dean. Second-site suppression of regulatory phosphorylation in *Escherichia coli* isocitrate dehydrogenase. Protein Sci 5:287–295, 1996.
11. RD Chen, A Greer, AM Dean. Redesigning secondary structure to invert coenzyme specificity in isopropylmalate dehydrogenase. Proc Natl Acad Sci (USA) 93:12171–12176, 1996.
12. RD Chen, A Greer, AM Dean. Structural constraints in protein engineering: the coenzyme specificity of *Escherichia coli* isocitrate dehydrogenase. Eur J Biochem 250:578–582, 1997.
13. JH Hurley, RD Chen, AM Dean. Determinants of cofactor specificity in isocitrate dehydrogenase: structure of an engineered NADP → NAD specificity-reversal mutant. Biochemistry 35:5670–5678, 1996.
14. RD Chen. A general strategy for enzyme engineering. Trends Biotechnol 17:344–345, 1999.
15. RD Chen, P Gadal. Structure, function and regulation of NAD- and NADP-isocitrate dehydrogenases in higher plants and in other organisms. Plant Physiol Biochem 28: 411–427, 1990.
16. JH Hurley, P Thorsness, V Ramalingham, N Helmers, DE Koshland Jr, RM Stroud. Structure of a bacterial enzyme regulated by phosphorylation, isocitrate dehydrogenase. Proc Natl Acad Sci (USA) 86:8635–8639, 1989.
17. JH Hurley, AM Dean, DE Koshland Jr, RM Stroud. Catalytic mechanism of $NADP^+$-dependent isocitrate dehydrogenase: implications from the structures of magnesium-isocitrate and $NADP^+$ complexes. Biochemistry 30:8671–8678, 1991.
18. BL Stoddard, AM Dean, DE Koshland Jr. Structure of isocitrate dehydrogenase with

isocitrate, nicotinamide adenine dinucleotide phosphate, and calcium at 2.5-Å resolution: a pseudo-Michaelis ternary complex. Biochemistry 32:9310–9316, 1993.

19. JM Bolduc, DH Dyer, WG Scott, P Singer, RM Sweet, DE Koshland Jr, BL Stoddard. Mutagenesis and Laue structures of enzyme intermediates: isocitrate dehydrogenase. Science 268:1312–1318, 1995.
20. AM Dean, DE Koshland Jr. The kinetic mechanism of *Escherichia coli* isocitrate dehydrogenase. Biochemistry 32:9302–9309, 1993.
21. AM Dean, AK Shiau, DE Koshland Jr. Determinants of performance in the isocitrate dehydrogenase of *Escherichia coli*. Protein Sci 5:341–347, 1995.
22. JH Hurley, AM Dean. Structure of 3-isopropylmalate dehydrogenase in complex with NAD^+: ligand-induced loop closing and mechanism for cofactor specificity. Structure 2:1007–1016, 1994.
23. K Imada, M Sato, N Tanaka, Y Katsube, Y Matsuura, T Oshima. Three-dimensional structure of a highly thermostable enzyme, 3-isopropylmalate dehydrogenase of *Thermus thermophilus* at 2.2 Å resolution. J Mol Biol 222:725–738, 1991.
24. AM Dean, L Dvorak. The role of glutamate 87 in the kinetic mechanism of *Thermus thermophilus* isopropylmalate dehydrogenase. Protein Sci 4:2156–2167, 1995.
25. JM Thornton. The future of bioinformatics. Trends Guide to Bioinformatics (TIBS suppl) 30–31, 1998.
26. SP Miller, RD Chen, EJ Karschnia, C Romfo, A Dean, DC LaPorte. Locations of the regulatory sites for isocitrate dehydrogenase kinase/phosphatase. J Biol Chem 275:833–839, 2000.

24

Proteomics: Chromatographic Fractionation Prior to Two-Dimensional Polyacrylamide Gel Electrophoresis for Enrichment of Low-Abundance Proteins to Facilitate Identification by Mass Spectrometric Methods

Srinivasan Krishnan, John E. Hale, and Gerald W. Becker
Eli Lilly and Company, Indianapolis, Indiana

I. INTRODUCTION

As the sequences of the genomes of several species are completed (www.tigr.org/tdb/mdb/mdb.html), interest is growing in the functional component of biological systems, the ''proteome.'' Proteome [1] refers to the complete protein expression profile of a biological system. Examination of the proteome will allow one to determine important proteins, enzymes, and pathways involved in disease states by comparing the proteome of normal and diseased tissues. One may determine alteration in protein expression of tissues or cells in response to stresses such as exposure to drugs or toxins. Thus the proteome is a rich source for identification of protein targets or therapeutics for drug development. Another important aspect of studying the proteome is the discovery of novel proteins.

One of the most useful techniques for visualization of the proteome is two-dimensional sodium dodecyl sulfate-polyacrylamide gel electrophoresis (2-D SDS-PAGE). This technique possesses unmatched resolving power for separation of proteins [2–4] and has been used extensively to analyze proteins [5–8], their regulation [9–18], and posttranslational modifications [19–22]. Several techniques have been used to identify proteins from various organisms following separation by 2-D PAGE. These techniques include amino acid analysis [6,23–26], Edman sequencing [27–30], immunological methods [31–34], and mass spectrometry [35–46]. Mass spectrometric technologies are particularly attractive because of their high sensitivity and adaptability to high-throughput analysis.

However, protein extracts from cells or biological fluids such as serum contain thousands of proteins in varying abundance, and only the most abundant protein species are visible in the 2-D gels of these samples. The minor protein components are either below detection levels or are obscured by the abundant species. For example, of the possible 1700 gene products in *Haemophilus influenzae*, only about 400 are visible on a Coomassie-stained 2-D gel of an extract of this organism [47]. In the case of *Saccharomyces cerevisiae*, even a 2-D gel image generated by [^{35}S]methionine-labeled polypeptides contains about 1200 spots, corresponding to roughly 20% of the gene products [48]. Wilkins et al. [49] have pointed out that this discrepancy arises due to two factors, poor electrophoretic behavior and low copy number. Using a graphical model, these authors predict that proteins present at less than 1000 copies per cell cannot be visualized on 2-D gels considering the current maximum 2-D gel loading capacity. Similar observations are made with many biological systems, where only a fraction of the proteins are visualized on a 2-D gel. Therefore the challenge is to enhance the sensitivity of detection of the low-abundance proteins. This chapter describes approaches that accomplish this enhancement in sensitivity and help to identify novel low-abundance proteins.

II. PROTEOMICS

A. Two-Dimensional SDS-PAGE

High-resolution 2-D gel electrophoresis [2,3] involves separation of proteins based on their isoelectric point (pI) in the first dimension and size in the second. Originally, the first dimension consisted of isoelectric focusing (IEF) tube gels with carrier ampholytes [2,3,50]. These gels were irreproducible, difficult to transfer to the PAGE gel, and were limited in the amount of protein that could be loaded. Recently, immobilized pH gradients (IPGs) have gained popularity [51–53]. These IEF strips are produced by copolymerization of ampholytes within the fibers of a polyacrylamide matrix on GelBond™ film. The IPGs are compatible with various additives used in 2-D gel electrophoresis, such as urea (8 M), and nonionic and zwitterionic detergents which improve the solubility and resolution of proteins. The high degree of reproducibility in spot position, ease of handling, and improved protein capacity have made this technique an improvement over tube gels and has increased the sensitivity of the 2-D gel technique. After isoelectric focusing on the IPG, the proteins are further resolved by applying the IPG strip to the top of an SDS-PAGE gel and transferring the proteins electrophoretically into the second-dimension gel. Following electrophoresis the proteins are visualized by a variety of staining techniques [54–59]. The most common stain used is Coomassie blue, which can detect as little as 50 ng of an individual protein. More sensitive stains such as silver may detect as little as 1 ng of an individual protein.

B. Identification of Protein Spots on 2-D Gels Using Mass Spectrometry

The method of choice for identification of proteins separated by 2-D PAGE is mass spectrometry (MS). Recent developments in mass spectrometry have enabled the rapid identification of proteins separated by 2-D PAGE with sensitivities routinely in the femtomole range [60,61] and, in some cases, the attomole range [62]. Two mass spectrometric methods utilized in the analysis of proteins and peptides are matrix-assisted laser desorption ionization–time-of-flight mass spectrometry (MALDI-Tof MS) and electrospray ionization [35–46]. The identification by these mass spectrometric techniques is done after proteolytic digestion of the protein in the gel, typically using the enzyme trypsin, and extraction of the resulting peptides.

The most widely used mass spectrometric identification procedure is MALDI-Tof analysis of the entire peptide mixture. Gas-phase matrix interaction with peptide ions in MALDI-Tof results in singly charged ions, giving a mass profile that is highly characteristic of the protein from which the peptides are derived. These peptide masses (actually protonated peptide molecular ions, MH^+) can be used to search databases (either protein or nucleic acid databases) to identify the proteins. The two most important factors in successfully identifying proteins by this approach are the number of matching peptide masses and the accuracy of the peptide mass determination.

In some cases the identification of proteins based on peptide masses is ambiguous or a suitable match cannot be found. Further information may be obtained by performing a collision-induced dissociation tandem mass spectrometry (CID MS/MS) experiment, in which the peptide molecular ions are fragmented in a collision cell and the resulting product ions are analyzed. These product ions often reveal sequence information that can then be used to identify the protein.

The sensitivity of these techniques is excellent when applied to the levels of protein present in spots from a 2-D gel. The most sensitive staining technique visualizes a nanogram of protein, a level well within the detection limits of mass spectrometric techniques. However, many proteins, such as cytokines, nuclear factors, transcription factors, etc., are present at levels far below these limits of detection and are not visible on gels using any of the available staining methods. Therefore, enrichment methods are critical for successfully applying the techniques of 2-D PAGE and mass spectrometry to interesting biological questions.

C. Prefractionation as an Approach to Enhancement of Sensitivity

One approach to detection of low-abundance proteins is to increase the amount of sample loaded on the IPG strips. While this increases the amount of low-abundance proteins on the gels, it also increases the high-abundance proteins. This causes problems with the running of the IPG strips and excessive streaking in the second dimension. Higher protein loads lead to spot overlapping and decreased resolution. In biological samples such as serum, where one or two species (albumin and immunoglobulins) predominate, isoelectric focusing encounters severe problems due to protein precipitation and gel expansion in those regions. Simply increasing total protein loads is not a solution to visualization of low-abundance proteins.

Fractionating the protein complex prior to running the 2-D gel can be used to accomplish two purposes, removal of the abundant species and enrichment of low-abundance species. A variety of high-resolution chromatographic modes for fractionation of proteins exists, and many of these are readily adaptable to 2-D gel electrophoresis. We have utilized ion-exchange chromatography as a method for separation of proteins in complex mixtures. By careful selection of buffer and pH, subsets of proteins in a mixture may be bound to ion-exchange resins. The bound proteins may be further fractionated by elution with salt gradients. Protein not bound in the first round of chromatography may be bound and fractionated by altering conditions and rechromatographing or by chromatography in a different mode. We have used a chromatographic prefractionation approach in the analysis of serum proteins. Serum contains two major protein fractions, albumin and immunoglobulins. In addition, serum contains thousands of proteins in very low abundance. The presence of huge quantities of a single protein, albumin, negates high loads of serum on IPG strips for isoelectric focusing. Serum loads above 100 μl on a 17- to 18-cm IPG strip cause severe problems in the isoelectric focusing and hence improper running of the strip. In addition, the presence of relatively high amounts of certain other proteins, such as transferrin, alpha 1-antitrypsin, etc., causes extensive spot overlapping. We have used anion-exchange and cation-exchange chromatographic procedures to enrich lower-abundance acidic and basic proteins from rat serum.

III. PROTEOMICS: GENERAL METHODS

A. Chromatography

Anion-exchange chromatography was performed at pH 4.5 in 50 mM sodium acetate buffer on a DEAE column (0.75 × 7.5 cm, Tosohaas). Rat serum (5 ml), dialyzed in the same buffer, was loaded at 1 ml/min and the column was washed with 20 ml of buffer followed by 10 ml of 0.1 M NaCl. Proteins were eluted with a linear gradient of NaCl (0.1–1.0 M) over the next 30 min and a final isocratic elution at 1.0 M NaCl for an additional 20 min. The elution was monitored at 280 nm, and 1-ml fractions were collected.

Cation-exchange chromatography was performed at pH 7.5 in 50 mM sodium phosphate buffer on a sulfopropyl column (0.75 × 7.5 cm, Tosohaas). The run conditions were maintained as above except in the present case the rat serum was dialyzed in 50 mM sodium phosphate buffer, pH 7.5.

B. Two-Dimensional Gel Electrophoresis

Chromatographic fractions were pooled, dialyzed extensively against water, and lyophilized. Prior to electrophoresis, samples were dissolved in approximately 0.4 ml of rehydration solution (0.3% dithiothreitol, 2% 3-[(3-cholamidopropyl)-dimethylammonio]-1-propanesulfonate (CHAPS), 8 M urea, 2% immobilines, pH 4–7, and bromophenol blue) and applied to immobilized pH gradient strips (Pharmacia, linear gradient, pH 4–7, 18 cm). The rehydration of the IPG strips was performed overnight in a rehydration tray (Pharmacia) with 400 μl of sample applied to each strip. The rehydrated IPG strips were focused for 50,000 V-hr with a slow voltage ramp for the first 4 hr reaching a maximum of 2500 V. Following focusing, the proteins on the strips were reduced by immersing the strips in a buffer (0.1 M Tris-HCl, pH 6.8, 6 M urea, 2 M thiourea, 1% SDS, 6.4 mM dithiothreitol (DTT), 30% glycerol, and a few grains of bromophenol blue) for 15 min at room temperature. Alkylation was achieved by incubating for 15 min at room temperature in a second solution containing 24 mM iodoacetamide instead of DTT. The IPG strips were then overlayed on SDS PAGE gels (11–18% acrylamide) using 1% agarose to hold them in place. The gels were subjected to electrophoresis at an initial low voltage (50 V) until the dye front had entered the gels and then at a higher voltage (200 V) overnight. The tank buffer (24 mM Tris, 200 mM glycine, and 0.1% SDS) was maintained at 5°C throughout the electrophoresis run. The gels were fixed by soaking in 50% ethanol, 2% phosphoric acid, and 48% water for 14 hr. Proteins were visualized by staining with colloidal Coomassie blue in Neuhoff concentrate (17% w/v ammonium sulfate, 34% methanol, and 3% phosphoric acid in water) for 2 days.

C. Protein Spot Digestion

Protein spots of interest were excised from the gels, suspended in 50% acetonitrile in 50 mM ammonium bicarbonate, and crushed using a plastic pestle. The Coomassie blue was extracted from the gel pieces by repeated washing with the same buffer. The pieces were finally suspended in a small volume of acetonitrile and dried completely using a concentrator (Savant). Protein in the gel pieces was proteolyzed by swelling the gel in a solution of trypsin (Promega, sequencing grade, 50 μg/ml) in 50 mM ammonium bicarbonate and incubating overnight at

37°C. Peptides were extracted from the gel three times using trifluoroacetic acid/acetonitrile/50 mM ammonium bicarbonate (5/45/50). The combined extracts for each digest were dried completely and washed with 200 μl water four times. Each extract was finally dissolved in 10 μl of 50% acetonitrile/1% acetic acid in water and used for further mass spectrometric analyses.

D. Mass Spectrometry of Digests

Peptide extracts prepared as described above were analyzed on a MALDI-Tof mass spectrometer (Micromass TofSpec E). A small portion of the extract was mixed with an equal volume of α-cyano-4-hydroxycinnamic acid (alpha CHC) matrix in methanol and loaded onto a MALDI plate. Calibration of the flight tube was performed with the internal standards angiotensin I and adrenocorticotrophic hormone (both from Sigma).

When identification of the protein could not be achieved using MALDI data alone, peptides were fragmented and the product ions analyzed. Peptide fragmentation was performed on a Micromass quadrupole time-of-flight (Q-Tof) mass spectrometer. The peptide extracts were desalted on a 0.5-ml C18 reverse-phase column and dried in a Speed Vac concentrator, then suspended in 10 μl of 50% acetonitrile/1% acetic acid in water. An infusion pump was used to introduce the sample into the Q-Tof and the peptide ions fragmented by varying the collision gas (nitrogen) energy. Information from the fragmentation pattern was used for protein identification.

E. Protein Identification

Proteins from the gels were identified by searching protein sequence databases with data derived from the mass spectrometric analyses described above. In most cases, positive identifications could be made by searching tryptic peptide masses derived from the MALDI-Tof experiments. Programs used to perform this type of search include PeptideSearch [63] and MS Fit from Protein Prospector (http://prospector.ucsf.edu). When a positive identification could not be made by searching the peptide masses, searches were carried out using peptide fragmentation data from the Q-Tof experiments. The programs used for these searches were PeptideSearch and MS-Tag from Protein Prospector.

IV. ENRICHMENT AND IDENTIFICATION OF LOW-ABUNDANCE SERUM PROTEINS

A representative separation of rat serum is shown in Fig. 1. It is readily apparent from the pattern that even though serum contains thousands of proteins, the 2-

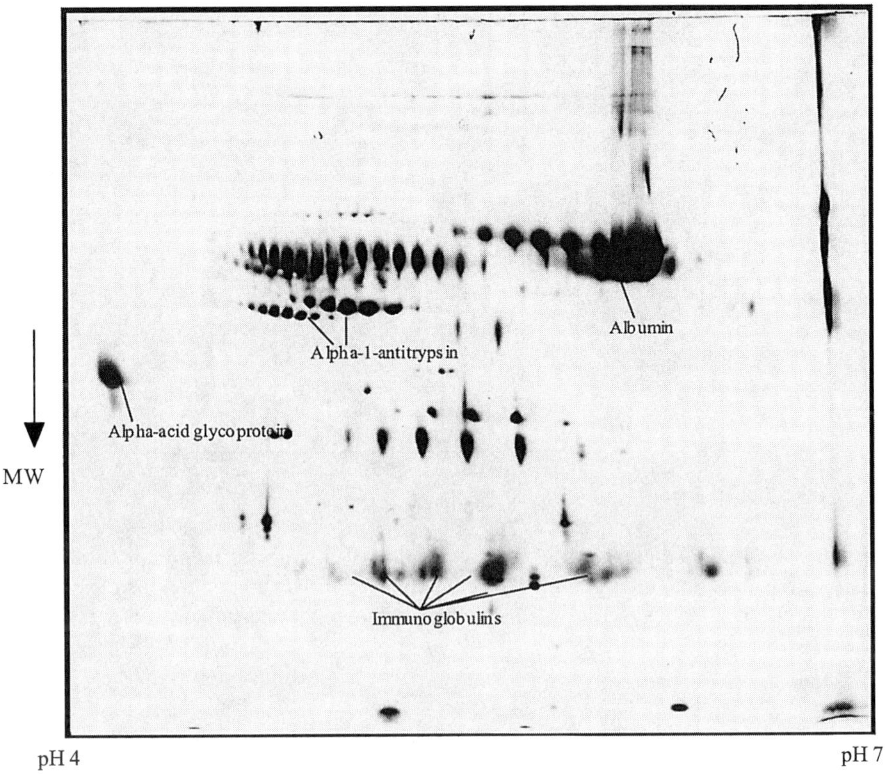

Figure 1 Reference 2-D PAGE map of rat serum. Unfractionated rat serum (30 μl) was analyzed by 2-D PAGE. The separation was based on a pH 4–7 linear IPG in the first dimension and 11–18% SDS-PAGE in the second dimension. Proteins were visualized by staining with colloidal Coomassie. Some of the major proteins of rat serum are labeled.

D gel pattern is dominated by a relatively small number of proteins. Our first step in enrichment of low-abundance proteins was anion-exchange chromatography. Figure 2 shows a typical anion-exchange separation of whole rat serum on a weak anion-exchange resin at relatively low pH. Retention of a small percentage of the total serum proteins was observed and almost all of the retained proteins eluted at salt concentrations lower than 0.4 M. Several chromatographic runs were made, indicating the high degree of reproducibility of this process. The chromatographic fractions were pooled as shown and were run on 2-D gels after extensive dialysis against water and lyophilization to dryness.

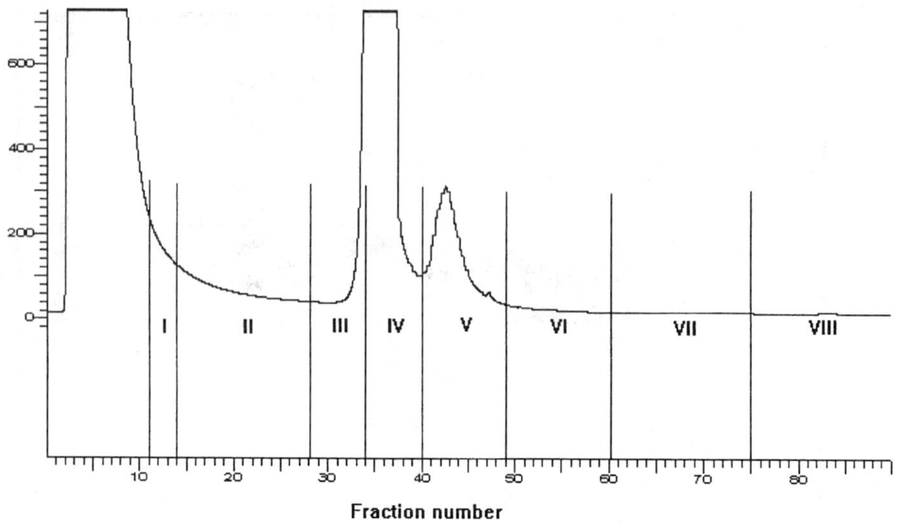

Figure 2 Chromatographic profile of separation of proteins in rat serum on a DEAE anion-exchange column at pH 4.5. The conditions of the separation are detailed in Sec. III. Individual fractions were pooled as indicated into eight major pools and analyzed by 2-D PAGE (Fig. 3).

The 2-D gels of three chromatographic portions from anion-exchange column separation of 10 ml of rat serum are shown in Fig. 3. We observed from the gel patterns that almost 90% of the albumin was removed from the fractions shown. In addition, almost all of the immunoglobulins were removed and the amount of alpha 1-antitrypsin was greatly reduced. The reduction in these major protein species has helped overcome the issues related to improper running of the gels, spot overlapping, and resolution. A comparison of the whole-rat serum gel and the chromatographic fractions highlights this point.

Several spots that were not visible on the stained gel of whole serum were now predominant. Although several protein spots not seen in the unfractionated rat serum gel were visible, many of these proved to be fragments of abundant proteins such as alpha 1-macroglobulin, albumin, alpha 1-antitrypsin, etc. The low-abundance proteins that were enriched are indicated by arrowheads. Many of these protein spots are at the acidic end of the gel, indicating that the anionic chromatographic fractionation not only accomplished removal of major protein species but also enriched several low-abundance acidic proteins. We next enriched basic proteins by cation exchange. From the chromatographic profile of 10 ml of whole serum separated on a strong cation-exchange resin at relatively

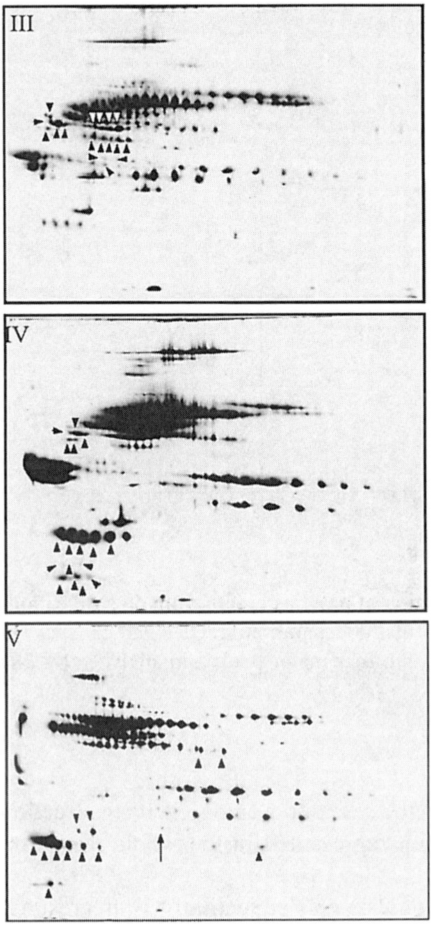

Figure 3 Analysis of selected fractions from DEAE column by 2-D PAGE. Fractions III, IV, and V from the anion-exchange separation of rat serum (Fig. 2) were analyzed by 2-D PAGE. Low-abundance proteins enriched by fractionation are noted with arrowheads.

high pH (Fig. 4), it can be seen that only a small portion of the proteins bound to this column. These proteins were also eluted with salt concentrations below 0.4 M. The 2-D gels of cation-exchange column fractions are shown in Figure 5. A striking feature seen in these gels is the complete removal of the abundant protein species in serum—albumin, immunoglobulins, and trypsin inhibitors—indicating the usefulness of this approach. The protein pattern is concentrated

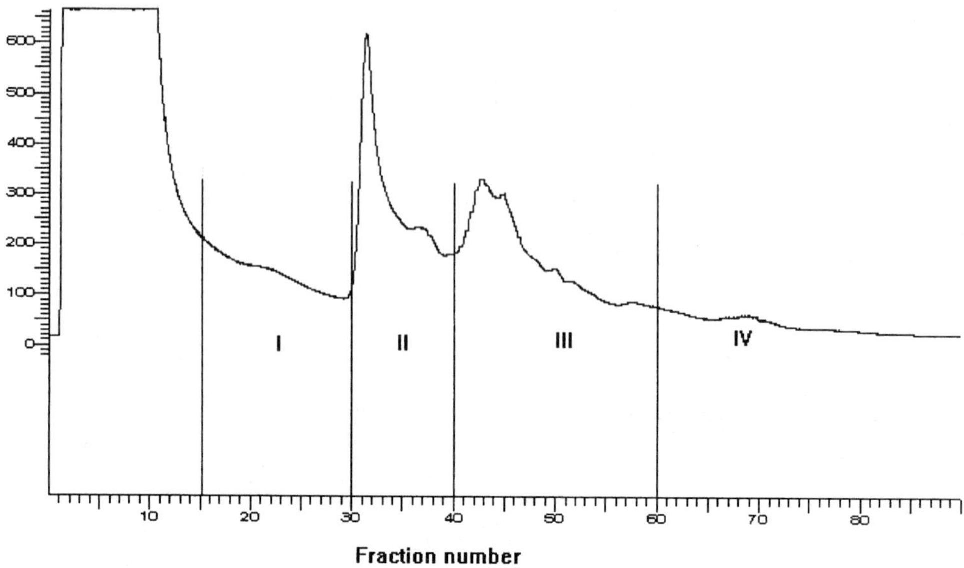

Figure 4 Chromatographic profile of separation of proteins in rat serum on a SP cation-exchange column at pH 7.5. The conditions of the separation are detailed in Sec. III. Individual fractions were pooled as indicated into four major pools and analyzed by 2-D PAGE (Fig. 5).

more toward the basic end of the gel. As in the case of the anion-exchange fraction gels, these gels reveal several proteins not represented in the whole-rat serum gel (indicated by the arrowheads.)

While enrichment of protein spots on 2-D gels is qualitatively interesting, the full power of proteomics may not be realized without identification of these proteins. Following digestion of the proteins in the gel spots with trypsin, the tryptic peptides were analyzed by two mass spectrometric techniques, MALDI-Tof and Q-Tof. A MALDI-Tof mass spectrum obtained from the tryptic digest of a spot (arrow in Fig. 5) from a 2-D gel of rat serum fractionated by cation-exchange chromatography is shown in Figure 6. Three ions in the spectrum, 842.50, 2211.14, and 2261.24, represent auto-catalytic fragments of trypsin. These ions, as well as the ions at 1296.69 (angiotensin D) and 2465.20 (ACTH), have been used to calibrate the mass analyzer. With careful calibration, accuracies better than 100 ppm are routinely achieved.

Thirteen ions were selected from the spectrum shown in Fig. 6 and used to search the SwissProt protein sequence database. The results of this search are shown in Table 1. The program MS-Fit, which is part of the Protein Prospector

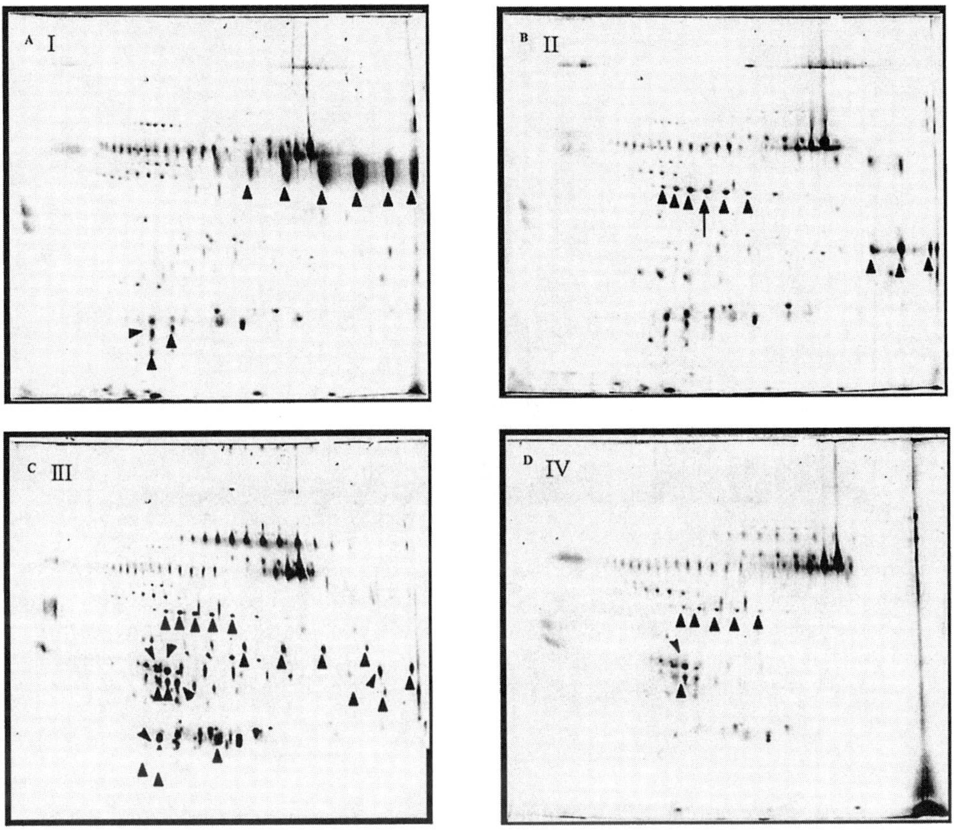

Figure 5 Analysis of selected fractions from SP column by 2-D PAGE. Fractions I through IV from the cation-exchange separation of rat serum (Fig. 4) were analyzed by 2-D PAGE. Low-abundance proteins enriched by fractionation are noted with arrowheads.

package of programs, was used to conduct the search. Eight of the *m/z* values searched matched within 0.1 dalton of the theoretical MH^+ of ions from the protein ''mouse pigment epithelium-derived factor precursor (PEDF),'' also known as stromal cell-derived factor 3 (SDF-3). Interestingly, the protein identified in the SwissProt database was a mouse protein, whereas the sample being analyzed was obtained from rat serum. A more careful examination of the databases revealed no entry for a rat homolog of this protein. This represents the first report that pigment epithelium-derived factor is found in the rat.

Occasionally, a protein was not identified by MALDI-MS methods. This may occur when more than one protein is present in a spot. Protein identification

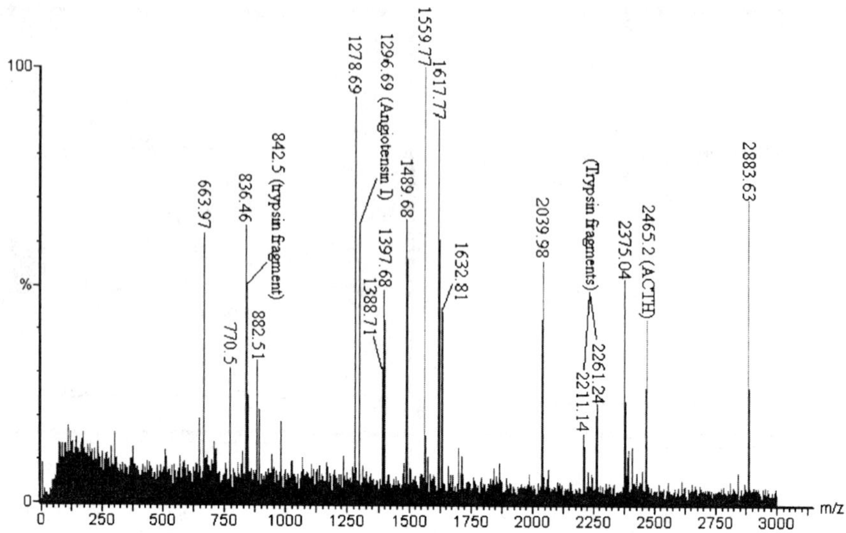

Figure 6 MALDI-Tof mass spectrum. A protein spot separated from rat serum by SP chromatography (indicated by arrow in Fig. 5, fraction II) was digested with trypsin as described in Sec. III and analyzed by MALDI-Tof mass spectrometry (TofSpec-E, Micromass). Ions generated as a result of autocatalytic cleavage of trypsin and two internal standards, angiotensin I and adrenocorticotrophic hormone, are indicated.

may still be accomplished by fragmentation of peptide ions by an MS/MS technique. This requires the use of mass spectrometers equipped with collision chamber (such as the Sciex API III triple quadrupole, the Micromass Q-Tof, etc.,) or an ion-trap type of instrument (Finnigan LCQ). We have used the Q-Tof using the general procedure as follows. The ions in the protein digest were first identified by an initial ion scan in the quadrupole of the instrument. The ionization potential was set such that doubly charged ions predominate in the spectra. An individual ion (known as the precursor ion) was selected and diverted into a collision chamber, where it was fragmented into a series of product ions. The most scissile bonds are the peptide bonds, fragmentation of which yields a series of product ions differing in mass by the mass of successive amino acid residues in the peptide sequence. A fragmentation obtained on a Q-Tof will generate b-ions (N-terminal fragments) and y-ions (C-terminal fragments) which are passed into the flight tube for mass analysis. Charge tends to be retained on the most basic residues (which are the C-terminal amino acids in tryptic peptides) and so y ions tend to be the most intense in MS/MS spectra. Interpretation of the MS/MS spectra may

Table 1 Search of a Protein Sequence Database With Ions Resulting from a MALDI-Tof Analysis of a Tryptic Digest of a Rat Serum Protein Enriched by Cation-Exchange Chromatography [a]

m/z submitted	MH^+ matched	Delta (Da)	Peptide sequence
770.50	770.45	0.0475	(R)ILTGNPR(V)
836.46	836.4379	0.0221	(K)SYGTRPR(I)
882.51	882.4797	0.0303	(K)LTQVEHR(A)
1278.69	1278.6694	0.0206	(R)DTDTGALLFIGR(I)
1397.68	1397.6953	−0.0153	(K)LQSLFESPDFSK(I)
1489.68	1489.6923	−0.0123	(K)TTLQDFHLDEDR(T)
1559.77	1559.7858	−0.0158	(K)LAAAVSNFGYDLYR(L)
1617.77	1617.7873	−0.0173	(R)KTTLQDFHLDEDR(T)

[a] The best match was to mouse pigment epithelium-derived factor precursor (PEDF) (stromal cell-derived factor 3) (SDF-3); 46204.3 Da, pI = 6.48 and accession # P97298.

Five unmatched masses: 663.9700 1388.7100 1632.8100 2039.9800 2375.0400. The database searched was the SwissProt database; cysteines were carbamidomethylated; the peptide masses were monoisotopic; the peptide mass error was <0.1 Da; the protein molecular weight range was 1,000–100,000 Da; one missed cleavage was allowed.

be accomplished using sequence tags generated by manual interpretation of the MS/MS spectrum. More automated programs such as Protein Prospector, PeptideSearch, or Sequest are capable of searching databases with MS/MS data in tabular format. One such fragmentation pattern is shown in Figure 7 for the parent ion 537.83 (+2 charge state) of a peptide obtained from the tryptic digest of the spot denoted by the arrow in Figure 3. Calculation of the differences in the y-ion series in the spectra between masses 414.32 and 860.55 yielded the sequence I/LVSF (read in C-terminal to N-terminal order). This partial sequence information along with the masses of the product ions encompassing the sequence (414.32 and 860.55 in this case) is termed a ''sequence tag,'' which was used to search a database. Such a search using the PeptideSearch program identified the peptide, K⟨TLFSVLPGLK⟩M, which is present in two isoforms of the inter-alpha-inhibitor 4 protein. Many of the theoretical MS/MS ions of this peptide can be identified in the actual MS/MS spectrum (Fig. 7). The sequence FSVL (read from N-terminal to C-terminal) matches the sequence determined from the y series in the MS/MS spectrum.

Using approaches described above, the low-molecular-weight spots that were enriched in the acidic region in Figure 3 have been characterized to be different peptides of rat kininogens from which bradykinin peptides originate. The higher-molecular-weight enriched proteins were identified to be several dif-

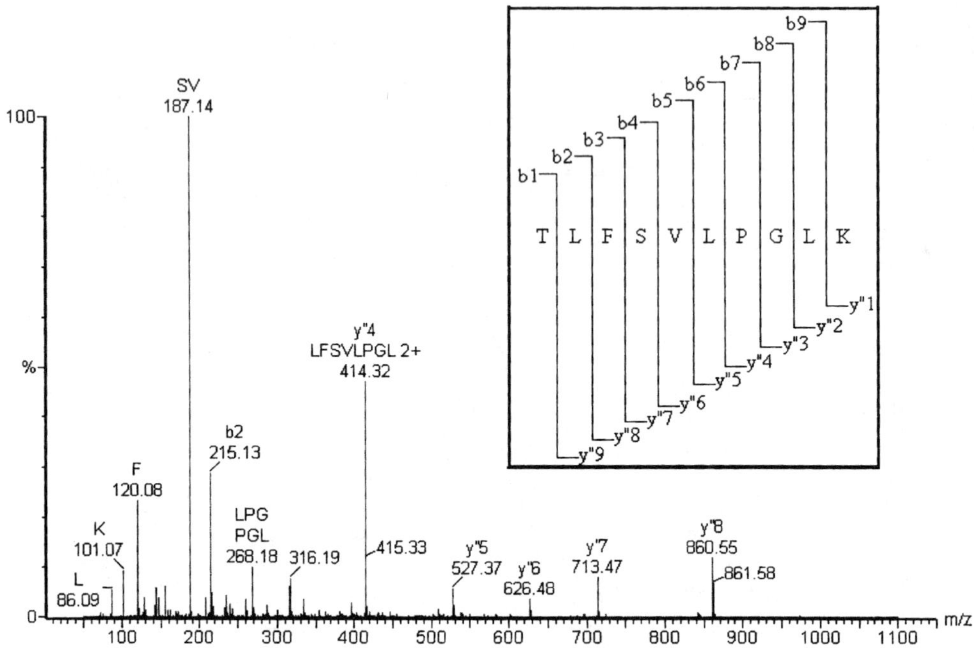

Figure 7 Q-Tof MS/MS spectrum. A protein spot separated from rat serum by DEAE chromatography (indicated by arrow in Fig. 3, fraction V) was digested with trypsin as described in Sec. III. A precursor ion ($m/z = 537.83^{2+}$) was fragmented and the resulting product ions analyzed by MS/MS (Q-Tof, Micromass).

ferent forms of serpins (Fig. 8A). In addition a novel rat form of the thrombospondin 1 protein was discovered in one of the 2-D gels (Fig. 8B). In the case of unfractionated serum this protein would not have been identified, due to the highly abundant immunoglobulins appearing in the same region and the relatively lower abundance of this protein in serum.

Cation exchange-enriched spots that were identified include several forms of the APO-E protein (Fig. 8C). The most striking feature was the enrichment of a low-abundance novel rat homolog of stromal cell derived factor 3 represented as a series of protein spots on the 2-D gel differing by charge (due to variations in the glycosylation state). The location of this protein spot on the gel indicates that this charge series would have overlapped with the alpha 1-antitrypsins and other serpins had the entire 10 ml of serum been run. We also found fragments of a stem-cell factor receptor precursor, indicating that this fractionation approach has enabled us to enhance the sensitivity of the 2-D gel approach.

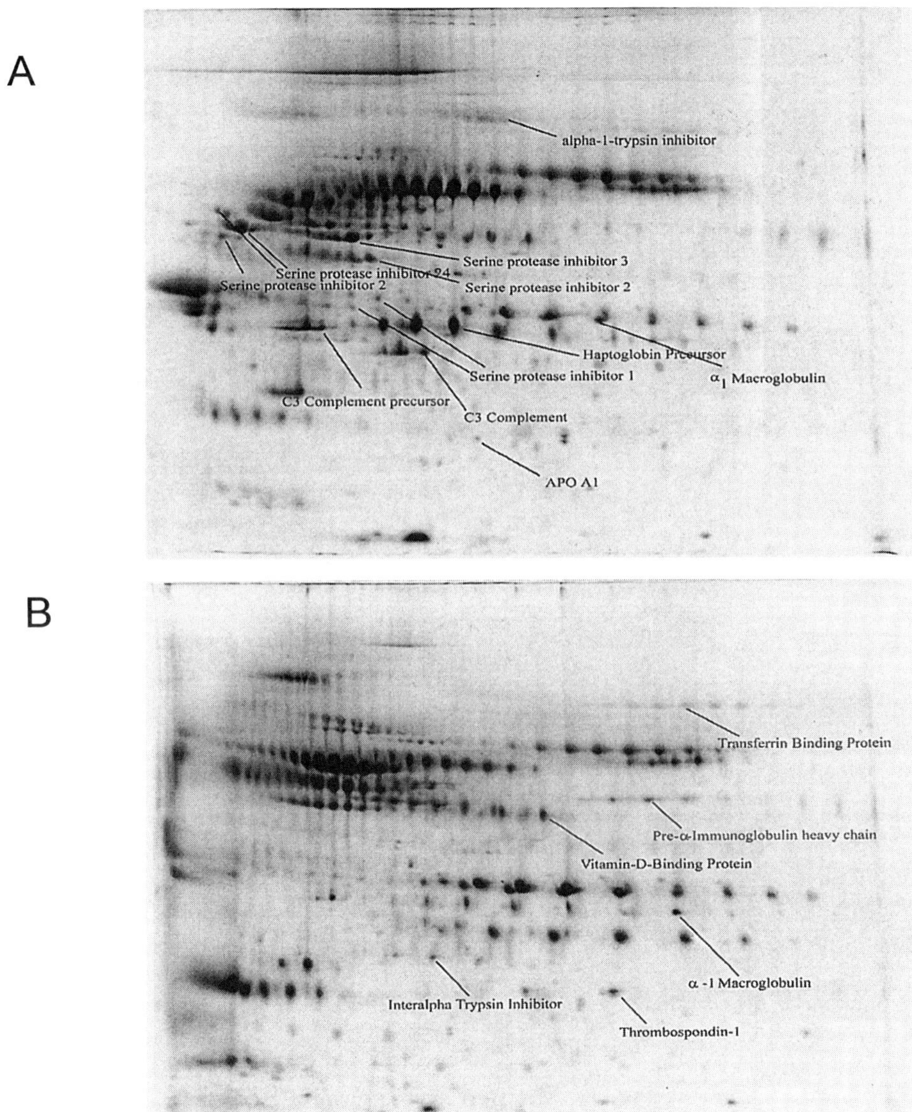

Figure 8 Summary of proteins identified from the anion-exchange and cation-exchange fractionations of rat serum. (A) 2-D gel of anion-exchange fraction III. (B) 2-D gel of anion-exchange fraction V. (C) 2-D gel of cation-exchange fraction III.

C

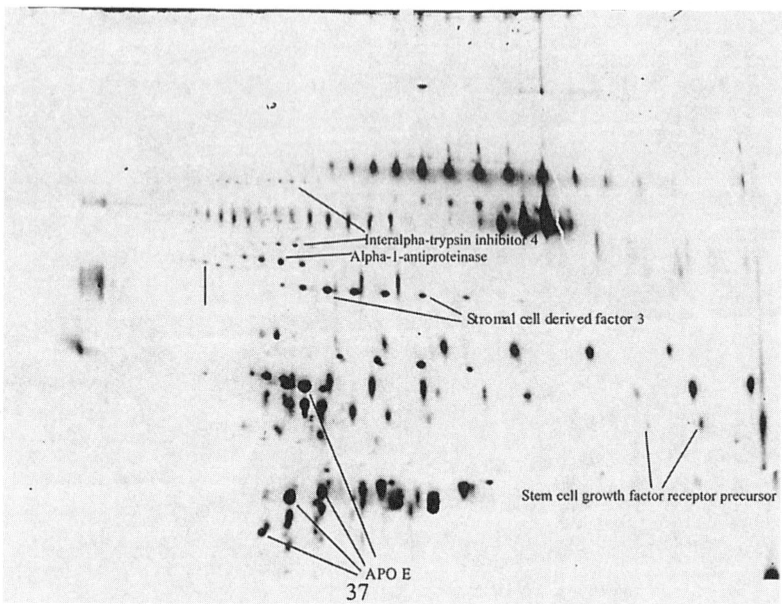

Figure 8 Continued.

V. USE OF OTHER PREFRACTIONATION METHODS

Examples have been reported in the literature in which other forms of chromatography have been used for prefractionation of sample prior to the use of 2-D gel. Fountoulakis et al. have used hydroxyapatite chromatography to enrich low-abundance proteins in *Escherichia coli* [64]. In this study, about 130 proteins not detected in the 2-D gel of the total extract were identified. The same group has reported the use of a heparin-affinity approach in the identification of low-abundance proteins in *Haemophilus influenzae* [65]. The availability of other affinity approaches to enrich protein classes of interest, such as metal-affinity chromatography used to enrich phospho-proteins, lectin chromatography to enrich glycoproteins, and antibody columns targeted to phosphorylated amino acids or certain protein domains, makes this technique a very versatile one. Enrichment and identification of low-abundance proteins in *H. influenzae* by hydrophobic interaction chromatography has been reported [66].

Different physical techniques can also be used for prefractionation of samples to be analyzed by 2-D gels. Corthals et al. have fractionated human serum by an electrokinetic technique [67]. Fountoulakis et al. have used chromatofocusing as a prefractionation step in the analysis of low-abundance proteins in *H.*

influenzae [68]. Ultrafiltration in conjunction with other chromatographic methods offers an attractive approach to enrich proteins of the desired size and characteristic, although nonspecific binding of proteins to the membrane limits the use of this method. Several reports have appeared in the literature recently using multidimensional chromatographic approaches in tandem with ESI mass spectrometry in the identification of proteins [69–71].

VI. FUTURE DIRECTIONS IN PROTEOMICS

Currently, efforts are being made by a number of companies in the automation of 2 D gel technology and protein identification by mass spectrometry. These advances will be critical to the success of characterizing complete proteomes of organisms. Automation will address the problems of separating and identifying large numbers of proteins, but issues in identification of low-abundance proteins will remain. We believe that prefractionation of proteins either by multidimensional chromatography or other techniques prior to 2D-gel electrophoresis will be needed in order to obtain complete characterization of the proteome. Already, some genomics companies have begun to address the proteome [72] and consider the problems associated with protein analysis. Fractionation of proteins and protein classes will play an important role in the ultimate success of proteome characterization.

REFERENCES

1. MR Wilkins, JC Sanchez, AA Gooley, RD Appel, I Humphery-Smith, DF Hochstrasser, KL Williams. Progress with proteome projects: why all proteins expressed by a genome should be identified and how to do it. Biotechnol Genet Eng Rev 13: 19–50, 1995.
2. J Klose. Protein mapping by combined isoelectric focusing and electrophoresis of mouse tissues. A novel approach to testing for induced point mutations in mammals. Humangenetik 26:231–243, 1975.
3. PH O'Farrell. High resolution two-dimensional electrophoresis of proteins. J Biol Chem 250:4007–4021, 1975.
4. P Jungblut, B Thiede, U Zimny-Arndt, E-C Muller, C Scheler, B Wittmann-Liebold, A Otto. Resolution power of two-dimensional electrophoresis and identification of proteins from gels. Electrophoresis 17:839–847, 1996.
5. H Langen, C Gray, D Roder, J-F Juranville, B Takacs, M Fountoulakis. From genome to proteome: protein map of *Haemophilus influenzae*. Electrophoresis 18: 1184–1192, 1997.
6. C Pasquali, S Frutiger, MR Wilkins, GJ Hughes, RD Appel, A Bairoch, D Schaller, J-C Sanchez, DF Hochstrasser. Two-dimensional gel electrophoresis of *Escherichia*

coli homogenates: the *Escherichia coli* SWISS-2DPAGE database. Electrophoresis 17:547–555, 1996.

7. WP Blackstock, MP Weir. Proteomics: quantitative and physical mapping of cellular proteins. Trends Biotechnol 17:121–127, 1999.
8. CA Mangolin, LMM Ottoboni, M de FPS Machado. Two-dimensional electrophoresis of *Cereus peruvianus* (Cactaceae) callus tissue proteins. Electrophoresis 20:626–629, 1999.
9. C Vasseur, J Labadie, M Hebraud. Differential protein expression by *Pseudomonas fragi* submitted to various stresses. Electrophoresis 20:2204–2213, 1999.
10. B Franzen, S Linder, AA Alaiya, E Eriksson, K Fujioka, A-C Bergman, H Jornvall, G Auer. Analysis of polypeptide expression in benign and malignant human breast lessions. Electrophoresis 18:582–587, 1997.
11. DH Geschwind, FR Thormodsson, S Hockfield. Changes in protein expression during neural development analyzed by two-dimensional electrophoresis. Electrophoresis 17:1677–1682, 1996.
12. L Bini, B Magi, B Marzocchi, C Cellesi, B Berti, R Raggiaschi, A Rossolini, V Pallini. Two-dimensional electrophoretic patterns of acute-phase human serum proteins in the course of bacterial and viral diseases. Electrophoresis 17:612–616, 1996.
13. L Martinelli, E Gianazza. Biochemical changes during regeneration of sunflower (*Helianthus annuus* L.). Electrophoresis 17:191–197, 1996.
14. FA Witzmann, CD Fultz, JC Lipscomb. Toxicant-induced alterations in two-dimensional electrophoretic patterns of hepatic and renal stress proteins. Electrophoresis 17:198–202, 1996.
15. J Weekes, CH Wheeler, JX Yan, J Weil, T Eschenhagen, G Scholtysik, MJ Dunn. Bovine dilated cardiomyopathy: proteomic analysis of an animal model of human dilated cardiomyopathy. Electrophoresis 20:898–906, 1999.
16. U Edvardsson, M Alexandersson, H Brockenhuus von Lowenhielm, A-C Nystrom, B Ljung, F Nilsson, B Dahllof. A proteome analysis of livers from obese (*ob/ob*) mice treated with the peroxisome proliferator WY14,643. Electrophoresis 20:935–942, 1999.
17. FA Witzmann, CD Fultz, RA Grant, LS Wright, SE Kornguth, FL Siegel. Regional protein alterations in rat kidneys induced by lead exposure. Electrophoresis 20:943–951, 1999.
18. T Rabilloud, P Vincens, D Asselineau, J-L Pennetier, M Darmon, P Tarroux. Computer analysis of two-dimensional electrophoresis gels as a tool in cell biology: study of the protein expression of human keratinocytes from normal to tumor cells. Cell Mol Biol 40:17–27, 1994.
19. MR Wilkins, E Gasteiger, AA Gooley, BR Herbert, MP Molloy, P-A Binz, K Ou, J-C Sanchez, A Bairoch, KL Williams, DF Hochstrasser. High-throughput mass spectrometric discovery of protein post-translational modifications. J Mol Biol 289: 645–657, 1999.
20. E Mortz, T Sareneva, S Haebel, I Julkunen, P Roepstroff. Mass spectrometric characterization of glycosylated interferon-γ variants separated by gel electrophoresis. Electrophoresis 17:925–931, 1996.
21. J Liao, N-O Ku, MB Omary. Two-dimensional gel analysis of glandular keratin intermediate filament phosphorylation. Electrophoresis 17:1671–1676, 1996.

22. NH Packer, A Pawlak, WC Kett, AA Gooley, JW Redmond, KL Williams. Proteome analysis of glycoforms: a review of strategies for the microcharacterization of glycoproteins separated by two-dimensional polyacrylamide gel electrophoresis. Electrophoresis 18:452–460, 1997.
23. O Golaz, MR Wilkins, J-C Sanchez, RD Appel, DF Hochstrasser, KL Williams. Identification of proteins by their amino acid composition: an evaluation of the method. Electrophoresis 17:573–579, 1996.
24. P Touzet, D de Vienne, J-C Huet, C Ouali, F Bouet, M Zivy. Amino acid analysis of proteins separated by two-dimensional gel electrophoresis protein database ''HEART-2DPAGE'': a practical approach. Electrophoresis 17:1393–1401, 1996.
25. A Galat, V Rioux. Convergence of amino acid compositions of certain groups of proteins aids in their identification on two-dimensional electrophoresis gels. Electrophoresis 18:443–451, 1997.
26. M Fountoulakis, J-F Juranville, P Berndt. Large-scale identification of proteins of *Haemophilus influenzae* by amino acid composition analysis. Electrophoresis 18: 2968–2977, 1997.
27. LB Epstein, DM Smith, NM Matsui, HM Tran, C Sullivan, I Raineri, AL Burlingame, KR Clauser, SC Hall, LE Andrews. Identification of cytokine-regulated proteins in normal and malignant cells by the combination of two-dimensional polyacrylamide gel electrophoresis, mass spectrometry, Edman degradation and immunoblotting and approaches to the analysis of their functional roles. Electrophoresis 17:1655–1670, 1996.
28. AP Teixeira-Gomes, A Cloeckaert, G Bezard, G Dubray, MS Zygmunt. Mapping and identification of *Brucella melitensis* proteins by two-dimensional electrophoresis and microsequencing. Electrophoresis 18:156–162, 1997.
29. L Bini, H Heid, S Liberatori, G Geier, V Pallini, R Zwilling. Two-dimensional gel electrophoresis of *Caenorhabditis elegans* homogenates and identification of protein spots by microsequencing. Electrophoresis 18:557–562, 1997.
30. M Luck, W Schroder, S Harnisch, K Thode, T Blunk, B-R Paulke, M Kresse, RH Muller. Identification of plasma proteins facilitated by enrichment on particulate surfaces: analysis by two-dimensional electrophoresis and N-terminal microsequencing. Electrophoresis 18:2961–2967, 1997.
31. D Lutomski, M Caron, J-D Cornillot, P Bourin, C Dupuy, M Pontet, D Bladier, R Joubert-Caron. Identification of different galectins by immunoblotting after two-dimensional polyacrylamide gel electrophoresis with immobilized pH gradients. Electrophoresis 17:600–606, 1996.
32. E Contiero, R Ferrari, GM Vaselli, M Folin. Apolipoprotein A1 isoforms in serum determined by isoelectric focusing and immunoblotting. Electrophoresis 18:122–126, 1997.
33. AP Teixeira-Gomes, A Cloeckaert, G Bezard, RA Bowden, G Dubray, MS Zygmunt. Identification and characterization of *Brucella ovis* immunogenic proteins using two-dimensional electrophoresis and immunoblotting. Electrophoresis 18:1491–1497, 1997.
34. M Sanchez-Campillo, L Bini, M Comanducci, R Raggiaschi, B Marzocchi, V Pallini, G Ratti. Identification of immunoreactive proteins of *Chlamydia trachomatis* by

Western blot analysis of a two-dimensional electrophoresis map with patient sera. Electrophoresis 20:2269–2279, 1999.
35. B Thiede, A Otto, U Zimny-Arndt, E-C Muller, P Jungblut. Identification of human myocardial proteins separated by two-dimensional electrophoresis with matrix-assisted laser desorption/ionization mass spectrometry. Electrophoresis 17:588–599, 1996.
36. M Schuhmacher, MO Glocker, M Wunderlin, M Przybylski. Direct isolation of proteins from sodium dodecyl sulfate-polyacrylamide gel electrophoresis and analysis by electrospray-ionization mass spectrometry. Electrophoresis 17:848–854, 1996.
37. K Gavaert, J-L Verschelde, M Puype, J Van Damme, M Goethals, S De Boeck, J Vandekerckhove. Structural analysis and identification of gel-purified proteins, available in the femtomole range, using a novel computer program for peptide sequence assignment, by matrix-assisted laser desorption ionization–reflectron time-of-flight–mass spectrometry. Electrophoresis 17:918–924, 1996.
38. KL O'Connell, JT Stults. Identification of mouse liver proteins on two-dimensional electrophoresis gels by matrix-assisted laser desorption/ionization mass spectrometry of *in situ* enzymatic digests. Electrophoresis 18:349–359, 1997.
39. D Figeys, R Aebersold. High sensitivity identification of proteins by electrospray ionization tandem mass spectrometry: Initial comparison between an ion trap mass spectrometer and a triple quadrupole mass spectrometer. Electrophoresis 18:360–368, 1997.
40. G Li, M Waltham, NL Anderson, E Unsworth, A Treston, JN Weinstein. Rapid mass spectrometric identification of proteins from two-dimensional polyacrylamide gels after in gel proteolytic digestion. Electrophoresis 18:391–402, 1997.
41. F Gharahdaghi, CR Weinberg, DA Meagher, BS Imai, SM Mische. Mass spectrometric identification of proteins from silver-stained polyacrylamide gel: a method for the removal of silver ions to enhance sensitivity. Electrophoresis 20:601–605, 1999.
42. E-C Muller, M Schumann, A Rickers, K Bommert, B Wittmann-Liebold, A Otto. Study of Burkitt lymphoma cell line proteins by high resolution two-dimensional gel electrophoresis and nanoelectrospray mass spectrometry. Electrophoresis 20: 320–330, 1999.
43. A Shevchenko, ON Jensen, AV Podtelejnikov, F Sagliocco, M Wilm, O Vorm, P Mortensen, A Shevchenko, H Boucherie, M Mann. Linking genome and proteome by mass spectrometry: large-scale identification of yeast proteins from two dimensional gels. Proc Natl Acad Sci (USA) 93:14440–14445, 1996.
44. JR Yates III, JK Eng, AL McCormack. Mining genomes: correlating tandem mass spectra of modified and unmodified peptides to sequences in nucleotide databases. Anal Chem 67:3202–3210, 1995.
45. JR Yates III, JK Eng, AL McCormack, D Schieltz. Method to correlate tandem mass spectra of modified peptides to amino acid sequences in the protein database. Anal Chem 67:1426–1436, 1995.
46. SD Patterson, R Aebersold. Mass spectrometric approaches for the identification of gel-separated proteins. Electrophoresis 16:1791–1814, 1995.

47. M Fountoulakis, H Langen, S Evers, C Gray, B Takacs. Two-dimensional map of *Haemophilus influenzae* following protein enrichment by heparin chromatography. Electrophoresis 18:1193–1202, 1997.
48. M Perrot, F Sagliocco, T Mini, C Monribot, U Schneider, A Shevchenko, M Mann, P Jeno, H Boucherie. Two-dimensional gel protein database of *Saccharomyces cerevisiae* (update 1999). Electrophoresis 20:2280–2298, 1999.
49. MR Wilkins, E Gasteiger, J-C Sanchez, A Bairoch, DF Hochstrasser. Two-dimensional gel electrophoresis for proteome projects: the effects of protein hydrophobicity and copy number. Electrophoresis 19:1501–1505, 1998.
50. PG Righetti. Isoelectric Focusing: Theory, Methodology and Applications. Amsterdam, Elsevier, 1983.
51. PG Righetti. Immobilized pH gradients: theory and methodology. Lab Tech Biochem Mol Biol 20, 1990.
52. PG Righetti, A Bossi. Isoelectric focusing in immobilized pH gradients: an update. J Chromatogr B 699:77–89, 1997.
53. A Gorg, C Obermaier, G Boguth, A Csordas, J-J Diaz, J-J Madjar. Very alkaline immobilized pH gradients for two-dimensional electrophoresis of ribosomal and nuclear proteins. Electrophoresis 18:328–337, 1997.
54. S Fazekas de St. Groth, RG Webster, A Datyner. Two new staining procedures for quantitative estimation of proteins on electrophoretic strips. Biochim Biophys Acta 71:377–391, 1963.
55. V Neuhoff, R Stamm, H Eibl. Clear background and highly sensitive protein staining with Coomassie blue dyes in polyacrylamide gels: a systematic analysis. Electrophoresis 6:427–448, 1985.
56. U Tessmer, R Dernick. Preparative separation of poliovirus structural polypeptides by sodium dodecyl sulfate-polyacrylamide gel electrophoresis, copper staining and electroelution, and induction of monospecific antisera. Electrophoresis 10:277–279, 1989.
57. C Fernandez-Paton, L Castellanos-Serra, P Rodriguez. Reverse staining of sodium dodecyl sulfate polyacrylamide gels by imidazole-zinc salts: sensitive detection of unmodified proteins. BioTechniques 12:564–573, 1992.
58. RK Hamby. A sensitive fluorescent gel stain. Am Biotech Lab 14:12, 1996.
59. CR Merril, D Goldman, SA Sedman, MH Ebert. Ultrasensitive stain for proteins in polyacrylamide gels show regional variation in cerebrospinal fluid proteins. Science 211:1437–1438, 1981.
60. M Wilm, A Shevchenko, T Houthaeve, S Breit, L Schweigerer, T Fotsis, M Mann. Femtomole sequencing of proteins from polyacrylamide gels by nano-electrospray mass spectrometry. Nature 379:466–469, 1996.
61. A Shevchenko, M Wilm, O Vorm, M Mann. Mass spectrometric sequencing of proteins from silver-stained polyacrylamide gels. Anal Chem 68:850–858, 1996.
62. IM Lazar, RS Ramsey, S Sundberg, JM Ramsey. Subattomole-sensitivity microchip nanoelectrospray source with time-of-flight mass spectrometry detection. Anal Chem 71:3627–3631, 1999.
63. M Mann, M Wilm. Error-tolerant identification of peptides in sequence databases by peptide sequence tags. Anal Chem 66:4390–4399, 1994.

64. M Fountoulakis, M-F Takacs, P Berndt, H Langen, B Takacs. Enrichment of low abundance proteins of *Escherichia coli* by hydroxyapatite chromatography. Electrophoresis 20:2181–2195, 1999.
65. M Fountoulakis, B Takacs. Design of protein purification pathways: application to the proteome of *Haemophilus influenzae* using heparin chromatography. Prot Express Purif 14:113–119, 1998.
66. M Fountoulakis, M-F Takacs, B Takacs. Enrichment of low-copy-number gene products by hydrophobic interaction chromatography. J Chromatogr A 833:157–168, 1999.
67. GL Corthals, MP Molloy, BR Herbert, KL Williams, AA Gooley. Prefractionation of protein samples prior to two-dimensional electrophoresis. Electrophoresis 18: 317–323, 1997.
68. M Fountoulakis, H Langen, C Gray, B Takacs. Enrichment and purification of proteins of *Haemophilus influenzae* by chromatofocusing. J Chromatogr 806:279–291, 1998.
69. GJ Opiteck, KC Lewis, JW Jorgenson. Comprehensive on-line LC/LC/MS of proteins. Anal Chem 69:1518–1524, 1997.
70. GJ Opiteck, JW Jorgenson. Two-dimensional SEC/RPLC coupled to mass spectrometry for the analysis of peptides. Anal Chem 69:2283–2291, 1997.
71. GJ Opiteck, SM Ramirez, JW Jorgenson, MA Moseley III. Comprehensive two-dimensional high-performance liquid chromatography for the isolation of overexpressed proteins and proteome mapping. Anal Biochem 258:349–361, 1998.
72. RF Service. Can Celera do it again? Science 287:2136–2138, 2000.

Index